Methods in Enzymology

Volume 98

BIOMEMBRANES

Part L

Membrane Biogenesis:

Processing and Recycling

METHODS IN ENZYMOLOGY

EDITORS-IN-CHIEF

Sidney P. Colowick   Nathan O. Kaplan

Methods in Enzymology

Volume 98

Biomembranes

Part L

Membrane Biogenesis: Processing and Recycling

EDITED BY

Sidney Fleischer
Becca Fleischer

DEPARTMENT OF MOLECULAR BIOLOGY
VANDERBILT UNIVERSITY
NASHVILLE, TENNESSEE

1983

ACADEMIC PRESS
A Subsidiary of Harcourt Brace Jovanovich, Publishers

New York London
Paris San Diego San Francisco São Paulo Sydney Tokyo Toronto

ACADEMIC PRESS, INC.
111 Fifth Avenue, New York, New York 10003

United Kingdom Edition published by
ACADEMIC PRESS, INC. (LONDON) LTD.
24/28 Oval Road, London NW1 7DX

Library of Congress Cataloging in Publication Data
Main entry under title:

Biomembranes.

(Methods in enzymology ; v. 31–)
Pt.H– edited by Lester Packer.
Includes bibliographical references and indexes.
Contents: pts. A-B. [without special title] -- C-E.
Biological oxidations -- [etc.] -- pt. L. Membrane
biogenesis.
1. Cell membranes. 2. Oxidation, Physiological.
3. Bioenergetics. 4. Nucleic acid synthesis.
I. Fleischer, Sidney. II. Packer, Lester. III. Series:
Methods in enzymology; v. 31, etc. [DNLM: 1. Cell
membrane. 2. Membranes--Enzymology. W1 ME9615K v. 31,
etc. / QH 601 B6192 1974]
QP601.M49 vol. 98 [QH601] 574.19'25s[574.87'5] 74-11352
ISBN 0–12–181998–1 (v.98)

PRINTED IN THE UNITED STATES OF AMERICA

83 84 85 86 9 8 7 6 5 4 3 2 1

Table of Contents

Section I. Specialized Methods

Section II. Importance of Glycosylation and Trimming; Targeting

Section III. Exocytosis

Section IV. Internalization of Plasma Membrane Components during Absorptive Endocytosis

A. Receptor-Mediated Uptake

B. Coated Vesicles

Section V. Recycling of Plasma Membrane Proteins

Section VI. Transcellular Transport

Section VII. Plasma Membrane Differentiation

Section VIII. Transfer of Phospholipids between Membranes

Addendum

Contributors to Volume 98

Article numbers are in parentheses following the names of contributors.
Affiliations listed are current.

PETER ARVAN (8), *Section of Cell Biology, Yale University School of Medicine, New Haven, Connecticut 06510*

ROOBIK AZARNIA (47), *Department of Physiology and Biophysics, University of Miami, Miami, Florida 33101*

JACQUES U. BAENZIGER (12), *Department of Pathology, Washington University School of Medicine, St. Louis, Missouri 63110*

P. F. BAKER (3), *Department of Physiology, King's College London, Strand, London WC2R 2LS, England*

WILLIAM E. BALCH (4), *Department of Biochemistry, Stanford University, Stanford, California 94305*

SANDIP K. BASU (19), *Department of Molecular Genetics, University of Texas Health Science Center at Dallas, Dallas, Texas 75235*

SUBAL BISHAYEE (49), *Department of Biochemistry and Biophysics, University of Pennsylvania School of Medicine, Philadelphia, Pennsylvania 19104*

GÜNTER BLOBEL (40), *Laboratory of Cell Biology, The Rockefeller University, New York, New York 10021*

BERNABÉ BLOJ (51), *Division of Nutritional Sciences and Section of Biochemistry, Molecular, and Cell Biology, Cornell University, Ithaca, New York 14853*

ENRIQUE BRANDAN (28), *Departamento Biologia Celular, Universidad Catolica, Cassilla 114-D, Santiago, Chile*

MICHAEL S. BROWN (19), *Department of Molecular Genetics, University of Texas Health Science Center at Dallas, Dallas, Texas 75235*

RICHARD S. CAMERON (8), *Section of Cell Biology, Yale University School of Medicine, New Haven, Connecticut 06510*

J. DAVID CASTLE (8), *Section of Cell Biology, Yale University School of Medicine, New Haven, Connecticut 06510*

MARIANNE CHAMBARD (42), *Laboratoire de Biochimie Médicale et INSERM U38, Faculté de Médicine, F 13385 Marseille Cedex 5, France*

A. M. CHAMBAUT-GUÉRIN (15), *Laboratoire de Biochimie des Transports Cellulaires, Institut de Biochimie, Université Paris Sud, 91405 Orsay Cedex, France*

ZANVIL A. COHN (35), *Department of Cellular Physiology and Immunology, The Rockefeller University, New York, New York 10021*

JERRY R. COLCA (16), *Department of Pathology, Washington University School of Medicine, St. Louis, Missouri 63110*

KIM E. CREEK (23), *The Edward Mallinckrodt Department of Pediatrics, Washington University School of Medicine, and Division of Medical Genetics, St. Louis Children's Hospital, St. Louis, Missouri 63178*

GERHARD DAHL (47), *Department of Physiology and Biophysics, University of Miami, Miami, Florida 33101*

JOHN L. DAISS (29), *Biosciences Division, Kodak Research Laboratories, Eastman Kodak Company, Rochester, New York 14650*

MANJUSRI DAS (49), *Department of Biochemistry and Biophysics, University of Pennsylvania School of Medicine, Philadelphia, Pennsylvania 19104*

RANDALL L. DIMOND (13), *Department of Bacteriology, University of Wisconsin, Madison, Wisconsin 53706*

JACK J. DISTLER (25), *Rackham Arthritis Research Unit and Department of Internal Medicine, University of Michigan, Ann Arbor, Michigan 48109*

WILLIAM A. DUNN (18), *Department of Cell Biology and Anatomy, The Johns Hopkins University School of Medicine, Baltimore, Maryland 21205*

WILLIAM G. DUNPHY (4), *Department of Biochemistry, Stanford University, Stanford California 94305*

HANSJORG EIBL (58), *Max-Planck-Institut*

für Biophysikalische Chemie, Karl-Friedrich-Bonhoeffer-Institut, Abteilung 13, D-3400 Göttingen-Nikolausberg, Federal Republic of Germany

ALAN D. ELBEIN (11), *Department of Biochemistry, University of Texas Health Science Center, San Antonio, Texas 78284*

MARILYN GIST FARQUHAR (1, 17) *Section of Cell Biology, Yale University School of Medicine, New Haven, Connecticut 06510*

JEFFREY FEINMAN (49), *Department of Biochemistry and Biophysics, University of Pennsylvania School of Medicine, Philadelphia, Pennsylvania 19104*

H. DAVID FISCHER (23), *Department of Biochemistry, Molecular Biology, and Cell Biology, Northwestern University, Evanston, Illinois 60201*

EDUARD A. M. FLEER (58), *Department of Molecular Biology, Vanderbilt University School of Medicine, Nashville, Tennessee 37232*

BECCA FLEISCHER (6, 28), *Department of Molecular Biology, Vanderbilt University, Nashville, Tennessee 37235*

SIDNEY FLEISCHER (28, 56, 58), *Department of Molecular Biology, Vanderbilt University, Nashville, Tennessee 37235*

ERIK FRIES (4), *The Wallenberg Laboratory, University of Uppsala, 22 Uppsala, Sweden*

JACQUELINE GABRION (42), *Laboratoire de Biochimie Médicale et INSERM U38, Faculté de Médicine, F 13385 Marseille Cedex 5, France*

CYNTHIA J. GALLOWAY (48), *Section of Cell Biology, Yale University School of Medicine, New Haven, Connecticut 06510*

NORTON B. GILULA (45), *Department of Cell Biology, Baylor College of Medicine, Texas Medical Center, Houston, Texas 77030*

PAUL GLEESON (10), *National Institute for Medical Research, The Ridgeway, London NW7 1AA, England*

JOSEPH L. GOLDSTEIN (19), *Department of Molecular Genetics, University of Texas Health Science Center at Dallas, Dallas, Texas 75235*

HARRY T. HAIGLER (22), *Department of Physiology and Biophysics, University of California, Irvine, California 92717*

ARI HELENIUS (20), *Section of Cell Biology, Yale University School of Medicine, New Haven, Connecticut 06510*

ELLIOT L. HERTZBERG (44), *Verna and Marrs McLean Department of Biochemistry, Baylor College of Medicine, Texas Medical Center, Houston, Texas 77030*

VOLKER HERZOG (17, 39), *Institut für Zellbiologie der Universität München, D-8000 München, Federal Republic of Germany*

ROBERT L. HILL (26), *Department of Biochemistry, Duke University Medical Center, Durham, North Carolina 27710*

G. HOLT (24), *Department of Physiology and Biophysics, Washington University School of Medicine, St. Louis, Missouri 63110*

ANN L. HUBBARD (18), *Department of Cell Biology and Anatomy, The Johns Hopkins University School of Medicine, Baltimore, Maryland 21205*

GEORGE W. JOURDIAN (25), *Rackham Arthritis Research Unit and Department of Internal Medicine, University of Michigan, Ann Arbor, Michigan 48109*

H. H. KAMP (52), *Laboratory of Biochemistry, State University Hospital, Utrecht, NL-3500 CG Utrecht, The Netherlands*

JAMES H. KEEN (31), *Fels Research Institute and Department of Biochemistry, Temple University School of Medicine, Philadelphia, Pennsylvania 19140*

GILBERT A. KELLER (32), *Department of Biology, University of California at San Diego, La Jolla, California 92037*

EUGENE P. KENNEDY (55), *Department of Biological Chemistry, Harvard Medical School, Boston, Massachusetts 02115*

DAVID A. KNECHT (13), *Department of Biology, University of California at San Diego, La Jolla, California 92037*

D. E. KNIGHT (3), *Department of Physiology, King's College London, Strand, London WC2R 2LS, England*

M. KONISH (24), *Department of Physiology and Biophysics, Washington University School of Medicine, St. Louis, Missouri 63110*

TETSURO KONO (38), *Department of Physiology, Vanderbilt University School of Medicine, Nashville, Tennessee 37232*

NIRMALA KOTAGAL (16), *Department of Pa-*

thology, Washington University School of Medicine, St. Louis, Missouri 63110

NALIN M. KUMAR (45), *Department of Cell Biology, Baylor College of Medicine, Texas Medical Center, Houston, Texas 77030*

PAUL E. LACY (16), *Department of Pathology, Washington University School of Medicine, St. Louis, Missouri 63110*

MARK A. LEHRMAN (26), *Department of Molecular Genetics, University of Texas Health Science Center at Dallas, Dallas, Texas 75235*

WILLIAM LENNARZ (9), *Department of Biochemistry and Molecular Biology, University of Texas System Cancer Center, M. D. Anderson Hospital and Tumor Institute, Texas Medical Center, Houston, Texas 77030*

DANIEL LOUVARD (33), *Laboratoire de Biologie des Membranes, Institut Pasteur, F-75724 Paris Cedex 15, France*

MICHAEL L. MCDANIEL (16), *Department of Pathology, Washington University School of Medicine, St. Louis, Missouri 63110*

J. OLIVER MCINTYRE (58), *Department of Molecular Biology, Vanderbilt University School of Medicine, Nashville, Tennessee 37232*

MARK MARSH (20), *Section of Cell Biology, Yale University School of Medicine, New Haven, Connecticut 06510*

KARL MATLIN (20), *European Molecular Biology Laboratory, 6900 Heidelberg, Federal Republic of Germany*

JEAN MAUCHAMP (42), *Laboratoire de Biochimie Médicale et INSERM U38, Faculté de Médicine, F 13385 Marseille Cedex 5, France*

JACOPO MELDOLESI (7), *Department of Pharmacology and CNR Center of Cytopharmacology, Università Degli Studi di Milano, 20129 Milano, Italy*

IRA MELLMAN (48), *Section of Cell Biology, Yale University School of Medicine, New Haven, Connecticut 06510*

RAYMOND J. METZ (57), *Mental Health Research Institute and Department of Biological Chemistry, University of Michigan, Ann Arbor, Michigan 48109*

HERMAN MICHAEL (49), *Department of Biochemistry and Biophysics, University of Pennsylvania School of Medicine, Philadelphia, Pennsylvania 19104*

ROBERT C. MIERENDORF, JR. (13), *Department of Bacteriology, University of Wisconsin, Madison, Wisconsin 53706*

DIANE C. MITCHELL (25), *Rackham Arthritis Research Unit and Department of Internal Medicine, University of Michigan, Ann Arbor, Michigan 48109*

KEITH E. MOSTOV (40), *Laboratory of Cell Biology, The Rockefeller University, New York, New York 10021*

P. MULLER (15), *Laboratoire de Biochimie des Transports Cellulaires, Institut de Biochimie, Université Paris Sud, 91405 Orsay Cedex, France*

WILLIAM A. MULLER (35), *Department of Cellular Physiology and Immunology, The Rockefeller University, New York, New York 10021*

SAROJA NARASIMHAN (10), *Department of Biochemistry, Hospital for Sick Children, Toronto, Ontario M5G 1X8, Canada*

BRUCE J. NICHOLSON (46), *Division of Biology, California Institute of Technology, Pasadena, California 91125*

PAULA NORTH (56), *Department of Physiology, Vanderbilt University School of Medicine, Nashville, Tennessee 37232*

IRA H. PASTAN (21), *Laboratory of Molecular Biology, National Cancer Institute, National Institutes of Health, Bethesda, Maryland 20205*

B. M. F. PEARSE (27), *Laboratory of Molecular Biology, Medical Research Council Center, University Medical School, Cambridge CB2 2QH, England*

MARK PITTENGER (49), *Department of Biochemistry and Biophysics, University of Pennsylvania School of Medicine, Philadelphia, Pennsylvania 19104*

B. J. H. M. POORTHUIS (53, 54), *Laboratory of Biochemistry, State University of Utrecht, NL-3584 CH Utrecht, The Netherlands*

S. PUSZKIN (30), *Department of Pathology, Mount Sinai School of Medicine, City University of New York, New York, New York 10029*

NORMAN S. RADIN (57), *Mental Health Research Institute and Department of Biolog-*

ical Chemistry, University of Michigan, Ann Arbor, Michigan 48109

COLVIN M. REDMAN (14), *Lindsley F. Kimball Research Institute, New York Blood Center, New York, New York 10021*

HUBERT REGGIO (33), *Institut de Cytologie et de Biologie Cellulaire, Université d'Aix-Marseille, Centre de Marseille-Luminy, F-13288 Marseille Cedex 2, France*

JEAN-PAUL REVEL (46), *Division of Biology, California Institute of Technology, Pasadena, California 91125*

ENRIQUE RODRIGUEZ-BOULAN (43), *Department of Pathology, Downstate Medical Center, State University of New York, Brooklyn, New York 11203*

GERHARD ROHR (41), *Medizinische Klinik, Philipps University, D-3550 Marburg, Federal Republic of Germany*

B. ROSSIGNOL (15), *Laboratoire de Biochimie des Transports Cellulaires, Institut de Biochimie, Université Paris Sud, 91405 Orsay Cedex, France*

THOMAS F. ROTH (29), *Department of Biological Sciences, University of Maryland, Baltimore County, Catonsville, Maryland 21228*

JAMES E. ROTHMAN (4), *Department of Biochemistry, Stanford University, Stanford, California 94305*

G. GARY SAHAGIAN (25), *Biochemistry Branch, National Institute of Arthritis, Diabetes, Digestive and Kidney Diseases, National Institutes of Health, Bethesda, Maryland 20205*

JEFFREY L. SALISBURY (32), *Department of Anatomy, Albert Einstein College of Medicine, Bronx, New York 10461*

HARRY SCHACHTER (10), *Department of Biochemistry, Hospital for Sick Children, Toronto, Ontario M5G 1X8, Canada*

GEORGE SCHEELE (2, 41), *Laboratory of Cell Biology, The Rockefeller University, New York, New York 10021*

W. SCHOOK (30), *Department of Pathology, Mount Sinai School of Medicine, City University of New York, New York, New York 10029*

V. SHEPHERD (24), *Department of Physiology and Biophysics, Washington University School of Medicine, St. Louis, Missouri 63110*

F. ANTHONY SIMION (28), *Department of Molecular Biology, Vanderbilt University, Nashville, Tennessee 37235*

KAI SIMONS (20), *European Molecular Biology Laboratory, 6900 Heidelberg, Federal Republic of Germany*

WILLIAM S. SLY (23), *The Edward Mallinckrodt Department of Pediatrics, Washington University School of Medicine, Division of Medical Genetics, St. Louis Children's Hospital, St. Louis, Missouri 63178*

P. STAHL (24), *Department of Physiology and Biophysics, Washington University School of Medicine, St. Louis, Missouri 63110*

RALPH M. STEINMAN (35), *Department of Cellular Physiology and Immunology, The Rockefeller University, New York, New York 10021*

ALAN MICHAEL TARTAKOFF (5), *Pathology Department, University of Geneva School of Medicine, 1211 Geneva 4, Switzerland*

T. TEERLINK (53), *Laboratory of Biochemistry, State University of Utrecht, NL-3584 CH Utrecht, The Netherlands*

LUTZ THILO (36), *National Chemical Research Laboratory, Council for Scientific and Industrial Research, Pretoria 0001, Republic of South Africa*

LENORE J. URBANI (4), *Department of Biochemistry, Stanford University, Stanford, California 94305*

GEORGE VELLA (10), *Department of Biochemistry, Hospital for Sick Children, Toronto, Ontario M5G 1X8, Canada*

BERNARD VERRIER (42), *Laboratoire de Biochimie Médicale et INSERM U38, Faculté de Médicine, F 13385 Marseille Cedex 5, France*

DENNIS R. VOELKER (55), *Department of Medicine, National Jewish Hospital, Denver, Colorado 80206*

GÜNTER VOGEL (37), *Bergische Universität, 5600 Wuppertal, Federal Republic of Germany*

RENATE M. WAGNER (25), *Rackham Arthritis Research Unit and Department of Internal Medicine, University of Michigan, Ann Arbor, Michigan 48109*

DORIS A. WALL (18), *Department of Cell Biology and Anatomy, The Johns Hopkins*

University School of Medicine, Baltimore, Maryland 21205

PAUL WEBSTER (33), *International Laboratory for Research on Animal Diseases, P. O. Box 30709, Nairobi, Kenya*

RUDOLF WERNER (47), *Department of Biochemistry, University of Miami, Miami, Florida 33101*

JAN WESTERMAN (52), *Laboratory of Biochemistry, State University of Utrecht, NL-3584 CH Utrecht, The Netherlands*

CHRISTOPHER C. WIDNELL (34), *Department of Anatomy and Cell Biology, University of Pittsburgh School of Medicine, Pittsburgh, Pennsylvania 15261*

EDWARD H. WILLIAMS (45), *Department of Anatomy, University of Virginia, Charlottesville, Virginia 22908*

MARK C. WILLINGHAM (21), *Laboratory of Molecular Biology, National Cancer Institute, National Institutes of Health, Bethesda, Maryland 20205*

DAVID WINEK (28), *Washington University Medical School, St. Louis, Missouri 63110*

K. W. A. WIRTZ (52, 53, 54), *Laboratory of Biochemistry, State University of Utrecht, NL-3584 CH Utrecht, The Netherlands*

DONALD B. ZILVERSMIT (50, 51), *Division of Nutritional Sciences and Section of Biochemistry, Molecular, and Cell Biology, Cornell University, Ithaca, New York 14853*

Preface

Volumes 96 to 98, Parts J, K, and L of the Biomembranes series, focus on methodology to study membrane biogenesis, assembly, targeting, and recycling. This field is one of the very exciting and active areas of research. Future volumes will deal with transport and other aspects of membrane function.

We were fortunate to have the advice and good counsel of our Advisory Board. Additional valuable input to this volume was obtained from Drs. Kai Simons, Marilyn Farquhar, Oscar Touster, William S. Sly, Norton Gilula, James D. Jamieson, and Stuart Kornfield. We were gratified by the enthusiasm and cooperation of the participants in the field whose contributions and suggestions have enriched and made possible these volumes. The friendly cooperation of the staff of Academic Press is gratefully acknowledged.

SIDNEY FLEISCHER
BECCA FLEISCHER

METHODS IN ENZYMOLOGY

EDITED BY

Sidney P. Colowick and Nathan O. Kaplan

VANDERBILT UNIVERSITY
SCHOOL OF MEDICINE
NASHVILLE, TENNESSEE

DEPARTMENT OF CHEMISTRY
UNIVERSITY OF CALIFORNIA
AT SAN DIEGO
LA JOLLA, CALIFORNIA

I. Preparation and Assay of Enzymes
II. Preparation and Assay of Enzymes
III. Preparation and Assay of Substrates
IV. Special Techniques for the Enzymologist
V. Preparation and Assay of Enzymes
VI. Preparation and Assay of Enzymes (*Continued*)
Preparation and Assay of Substrates
Special Techniques
VII. Cumulative Subject Index

METHODS IN ENZYMOLOGY

EDITORS-IN-CHIEF

Sidney P. Colowick Nathan O. Kaplan

VOLUME VIII. Complex Carbohydrates
Edited by ELIZABETH F. NEUFELD AND VICTOR GINSBURG

VOLUME IX. Carbohydrate Metabolism
Edited by WILLIS A. WOOD

VOLUME X. Oxidation and Phosphorylation
Edited by RONALD W. ESTABROOK AND MAYNARD E. PULLMAN

VOLUME XI. Enzyme Structure
Edited by C. H. W. HIRS

VOLUME XII. Nucleic Acids (Parts A and B)
Edited by LAWRENCE GROSSMAN AND KIVIE MOLDAVE

VOLUME XIII. Citric Acid Cycle
Edited by J. M. LOWENSTEIN

VOLUME XIV. Lipids
Edited by J. M. LOWENSTEIN

VOLUME XV. Steroids and Terpenoids
Edited by RAYMOND B. CLAYTON

VOLUME XVI. Fast Reactions
Edited by KENNETH KUSTIN

VOLUME XVII. Metabolism of Amino Acids and Amines (Parts A and B)
Edited by HERBERT TABOR AND CELIA WHITE TABOR

VOLUME XVIII. Vitamins and Coenzymes (Parts A, B, and C)
Edited by DONALD B. MCCORMICK AND LEMUEL D. WRIGHT

Volume XIX. Proteolytic Enzymes
Edited by Gertrude E. Perlmann and Laszlo Lorand

Volume XX. Nucleic Acids and Protein Synthesis (Part C)
Edited by Kivie Moldave and Lawrence Grossman

Volume XXI. Nucleic Acids (Part D)
Edited by Lawrence Grossman and Kivie Moldave

Volume XXII. Enzyme Purification and Related Techniques
Edited by William B. Jakoby

Volume XXIII. Photosynthesis (Part A)
Edited by Anthony San Pietro

Volume XXIV. Photosynthesis and Nitrogen Fixation (Part B)
Edited by Anthony San Pietro

Volume XXV. Enzyme Structure (Part B)
Edited by C. H. W. Hirs and Serge N. Timasheff

Volume XXVI. Enzyme Structure (Part C)
Edited by C. H. W. Hirs and Serge N. Timasheff

Volume XXVII. Enzyme Structure (Part D)
Edited by C. H. W. Hirs and Serge N. Timasheff

Volume XXVIII. Complex Carbohydrates (Part B)
Edited by Victor Ginsburg

Volume XXIX. Nucleic Acids and Protein Synthesis (Part E)
Edited by Lawrence Grossman and Kivie Moldave

Volume XXX. Nucleic Acids and Protein Synthesis (Part F)
Edited by Kivie Moldave and Lawrence Grossman

Volume XXXI. Biomembranes (Part A)
Edited by Sidney Fleischer and Lester Packer

Volume XXXII. Biomembranes (Part B)
Edited by Sidney Fleischer and Lester Packer

VOLUME XXXIII. Cumulative Subject Index Volumes I–XXX
Edited by MARTHA G. DENNIS AND EDWARD A. DENNIS

VOLUME XXXIV. Affinity Techniques (Enzyme Purification: Part B)
Edited by WILLIAM B. JAKOBY AND MEIR WILCHEK

VOLUME XXXV. Lipids (Part B)
Edited by JOHN M. LOWENSTEIN

VOLUME XXXVI. Hormone Action (Part A: Steroid Hormones)
Edited by BERT W. O'MALLEY AND JOEL G. HARDMAN

VOLUME XXXVII. Hormone Action (Part B: Peptide Hormones)
Edited by BERT W. O'MALLEY AND JOEL G. HARDMAN

VOLUME XXXVIII. Hormone Action (Part C: Cyclic Nucleotides)
Edited by JOEL G. HARDMAN AND BERT W. O'MALLEY

VOLUME XXXIX. Hormone Action (Part D: Isolated Cells, Tissues, and Organ Systems)
Edited by JOEL G. HARDMAN AND BERT W. O'MALLEY

VOLUME XL. Hormone Action (Part E: Nuclear Structure and Function)
Edited by BERT W. O'MALLEY AND JOEL G. HARDMAN

VOLUME XLI. Carbohydrate Metabolism (Part B)
Edited by W. A. WOOD

VOLUME XLII. Carbohydrate Metabolism (Part C)
Edited by W. A. WOOD

VOLUME XLIII. Antibiotics
Edited by JOHN H. HASH

VOLUME XLIV. Immobilized Enzymes
Edited by KLAUS MOSBACH

VOLUME XLV. Proteolytic Enzymes (Part B)
Edited by LASZLO LORAND

VOLUME XLVI. Affinity Labeling
Edited by WILLIAM B. JAKOBY AND MEIR WILCHEK

VOLUME XLVII. Enzyme Structure (Part E)
Edited by C. H. W. HIRS AND SERGE N. TIMASHEFF

VOLUME XLVIII. Enzyme Structure (Part F)
Edited by C. H. W. HIRS AND SERGE N. TIMASHEFF

VOLUME XLIX. Enzyme Structure (Part G)
Edited by C. H. W. HIRS AND SERGE N. TIMASHEFF

VOLUME L. Complex Carbohydrates (Part C)
Edited by VICTOR GINSBURG

VOLUME LI. Purine and Pyrimidine Nucleotide Metabolism
Edited by PATRICIA A. HOFFEE AND MARY ELLEN JONES

VOLUME LII. Biomembranes (Part C: Biological Oxidations)
Edited by SIDNEY FLEISCHER AND LESTER PACKER

VOLUME LIII. Biomembranes (Part D: Biological Oxidations)
Edited by SIDNEY FLEISCHER AND LESTER PACKER

VOLUME LIV. Biomembranes (Part E: Biological Oxidations)
Edited by SIDNEY FLEISCHER AND LESTER PACKER

VOLUME LV. Biomembranes (Part F: Bioenergetics)
Edited by SIDNEY FLEISCHER AND LESTER PACKER

VOLUME LVI. Biomembranes (Part G: Bioenergetics)
Edited by SIDNEY FLEISCHER AND LESTER PACKER

VOLUME LVII. Bioluminescence and Chemiluminescence
Edited by MARLENE A. DELUCA

VOLUME LVIII. Cell Culture
Edited by WILLIAM B. JAKOBY AND IRA H. PASTAN

VOLUME LIX. Nucleic Acids and Protein Synthesis (Part G)
Edited by KIVIE MOLDAVE AND LAWRENCE GROSSMAN

VOLUME LX. Nucleic Acids and Protein Synthesis (Part H)
Edited by KIVIE MOLDAVE AND LAWRENCE GROSSMAN

VOLUME 61. Enzyme Structure (Part H)
Edited by C. H. W. HIRS AND SERGE N. TIMASHEFF

VOLUME 62. Vitamins and Coenzymes (Part D)
Edited by DONALD B. MCCORMICK AND LEMUEL D. WRIGHT

VOLUME 63. Enzyme Kinetics and Mechanism (Part A: Initial Rate and Inhibitor Methods)
Edited by DANIEL L. PURICH

VOLUME 64. Enzyme Kinetics and Mechanism (Part B: Isotopic Probes and Complex Enzyme Systems)
Edited by DANIEL L. PURICH

VOLUME 65. Nucleic Acids (Part I)
Edited by LAWRENCE GROSSMAN AND KIVIE MOLDAVE

VOLUME 66. Vitamins and Coenzymes (Part E)
Edited by DONALD B. MCCORMICK AND LEMUEL D. WRIGHT

VOLUME 67. Vitamins and Coenzymes (Part F)
Edited by DONALD B. MCCORMICK AND LEMUEL D. WRIGHT

VOLUME 68. Recombinant DNA
Edited by RAY WU

VOLUME 69. Photosynthesis and Nitrogen Fixation (Part C)
Edited by ANTHONY SAN PIETRO

VOLUME 70. Immunochemical Techniques (Part A)
Edited by HELEN VAN VUNAKIS AND JOHN J. LANGONE

VOLUME 71. Lipids (Part C)
Edited by JOHN M. LOWENSTEIN

VOLUME 72. Lipids (Part D)
Edited by JOHN M. LOWENSTEIN

VOLUME 73. Immunochemical Techniques (Part B)
Edited by JOHN J. LANGONE AND HELEN VAN VUNAKIS

VOLUME 74. Immunochemical Techniques (Part C)
Edited by JOHN J. LANGONE AND HELEN VAN VUNAKIS

VOLUME 75. Cumulative Subject Index Volumes XXXI, XXXII, and XXXIV–LX
Edited by MARTHA G. DENNIS AND EDWARD A. DENNIS

VOLUME 76. Hemoglobins
Edited by ERALDO ANTONINI, LUIGI ROSSI-BERNARDI, AND EMILIA CHIANCONE

VOLUME 77. Detoxication and Drug Metabolism
Edited by WILLIAM B. JAKOBY

VOLUME 78. Interferons (Part A)
Edited by SIDNEY PESTKA

VOLUME 79. Interferons (Part B)
Edited by SIDNEY PESTKA

VOLUME 80. Proteolytic Enzymes (Part C)
Edited by LASZLO LORAND

VOLUME 81. Biomembranes (Part H: Visual Pigments and Purple Membranes, I)
Edited by LESTER PACKER

VOLUME 82. Structural and Contractile Proteins (Part A: Extracellular Matrix)
Edited by LEON W. CUNNINGHAM AND DIXIE W. FREDERIKSEN

VOLUME 83. Complex Carbohydrates (Part D)
Edited by VICTOR GINSBURG

VOLUME 84. Immunochemical Techniques (Part D: Selected Immunoassays)
Edited by JOHN J. LANGONE AND HELEN VAN VUNAKIS

VOLUME 85. Structural and Contractile Proteins (Part B: The Contractile Apparatus and the Cytoskeleton)
Edited by DIXIE W. FREDERIKSEN AND LEON W. CUNNINGHAM

VOLUME 86. Prostaglandins and Arachidonate Metabolites
Edited by WILLIAM E. M. LANDS AND WILLIAM L. SMITH

VOLUME 87. Enzyme Kinetics and Mechanism (Part C: Intermediates, Stereochemistry, and Rate Studies)
Edited by DANIEL L. PURICH

VOLUME 88. Biomembranes (Part I: Visual Pigments and Purple Membranes, II)
Edited by LESTER PACKER

VOLUME 89. Carbohydrate Metabolism (Part D)
Edited by WILLIS A. WOOD

VOLUME 90. Carbohydrate Metabolism (Part E)
Edited by WILLIS A. WOOD

VOLUME 91. Enzyme Structure (Part I)
Edited by C. H. W. HIRS AND SERGE N. TIMASHEFF

VOLUME 92. Immunochemical Techniques (Part E: Monoclonal Antibodies and General Immunoassay Methods)
Edited by JOHN J. LANGONE AND HELEN VAN VUNAKIS

VOLUME 93. Immunochemical Techniques (Part F: Conventional Antibodies, Fc Receptors, and Cytotoxicity)
Edited by JOHN J. LANGONE AND HELEN VAN VUNAKIS

VOLUME 94. Polyamines
Edited by HERBERT TABOR AND CELIA WHITE TABOR

VOLUME 95. Cumulative Subject Index Volumes 61–74, 76–80 (in preparation)
Edited by MARTHA G. DENNIS AND EDWARD A. DENNIS

VOLUME 96. Biomembranes [Part J: Membrane Biogenesis: Assembly and Targeting (General Methods, Eukaryotes)]
Edited by SIDNEY FLEISCHER AND BECCA FLEISCHER

VOLUME 97. Biomembranes [Part K: Membrane Biogenesis: Assembly and Targeting (Prokaryotes, Mitochondria, and Chloroplasts)]
Edited by SIDNEY FLEISCHER AND BECCA FLEISCHER

VOLUME 98. Biomembranes [Part L: Membrane Biogenesis (Processing and Recycling)]
Edited by SIDNEY FLEISCHER AND BECCA FLEISCHER

VOLUME 99. Hormone Action (Part F: Protein Kinases)
Edited by JACKIE D. CORBIN AND JOEL G. HARDMAN

VOLUME 100. Recombinant DNA (Part B)
Edited by RAY WU, LAWRENCE GROSSMAN, AND KIVIE MOLDAVE

VOLUME 101. Recombinant DNA (Part C)
Edited by RAY WU, LAWRENCE GROSSMAN, AND KIVIE MOLDAVE

VOLUME 102. Hormone Action (Part G: Calmodulin and Calcium-Binding Proteins)
Edited by ANTHONY R. MEANS AND BERT W. O'MALLEY

VOLUME 103. Hormone Action (Part H: Neuroendocrine Peptides) (in preparation)
Edited by P. MICHAEL CONN

VOLUME 104. Enzyme Purification and Related Techniques (Part C) (in preparation)
Edited by WILLIAM B. JAKOBY

VOLUME 105. Oxygen Radicals in Biological Systems (in preparation)
Edited by LESTER PACKER

Methods in Enzymology

Volume 98

BIOMEMBRANES

Part L

Membrane Biogenesis:

Processing and Recycling

[1] Intracellular Membrane Traffic: Pathways, Carriers, and Sorting Devices*

By MARILYN GIST FARQUHAR

A major development during the past few years has been the gradual realization of the magnitude of intracellular membrane traffic connected with such diverse cell processes as protein secretion, biogenesis of membrane constituents, production of lysosomal enzymes, endocytosis, transepithelial transport of immunoglobulins, and transendothelial transport of macromolecules. It has become clear that all these processes are mediated by vesicular traffic and that the cells economize on the vesicular containers used in these operations by reutilizing or recycling them rather than synthesizing them anew.

This brief overview will focus on outlining what is known about the pathways available, briefly summarizing what little we know about the carriers used and the sites and mechanisms for sorting and directing such traffic.

The Exocytosis Pathway

A well-studied route of membrane traffic is the exocytosis pathway, which is now known to be the general mode for release of secretory products from exocrine, endocrine (including neuroendocrine), and immunoglobulin-secreting cells[1] (route 1 in Fig. 1). According to many[2-4] it is also the mode of release of neurotransmitters in neurons. In glandular cells, exocytosis is clearly one of the predominant pathways, because secretory granule membranes continually fuse with the plasmalemma and discharge their contents. Besides its well-known role in delivery of secretory products, exocytosis is also believed to be the route used for the delivery of intrinsic membrane components, primarily glycoproteins, to the plasmalemma (routes 2 and 3 in Fig. 1). There is evidence that plasmalemmal glycoproteins follow the same intracellular route as secretory proteins, being glycosylated in the Golgi complex and ferried to the cell surface via vesicles that

* This work was supported by Grant AM 17780 from the National Institute of Arthritis, Diabetes, Digestive and Kidney Diseases.

[1] G. E. Palade, *Science* **189,** 347 (1975).

[2] J. E. Heuser and T. S. Reese, *J. Cell Biol.* **88,** 564 (1981).

[3] J. E. Heuser, *in* "Transport of Macromolecules in Cellular Systems" (S. C. Silverstein, ed.), p. 445. Dahlem Konferenzen, Berlin, 1978.

[4] B. Ceccarelli and W. B. Hurlbut, *J. Cell Biol.* **87,** 297 (1980).

ISBN 0-12-181998-1

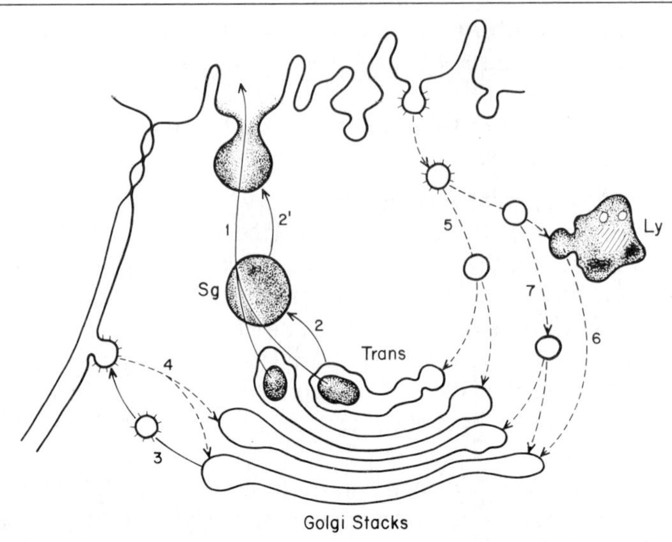

FIG. 1. The exocytosis (1–3) and plasmalemma to Golgi (4–7) pathways. See text for details. From Farquhar.[66]

fuse with the plasmalemma.[5-8] In polarized glandular cells and in other cell types that are known to possess plasmalemmal domains (e.g., apical vs basolateral) of a different composition, there must also be a mechanism for delivery of surface membrane constituents to these different regions of the cell surface. In the case of the apical domain of glandular cells, delivery could be via secretory granule membranes (route 2–2′ in Fig. 1) or via another population of vesicles, and it is not clear which is the case. Evidence has been presented[9] that in a pituitary tumor cell line the secretory product (ACTH) and an endogenous murine leukemia viral membrane glycoprotein (gp 70) are transported in separate membrane containers; however, it remains to be seen whether or not the same applies to the cell's own membrane proteins.

In the case of the basolateral plasmalemmal domain beyond the occluding junctions, which in polarized cells has a different composition than the apical domain, a separate pathway for delivery of membrane constituents seems likely (route 3 in Fig. 1). Studies in Madin-Darby canine kidney (MDCK) cells[10] indicate that viral envelope glycoproteins (presumed to

[5] J. E. Michaels and C. P. Leblond, *J. Microsc. Biol. Cell.* **25,** 243 (1976).

[6] J. E. Rothman, H. Bursztyn-Pettegrew, and R. E. Fine, *J. Cell Biol.* **86,** 162 (1980).

[7] J. E. Bergmann, K. T. Tokuyasu, and S. J. Singer, *Proc. Natl. Acad. Sci. U.S.A.* **78,** 1746 (1981).

[8] J. Green, G. Griffiths, D. Louvard, P. Quinn, and G. Warren, *J. Mol. Biol.* **152,** 663 (1981).

[9] B. Gumbiner and R. B. Kelly, *Cell* **28,** 51 (1982).

[10] M. J. Rinder, I. E. Ivanov, E. J. Rodriguez-Boulan, and D. D. Sabatini, *Membr. Recycling Ciba Found. Symp.* **92,** 184 (1982).

mimic the behavior of native membrane proteins) are selectively inserted into different regions of the cell surface; influenza and Simian virus membrane proteins bud from the apical surface, whereas VSV "G" protein buds from the basolateral domain of the plasmalemma, but the precise routes taken to arrive at specific locations on the plasmalemma are not known. All these glycoproteins traverse the same Golgi apparatus; therefore, the critical sorting steps must take place during or after passage of the glycoproteins through this organelle.

Pathway from the Plasmalemma to the Golgi Complex

Another route for which evidence has been obtained is that from the plasmalemma to the Golgi complex (routes 5–7 in Fig. 1). This appears to be a major pathway in secretory cells producing proteins for export, and indirect evidence (to be summarized) suggests that a large fraction of this traffic is connected with the retrieval and recycling of secretory granule membranes. The existence of such a plasmalemma to Golgi route has been demonstrated in cells from the parotid and lacrimal glands,[11] endocrine cells of the anterior pituitary gland[12] and pancreas,[13] thyroid epithelium,[14] the exocrine pancreas,[15] and immunoglobulin-secreting cells.[16,17] In all these cases it was found that, after exocytosis, surface membrane recovered by endocytosis fuses—directly or indirectly—with multiple stacked Golgi cisternae and condensing granules as well as with lysosomes. The most likely explanation for the bulk of this traffic in secretory cells is that it represents the recovery of granule membranes to be reutilized in the packaging of secretory granules, i.e., that it represents recycling granule membrane. This assumption is supported by the observations that (*a*) this pathway has been detected primarily in cells producing proteins for export, where such traffic could be expected to be heaviest; (*b*) traffic is most abundant in secretory cells actively concentrating and packaging secretory granules and is minimal or absent in secretory cells that are quiescent or not actively involved in packaging; (*c*) the traffic is heaviest to trans Golgi components, which is where packaging of secretory products usually takes place[12,18,19]; and (*d*) recent turnover data (see Farquhar and Palade[19]) indicate that proteins of

[11] V. Herzog and M. G. Farquhar, *Proc. Natl. Acad. Sci. U.S.A.* **74** (1977).
[12] M. G. Farquhar, *J. Cell Biol.* **77,** R35 (1978).
[13] L. Orci, A. Perrelet, and P. Gorden, *Recent Prog. Hormone Res* **34,** 95 (1978).
[14] V. Herzog and F. Miller, *Eur. J. Cell Biol.* **19,** 203 (1979).
[15] V Herzog and H. Reggio, *Eur. J. Cell Biol.* **21,** 11 (1980).
[16] P. D. Ottosen, P. J. Courtoy, and M. G. Farquhar, *J. Exp. Med.* **152,** 1 (1980).
[17] M. G. Farquhar, *Methods Cell Biol.* **23,** 399 (1981).
[18] M. G. Farquhar, *Membr. Recycling Ciba Found. Symp.* **92,** 157 (1982).
[19] M. G. Farquhar and G. E. Palade, *J. Cell Biol.* **91,** 77s (1981).

the membranes of secretory granules turnover at a much slower rate than those of their contents.

It should be noted that the plasmalemma to Golgi traffic has also been detected in other cell types (e.g., neurons[20] and macrophages[21]) in which its coupling to exocytosis (secretion) is not so evident.

The question still remains whether the route taken by recovered membrane to reach the Golgi complex in secretory cells is direct (route 5 in Fig. 1) or indirect, involving a stopover in lysosomes (route 6 in Fig. 1), or in some intermediate compartment (route 7 in Fig. 1). In studies in which electron-dense tracers have been used to investigate this problem, the tracers were usually found in lysosomes as well as in Golgi cisternae.[11,12,14,16] However, the available evidence suggests that both options may be used (in different cell types) because in thyroid epithelial cells tracer (cationized ferritin) appears in lysosomes prior to delivery to the Golgi complex, whereas in myeloma cells, when membrane was traced using a covalent labeling procedure (lactoperoxidase-mediated radioiodination) it was found that the bulk of the membrane traffic to the Golgi bypasses lysosomes.[22] In the latter case, however, a stopover at an as yet unidentified intermediate compartment, where sorting of membrane and contents[16] could take place, cannot be ruled out.

The Endocytic Pathway from the Plasmalemma to Lysosomes

Another familiar route of membrane traffic is that from the plasmalemma to lysosomes (route 1 in Fig. 2). Up until a few years ago, endocytosis was virtually synonymous with incorporation into lysosomes, because this was the only route that had been well characterized,[23,24] the most extensively studied endocytic events being pinocytosis and phagocytosis in macrophages.[24,25] According to the original lysosome concept, it was assumed that whatever entered the cell by endocytosis was degraded within lysosomes. We now know that this is *not* the case, in that the fate of the membrane and the content of the endocytic vesicle differ: the content is delivered to lysosomes and, in the majority of cases studied, degraded down to its constituent molecules,[23] whereas the bulk of the membrane returns to the

[20] N. K. Gonatas, *J. Neuropathol. Exp. Neurol.* **41,** 6 (1982).

[21] J. Thyberg, J. Nilsson, and D. Hellgren, *Eur. J. Cell Biol.* **23,** 85 (1980).

[22] P. Wilson, D. Sharkey, N. Haynes, P. J. Courtoy, and M. G. Farquhar, *J. Cell Biol.* **91,** 417a (1981).

[23] C. de Duve and R. Wattiaux, *Annu. Rev. Physiol.* **28,** 435 (1966).

[24] S. C. Silverstein, R. M. Steinman, and Z. A. Cohn, *Annu, Rev. Biochem.* **446,** 669 (1977).

[25] R. M. Steinman, J. M. Silver, and Z. A. Cohn, *in* "Transport of Macromolecules in Cellular Systems" (S. C. Silverstein, ed.), p. 181. Dahlem Konferenzen, Berlin, 1978.

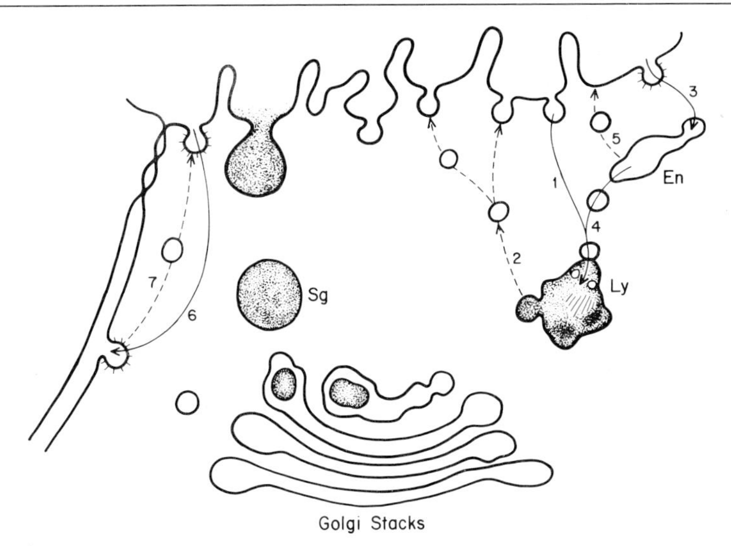

FIG. 2. The lysosomal (1–5) and transcellular (6–7) routes of membrane traffic. See text for details. From Farquhar.[66]

cell surface and is reutilized or recycled (route 2 in Fig. 2).[24–27] Morphometric data [obtained using horseradish peroxidase (HRP) as a marker] have provided the only quantitation available on the magnitude of endocytosis and (by implication) of recycling membrane traffic. A macrophage internalizes the equivalent of its cell membrane every 20 min and a fibroblast does the same every 60 min.[25,27] Such data emphasize the necessity for membrane reutilization or recycling, because it would be quite impractical for the cell to synthesize that amount of membrane anew with each endocytic event.

It has been shown that a number of ligands of great physiological interest, such as low-density lipoproteins (LDL),[28,29] peptide hormones,[30] and transport proteins, e.g., transferrin,[31] enter the cell by receptor-mediated endocytosis[28]; i.e., they bind to specific receptors on the cell membrane and are internalized by concentration in clathrin-coated pits and vesicles[31a] (route 3 in Fig. 2). In the majority of cases studied so far, for which

[26] Y.-J. Schneider, P. Tulkens, C. de Duve, and A. Trouet, *J. Cell Biol.* **82,** 466 (1979).

[27] R. M. Steinman, I. S. Mellman, W. A. Muller, and Z. A. Cohn, *J. Cell Biol.* **96,** 1 (1983).

[28] J. L. Goldstein, R. G. W. Anderson, and M. S. Brown, *Nature (London)* **279,** 679 (1979).

[29] J. L. Goldstein, R. G. W. Anderson, and M. S. Brown, *Membr. Recycling Ciba Found. Symp.* **92,** 77 (1982).

[30] A. C. King and P. Cuatrecasas, *N. Engl. J. Med.* **305,** 77 (1981).

[31] E. Regoeczi, P. A. Chindemi, M. T. DeBanne, and P. A. Charlwood, *Proc. Natl. Acad. Sci. U.S.A.* **79,** 2226 (1982).

[31a] The coat has been shown to be composed of several proteins (180, 100, 50, 36, and 33 kilodaltons) of which the predominant one is the 180 kilodalton protein clathrin.[31b]

the prototype is LDL, the internalized ligand has been shown to be delivered to lysosomes (route 4 in Fig. 2), but delivery to other destinations (e.g., the Golgi complex[32] or GERL[33]) has also been documented.

According to the classical lysosome concept[23] the endosome, i.e., the pinosome or phagosome, fuses directly with a primary or secondary lysosome (route 1 in Fig. 2). Evidence has been obtained on several systems (fibroblasts in culture[34] and hepatocytes[35]) indicating that the incoming vesicles fuse with an intermediate or prelysosomal compartment prior to delivery of the internalized ligand (α_2-macroglobulin and asialoglycoproteins, respectively) to lysosomes (route 3 in Fig. 2). There is now both morphological[34-36] and biochemical (cell fractionation)[36] evidence for the existence of a distinctive intermediate endosome or prelysosomal compartment, which has been called a "receptosome"[34] or a "sorting vesicle."[29]

The existence of such a prelysosomal or sorting compartment represents a useful device for the cell because it provides a means by which surface membrane constituents such as LDL[29] and asialoglycoprotein[37] receptors can recycle directly back to the cell surface and escape lysosomal digestion —the ligand can be delivered to the lysosome (route 4 in Fig. 2), and the receptor can return to the cell surface (route 5 in Fig. 2) or be redistributed intracellularly. This requires that the ligand dissociate from its receptor in transit through the receptosome or sorting vesicle. One mechanism to explain how this takes place has been provided by the finding[38,39] that this intermediate compartment (which lacks lysosomal enzymes) has a low pH (~ 5.0). The low pH would facilitate dissociation of many ligands from their receptor because a number of ligands (e.g., LDL,[29,40] asialoglycoproteins,[35,37] EGF,[41] and insulin[32]) undergo rapid dissociation from their receptors when the pH drops below 5.5. Thus, the intermediate endosome compartment may serve as an "acid wash,"[42] which promotes ligand–re-

[31b] B. M. F. Pearse, *Membr. Recycling Ciba Found. Symp.* **92,** 246 (1982).

[32] B. I. Poser, J. J. M. Bergeron, Z. Josefsberg, M. N. Khan, R. J. Khan, B. A. Patel, A. Sikstrom, and A. K. Verma, *Recent Prog. Hormone Res.* **37,** 539 (1981).

[33] M. C. Willingham and I. H. Pastan, *J. Cell Biol.* **94,** 207 (1982).

[34] I. H. Pastan and M. C. Willingham, *Science* **214,** 504 (1981).

[35] D. A. Wall, G. Wilson, and A. L. Hubbard, *Cell* **21,** 79 (1980).

[36] P. J. Courtoy, J. Quintart, J. N. Limet, C. De Roe, and P. Baudhuin, *J. Cell Biol.* **95,** 425a (1982).

[37] A. L. Schwartz, S. E. Fridovich, and H. F. Lodish, *J. Biol. Chem,* **257,** 4230 (1982).

[38] B. Tycko and F. R. Maxfield, *Cell* **28,** 643 (1982).

[39] F. R. Maxfield, *J. Cell Biol.* **95,** 676 (1982).

[40] R. G. W. Anderson, M. S. Brown, U. Beisiegel, and J. L. Goldstein, *J. Cell Biol.* **93,** 523 (1982).

[41] H. T. Haigler, F. R. Maxfield, M. C. Willingham, and I. Pastan, *J. Biol. Chem.* **255,** 1239 (1980).

[42] G. E. Palade, *Membr. Recycling Ciba Found. Symp.* **92,** 293 (1982).

ceptor dissociation and at the same time cleanses internalized plasma membrane of extraneous molecules taken up from the extracellular environment nonspecifically (e.g., by electrostatic interaction) before it is recycled back to the plasmalemma.[42a]

The Transcellular Route

It is now quite clear that not all membrane internalized by endocytosis is destined to fuse with lysosomes. Although the lysosomal route is the predominant one in cultured cells and in phagocytes, where it is most extensively studied, it is clear that other pathways exist and may even predominate in other cell types (e.g., plasmalemma to Golgi traffic in secretory cells). The *transcellular route,* whereby membrane is incorporated by endocytosis, moves across the cell and fuses with the opposite cell surface by exocytosis (routes 6 and 7 in Fig. 2), represents another well-documented route of membrane traffic. It is the main route of vesicular traffic in endothelial cells, especially in capillaries with a continuous endothelium of the type found in skeletal and cardiac muscle. Using electron-dense tracers as content markers, it has been shown[45] that there is a shuttle mechanism whereby vesicles pick up material on the luminal front of the endothelium and release it on the adventitial front (route 6 in Fig. 2). It is implied that those on the adventitial front similarly serve as shuttles to ferry material back to the lumen (route 7 in Fig. 2). In endothelial cells, lysosomal and Golgi traffic is minimal, and the transcellular route is highly amplified. The cell geometry (flattened cells with attenuated peripheral cytoplasm) facilitates this type of transcellular transport by shortening the transit pathway.

A similar transcellular pathway involving a longer journey is used for immunoglobulin transport across epithelial cells, the best documented example being the route taken by absorbed IgG in the newborn rat intestine[46,47]: maternal IgG present in milk binds to Fc receptors on the brush border membranes at the luminal cell front, is internalized by receptor-mediated endocytosis in coated vesicles, and is transported across the cell to the lateral surface, where it is released into the intercellular spaces at the

[42a] It has recently been shown [Galloway *et al., Proc. Natl Acad. Sci. U. S. A.* **80,** 3334 (1983)] that the membrane of endosomes, like that of lysosomes,[43,44] contains an ATP-driven proton pump believed to be responsible for maintaining the low pH.

[43] S. Ohkuma, Y. Moriyama, and T. Takano, *Proc. Natl. Acad. Sci. U.S.A.* **79,** 2758 (1982).

[44] D. L. Schneider, *J. Biol Chem.* **256,** 3858 (1981).

[45] G. E. Palade, M. Simionescu, and N. Simionescu, *Acta Physiol. Scand. Suppl.* **463,** 11 (1979).

[46] R. Rodewald, *J. Cell Biol.* **85,** 18 (1980).

[47] R. Rodewald and D. R. Abrahamson, *Membr. Recycling Ciba Found. Symp.* **92,** 209 (1982).

vascular cell front (route 6 in Fig. 2). In this case, binding of IgG to its receptor is highest at pH 6.0–6.5,[47] the usual pH of the intestinal luminal fluid, and is less at neutral pH (~7.4), which presumably is the situation along the intercellular spaces at the basolateral cell front. A similar transcellular route apparently operates in the reverse direction for the transepithelial transport of IgA[48,49]: e.g., in the hepatocyte[48] polymeric IgA binds to its receptor protein ("secretory component") at the vascular front and is transported across the cell to the bile capillary front where IgA is discharged intact (along with a proteolytic fragment of the secretory component) into the bile (route 7 in Fig. 2).

In the newborn rat intestine there is evidence that the pathway taken by IgG is indirect, because at least some of the transporting vesicles make a stopover en route in an intermediate compartment, where sorting of membrane and contents takes place.[47] When HRP (a fluid-phase or content marker), and ferritin-conjugated IgG or cationized ferritin (which bind to the membrane) were given together, they were internalized—often in the same vesicle—but HRP was delivered to lysosomes and the IgG was transported transcellularly and discharged at the vascular cell front. These results suggest that the fluid contents, but not the membrane-bound IgG, are released during the intracellular stopover.

The examples just cited serve to demonstrate that—whether direct or indirect—a pathway exists whereby macromolecules can be transported across cells intact, bypassing lysosomes and thereby avoiding intralysosomal digestion. They also provide another example of how pH modulations are used to enhance binding and dissociation of a ligand from its receptor.

Intracellular Biosynthetic Pathways: Endoplasmic Reticulum (ER) to Golgi and Golgi to Lysosomes

There is considerable intracellular biosynthetic traffic connected with the synthesis of secretory products, lysosomal enzymes, and membrane glycoproteins (Fig. 3). Evidence has been summarized[50] that all three types of products when synthesized contain signal sequences, and accordingly are sequestered in the rough ER (through ribosomes that attach to the rough ER). They are then transported to the Golgi complex via small vesicles (routes 1 and 2 in Fig. 3). Somewhere in the Golgi complex these products are sorted (Fig. 4), and thereafter follow different routes: the secretory products are concentrated into secretory granules or vacuoles discharged by exocytosis (route 1 in Fig. 1), membrane proteins are delivered to the

[48] R. H. Renston, A. L. Jones, W. D. Christiansen, and G. Hradek, *Science* **208,** 1276 (1980).
[49] H. Nagura, P. K. Nakane, and W. R. Brown, *J. Immunol.* **123,** 2359 (1979).
[50] G. E. Palade, this series, Vol. 96.

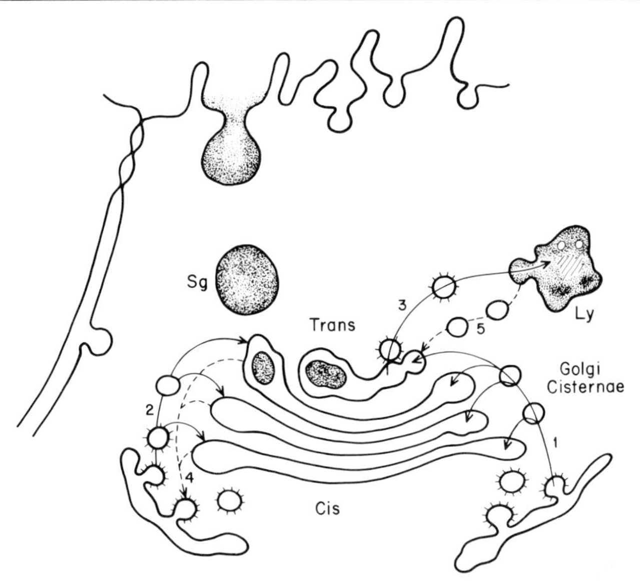

FIG. 3. The biosynthetic pathways for transport between the ER and the Golgi complex (1–3) and the Golgi complex and lysosomes (4, 5). From Farquhar.[66]

appropriate membrane compartments (routes 2 and 3 in Fig. 1), and lysosomal enzymes are delivered to lysosomes (route 3 in Fig. 1). From what we know about the cell's tendency to conserve its membrane containers, it is reasonable to assume that there is recycling of the membranes of the vesicles involved in transport from the ER to the Golgi complex as well as those involved in transport from the Golgi complex to lysosomes (routes 4 and 5 in Fig. 3). However, at present there is no direct evidence on this point because these are difficult membranes to label and therefore difficult problems to study.

Coated Vesicle Carriers

Perusal of the diagrams (Figs. 1–3) indicates that coated vesicles are involved in each of the transport operations depicted. Besides their well-known role in receptor-mediated endocytosis,[28,29] clathrin-coated vesicles[31b] have been implicated in (*a*) transport of secretory products[1] and membrane proteins[6] from the ER to the Golgi complex; (*b*) transport of lysosomal enzymes from the Golgi complex to lysosomes[51–53]; (*c*) transport of mem-

[51] D. S. Friend and M. G. Farquhar, *J. Cell Biol*, **35**, 357 (1967).

[52] E. Holtzman, A. B. Novikoff, and H. Villaverde, *J. Cell Biol.* **33**, 419 (1967).

[53] B. A. Nichols, D. F. Bainton, and M. G. Farquhar, *J. Cell Biol.* **50**, 498 (1971).

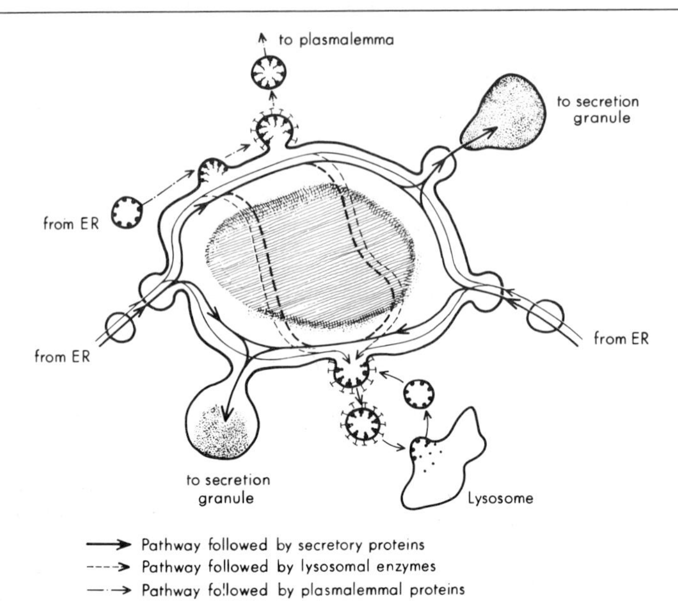

FIG. 4. Diagram of a Golgi cisterna viewed en face showing the presumed routing of the biosynthetic traffic of membranes and secretory products along its dilated rims. Four types of traffic are depicted: ER → Golgi; Golgi → lysosomes; Golgi → condensing granules or vacuoles; and Golgi → plasmalemma. In all cases, transport is assumed to be effected by vesicular carriers that must possess specific receptors for transported species on their inner (cisternal) surfaces. In only one case (Golgi → lysosomes) is the specific recognition marker (mannose 6-phosphate) known. The large dots attached to the membrane represent the receptor, and the small dots respresent the lysosomal enzymes. Most of the traffic is assumed to move along the dilated periphery of the cisterna (solid lines) rather than through its flattened central region (dashed lines, shaded area). From Farquhar and Palade.[19]

brane components from the Golgi to the cell surface[6] (see, however, Wehland *et al.*[54]); (*d*) recovery of membrane from the cell surface[55] and its recycling to the Golgi complex[11,16,17]; and (*e*) transcellular transport of immunoglobulins.[47] The only pathways in which coated vesicles have *not* yet been implicated are in transcellular transport across endothelial cells[45] and in fluid-phase pinocytosis in macrophages.[25]

This means that there must be multiple populations of coated vesicles with different transport functions in any given cell. Accordingly, it becomes of interest to isolate these vesicles and attempt to separate them into subpopulations. The original sucrose gradient centrifugation procedure, introduced by Pearse[56] for coated vesicle isolation, yielded preparations still detectably contaminated by other, mostly smooth, membranes. Improvements in isolation procedures involving alternative gradient materials,[56]

[54] J. Wehland, M. C. Willingham, M. G. Gallo, and I. Pastan, *Cell* **28,** 831 (1982).

[55] W. W. Douglas, *Methods Cell Biol.* **23,** 483 (1981).

[56] B. M. F. Pearse, *J. Mol. Biol.* **126,** 803 (1978).

permeation chromatography,[57] or immunoaffinity absorption (onto *Staphylococcus aureus* cells coated with anti-clathrin antibodies[58]) yield more highly purified coated vesicle fractions. Using one of these procedures in combination with a second step designed to sort coated vesicles (e.g., according to difference in size[57] or exposed membrane constituents[59]), it may be possible in the future to isolate and to characterize subfractions of coated vesicles involved in specific transport functions by selectively extracting the vesicle coat, contents, and membrane components. Identification of the contents will be useful for defining the transport functions of a given vesicle population, and analysis of the coat proteins will yield information as to the extent to which coat proteins are similar or different among the different subpopulations. Analysis of the membrane constituents will be particularly useful because to date virtually nothing is known about them, and yet it is assumed that coated vesicles possess membrane proteins whose ectodomains (facing the vesicle contents) contain specific receptors for the transported species and whose endodomains (facing the cytoplasmic matrix) possess appropriate recognition signals for receiving compartments with which they fuse. A variety of receptors have been identified in coated vesicles in various cell types, based on binding of the appropriate ligands (e.g., LDL,[29] transferrin,[31b] IgG,[31b] α_2-macroglobulin,[34] and peptide hormones[31b,34,59a]) but nothing is known at the present concerning the nature of the signals for recognition of the receiving compartments.

Sites and Mechanisms of Sorting

It is becoming increasingly evident that during these various vesicular transport operations the cell is capable of considerable sorting—i.e., sorting of membrane constituents from fluid contents, sorting of receptors (membrane components) from ligands, as well as sorting of membrane components from one another. The sites identified so far where such sorting can take place include the plasmalemma, the Golgi complex, the endosomes (receptosomes or sorting vesicles), of which the most familiar is the plasmalemma, where selective binding and internalization of both receptor and ligands is well documented. It has also become clear that the Golgi complex is the main intracellular site where membrane and product traffic converges and where sorting takes place. Indeed, some of the main functions of the Golgi complex are to sort secretory proteins, lysosomal enzymes, certain membrane proteins (mainly glycoproteins) during biogenesis (Fig. 4), and recycling membranes in secretory cells and to direct them to their correct

[57] S. R. Pfeffer and R. B. Kelly, *J. Cell Biol.* **91,** 385 (1981).
[58] E. M. Merisko, M. G. Farquhar, and G. E. Palade, *J. Cell Biol.* **92,** 846 (1982).
[59] S. R. Pfeffer and R. B. Kelly, *J. Cell Biol.* **95,** 271a (1982).
[59a] For a complete list see Goldstein *et al.*[29]

intracellular or extracellular destinations. Little is known at present concerning how the sorting of these products is done or where in the Golgi complex these events occur. The available evidence (summarized by Farquhar and Palade[19]) indicates that the Golgi complex consists of an unknown number of subcompartments of different morphology, membrane composition, and function,[19,60] with transport between the various subcompartments presumed to be effected by specific vesicular carriers which have specific receptors for transported species on their inner (cisternal) surfaces and appropriate recognition signals for the receiving compartment on their outer surfaces. In the case of the Golgi, only one type of specific receptor is known—that for the sorting of lysosomal enzymes. There is evidence[61,62] that phosphomannosyl groups (mannose 6-phosphate) constitute at least one of the recognition markers that lead to the receptor-mediated sorting of lysosomal enzymes and their subsequent delivery to lysosomes (Fig. 4) in the same manner as peptide hormone or LDL receptors mediate the uptake of their specific ligand at the plasmalemma.

Apart from the Golgi elements, there appear to be an unknown number of prelysosomal sites involved in the sorting of receptors from ligands (e.g., LDL receptors and LDL[29] or asialoglycoprotein receptors from desialylated glycoproteins[35]) and sorting of different internalized ligands (e.g., IgG from HRP,[47] IgA from desialylated glycoproteins[36]) takes place. The only mechanism known so far that affects the sorting operation is the pH of the receiving compartment; however, as pointed out by Palade,[42] pH cannot be the only mechanism involved since electrostatic interactions lack the necessary specificity. It seems more likely that specificity is accomplished by appropriate interactions among membrane constituents. For example, binding of ligand to receptors may trigger a conformational change in a membrane protein that triggers membrane fission or exposes, in the cytoplasmic domain, a sequence that recognizes the appropriate receiving compartment and promotes fusion.[63,64] In such cases, pH may serve to modulate ligand binding or trigger (as in the case of viral entry[65]) the necessary fusion event.[66]

Summary and Conclusions

Multiple pathways of intracellular membrane traffic have been detected

[60] J. E. Rothman, *Membr. Recycling Ciba Found. Symp.* **92,** 120 (1982).

[61] W. S. Sly, H. S. Fischer, A. Gonzalez-Noriega, J. H. Grubb, and M. Natowicz, *Methods Cell Biol.* **23,** 191 (1981).

[62] S. Kornfeld, M. L. Reitman, A. Varki, D. Goldberg, and C. A. Gabel, *Membr. Recycling Ciba Found. Symp.* **92,** 138 (1982).

[63] G. Blobel, *Proc. Natl. Acad. Sci. U.S.A.* **77,** 1496 (1980).

[64] D. D. Sabatini, G. Keibich, T. Morimoto, and M. Adesnik, *J. Cell Biol.* **92,** 1 (1982).

[65] A. Helenius and M. Marsh, *Membr. Recycling Ciba Found. Symp.* **92,** 59 (1982).

[66] M. G. Farquhar, *Fed. Proc., Fed. Am. Soc. Exp. Biol.* **42,** 2407 (1983).

in various cell types. The major established routes are (*a*) *the exocytosis pathway,* utilized in secretory cells for the discharge of secretory products, and which is also believed to be used for delivery of intrinsic membrane glycoproteins in all cell types; (*b*) *the plasmalemma to Golgi route,* also highly developed in secretory cells, which is believed to be utilized for the recovery and recycling of the membranes of containers used in packaging of secretory products (i.e., secretory granules or vesicles); (*c*) *the lysosomal pathway,* which is available in all cells but is the major route utilized in phagocytic cells; (*d*) *the transcellular route,* which represents the major type of traffic encountered in nonfenestrated, capillary endothelial cells and also appears to be the preferred route for the transport of immunoglobulins (intact) across cells; and (*e*) *the biosynthetic pathways* used for transport of secretory products, lysosomal enzymes, and membrane proteins from the ER to the Golgi complex and for transport of lysosomal enzymes from the Golgi complex to lysosomes in all cell types.

It has become clear that cells repeatedly reutilize or recycle the vesicular membranes involved in carrying out these various transport operations. Clathrin-coated vesicles have been found to be involved in transport along all the routes detected so far, suggesting that there are multiple populations of coated vesicles with different transport functions in every cell.

It has become clear that considerable sorting of membrane constituents and ligands takes place at the plasmalemma (receptor-mediated uptake), in the Golgi complex, and in endosomes. The Golgi complex is the intracellular site where much of the biosynthetic and recycling membrane traffic converges and where products are sorted and directed to their correct destinations.

In summary, we have become aware of the existence of multiple pathways of membrane traffic and of the extensive reutilization or recycling of membranes that occurs in cells. The basic pathways are similar in all cells except that some are emphasized or deemphasized according to the predominant function and organization of a given cell type. What now remains to be done is to determine how these transporting membranes and the membranes of the receiving compartments are constructed, how their specific interactions are controlled, and how individual cell types utilize these pathways to carry out their specific functions. The resolution of these problems will require a multifaceted approach involving use of electron-dense tracers, biosynthetic and covalent labels to follow membrane traffic, isolation and characterization of the vesicular carriers and receiving compartments, immunocytochemical localization of membrane constituents and ligands in specific cell structures, and purification and analysis of the membrane constituents involved as receptors or mediators of specific transport functions.

Section I

Specialized Methods

[2] Pancreatic Lobules in the *in Vitro* Study of Pancreatic Acinar Cell Function

By GEORGE SCHEELE

In the past 15 years, *in vitro* preparations that have been developed to study intact pancreatic acinar cells have included pancreatic slices,[1] pieces,[2] lobules,[3] dissociated single cells,[4] and dispersed acini.[5] Compared to pancreatic lobules, slices and pieces show considerably increased levels of mechanical damage to cells. Compared to dissociated acini, dissociated single cells show increased levels of damage due to the proteases, phospholipases, chelators, and mechanical shearing forces required for further tissue dissociation. When properly prepared, pancreatic lobules and dissociated acini show similar *in vitro* responses to varying levels of exogenously added secretagogues. The use of pancreatic lobules has two advantages over the use of dispersed acini: (*a*) ease of preparation; and (*b*) omission of degradative enzymes and chelators during tissue preparation. The use of dispersed acini, however, shows two advantages over the use of lobules: (*a*) increased access of labeled probes to the plasma membrane, and (*b*) apparent removal of nerve endings from isolated acini.

Methods for the preparation and study of pancreatic lobules are presented in this chapter.

Preparation of Pancreatic Lobules

Pancreatic tissue excised from a small animal (e.g., mouse, rat, guinea pig, or rabbit) is placed in a petri dish containing ice-cold Krebs–Ringer bicarbonate (KRB) buffer (125 m*M* NaCl, 5 m*M* KCl, 2.53 m*M* $CaCl_2$, 1.16 m*M* $MgSO_4$, 25 m*M* $NaHCO_3$). A 10-ml syringe with a No. 26 needle, 1 in. long, is used to inject KRB buffer directly into the gland at random sites. Injected buffer dissects along the connective tissue planes and separates macroscopic lobules (Fig. 1a), which are then individually removed by simple excision using a pair of small curved ophthalmic scissors. This procedure minimizes damage to acinar cells, since most of the surgical trauma is limited to ducts and vessels. The excised lobules preserve the overall acinar architecture of the tissue (Fig. 1b), and their small size

[1] J. D. Jamieson and G. E. Palade, *J. Cell Biol.* **34,** 57 (1967).

[2] L. Benz, B. Eckstein, E. K. Mathews, and J. A. Williams, *Br. J. Pharmacol.* **46,** 66 (1972).

[3] G. A. Scheele and G. E. Palade, *J. Biol. Chem.* **250,** 2660 (1975).

[4] A. Amsterdam and J. D. Jamieson, *J. Cell Biol.* **63,** 1057 (1974).

[5] G. S. Schultz, M. P. Sarras, G. R. Gunther, B. E. Hull, H. A. Alicea, F. S. Gorelick, and J. D. Jamieson, *Exp. Cell Res.* **130,** 49 (1980).

ISBN 0-12-181998-1

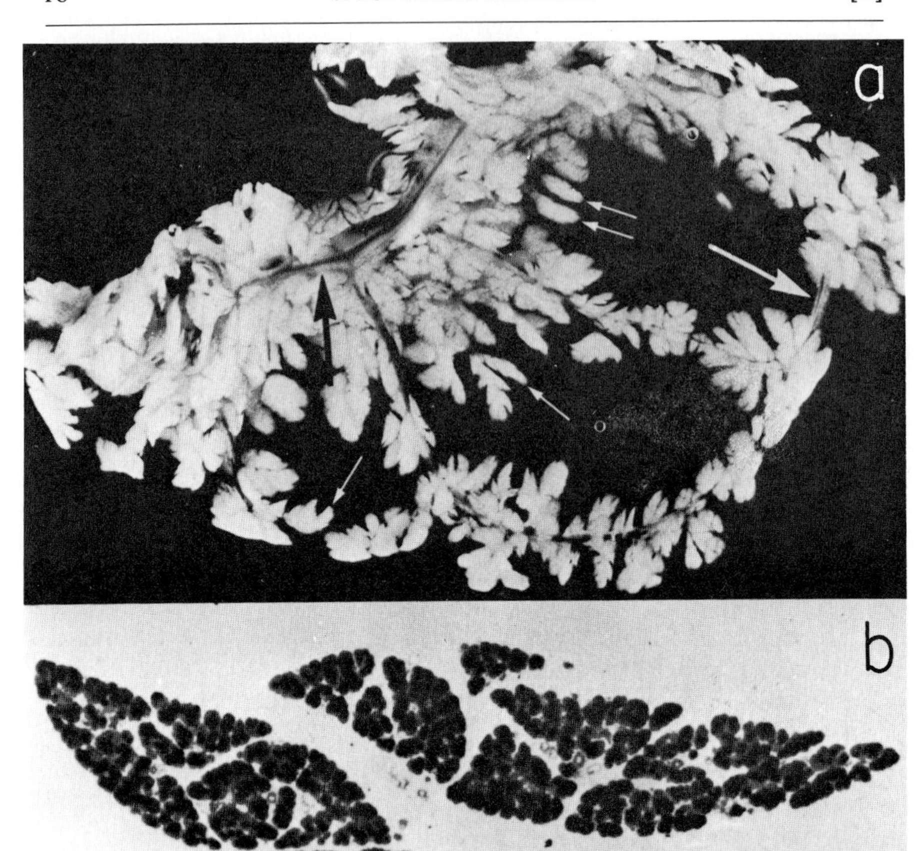

FIG. 1. (a) Guinea pig pancreas in which "macroscopic" lobules have been spread out by injecting incubation medium (Krebs–Ringer bicarbonate) into the gland. The actual length of the distended pancreas was ~6 cm. Long arrows mark large interlobular ducts. Short arrows indicate well-separated lobules that are removed by excision with fine scissors for use in analytical *in vitro* incubation studies. For preparative studies larger segments of tissue, containing four to eight macroscopic lobules, can be taken. (b) Low-power light micrograph of two lobules used in the analytical *in vitro* incubation system. Black dots represent groups of zymogen granules located near central lumens in pancreatic acini. Note the structural integrity of the lobules. Transverse section, ×90. Taken from Scheele and Palade.[3]

($\leqslant 2.0 \times 1.0 \times 0.5$ mm) allows for easy penetration of oxygen and solutes from the incubation medium. Individual lobules from the guinea pig pancreas are approximately 6 mg wet weight, contain ~16.6 μg of DNA, 0.5 mg of protein, and $\sim 1.5 \times 10^6$ cells. Lobules prepared from pancreatic tissue taken from cats, dogs, and humans cannot be used for *in vitro* studies because they exceed the size limits required for efficient exchange of gas and solutes. In these larger glands one is restricted to the use of pancreatic tissue slices (or dissociated acini).

Lobules are incubated in either (*a*) KRB medium gassed with 95% O_2 – 5% CO_2 and supplemented with 1 mg of glucose and 100 μg of bovine serum albumin per milliliter and with essential and nonessential amino acids at 0.2 m*M* each; or (*b*) Krebs–Ringer HEPES medium (phosphate and bicarbonate replaced by 25m*M* HEPES, pH 7.4) gassed with 100% O_2 and supplemented as above with glucose, albumin, and amino acids. The HEPES-buffered medium has two advantages: its pH value is less sensitive to temperature changes and gassing conditions; and divalent cations, particularly Ca^{2+}, can be elevated above physiological levels without precipitation of calcium salts. Guinea pig pancreatic lobules are stable in these media without trypsin inhibitors. However, since exocrine proteins from the mouse, rat, and rabbit show a tendency toward autoactivation, pancreatic lobules prepared from these species are incubated in medium supplemented (per milliliter) with 20 μg of soybean trypsin inhibitor (STI; Worthington Corp., Freehold, New Jersey,) and 100 KIU of Trasylol (FBA Pharmaceuticals, New York, New York).

Applications for Study of Pancreatic Lobules

Preparative Isolation of Pancreatic Exocrine Proteins

Purified preparations of secretory proteins can be extracted from a crude fraction of zymogen granules or obtained physiologically after carbamylcholine-induced secretion of proteins into the incubation medium.

Zymogen Granule Lysate. Lobules prepared from one pancreatic gland (~1 g of tissue wet weight) are homogenized in 10 ml of 0.3 *M* sucrose containing 20 μg of soybean trypsin inhibitor and 100 KIU of Trasylol per milliliter, 1 m*M* DFP (Sigma Chemical Co., St. Louis, Missouri) and 1 m*M* benzamidine (Aldrich Chemical Co., Milwaukee, Wisconsin). Nuclei and cell debris are removed by sedimentation for 10 min at 3° and 600 *g*. A crude zymogen granule pellet is obtained from the postnuclear supernatant by sedimentation for 10 min at 3° and 1000 *g*.[6] The granule pellet is lysed either in a 1.0-ml solution containing 1% Triton X-100, 25 m*M* Tris–HC1 (pH 9.0), 20 μg of STI and 100 KIU of Trasylol per milliliter, 1 m*M* DFP, and 1 m*M* benzamidine or in a 1.0-ml solution containing 0.2 *M* $NaHCO_3$ (pH 8.5) and the protease inhibitors indicated above. The majority of membranes are removed by sedimentation in a Brinkmann microcentrifuge for 15 min at 3° and 8000 *g*.

Secreted Proteins. Pancreatic lobules from one gland (~1 g of tissue) are

[6] A. M. Tartakoff and J. D. Jamieson, this series, Vol. 31, p. 57.

incubated under physiological conditions for 3 hr in the presence of 10^{-5} *M* carbamylcholine (Sigma). After this incubation period, the medium is decanted and supplemented with 2 μg of soybean trypsin inhibitor and 10 KIU of Trasylol per milliliter, 1 m*M* diisopropyl fluorophosphate, and 1 m*M* benzamidine. Contaminating cells and loose debris are removed by sedimentation for 30 min at 105,000 *g* and 3°.

Secretory proteins obtained either from the incubation medium (1–2 mg/ml) or a zymogen granule lysate (5–10 mg/ml) are aliquoted in samples of 50–100 μl and stored at −80° after rapid freezing in liquid N_2. Samples are thawed and used only once, since repeated freezing and thawing can result in autoactivation and degradation of samples.

Preparative Isolation of Endogenously Labeled Radioactive Proteins

Radioactive proteins can be endogenously labeled by incubation of pancreatic lobules under physiological conditions for 3 hr in the presence of radioactive amino acids and the remaining complement of unlabeled amino acids at 0.2 m*M* each. Highest levels of specific radioactivity can be obtained using [^{35}S]methionine or [^{35}S]cysteine. Approximately 95% of radioactivity incorporated into pancreatic lobules appears within secretory proteins.[7] Labeled secretory proteins can be isolated from the incubation medium or a zymogen granule extract as described above.

Analysis of mRNA-Directed Protein Synthesis

The high resolution achieved in the separation of exocrine pancreatic proteins by two-demensional isoelectric focusing/sodium dodecyl sulfate (IEF/SDS)-gel electrophoresis allows the precise measurement of biosynthetic rates of individual secretory products. We have made such measurements in the following manner. Guinea pig pancreatic lobules are incubated with [^{35}S]methionine for 15 min, and the incorporation of radioactive methionine is terminated by rapid freezing in liquid nitrogen. The tissue is then homogenized in 1% Triton X-100 and 25 m*M* Tris-HC1 (pH 9.0), and proteins contained in tissue homogenates are separated by two-dimensional IEF/SDS-gel electrophoresis. Coomassie blue-stained spots characterized as secretory proteins and unstained regions of the gel containing short-lived precursor forms (identified previously using fluorography) are removed from the second-dimension gel and dissolved in hydrogen peroxide, and solubilized radioactivity is quantitated by liquid scintillation spectrometry. These methods have allowed us to measure changes in bio-

[7] G. A. Scheele, G. E. Palade, and A. M. Tartakoff, *J. Cell Biol.* **78,** 110 (1978).

synthetic rates of individual exocrine proteins in the presence of hormonal stimulation.[8]

Analysis of the Pathway for Intracellular Transport of Secretory Proteins

In association with cell fractionation techniques and two-dimensional gel separation of proteins, pulse-chase studies carried out on pancreatic lobules allow a thorough analysis of the movements of individual secretory proteins through intracellular compartments and their eventual discharge into the extracellular medium. Previous studies utilizing [^{3}H]leucine-labeled pulse-chase studies and cell fractionation techniques were unable, with certainty, to determine the route of intracellular transport for these proteins, largely owing to redistribution artifacts associated with tissue homogenization. Scheele *et al.*[7] used guinea pig pancreatic lobules pulse-labeled with ^{14}C amino acids and chased for varying periods of time with ^{12}C amino acids, and they performed tissue homogenization in the presence of tracer amounts of ^{3}H-labeled pancreatic proteins. Analysis of ^{3}H/^{14}C ratios in two-dimensional gel spots derived from individual cell fractions provided information on the extent of leakage of secretory proteins from individual membrane-enclosed compartments along the secretory pathway and the extent of redistribution of leaked molecules to membrane surfaces by nonspecific adsorption. These studies indicated that the appearance of exportable proteins in the postmicrosomal supernatant fraction could be accounted for by the uniform leakage of proteins during tissue homogenization and the preferential binding of positively charged molecules to negatively charged surfaces of membrane-bound organelles. Figure 2 shows the extent to which individual radioactively labeled exocrine proteins, when introduced into the homogenizing medium, will adsorb to the surfaces of subcellular organelles. The data indicate, for the majority of enzymes and zymogens, that adsorption to membrane surfaces is directly dependent on the isoelectric point of the protein. The majority of proteins adsorbed to individual subcellular fractions could be removed with a 100 m*M* KCl wash. Figure 3 shows in schematic form the extent of leakage and adsorption artifacts that occur when pulse-labeled proteins are located in the rough endoplasmic reticulum (RER) (Fig. 3b) and in zymogen granules (Fig. 3c). Figure 4 shows the improvement observed in intracellular transport kinetics when the data are corrected for redistribution artifacts.

These studies provided important evidence in support of the hypothesis that secretory proteins remain segregated within a series of interconnectable membrane-bound compartments during their transport through the cell. They also provided the first clear demonstration, by cell fractionation

[8] G. A. Scheele, *Methods Cell Biol.* **23,** 345 (1981).

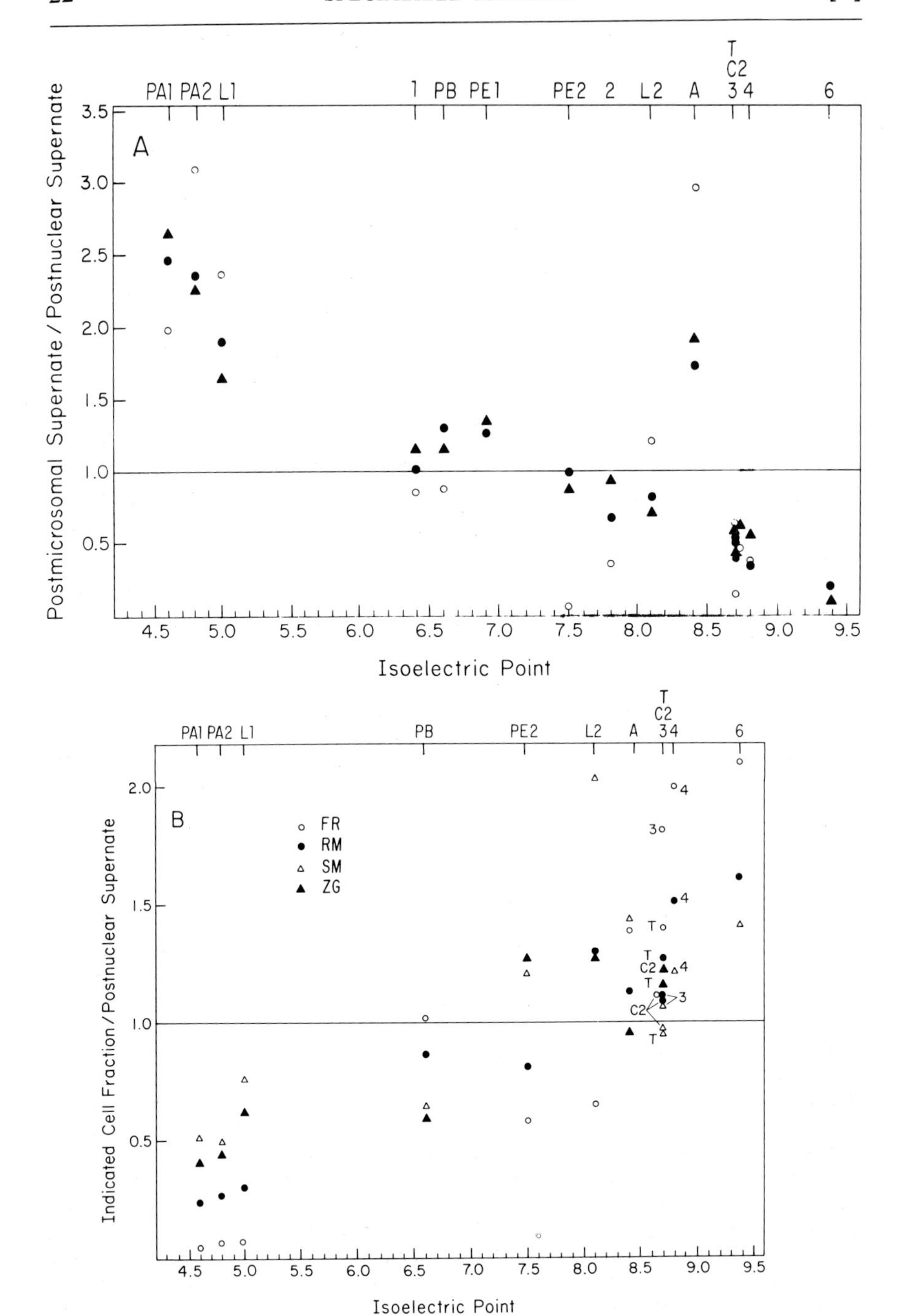

PA1 PA2 L1
1 PB PE1
PE2 2 L2 A
T
C2
3 4
6
A
Postmicrosomal Supernate / Postnuclear Supernate
Isoelectric Point
PA1 PA2 L1
PB
PE2
L2 A
6
B
FR
RM
SM
ZG
Indicated Cell Fraction / Postnuclear Supernate
Isoelectric Point

techniques, that exocrine pancreatic proteins are transported successively from the RER to the Golgi apparatus and from there to zymogen granules.

Analysis of Posttranslational Modification of Pancreatic Gene Products

Since there exist multiple forms of enzymes and zymogens among pancreatic proteins, it is necessary to determine which forms represent separate gene products and which represent modified forms of single gene products. Characterizations of this type are vital to the quantitation of mRNA-directed protein synthesis described above (see Analysis of mRNA-Directed Protein Synthesis). Pancreatic lobules are given a 10-min pulse with [^{35}S]methionine followed by a varying chase period with [^{32}S]methionine. Chase intervals are terminated by rapid freezing in liquid nitrogen, lobules are homogenized in 1% Triton X-100 and 25 m*M* Tris-HCl, pH 9.0, and radioactive proteins contained in tissue homogenates are analyzed by fluorography after their separation by two-dimensional IEF/SDS–gel electrophoresis. Radioactive proteins that change positions in the two-dimensional pattern of protein spots after varying intervals of chase represent posttranslational modifications. Among guinea pig pancreatic proteins isolated from pancreatic lobules, both short-lived and long-lived intermediate forms have been identified.[8]

Secretagogue-Induced Discharge of Exocrine Proteins

Functional discharge of exocrine proteins is studied by incubation of 5–10 pancreatic lobules in 5 ml of KRB solution, supplemented as de-

FIG. 2. Differential adsorption of exogenous ^{3}H-labeled exocrine pancreatic proteins on subcellular fractions. Radioactive proteins associated with individual fractions from a typical cell fractionation experiment were separated by two-dimensional isoelectric focusing/SDS–gel electrophoresis, and the distribution of radioactivity among protein spots was expressed relative to the distribution of radioactivity found in the postnuclear supernate. The lower abscissa indicates the isoelectric point of individual exocrine proteins. The identity of these spots is given in the upper abscissa as follows: PA, procarboxypeptidase A; L, lipase; PB, procarboxypeptidase B; PE, proelastase; A, amylase; T, trypsinogen; C, chymotrypsinogen. Spots 3 and/or 4 are ribonuclease. Spots 1, 2, and 6 are unidentified. Individual enzyme forms are numbered as given by G. A. Scheele [*J. Biol. Chem.* **250,** 5375 (1975)]. (A) Comparison of individual exocrine proteins in the postmicrosomal supernate (PMS) with those in the postnuclear supernate. The symbols are as follows: ●, ^{3}H pancreatic proteins (exogenous label) in pancreatic PMS; ▲, ^{14}C pancreatic proteins (endogenous label, 10 min pulse, 10 min chase) in pancreatic PMS; ○, ^{3}H pancreatic proteins (exogenous label) in liver PMS. (B) Comparison of the distribution of individual proteins in the exogenous tracer in four pelletable fractions with the distribution of those in the postnuclear supernatant fraction. The symbols represent the following cell fractions; ○, free ribosomes; ●, rough microsomes; △, smooth microsomes; ▲, zymogen granules. Taken from Scheele *et al.*[7]

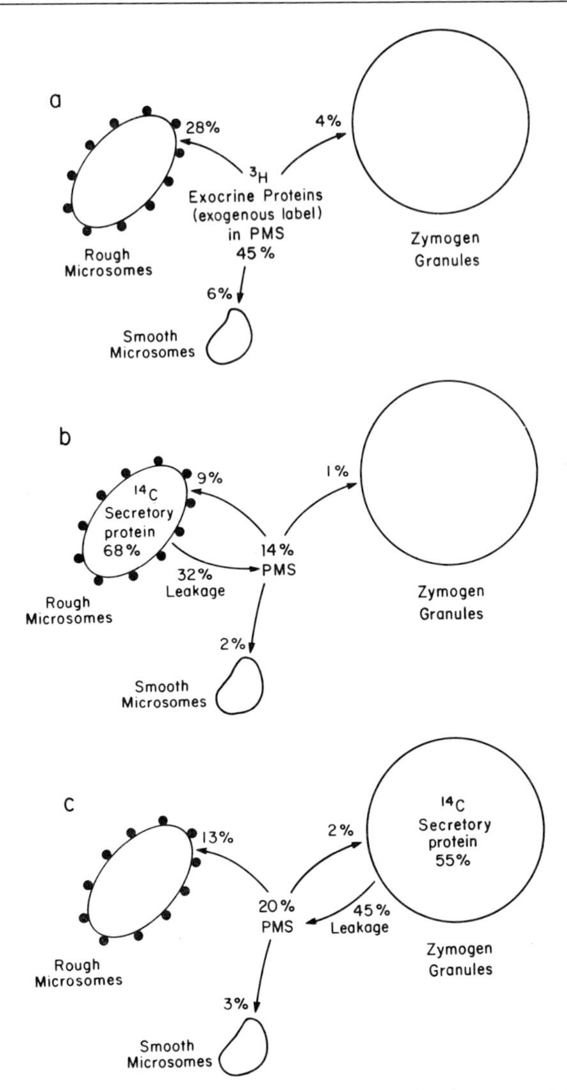

FIG. 3. Diagrammatic representation of leakage and relocation by adsorption of exocrine proteins during homogenization and subcellular fractionation of guinea pig pancreatic lobules. (a) Distribution of exogenously labeled exocrine proteins (added to the homogenizing medium) among each of the subcellular fractions derived from the secretory pathway. (b) Leakage of exocrine proteins from the rough endoplasmic reticulum during homogenization and their relocation by adsorption to subcellular fractions derived from the secretory pathway. (c) Leakage of exocrine proteins from zymogen granules and relocation as described in (b). Each cell fraction is represented diagrammatically by its major component. Data for nuclear and mitochondrial fractions are not included; this explains the apparent underrecoveries. Taken from Scheele *et al.*[7]

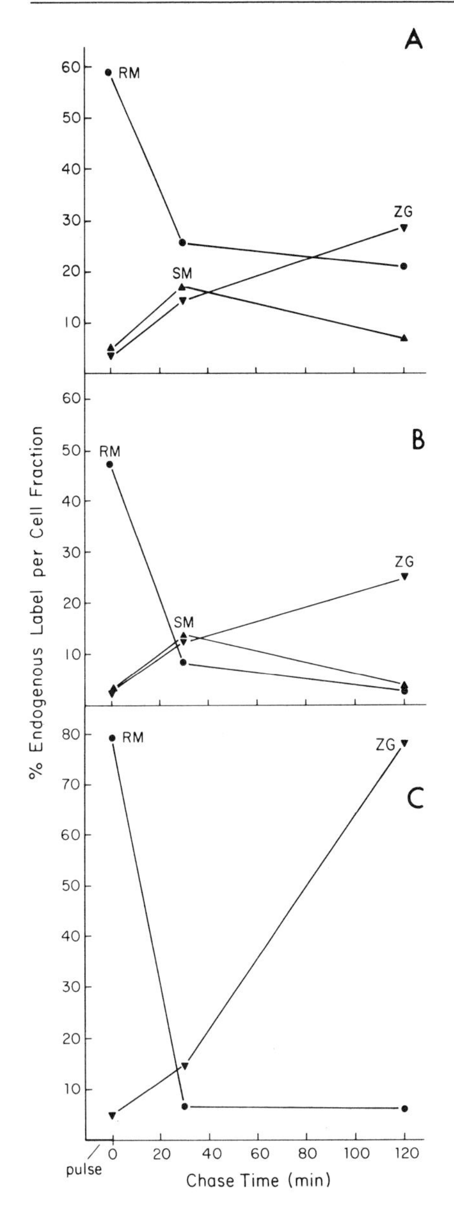

FIG. 4. Intracellular transport kinetics for pulse-labeled proteins in the exocrine pancreas presented (A) as raw data (uncorrected) (B) after correction for adsorption artifact, and (C) after correction for both adsorption and leakage artifacts. A curve for the smooth microsomal fraction (SM) cannot be given in (C) because data on leakage from this fraction are not available at present. RM, rough microsomal fraction; ZG, zymogen granules. Taken from Scheele *et al.*[7]

scribed above with glucose, amino acids, and albumin, and in the presence or the absence of known secretagogues and experimental agents. The addition of a circular piece of nylon mesh [Nitex, 118 μm (Tetko Co., Elmsford, New York)] cut 1.5–2.0 cm in diameter allows rapid transfer of lobules from one flask to another, since lobules adhere to the mesh. Discharge of exocrine proteins can be quantified by two methods:

Method a. Radioactive Protein Assay. Secretory proteins are labeled by the following pulse–chase protocol. Pancreatic lobules are pulsed with 0.44 μM L-[4, 5-^{3}H]leucine (20 μCi/ml) for 15 min, during which time approximately 95% of trichloroacetic acid-insoluble radioactivity is incorporated into secretory proteins. Pancreatic lobules are then briefly washed in KRB solution and incubated in a second flask containing KRB solution with physiological quantities of nonradioactive essential and nonessential amino acids for 90 min; during this time the majority of radiolabeled secretory proteins are transported to the zymogen granule pool. The lobules are then transferred to the experimental flask, which contains additional chase medium (5 ml) with the appropriate experimental agents under study. Sequential aliquots, usually 100 μl, are withdrawn during the incubation period without readdition of fresh medium. After the incubation period, pancreatic lobules containing undischarged protein are homogenized in 5 ml of 1% Triton X-100 and 25 mM Tris-HCl, pH 9.0. Aliquots (100 μl) of medium and tissue homogenate (250–4000 cpm per aliquot) are applied to Whatman 3MM filter disks, immersed in ice-cold 10% trichloroacetic acid and processed for trichloroacetic acid-insoluble radioactivity by the procedure of Mans and Novelli.[9] The data are expressed as cumulative discharge of labeled protein into the incubation medium as percentage of total labeled protein (tissue plus medium) as a function of time. In control studies, data derived from this pulse–chase protocol closely agreed with those derived from assay of amylase.

Method b. Assay for Enzyme and Potential Enzyme Activity. Quantitation of secretory activity can also be achieved by measuring pancreatic enzymes (amylase, lipase, and ribonuclease) or zymogens (trypsinogen, after its activation by enterokinase or chymotrypsinogen; procarboxypeptidase A and procarboxypeptidase B, after their activation by trypsin). Optimal conditions for activation of guinea pig pancreatic zymogens, including time, temperature, and protein (zymogen) to activator ratios, are given by Scheele and Palade.[3] Using these conditions, we obtained linear relationships between the amounts of protein-activated and enzyme activity elicited by activation. Assay conditions for measurement of resulting enzyme activities are also summarized.[3] Triton X-100 is added to tissue samples to ensure the release of enzymes and zymogens from membrane-bound compartments. All samples of homogenate and incubation medium are therefore adjusted

[9] R. J. Mans and G. D. Novelli, *Arch. Biochem. Biophys.* **94**, 48 (1961).

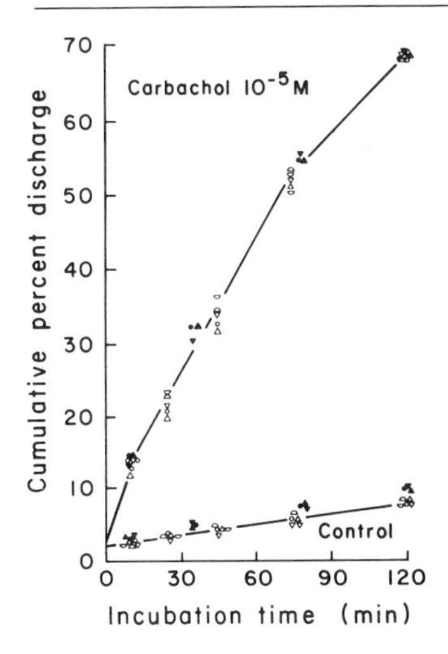

FIG. 5. Kinetics of discharge of seven secretory protein activities from guinea pig pancreatic lobules incubated in the presence and in the absence of 10^{-5} *M* carbamylcholine. Eight lobules were incubated for 2 hr in 5 ml of incubation medium. In a first experiment (open symbols) discharge of amylase activity (○) was compared to discharge of chymotrypsinogen (△), trypsinogen (▽), procarboxypeptidase A (⌓), and procarboxypeptidase B (◡). The values given for the four zymogens represent the corresponding enzyme activities assayed after activation. In a second experiment (filled symbols) discharge of amylase activity (●) was compared to that of lipase (▲) and ribonuclease (▼) activity. The results are expressed as percentage of activities released into the medium at each time point relative to the sum of activities retained in the tissue and discharged into the medium at the end of the incubation period. Note that under these conditions the seven activities were discharged in synchrony and in constant proportions during rest or secretagogue stimulation. Taken from Scheele and Palade.[3]

to 0.1% Triton X-100 (v/v) prior to enzyme assay or activation. Data are expressed as described for discharge of radioactively labeled proteins. Figure 5 shows the kinetics of discharge of three enzymes (amylase, lipase, and ribonuclease) and four zymogens (trypsinogen, chymotrypsinogen, procarboxypeptidase A, and procarboxypeptidase B) from pancreatic lobules incubated *in vitro* for 2 hr in the presence and in the absence of optimal doses of carbamylcholine.

Two-dimensional IEF/SDS–gel electrophoresis can be used also to quantitate exocrine proteins discharged from pancreatic lobules into the incubation medium under conditions of rest and secretagogue stimulation. Proteins separated by the two-dimensional procedure are quantitated after Coomassie Blue staining by two-dimensional spectrophotometric scanning and computer analysis of the scanning data.[8] Scanning was performed on photographic reproductions of stained gels with an Optronics two-dimensional gel scanner with optical density measurements taken at 100-μm intervals. Measurements recorded on magnetic tape were analyzed on a Digital Equipment Corporation PDP-11/70 computer using programs that determined the cumulative densities within and fractional densities among the two-dimensional spots.

By use of the methods described in this section, we have (*a*) measured the effects of secretagogues,[3] ions,[10,11] cyclic nucleotides,[12] and ATP stores[3] on

[10] G. A. Scheele and A. Haymovits, *J. Biol. Chem.* **254,** 10346 (1979).

[11] G. A. Scheele and A. Haymovits, *J. Biol. Chem.* **255,** 4918 (1980).

[12] A. Haymovits and G. A. Scheele, *Proc. Natl. Acad. Sci. U.S.A.* **73,** 156 (1976).

protein discharge; (*b*) defined the role of intracellular and extracellular calcium during stimulus–secretion coupling[10]; (*c*) identified changes in cellular cyclic nucleotide levels during cholinergic and hormonal stimulation[12]; and (*d*) outlined mechanisms for the appearance of both parallel and nonparallel discharge of secretory proteins in the exocrine pancreas.[8,13]

[13] G. A. Scheele, *Am. J. Physiol.* **238,** 6467 (1980).

[3] High-Voltage Techniques for Gaining Access to the Interior of Cells: Application to the Study of Exocytosis and Membrane Turnover

By P. F. BAKER and D. E. KNIGHT

The plasma membrane barrier greatly hinders experimental manipulation of the cytosolic environment in which all intracellular events take place. Various techniques are available to overcome this problem. These can be divided very roughly into two groups: those that permit the introduction of substances into cells, but not their removal, and those that permit more complete control over the intracellular environment. The first group includes both microinjection and fusion with a suitable membrane-bound carrier, and the second group includes internal perfusion and dialysis, partial or complete destruction of the plasma membrane by detergents, and more restricted breakdown of the plasma membrane permeability barrier by treatment with complement or brief exposure either to hypotonic solutions or to intense electric fields. Each has its advantages and disadvantages. The techniques of perfusion and dialysis offer the greatest experimental control but can be applied only to single relatively large cells. For smaller cells most of the available methods involve exposure to foreign chemicals (detergents, complement) with all the uncertainties that attend the use of unphysiological and often highly reactive reagents. Exposure to brief, intense electric fields provides a simple, rapid, and chemically clean method for gaining access to the interior of cells. Under certain conditions high-voltage techniques can be used to render the plasma membrane transiently permeable, whereas under other conditions the plasma membrane remains permeable for prolonged periods. This chapter describes the main features of the high-voltage technique and its application in particular to the problem of membrane turnover by exocytosis and endocytosis.

METHODS IN ENZYMOLOGY, VOL. 98

ISBN 0-12-181998-1

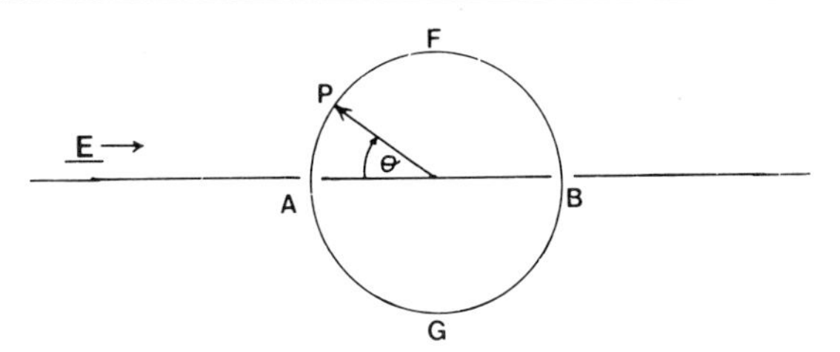

FIG. 1. A description is given in the text.

Effect of an Electric Field on a Cell Suspension

If two flat parallel metal plates are positioned facing each other 1 cm apart and at a potential difference of, say, 1000 V, anything placed in between them is said to experience an electric field of 1000 V/cm. When a cell, which may be considered as a hollow thin-walled sphere, is exposed to such an electric field, a voltage will be imposed across its membrane. The magnitude of this potential depends on the intensity of the field [E (V/cm)] and the size of the cell [radius r (cm)] and will vary around the cell's circumference. With reference to Fig. 1 the voltage at point P (V_p) across the membrane of a spherical cell is given by the simplified equation

$$V_p = C \cdot r \cdot E \cdot \cos\theta \text{ volts}$$

where the value of the constant C depends on the relative conductivities of the extracellular fluid, the cytosol, and the membrane. For most cells the conductivities of the extra- and intracellular fluids are much greater than that of the membrane and the dimensions of the cell are much greater than the membrane thickness. Under these conditions the constant term C approaches a value of 1.5, and the equation becomes

$$V_p = 1.5r \cdot E \cdot \cos\theta \text{ volts} \tag{1}$$

Clearly the potential difference imposed across the membrane will be a maximum at points A and B in Fig. 1 corresponding to $\cos\theta = \pm 1$ and will decrease around the cell to positions F and G where the electric field has no influence on the membrane ($\cos\theta = 0$). If, therefore, a cell 20 μm in diameter is placed in a field of, say, 2000 V/cm, a transmembrane potential of 3 V would occur at opposite ends of the cell in line with the electric field, i.e., at points A and B. It is this large potential that is thought to cause the membrane to break down and become permeable to solutes.

Studies on nerve, muscle, and artificial membranes suggest that a potential difference of 0.3 V is sufficient to bring about membrane breakdown insofar as the membrane exhibits a greatly reduced resistance. However, other studies using cell suspensions, artificial membranes, and plant cells[1-4] indicate strongly that a transmembrane voltage of about 1 V is necessary to cause a marked increase in membrane permeability. It is probable that this rather wide range of voltages reflects different criteria upon which breakdown is assumed to have occurred, but the possibility of different mechanisms cannot be excluded.[5,6] Since changes in permeability to solutes of low molecular mass seem to be associated with membrane voltages of about 1 V, this figure can be used as a guide to calculate from Eq. (1) the field strength required to gain access to the cytosol.

Figure 2 describes the maximum voltage imposed across the membranes of various-sized cells and organelles as a result of exposing them to electric fields. As expected from Eq. (1), the smaller the cell, the larger the field required to charge up the membrane to any given voltage. Figure 2 also includes some particularly instructive data relating to adrenal medullary cells. It compares the influence of the applied field on the membrane potential across the plasma membrane (cell diameter, 20 μm) and on that across the membranes of two groups of intracellular organelles, the chromaffin granules (approximate diameter, 0.2 μm), which contain the secretory product catecholamine, and the mitochondria (approximate diameter, 1 μm). When a cell suspension is exposed to a field of 2000 V/cm, the plasma membrane should experience a transmembrane potential of 3 V, which ought to bring about breakdown, whereas the potential experienced by the membranes of the chromaffin granules and mitochondria should be only 30 and 150 mV, respectively, which is insufficient to bring about breakdown.

The dependence of breakdown in an intense electric field on particle size is an exciting and potentially very useful aspect of the high-voltage technique, as it predicts that the cytosol can be accessed without disturbing the structure and function of the intracellular organelles. This has been demonstrated in such diverse cells as adrenal medullary cells,[3,7] sea urchin eggs,[8] and platelets,[9] where access has been achieved without altering the secretory mechanism or calcium-sequestering ability of the organelles.

[1] A. J. H. Sale and W. A. Hamilton, *Biochim. Biophys. Acta* **163,** 37 (1968).
[2] U. Zimmerman, P. Scheurich, G. Pilwat, and R. Benz, *Angew. Chem.* **20,** 325 (1981).
[3] P. F. Baker and D. E. Knight, *Nature, (London)* **276,** 620 (1978).
[4] D. E. Knight, *in* "Techniques in Cell Physiology" (P. F. Baker, ed.), p. 113. Elsevier/North Holland, Amsterdam, 1981.
[5] J. M. Crowley, *Biophys. J.* **13,** 711 (1975).
[6] R. Benz, F. Beckers, and U. Zimmerman, *J. Membr. Biol.* **48,** 181 (1979).
[7] P. F. Baker and D. E. Knight, *Philos. Trans. R. Soc. London Ser. B* **296,** 83 (1981).
[8] P. F. Baker, D. E. Knight, and M. Whitaker, *Proc. R. Soc, London, Ser. B* **207,** 149 (1980).
[9] D. E. Knight and M. C. Scrutton, *Thromb. Res.* **20,** 437 (1980).

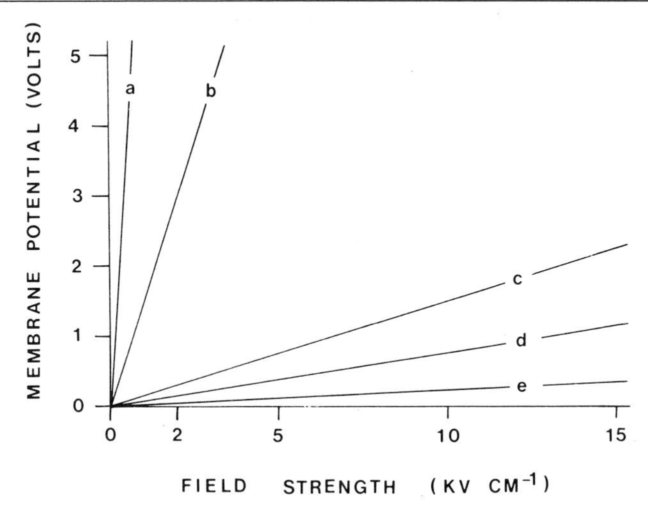

FIG. 2. The expected membrane potential [from Eq. (1)] across (a) sea urchin eggs of 100 μm diameter, (b) adrenal medullary cells of 20 μm diameter, (c) platelets of 2 μm diameter, (d) mitochondria of 1 μm diameter, and (e) chromaffin granules of 0.2 μm diameter as a result of exposure to various field strengths.

Methodology

Two methods have been used routinely to subject isolated cells to electric fields: (a) passing cells through a Coulter counter[2] and (b) discharging a capacitor through a cell suspension. The main disadvantage of using the Coulter-counter method is the length of time taken to harvest a worthwhile number of "leaky" cells, and for this reason we routinely use the second method, which is described below. Our experimental setup is illustrated in Fig. 3.

The capacitor is charged to a known voltage, isolated from the power supply, and discharged through a cell suspension between metal electrodes. The electric field in the cell suspension will decay exponentially with a time constant determined by the product of the capacitance and the resistance of the cell suspension. A 2-μF capacitor discharged across electrodes 1 cm apart and containing 1 ml of cell suspension having a resistance about 100 Ω will produce a field through the suspension decaying with a time constant of 200 μsec. The duration of the field may be increased by increasing either the capacitance or the resistance of the suspending medium. Partial replacement of the ions in the suspending medium by sucrose is an easy way to increase the resistance. When a series of discharges is given, the reversing switch (Fig. 3) should reduce the effects of electrode polarization. In practice, however, for short pulse durations and large voltages the polarization effect is negligible.

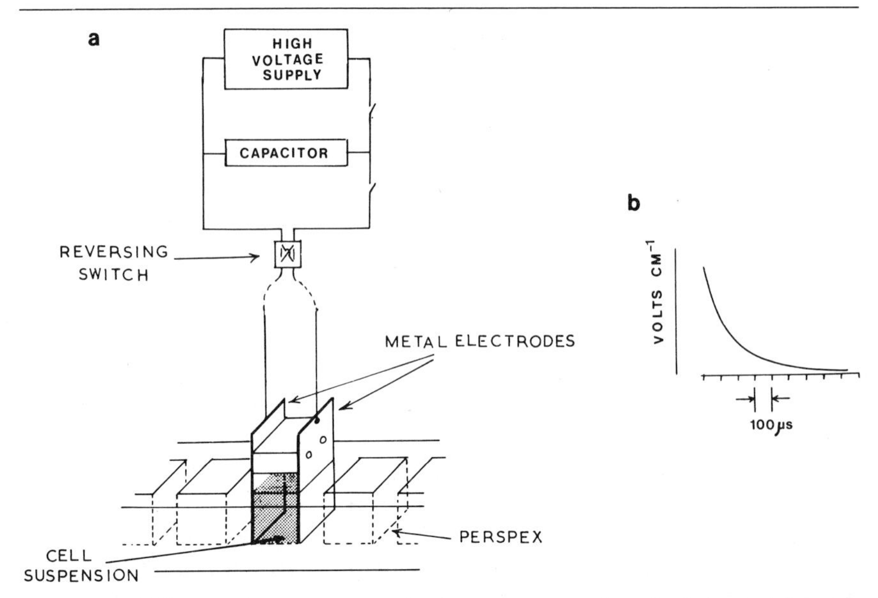

FIG. 3. (a) The experimental setup used to expose a suspension of cells to a brief electric field. (b) The time course of the electric field through the cell suspension made up in a glutamate solution as described in Table I when the applied high voltage is 2 kV and the capacitor 2 μF. The metal electrodes were stainless steel.

The following points must also be borne in mind (for further details, see Knight[4]).

1. The duration of exposure to the field must be long enough for the membrane to charge up. In practice this is a few microseconds at most. To some extent the choice of strength and duration of the applied field will depend on whether the experimental protocol requires the cells to reseal or remain "leaky." In general, resealing becomes less likely as the strength and duration of the field is increased; but resealing is much more common in some cell types than in others, and it is not possible to generalize. Thus red cells, macrophages, and HeLa cells reseal more readily then adrenal medullary cells at 37° (although they remain permeable at 0°).

2. The discharge of the capacitor will cause the cell suspension to heat up. For 2000 V/cm, the heating is approximately 1° per discharge.

3. Increasing the number of discharges does not necessarily increase the number of effective pores in the plasma membrane, because once the plasma membrane has been breached it becomes progressively more difficult to charge up the remaining intact portions of membrane. In practice, the maximum number of pores that can be produced is probably between 6 and 10. The effective diameter of the "pores" can be calculated from the rates of uptake or loss of suitable markers.[4,10] In the specific case of bovine

[10] D. E. Knight and P. F. Baker, *J. Membr. Biol.* **68,** 107 (1982).

adrenal medullary cells exposed to 2000 V/cm and assuming two pores per discharge, the pores have an effective diameter of 4 nm.

Choice of Medium in Which Cells Are Suspended prior to Exposure to Intense Electric Fields

Once the plasma membrane becomes permeable, the cytosolic components of low molecular mass will tend to exchange with those present in the suspension medium. This raises an absolutely fundamental problem. What artificial medium will preserve intracellular function? It seems likely that this may well depend on the cytosolic function under investigation, but we have been surprised at how difficult it is to find a fully satisfactory medium. Table I describes the medium that we have found to be best for preserving calcium-dependent membrane turnover in "leaky" bovine adrenal medullary cells; even in this medium, however, calcium can release only 30% of the total catecholamine, and we suspect that our medium lacks factors essential for the mobilization of the rest of the catecholamine.

The inclusion of MgATP, the maintenance of an ionized Ca less than 10^{-7} *M*, and a low concentration of chloride are all important features of this medium. The same features have proved to be of importance for the preservation of cortical granule exocytosis in sea urchin eggs,[8] serotonin release from platelets,[9] and particle movement in crab axons.[11,12]

Exocytosis and Endocytosis in Bovine Adrenal Medullary Cells

Cells obtained by enzyme digestion of the bovine adrenal medulla are approximately 20 μm in diameter, and cell suspensions (up to 10^7 cells per milliliter) can be rendered permeable by one or more exposures to 2000 V/cm, $\tau = 200$ μsec. Once rendered permeable, adrenal medullary cells are quite robust and can be handled in much the same way as intact cells. As already mentioned, the cells behave as though each discharge produces two "pores" of effective diameter 4 nm; once formed, these pores remain patent for up to 2 hr. Despite the presence of these pores less than 1% of the total cellular catecholamine leaks out of the permeable cells, which shows both that the bulk of the catecholamine is within organelles and that the integrity of these organelles is not affected by exposure to a field of 2000 V/cm. Release of catecholamine can be evoked by raising the ionized Ca into the micromolar range of concentrations, and in these circumstances catecholamine is released along with both dopamine β-monooxygenase (EC 1.14.17.1) and Met-enkephalin—two other components of the chromaffin gran-

[11] R. J. Adams, *Nature* (*London*) **297,** 327 (1982).
[12] R. J. Adams, P. F. Baker, and D. Bray, *J. Physiol.* (*London*) **326,** 7P (1982).

TABLE I
COMPOSITION OF MEDIUM IN WHICH CELLS ARE RENDERED "LEAKY"[a]

Medium	Concentration	Comments
Potassium glutamate	140 mM	K can be replaced by Na, but the nature of the major anion is important
MgATP	5.0 mM	Essential
PIPES, pH 6.6	20.0 mM	pH can be varied from 6.0 to 8.0
EGTA[b]	0.4 mM	Can be increased to at least 50 mM without altering the results
Free Mg^{2+}	2.0 mM	—
Ionized Ca	<100 nM	—
Glucose	5.0 mM	Not essential

[a] This medium has proved to be entirely satisfactory for adrenal medullary cells, but in order to prevent progressive lysis of other cells, it may be necessary to add a high MW substance (e.g., dextran; M_r 10,000) to offset the intracellular colloid osmotic pressure.

[b] Ethylene glycol bis (β-aminoethyl ether)-N,N'-tetraacetic acid.

ule core. The possibility that the chromaffin granules are lysing within the cell and their contents leaking out through the pores in the plasma membrane is rendered extremely unlikely because of the similar time courses of release of all granular contents, despite their very different molecular masses.[7,10] Examination of these leaky cells in the electron microscope reveals a quite normal ultrastructure when the ionized Ca is less than 10^{-7} M; but after exposure to micromolar concentrations of Ca, dense-cored granules are replaced by larger vacuolar structures, which suggests that Ca evokes both exocytosis and membrane retrieval by endocytosis, i.e., a complete cycle of membrane turnover.[7,13]

This suggests that leaky cells may be excellent preparations in which to study the requirements of membrane turnover; but until methods are available for distinguishing the exocytotic and endocytotic phases of this process, we cannot be sure where in the cycle agents that affect catecholamine release are exerting their effects. Table II and Fig. 4 summarize some of the main features of Ca-induced release of catecholamine from leaky bovine adrenal medullary cells (see also Baker and Knight[7,10]). Secretion requires MgATP (apparent K_m, 1 mM), is half-maximally activated by approximately 1 μM Ca, and is inhibited by chloride and similar halides. Secretion is also inhibited by trifluoperazine, by an increase in the osmotic pressure, by detergents, by the sulfhydryl reagent N-ethylmaleimide, and by high concentrations of the uncoupler carbonyl cyanide p-trifluoromethoxyphenylhydrazone (FCCP). It is of particular interest that secretion from leaky cells is blocked by detergents (Fig. 5) at concentrations at the very bottom end of

[13] P. F. Baker, D. E. Knight, and C. S. Roberts, *J. Physiol.* (*London*) **326,** 6P (1981).

TABLE II
PROPERTIES OF CALCIUM-DEPENDENT CATECHOLAMINE RELEASE FROM "LEAKY" ADRENAL MEDULLARY CELLS

1. Activation is half-maximal at an ionized Ca concentration of 1 μM
2. Requirement for MgATP is very specific. Half-maximal activation requires 1 mM
3. Release is unaffected by:
   a. Agonists and antagonists of acetylcholine receptors
   b. Ca channel blocker D600 (100 μM)
   c. Agents that bind to tubulin (colchicine, vinblastine, 100 μM)
   d. Cytochalasin B (1 mM)
   e. Inhibitors of anion permeability (SITS,[a] DIDS,[b] 100 μM)
   f. Protease inhibitor TLCK[c] (1 mM)
   g. Cyclic nucleotides (cyclic AMP, cyclic GMP, 1 mM)
   h. *S*-Adenosylmethionine (5 mM)
   i. Phalloidin (1 mM)
   j. Vanadate (1 mM)
   k. Leu- and Met-enkephalins, substance P (100 μM)
   l. Somatostatin (1 μM)
   m. NH_4Cl (30 mM)
4. Release is inhibited by
   a. Chaotropic anions: SCN > Br > Cl
   b. Detergents (complete inhibition after 10-min incubation with 10 μg/ml of digitonin, Brij 58, or saponin)
   c. Trifluoperazine (complete inhibition with 20 μg/ml)
   d. High Mg concentration: small increase in apparent K_m for Ca accompanies large reduction in V_{max}
   e. High osmotic pressure: large reduction in V_{max} but no significant changes in the affinity for Ca
   f. Carbonyl cyanide *p*-trifluoromethoxyphenylhydrazone (FCCP) (45% inhibition by 10 μM)
   g. *N*-Ethylmaleimide (1 mM)

[a] SITS, 4-acetamide-4′-isothiocyanostilbine-2,2′-disulfonic acid.
[b] 4,4′-Diisothiocyano-2,2′-stilbenedisulfonic acid.
[c] Tosyllysine chloromethyl ketone.

the range used in other systems to gain access to the cell interior.[14-16] This provides further evidence in favor of the high-voltage technique over the use of detergents as a means of gaining access to the cell interior.

The two main disadvantages of the high-voltage technique are that accessibility is largely restricted to small molecules and that even for these molecules the rate of change in intracellular solute composition is dependent on the diameter of the cell and the number and size of the pores that have been created by exposure to the electric field. Diffusional equilibrium will be fastest in very small cells, such as platelets: in adrenal medullary cells it may take a few minutes even for quite small molecules. The problem of

[14] G. L. Becker, G. Fiskum, and A. L. Lehninger, *J. Biol. Chem.* **255,** 9009 (1980).
[15] E. Murphy, K. Coll, T.L. Rich, and J. R. Williamson, *J. Biol. Chem.* **255,** 6600 (1980).
[16] W. Z. Cande, K. McDonald, and R. L. Meeusen, *J. Cell Biol.* **88,** 618 (1981).

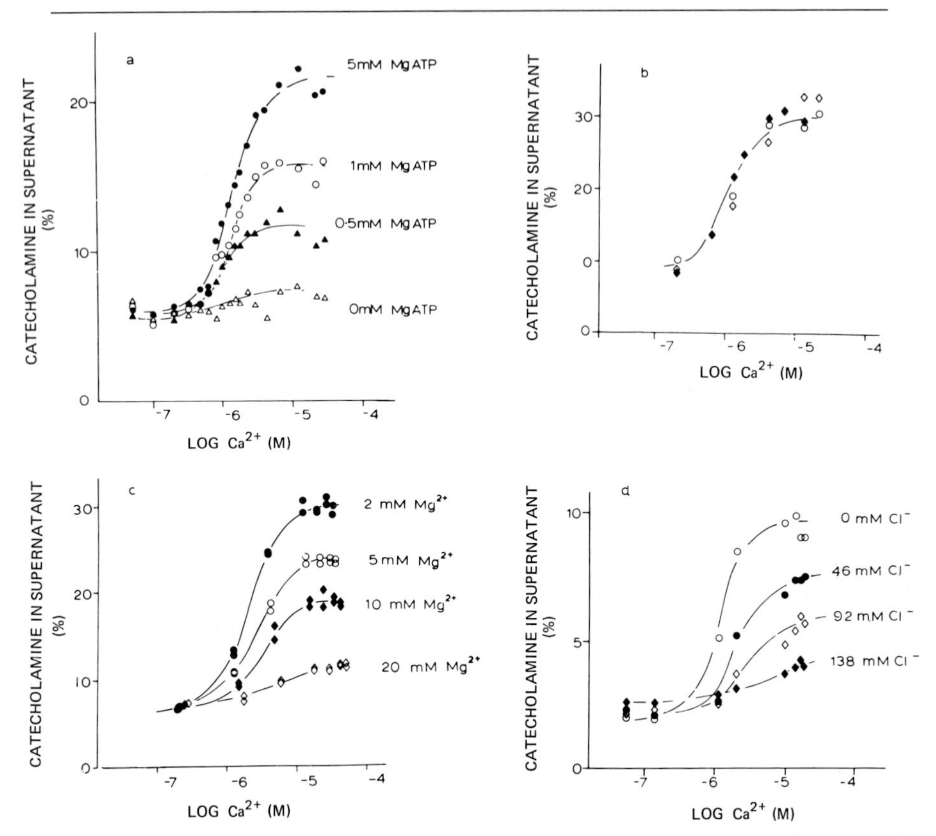

FIG. 4. (a) Dependence of catecholamine release on Mg-ATP. Cells in K glutamate medium (Table I) lacking Mg-ATP were rendered leaky by 10 discharges of 2 kV cm^{-1} ($\tau = 200$ μsec) and immediately diluted into similar solutions containing various concentrations of Mg-ATP. After 10 min the cells were challenged with 10 m*M* Ca EGTA buffers and the catecholamine in the supernatant was determined 15 min later. Temperature 37°. (b) Effect of Na^+,K^+ and low ionic strength on Ca-sensitive release of catecholamine. Isosmotic replacement of K glutamate by Na glutamate or sucrose. Cells in a K glutamate solution (Table I) were rendered leaky by 10 exposures to 2 kV cm^{-1} ($\tau = 200$ μsec) and immediately diluted into either an identical K solution (◇), or into one in which the K had been replaced by Na to give a final concentration of 120 m*M* Na^+ (○), or into one in which the K glutamate had been replaced by sucrose to give a final concentration of 300 m*M* (●); after 3 min the cells were challenged with 10 m*M* Ca EGTA buffer. The catecholamine in the supernatant was determined 15 min later. Temperature 37°. (c) Effect of Mg acetate on Ca-sensitive release of catecholamine. Cells on a K glutamate solution (Table I) were rendered "leaky" by 10 exposures of 2 kV cm^{-1} ($\tau = 200$ μsec) and immediately transferred to similar solutions containing various concentrations of free Mg^{2+} (Mg acetate used). After 5-min incubation the cells were challenged with 10 m*M* Ca EGTA buffer and the catecholamine was determined 10 min later. Temperature 37°. (d) Ca activation curves at different chloride concentrations. Cells in a K glutamate medium (Table I) (ca. 10 n*M* Ca^{2+}) were rendered permeable by 10 exposures to 2 kV cm^{-1} ($\tau = 200$ μsec) and then injected into similar solutions in which the glutamate had been wholly or partly replaced by chloride. After 5 min the cells were challenged with 10 m*M* Ca EGTA buffer and the catecholamine in the supernatant was determined 10 min later. Temperature 37°.

diffusional equilibrium can be overcome to some extent by equilibrating the leaky cells at 0° and then initiating secretion by imposing a sudden temperature jump in which the preparation is warmed rapidly to 37°.

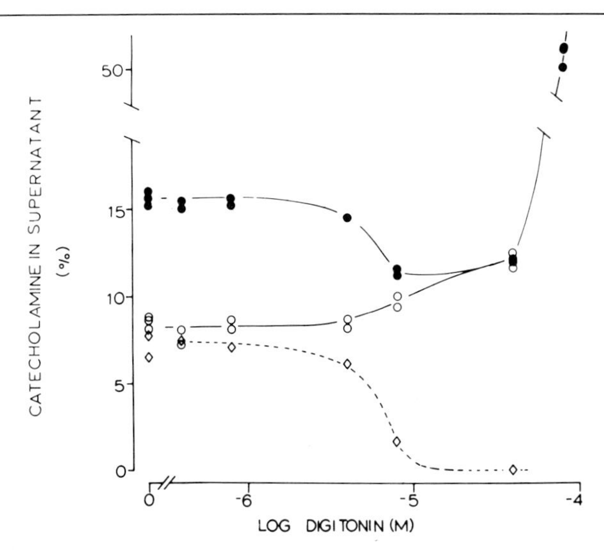

FIG. 5. Effect of digitonin on secretion. Cells (less than 10^4/ml) in K glutamate medium (Table I) were rendered "leaky" by 10 exposures to 2000 V cm^{-1} ($\tau = 200\ \mu sec$) and incubated for 3 min with various concentrations of digitonin before being challenged with 10 μM Ca^{2+} and the catecholamine in the supernatant was determined 15 min later. ●, Cells challenged with 10 μM Ca^{2+}; ○, cells held at 10 nM Ca^{2+}; ◇- - -◇, difference curve, Ca-dependent release. Results similar to these were also obtained with saponin and Brij 58. Temperature 37°.

Acknowledgment

This work was supported by grants from the Medical Research Council of Great Britain.

[4] Transport-Coupled Oligosaccharide Processing in a Cell-Free System

By WILLIAM E. BALCH, ERIK FRIES, WILLIAM H. DUNPHY, LENORE J. URBANI, and JAMES E. ROTHMAN

An understanding of the biochemical basis of the intracellular transport of proteins among organelles of eukaryotic cells will require reconstitution of segments of these pathways in cell-free systems and then fractionation into their component parts. A large body of biochemical evidence (see reviews[1-6]) has established that glycoproteins maturing from their site of synthesis in the rough endoplasmic reticulum (RER) to the various intracellular organelles or cell surface undergo a series of modifications: glycosylations,[3] sulfation,[7] phosphorylation,[8,9] proteolytic cleavage,[10] and addition of fatty acids.[11] The intracellular passage of the membrane glycoprotein (G protein) in vesicular stomatitis virus (VSV)-infected cells provides an experimental system[12] to utilize such sequential alterations in protein structure as

ISBN 0-12-181998-1

landmarks with which to delineate segments of the transport pathway for biochemical studies.

G protein, the major glycoprotein produced in VSV-infected cells, is transported to the cell surface via the Golgi complex.[3, 14–24a] The small genome of VSV,[12] in contrast to the genetic complexity of intracellular transport,[13] ensures that the maturation of G follows pathways provided by the host. G protein inserted cotranslationally[15] into the RER acquires two "high-mannose" asparagine-linked oligosaccharides of the form [-Asn-$(GlcNAc)_2Man_9Glc_3$]. These oligosaccharides are processed during intracellular transport in a series of steps by Golgi-associated mannosidases and glycosyltransferases to yield the mature "complex" [-Asn-$(GlcNAc)_2$ $Man_3(GlcNAc\text{-}Gal\text{-}SA)_2$] forms that are found attached to G protein in plasma membranes and budded virions.[20–22, 24]

These processing steps can be used as the basis for assays that measure transport of G protein between organelle membranes *in vitro*. Membrane fractions containing G protein ("donor" membranes) can be prepared from a VSV-infected mutant cell line defective in a stage of oligosaccharide processing, and incubated with membranes from uninfected wild-type cells

---

[1] M. G. Farquhar and G. E. Palade, *J. Cell Biol.* **91,** 77s (1981).

[2] M. Farquhar, *in* "Transport of Macromolecules in Cellular Systems" (S. Silverstein, ed.), p. 341. Abakon Verlagsgesellschaft, Berlin, 1978.

[3] W. Lennarz, ed. "The Biochemistry of Glycoproteins and Proteoglycans." Plenum, New York, 1980.

[4] C. Leblond and G. Bennett, *in* "International Cell Biology" (B. Brinkley and K. Porter, eds.), p. 267. Rockefeller Univ. Press, New York, 1977.

[5] A. Tartakoff, *Int. Rev. Exp. Pathol.* **22,** 228 (1980).

[6] J. E. Rothman, *Science* **213,** 1212 (1981).

[7] R. Young, *J. Cell Biol.* **57,** 175 (1973).

[8] A. Kaplan, D. T. Achord, and W. S. Sly, *Proc. Natl. Acad. Sci. U.S.A.* **74,** 2026 (1977).

[9] I. Tabas and S. Kornfeld, *J. Biol. Chem.* **255,** 6633 (1980).

[10] B. Gumbiner and R. Kelly, *Proc. Natl. Acad. Sci. U.S.A.* **78,** 318 (1981).

[11] M. Schmidt and M. Schlesinger, *J. Biol. Chem.* **255,** 3334 (1980).

[12] J. Lenard, *Annu. Rev. Biophys. Bioeng.* **7,** 139 (1978).

[13] P. Novick, C. Field, and R. Schekman, *Cell* **21,** 205 (1980).

[14] L. A. Hunt and D. F. Summers, *J. Virol.* **20,** 646 (1976).

[15] F. N. Katz and H. F. Lodish, *J. Cell Biol.* **80,** 416 (1979).

[16] D. M. Knipe, D. Baltimore, and H. F. Lodish, *J. Virol.* **21,** 1128 (1977).

[17] D. M. Knipe, H. F. Lodish, and D. Baltimore, *J. Virol.* **21,** 1121 (1977).

[18] S. Kornfeld, E. Li, and I. Tabas, *J. Biol. Chem.* **253,** 7771 (1978).

[19] C. L. Reading, E. E. Penhoet, and C. E. Ballou, *J. Biol. Chem.* **253,** 5600 (1978).

[20] P. W. Robbins, S. C. Hubbard, S. J. Turco, and D. F. Wirth, *Cell* **12,** 893 (1977).

[21] I. Tabas and S. Kornfeld, *J. Biol. Chem.* **253,** 7779 (1978).

[22] S. Narasimhan, P. Stanley, and H. Schachter, *J. Biol. Chem.* **252,** 3926 (1977).

[23] I. Tabas and S. Kornfeld, *J. Biol. Chem.* **254,** 11655 (1979).

[24] S. Hubbard and R. Ivatt, *Annu. Rev. Biochem.* **50,** 555 (1981).

[24a] J. Bergmann, K. Tokuyasu, and S. J. Singer, *Proc. Natl. Acad. Sci. U.S.A.* **78,** 1746 (1981).

("acceptor" membranes) housing the enzyme missing from the mutant. For example, the incomplete oligosaccharide chains of G protein present in membranes prepared from a VSV-infected mutant CHO cell line, clone 15B (missing a Golgi-associated glycosyltransferase[25]), are in fact completed *in vitro* during incubations with Golgi-containing membranes of the wild-type CHO cell line and other necessary components.[26] The appearance of G protein with the completed, complex oligosaccharide structure provides evidence for the transport of G from the donor membrane to the acceptor, Golgi membrane in this cell-free system. This type of measure of transport-coupled processing provides an assay that may help to define the functional components required for intercompartmental flow in the cell.

Work in our laboratory has established this approach. We have provided evidence for a transport of G protein *in vitro* between distinct compartments of the Golgi apparatus.[26-29] Transport was dependent on the presence of ATP as well as the addition of a cytosol fraction (i.e., supernatant from a high-speed centrifugation of cell homogenate). The assays used for *in vitro* reconstitution of transport of G protein between two compartments of the Golgi are presented in detail below.

Preparation of Components for Assay of Transport-Coupled Oligosaccharide Processing

Materials. [^{35}S]Methionine ([^{35}S]Met; 1000–1200 Ci/mmol) and uridine diphosphate-*N*-acetyl-D-glucosamine [glucosamine-6-^{3}H(N)-] (UDP-[^{3}H]GlcNAc; 24 Ci/mmol) were purchased from New England Nuclear (Boston, Massachusetts). Endo-β-*N*-acetylglucosaminidase H (Endo H), purified according to Tarentino and Maley,[30] was purchased from F. Maley Health Research Institute (Tower Building, 13th floor, Empire Street Plaza, Albany, New York 12237). Creatine phosphokinase (CPK), uridine diphosphate-*N*-acetyl-D-glucosamine (UDP-GlcNAc), and uridine diphosphate-galactose (UDP-Gal) were obtained from Sigma Chemical Co. (St. Louis, Missouri).

Cells and Virus. A wild-type cell line of Chinese hamster ovary (CHO) cells (originally obtained from H. Lodish, Massachusetts Institute of Technology) was maintained in a suspension in α-MEM (Grand Island Biological Co., Grand Island, New York) containing 7.5% fetal calf serum[17] (com-

[25] C. Gottlieb, J. Baenziger, and S. Kornfeld, *J. Biol. Chem.* **250,** 3303 (1975).
[26] E. Fries and J. E. Rothman, *Proc. Natl. Acad. Sci. U.S.A.* **77,** 3870 (1980).
[27] J. E. Rothman and E. Fries, *J. Cell Biol.* **89,** 162 (1981).
[28] E. Fries and J. E. Rothman, *J. Cell Biol.* **90,** 697 (1981).
[29] W. G. Dunphy, E. Fries, L. J. Urbani, and J. E. Rothman, *Proc. Natl. Acad. Sci. U.S.A.* **78,** 7453 (1981).
[30] A. L. Tarentino and F. Maley, *J. Biol. Chem.* **249,** 811 (1974).

plete growth medium). The CHO cell mutant clone 15B (obtained from S. Kornfeld, Washington University, St. Louis, Missouri[25]) was grown in monolayers in the same medium. The stock of vesicular stomatitis virus (VSV) (Indiana strain) was prepared as described[17] using a baby hamster kidney cell line (BHK) (obtained from R. R. Wagner, University of Virginia School of Medicine, Charlottesville, Virginia). The virus titer was typically 2×10^9 plaque-forming units (PFU)/ml. Intact virions for preparation of anti-G protein antiserum were purified from culture medium at 48 hr after infection (multiplicity of infection was 0.1–0.2 PFU/cell) by centrifugation for 90 min at 25,000 rpm (90,000 *g*) in an SW27 rotor (Beckman Instruments, Palo Alto, California). The pellet was resuspended in 50 m*M* Tris (pH 7.5), 250 m*M* NaCl, 0.5 m*M* Na_2EDTA and layered on a 15 to 40% sucrose gradient in the same buffer, and centrifuged for 60 min at 24,000 rpm (80,000 *g*) in an SW27 rotor. The turbid band corresponding to virions (20–25% sucrose) was harvested and diluted in 10 m*M* Tris (pH 7.4), 1 m*M* Na_2EDTA and pelleted for 90 min at 25,000 rpm in an SW27 rotor. The pellet was resuspended in phosphate-buffered saline (see below), frozen in liquid N_2, and stored at −80°.

Preparation of Donor and Acceptor Fractions

Infection of 15B Cells. Twenty plates (10 cm in diameter) of densely confluent cells (2×10^7 cells per plate) were infected with 5–10 PFU of VSV per cell in serum-free growth medium (2 ml per plate) containing actinomycin D at 5 μg/ml and 20 m*M* HEPES–NaOH, pH 7.3. At 1 hr after the start of infection, 8 ml of complete growth medium were added to each plate. At 4 hr, cells were removed from the plates by trypsinization by the following procedure. The medium was aspirated, and each plate was rinsed with 5 ml of phosphate-buffered saline (P_i–NaCl) (per liter: 0.20 g of KCl, 0.20 g of KH_2PO_4, 8.0 g of NaCl, 1.15 g of Na_2PO_4, pH 7.4) and then rinsed quickly with 3 ml of P_i–NaCl containing 0.05% trypsin and 0.02% Na_2EDTA. After 5–10 min at room temperature, cells in each plate used to prepare unlabeled extracts were suspended by addition of 2 ml of ice-cold complete medium and pelleted once (600 *g* for 5 min at 4°). Cells used to prepare [^{35}S]Met-labeled extracts were suspended with 2 ml ice-cold P_i–NaCl and pelleted. The pellet was then suspended for labeling in 40 ml of a modified methionine-free Joklik's minimal essential medium (JMEM) (Grand Island Biological Co.) containing 6.8 g of NaCl, 0.4 g of KCl, 0.2 g of $MgCl_2 \cdot 6H_2O$, 1.5 g of $Na_2HPO_4 \cdot H_2O$, 0.5 g of $NaHCO_3$, 4.76 g of HEPES, 20 ml of 50 × MEM essential amino acids minus methionine (prepared according to the recipe provided in the Grand Island Biological Co. catalog except that methionine was omitted), 10 ml of 100 × MEM vitamins (Grand Island Biological Co.),

2 ml of 0.5% phenol red, H_2O to 500 ml, and titrated to a final pH of 7.3 with NaOH. The final concentration of cells was about 10^7 per milliliter.

Labeling of Cells. Cells suspended for labeling (normally 10 ml per condition) were equilibrated for 5 to 10 min at 37° prior to addition of 0.5–1.0 mCi of [^{35}S]Met per 10 ml of cell suspension. Cell suspensions were continuously and gently stirred during all incubations. Chase conditions (where applicable, see Fries and Rothman[28]) were obtained by the addition of unlabeled methionine to give a final concentration of 2.5 m*M*. Chase was terminated by mixing 10 ml of the cell suspension with 30 ml of ice-cold P_i–NaCl containing 0.01% $CaCl_2$ and 0.006% $MgSO_4$. The cells were pelleted and then washed once in ice-cold P_i–NaCl.

Lysis of Cells and Preparation of Postnuclear Supernatant (PNS). It is important that all subsequent manipulations be at 0–4°, as fractions rapidly lose activity at elevated temperatures. Unlabeled or labeled cells were washed once with P_i–NaCl and suspended in 10 ml of a "swelling buffer" containing 15 m*M* KCl, 1.5 m*M* magnesium acetate, 1 m*M* dithiothreitol, 10 m*M* HEPES–KOH at pH 7.3. After 5 min on ice to permit swelling followed by centrifugation (600 *g* for 3 min), the supernatant was removed until a volume three times that of the swollen cell pellet remained. Then, a homogenate was made with 20 strokes of a 7-ml tight-fitting Dounce homogenizer (Wheaton, Millville, New Jersey). One-tenth volume of a 10 × concentrated "extract buffer" [consisting of 300 m*M* HEPES (pH 7.5), 600 m*M* KCl, 30 m*M* magnesium acetate] was added to yield the following final concentrations: 45 m*M* HEPES–KOH at pH 7.3, 75 m*M* KCl, 4.5 m*M* magnesium acetate, and 1 m*M* dithiothreitol. The homogenate was centrifuged at 600 *g* for 5 min. The resulting postnuclear supernatant (PNS) was removed and used at once or frozen in liquid N_2 and stored at −80° (with essentially identical results). Frozen PNS was stable for several months. Prior to assay, frozen extracts should be thawed rapidly at 37° and maintained on ice. Typically, twenty 10-cm plates yielded 3 ml of PNS (protein concentration, 7–10 mg/ml). For PNS fractions of uninfected 15B and of wild-type CHO cells, 10^8 cells were swollen, lysed, and processed as described above.

Preparation of Anti-G Protein Antiserum. Antibody directed against VSV G protein was made according to the procedure of Miller *et al.*[31] Vesicular stomatitis virions (10 mg of protein) were suspended in 3 ml of 10 m*M* potassium phosphate buffer at pH 8.0 containing 0.05% β-octylglucoside and stirred vigorously at 20° for 45 min. The suspension was centrifuged at 180,000 *g* for 60 min at 4°. The supernatant was dialyzed for 36 hr against three changes of P_i–NaCl. The resulting vesicles (containing princi-

[31] D. K. Miller, B. I. Feuer, R. Vanderoef, and J. Lenard, *J. Cell Biol.* **84,** 421 (1980).

pally G protein) were pelleted at 180,000 *g* for 60 min at 4°. Vesicles were resuspended in 1 ml of P_i–NaCl to a final concentration of 1 mg of protein per milliliter. Then, 0.5 mg of G protein in vesicles (0.5 ml) was mixed on ice with 2.5 ml of P_i–NaCl, 1.6 ml 10% $KAl(SO_4)_2 \cdot 12H_2O$, and neutralized to pH 7.0 by addition of 1 *N* NaOH. The precipitate was centrifuged at 5000 *g* for 5 min at 4°, and the pellet was resuspended to a final volume of 2 ml of P_i–NaCl, and injected into a rabbit intramuscularly. Immunoprecipitating, rabbit anti-G protein antiserum was obtained within 3 weeks after a single boost of 0.5 mg of G protein given at 2 weeks.

Assay of Transport-Coupled Oligosaccharide Processing: Resistance to Endo H

Principle

As the oligosaccharides of G protein are sequentially processed in the Golgi from the "high" mannose [-Asn-$(GlcNAc)_2Man_9Glc_3$] form found in the RER to the mature "complex" form [-Asn-$(GlcNAc)_2Man_3$(GlcNAc-Gal-SA)$_2$] the oligosaccharides become resistant to cleavage by *β-N*-endoglycosidase H (Endo H). Endo H will cleave mannose-rich precursors of G, but will not attack the Golgi-processed oligosaccharide.[20] For the assay, a PNS is prepared from VSV-infected CHO cells of a mutant line (clone 15B) that specifically lacks UDP-GlcNAc glycosyltransferase I,[25] an enzyme found in the Golgi complex.[3,21] Within this processing pathway, the action of UDP-GlcNAc glycosyltransferase I is required before Endo H resistance can occur. G is transported normally within infected 15B cells,[32] but it always remains sensitive to Endo H as a result of the transferase I deficiency. When the PNS fraction of [^{35}S]Met-labeled, VSV-infected, clone 15B cells (donor membranes) are incubated with the PNS of uninfected wild-type CHO cells (acceptor membranes), the appearance of the Endo H-resistant form of G *in vitro* demonstrates the processing of G initially present in 15B cell membranes by transferase I present in the Golgi membranes derived from the wild-type cell. The assay yields two principal species of G protein differing in molecular weight after treatment with Endo H. They are quantitated using SDS–gel electrophoresis.[33] The Endo H-resistant forms retain their original apparent molecular weights after incubation with Endo H. The Endo H-sensitive form electrophoreses faster because it loses apparent molecular weight when the oligosaccharide is cleaved by Endo H.

[32] G. Schlessinger, C. Gottlieb, P. Fiel, N. Gelb, and S. Kornfeld, *J. Virol.* **17,** 239 (1976).
[33] U. K. Laemmli, *Nature (London)* **227,** 680 (1970).

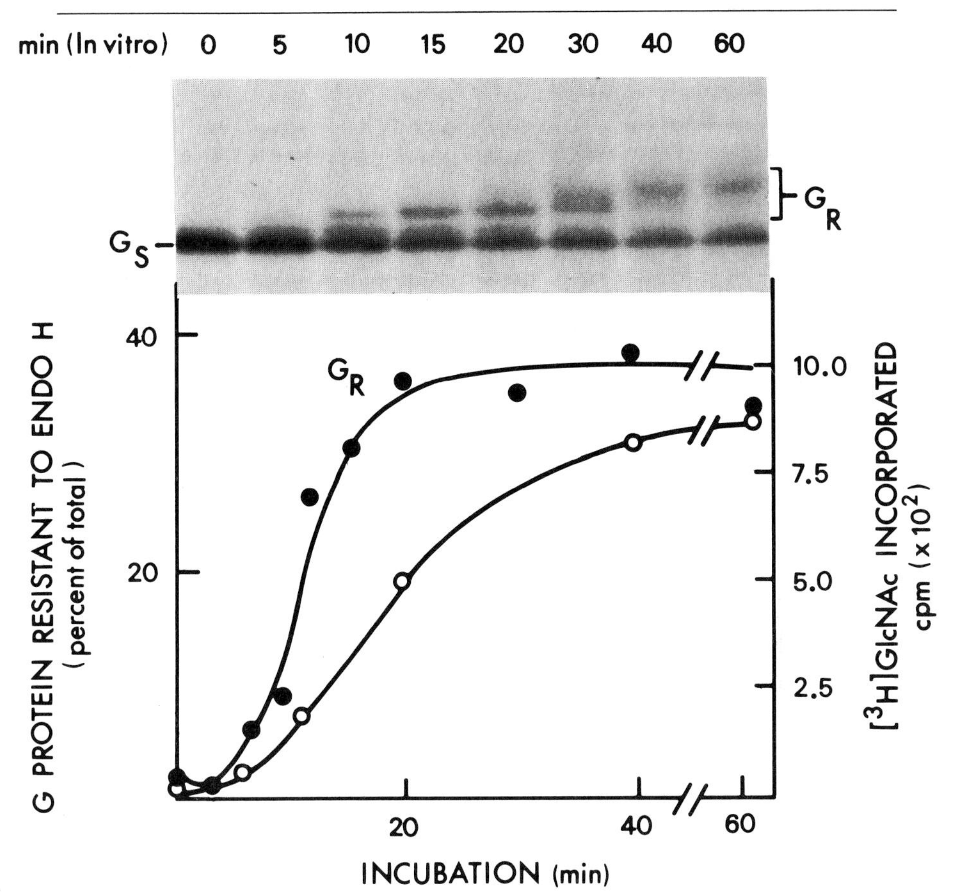

FIG. 1. Time course of the conversion of G protein to the Endo H-resistant form (●) or the incorporation of [^{3}H]GlcNAc (○) *in vitro*. ●, Postnuclear supernatant of VSV-infected 15B cells (labeled for 5 min with [^{35}S]Met and chased for 10 min[28]) was incubated with postnuclear supernatant of wild-type cells under standard conditions. Aliquots were taken at the time indicated, treated with Endo H, and analyzed by gel electrophoresis. Shown on top is an autoradiograph of the portion of the gel containing G protein. Below, the percentage of G protein converted to Endo H-resistant forms was determined by integration of densitometer tracings of the autoradiograph. ○, Postnuclear supernatant of vesicular stomatitis virus-infected 15B cells was incubated with postnuclear supernatant of wild-type cells under standard conditions in the presence of UDP-[^{3}H]GlcNAc. Aliquots were taken at the time indicated and mixed with detergent buffer to terminate the reaction. G protein was immunoprecipitated, filtered, and counted as described in the text.

Assay

For standard assays, 10 μl (50–100 μg) of [^{35}S]Met-labeled PNS of VSV-infected clone 15B cells are mixed on ice in a 10 × 75 mm glass disposable tube with 10 μl (50–100 μg) of PNS from uninfected, unlabeled wild-type CHO cells. To the PNS mix was added 10 μl of an incubation cocktail made by mixing 40 μl of a 10× concentrated stock of "incubation buffer" [450 m*M* HEPES–KOH (pH 6.8), 750 m*M* KCl, 45 m*M* magnesium acetate, and 10 m*M* dithiothreitol], 100 μl of 40 m*M* ATP (Na form, titrated to pH 7.0 with dilute NaOH), 100 μl of 200 m*M* creatine phosphate (Na form), 75 μl each of 20 m*M* UDP-GlcNAc and UDP-Gal (Sigma Chemical Company, St. Louis, Missouri), and 10 μl of rabbit skeletal muscle creatine phosphokinase (Sigma) at 1600 IU/ml. All assays are brought to a final volume of 50 μl (usually with 1× incubation buffer) to yield the following final concentrations: 1.5 m*M* ATP, 10 m*M* creatine phosphate

REQUIREMENTS FOR TRANSPORT-COUPLED OLIGOSACCHARIDE PROCESSING

Incubation conditions	Endo H-resistant G protein	[^{3}H]GlcNAc incorporated
1. Complete[a]	(1)	(1)
2. −ATP, −creatine phosphate (CP), −creatine phosphokinase (CPK), +ADP (3 m*M*)	0.54	—
3. −ATP, −CP, −CPK, + hexokinase, + glucose	<0.1	<0.05
4. −UDP-GlcNAc[b]	0.13	—
5. +UDP-GlcNAc (10 μ*M*)	—	0.1
6. −UDP-Gal	1.46	—
7. +UDP-Gal (10 μ*M*)	—	1.06
8. −Cytosol	0.15	<0.05
9. −Membranes from wild-type PNS, +membranes from noninfected 15B	<0.1	<0.05
10. Preimmune serum replacing anti-G serum	—	<0.05

[a] Complete incubation mix was as described in the text for each assay. Membranes were pelleted for 5 min at 100,000 *g* in a Beckman Airfuge (Beckman Instruments, Palo Alto, California) and resuspended in extract buffer to the original volume. The supernatant (cytosol) was dialyzed for 16 hr against extract buffer prior to use. For demonstration of ATP dependence (line 3), the assay was supplemented with 1 μl of a hexokinase solution (1400 IU/ml) prepared by dissolving centrifuged crystals of an ammonium sulfate precipitate yeast hexokinase (Boehringer Mannheim Biochemicals, Indianapolis, Indiana) in water, and 1 μl of a 0.25 *M* solution of glucose. Results are expressed relative to the value obtained for complete incubation mix [taken as unity (1)], typically 40–45% Endo H-resistant, and about 1000 cpm of [^{3}H]GlcNAc for the two assays.

[b] GlcNAc = *N*-acetyl-D-glucosamine.

(Na form), 4 IU of rabbit muscle creatine phosphokinase, 0.75 mM each UDP-GlcNAc and UDP-Gal, 1.0 mM dithiothreitol, 4.5 mM magnesium acetate, 75 mM KCl, and 45 mM HEPES–KOH (pH 6.8). The final pH of the incubation mixture was ~6.8–7.0. The incubation was initiated by transfer to 37° for 20–60 min.

Analysis with Endo H

Fifteen-microliter portions of incubations were mixed in 1.5-ml microfuge tubes with 15 μl of 0.1 M Tris-HCl (pH 6.8) containing 2% $NaDodSO_4$ and 30 mM dithiothreitol, and then boiled for 3 min. Then 20 μl of 0.3 M sodium citrate buffer (pH 5.5), containing 0.1% $NaDodSO_4$ and 75 ng of Endo H per milliliter, was added. After 12–18 hr at 37°, proteins were precipitated by adding 0.3 ml of 10% trichloroacetic acid, and pelleted for 2 min in the Eppendorf microfuge (Brinkmann Instruments, Inc., Westbury, New York). The pellet was dissolved in 50 μl of electrophoresis sample buffer.[33] Then, 5 μl of 1 M Tris free base was added; after boiling, the sample was analyzed by electrophoresis in a 10% polyacrylamide gel according to Laemmli.[33] The separation between Endo H-sensitive and -resistant forms of G protein is enhanced by extended electrophoresis. Dried gels were autoradiographed after fluorographic enhancement (En^{3}Hance; New England Nuclear, Boston, Massachusetts) and exposed for 1–7 days. Presented are a typical time course for *in vitro* transport-coupled oligosaccharide processing (Fig. 1) and dependence on assay components (see the table).

Assay of Transport-Coupled Oligosaccharide Processing: UDP-[^{3}H]GlcNAc Incorporation

Principle

Donor membrane fractions, prepared from unlabeled VSV-infected clone 15B cells (lacking UDP-GlcNAc glycosyltransferase I) are incubated with acceptor membrane fractions prepared from the uninfected wild-type CHO cell line and UDP-[^{3}H]GlcNAc. [^{3}H]GlcNAc should be incorporated into G protein only following transport to the wild-type Golgi (housing transferase I) *in vitro*. Incorporation into G protein is measured directly by immunoprecipitation.

Assay

Standard incubations contained 10 μl of unlabeled PNS from VSV-infected 15B cells, and 10 μl of PNS from uninfected, unlabeled wild-type cells in a final volume of 50 μl. Other components were identical to those described in the Endo H-resistant assay with the exception that UDP-GlcNAc and UDP-Gal were not included in the incubation cocktail and were replaced by 150 μl of H_2O. The assay was initiated by the addition of 0.5 μCi UDP-[^{3}H]GlcNAc (prepared by evaporation to dryness of an aliquot of the ethanolic stock of sugar nucleotide with a gentle stream of N_2, and dissolving in water at 0.1 mCi/ml) and transfer to 37° for 20–60 min before immunoprecipitation.

Immunoprecipitation and Filter Binding

The reaction was terminated and membranes solubilized by addition of 50 μl of a detergent buffer containing 50 m*M* Tris–HCl (pH 7.5), 250 m*M* NaCl, 5 m*M* Na_2EDTA, 1% Triton X-100, and 1% sodium cholate. Then 10–15 μl of rabbit anti-G protein antiserum (or, the same volume of preimmune serum for control incubations) were added. After 45 min at 37° or an overnight incubation at 0–4°, the immunoprecipitate was collected by Millipore filtration (type HA filters, 0.45 μm; Millipore Corporation, Bedford, Massachusetts). To do this, the filters were first rinsed with 3 ml of a "washing buffer" [0.05 *M* Tris-HCl (pH 7.5), 250 m*M* NaCl, 5 m*M* Na_2EDTA, 1% Triton X-100]. Then the immunoprecipitated incubation mix was diluted with 3 ml of the washing buffer and filtered. Filters were then washed with three 3-ml portions of washing buffer, dried, and counted in 10 ml of Aquasol 2 (New England Nuclear, Boston, Massachusetts). It is probably important that filters be processed one at a time to avoid extensive air drying in between washes. The results of a typical assay (Fig. 1) and dependence on assay components (see the table) are illustrated.

Concluding Remarks

Two methods for measuring transport-coupled oligosaccharide processing *in vitro* have been presented. They differ in that in the first case the polypeptide backbone of G protein is labeled (with [^{35}S]Met) and oligosaccharide processing is inferred from Endo H resistance, whereas the second method follows glycosylation of G directly via incorporation of the [^{3}H]GlcNAc into unlabeled G proteins. The assay based on the conferral of resistance to Endo H provides information regarding the location and nature of the donor[28, 29] but requires a minimum of 2–3 days postincubation processing to complete the analysis. The assay based on the incorporation of [^{3}H]GlcNAc into G protein from its sugar–nucleotide precursor is simpler, far more rapid, and quantitative, making it the assay of choice to be employed for purification and characterization of components required for G protein transport.

Both assays yield similar results (see the table) with respect to ATP dependence (lines 2 and 3) and the requirement for a cytosol fraction (line 8). An essential control is that fractions prepared from the uninfected 15B cells do not substitute for wild-type membranes as acceptors (see the table), demonstrating that the oligosaccharide processing observed *in vitro* follows the same enzymic pathway used *in vivo*. Under suitable conditions, processing is linear with respect to time and to concentrations of cytosol, donor, and acceptor membranes. The wild-type "acceptor" membranes may be replaced by extensively purified rat liver Golgi.[27] Donor fractions are especially sensitive to manipulation. For example, it is important to maintain the PNS on ice prior to assay. Additionally, resuspension of donor membranes from tightly packed pellets (e.g., 180,000 *g* for 60 min) results in marked loss of activity. However, a brief centrifugation (100,000 *g* for 5 min) in a Beckman Airfuge (Beckman Instruments, Palo Alto, California) yields efficient recovery of activity. This step is essential for demonstrating the cytosol dependence (see the table).

The principles embodied in the design of these assays are illustrative of a broad class of transport-coupled processing assays that are possible. Considering the extensive number of covalent modifications that occur as proteins are passed between organelles, the careful choice of substrates for *in vivo*[29] or *in vitro* labeling, and the use of different cell[3] and virus[34] mutants defective in stages of processing should allow the construction of a variety of assays that can divide the transport pathway into distinct segments. Each may be amenable to dissection by fractionation, providing an experimental approach for defining the biochemical basis of intracellular transport.

[34] A. Zilberstein, M. D. Snider, M. Porter, and H. F. Lodish, *Cell* **21**, 417 (1980).

[5] Perturbation of the Structure and Function of the Golgi Complex by Monovalent Carboxylic Ionophores

By ALAN MICHAEL TARTAKOFF

A convergence of research on the Golgi complex has reaffirmed its pivotal role in protein and glycoprotein intracellular transport and processing and has made it clear that the Golgi complex serves as a crossroads for vesicular traffic in the cytoplasm. Nevertheless, understanding of how the Golgi complex executes these important operations is far from complete. Subcellular fractionation has been able to shed only a limited light on these events. For these reasons the development of a pharmacological approach to the structure and function of the Golgi complex has assumed a particular

METHODS IN ENZYMOLOGY, VOL. 98

ISBN 0-12-181998-1

importance. For background on the basic structure and identified functions of the Golgi complex, the reader is refered to several reviews.[1-4]

Perturbation of Structure and Function by Monovalent Ionophores

Background

The decision to explore the effects of monovalent carboxylic ionophores on intracellular transport (ICT) was based on the supposition that cells should tolerate imposed alterations in Na^+/K^+ balance, since such alterations are commonplace in normal cell physiology. Furthermore, several reports suggested that shifts in Na^+/K^+ balance[5] or addition of the relatively nonspecific ionophore X 537A[6] caused dilation of Golgi cisternae or related structures of several types of cells.

The ionophore that has proved to be most useful is monensin, a heterocyclic metabolite of *Streptomyces cinnamonensis*.[7,8] Available preparations are in fact a mixture of at least four components. The covalent[9] and crystal[10,11] structures of monensin A have been solved, and the molecule can be easily recognized as a member of the class of carboxylic ionophores, other well-characterized members of which are nigericin, dianemycin, X 537A, and A 23187.[12] Solution studies[7,13] show that monensin forms complexes with a range of ions but the only such ions of physiological significance are Na^+ and K^+. The K_d for Na^+ is $6 \times 10^{-5} M$ and the K_d for K^+ is $4 \times 10^{-2} M$. The crystallographic studies[10,11] show that monensin assumes a roughly cyclic structure with the metal ion ringed by six liganding oxygens. Unlike nigericin, the carboxyl group of monensin does not participate directly in complexation.

An extensive literature,[12] largely making use of artificial membranes,

[1] A. Tartakoff, *Int. Rev. Exp. Pathol.* **22,** 227 (1980); A. Tartakoff, *Int. Rev. Cytol.* **85,** 221 (1983).
[2] M. Farquhar, *Life Sci. Res. Rep.* **11,** 341 (1978).
[3] C. Leblond and G. Bennett, *Int. Cell Biol., Papers Congr. 1st 1976* p. 326 (1977).
[4] G. Whaley, *Springer Cell Biol. Monogr.* **2** (1975).
[5] W. Whetsell and R. Bunge, J. Cell Biol. **42,** 490 (1969).
[6] A. Somlyo, R. Garfield, S. Chacko, and A. Somlyo, *J. Cell Biol.* **66** 425 (1975).
[7] B. Pressman, *Fed. Proc., Fed. Am. Soc. Exp. Biol.* **27,** 1283 (1968).
[8] B. Pressman, *Fed. Proc., Fed. Am. Soc. Exp. Biol.* **32,** 1698 (1973).
[9] A. Agtarap, J. Chamberlain, M. Pinkerton, and L. Steinrauf, *J. Am. Chem. Soc.* **89,** 5737 (1967).
[10] L. Steinrauf, E. Czerwinski, and M. Pinkerton, *Biochem. Biophys. Res. Commun.* **45,** 1279 (1971).
[11] W. Lutz, F. Winkler, and J. Dunitz, *Helv. Chim. Acta* **54,** 1103 (1971).
[12] B. Pressman, *Annu. Rev. Biochem.* **45,** 501 (1976); B. Pressman and M. Fahim, *Annu. Rev. Pharmacol. Toxicol.* **22,** 465 (1982).
[13] W. Simon, W. Morf, and P. Meier, *Struct. Bonding (Berlin)*, **16,** 113 (1973).

indicates that monensin inserts into lipid bilayers and can complex a given ion at one membrane interface and diffuse to the opposite face of the bilayer, where the ion can be released. For a return voyage the ionophore may either carry another metal ion or become electroneutral by protonation of its carboxyl ($pK_a = 7.95$).[8] Such electroneutral transport is sensitive to the viscosity of the bilayer and is profoundly slowed at reduced temperature.[14]

To the extent that monensin serves exclusively as a sodium ionophore, it would be expected to raise cellular sodium levels and possibly raise intracellular pH. In reality, the rapid increase in cellular sodium is largely balanced by potassium loss,[15,16] and the only documented pH changes are minor and transient.[17] A secondary consequence of partial Na^+/K^+ equilibration is hyperpolarization of the cell membrane due to stimulation of the cell-surface Na^+,K^+-ATPase.[17,18] There has been no systematic evaluation of the influence of monensin on other ions; however, in plasma cells, monensin does not influence the rate of calcium efflux.[15]

The closely related ionophore nigericin[7,12] has a much higher K^+ affinity and is known as an uncoupler of oxidative phosphorylation.[19] When K^+ is present it abolishes ion gradients across the inner mitochondrial membrane that are essential for coupled electron flow. Monensin-treated cells have greatly reduced K^+ and their mitochondria contract,[15] suggesting extensive cation efflux; yet in many cell types a number of biosynthetic activities proceed normally. For example, incorporation of radioactive amino acids into newly synthesized proteins proceeds almost unimpaired.[15,20] This certification of the adequacy of energy reserves is of particular importance, since ongoing ATP production is known to be necessary for a pre-Golgi step in ICT—exit from the rough endoplasmic reticulum (RER).[1]

Morphological Consequences

The reported morphological consequences of 1 to 4 hr treatment with 0.05 – 1 μM monensin are similar for all cells and are remarkably selective in the sense that only the Golgi region and mitochondria are altered (Fig. 1).[15]

First visible changes occur rapidly. The Golgi cisternae of murine plasma cells begin to dilate after treatment for 30 – 60 sec with 1 μM

[14] S. Hladsky, *Curr. Top. Membr. Transp.* **12,** 53 (1979).
[15] A. Tartakoff and P. Vassalli, *J. Exp. Med.* **146,** 1332 (1977).
[16] M. Feinstein, E. Henderson, and R. Sha'afi, *Biochim. Biophys. Acta* **468** 284 (1977).
[17] D. Lichtenstein, K. Dunlop, H. Kaback, and A. Blume, *Proc. Natl. Acad. Sci. U.S.A.* **76,** 2580 (1979).
[18] S. Mendoza, N. Wiggelsworth, P. Pohjanpelto, and E. Rozengurt, *J. Cell. Physiol.* **103,** 17 (1980).
[19] S. Ferguson, S. Estrada-O., and H. Lardy, *J. Biol. Chem.* **246,** 5645 (1971).
[20] N. Uchida, H. Smilowitz, and M. Tanzer, *Proc. Natl. Acad. Sci. U.S.A.* **76,** 1868 (1979).

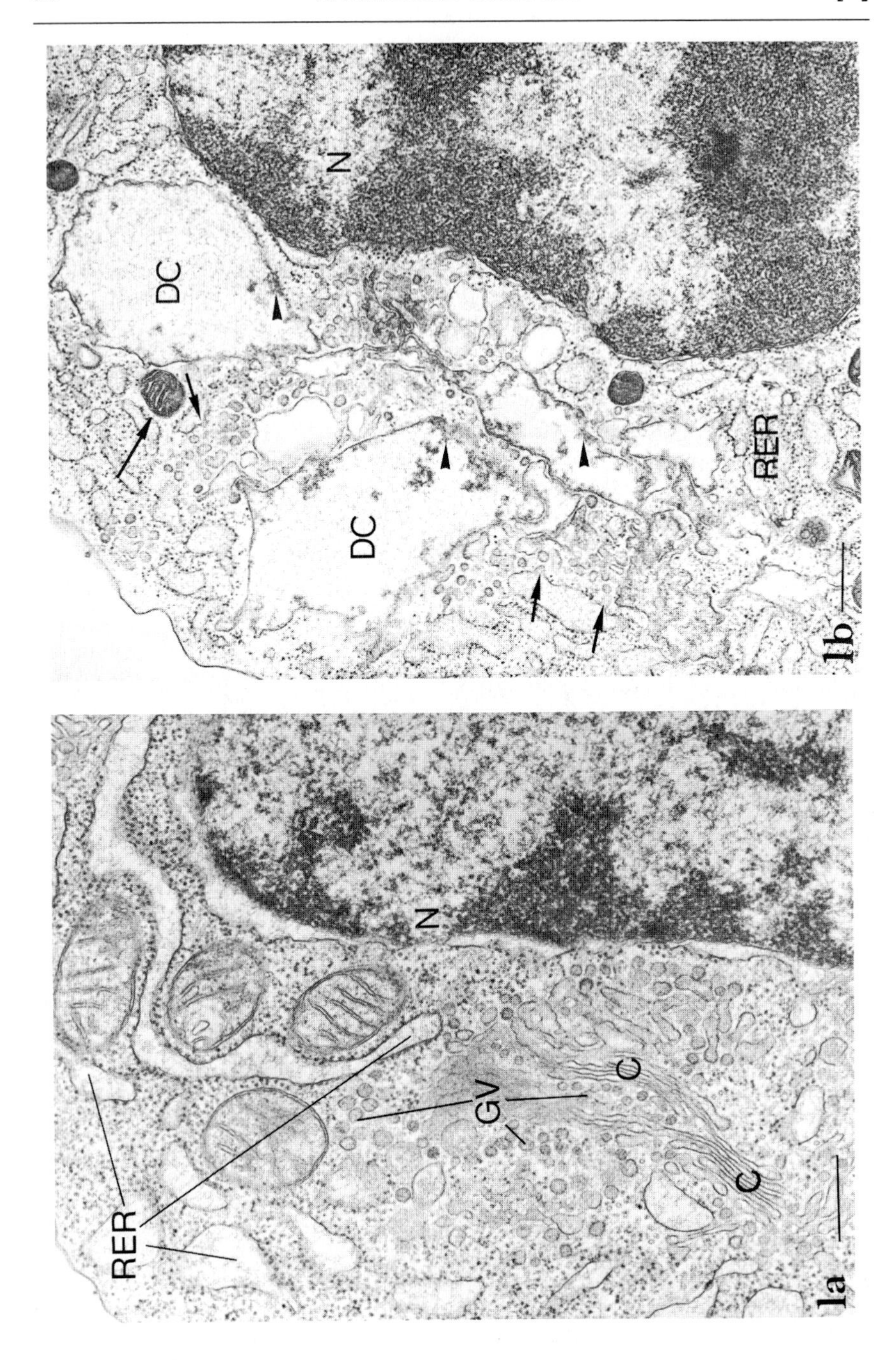
N
DC
DC
RER
1b
N
RER
GV
C
C
1a

monensin at 37°, and by 5 min there are essentially no remaining compressed cisternae.[21] Mitochondrial contraction proceeds simultaneously. Examination of the early stages of dilation does not show an initial effect restricted to any subregion of the Golgi stack. The relatively uniform response of all cisternae is also suggested by the histochemical detection of thiamin pyrophosphatase in monensin-treated plasma cells, pancreatic acinar cells, and fibroblasts.[21,22] The product of this enzyme is customarily restricted to a few distal cisternae. After monensin treatment a minor proportion of the dilated elements are histochemically positive. Another characteristic of the dilated structures is that they tend to adhere one to the next along an extended portion of their perimeter. The cause of this affinity, like the close apposition of unperturbed Golgi cisternae, is not known.

Effects on Protein Intracellular Transport

Intracellular transport of secretory proteins is effectively slowed by monensin. For example, in [^{35}S]methionine pulse–chase protocols, the half-time for secretion of newly synthesized IgM or IgA from murine plasma cells is lengthened by at least an order of magnitude by 1 μM monensin.[15] Comparable observations have been made for exocrine pancreatic[23] and macrophage[23] secretory products, procollagen[20,23] and fibronectin[20] secretion by fibroblasts, acetylcholinesterase secretion by cultured muscle cells, transferrin secretion by a hepatoma,[24] and transport of numerous proteins along axons of nerve cells.[46] No mammalian cell is known to resist this inhibitory effect.

The ICT of several integral cell surface glycoproteins is also greatly retarded by monensin. The proteins investigated are the envelope glycoproteins of vesicular stomatitis virus (VSV),[24,25] Sindbis,[25] and Semliki

[21] A. Tartakoff and J. Reigner, unpublished observations (1979–1981).

[22] G. Griffiths, P. Quinn, and G. Warren, *J. Cell Biol.* **96,** 835 (1983).

[23] A. Tartakoff and P. Vassalli, *J. Cell Biol.* **79,** 694 (1978).

[24] G. Strous and H. Lodish, *Cell* **22,** 709 (1980).

[25] D. Johnson and M. Schlesinger, *Virology* **103,** 407 (1980).

FIG. 1a. Thin section of control mouse plasma cell. Note the compressed Golgi cisternae and associated smooth-surfaced vesicles. C, Golgi cisternae; GV, Golgi vesicles; N, Nucleus, RER, rough endoplasmic reticulum. Bar = 0.5 μM

FIG. 1b. Thin section of a plasma cell after 1 hr at 37° in the presence of 1 μM monensin. Most aspects of cell morphology are unchanged, but in the place of compressed Golgi cisternae are large dilated sacs with a precipitate content. The mitochondria are condensed. DC, dilated cisterna; arrow heads, precipitate content of the sacs; short arrows, transitional elements of the RER; long arrow, a condensed mitochondrion. Bar = 0.5 μM.

Forest[26,27] viruses in infected fibroblasts, the acetylcholine receptor[28] of cultured muscle cells, and the heavy chains of surface IgM and H2 antigens[29] of lymphocytes. Again, no exceptions are known.

The Site of Arrest Due to Monensin

The Cytologic Site. The cytologic site of ICT arrest has been examined by autoradiography, subcellular fractionation, and immunocytochemistry.

The autoradiographic data pertain to a mixed population of IgM-secreting mouse plasma cells[15] and the exocrine pancreas.[23] In both cases, pulse–chase experiments employing radioactive amino acids show that newly synthesized protein accumulates within the dilated Golgi elements when monensin is present. Some of these elements actually contain a visible precipitate (Fig. 2). The subcellular fractionation data for the exocrine pancreas[23] also clearly indicate a site of arrest within the Golgi complex: the rate of drainage of pulse-labeled protein from the rough microsomal fraction is normal, and radioactive protein overaccumulates in the smooth microsomal fraction and fails even after several hours to gain access to the zymogen granule fraction. Under the conditions studied, monensin can also be shown to be without effect on the discharge of granule content, under both control and secretagogue-stimulated conditions. Immunocytochemical data implicating the Golgi complex have been obtained at both the light microscopic[26,30] and electron microscopic levels.[31]

The Biochemical Site. The site of arrest due to monensin can be described with reference to posttranslational Golgi-associated processing events. For example, in the case of newly synthesized glycoproteins bearing asparagine-linked complex oligosaccharides one can inquire whether, in the presence of monensin, they acquire their terminal sugars. This question has a particular significance, since several lines of investigation[1] indicate that terminal sugar addition occurs only within the more distal portion of the Golgi complex; i.e., the degree of maturation of the oligosaccharides would be diagnostic of the precise cytologic site of arrest within the Golgi complex.

Ideally, two "sequential" protocols should be employed in order to judge whether a given event occurs proximal or distal to the monensin site. With respect to the addition of galactose residues to secretory IgM,[32] for example,

[26] L. Kaariainen, K. Hashimoto, J. Saraste, and K. Penttinen, *J. Cell Biol.* **87,** 783 (1980).

[27] J. Saraste and K. Hedman, *J. Cell Biol.* (in press).

[28] R. Rotundo, and D. Fambrough, *Cell,* **22,** 595 (1980).

[29] A. Tartakoff, D. Hoessli, and P. Vassalli, *J. Mol. Biol.* **150,** 525 (1981).

[30] P. Ledger, N. Uchida, and M. Tanzer, *J. Cell Biol.* **87,** 663 (1980).

[31] J. Roth and A. Tartakoff, unpublished observations, 1980.

[32] A. Tartakoff and P. Vassalli, *J. Cell Biol.* **83,** 284 (1979).

the first protocol involves pulse labeling with [^{3}H]galactose followed by a chase interval in the absence or the presence of monensin. The efficiency of transport of labeled Ig (in this case, secretion) is then evaluated, and it is observed that the secretion of [^{3}H]galactose-labeled Ig is little slowed by monensin—unlike Ig labeled with [^{35}S]methionine or [^{3}H]mannose. In the second protocol, cells are pretreated with monensin so as to impose a block in ICT and to allow compartments distal to the block to drain. Cells are then challenged with a brief [^{3}H]galactose pulse to assess whether Ig is still present in the compartment where galactose addition normally takes place and, therefore, can be labeled. In this protocol, [^{3}H]galactose incorporation into Ig is greatly slowed relative to nonpretreated controls. Nevertheless, [^{3}H]galactose continues to be added to proteins other than Ig. Thus, as a first approximation one can conclude that the sugar transferase itself remains active and that the reduction in labeling of Ig is to be ascribed to topographic, not enzymologic, considerations.

Both protocols therefore suggest that the addition of galactose occurs distal to the monensin site. Comparable data are available for fucose.[32]

In the case of another secretory glycoprotein, transferrin, [^{35}S]methionine pulse–chase protocols show that in the presence of monensin this protein remains sensitive to endo-*β*-*N*-acetylglucosaminidase H (Endo H).[24] The retention of Endo H sensitivity means that oligosaccharide processing has not reached the $Man_3GlcNAc_2$ core.[33] In the conversion from $Man_9GlcNAc_2$ it is thought that a single mannose residue is removed in the RER[34] and that two Golgi mannosidases complete the trimming along with addition of a single mannose-linked *N*-acetylglucosamine residue. It is not known which of these processing activities occur in monensin-treated cells.

The data for membrane glycoproteins are not so coherent. In the case of surface IgM and H2, it is clear that at least sialic acid is not added in the presence of monensin[29]; however, for several viral envelope glycoproteins, resistance to Endo H can be acquired.[24,25] In fact, such observations are not surprising since (*a*) the ICT block due to monensin, though major, is incomplete; and (*b*) the G protein of VSV, for example, is normally transported much more rapidly than certain of the secretory glycoproteins in question.[24]

[^{35}S]methionine pulse–chase experiments making use of a VSV-infected rat myeloma do in fact document an analogous slowing of acquisition of Endo H resistance for both Ig heavy chains and the G protein when appropriate, *early,* chase time points are examined.[21]

[33] H. Schachter, *in* "Biochemistry of Glycoproteins and Proteoglycans" (W. Lennarz, ed.), p. 85. Plenum, New York, 1980.

[34] D. Godelaine, M. Spiro, and R. Spiro, *J. Biol. Chem.* **256,** 10161 (1981).

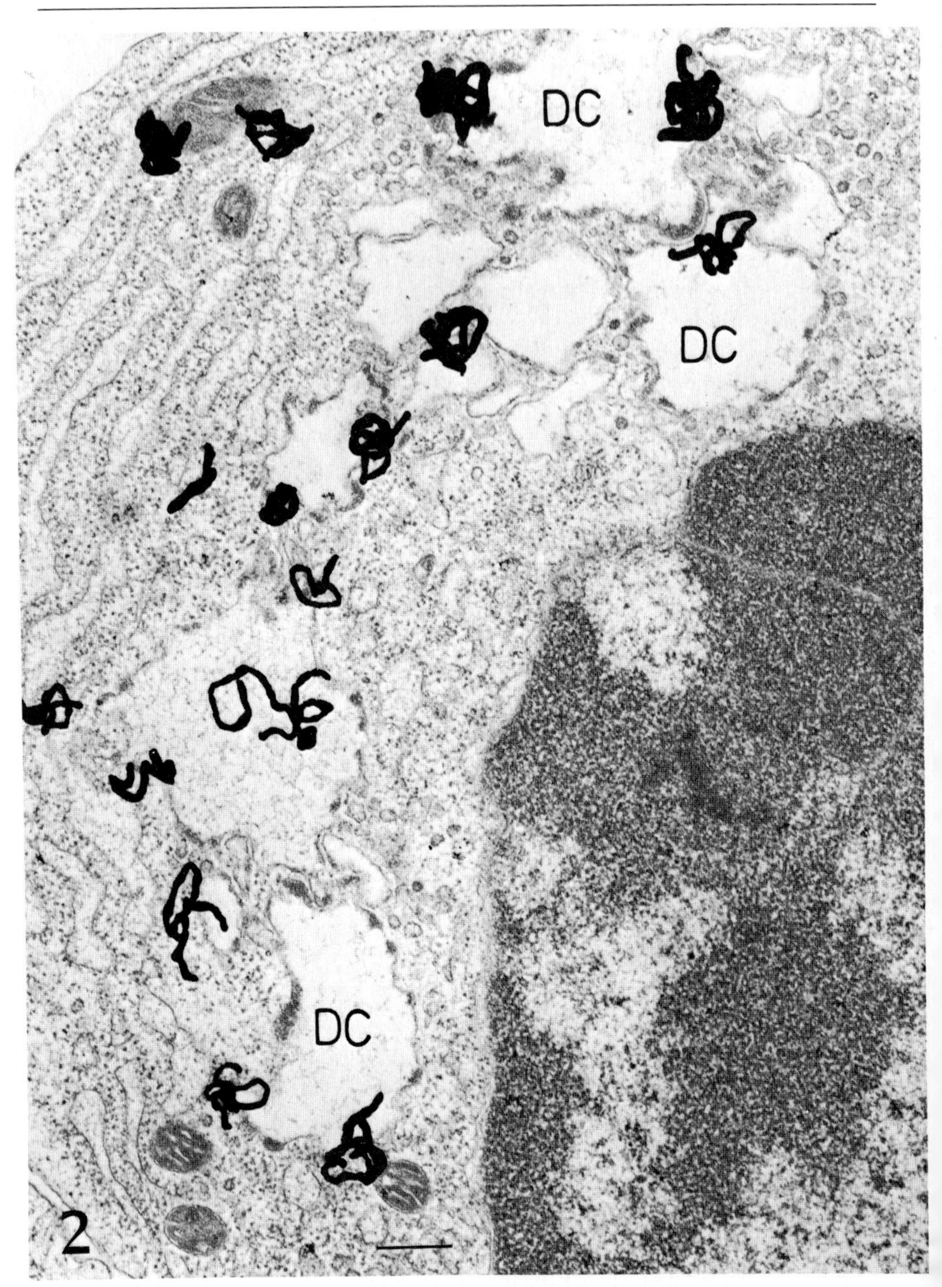

FIG. 2. Autoradiogram of a mouse plasma cell pulse-labeled for 5 min with [^{3}H]leucine and chased for 90 min in the presence of 1 μM monensin.[15] In this extreme example one sees the close association between autoradiographic grains and the dilated cisternae (DC). The content of these sacs is especially abundant and coherent. Bar = 0.5 μM.

Thus, the Golgi site at which monensin slows ICT is proximal to the activity of the enzymes conferring Endo H resistance on both secretory and membrane proteins.

Several additional Golgi-associated activities can be positioned proximal to the monensin site, since they persist in monensin-treated cells. As the available data are not based on a pair of "sequential" protocols of the sort outlined above for secretory IgM, these assignments should be considered provisional. For example, lipid addition to the envelope glycoproteins of VSV and Sindbis virus persists,[25] as does the phosphorylation of selected lysosomal enzymes.[35] The interpretation of any inhibition of isotope incorporation under such conditions is, however, of uncertain significance, since monensin treatment may alter cellular pools of key isotopically labeled intermediates. Both monensin and the less specific ionophore × 537A have been reported to inhibit addition of sulfate to the proteoglycans of chondrocytes and to have a lesser effect on [^{3}H]glucosamine incorporation.[6,36] A histochemical approach has been utilized to attempt to establish whether the site of arrest of ICT is proximal to those distal Golgi cisternae that contain thiamin pyrophosphatase (TPPase).[22] Fibroblasts were infected with Semliki Forest virus and incubated for several hours with monensin. Unlike controls, where nucleocapsids are found associated with and budding across the plasma membrane, in the presence of monensin the nucleocapsids are found to rim, and to bud into, certain of the dilated Golgi structures.[25] In contrast to nearby dilated cisternae, those bearing nucleocapsids are negative for TPPase. Thus it can be concluded that the monensin site is proximal to TPPase.[36a]

These several observations have been put together in a confined function model of the Golgi complex, as illustrated in Fig. 3.

The sensitivity of the intracellular transport of a given protein to monensin is clearly diagnostic of the path the protein traverses. For this reason, several authors have attempted to distinguish distinct pathways of transport occurring within one and the same cell by making use of ionophores. For example, there is a report that the ICT of the acetylcholine receptor (a plasma membrane protein) is much less sensitive to inhibition by nigericin than the ICT of secretory acetylcholinesterase.[37] A reinvestigation of the same events has shown that the apparently lesser sensitivity of the receptor actually reflects a secondary effect of monensin, nigericin, and × 537A: they

[35] A. Hasilik, unpublished observations, 1981.

[36] K. Tajiri, N. Uchida, and M. Tanzer, *J. Biol. Chem.* **255**, 6036 (1980).

[36a] Strictly speaking, since there are known to be TPPase-negative cisternae known as "GERL"[1] distal to the Golgi stack, the nucleocapsids might be adhering to these structures. There are no published data on the sensitivity of the GERL to dilation by monensin.

[37] H. Smilowitz, *Cell* **19**, 237 (1980).

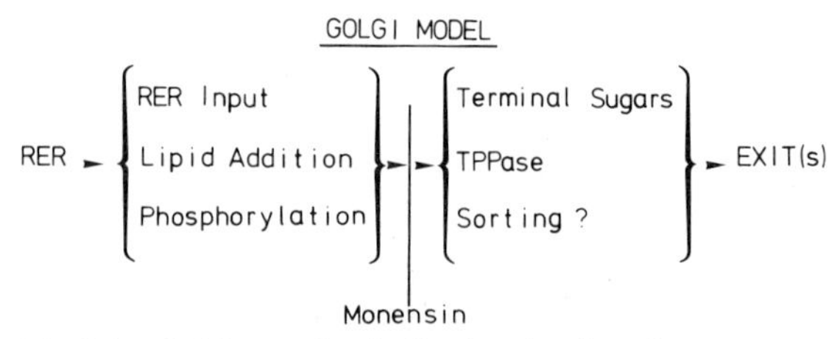

FIG. 3. Model of the Golgi complex indicating the site of transport arrest imposed by monensin. Selected activities are assigned to the relatively proximal (left) or distal (right) subcompartments for the reasons given in the text.

slow the degradation of the acetylcholine receptor and possible other proteins by slowing their internalization.[28]

A second apparent differential effect of monensin on ICT pertains to the infection of monolayers of Madin-Darby canine kidney (MDCK) cells by two viruses that bud from distinct cell surface domains: VSV, which buds at the basolateral surface, and influenza virus, which buds at the apical surface. Only VSV replication is reported to be blocked by monensin.[38] It is not clear to what extent these observations pertain directly to the ICT of the respective envelope glycoproteins of the two viruses. The selection of an MDCK cell host may not be fortuitous, since these cells are professional ion transporters and therefore might be influenced somewhat differently by the ionophore than, for example, a plasma cell.

Reversibility[21]

In murine plasma cells, total reversibility of both the morphologic and biochemical lesions due to monensin can be achieved, but, for the reversibility to be complete within several hours, the conditions of monensin treatment must be modest. If cells are exposed to 0.05 μM monensin for 1 hr at 37° and then washed and returned to culture, they can be challenged in a [^{35}S]methionine pulse–chase protocol at once or after various recovery intervals. As shown in Fig. 4, overnight reculture confers upon the cells the ability to secrete Ig whose heavy chains have a gel mobility diagnostic of the presence of mature (sialic acid-containing) oligosaccharides. The kinetics of secretion of this IgM are as rapid as for control cells. If, however, the dose of monensin is higher or the period of reculture is reduced to a few hours, secretion is slow and the limited quantity of Ig secreted contains immature oligosaccharides. The mechanism of secretion of this Ig is not known, but it does not reflect any increase in the proportion of dead cells by comparison with untreated controls.

Using the low dose of monensin, overnight reculture also allows the morphology of the Golgi complex to return to normal. There is no obvious

[38] F. Alonso and R. Compans, *J. Cell Biol.* **89**, 700 (1981).

increase in the number of lysosomes in such cultured cells. Since the return to normal morphology can also be achieved in the presence of 5 μg of cycloheximide per milliliter, one has the impression that those Golgi elements that dilate ultimately collapse to their initial form. A "new" Golgi complex does not have to be generated.

A number of attempts have been made to speed up the biochemical and morphological reversibility. On the supposition that the long period normally required reflects the high affinity of the lipophilic ionophore for membranes, defatted albumin or liposomes have been included during the reculture interval; however, no obvious speeding of recovery was observed.

Mechanism of Perturbation

Two lines of evidence suggest that the impact of monensin on Golgi structure and function is a direct consequence of the shift in cytoplasmic Na^+/K^+ balance: (*a*) a partial Na^+/K^+ equilibration is seen; and (*b*) ouabain or incubation in K^+-free medium can simulate monensin's effect on plasma cells.[15] Nevertheless, it is certain that many secondary consequences follow upon Na^+/K^+ shifts, and their significance remains to be evaluated. One would expect, for example, that the impact of monensin on a given cell should reflect the activity of its surface Na^+,K^+-ATPase, which, when activated (e.g., by Na^+ influx) both hyperpolarizes the cell and, possibly, reduces available ATP stores.

As a first approximation, one can attribute the dilation of Golgi elements to a relaxation of whatever (unidentified) constraints normally endow cisternae with their compressed shape; however, vesicle–cisterna or cisterna–cisterna fusion may further contribute to dilation. It remains altogether unclear whether there is a causal relation between dilation and the interruption of ICT. It is conceivable that either of the events might be responsible for the other.

The very different morphologic and functional responses of the RER and the Golgi complex point to a hitherto unsuspected and surely major difference in the ion permeabilities of these two classes of membranes. The only literature on Golgi ion pumps pertains to calcium uptake.[39,40]

Other Applications of Monensin

Modest or high doses of monensin interrupt fibroblast receptor-mediated endocytosis.[41,42] In the case of low-density lipoprotein receptors,[42]

[39] M. Neville, F. Selker, K. Semple, and C. Watters, *J. Membr. Biol.* **61,** 97 (1981).

[40] R. Freedman, J. MacLaughlin, and M. Weiser, *Arch. Biochem. Biophys.* **206,** 233 (1981).

[41] G. Vladutiu and M. Rattazzi, *Biochem. J.* **192,** 813 (1980).

[42] S. Basu, J. Goldstein, R. Anderson, and M. Brown, *Cell* **24,** 493 (1981).

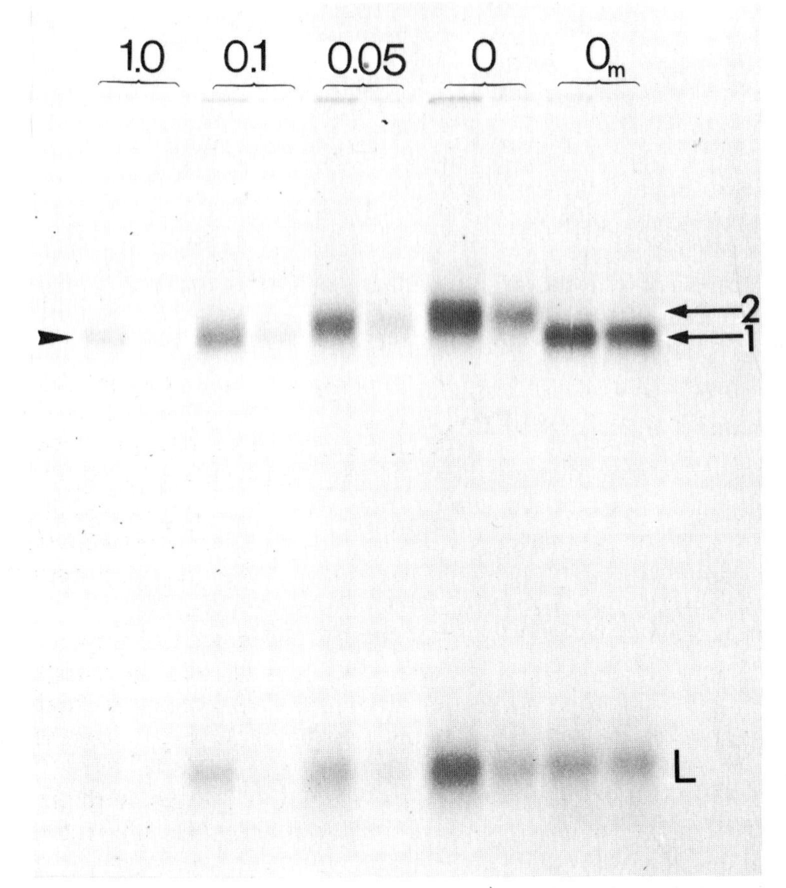

FIG. 4. Studies of reversibility of monensin's inhibition of IgM intracellular transport. Evaluation of the kinetics of IgM secretion and the degree of completion of μ-chain oligosaccharides. Cultured mouse plasmablasts were exposed for 1 hr at 37° to 0, 0.05, 0.1, or 1 μM monensin, washed, and recultured for 15 hr. The cultures were then washed, pulse-labeled for 5 min with [^{35}S]methionine, and chased in the absence or the presence (O_m) of 1 μM monensin for 90 min. These first culture supernatants were collected as well as a second 90-min sample of secretion. [^{35}S]immunoglobulin was recovered by immunoprecipitation from each pair of secretion samples, reduced, and analyzed on a 10 to 15% gradient sodium dodecyl sulfate (SDS) gel that was dried and autoradiographed. The heavy chains (μ) are of mobility 1 (also arrowhead on left) or 2. Earlier studies[32] have shown that neuraminidase converts mobility 2 to 1. Light chains are labeled L. The unperturbed control (O) secretes μ chains of mobility 2, and the amount secreted during the first interval greatly exceeds that secreted during the second. As expected, the O_m sample (no pretreatment, 1 μM monensin present during the chase) secretes much less than the control (O) and the distribution of [^{35}S]Ig between the two chase intervals is about equal. Unlike the control, after the second chase interval the remaining amount of intracellular [^{35}S]Ig exceeds that secreted during either chase interval (not shown). In the presence of monensin (O_m) all released μ chains are of mobility 1. In all the samples pretreated with monensin (1, 0.1, 0.05 μM) secretion is rapid (i.e., more [^{35}S]Ig is released during the first chase interval than during the second, and the cumulative 180-min percentage of A secretion of

when the ionophore is present, the receptor can function in a single cycle of ligand internalization to lysosomes.

Monensin has also been reported to cause an increase in miniendplate potential frequency at neuromuscular junctions,[43] to trigger exocytosis[44] and sperm maturation,[45] and to block axonal flow.[46] It is used in an experimental setting to increase the contractility of heart muscle,[12] and has a large veterinary significance in treatment of the parasitic protozoa of poultry[12] and in increasing weight gain by cattle and other farm animals.[12]

Other Golgi Perturbants

1. Uncouplers of oxidative phosphorylation or nitrogen effectively block exit of secretory proteins from the RER. A brief provocative electron microscopic study[47] suggests, that such arrest is accompanied by an accumulation of clathrin-like "shed coats" in the Golgi region. This "Golgi perturbant" has been used to demonstrate that asparagine-linked oligosaccharides of thyroglobulin lose one mannose and three glucose residues[34] while still in the RER and that murine IgM attains its pentameric structure ($M_r \sim 10^6$) before exit from the RER.[32]

2. An abundant literature shows that agents known to disrupt microtubules fragment, disorganize, and dilate the Golgi complex. In parallel, ICT of many (but not all) secretory proteins is slowed.[1] One study reports that the secretory glycoproteins caused to overaccumulate have acquired their terminal sugars.[48]

3. Certain anesthetics[49] or calcium withdrawal[15] also perturb the structure and function of the Golgi Complex. These agents have not been studied in depth.

Acknowledgment

This work was supported by Grant 3.059-0.81 from the Swiss National Science Foundation.

[43] H. Meiri, S. Erulkar, T. Lerman, and R. Rahamimoff, *Brain Res.* **204,** 204 (1981).
[44] E. Watson, C. Farnham, J. R. Friedman, and W. Farnham, *Am. J. Physiol.* **240,** C189 (1981).
[45] G. Nelson and S. Ward, *Cell* **19,** 457 (1980).
[46] R. Hammerschlag, G. Stone, J. Lindsey, and M. Ellisman, *J. Cell Biol.* **93,** 568 (1982).
[47] G. Palade and M. Fletcher, *J. Cell Biol.* **75,** 371a (1977).
[48] D. Banerjee, C. Manning, and C. Redman, *J. Biol. Chem.* **251,** 3887 (1976).
[49] J. Eichhorn and B. Peterkofsky, *J. Cell Biol.* **81,** 26 (1979).

[^{35}S]Ig is greater than 80%); however, the mobility of the μ chains depends on the concentration of monensin employed. Furthermore, although cell viability is constant, samples pretreated with a high dose of monensin incorporate relatively little radioactivity into IgM.

[6] Isolation of Golgi Apparatus from Rat Liver Using D_2O – Sucrose Gradients

By BECCA FLEISCHER

A number of methods are now available for the isolation of Golgi apparatus from rat liver.[1-4] These methods have in common a purification step involving either flotation or sedimentation in sucrose gradients followed by dilution of the purified fraction with sucrose of much lower osmolarity. Thus, the Golgi apparatus undergoes a significant osmotic shock during the preparation and is generally recovered as vesicles. In our preparations, we have shown the vesicles to be about 90% oriented as in the intact cells, with the cytoplasmic face toward the medium.[5]

In the course of studies on the nucleotide content of Golgi apparatus in which we wished to avoid osmotic shock as much as possible, we investigated the use of D_2O – sucrose gradients for the preparation of Golgi from rat liver.[6] The use of D_2O allowed us to use a shallower gradient of sucrose for the isolation. It also allowed us to avoid osmotic shock in the last step of the preparation, since we could dilute the Golgi fractions recovered from the gradient with H_2O – sucrose of only slightly less osmolarity than that of the fractions themselves. Golgi fractions prepared in D_2O – sucrose gradients retain higher levels of nucleotides than those obtained in H_2O – sucrose gradients.[6] This finding supports our hypothesis that osmotic shock is detrimental to the preservation of the Golgi membrane.

For comparison, preparation of Golgi apparatus in H_2O – sucrose gradients in a manner similar to the preparation in D_2O – sucrose gradients is also described. This is a modification of method B described previously.[2]

Preparation of Golgi Apparatus

Male Holtzman rats, 250 – 300 g, fed *ad libitum,* are used. They are killed by decapitation and exsanguinated. The livers are removed and placed in ice-cold 0.25 *M* sucrose. They are blotted dry and weighed before homogenization. All sucrose solutions are prepared using a special grade of sucrose

[1] D. J. Morré, this series, Vol. 22, p. 130.
[2] B. Fleischer, this series, Vol. 31, p. 180.
[3] D. E. Leelavathi, L. W. Estes, D. S. Feingold, and B. Lombardi, *Biochim. Biophys. Acta* **211,** 124 (1970).
[4] J. H. Ehrenreich, J. J. M. Bergeron, P. Siekevitz, and G. E. Palade, *J. Cell Biol.* **59,** 45 (1973).
[5] B. Fleischer, *J. Cell Biol.* **89,** 246 (1981).
[6] B. Fleischer, *Arch. Biochem. Biophys.* **212,** 602 (1981).

ISBN 0-12-181998-1

for density gradients (EM Laboratories, Associate of E. Merck, Darmstadt, 500 Executive Blvd., Elmsford, New York). Unbuffered solutions of sucrose in H_2O used in the step gradients or for final suspension of the Golgi fractions are neutralized to pH 7.0 ± 0.1 with KOH before use. The percentage of sucrose (w/w) for H_2O–sucrose solutions is adjusted finally to the desired concentration ± 0.1% using a Bausch and Lomb Abbe 3L refractometer at room temperature. These solutions are prepared immediately before use, the D_2O (99.8%) is obtained from Bio-Rad Laboratories (Richmond, California). D_2O–sucrose solutions are prepared on a weight/weight basis, need not be neutralized, and can be stored frozen before use. All centrifugations are carried out at 6° in a Spinco L8-70 ultracentrifuge unless otherwise specified. Centrifugation times are given as settings on the time dial. All solutions are used at refrigerator temperature (6°), and all steps after removal of the livers are carried out in a cold room at 6° or in an ice bath.

Procedure A. H_2O–Sucrose Gradient

This is a modification of method B described previously.[2] Rat livers, obtained as described above, are minced slightly with scissors and homogenized in 3 volumes of 52% sucrose containing 0.1 *M* sodium phosphate, pH 7.10 ± 0.05, using a Potter–Elvehjem homogenizer (50 ml) and three full strokes of a Teflon pestle with a clearance of 0.026 in. at 1000 rpm. The homogenization is repeated as before with a pestle having a clearance of 0.018 in. The final pH of the homogenate should be 7.1–7.2. The homogenate is filtered through two double layers of a loose-mesh cheesecloth to remove connective tissue. Filtration of the viscous homogenate is aided by stirring the liquid on the cheesecloth with a thick, blunt stirring rod. The filtrate is adjusted to 43.7% sucrose by adding 0.25 *M* sucrose as necessary. Fifteen milliliters of the homogenate is carefully pipetted into the bottom of a Spinco type SW 25.2 centrifuge tube. Cellulose nitrate tubes are preferred so that the interfaces are clearly visible. The homogenate is carefully overlaid sequentially with 10 ml of 38.7%, 10 ml of 36%, 10 ml of 33%, and 12 ml of 28% sucrose. This is best done using 10-ml pipettes and running the solutions down the side of the tube. The gradient tubes are centrifuged for 55 min at 24,000 rpm (100,000 *g*) using a precooled Spinco SW 25.2 rotor. Fractions are collected separately from the 28/33% and 33/36% sucrose interfaces, diluted 50% with deionized H_2O, and centrifuged for 10 min at 13,000 rpm (16,000 *g*) in a J20 rotor using a J21 Beckman centrifuge. The Golgi fractions are recovered from the supernatants by centrifugation at 40,000 rpm (180,000 *g*) in a Spinco type 42.1 rotor for 60 min. The pellets are suspended finally in 0.25 *M* sucrose using a Potter–Elvehjem-type homogenizer, 1 ml, with a hand-driven Teflon pestle. For 14 preparations, the yield of purified Golgi was 0.30 ± 0.19 mg of protein per gram weight of liver, and the galactosyltransferase activity averaged 780 ± 240 nmol of galactose transferred per milligram of protein per hour.

Procedure B. D_2O–Sucrose Gradient

Rat livers, obtained as described above, are minced slightly with scissors and homogenized in 3 volumes of D_2O containing 26.3% (w/w) sucrose and 0.1 *M* sodium phosphate, pH 7.1, as described in procedure A and filtered through four layers of cheesecloth. The final sucrose concentration of the homogenate is calculated as 20.6% (w/w), since 1 g of liver is added to 3.6 g (3.0 ml) of homogenizing medium, diluting the sucrose concentration of the homogenizing medium by a factor of 3.6/4.6 on a weight basis. Fifteen milliliters of the homogenate are carefully pipetted into the bottom of a Spinco type 25.2 centrifuge tube and overlaid as in procedure A, above, with 20 ml of D_2O containing 16.3% sucrose, 10 ml of D_2O containing 13.1% sucrose, and 12 ml of D_2O containing 9.6% sucrose. The gradient tubes are centrifuged for 55 min at 24,000 rpm (100,000 *g*) using a precoiled Spinco SW 25.2 rotor. Fractions are collected from the 9.6/13.1% and 13.1/16.3% sucrose interfaces using Pasteur pipettes. The fractions are combined, diluted 50% with 0.25 *M* sucrose, and centrifuged at 40,000 rpm (180,000 *g*) in a Spinco type 42.1 rotor for 60 min. The pellet is suspended finally in 0.25 *M* sucrose in H_2O as described in A, above. For 12 preparations, the yield of purified Golgi was 0.32 ± 0.18 mg of protein per gram wet weight of liver, and the galactosyltransferase activity[7] averaged 1000 ± 180 nmol of galactose transferred per milligram of protein per hour.

Comments

Figures 1A and B and Table I summarize the preparation and enzymic properties of fractions obtained from rat liver homogenates using H_2O–sucrose or D_2O–sucrose step gradients. As we have shown previously,[2] Golgi-rich fractions from this source are characterized by high specific activities of galactosyltransferase and low specific activities of glucose-6-phosphatase, 5′-nucleotidase, and succinate–cytochrome *c* reductase. In liver, the latter activities are largely characteristic of endoplasmic reticulum, plasma membrane, and mitochondria, respectively.

When prepared using H_2O–sucrose gradients, the Golgi is generally recovered at the 28/33% sucrose interface. Quite often, however, the fraction at the 33/36% sucrose interface is also mainly Golgi in origin, as judged by its galactosyltransferase activity. We therefore routinely collect and assay both fractions and combine them after analysis when fraction 2 is sufficiently pure.

[7] B. Fleischer and M. Smigel, *J. Biol. Chem.* **252,** 1632 (1978).

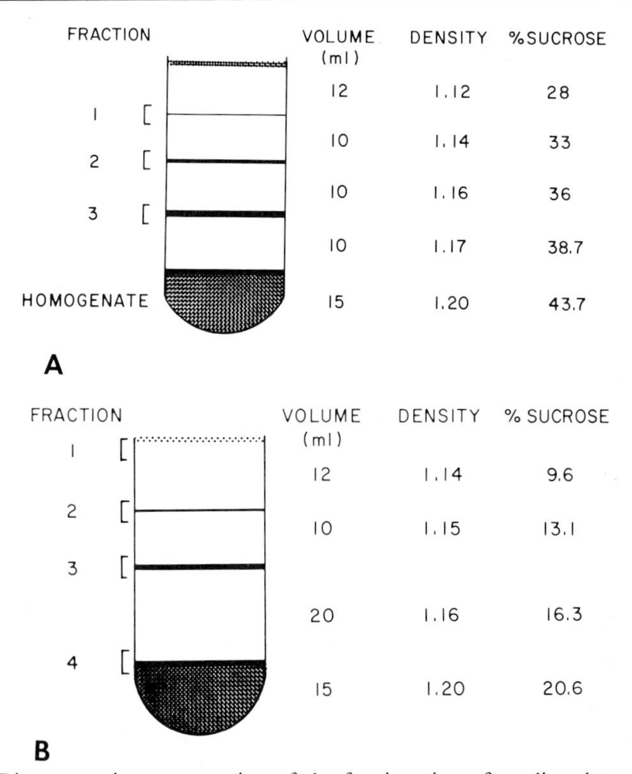

FIG. 1. Diagrammatic representation of the fractionation of rat liver homogenate on sucrose step gradients. The crosshatched area represents the initial homogenate. (A) H_2O–sucrose step gradient (procedure A). (B) D_2O–sucrose step gradient (procedure B). Analysis of the fractions from the interfaces of both gradients is shown in Table I.

When procedure A is used, a variable amount of white flocculent material is often seen at each interface. This contaminant is largely removed by centrifugation of the diluted fractions at 13,000 *g* for 10 min. This contaminant usually does not appear in procedure B, so that this step can be omitted.

When prepared using D_2O–sucrose gradients, Golgi is recovered in both the 9.6/13.1% and 13.1/16.3% sucrose interfaces. This is higher density than the Golgi fractions obtained in H_2O–sucrose gradients. This is probably due to replacement of H_2O in the lumen of the Golgi by D_2O during the isolation procedure. It is important to remove most of the D_2O from the Golgi before carrying out the galactosyltransferase assay, as D_2O inhibits galactosyltransferase activity.[8]

[8] F. A. Simion and B. Fleischer, unpublished observations.

TABLE I

DISTRIBUTION OF GALACTOSYLTRANSFERASE AND OTHER MARKER ENZYMES AFTER CENTRIFUGATION OF RAT LIVER HOMOGENATE IN H_2O–SUCROSE OR D_2O–SUCROSE STEP GRADIENTS[a]

Fraction	Total protein (mg)	Phosphorus: Protein (μg/mg)	Galactosyltransferase (nmol/hr × mg Pr)	NADH–cytochrome *c* reductase (μmol/min × mg Pr)	Succinate–cytochrome *c* reductase (μmol/min × mg Pr)	5′-Nucleotidase (μmol/min × mg Pr)	Glu-6-phosphatase (μmol/min × mg Pr)
A. H_2O–sucrose (as in Fig. 1A)							
Homogenate	1787	—	11.7	0.385	0.127	0.016	0.075
1	1.6	29.0	599	0.376	0.008	0.100	0.035
2	1.5	28.3	314	0.376	0.010	0.145	0.070
3	3.0	33.5	120	0.994	0.050	0.150	0.115
B. D_2O–sucrose (as in Fig. 1B)							
Homogenate	1640	—	13.7	0.487	0.154	0.020	0.112
1	0.4	27.2	419	0.473	0.010	0.107	—
2	1.1	28.5	1175	0.291	0.008	0.063	0.022
3	1.7	30.9	1324	0.586	0.010	0.061	0.036
4	5.6	36.0	60	1.65	0.061	0.064	0.089

[a] Galactosyltransferase and glucose-6-phosphatase assays were carried out at 37°; all others at 32°. Galactosyltransferase was assayed in the presence of 0.6 mg Triton X-100 per milliliter using 1 m*M* GlcNAc as acceptor.[7] NADH–cytochrome *c* reductase, succinate–cytochrome *c* reductase, and glucose-6-phosphatase assays were carried out as described previously.[2] 5′-Nucleotidase assays were carried out using [^{14}C]AMP.[6] Protein was determined by the method of O. H. Lowry, N. J. Rosebrough, A. L. Farr, and R. J. Randall [*J. Biol. Chem.* **193,** 265 (1951)] using bovine serum albumin as a standard.[6] Phosphorus was determined by the method of P. S. Chen, T. Y. Toribara, and H. Warren [*Anal. Chem.* **28,** 1756 (1956)]. Pr = protein.

In general, the preparations from D_2O–sucrose gradients have a higher galactosyltransferase activity than comparable fractions from H_2O–sucrose gradients and the yield is better. In addition, the fractions appear to be more intact (Table II). Both galactyosyl- and sialyltransferase activities are stimulated to a greater extent by treatment with Triton X-100 in the Golgi preparation made in D_2O–sucrose than in those made in H_2O–sucrose. This is true with substrates of small molecular weight as well as with protein substrates.

Morphologically, the Golgi apparatus fractions prepared using D_2O–sucrose step gradients appear virtually intact when the fractions are fixed as they are isolated from the gradients before dilution and centrifugation to recover the fractions in a concentrated form (Fig. 2A). They are similar in morphology to the equivalent fractions prepared in H_2O–sucrose gradients that we have described previously.[2] After dilution with 0.25 *M* sucrose in H_2O, they remain intact. Centrifugation and resuspension using gentle homogenization disrupts the characteristic Golgi morphology, however, and produces a heterogeneous population of filled and empty vesicles, ringlike structures, and unstacked cisternae (Fig. 2B). Attemtps to preserve the structure during the final centrifugation step by sedimenting on a cushion have not been entirely successful. The cisternae are better preserved but are largely unstacked. It is likely that the extra shear resulting from recentrifugation, as well as the loss of some components necessary to maintain the integrated structure, is involved in the partial fragmentation of the structure during the final steps of the isolation.

TABLE II

EFFECT OF TRITON X-100 ON ACCESSIBILITY OF SUBSTRATES TO GLYCOSYLTRANSFERASES OF RAT LIVER GOLGI

	Specific activity[a]					
	Prepared in H_2O–sucrose			Prepared in D_2O–sucrose		
Triton X-100	–	+	+/–	–	+	+/–
UDPGal → GlcNAc	348	600	1.7	253	1121	4.5
UDPGal → ovalbumin	22	143	6.4	6	283	46
CMP-NAN → lactose	30	123	4.1	64	387	5.7
CMP-NAN → transferrin	84	1106	13.2	39	1804	46

[a] Expressed as nanomoles of galactose or sialic acid transferred per milligram of protein per hour at 37°. Incubations were carried out with the substrate indicated for 5 min at 37°.[7] Triton X-100, when present, was added to a final concentration of 0.6% (w/v). The column marked +/– is the ratio of specific activity found in the presence of Triton X-100 (+) to that found in the absence of detergent (–).

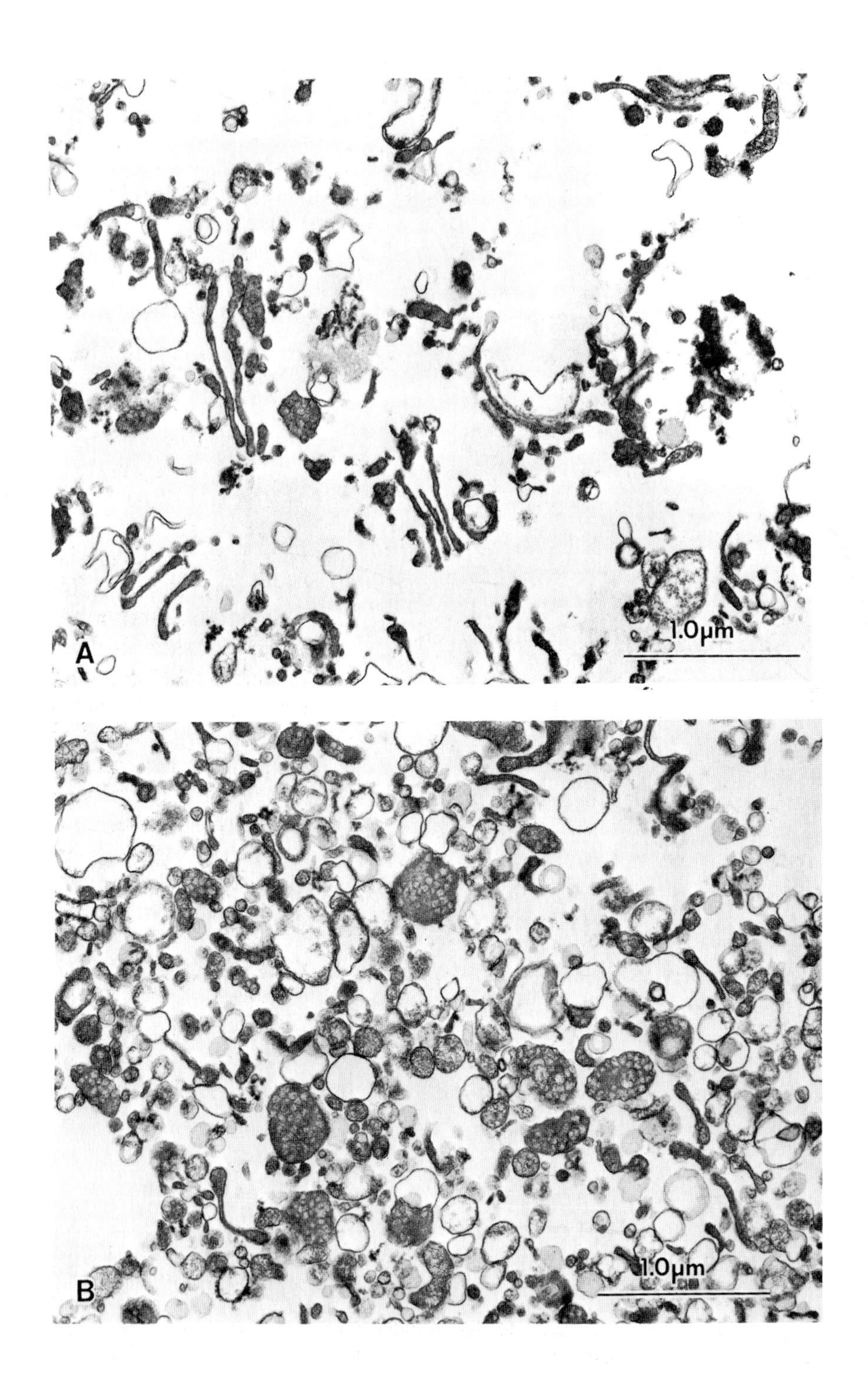
1.0µm
A
1.0µm
B

Acknowledgments

The author would like to thank Akitsugu Saito for the electron micrographs used in this paper and Kathryn Dewey for technical assistance. This work was supported in part by NIH Grant AM-14632.

[7] Membranes of Pancreatic Zymogen Granules

By JACOPO MELDOLESI

Zymogen granules (ZG) of pancreatic acinar cells are large (~ 1 μm in diameter) spherical cytoplasmic organelles, where digestive enzymes (zymogens) are stored in concentrated form prior to their release to the extracellular space. The number of ZG per cell varies in different species and fluctuates depending on the feeding state of the animal. In starved guinea pigs, the ZG compartment accounts for ~ 6.4% of the volume[1] and ~ 30% of the protein[2] of acinar cells. However, owing to the large size of granules, its cumulative surface area (~ 2.6% of total cellular membranes[1]) is much smaller than that of the other cytoplasmic membrane-bound compartments. In conventional thin-section electron microscopy, the ZG membrane reveals no distinctive features. In freeze-fracture, however, it is characterized by a very low density (number per unit area) of intramembrane particles, most of which are attached to the protoplasmic fracture face.[3] Zymogen granules arise in the Golgi area by concentration of secretory zymogens within immature granules (called the condensing vacuoles) formed by the fusion of multiple trans Golgi vesicles. Mature ZG move to the apical portion of the cytoplasm. Discharge occurs by exocytosis: the ZG membrane and the lumenal portion of the plasmalemma undergo focal fusion followed by collapse of the fused area. Thus, digestive zymogens gain access to the acinar lumen, while the granule membrane is incorporated at the cell surface. Discharge of pancreatic zymogens is a regulated process; it occurs at a low rate during starvation and is greatly accelerated by either feeding or acinar cell stimulation with peptide hormones (cholecystokinin,

[1] R. P. Bolender, *J. Cell Biol.* **61,** 269 (1974).
[2] J. Meldolesi, J. D. Jamieson, and G. E. Palade, *J. Cell Biol.* **49,** 109 (1971).
[3] P. De Camilli, D. Peluchetti, and J. Meldolesi, *Nature* (*London*) **248,** 245 (1974).

FIG. 2. Electron micrographs of the Golgi fraction isolated by procedure B using D_2O–sucrose gradients. Samples were fixed in suspension in 2.5% glutaraldehyde as described previously[2] (×25,000). (A) Fractions 2 + 3 were fixed as recovered from the gradients. (B) Final Golgi sample was fixed after diluting Golgi fractions 2 + 3 as in (A) and recovering the Golgi by ultracentrifugation and gentle rehomogenization in 0.25 *M* sucrose in H_2O.

ISBN 0-12-181998-1

bombesin), neurotransmitters (acetylcholine) and their analogs, *in vivo* as well as *in vitro.* No function, except storage and discharge, and no membrane enzyme activity has been unambiguously shown to be associated with pancreatic ZG.

Zymogen granule membranes are not isolated directly from tissue homogenates but are obtained by subfractionation of purified granules. Most studies carried out during the last decade have employed, sometimes with modifications, the isolation procedure[2] that I developed in collaboration with J. D. Jamieson and G. E. Palade, following original observations of L. E. Hokin.[4] A critical account of this technique is given below.

Isolation of Zymogen Granules

For the purification of ZG we took advantage of the fact that these organelles, when centrifuged into a pellet, tend to stick tightly to each other at the bottom of tubes, and thus to separate from other sedimented organelles, which are recovered in more superficial layers of the pellet. Under carefully controlled centrifugation conditions, and provided that the ZG concentration in the starting preparation is large, a highly purified fraction can be obtained.

Pancreata, rapidly removed from guinea pigs starved for 18 hr, are chilled by dipping into ice-cold, 0.3 *M*, unbuffered sucrose. The rest of the isolation is carried out at 0–4°. The glands are trimmed free of fat and mesentery, minced (with scissors or razor blades), and homogenized in 10 volumes of the same sucrose solution. Originally, three up-and-down strokes in a loose-fitting Teflon–glass, Brendler-type homogenizer (A. H. Thomas, Philadelphia, type C), operated at 3000 rpm, were applied. Later, however, other homogenization protocols were employed as profitably, provided that harsh shearing was avoided. Homogenates are not filtered before centrifugation; actually, the presence of tissue debris is essential at this stage. Six-milliliter aliquots distributed in 12-ml conical glass tubes are centrifuged at 1000 g_{max} in a Model K International centrifuge (or similar machine equipped with a swinging-bucket rotor); the supernatants are discarded. The pellets contain the bulk of ZG present in the homogenate. Although some of these granules are clumped into small white masses, most of them remain separate, interspersed in the tridimensional network formed by collagen fibers, large tissue and cell debris, nuclei, and membrane sheets. This pellet is overlaid with 5 ml of 0.3 *M* sucrose and resuspended manually using a small Teflon rod. Special care is applied to crush the small white masses rich in ZG, that might have remained attached to the glass wall. Tubes are then centrifuged at very low speed (180 g_{max}, 12 min; same

[4] L. E. Hokin, *Biochim. Biophys. Acta* **18**, 379 (1955).

centrifuge), enough to sediment intact cells, nuclei, and large debris and leave in suspension most ZG and small contaminants (primarily mitochondria). The resulting supernatants (~ 3 ml per tube) are aspirated and filtered through 110-mesh nylon gauze into 15-ml, round-bottomed Corex tubes, which are centrifuged at 1000 g_{max} for 3.5 min in an International centrifuge. Under these conditions many ZG pack tightly into thin, round white pellets, covered by loose tan layers rich in mitochondria. These layers are easily removed be gently streaming 1 ml of 0.3 *M* sucrose along the tube wall, followed by careful swirling and aspiration (repeated twice). Recovery of ZG can be increased by ~ 25% by resuspending the pellet of the 180 *g* centrifugation in 3 ml of 0.3 *M* sucrose and repeating the isolation procedure.

The pellets thus obtained were systematically monitored by electron microscopy and found to be composed almost exclusively of well-preserved ZG (90–95%) and condensing vacuoles (2–5%). Rare mitochondria (~ 2%) and occasional rough microsomes (~ 0.5%) were the only observed contaminants. The high purity of the ZG fraction was confirmed by extensive biochemical and marker enzyme characterization, part of which is summarized in the table.

Subfractionation of Zymogen Granules

Like all other secretory granules, ZG consist of two components: limiting membrane and segregated content. The latter is accounted for essentially by soluble proteins, the digestive enzymes, copacked in a highly concentrated form. Thus, isolation of the membrane can be obtained easily by centrifugation after solubilization of the content. With other types of granules, such a solubilization has been successfully obtained by hypotonic shock. Zymogen granules, however, are stable even in distilled water, provided that the pH < 7, but are lysed readily when exposed to slightly alkaline buffers. Thus, suspensions of ZG in saline or distilled water (protein concentration: 5–10 mg/ml) are cleared by dilution with 2.5 volumes of 0.2 *M* $NaCO_3$, pH 7.8, or of other buffers, such as Tris and HEPES. With guinea pig ZG, clearing is very rapid, so that shearing or incubation of the diluted preparations in the cold is not necessary. However, a 20-min incubation in the cold has been recommended by others working with rat ZG.[5] Centrifugation at high speed (200,000 g_{max}, 30 min) of the cleared preparations yields supernatants (digestive zymogens), and pellets that contain ZG ghosts, a few partially disrupted ZG as well as contaminants. Although the latter constitute only a minute component (2–3%) of the intact ZG preparations, they

[5] R. A. Ronzio, K. E. Kronquist, D. S. Lewis, R. J. MacDonald, S. H. Mohrlok, and J. J. O'Donnel, Jr., *Biochim. Biophys, Acta* **508,** 65 (1978).

CHEMISTRY AND ENZYME ACTIVITIES OF GUINEA PIG PANCREATIC HOMOGENATES, ZYMOGEN GRANULES (ZG), AND ZG SUBFRACTIONS[a,b]

Fraction	Protein (mg/g tissue wet wt)	PLP (μg/mg protein)	α-Amylase[c] (mg/mg protein)	Chymotrypsin[c] (μmol/mg protein)	Cytochrome oxidase[c]	NADH[c] + cytochrome c reductase	NADPH[c] + cytochrome c reductase
						(nmol/mg PLP)	
Homogenate	144.0	77.1	5.0	4.8	210.0	550.0	32.0
ZG	3.8	13.5	14.5	20.1	195.0	85.0	0
Lysed ZG pellet	0.23	175.5	2.9	3.6	190.5	80.5	0
ZG membranes	0.035	580.0	0.0	0.0	97.0	32.8	0

[a] Data are from Meldolesi *et al.*[2,11] and unpublished results.

[b] The figures represent: α-amylase, milligrams of maltose released per minute at 30°; chymotrypsin, micromoles of BTEE split per minute at 25°; cytochrome oxidase, NADH- and NADPH-cytochrome *c* reductase, nanomoles of cytochrome *c* oxidized or reduced per minute at 25°.

[c] Specific activity on a phospholipid (PLP) basis of cytochrome oxidase and NADH-cytochrome *c* reductase in purified mitochondrial fraction was 1789 and 110; of NADH- and NADPH-cytochrome *c* reductase of purified rough microsomal fraction, 181.0 and 27.7.

are greatly enriched in the membrane pellets as a consequence of the extraction of digestive enzymes (which account for over 99% of the protein in intact ZG). In order to obtain more highly purified ZG membranes, a further step is therefore necessary. The pellets are resuspended in a small volume (2–3 ml) of 1 *M* sucrose (which has a density intermediate between that of ZG membranes and contaminants), transferred to centrifuge tubes and overlaid with 0.3 *M* sucrose. After centrifugation at 100,000 *g* for 60 min (Spinco SW 50.1 or similar rotors), the ZG membranes are recovered as a band at the interface, and the bulk of the contaminants sediment in a pellet.

As already mentioned, the isolation procedure that I have described was developed originally for guinea pig ZG membranes. Later, however, it was used by us and others with other species as well: rat, mouse, dog, calf, pig. In general, the results were similar, but a few differences (indicated below) also emerged. The procedure is satisfactory especially for the purity of the membrane fraction obtained, inasmuch as the contamination by other membranes does not exceed 10% (see the table). However, the procedure suffers from two major limitations: low recovery (~ 10%) and artifactual adsorption (or incomplete solubilization) of soluble zymogens onto the isolated membranes.

Low recovery becomes a problem especially when isolation is attempted from pancreatic tissue poor in ZG, namely, after animal feeding or pharmacological stimulation. Under these conditions, or for studies necessitating large quantities of membranes, a new procedure for ZG isolation proposed by Brockmeyer and Palade, and so far published only in abstract form,[6] might be appropriate. In this procedure rat pancreatic postnuclear supernatants are loaded onto a linear sucrose (10 to 44%) gradient containing 5% Ficoll 400 and 1 m*M* EDTA, in a Beckman zonal rotor Ti 15. After centrifugation, 50–60% of ZG originally present in the homogenate are said to separate in a band, with a purity comparable to that obtained with our procedure. This interesting result seems to depend primarily on the presence of Ficoll throughout the gradient, inasmuch as centrifugation on pure sucrose gradients yields fractions heavily contaminated with mitochondria.

Adsorption of soluble zymogens might also be a problem, especially for studies of membrane biogenesis. In fact, secretory proteins turn over at rates that are one to two orders of magnitude faster than membrane proteins, so that under certain conditions, even if present in minute amounts, they can account for the bulk of the radioactivity recovered in the membrane fraction.[7] The extent of the adsorption phenomenon seems to vary in different species. In fact, rat ZG membranes, when analyzed enzymically or by

[6] T. F. Brockmeyer and G. E. Palade, *J. Cell Biol.* **83,** 272a (1979).

[7] J. Meldolesi, *J. Cell Biol.* **61,** 1 (1974).

SDS–polyacrylamide gel electrophoresis, show measurable contamination by content enzymes, such as amylase and trypsinogen,[5] whereas in the guinea pig the contamination, although present, is below detection with these techniques.[7] To deal with the adsorption, additional washes of the isolated membranes have been proposed: with 0.25 *M* NaBr plus sonication[5]: with 0.1 M Na_2CO_3, pH 11.2.[6] This last treatment, especially, has been reported to remove the adsorbed zymogens very efficiently. However, the possibility that, in addition, exposure to high pH causes disruption, or partial extraction, of the membranes has not yet been explored in detail.

Characterization of ZG Membranes

Zymogen granule membranes, isolated from various animal species, have been analyzed in various laboratories. Results on lipids appear to be in good agreement.[2,5,8,9] The phospholipid (PLP) -to-protein ratio (w/w) is of the order of 0.6–0.8, and the major phospholipids are phosphatidylcholine and phosphatidylethanolamine. Analogously to the plasma membrane, ZG membranes are rich in sphingomyelin ($\sim$ 23%) and free cholesterol (0.55 mol per mole of PLP), and contain glycolipids. Phosphatidylserine and phosphatidylinositol are also present. Moreover, a large proportion ($\sim$ 40%) of fatty acids in PLP are saturated.[9] Both in the guinea pig and in the pig, considerable concentrations of lysophosphatidylcholine, -ethanolamine, and -serine were found in ZG membranes. However, these lyso-PLP are not genuine membrane components, but arise during isolation by hydrolysis of the corresponding diacyl-PLP.[8,9] In the guinea pig, the enzyme responsible for such an artifact is probably lipase.[8,10] Another artifact that occurs during isolation is accumulation of large quantities of free fatty acids that are digestion products of triglycerides released during homogenization from peripancreatic fat cells.[8] The high concentrations of cholesterol and saturated fatty acids suggest that ZG membranes have a low permeability, as is to be expected for a membrane destined to be temporarily inserted at the surface of the cell.

Zymogen granule membranes are extraordinarily poor in enzymes.[11] Only a Mg^{2+}-ATPase was found to exhibit a considerable activity, which, however, was certainly contributed, at least in large part, by contaminating mitochondrial membrane fragments.[11] Enzymes typical of endoplasmic reticulum and Golgi membranes (including sugar transferases[5]) were not

[8] J. Meldolesi, J. D. Jamieson, and G. E. Palade, *J. Cell Biol.* **49,** 130 (1971).

[9] W. J. Rutten, J. J. H. H. M. De Pont, S. L. Bonting, and P. J. M. Daemen, *Eur. J. Biochem.* **54,** 259 (1975).

[10] S. Durand, F. Clemente, J. P. Thouvenot, J. Fauvel, and L. Douste-Blazy, *Biochimie* **60,** 1215 (1978).

[11] J. Meldolesi, J. D. Jamieson, and G. E. Palade, *J. Cell Biol.* **49,** 150 (1971).

found, and 5′-nucleotidase and other plasmalemma enzymes show low activity (one-fifth to one-tenth of the plasmalemma level[11]). Moreover, the distribution of these enzymes can be nonhomogeneous in the granule population. Thus, acid phosphatase is present in condensing vacuoles but is lacking in mature ZG.[12]

Interesting data emerged from SDS–polyacrylamide gel electrophoresis. Compared to the other membranes of acinar cells, ZG membranes appear to be extraordinarily simple; i.e., they are composed of only a few (10–20) polypeptides, instead of several hundreds. In the rat and dog, the major component (~40%) is a glycoprotein (M_r 70,000–80,000) named GP2, which is exposed at the inner surface of the membrane.[5,6] In other species, especially the guinea pig,[13] GP2 is present, but in a much smaller amount. The existence of this well-identified component, present in large amount and endowed in addition with good antigenicity (see below), initially attracted a great deal of interest. More recently, however, biochemical and immunochemical studies revealed that a glycoprotein with a slightly lower molecular weight but otherwise indistinguishable (immunologically and by fingerprinting) from GP2 is present in the ZG content and is released to the pancreatic juice.[14,15]

Thus, at the moment the membrane status of GP2 is in question. In fact, it might well be only a secretory protein capable of adsorbing to the inner face of the ZG membrane, or, alternatively, a protein with a dual (membrane-content) localization,[14] as is recognized for dopamine β-hydroxylase in chromaffin granules.[16] Two minor glycoproteins (GP1, possibly exposed at the cytoplasmic surface, and GP3, exposed at the internal face) have been described in rat ZG membranes.[5] However, little is known about them except their apparent molecular weights. In all animal species, one or a few nonglycosylated peptides with $M_r \sim 15{,}000$ have been observed. The possibility that these peptides are digestion artifacts has been considered.[5] However, in the guinea pig no change in the low-molecular-weight peptide was observed when the membranes were isolated in the presence of protease inhibitors.[13]

Finally, sulfated macromolecules (glycosaminoglycans and glycopeptides), as yet characterized only in small part, have been described in ZG, where they could be associated with the inner face of the membrane. As in the case of GP2, however, their status (truly membrane-bound or adsorbed) is still uncertain.

[12] A. B. Novikoff, M. Mori, N. Quintana, and A. Yam, *J. Cell Biol.* **75,** 148 (1977).
[13] J. Meldolesi and D. Cova, *J. Cell Biol.* **55,** 1 (1972).
[14] R. C. T. Scheffer, G. Poort, and J. W. Slot, *Eur. J. Cell Biol.* **23,** 122 (1980).
[15] H. J. Geuze, K. W. Slot, P. van der Lay, and R. C. T. Scheffer, *J. Cell Biol.* **89,** 653 (1981).
[16] H. Winkler and E. Westhead, *Neuroscience* **5,** 1803 (1980).

Exocytosis and Membrane Biogenesis

The ZG membrane fraction might be a useful tool with which to investigate two problems of paramount importance in cell biology: exocytosis and the biogenesis and recycling of membranes. With regard to exocytosis, a reasonable hypothesis is that the rate changes of the process are due to molecular change(s) of specific components in the membranes involved: ZG and/or lumenal plasma membranes. In the ZG membrane, however, no such changes have been identified as yet. Thus, the hydrolysis of phosphatidylinositol, which occurs in acinar cell membranes after stimulation, can be dissociated from the exocytotic response,[17] and the accumulation of lysoPLP is most probably an artifact (see above). Moreover, the phosphorylations of ZG membrane proteins described so far have doubtful physiological meaning and are not correlated with zymogen discharge.[18] Ca^{2+}-dependent binding of calmodulin to ZG membranes has been described, but the responsible membrane proteins could not be identified.[19]

With regard to membrane biogenesis, our own studies demonstrated that the various proteins of ZG membranes turn over asynchronously, at rates that are much slower than those of secretory zymogens ($t_{\frac{1}{2}}$ of the order of days instead of hours[7]). Thus, ZG membranes are not destroyed after insertion at the cell surface during exocytosis but are recycled to the cytoplasm, reused in several secretory cycles, and replaced not in bulk, but molecule by molecule, as is the case for other cellular membranes. The mechanisms underlying the process of recycling are as yet incompletely understood. In fact, taking into account the marked differences in composition between ZG membranes, on the one hand, and the membranes existing at both terminals of the pathway traveled by ZG, i.e., the Golgi and plasma membranes, it is necessary to postulate not only that recycling matches in quantity the membrane additions and removals occurring in relation to zymogen transport, but also that the recycled membranes are identical in composition to ZG membranes. Tracer studies by Herzog and Reggio[20] have lead to the identification of a pathway of membrane traffic initiated at the luminal surface by coated pits and vesicles. These vesicles, after shedding their coat, move to the Golgi area and fuse with trans cisternae and condensing vacuoles. On the other hand, formation of coated vesicles does not occur at random over the entire lumenal surface, but is more concentrated at the level of large infoldings, which represent previously fused granules.[21] Thus, the ZG membrane inserted at the cell surface might be

[17] R. H. Michell, *Biochim. Biophys, Acta* **415,** 81 (1975).

[18] D. S. Lewis and R. A. Ronzio, *Biochim. Biophys, Acta* **583,** 422 (1979).

[19] D. Bartelt and G. Scheele, *J. Cell Biol.* **91,** 398a (1981).

[20] V. Herzog and H. Reggio, *Eur. J. Cell Biol.* **21,** 141 (1980).

[21] J. Meldolesi and B. Ceccarelli, *Proc. R. Soc. London Ser. B* **296,** 55 (1981).

directly recaptured shortly after exocytosis. Moreover, the molecular intermixing that might possibly have occurred after fusion between ZG and plasma membranes could be eliminated by the coated vesicle mechanism, whose molecular filtration function has been recognized.[22] Since coated pits and vesicles have been found also in the Golgi area, in apposition or continuity to cisternae and condensing vacuoles, it is possible that this mechanism operates also at the proximal terminal of the ZG pathway, to sort out patches of ZG membrane from Golgi membrane components.

The scheme of membrane recycling that we have briefly summarized is certainly compatible with the available evidence. However, at the present time it should be considered only hypothetical. Direct investigations in this field would require the study of specific ZG membrane proteins, to be identified in intact cells and isolated in pure form from homogenates or cell fractions. Taking into account the low concentration of these proteins, these studies could be carried out only by using specific antibodies (immunochemistry and immunocytochemistry at the electron microscope level). Unfortunately, however, the only available antibodies (obtained in the rabbit by injection of either the pure protein or entire ZG membranes) are directed against GP2,[6,15,16] which, because of its yet dubious association with the membrane, and concomitant localization in the ZG content, does not appear to be suitable for these studies.

[22] B. M. F. Pierce and M. S. Bretscher, *Anna. Rev. Biochem.* **50,** 85 (1981).

[8] Secretory Membranes of the Rat Parotid Gland: Preparation and Comparative Characterization[1]

By PETER ARVAN, RICHARD S. CAMERON, and J. DAVID CASTLE

Exocrine secretory glands consist of functionally polarized epithelial cells having an organization such that synthesis of macromolecules (especially proteins) destined for export begins near the base of the cell, packaging into membrane-bounded containers and intracellular storage as secretion granules occur near the apex of the cell, and discharge takes place selectively at the apical cellular surface. This region, designated the apical plasma membrane, represents a discrete fraction of the cell surface that is segregated from the basolateral domain by junctional complexes. Release of secretory products by exocytosis involves the specific interaction of secretion granule and apical plasma membranes, focal rearrangement of bilayer structures in

[1] These studies were supported in part by USPHS Grants GM 26524 and AM 29868.

ISBN 0-12-181998-1

regions of contact, and eventual coalescence of the two membranes with direct extracellular deposition of secretory products. Although exocytosis may be initiated by the fusion of these two membrane types, secreting cells of many tissues (including nonexocrine varieties) can exhibit compound exocytosis in which further release takes place by fusion of granules with the membranes of previously discharged granules that are already continuous with the apical plasma membrane. In this case, granule membranes are equivalent to plasma membranes as acceptors for further granule discharge. Consequently, granule and apical plasma membranes may contain common constituents that facilitate the specific events of the release process in stimulated cells.

As initial steps in undertaking a comparative compositional analysis of membranes that serve as partners in exocytosis, we have obtained purified fractions of secretion granule membranes and plasma membranes containing the apical domain from rat parotid gland. This tissue was chosen for study because a single secretory cell type constitutes 85–90% of the cellular volume of the gland and because it is known to contain especially low levels of lipolytic and proteolytic activities[2] that potentially could elicit alterations of membrane components during isolation.

Preparation of a Plasma Membrane Fraction Containing the Apical Domain

Since enzymic marker activities exclusive to the apical surface of acinar cells of parotid are not known, we sought an isolation procedure that could at least partially preserve morphologically recognizable regions constituting the apical surface. Two important features of the procedure developed are that (*a*) it selects for large sheets of membrane and (*b*) very hypoosmotic media are used at the outset, as in the case of plasma membrane purification from rat liver,[3] so that osmotically sensitive organelles (especially secretion granules and mitochondria) are either damaged or disrupted during homogenization. A flow diagram of the procedure is shown in Fig. 1.

Reagents

Medium A: 0.5 m*M* $MgCl_2$, 1 m*M* $NaHCO_3$, pH 7.4

Medium B: 0.5 m*M* $MgCl_2$, 1 m*M* $NaHCO_3$, 0.7 m*M* EDTA, pH 7.4

Medium C: 0.5 m*M* $MgCl_2$, 1 m*M* $NaHCO_3$, 1.7 m*M* EDTA, pH 7.4

Procedure. The plasma membrane preparation is carried out entirely at 4°. Parotid tissue (5.5 g), obtained from 16 male Sprague–Dawley rats (100–125 g) starved overnight, is cleaned of connective tissue, minced with

[2] M. Schramm and D. Danon, *Biochim. Biophys. Acta* **50,** 102 (1961).

[3] T. K. Ray, *Biochim. Biophys. Acta* **196,** 1 (1970).

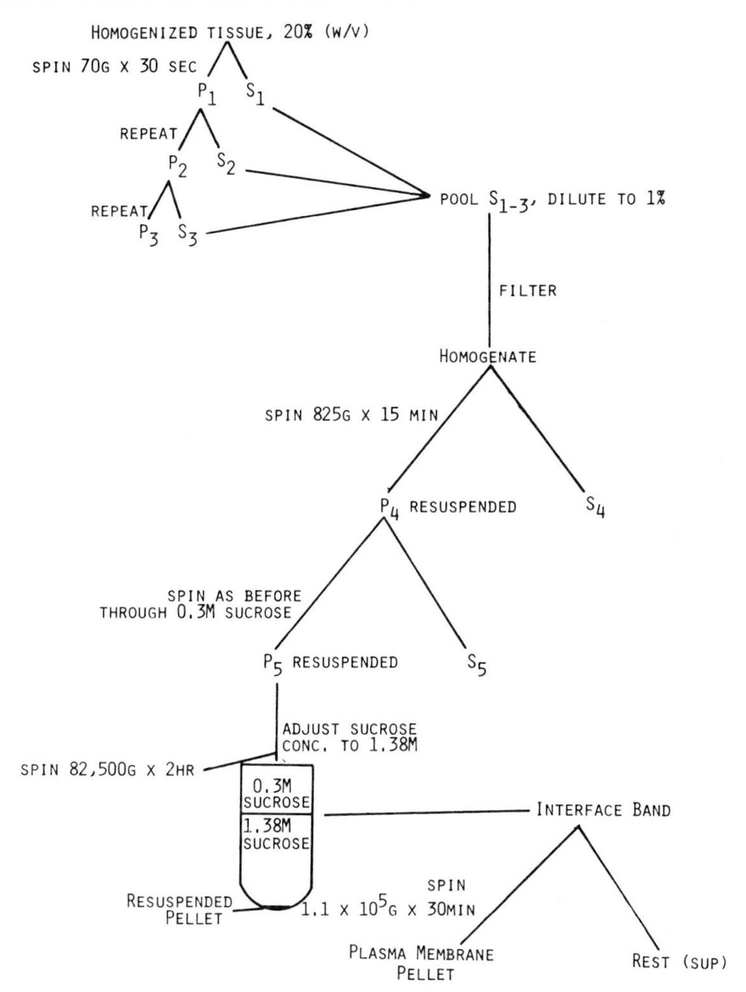

FIG. 1. Flow diagram of plasma membrane preparation.

razor blades, and homogenized (15 sec at 1900 rpm with a Polytron homogenizer followed by three strokes with a Brendler Teflon pestle homogenizer at 1300 rpm) in medium A at 20% w/v. The resulting suspension is spun at 70 *g* for 0.5 min to yield the first supernatant and a pellet that is rehomogenized with the Brendler homogenizer in the same volume of medium A. The procedure is repeated again, the final pellet is discarded, and the three supernatants are pooled and diluted to 1% (original tissue weight per volume) with medium B. Dilution and net EDTA serve to reduce organelle aggregation. The suspension is filtered through four layers of cheesecloth plus one layer of 105-μm nylon mesh to produce the homogenate, the initial fraction on which recoveries of enzyme activities are based. Differential centrifugation of the homogenate (825 $g_{av} \times 15$ min) is used to obtain the fourth supernatant (S_4, saved for assays) and the fourth pellet (P_4) consisting of nuclei, acinar debris (especially basement membrane-containing elements), aggregated microsomes, and plasma membrane sheets. P_4 is resuspended (three strokes in a Dounce homogenizer with tight pestle), diluted to one-half the volume of S_{1-3} with medium B, layered above 5-ml 0.3 *M* sucrose in medium B in 30-ml Corex (Corning Glass) tubes, and subjected to recentrifugation under identical conditions. The 0.3 *M* sucrose layer minimizes contamination of the pellet (P_5) by microsomal elements retained in the supernatant (S_5). P_5 is resuspended in ~ 4 ml of sucrose concentration of 1.38 *M*, and diluted to 125 ml using 1.38 *M* sucrose in medium B. The diluted suspension is loaded into centrifuge tubes, overlaid with 0.3 *M* sucrose in medium B, and subjected to centrifugation ($9.9 \times 10^6 g_{av} \times$ min) in a Beckman SW 27 rotor in order to float plasma membrane sheets to the 0.3–1.38 *M* interface away from the other particulates noted in P_4 (and P_5) that pellet under these conditions. The interface band collected in a small volume is supplemented with EDTA to give a net concentration of 1.2 m*M*, diluted to 0.35 *M* sucrose with medium C, thoroughly dispersed with three strokes in a tight-fitting Dounce homogenizer (to reduce membrane aggregation), and subjected to centrifugation ($3.3 \times 10^6 g_{av} \times$ min) in a Beckman SW 41. Plasma membrane pellets and the overlying supernatant are used for assays.

Electron Microscopic Characterization of the Plasma Membrane Fraction

Since morphological appearance constitutes an important initial evaluation for the purity and integrity of isolated membranes containing distinct regions, aliquots of the fraction are fixed in aldehydes,[4] postfixed in 1% osmium tetroxide, and prepared for microscopy according to standard procedures. Figure 2 shows a representative low-power electron micrograph of the plasmalemmal fraction. The fraction predominantly consists of large membrane sheets that show in part regions retaining the organization of the apical surface *in situ.* Apical membranes are joined to one another by elements of the junctional complex, often forming closed profiles, the former cytoplasmic aspects of which are oriented outwardly. Extracted or collapsed microvilli invaginate from the surface into the space bounded by the apical membranes; basolateral membranes extend beyond the junctional complexes either to contact other apical compartments or to terminate as free ends. Frequently several lumenal profiles, mostly with cross-sec-

[4] R. C. Graham and M. J. Karnovsky, *J. Histochem. Cytochem.* **14,** 291 (1966).

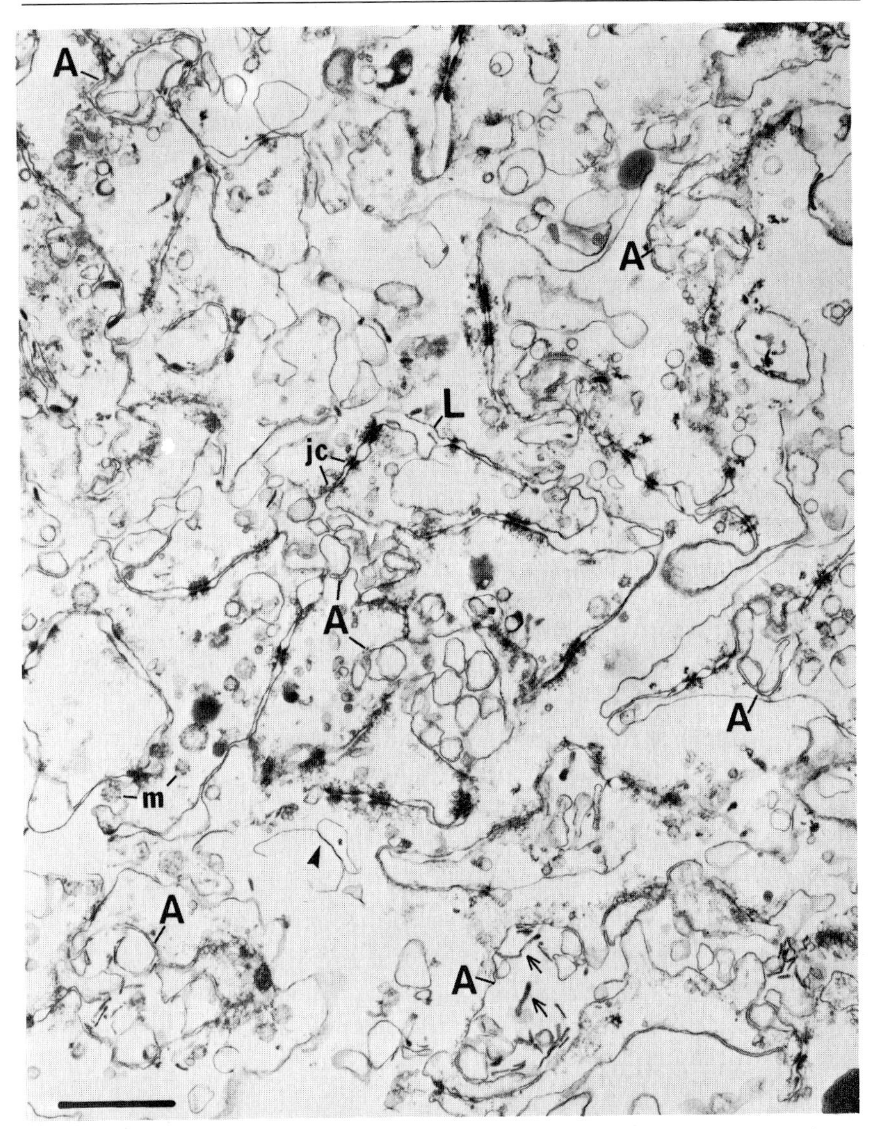

FIG. 2. Representative electron micrograph of the purified parotid plasma membrane fraction. Individual plasmalemmal sheets are quite extended and contorted and contain multiple profiles of apical membrane (A) enclosing the former lumenal space. Junctional complexes (jc) adjoin the apical surfaces of neighboring cells, and the contiguous lateral membrane (L) is often multiply studded with desmosomes. Lumenal compartments contain remnants of microvilli often appearing as vesicular profiles (in favorable sections, continuous with the apical surface) and less frequently as apparently highly flattened vesicles (arrows at lower right). In addition, the fraction can be seen to contain unidentified smooth membrane elements (arrowhead) possibly deriving from the plasma membrane sheets, microsomes (m), and filamentous material especially in the vicinity of junctional elements and the apical surface. Bar = 1 μm.

tional diameters comparable to those of secretion granules, are observed in individual sheets of plasma membrane. Both the small diameters and multiplicity of profiles suggest that the membranes originate from acinar secretory cells and that the organization corresponds to that of the tubular secretory canaliculi constituting the luminal surface *in situ*.[5] This interpretation is favored further since an estimated 85–90% of the gland volume is composed of acinar cells.

Morphological characterization additionally indicates that the plasmalemmal fraction is contaminated by low levels of rough microsomes and ribosomal aggregates; occasionally, damaged mitochondria and, rarely, centrioles and extracted nuclei are observed in the fraction. Cytoplasmic filaments, emanating from elements of the junctional complex, frequently can be seen in association with the membrane sheets.

Plasma Membrane Yield and Analysis for Organelle Contaminants Using Marker Enzyme Activities

Typically ~0.15 mg of protein is recovered in the purified plasma membrane fraction per gram wet weight of tissue. This yield is unusually low in comparison to that of preparations from liver, where 5–10 times as much protein is obtained from a comparable amount of tissue.[6] The reduced yield probably reflects a decreased amount of large membrane sheets as a result of the vigorous homogenization required to disrupt connective tissue and basement membranes surrounding acini.

The low level of contamination of the plasma membrane fraction observed by electron microscopy is confirmed by assays of enzymes considered to mark specific organelles selectively. Membrane-associated enzymes are assayed except for secretion granules and lysosomes, where unique markers have not been identified. In these cases it is possible to examine only the extent of contamination by species present in the organelle content.

Table I presents, for the purified plasmalemmal fraction, data that have been extracted from comprehensive studies of activity distributions throughout the fractionation procedure. Generally, recoveries of enzyme activities at each step of fractionation are very good; the exception is UDPgalactosyltransferase (EC 2.4.1.22), where recovery of activity at the final step of plasma membrane purification is incomplete. As can be seen in Table I, 0.4% or less of total homogenate activity for each marker is recovered in the final plasma membrane fraction, with at most a low level of enrichment for any contaminant.

[5] J. F. Parks, *Am. J. Anat.* **108,** 303 (1961).

[6] A. L. Hubbard, D. A. Wall, and A. Ma, *J. Cell Biol.* **96,** 217 (1983).

TABLE I
ORGANELLE CONTAMINANT ANALYSIS OF THE PLASMALEMMAL FRACTION[a]

Enzyme	Organelle	% Recovery[b]	Relative specific activity[c]
Cytochrome *c* oxidase[d]	Mitochondria	0.27	1.17
NADH–cytochrome *c* reductase[e] (rotenone-insensitive)	Microsomes	0.40	1.76
UDPgalactosyltransferase[f]	Golgi complex	0.39	1.70
β-*N*-Acetyl-D-glucosaminidase[g]	Lysosomal content	0.29	1.26
α-Amylase[h]	Granule content	0.008	0.04

[a] Values presented are averages for a minimum of three experiments.

[b] Percentage recovery is based on 100% activity in the homogenate.

[c] Relative specific activity = specific activity ratio, plasmalemmal fraction: homogenate. Protein was assayed according to Markwell *et al.*[12]

[d] Cytochrome *c* oxidase [T. J. Peters, M. Muller, and C. de Duve, *J. Exp. Med.* **136,** 1117 (1972)]. The first-order rate constant is used as a measure of activity [E. E. Max, D. B. P. Goodman, and H. Rasmussen, *Biochim. Biophys. Acta* **511,** 224 (1978)]. Analysis of amine oxidase indicates a distribution very similar to that reported for cytochrome *c* oxidase.

[e] G. L. Sottocasa, B. Kuylenstierna, L. Ernster, and A. Bergstrand, *J. Cell Biol.* **32,** 415 (1967). This activity is indicated as a microsomal marker, although in other systems it has been shown to mark mitochondrial outer membranes, Golgi membranes [N. Borgese and J. Meldolesi, *J. Cell Biol.* **85,** 501 (1980)], and even plasma membranes [E. D. Jarasch, J. Kartenbeck, G. Bruder, D. J. Morré, and W. W. Franke, *J. Cell Biol.* **80,** 37 (1979)]. Thus the recovery and purification probably reflect the sum of contamination by a variety of organelles.

[f] UDPgalactosyltransferase: B. Fleischer, this series, Vol. 31, p. 180. The assay uses [^{3}H]UDPgalactose in the presence of 2 m*M* ATP and 2 mg of asialogalactofetuin per milliliter.

[g] β-*N*-Acetyl-D-glucosaminidase [J. Findlay, G. A. Levvy, and C. A. Marsh, *Biochem. J.* **669,** 467 (1958)].

[h] α-Amylase. P. Bernfeld, this series, Vol. 1, p. 149.

Preparation of Secretion Granule Membranes

Precursor Fraction: Secretion Granules. As starting material for granule membrane purification, secretion granules have been isolated by a procedure that capitalizes on their high density in hyperosmotic sucrose solutions. Routinely, 4.5–5.0 g of tissue (corresponding to the parotid glands of 14–16 rats starved overnight) thoroughly cleaned of lymph nodes and surrounding connective tissue are used in a procedure modified from that developed by Castle *et al.*[7] for purification of granules from rabbit parotid glands. Processing involved the following steps carried out at 4°: (*a*) mincing of tissue with razor blades and homogenization—10 sec with a Polyton (1900 rpm) followed by five strokes (1300 rpm) with a Brendler homogenizer—in 0.3 *M*

[7] J. D. Castle, J. D. Jamieson, and G. E. Palade, *J. Cell Biol.* **64,** 182 (1975).

sucrose, 10 mM imidazole, 0.5 mM $MgCl_2$, pH 7.1; (*b*) sedimentation (70 g $\times$ 30 sec) of large particulates and their rehomogenization (Brendler mortar and pestle only) in the same volume to yield a total homogenate, 5% tissue weight per volume; (*c*) low-speed centrifugation (7000 $g_{av} \times$ min) to remove nuclei and cellular debris; (*d*) filtration of the supernatant through 20-μm nylon mesh and addition of EDTA to a final concentration of 5 mM; (*e*) recentrifugation (1.3×10^5 $g_{av} \times$ min) of the supernatant on a sucrose step gradient having two layers containing 1.0 M and 2.0 M sucrose, respectively, each supplemented with 4% Ficoll 400, 10 mM imidazole, and 5 mM EDTA; (*f*) collection of a crude granule fraction at the 1.0/2.0 M sucrose interface and adjustment of the suspension to contain 1.5 M sucrose, 4% Ficoll 400, 10 mM imidazole, 5 mM EDTA; (*g*) centrifugation (1.24×10^7 $g_{av} \times$ min) in a sucrose step gradient having underlayers of 1.55 M and 2.0 M sucrose and an overlayer of 0.8 M sucrose (each containing the same additives as the crude granule load); (*h*) collection of the purified granules from the 1.55–2.0 M interface. Using the secretory protein α-amylase as an approximation of a granule-specific marker for parotid tissue, we routinely recover ~22% of the total activity of the homogenate in the purified fraction. Mitochondria constitute the only significant organelle contaminant both morphologically and biochemically; 1.7% of the amine oxidase (flavin-containing) activity (an outer membrane marker[8]) of the homogenate is located in the granule fraction.[9]

To prepare purified granules for lysis, sucrose is diluted to 0.9 M using 0.2 M sucrose, 10 mM imidazole, and 5 mM EDTA, and the granules are pelleted by centrifugation (1.6×10^6 $g_{av} \times$ min). The dilution typically results in a 15–20% loss of granules by lysis, which is thus far unavoidable.

Reagents for Granule Lysis and Membrane Purification

Lysis medium: 0.19 M KCl, 10 mM imidazole, 5 mM EDTA, pH 7.1
Buffered sucroses: 2.0 M, 0.9 M sucrose each containing 0.19 M KCl, 10 mM imidazole, 5 mM EDTA, pH 7.1

Procedure. The protocol developed for lysis takes advantage of the known lability of parotid secretion granules in KCl-containing media.[10] Purified granules are resuspended and diluted to 30 ml in lysis medium and maintained at 0° for 12 hr; clearing of the suspension indicates extensive lysis. In order to separate the low-density granule membranes from organelle contaminants and soluble content proteins, the lysate is adjusted to a

[8] M. R. Castro Costa, S. Edelstein, C. M. Castiglione, H. Chao, and X. O. Breakfield, *Biochem. Genet.* **18,** 577 (1980).

[9] We are currently modifying the granule purification procedure to reduce the level of mitochondrial contamination further.

[10] M. Schramm, R. Ben-Zvi, and A. Bdolah, *Biochem. Biophys. Res. Commun.* **18,** 446 (1965).

final sucrose concentration of 1 *M* (using buffered 2 *M* sucrose), overlaid with buffered 0.9 *M* sucrose and lysis medium, respectively, and subjected to centrifugation ($1.5 \times 10^7\ g_{av} \times$ min) in a Beckman SW 27.1 rotor. Granule membranes band at the lysis medium–0.9 *M* interface, whereas residual contaminants (mitochondria) and unlysed granules sediment into a pellet. The membranes are collected, diluted to 0.07 *M* sucrose with lysis medium, and pelleted by centrifugation ($1.5 \times 10^7\ g_{av} \times$ min). Resuspended membranes are used for assays and morphology.

Electron Microscopic Observation of the Granule Membrane Fraction

Processing of secretion granule membranes for electron microscopy employs the same procedure already outlined for plasma membranes. The fraction is very homogeneous in appearance, consisting of closed, smooth-surfaced vesicles 0.3–0.8 μm in diameter with no evidence of residual granule content (see Fig. 3). The diameters are smaller than those of intact granules (~ 1 μm) indicating that the vesicles represent fragments of the original membranes.

Estimation of Mitochondrial and Soluble Secretory Contaminants

Since mitochondria constitute the principal organelle contaminant of the granule fraction, the distribution of amine oxidase activity is followed during purification of granule membranes. With 90–100% recovery of total activity, 3–4% of that originally present in the granule lysate is detected in the membrane fraction. On the basis of the specific activity (units of amine oxidase per micromole of lipid phosphorus[11] and per milligram of protein[12]) of a purified parotid mitochondrial fraction, we estimate that a maximum[13] of 1% of the lipid and 3% of the protein of the granule membrane fraction could be contributed by mitochondrial contamination.

Contamination of granule membranes by secretory proteins is estimated by two means: (*a*) by following routinely the distribution of α-amylase during subfractionation of granule lysates; (*b*) for a more comprehensive examination, by adding biosynthetically labeled rat parotid secretory proteins (prepared in the manner of Castle *et al.*[7])to granules during lysis, since this reagent is a more representative marker of total granule content.

[11] G. R. Bartlett. *J. Biol. Chem.* **234,** 446 (1959).

[12] M. A. K. Markwell, S. M. Hass, L. L. Bieber, and N. E. Tolbert, *Anal. Biochem.* **37,** 206 (1978).

[13] The estimated levels of contamination constitute upper limits because the most probable contaminants of the granule membrane fraction are outer mitochondrial membranes rather than intact mitochondria. The former will have much higher amine oxidase specific activities, hence will contribute less lipid and protein per unit of activity as contaminants.

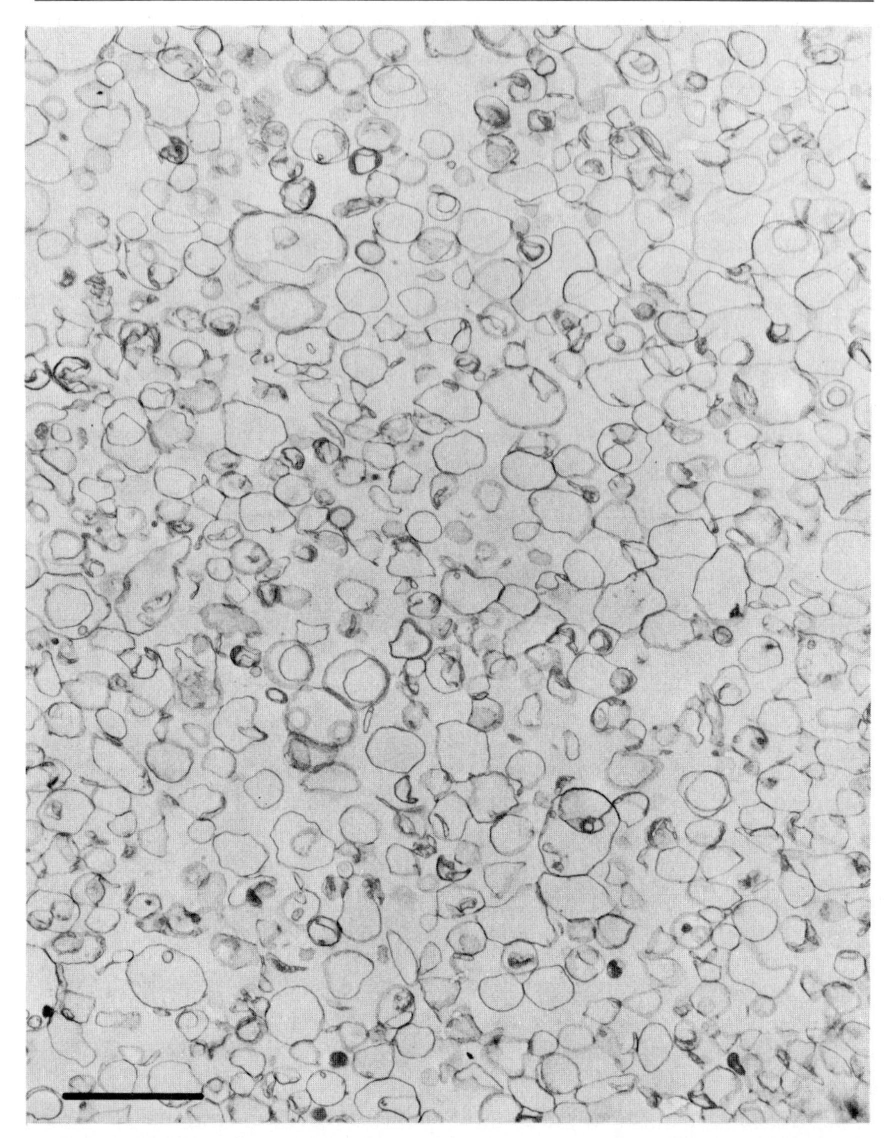

FIG. 3. Electron micrograph of the secretion granule membrane fraction. Membranes appear as smooth-surfaced, mostly closed vesicles, in some cases multivesicular, apparently owing to resealing of larger membrane pieces around smaller membrane fragments. Bar = 1 μm.

Radioactive protein found in the final fraction (detected by scintillation counting and by fluorography[14] of ^3H-labeled polypeptides resolved in SDS–gel electrophoretograms) is thought to compete and hence is used to approximate the extent of trapping and/or adsorption of residual content to the purified granule membranes. In both cases the membranes are freed of more than 99.99% of the contaminant markers. Although these levels of contaminant removal suggest that the granule membranes are highly purified, residual amylase in the membrane fraction is estimated to represent ~5% of the total protein of this fraction (based on the specific activity of purified rat parotid amylase: 2780 units per milligram of protein[15]). Consequently, we estimate that, taken together, mitochondrial protein and residual secretory protein constitute at most 10–15% of the protein of the granule membrane fraction.

Enzyme Activities Associated with the Plasmalemma and Secretion Granule Membranes

Assays for several enzymes known to mark plasma membranes of other cell types indicate that our parotid plasmalemmal fraction is not significantly enriched in alkaline phosphatase and leucyl aminopeptidase (1.4-fold and 1.2-fold higher specific activities than the homogenate, respectively) and is only modestly enriched in 5′-nucleotidase (3.8-fold over the homogenate). As shown in Table II, two enzymes, sodium–potassium adenosine triphosphatase (Na^+,K^+-ATPase) and γ-glutamyltransferase (γ-glutamyltranspeptidase; GGTPase) exhibit large enrichments. Although distributions of these markers are discussed in detail elsewhere,[16] several features are worthy of mention here.

1. The low enrichments for alkaline phosphatase and 5′-nucleotidase favor an origin of the plasma membrane sheets in acinar cells, since histochemical studies have suggested that these activities are associated with other parotid cell types.[17, 18] Consequently, these markers are of little use where the isolation of membranes involved in exocytosis in parotid tissue is intended; morphological observations provide the key means of analysis for apical secretory surface.

2. Na^+,K^+-ATPase is a classical marker of the basolateral domain in polarized epithelia. Although the specific activity increases substantially in discontinuous gradient centrifugation, ~70% of the activity of P_5 pellets

[14] R. A. Laskey and A. D. Mills, *Eur. J. Biochem.* **56,** 335 (1975).
[15] T. G. Sanders and W. J. Rutter, *Biochemistry* **11,** 130 (1972).
[16] P. Arvan and J. D. Castle, *J. Cell Biol.* **95,** 8 (1982).
[17] J. R. Garrett and P. A. Parsons, *Histochem. J.* **5,** 463 (1973).
[18] S. Yamashina and K. Kawai, *Histochemistry* **60,** 255 (1979).

TABLE II
ENZYMES MARKING PAROTID PLASMALEMMA AND SECRETION GRANULE MEMBRANES[a]

Enzyme	Fraction	Recovery (%)	Relative specific activity	Specific activity (per μmol lipid phosphate)
Na^+,K^+-ATPase[b]	Homogenate	100	1.0	—
	Plasma membranes	4.6	20.1	—
	Homogenate	100	1.0	—
	Secretion granules (23% of total amylase)	0.4	0.07	—
	Granule membranes	ND[c]	—	—
GGTPase[d]	Homogenate	100	1.0	—
	Plasma membranes	5.9	25.6	0.68
	Homogenate	100	1.0	—
	Secretion granules (23% of total amylase)	4.4	0.3	—
	Granule membranes	2.8	30.6	0.21

[a] With the exception of data reported for Na^+,K^+-ATPase recovery in granules, which represent the average of two experiments, all data reported constitute the averages for at least six different experiments.

[b] Na^+,K^+-ATPase has been shown to be equivalent in rat parotid tissue to potassium-stimulated *p*-nitrophenylphosphatase and is quantitated by measuring *p*-nitrophenol formation in an assay mixture containing 5 m*M* *p*-nitrophenylphosphate, 0.1 *M* imidazole, pH 7.5, 0.01 *M* $MgCl_2$, 0.75 m*M* EGTA, 0.01 *M* NaCl, and 0.01 *M* either KCl or choline chloride.

[c] ND, not detected.

[d] γ-Glutamyltransferase [S. S. Tate and A. Meister, *J. Biol. Chem.* **249,** 7593 (1974)].

rather than floats to the interface band. This activity is likely to be associated with large plasma membrane-containing structures intimately contacting basement membranes. We are anticipating that for the purified fraction the Na^+,K^+-ATPase activity resides in the lateral membrane segments emanating from the junctional complexes in the plasmalemmal sheets.

3. γ-Glutamyltransferase has been shown to be highly enriched on the free surface of epithelial cells, notably those specialized for secretion and absorption.[19] We favor the notion of a similar localization in parotid acinar cells, especially since, in contrast to Na^+,K^+-ATPase, ~70% of the GGTPase activity of P_5 is found in the interface band. Consequently, these two enzymes may constitute markers for distinct plasmalemmal domains in parotid acinar cells and in the isolated membrane sheets.

4. γ-Glutamyltransferase is present in secretion granule fractions at an average yield (six separate preparations) of 4.4% of the total homogenate

[19] A. Meister and S. S. Tate, *Annu. Rev. Biochem.* **45** 559 (1976).

activity. We estimate that ~ 17% of the total tissue GGTPase is associated with secretion granules according to the assumption that amylase approximates a granule-specific marker and that we recover about one-quarter (23%) of the total tissue amylase activity in the purified granule fraction. Since granule-associated protein represents a large percentage of the total protein of parotid, GGTPase is actually depurified (threefold in Table II) in preparing a secretion granule fraction. The recovery of GGTPase activity in the steps following granule lysis is 98%. After taking into account the loss of purified granules during sucrose dilution and centrifugation prior to lysis, we find that 80% of the GGTPase activity of the granule fraction is recovered in the purified membranes. Most of the remaining activity is found in the load and pellet fractions of the discontinuous gradient.

5. The presence of GGTPase activity in granule membranes (and the apparent absence of Na^+,K^+-ATPase) suggests that there is a compositional overlap of membranes that serve as fusion partners during exocytosis. Further, the overlap is qualitative but apparently not quantitative, since the specific activity of GGTPase relative to lipid phosphorus (as a reliable normalization for this membrane-associated enzyme) for the granule membrane fraction is 30% that of the plasmalemmal fraction. Considering that the plasma membrane-associated activity may be concentrated in regions comprising the apical surface, the quantitative difference in surface concentration may be as much as 10-fold.[20]

These studies provide the rationale for a detailed immunocytochemical study of GGTPase localization. Further, they form the basis for future detailed investigations of compositional similarities and differences between membranes that attain functional continuity and of the resulting implications to the mechanisms of both exocytosis and membrane recycling.

[20] At this point 10-fold clearly represents an estimate for two reasons: (1) The value is based on data obtained with purified fractions, rough microsomes; and mitochondria (which are low-level contaminants of the plasmalemmal fraction) contain very low activities of GGTPase. Hence their contribution of lipid phosphorus exceeds their contribution of GGTPase to the plasmalemmal fraction. (2) At present we do not have quantitative estimates of the fraction of surface area of isolated plasma membrane that constitutes apical region.

Section II

Importance of Glycosylation and Trimming; Targeting

[9] Overview: Role of Intracellular Membrane Systems in Glycosylation of Proteins

By WILLIAM LENNARZ

All cells are involved in the synthesis of glycoproteins destined to become components of the plasma membrane. In addition, many cell types commit a significant portion of their protein biosynthetic activity to the synthesis of glycoproteins that are secreted and others that are packaged in lysosomes. The synthesis of these three classes of glycoproteins is a highly segregated process that occurs within an intracellular membrane system composed of the endoplasmic reticulum, transfer vesicles, Golgi complex, and secretory vesicles. During their translation, glycosylation, and processing, the glycoproteins are completely isolated from the cytoplasm and travel to the cell surface (in the case of membrane glycoproteins), to the extracellular environment (in the case of secretory glycoproteins), or to the lysosomes (in the case of lysosomal enzymes) as part of, or within, these membrane compartments. The two common classes of glycoproteins to be considered in this overview are those containing N-glycosidically linked oligosaccharide chains attached to asparagine residues and O-glycosidically linked oligosaccharide chains linked to serine or threonine residues in the polypeptide. Synthesis of the O-linked chains found in proteoglycans and in collagen will not be discussed.

It has been known for many years that the polysomes containing mRNA species that direct the synthesis of secretory proteins are found in association with the endoplasmic reticulum. It is now clear that this association is the result of synthesis of a signal peptide at the amino terminus of nascent chains of membrane and secretory proteins that directs their insertion into the rough endoplasmic reticulum (RER).[1] Upon elongation and completion of the polypeptide chain, the secretory proteins are deposited in the lumen of the RER, whereas the plasma membrane proteins remain associated with the membrane of the endoplasmic reticulum.

The relationship between the translation of polypeptides destined to become glycoproteins and their glycosylation is shown schematically in Fig. 1. There is considerable evidence that N-glycosylation of asparagine residues occurs in concert with polypeptide synthesis; i.e., it is a cotranslational process. In contrast, most of the available evidence indicates that O-glycosylation of serine (or threonine) residues is a posttranslational process and

[1] D. Sabatini, G. Kreibich, T. Morimoto, and M. Adesnick, *J. Cell Biol.* **92,** 1 (1981).

ISBN 0-12-181998-1

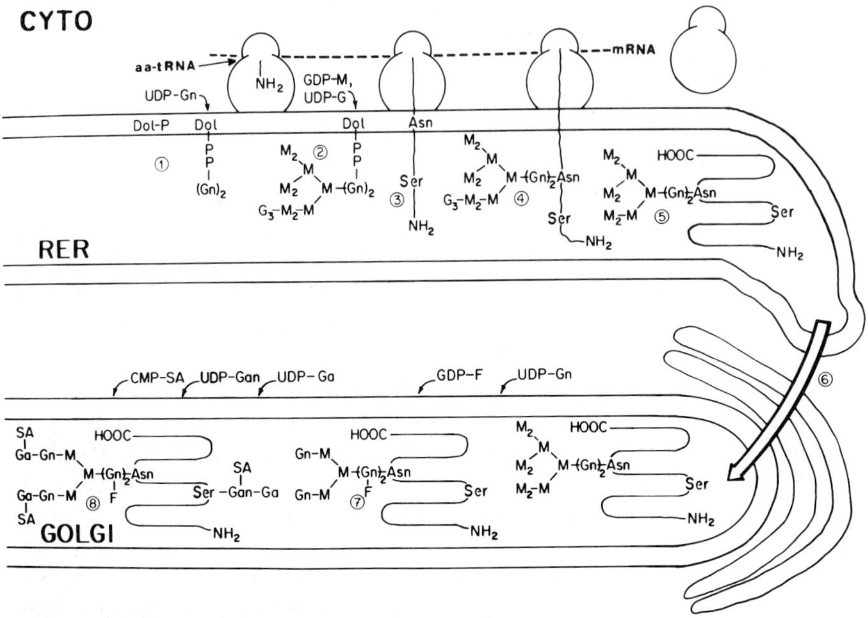

FIG. 1. Diagram outlining some of the steps in the assembly and processing of N- and O-linked oligosaccharide chains on glycoproteins. Adapted from Hanover and Lennarz.[2] Gn, *N*-acetylglucosamine; M, mannose; G, glucose; F, fucose; Ga, galactose; Gan, galactosamine; SA, sialic acid; Dol, dolichol.

that it occurs at a later stage in intracellular routing in the Golgi complex.[2,3] In Fig. 1, for simplicity a single polypeptide chain, containing both a N- and a O-linked oligosaccharide chain is shown, and the key steps in synthesis, attachment of each type of chain to the polypeptide, and processing are denoted by the numerals 1 through 8.

The initial precursors to the saccharide units of oligosaccharide chains are the sugar nucleotides, which are made by enzymes in the cytoplasm. However, the assembly as well as the processing of the oligosaccharide is catalyzed by membrane-associated enzymes. Furthermore, in the case of the dolichylphosphate-mediated synthetic reactions it is clear that these enzymes are restricted to the endoplasmic reticulum; little activity is found in the Golgi complex.[4] In the initial step in synthesis of N-glycosidically linked chains, two *N*-acetylglucosamine units are transferred to dolichylphosphate,

[2] J. Hanover and W. J. Lennarz, *Arch. Biochem. Biophys.* **211,** 1 (1981).

[3] D. Struck and W. J. Lennarz *in* "The Biochemistry of Glycoproteins and Proteoglycans" (W. J. Lennarz, ed.), p. 35. Plenum, New York, 1980.

[4] A. M. Ravoet, A. Amar-Costesec, D. Godelaine, and H. Beaufay, *J. Cell Biol.* **91,** 679 (1981).

a long-chain, phosphorylated polyisoprenol that serves as an anchor for the preassembly of the entire oligosaccharide chain (step 1).[3] The transfer of the first *N*-acetylglucosamine units is inhibited by the drug tunicamycin, which has proved to be very useful in studies on glycoprotein synthesis and function.[4a] After the second *N*-acetylglucosamine unit is added, the resulting disaccharide lipid (chitobiosyl-lipid) is known to be asymmetrically situated on the membrane of the endoplasmic reticulum, with the disaccharide unit facing the lumen. Since both *in vitro* and *in vivo* experiments indicate that UDP-*N*-acetylglucosamine does not enter the lumen and that chitobiosyl-lipid does not undergo transmembrane "flip-flop," it appears that this asymmetric orientation is introduced during the synthesis of the chitobiosyl-lipid.[2]

Subsequent elongation of the chitobiosyl-lipid to produce a "mature" oligosaccharide chain, still attached to dolichylpyrophosphate is a complex process (step 2). It involves not only GDPmannose, but also mannosylphosphoryldolichol and glucosylphosphoryldolichol as direct glycosyl donors, as outlined in Fig. 2. As noted, the enzymes involved in synthesis of these dolichol-linked monosaccharides from the appropriate sugar nucleotide and dolichylphosphate are specifically associated with the endoplasmic reticulum. As shown, the inner five mannosyl units come directly from GDPmannose whereas the outer four mannosyl units are derived from mannosylphosphoryldolichol.[5] The source of the glucose units is glucosylphosphoryldolichol. Since it appears that GDPmannose (like UDP-*N*-acetylglucosamine) is not permeable to the membrane,[6] this process probably involves some type of vectorial enzyme-mediated transfer of the mannosyl unit from the nucleotides to the growing oligosaccharide chains. The completed oligosaccharide chains attached to dolichylpyrophosphate face the lumen.[6a] This finding is consistent with the observation that oligosaccharide released from the lipid, probably by enzymic hydrolysis, is situated in the lumen.[6] Therefore it appears highly unlikely that the growing oligosaccharide–lipid undergoes flip-flop and chain elongation on the cytoplasmic face of the endoplasmic reticulum. Rather, it appears to retain the topological orientation of the chitobiosyl-lipid from which it was derived.

The next step is transfer of the completed oligosaccharide chain to the growing polypeptide chain (Fig. 1, steps 3 and 4). Not all asparagine residues in a given polypeptide serve as acceptors. Studies from a number of laboratories using unfolded proteins and simple peptides as substrates for oligosaccharyl transferase indicate that the sequence N----Asn-X-Ser/Thr----C is

[4a] A. D. Elbein, this volume [11].

[5] J. T. Rearick, K. Fujimoto, and S. Kornfeld, *J. Biol. Chem.* **256,** 3762 (1981).

[6] J. Hanover and W. J. Lennarz, *J. Biol. Chem.* **257,** 2787 (1982).

[6a] M. D. Snider and P. W. Robbins. *J. Biol. Chem.* **257,** 6796 (1982).

```
              Dol-PP-Gn2
        GDP-M→ |                M-M-M
               ↓               /
              Dol-PP-Gn2-M
               |               \
               |                M
GDP-M→M-P-Dol→ | ←G-P-Dol←UDP-G
               ↓                M-M-M-G-G-G
                               /
              Dol-PP-Gn2-M      M-M
                               \ /
                                M
                                 \
                                  M-M
```

FIG. 2. Participation of both nucleotide- and dolichylphosphate-linked sugars in synthesis of the dolichylpyrophosphate-linked oligosaccharide.

required.[3,7,8] Although it is clear that in small peptides this is both a necessary and a sufficient requirement for glycosylation, this does not appear to be true with large, native polypeptides. For example, in the case of ovalbumin the polypeptide chain contains two potentially glycosylatable sites, -Asn$_{293}$-Leu-Thr- and -Asn$_{311}$-Leu-Ser-, but *in vivo* only Asn$_{293}$ is found to be glycosylated. However, a peptide containing the second site, as well as the complete protein already glycosylated at Asn$_{293}$, can be glycosylated *in vitro* if it is first denatured.[9,10] Therefore it appears that in intact, native polypeptide chains conformational factors as well as the primary structure determine which -Asn-X-Ser/Thr- sites are glycosylated.

If, indeed, the oligosaccharide–lipid has the topological orientation shown, during translation the asparagine residue to be glycosylated must, at the very least, be located near the luminal face of the RER (step 3). For this to be possible, the polypeptide chain must be sufficiently long at the time of oligosaccharide transfer to transmit both the ribosomal cleft and most of the membrane bilayer. Experiments examining the glycosylation of nascent chains of ovalbumin indicate that this is the case; at least 50 amino acid residues are added after attachment of the asparagine residue to the growing ovalbumin chain before it is glycosylated.[9]

The function of the glucose residues added during the assembly of the oligosaccharide–lipid is unclear. Several *in vitro* studies have shown that the glycoslated oligosaccharide–lipid is a better substrate for transfer of the protein than unglycosylated or incompletely mannosylated oligosaccharide–lipid.[3,11,12] One possibility is that glucose addition serves as a signal for completion and therefore transfer. Such a signal would minimize transfer of

[7] C. Ronin, C. Granier, C. Coseti, S. Bouchilloux, and J. V. Rietschoten, *Eur. J. Biochem.* **118,** 159 (1981).

[8] C. Palmarczyk, G. Lehle, T. Mankowski, T. Chajnacki, and W. Tanner, *Eur. J. Biochem.* **105,** 517 (1980).

[9] C. Glabe, J. Hanover, and W. J. Lennarz, *J. Biol. Chem.* **255,** 9236 (1980).

[10] D. Pless and W. J. Lennarz, *Proc. Natl. Acad. Sci. U.S.A.* **74,** 134 (1977).

[11] S. J. Turco, B. Stetson, and P. W. Robbins, *Proc. Natl. Acad. Sci. U.S.A.* **74,** 4411 (1977).

[12] C. A. Murphy, and R. C. Spiro, *J. Biol. Chem.* **256,** 7487 (1981).

only partially completed oligosaccharide chains. However, the finding that *in vivo* transfer of mannose-poor chains to the polypeptide occurs in mutant cells that cannot add the terminal mannose units makes this idea less attractive.[13] In any case, it is clear that after transfer of the glucose-containing oligosaccharide to the protein the glycose units are excised (step 5). At least two membrane-associated glucosidases are involved.[14-16]

The next step is movement of the glycoproteins to the Golgi complex (step 6). The movement process is not well understood, but presumably is brought about by pinching off vesicles that bear the secretory glycoproteins in their lumen or the membrane glycoproteins in their membranes, or both. Subsequent fusion of these transfer vesicles, presumably at the cis face of the Golgi complex, results in deposition of the secretory protein in the lumen and the membrane protein in the membrane of the Golgi complex. This and other related processes in the Golgi complex are discussed in more detail elsewhere in this volume.[16a] After complete removal of the glucose residues and perhaps trimming of a few mannose residues, depending on the particular glycoprotein, the oligosaccharide is complete if the mature glycoprotein is destined to be of the polymannose type. If it is destined to be of the complex type, once in the Golgi complex a series of posttranslational modifications in the N-linked oligosaccharide chain occurs. The initial event is mannosidase-mediated removal of all but three of the innermost mannose residues (step 7). This is a somewhat more complex process than shown, since only after one branch of the chain has been completely trimmed and one *N*-acetylglucosamine residue has been added can the other branch be degraded.[17] Also, only after at least one *N*-acetylglucosamine residue has been added can the fucose unit be added.[17] At the present time it is unclear why the oligosaccharide chains of glycoproteins that, in their mature form, retain polymannose chains are not processed by these enzymes in the Golgi complex. One possibility is that the polymannose chains in these proteins, because of conformational factors, are not accessible to the processing mannosidases.

The final stage in maturation of the complex-type chains is addition of galactose and sialic acid units (step 8). These two glycose additions, as well as the earlier two that occur in the Golgi complex, utilize sugar nucleotides as the direct glycosyl donors; there is no evidence for an intermediate role of

[13] S. Kornfeld, W. Gregory, and A. Chapman, *J. Biol. Chem.* **254,** 11644 (1979).
[14] R. A. Ugalde, R. J. Staneloni, and L. F. Leloir, *FEBS Lett.* **91,** 209 (1978).
[15] L. S. Grinna and P. W. Ribbins, *J. Biol Chem.* **254,** 8814 (1979).
[16] J. J. Elting, W. W. Chen, and W. J. Lennarz, *J. Biol. Chem.* **255,** 2325 (1980).
[16a] A. M. Tartakoff, this volume [5].
[17] H. Schachter and S. Roseman, *in* "The Biochemistry of Glycoproteins and Proteoglycans" (W. J. Lennarz, ed.), p. 85. Plenum, New York 1980.

dolichylphosphate, as is the case in the RER. Another difference is that, unlike the RER, the membrane of Golgi complex appears to be permeable to at least some of the sugar nucleotides.[18, 19] Thus, the assembly, processing, and terminal glycosylation steps appear to be spatially and temporally segregated, the former occurring in the endoplasmic reticulum and the latter two in the Golgi complex. In addition, there is evidence that within the Golgi complex itself there may be segregation, with the earlier mannosidase reactions occurring in the cis portion of the Golgi and the later terminal glycosylation reactions occurring in the trans region.[20] This is discussed in more detail elsewhere in this volume.[20a] Finally, it should be noted that, in addition to these processing reactions in the Golgi complex, the oligosaccharide chains of glycoproteins that are destined to be lysosomal enzymes are modified by a phosphorylation process,[21-23] the end product of which is the introduction of a signal for routing to the lysosomes. This process is discussed elsewhere.[23a]

The process involving addition of *N*-acetylgalactosamine to form the key linkage found in oligosaccharide chains O-linked to serine (or threonine) residues (step 8) has not been studied as extensively as the N-glycosylation process. Although some evidence for cotranslational glycosylation has been reported, most findings indicate that it is a posttranslational process occurring in the Golgi complex. *In vitro* studies on the characteristics of the two processes studied in membranes from the same source indicate that, unlike N-glycosylation, O-glycosylation of peptide acceptors is unaffected by tunicamycin or dolichylphosphate, and that the enzyme activity is enriched in smooth membranes (Golgi complex plus smooth endoplasmic reticulum) rather than in the RER.[24] In addition, *in vivo* kinetic studies of the N- and O-glycosylation of a protein (human choriogonadotropin) that contains within its polypeptide chain both N- and O-linked oligosaccharide chains are consistent only with postttranslational addition of the O-linked chains.[25] Finally, completion of O-linked chains after attachment of the *N*-acetylgalactosamine unit involves, in the case shown, addition of sialic acid and galactose units. It should be mentioned that conformational factors may

[18] D. J. Carey and C. B. Hirschberg, *J. Biol. Chem.* **256,** 989 (1981).
[19] B. Fleischer, *J. Cell Biol.* **89,** 246 (1981).
[20] W. G. Dunphy, E. Fries, L. J. Urbani, and J. E. Rothman, *Proc. Natl. Acad. Sci. U.S.A.* **78,** 7453 (1981).
[20a] W. E. Balch, E. Fries, W. H. Dunphy, L. J. Urbani, and J. E. Rothman, this volume [4].
[21] A. Varki and S. Kornfeld, *J. Biol. Chem.* **255,** 8398 (1980).
[22] A. Waheed, R. Pohlman, A. Hasilik, and K. von Figura, *J. Biol. Chem.* **256,** 4150 (1981).
[23] D.E.Goldberg and S. Kornfeld, *J. Biol. Chem.* **256,** 13060 (1981).
[23a] K. E. Creek, H. D. Fischer, and W. S. Sly, this volume [23].
[24] J. Hanover. W. J. Lennarz, and J. D. Young, *J. Biol. Chem.* **255,** 6713 (1980).
[25] J. Hanover, J. Elting, G. R. Mintz, and W. J. Lennarz, *J. Biol. Chem.* **257,** 10172 (1982).

play a very important role in determining which serine residues in a polypeptide act as substrate for the *N*-acetylgalactosaminyl transferase. The only discernible primary structural requirement of serine-containing peptides is the presence of several proline residues.[26] Although the precise position of these proline residue in relation to the serine residue may not be crucial, it seems likely that their presence dramatically alters the conformation of the polypeptide.

It is of interest to contrast N- and O-glycosylation. First, the former is a cotranslational process that occurs in the RER, whereas the latter appears to be a posttranslational process occurring in the Golgi complex. Second, the former involves dolichylphosphate as an "anchor" on which the entire oligosaccharide chain is preassembled, whereas the latter involves simple, single glycose unit addition directly from the apparent sugar nucleotide. Given the first set of differences, the second set may be rationalized in terms of the differences in the dynamics of the two processes. Clearly, addition of nine sugars to an asparagine site in a rapidly growing polypeptide chain can be carried out with more fidelity if the entire oligosaccharide chain is "prepackaged." The necessity for such preassembly and rapid transfer may not exist after the polypeptide chain has been completed and deposited in the Golgi complex.

The final step in externalization of secretory and membrane proteins involves packaging in vesicles that arise by pinching off vesicles, possibly from the trans face of the Golgi complex. At the present time it is not clear if the same vesicle carries both the secretory and the plasma membrane glycoproteins, or if specialized vesicles exist. This subject is discussed in this volume.[27] In either case, the ultimate step is fusion of these vesicles with the plasma membrane, resulting in secretion of the secretory glycoprotein and incorporation of the membrane glycoprotein into the plasma membrane. Thus, from the very onset of their synthesis in the RER from cytoplasmic precursors, until they leave the cell, enter the lysosomes or become incorporated into the plasma membrane, glycoproteins exist within a highly organized membrane system. This membrane system contains a large battery of enzymes (at least 25 for formation of a glycoprotein with a complex-type chain), segregated in either the endoplasmic reticulum or the Golgi complex, that act in a sequential fashion to catalyze the synthesis and processing of oligosaccharide chains of glycoproteins.

[26] J. P. Briand, S. P. Andrews, E. Cahill, N. A. Conway, and J. D. Young, *J. Biol. Chem.* **256,** 12205 (1981).

[27] M. Farquhar, this volume [1].

[10] Glycosyltransferases Involved in Elongation of N-Glycosidically Linked Oligosaccharides of the Complex or *N*-Acetyllactosamine Type

By HARRY SCHACHTER, SAROJA NARASIMHAN, PAUL GLEESON, and GEORGE VELLA

Protein-bound oligosaccharides are classified according to the covalent linkage between amino acid and carbohydrate. The major linkages in avian and mammalian glycoproteins are the N-glycosidic linkage between Asn and GlcNAc and three types of O-glycosidic linkage, Ser(Thr)-GalNAc, Ser-xylose, and hydroxylysine-Gal.[1] The synthesis of the Asn-GlcNAc linkage is presently believed to be carried out by an oligosaccharide transferase in the rough endoplasmic reticulum that catalyzes the transfer of oligosaccharide from dolicholpyrophosphate oligosaccharide either to nascent ribosome-bound peptide or to postribosomal peptide.[2] In several tissues it has been shown that the initial peptide-bound oligosaccharide contains three Glc, nine Man, and two GlcNAc residues,[3] but this structure has not been established for all tissues. The peptide-bound oligosaccharide is then processed, first by removal of all three Glc residues in the rough endoplasmic reticulum and then by removal of 0 to 6 of the nine Man residues, primarily in the Golgi apparatus.[4] The number of Man residues removed varies from one oligosaccharide to the other. If zero to four Man residues are removed, the result is a high-mannose type of oligosaccharide. The smallest high-mannose oligosaccharide is the 5-Man structure, $(Man)_5(GlcNAc)_2$-Asn-peptide shown at the top of Fig. 1. This structure is the starting point for synthesis of *N*-glycosyl oligosaccharides of the complex or *N*-acetyllactosamine type.

Glycoproteins contain high-mannose oligosaccharides, complex oligosaccharides, or both (e.g., calf thyroglobulin). The biosynthetic factors that control this diversity are not known, but it is probable that the amino acid sequence near the glycosylation site is involved. The $(Man)_5(GlcNAc)_2$-Asn intermediate is the starting point for all complex *N*-glycosyl oligosaccharides (Fig. 1). A series of Golgi apparatus-localized glycosyltransferases and

[1] J. Montreuil, *Adv. Carbohydr. Chem. Biochem.* **37,** 157 (1980).

[2] D. K. Struck and W. J. Lennarz, *in* "The Biochemistry of Glycoproteins and Proteoglycans" (W. J. Lennarz, ed.), p. 35. Plenum, New York, 1980.

[3] R. Kornfeld, and S. Kornfeld, *in* "The Biochemistry of Glycoproteins and Proteoglycans" (W. J. Lennarz, ed.), p. 1. Plenum, New York, 1980.

[4] I. Tabas and S. Kornfeld, this series, Vol. 83, p. 416.

ISBN 0-12-181998-1

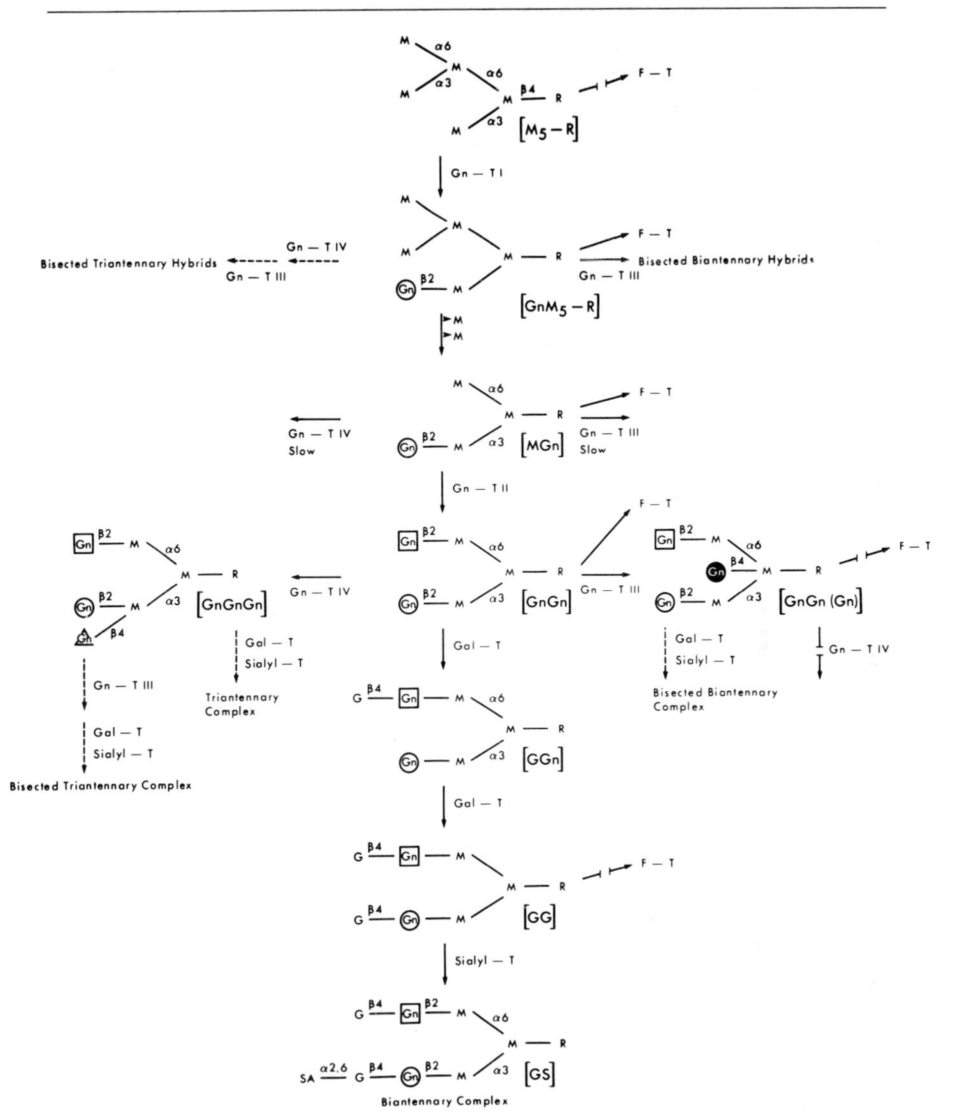

FIG. 1. Scheme showing the biosynthesis of the various classes of complex *N*-glycosyl oligosaccharides. These reactions are believed to occur in the Golgi apparatus and begin with the appearance of the 5-Man intermediate shown at the top of the scheme. The abbreviations are as follows: M, D-mannose (Man); Gn, *N*-acetyl-D-glucosamine (GlcNAc); R, GlcNAcβ1-4GlcNAc-Asn-X; F, L-fucose (Fuc); T, transferase; G, D-galactose (Gal); SA, sialic acid. Glycopeptides carrying complex oligosaccharides (3 Man residues only) are named according to the sugars present at the nonreducing termini of the antennae, the Man-α1-6- arm being named first; a bisecting GlcNAc is indicated by the symbol (Gn). Continuous arrows indicate reaction paths established *in vitro;* dashed arrows indicate postulated pathways, and interrupted arrows indicate non-allowed paths.

α-mannosidases act on this compound to produce several classes of complex structures.[5]

Bi-, tri-, and tetraantennary complex oligosaccharides contain a core with only three Man residues, $(Man)_3(GlcNAc)_2$-Asn, on which are attached two, three, or four antennae. An antenna can be incomplete (either a single GlcNAc, or a Gal-GlcNAc disaccharide) or complete (a sialyl-Gal-GlcNAc trisaccharide). The linkage between Gal and GlcNAc is usually β1-4 (hence the name *N*-acetyllactosamine) but is occasionally β1-3. The linkage between sialic acid and Gal is either α2-3 or α2-6. Bisected complex oligosaccharides are similar except for an additional GlcNAc residue linked β1-4 to the β-linked Man residue of the core; this GlcNAc bisects the antennae on the two arms of the core (Fig. 1), e.g., in ovotransferrin.[6] Hybrid oligosaccharides contain either one or two Man residues attached to the Man-α1-6 arm of the core and one or two antennae (usually incomplete) on the Man-α1-3 arm of the core. The structure is thus a hybrid in the sense of having a high-mannose structure on one arm of the core and a complex structure on the other arm. Hybrid oligosaccharides are usually isolated as bisected structures, e.g., from ovalbumin.[7] Nonbisected hybrid oligosaccharides have, however, been isolated from rhodopsin.[8,9] Structures representative of these oligosaccharide classes are shown in Fig. 1.

The key enzyme in the initiation of complex oligosaccharide synthesis is called UDP-GlcNAc: α-D-mannoside β1-2-*N*-acetylglucosaminyltransferase I (GlcNAc-transferase I). GlcNAc-transferase I adds a GlcNAc in β1-2 linkage to the Man-α1-3 arm of the 5-Man intermediate[10] (Fig. 1). This enzyme must act before the last two Man residues are removed by highly specific α-mannosidases[11,12] to yield the $(Man)_3(GlcNAc)_2$-Asn core present in all complex oligosaccharides. Hybrid structures result if, for some reason, α-mannosidase action is either slow or absent. Hen oviduct is rich in UDP-GlcNAc:glycopeptide β1-4-*N*-acetylglucosaminyltransferase III (GlcNAc-transferase III), which adds a bisecting GlcNAc residue.[13,14] This

[5] J. P. Carver and A. A. Grey, *Biochemistry* **20,** 6607 (1981).

[6] L. Dorland, J. Haverkamp, J. F. G. Vliegenthart, G. Spik, B. Fournet, and J. Montreuil, *Eur. J. Biochem.* **100,** 569 (1979).

[7] T. Tai, K. Yamashita, S. Ito, and A. Kobata, *J. Biol. Chem.* **252,** 6687 (1977).

[8] C.-J. Liang, K. Yamashita, C. G. Muellenberg, H. Shichi, and A. Kobata, *J. Biol. Chem.* **254,** 6414 (1979).

[9] M. N. Fukuda, D. S. Papermaster, and P. A. Hargrave, *J. Biol. Chem.* **254,** 8201 (1979).

[10] N. Harpaz and H. Schachter, *J. Biol. Chem.* **255,** 4885 (1980).

[11] N. Harpaz and H. Schachter, *J. Biol. Chem.* **255,** 4894 (1980).

[12] I. Tabas and S. Kornfeld, *J. Biol. Chem.* **253,** 7779 (1978).

[13] S. Narasimhan, D. Tsai, and H. Schachter, *Fed. Proc., Fed. Am. Soc. Exp. Biol.* **40,** 1597 (1981).

[14] S. Narasimhan, *J. Biol. Chem.* **257,** 10235 (1982).

residue inhibits α-mannosidase action,[11] thereby routing the synthetic pathway toward hybrid synthesis (Fig. 1). The reason for the presence of nonbisected hybrid structures in rhodopsin is not understood; it appears that α-mannosidase action is slow in this system even in the absence of a bisecting GlcNAc residue.

The appearance of GlcNAcβ1-2Manα1-3(Manα1-6)Manβ1-4GlcNAcβ1-4GlcNAc-Asn (structure MGn, Fig. 1) allows the action of UDP-GlcNAc:α-D-mannoside β1-2-*N*-acetylglucosaminyltransferase II (GlcNAc-transferase II), which adds the first GlcNAc of the second antenna.[10] We have also described in hen oviduct the presence of UDP-GlcNAc:β-*N*-acetylglucosaminide β1-4-*N*-acetylglucosaminyltransferase IV (GlcNAc-transferase IV), which adds GlcNAc in β1-4 linkage to the Man-α1-3 arm of the core to start the third antenna.[15] The enzyme that initiates the fourth antenna (GlcNAc in β1-6 linkage to the Man-α1-6 arm of the core) has not as yet been demonstrated *in vitro.*

Some complex oligosaccharides have a fucose linked α1-6 to the Asn-linked GlcNAc residue, and others have a fucose linked α1-3 to a peripheral GlcNAc residue in one of the antennae. Fucosyltransferases for both of these reactions have been demonstated *in vitro.*[16-18] Galactosyltransferases that incorporate Gal in β1-4[19,20] and β1-3[21] linkages to terminal GlcNAc residues have been described. Finally, sialyltransferases that incorporate sialic acid in α2-6[22,23] and α2-3[24,25] linkages to Galβ1-4GlcNAc termini have also been demonstrated *in vitro.*

All these reactions are believed to occur in the Golgi apparatus, although not all of the enzyme activities have been tested for enrichment in Golgi-rich membrane fractions. The Golgi apparatus appears to be the cellular site where not only *N*-glycosyl oligosaccharides receive their more terminal sugars, but also Ser(Thr)-GalNAc oligosaccharides (mucins and blood

[15] P. Gleeson, G. Vella, S. Narasimhan, and H. Schachter *Fed. Proc., Fed. Am. Soc. Exp. Biol.* **41,** 1147 (1982).
[16] G. D. Longmore and H. Schachter, *Carbohydr. Res.* **100,** 365 (1982).
[17] J.-P. Prieels, D. Monnom, M. Dolmans, T. A. Beyer, and R. L. Hill, *J. Biol. Chem.* **256,** 10456 (1981).
[18] P. H. Johnson, A. D. Yates, and W. M. Watkins, *Biochem. Biophys. Res. Commun.* **100,** 1611 (1981).
[19] R. Barker, K. W. Olsen, J. H. Shaper, and R. L. Hill, *J. Biol. Chem.* **247,** 7135 (1972).
[20] C. R. Geren, S. C. Magee, and K. E. Ebner, *Arch. Biochem. Biophys.* **172,** 149 (1976).
[21] B. T. Sheares, J. T. Y. Lau, and D. M. Carlson, *J. Biol. Chem.* **257,** 599 (1982).
[22] J. C. Paulson, W. E. Beranek, and R. L. Hill, *J. Biol. Chem.* **252,** 2356 (1977).
[23] J. C. Paulson, J. I. Rearick, and R. L. Hill, *J. Biol. Chem.* **252,** 2363 (1977).
[24] D. H. Van den Eijnden and W. E. C. M. Schiphorst, *J. Biol. Chem.* **256,** 3159 (1981).
[25] J. Weinstein, U. de Souza-e-Silva, and J. C. Paulson, *Fed. Proc., Fed. Am. Soc. Exp. Biol.* **41,** 663 (1982).

group glycoproteins), xylosyl-serine oligosaccharides (glycosaminoglycans), and hydroxylysine-galactose oligosaccharides (collagens and basement membranes). This chapter will be limited to the glycosyltransferases that carry out the elongation of complex *N*-glycosyl oligosaccharides.

Several reviews are available on these glycosyltransferases[26-28] as well as on the glycosyltransferases acting on mucins,[26-28] human and porcine blood group substances,[26-30] glycosaminoglycans,[28,31,32] and collagens.[28,33]

Preparation of Glycosyltransferase Substrates

Glycosyltransferases require two substrates, a glycosyl donor and a glycosyl acceptor. The accurate measurement of enzyme activities requires that both these substrates be present in the incubation at or near saturation levels. The glycosyl donors for the transferases to be discussed in this chapter are all nucleotide–sugars and all the enzyme assays to be described are radiochemical assays that measure the transfer of radioactive sugar from donor to acceptor. The various assays require four nucleotide–sugars, CMP-*N*-acetylneuraminic acid (CMP-NeuNAc), UDP-α-Gal, UDP-α-GlcNAc, and GDP-β-Fuc, all of which are now commercially available in various radiochemical forms. The nucleotide–sugar specific activities commonly used for assaying glycosyltransferases in our laboratory range between 1000 and 10,000 dpm/nmol, depending on the level of enzyme activity in the tissue being examined. If high-specific-activity nucleotide–sugar is used, there is a tendency to reduce the concentration well below saturation to keep the costs at a reasonable level. Such an assay may be satisfactory for some purposes, e.g., enzyme purification, but should be avoided for highly quantitative work, such as enzyme kinetics or the use of enzyme markers to assess the purity of subcellular fractions. Nonradioactive UDP-Gal and UDP-GlcNAc are commercially available and can be used to dilute the radioactive material to an appropriate level. Nonradioactive

[26] H. Schachter and S. Roseman, *in* "The Biochemistry of Glycoproteins and Proteoglycans" (W. J. Lennarz, ed.), p. 85. Plenum, New York, 1980.

[27] H. Schachter, *in* "The Glycoconjugates" (M. Horowitz and W. Pigman, eds.), Vol. 2, p. 87. Academic Press, New York, 1978.

[28] T. A. Beyer, J. E. Sadler, J. I. Rearick, J. C. Paulson, and R. L. Hill, *Adv. Enzymol.* **52,** 23 (1981).

[29] H. Schachter and C. A. Tilley, *Int. Rev. Biochem.* **16,** 209 (1978).

[30] W. M. Watkins, *Adv. Human Genet.* **10,** 1 (1980).

[31] L. Roden, J. R. Baker, J. A. Cifonelli, and M. B. Mathews, this series, Vol. 28, p. 73.

[32] L. Roden, J. R. Baker, T. Helting, N. B. Schwartz, A. C. Stoolmiller, S. Yamagata, and T. Yamagata, this series, Vol. 28, p. 638.

[33] N. Sharon, *in* "Complex Carbohydrates," p. 78. Addison–Wesley, Reading, Massachusetts, 1975.

CMP-NeuNAc has become commercially available (Sigma). Also, New England Nuclear sells a very low-specific-activity CMP-NeuNAc that is suitable for sialyltransferase assays. Nonradioactive GDP-Fuc is not commercially available, and the radioactive material being sold at present has a much higher specific activity than is suitable for fucosyltransferase assays. We make our own nonradioactive GDP-Fuc, and this preparation is described below.

An even more difficult problem is the preparation of suitable glycosyl acceptors. The early work in this field made use of high-molecular-weight glycoprotein derivatives. These materials are relatively simple to prepare and can be used in certain circumstances. However, as will be evident later in the discussion, glycoprotein acceptors are often too heterogeneous and thus may detect more than one transferase. We now use glycopeptides of high purity in most of our glycosyltransferase work. The preparation of suitable glycosyltransferase acceptors is described below.

Preparation of GDP-Fucose

The synthesis of GDP-Fuc has been described in this series.[34,35] The synthesis has been difficult for others to repeat, and suggestions for modifications are therefore given here.

Preparation of Fucokinase. The first step in the procedure is the preparation of fucokinase from pig liver.[35] We have on several occasions successfully prepared fucokinase to the 30% ammonium sulfate step as previously described.[35] The pellet from 250 g of pig liver is dissolved in 15–25 ml of 0.03 *M* sodium phosphate buffer at pH 7.4 and used for the preparation of β-L-fucose 1-phosphate. Several laboratories have reported difficulties in repeating our procedure. The problem appears to be the protamine sulfate step. There is apparently great variation between different batches of protamine sulfate. Yurchenco and Atkinson[36] have modified the procedure to avoid the use of protamine sulfate. They used a 30% ammonium sulfate precipitation step followed by gel filtration on Sephadex G-100 and ion-exchange chromatography on DEAE-cellulose. Their final fucokinase preparation was obtained in high yield and at least 1000-fold purified. The enzyme is stable indefinitely if stored in liquid nitrogen.

Preparation of β-L-Fucose 1-Phosphate. When an active fucokinase preparation is available, it is incubated with L-fucose, ATP, Mg^{2+}, fluoride, and buffer, as previously described.[34] It is useful to add some radioactive

[34] H. Schachter, H. Ishihara, and E. C. Heath, this series, Vol. 28, p. 285.
[35] H. Ishihara, H. Schachter, and E. C. Heath, this series, Vol. 28, p. 399.
[36] P. D. Yurchenco and P. A. Atkinson, *Biochemistry* **14,** 3107 (1975).

L-fucose so that fucose 1-phosphate and GDP-fucose can be readily located on columns during purification. Highly active kinase preparations will convert fucose to fucose 1-phosphate in 100% yield. We have allowed the incubation to run for as long as 21 hr, but shorter incubation times can be used. Protein was removed by ethanol precipitation, and fucose 1-phosphate was purified by ion-exchange chromatography on a column of Bio-Rad AG 1-X2 (100–200 mesh) in the biocarbonate form, as previously described.[34] Free fucose comes through the column unretarded, followed by L-fuconate (formed if the fucokinase preparation has some residual L-fucose dehydrogenase activity), and L-fucose 1-phosphate elutes between 0.3 and 0.4 *M* ammonium bicarbonate. The L-fucose 1-phosphate can be purified by barium precipitation as previously described.[34] A more convenient procedure is to concentrate the sample from the AG 1-X2 column by flash evaporation and subject it to gel filtration on a column (4 × 105 cm) of Sephadex G-10 equilibrated in 0.05 *M* triethylamine bicarbonate. The fucose 1-phosphate peak is flash evaporated to dryness to remove the volatile triethylamine bicarbonate. This preparation can be used for GDP-L-fucose synthesis.

Preparation of Fucose-1-phosphate Guanylyltransferase. The purification of fucose-1-phosphate guanylyltransferase from pig liver has been described.[37] We have used a relatively crude preparation of this enzyme for the synthesis of GDP-L-fucose.[34] The purified enzyme is unstable,[37] losing 80–85% of its activity at 0° in 24 hr. Storage in 10 m*M* dithioerythritol reduces loss to about 35% in 24 hr. It is therefore advisable to work at 4° as rapidly as possible during the preparation of this enzyme. A 30–50% ammonium sulfate cut is prepared from fresh pig liver as previously described.[34,37] It is not wise to use the 30% ammonium sulfate supernatant remaining from the fucokinase preparation, since presumably this supernatant has been stored for some time. The pellet from 250 g of pig liver is dissolved in 10–15 ml of 0.03 *M* sodium phosphate, pH 7.4, containing 1 m*M* dithiothreitol, and passed through a Sephadex G-100 column (4.5 × 90 cm) equilibrated with the same buffer. The enzyme peak appears at an elution volume of 700–800 ml. The enzyme is detected by incubating 0.025-ml aliquots from the column fractions in a total volume of 0.050 ml containing 0.1 *M* Tris-HCl, pH 8.0, 50 m*M* potassium fluoride, 25 m*M* magnesium chloride, 2.5 m*M* GTP, and 1.5 m*M* radioactive L-fucose 1-phosphate (preferably ^{14}C at about 500 dpm/nmol) at 37° for 1 hr. GDP-fucose is separated from fucose 1-phosphate by high-voltage paper electrophoresis at pH 6.5. Only those enzyme fractions showing extensive conversion (over 70%) of fucose 1-phosphate to GDP-fucose are pooled.

[37] H. Ishihara and E. C. Heath, this series, Vol. 28, p. 403.

Preparation of GDP-L-Fucose. The pyrophosphorylase must be used as soon as possible to prepare GDP-L-fucose. Incubation conditions are as previously reported[34] except that fucose 1-phosphate concentration can be as high as 0.5 m*M*. We have achieved conversions of 88% in 5 hr at 37%, but there is considerable variation depending on the quality of the enzyme preparation. Two volumes of 95% ethanol are added to stop the reaction, and GDP-fucose is purified by AG 1-X2 column chromatography and desalted on a Sephadex G-10 column as previously described.[34]

GDP-L-fucose can also be synthesized chemically in an overall yield of 40% from L-fucose.[38] Fucose is first converted to β-L-fucose 1-phosphate, which is then allowed to react with either guanosine 5′-phosphoric di-*n*-butylphosphinothioic anhydride or GMP morpholidate to form GDP-L-fucose.

Preparation of Glycosyl Acceptors

All the transferases to be described in this chapter can utilize both glycoproteins and low-molecular-weight glycopeptides and oligosaccharides as acceptors. Glycoproteins usually present a heterogeneous mixture of oligosaccharides to the enzyme, and product isolation and identification are difficult with these materials. The advantage of glycoprotein acceptors is that it is possible to use gel filtration or acid-insoluble radioactivity to measure product formation. The preparation of both high- and low-molecular-weight acceptors will be discussed.

Glycoprotein Acceptors. Any glycoprotein containing at least 1 mol of *N*-glycosyl bi-, tri-, or tetraantennary oligosaccharide (*N*-acetyllactosamine type) per 40,000 to 80,000 molecular weight (M_r) is in theory a potential acceptor. This laboratory has used human α_1-acid glycoprotein because it is very rich in oligosaccharides (1 mol per 9000 M_r). This material is not available commercially in large amounts and is prepared in this laboratory from Cohn fraction V supernatant. Fetuin (fetal calf serum) is commonly used because of its commercial availability. Since fetuin contains both *O*- and *N*-glycosyl oligosaccharides, it is not wise to use this protein in assaying glycosyltransferses in relatively crude preparations. Human transferrin is commercially available and contains about 80% biantennary oligosaccharides at a concentration of 1 mol per 38,000 M_r. Although relatively expensive, human transferrin is very well characterized and therefore suitable as a glycosyl acceptor. Other plasma glycoproteins, e.g., human fibrinogen (reported to have 1 mol of biantennary oligosaccharide per 80,000 M_r), may also become suitable as their oligosaccharide fine structures are determined.

[38] H. A. Nunez, J. V. O'Connor, P. R. Rosevear, and R. Barker, *Can. J. Chem.* **59,** 2086 (1981).

Sialic acid must be removed from the glycoprotein to generate the terminal Galβ1-4GlcNAc (*N*-acetyllactosamine) disaccharides needed for assay of sialyltransferase activity. This is most readily achieved by mild acid hydrolysis (0.1 *N* sulfuric acid at 80° for 1 hr) followed by extensive dialysis against water. A milder method of sialic acid removal is treatment with one of the commercially available bacterial (e.g., *Clostridium perfringens*[39]) *N*-acetylneuraminidases. Neuraminidase activity must be removed either by heat inactivation or column chromatography. An insoluble preparation of agarose-bound neuraminidase is available commercially (Sigma) and can be removed by simple centrifugation or filtration. Release of free sialic acid can be followed by the thiobarbiturate assay.[40]

Terminal galactose can be removed from asialoglycoprotein to generate an acceptor for galactosyltransferase either by a single Smith degradation step[41] or by treatment with a potent β-galactosidase free of β-*N*-acetylglucosaminidase activity. Jack bean β-galactosidase[42] (Sigma) is used for this purpose in our laboratory, and the release of free Gal is followed by a spectrophotometric assay using galactose dehydrogenase.[43]

In our earlier work on *N*-acetylglucosaminyltransferases, we used as acceptor asialoagalacto-human α_1-acid glycoprotein that had been treated with β-*N*-acetylglucosaminidase from *Clostridium perfringens*[44] (fraction Seph-Gm 1-DE-Gm 2). Extensive digestion with this enzyme at 37° for 48–72 hr removed about 15 mol of GlcNAc per mole of glycoprotein as determined by a colorimetric assay for GlcNAc.[45] The digest was passed through a Sephadex G-100 column to remove glycosidases (running at the void volume) from glycoprotein that is retarded by the resin. More recent work has shown that this acceptor assays at least two, and possibly more, *N*-acetylglucosaminyltransferases. We therefore now use specific glycopeptide acceptors in most of our work with these enzymes. The glycoprotein acceptor is useful in rapid assays of column fractions during enzyme purification work.

Glycopeptide Acceptors. Most acceptors will detect more than one glycosyltransferase; e.g., asialoglycoprotein with an *N*-acetyllactosamine terminus will detect α2-3- and α2-6-sialyltransferases as well as α1-3-fucosyltransferase. An ideal acceptor is specific for a single glycosyltransferase. Homogeneous low-molecular-weight glycopeptides or oligosaccharides or

[39] J. T. Cassidy, G. W. Jourdian, and S. Roseman, this series, Vol. 8, p. 680.
[40] L. Warren, *J. Biol. Chem.* **234,** 1971 (1959).
[41] R. G. Spiro, *J. Biol. Chem.* **239,** 567 (1964).
[42] Y.-T. Li and S.-C. Li, this series, Vol. 28, p. 702.
[43] H. Schachter, this seies, Vol. 41, p. 3.
[44] E. J. McGuire, S. Chipowsky, and S. Roseman, this series, Vol. 28, p. 755.
[45] C. T. Spivak and S. Roseman, *J. Am. Chem. Soc.* **81,** 2403 (1959).

monosaccharide glycosides are more likely to be specific acceptors than glycoproteins carrying a variety of oligosaccharide structures. Three glycopeptides (structures and nomenclature are given in Fig. 1 and its legend) are required for the assays described in this chapter. Glycopeptide M_5-R is prepared from ovalbumin and is a specific acceptor for GlcNAc-transferase I. Glycopeptide MGn is prepared from human myeloma IgG and is used to assay GlcNAc-transferase II. MGn is also an excellent acceptor for α1-6-fucosyltransferase and galactosyltransferase and is a poor acceptor for both GlcNAc-transferases III and IV (Fig. 1). Glycopeptide GnGn is prepared from IgG or transferrin and is an excellent acceptor for α1-6-fucosyltransferase, galactosyltransferase, and both GlcNAc-transferases III and IV (Fig. 1). When the same glycosyl donor and glycosyl acceptor detect more than one enzyme, as for example in the assay of GlcNAc-transferases III and IV using glycopeptide GnGn, it is necessary to separate the various products prior to counting. This is more readily achieved with low-molecular-weight acceptors than with glycoproteins, as will be demonstrated by the GlcNAc-transferase III and IV assays described below.

Preparation of Glycopeptide M_5-R. Ovalbumin is digested with Pronase,[46] and glycopeptides are obtained by gel filtration on Sephadex G-25 equilibrated in water. Digestion and gel filtration are carried out at least twice to ensure adequate proteolysis. The glycopeptides from 5 to 10 g of ovalbumin are fractionated on a column (2.5 × 90 cm) of AG 50W-X2 (200–400 mesh) equilibrated with sodium acetate buffer at pH 2.6, 1 m*M* in Na^+. It is important to get the pH of the column down to pH 2.6, and this can best be achieved by starting with the resin in the H^+ form. The column can be run at 4° or at room temperature. Five major and one minor asparaginyl oligosaccharides are obtained[46–48]; they are named either A to F or I to VI. Peak V is mainly M_5-R (Fig. 1). It can be further purified by preparative high-voltage paper electrophoresis in 1% borate.[49]

Preparation of Glycopeptide MGn. This glycopeptide (Fig. 1) is prepared by Pronase digestion (at least twice) of IgG from the serum of a patient (Tem) with multiple myeloma.[49,50] The glycopeptides are purified by gel filtration on Sephadex G-25 and G-75 followed by passage through an AG 50W-X2 column exactly as described above for ovalbumin glycopeptide

[46] C.-C. Huang, H. E. Mayer, Jr., and R. Montgomery, *Carbohydr. Res.* **13,** 127 (1970).

[47] T. Tai, K. Yamashita, M. Ogata-Arakawa, N. Koide, T. Muramatsu, S. Iwashita, Y. Inoue, and A. Kobata, *J. Biol. Chem.* **250,** 8569 (1975).

[48] P. H. Atkinson, A. A. Grey, J. P. Carver, J. Hakimi, and C. Ceccarini, *Biochemistry* **20,** 3979 (1981).

[49] S. Narasimhan, N. Harpaz, G. Longmore, J. P. Carver, A. A. Grey, and H. Schachter, *J. Biol. Chem.* **255,** 4876 (1980).

[50] S. Narasimhan, J. R. Wilson, E. Martin, and H. Schachter, *Can. J. Biochem.* **57,** 83 (1979).

M_5-R. The glycopeptide which passes through the column completely unretarded has a biantennary *N*-glycosyl oligosaccharide with sialylα2-6Galβ1-4GlcNAcβ1-2Manα1-3- and Galβ1-4GlcNAcβ1-2Manα1-6- as the two antennae. This glycopeptide is sequentially digested with a mixture of β-galactosidase and β-*N*-acetylglucosaminidase from *Clostridium perfringens* (fraction Seph-Gm 1-DE-Gm 2[44]), neuraminidase, and jack bean β-galactosidase. The mixture of β-galactosidase and β-*N*-acetylglucosaminidase must be inactivated before addition of neuraminidase. Glycopeptide MGn is purified by preparative high-voltage paper electrophoresis in 1% borate.[49]

It may be difficult to obtain a human myeloma IgG with the appropriate asymmetric oligosaccharide. An alternative procedure[51] is the preparation of the glycopeptide Manα1-3(Manα1-6)Manβ1-4GlcNAcβ1-4GlcNAcAsn by sequential digestion of a biantennary glycopeptide with neuraminidase, β-galactosidase, and β-*N*-acetylglucosaminidase. Transferrin is a good source for such a preparation (see below). This glycopeptide can then be incubated with GlcNAc-transferase I from bovine colostrum[10] or from mammalian liver[51] to generate glycopeptide MGn.

Preparation of Glycopeptide GnGn. Human transferrin is a good commercial source for preparation of GnGn (see Fig. 1 for structure). In a typical preparation, 4 g of transferrin are digested in 80 ml of 0.1 *M* Tris-HCl, pH 8.0, 5 m*M* $CaCl_2$, and 0.02% sodium azide containing 100 mg of Pronase (33 mg at zero time and 33 mg each at 12 and 24 hr) at 37° for a total of 48 hr. The digest is passed through a Sephadex G-25 column (10 × 24 cm) equilibrated in water containing 0.01% azide. The glycopeptide fraction is detected by a hexose assay.[52] Pronase digestion and Sephadex gel filtration are repeated once more and the glycopeptide fraction is concentrated to a 5.0-ml solution. A column (2 × 40 cm) of concanavalin A–Sepharose 4B (Sigma) is equilibrated with 5 m*M* sodium acetate, pH 5.2, containing 1 m*M* each of $MnCl_2$, $MgCl_2$, and $CaCl_2$ and 0.02% sodium azide. Concentrated glycopeptide solution (1.0 ml containing about 16 μmol of glycopeptide) is passed through the column at room temperature. The column is washed with 300 ml of the above buffer to elute triantennary glycopeptide (detected by a hexose assay[52]) followed by 15 m*M* α-methylglucoside in buffer to elute biantennary glycopeptide (detected by a sialic acid assay[53]). The biantennary fraction is concentrated and passed through a BioGel P-2 column (2.5 × 46 cm) equilibrated in water to remove α-methylglucoside, through a Sephadex G-75 column (1 × 15 cm) equilibrated in 0.01 *M* Tris-HCl, pH

[51] C. L. Oppenheimer, A. E. Eckhardt, and R. L. Hill, *J. Biol. Chem.* **256,** 11477 (1981).
[52] M. Dubois, K. A. Gilles, J. K. Hamilton, P. A. Rebers, and F. Smith *Anal. Chem.* **28,** 350 (1956).
[53] L. Svennerholm *Acta Chem. Scand.* **12,** 547 (1958).

7.5, containing 0.1 *M* NaCl, to remove concanavalin A, which tends to bleed from the column, and finally through another P-2 column to remove salts. The glycopeptide is hydrolyzed to remove sialic acid (0.1 *N* acid at 80° for 1 hr), acid is removed by gel filtration through BioGel P-2, and the glycopeptide is treated with jack bean β-galactosidase[42] to yield glycopeptide GnGn. Removal of galactose is followed by an enzymic assay.[43] The glycopeptide is purified by preparative high-voltage paper electrophoresis in 1% borate.[49]

Glycosyltransferase Assays

All the assays described in this chapter involve the measurement of transfer of radioactive sugar from donor to acceptor. The assays vary only in the method used to isolate the product from the other radioactive substances, such as the glycosyl donor, free sugar, and free sugar phosphate.

Assay A. High-voltage paper electrophoresis[54] in 1% sodium tetraborate, pH 9.0, is a relatively lengthy assay and not useful for monitoring columns during glycosyltransferase purification. However, it is a very reliable and accurate assay and can be used on crude enzyme preparations and with both glycoprotein and low-molecular-weight acceptors. Sialic acids, hexoses, methylpentoses, sugar phosphates, and nucleotide-sugars migrate away from the product, which remains near the origin. *N*-Acetylhexosamines do not bind borate well and must be washed away from the origin by descending chromatography with 80% ethanol.

Assay B. High-molecular-weight product can be precipitated[55,56] by addition of 1 volume of ice-cold 10% trichloroacetic acid-2% phosphotungstic acid followed by filtration on a glass-fiber filter. The filter is washed with 30 ml of 5% trichloroacetic acid-1% phosphotungstic acid and then by 10 ml of chloroform-methanol-ether (2 : 1 : 1, v/v/v). The filters are then dried and counted. Precipitated product can be collected and washed by centrifugation instead of filtration. A very rapid variant of this assay method involves spotting the incubation on 2.5-cm Whatman No. 1 paper disks, the disks are dried under a heat lamp, washed repeatedly with 10% trichloroacetic acid, ethanol-ether (2 : 1), and ether, dried, and counted.[57,58]

Assay C. Nucleotide-sugars and sugar phosphates can be removed from the enzyme digest by passage through small columns (0.5 × 5 cm) of AG

[54] S. Roseman, D. M. Carlson, G. W. Jourdian, E. J. McGuire, B. Kaufman, S. Basu, and B. Bartholomew, this series, Vol. 8, p. 354.

[55] P. J. Letts, M. L. Meistrich, W. R. Bruce, and H. Schachter, *Biochim. Biophys. Acta* **343**, 192 (1974).

[56] P. J. Letts, L. Pinteric, and H. Schachter, *Biochim. Biophys. Acta* **372**, 304 (1974).

[57] A. Baxter and J. P. Durham, *Anal. Biochem.* **98**, 95 (1979).

[58] J. P. Durham, D. Gillies, A. Baxter, and R. O. Lopez-Solis, *Clin. Chim. Acta* **95**, 425 (1979).

1-X8, chloride form, equilibrated in water.[10,59] The columns are washed with 2.0 ml of water, and the entire effluent is counted. The effluent contains not only product but also free radioactive sugar. The amount of free sugar formed must be determined by appropriate control incubations and subtracted from the assay. The assay is fast and can be used with both high- and low-molecular-weight products but may be misleading with crude enzyme preparations, which give high free sugar values.

Assay D. A convenient and rapid assay that can be used only with high-molecular-weight products is gel filtration of the enzyme digest on small columns (0.8 × 10 cm) of Sephadex G-50 (fine) equilibrated in 0.02*M* Tris-HCl, pH 7.5, containing 0.2 *M* NaCl and 0.01% sodium azide.[60] The product elutes ahead of low-molecular-weight radioactive contaminants. The column can be used repeatedly if thoroughly washed between assays.

Definition of Enzyme Unit. Throughout this chapter an enzyme unit is defined as the amount of enzyme that transfers 1 μmol of radioactive sugar to acceptor per minute in the presence of saturating concentrations of both substrates.

UDP-GlcNAc: α-D-Mannoside β1-2-GlcNAc-Transferase I

Johnston *et al.*[61] first reported the presence in goat colostrum of a GlcNAc-transferase acting on α_1-acid glycoprotein pretreated with neuraminidase, β-galactosidase, and β-*N*-acetylglucosaminidase. A similar enzyme has been found in many rat, pig, guinea pig, and human tissues[26,27] and has been localized to the Golgi apparatus in rat liver.[62] Studies on lectin-resistant mutants of Chinese hamster ovary cells[63–65] and baby hamster kidney cells[66] indicated that the high-molecular-weight acceptor detects at least two different GlcNAc-transferases. One of these enzymes, UDP-GlcNAc: α-D-mannoside β1-2-GlcNAc-transferase I, attaches GlcNAc in β1-2 linkage to the Manα1-3- terminus of the trimannosyl core of glycopeptide M_5-R, as shown in Fig. 1. This enzyme has been purified from bovine

[59] B. Fleischer, S. Fleischer, and H. Ozawa, *J. Cell Biol.* **43,** 59 (1969).

[60] M. Schwyzer and R. L. Hill, *J. Biol. Chem.* **252,** 2338 (1977).

[61] I. R. Johnston, E. J. McGuire, G. W. Jourdian, and S. Roseman, *J. Biol. Chem.* **241,** 5735 (1966).

[62] J. R. Munro, S. Narasimhan, S. Wetmore, J. R. Riordan, and H. Schachter, *Arch. Biochem. Biophys.* **169,** 269 (1975).

[63] P. Stanley, S. Narasimhan, L. Siminovitch, and H. Schachter, *Proc. Natl. Acad. Sci. U.S.A.* **72,** 3323 (1975).

[64] S. Narasimhan, P. Stanley, and H. Schachter, *J. Biol. Chem.* **252,** 3926 (1977).

[65] C. Gottlieb, J. Baenziger, and S. Kornfeld. *J. Biol. Chem.* **250,** 3303 (1975).

[66] P. Vischer and R. C. Hughes, *Eur. J. Biochem.* **117,** 275 (1981).

colostrum,[10] rabbit liver,[67] pig liver,[51] and pig tracheal mucosa.[51,68] The bovine colostrum procedure is described below.

The reaction catalyzed is of the following form and is shown in Fig. 1.

$$\text{UDP-GlcNAc} + \text{Man}\alpha\text{1-3(R}_1\text{-)Man}\beta\text{1-4GlcNAc-R}_2 \rightarrow \text{GlcNAc}\beta\text{1-2Man}\alpha\text{1-3(R}_1\text{-)Man}\beta\text{1-4GlcNAc-R}_2 + \text{UDP}$$

where R_1 and R_2 are defined below.

Assay Method

Principle. The assay measures the transfer of radioactive GlcNAc from UDP-GlcNAc to glycopeptide M_5-R (see Fig. 1 for structure).

Reagents

MES [2-(*N*-morpholino)ethanesulfonic acid] buffer, 0.5 *M*, pH 6.0
$MnCl_2$, 0.2 *M*
UDP-*N*-acetyl-D-[1-^{14}C]glucosamine (4×10^6 dpm/μmol), 5 m*M*
Tritiated UDP-GlcNAc; can be used at a specific activity of about 1×10^7 dpm/μmol
Bovine serum albumin, previously heated at 65° for 15 min, 2% (w/v) in water
Ovalbumin glycopeptide V (M_5-R), 10 m*M*
Triton X-100, 2% (v/v)
2% sodium tetraborate–0.25 *M* disodium EDTA
0.02 *M* sodium tetraborate–1 m*M* disodium EDTA

Assay Procedure. The standard incubation contains the following in a final volume of 0.050 ml: 10 μl of 0.5 *M* MES, pH 6.0 (5 μmol), 5 μl of 0.2 *M* $MnCl_2$ (1 μmol), 5 μl of 5 m*M* UDP-[1-^{14}C] GlcNAc (0.025 μmol), 5 μl of 2% bovine serum albumin (preheated to inactivate pyrophosphatases, 0.1 mg), 5 μl of 10 m*M* ovalbumin glycopeptide V (0.050 μmol), and enzyme to 0.050 ml. Triton X-100 is not needed when working with bovine colostrum. However, GlcNAc-transferases within cells are strongly bound to the membranes of the Golgi apparatus, and it is necessary to add 5 μl of 2% Triton X-100 (0.1 μl) to the incubation.[69] After 1 hr at 37°, the incubation is stopped by addition of 10 μl of 2% sodium tetraborate–0.25 *M* disodium EDTA followed by high-voltage paper electrophoresis in 1% sodium tetraborate at pH 9.0 for 1 hr at 2 kv. The paper is dried and subjected to descending paper chromatography in 80% ethanol for 18 hr to remove free radioactive GlcNAc at the origin (assay method A, above). The

[67] C. L. Oppenheimer and R. L. Hill, *J. Biol. Chem.* **256,** 799 (1981).
[68] J. Mendicino, E. V. Chandrasekaran, K. R. Anumula, and M. Davila, *Biochemistry* **20,** 967 (1981).
[69] H. Schachter, I. Jabbal, R. L. Hudgin, L. Pinteric, E. J. McGuire, and S. Roseman, *J. Biol. Chem.* **245,** 1090 (1970).

product remains at the origin and is counted. For more rapid assay of column fractions during purification, assay method C is used. The reaction is stopped by addition of 0.5 ml of 0.02 *M* sodium tetraborate–1 m*M* disodium EDTA and passed through a column (0.5 × 5.0 cm) of AG 1-X8, chloride form, equilibrated with water. The column is washed with 2.0 ml of water, and the effluent containing the product as well as free radioactive GlcNAc is counted. Control incubations lacking acceptor should be carried out routinely, especially when using assay method C.

For rapid assays during enzyme purification it is possible to use glycoproteins, such as α_1-acid glycoprotein or transferrin, that have been pretreated to remove terminal sialyl, galactosyl, and *N*-acetylglucosaminyl residues (see above). This acceptor allows the use of either assay B (precipitation of product) or assay D (Sephadex G-50 gel filtration). However, it must be remembered that both GlcNAc-transferases I and II (and possibly other GlcNAc-transferases) may be assayed by the use of high-molecular-weight acceptors (see section on GlcNAc-transferase II for further discussion of this point).

Purification of GlcNAc-Transferase I from Bovine Colostrum[10]

Ammonium Sulfate Fractionation. All steps are carried out at 4°. Bovine colostrum (277 g) is defatted by centrifugation at 15,000 g_{max} for 20 min, diluted with 3 volumes of cold-distilled water, and fractionated by addition of solid ammonium sulfate. Ammonium sulfate precipitates are collected by centrifugation at 15,000 g_{max} for 20 min. The pellet formed between 30 and 50% saturation is dissolved in 300 ml of cold 0.1 m*M* 6-aminocaproic acid and dialyzed extensively against 25 m*M* sodium cacodylate, pH 6.0, 5 m*M* $MnCl_2$, and 0.1 m*M* 6-aminocaproic acid (Buffer C).

CM-Sephadex. The solution is clarified by centrifugation as above, and the supernatant is applied to a column (3 × 57 cm) of CM-Sephadex C-50 (Pharmacia) equilibrated in 25 m*M* sodium cacodylate, pH 6.0, 5 m*M* $MnCl_2$, and 0.1 m*M* 6-aminocaproic acid (buffer C). The column is eluted stepwise with 1500 ml of buffer C, 600 ml of 0.1 *M* NaCl in buffer C, 900 ml of 0.15 *M* NaCl in buffer C, and 600 ml of 0.30 *M* NaCl in buffer C. GlcNAc-transferase I is located in the 0.3 *M* NaCl eluate. Fractions are pooled and dialyzed extensively against 25 m*M* sodium cacodylate, pH 6.8, 5 m*M* $MnCl_2$, and 0.1 m*M* 6-aminocaproic acid (buffer A).

UDP-Hexanolamine–Agarose. UDP-hexanolamine is commercially available from Sigma and can be coupled to Sepharose 4B by the cyanogen bromide method or by other coupling methods. A 70-ml aliquot of the dialyzed fraction eluted from CM-Sephadex by 0.3 *M* NaCl (containing about 60 milliunits of enzyme) is applied to a UDP-hexanolamine–agarose column (1.3 ×5 cm) equilibrated in buffer A. The column is washed with 25 mM sodium cacodylate, pH 6.8, containing 5 mM $MnCl_2$, but no 6-aminocaproic acid (buffer B), until the absorbance of the eluate at 280 nm is 0.07. GlcNAc-transferase I is very labile at low protein concentration (50% loss of activity per day at 4°), and attempts to obtain more highly purified enzyme by extensive washing of the UDP-hexanolamine–agarose column leads to complete loss of enzymic activity. GlcNAc-transferase I binds to the column and is eluted with 30 ml of 3 mM UDP-GlcNAc in buffer B. Enzyme-containing fractions are pooled, and bovine serum albumin (pretreated at 65° for 15 min) is added to a final concentration of 1 mg/ml. This preparation can be stored for at least 1 month at 4° and for longer periods at −20° without loss of activity. The final enzyme preparation has a specific activity of about 0.1 unit/mg, representing a purification of 11,000-fold. The yield is about 30%. The enzyme preparation is dialyzed against buffer B to remove UDP-GlcNAc prior to kinetic studies.

Rabbit Liver Enzyme.[67] GlcNAc-transferase I has been purified 7000-fold from rabbit liver in 36% yield by Triton X-100 extraction of an acetone powder followed by chromatography on UDP-hexanolamine–agarose. Two UDP-hexanolamine–agarose columns were run, enzyme being eluted from the first column by NaCl and from the second column by UDP. The final specific activity of the preparation is 2.5 units/mg, suggesting that the rabbit preparation may be 25 times purer than the bovine colostrum preparation, although there may be a species difference.

Pig Liver Enzyme.[51] GlcNAc-transferase I has been purified from pig liver by the same procedure used with rabbit liver except that three UDP-hexanolamine–agarose steps were used. The enzyme was purified 4800-fold at a yield of 10% and had a specific activity of 4.8 units/mg.

Properties of GlcNAc-Transferase I

Stability. The enzyme preparation is stable for several months at −20° provided bovine serum albumin is present at a concentration of 1 mg/ml.

Kinetic Parameters. The K_m values for both substrates, the optimum Mn^{2+} concentrations, and the pH optima for the different GlcNAc-transferase I preparations are shown in Table I. The enzyme is inactive if treated with EDTA and Mn^{2+} is the most effective cation activator. Less effective are Mg^{2+}, Co^{2+}, and Cd^{2+}.

Enzyme Purity. The purified bovine colostrum enzyme shows two major bands on SDS–polyacrylamide gel electrophoresis. However, the low specific activity (0.1 unit/mg) relative to the purified rabbit (2.5 units/mg) and pig (4.8 units/mg) liver enzymes indicates that the bovine colostrum preparation may not be pure. The bovine colostrum enzyme retains some activity toward glycopeptide MGn (see Fig. 1 for structure), the substrate for GlcNAc-transferase II. The fact that the K_m of glycopeptide MGn with

TABLE I
KINETIC PARAMETERS OF GLYCOSYLTRANSFERASES[a]

Enzyme	Source[b]	Substrates[a]	K_m (mM)	pH optimum	Optimum concentration of Mn^{2+} (mM)
GlcNAc-transferase I	Bovine colostrum[10]	UDP-GlcNAc	0.1	6.0	18–26
		M_5-R	0.12		
	Rabbit liver[67]	M_5-R	0.45	6.3	ND[c]
	Pig liver[51]	M_5-R	0.39	ND	ND
GlcNAc-transferase II	Bovine colostrum[10]	MGn	0.1	ND	ND
	Pig liver[51]	MGn	0.2	6.0–6.5	ND
GlcNAc-transferase III	Hen oviduct[13,14]	UDP-GlcNAc	1.1	6.0–7.0	12
		GnGn	0.23		
β1-4-Gal-transferase	Bovine milk[19,20]	UDP-Gal	0.1	7.5–9.0	13
		GlcNAc	6–8		
α2-6-Sialyl-transferase	Bovine colostrum[23]	CMP-NeuNAc	0.17	6.4–7.2	None[e]
		Asialo-AGP[d]	1.6		
α1-6-Fucosyl-transferase	Pig liver[16,80]	GnGn	0.08	ND	ND
		MGn	0.6		
		Asialo, agalacto-AGP[d]	17.0		
α1-3/4-Fucosyl-transferase	Human milk[17]	GDP-Fuc	0.005–0.013	7.0–7.8	5
		Galβ1-3GlcNAc	1.9		
		Galβ1-4GlcNAc	1.6		
		Galβ1-4Glc	59		

[a] Glycopeptide structures are shown in Fig. 1.
[b] Superscript numbers refer to text footnotes.
[c] ND, not done or not reported.
[d] AGP, α_1-acid glycoprotein.
[e] Metal is not essential for activity.

GlcNAc-transferase I is 100 times higher than the K_m of MGn with GlcNAc-transferase II[10] suggests that the bovine colostrum GlcNAc-transferase I has a low activity toward glycopeptide MGn. However, the highly purified rabbit and pig liver enzymes have no trace of activity toward MGn. The bovine colostrum enzyme must be purified to homogeneity to settle this discrepancy.

Substrate Specificity. Purified GlcNAc-transferase I acts on acceptors with the general formula Manα1-3(R_1-)Manβ1-4GlcNAc-R_2. A relatively poor acceptor with a K_m of 7.4 mM is the trisaccharide in which R_1 and R_2 are both H. Good acceptors are glycopeptides in which R_2 is -β1-4(Fucα1-6)GlcNAc-Asn-X and R_1 is Manα1-6- (K_m = 0.20 mM) or in which R_2 is -β1-4GlcNAc-Asn-X and R_1 is Manα1-6-, or Manα1-3Manα1-6-, or Manα1-6(Manα1-3)Manα1-6- (glycopeptide M_5-R, the standard acceptor with a K_m = 0.12 mM). The enzyme also works with the analogous oligosaccharides in which R_2 is H and R_1 is either Manα1-6- or Manα1-6(Manα1-3)Manα1-6-.[51,64,67] The bovine colostrum enzyme has a low activity toward glycopeptide MGn (see discussion above), but this may be due to residual GlcNAc-transferase II activity. This activity is at any rate too low to be of physiological importance. High-mannose glycopeptides in which the Manα1-3- terminus of the core is substituted with a Manα1-2- residue are not acceptors. The physiological role of GlcNAc-transferase I is the initiation of the first antenna or branch (Fig. 1). The role of the polypeptide sequence in the control of GlcNAc-transferase activity remains an unexplored problem.

UDP-GlcNAc:α-D-Mannoside β1-2-GlcNAc-Transferase II

The study of lectin-resistant mutants of Chinese hamster ovary cells[63-65] and baby hamster kidney cells[66] deficient in GlcNAc-transferase I indicated the presence of a second UDP-GlcNAc:α-D-mannoside β1-2-GlcNAc-transferase designated GlcNAc-transferase II.[63,64] The function of GlcNAc-transferase II is the initiation of the second antenna or branch by addition of GlcNAc in β1-2- linkage to the Manα1-6- terminus of glycopeptide MGn (Fig. 1). The enzyme has not been purified to homogeneity, but partially purified preparations have been obtained from bovine colostrum,[10] pig liver,[51] pig trachea,[51,68] and hen oviduct.[70] The reaction catalyzed is as follows (Fig. 1):

UDP-GlcNAc + Manα1-6(GlcNAcβ1-2Manα1-3)Manβ1-4GlcNAc-R $\rightarrow$
GlcNAcβ1-2Manα1-6(GlcNAcβ1-2Manα1-3)Manβ1-4GlcNAc-R + UDP

[70] G. Vella and H. Schachter, unpublished data.

Assay Method

Principle. The assay measures the transfer of radioactive GlcNAc from UDP-GlcNAc to glycopeptide MGn (see Fig. 1 for structure).

Reagents. The same reagents are used as in the assay of GlcNAc-transferase I.

Assay Procedure. The incubation is identical to that given for GlcNAc-transferase I except that ovalbumin glycopeptide V is replaced with 5 μl of 10 m*M* glycopeptide MGn (0.050 μmol). The incubation is stopped and product formation is determined by either assay A or C as described for GlcNAc-transferase I.

Both GlcNAc-transferases I and II are assayed when α_1-acid glycoprotein pretreated with neuraminidase, β-galactosidase, and β-*N*-acetylglucosaminidase is used as acceptor.[63,64] The reason for this is that removal of terminal GlcNAc residues from tri- and tetraantennary structures by *Clostridium perfringens* β-*N*-acetylglucosaminidase is incomplete, yielding oligosaccharides that act as acceptors for GlcNAc-transferase II. Any glycoprotein with tri- and tetraantennary oligosaccharides may therefore behave in this manner.

Partial Purification of GlcNAc-Transferase II[10]

Bovine colostrum GlcNAc-transferase II appears in the 0.1 *M* NaCl eluate of the CM-Sephadex column described above for the purification of GlcNAc-transferase I. Although this fraction is enriched only fourfold in GlcNAc-transferase II, it is free of GlcNAc-transferase I. The preparation also contains a large amount of UDP-Gal:GlcNAc galactosyltransferase activity, which is readily removed by passage through UDP-hexanolamine–agarose because galactosyltransferase adheres to this column whereas GlcNAc-transferase II does not. The specific activity of the preparation is about 13 μunit per milligram of protein.

Properties of GlcNAc-Transferase II

Stability. The partially purified bovine colostrum enzyme is stable for many months at 4°. However, Triton X-100 extracts of pig liver or hen oviduct membrane preparations lose about 50% of their activity over several days.

Kinetic Parameters. The kinetic parameters for the enzyme are given in Table I.

Substrate Specificity. The only substrate effective with this enzyme is glycopeptide MGn (Fig. 1). The enzyme does not work on any substrate of the form Manα1-3(R_1-)Manβ1-4GlcNAc-R_2. The presence of the GlcNAcβ1-2Manα1-3- sequence is essential, indicating that GlcNAc-transferase I must act before GlcNAc-transferase II can initiate synthesis of the second antenna as shown in Fig. 1. It is interesting that the structure Manα1-6(GlcNAcβ1-2Manα1-3)(GlcNAcβ1-4)-Manβ1-4GlcNAc(Fucα1-6)GlcNAc-Asn-X is not a substrate for GlcNAc-transferase II.[10] This glycopeptide differs from the substrate only by an additional GlcNAc residue linked β1-4- to the β-linked Man. This residue is called a "bisecting" GlcNAc residue (see Fig. 1). Its presence prevents not only the action of GlcNAc-transferase II, but also that of GlcNAc-transferase I-dependent α-mannosidase,[11] GlcNAc-transferase IV, and α1-6-fucosyltransferase (see below).

UDP-GlcNAc: Glycopeptide GnGn β1-4-GlcNAc-Transferase III[13,14]

This enzyme has been reported in hen oviduct membranes. It attaches a bisecting GlcNAc residue to glycopeptide GnGn (see Fig. 1 for the reaction catalyzed). A similar reaction also occurs in hen oviduct membranes at the 5-Man stage (Fig. 1), but it is not yet known whether the two activities are catalyzed by the same enzyme. The enzyme has not been purified.

Assay Method

Principle The assay measures the transfer of radioactive GlcNAc from UDP-GlcNAc to glycopeptide GnGn (structure shown in Fig. 1).

Reagents

- MES buffer, 0.5 *M*, pH 5.7, containing 50 m*M* $MnCl_2$
- UDP-*N*-acetyl-D-[U-^{14}C]glucosamine (4×10^6 cpm/μmol), 5 m*M*
- GlcNAc, 1.6 *M*
- Triton X-100, 5% (v/v)
- Glycopeptide GnGn, 4 m*M*
- 2% Sodium tetraborate–0.25 *M* disodium EDTA
- Tris-HCl buffer, pH 7.5, 0.01 *M*, containing 0.1 *M* NaCl
- Methyl α-D-glucopyranoside, 0.1 *M*

Assay Procedure. Five microliters of 4 m*M* glycopeptide GnGn (0.020 μmol) are dried in a small test tube. The following are added: 10 μl 0.5 *M* MES, pH 5.7, containing 50 m*M* $MnCl_2$ (5 μmol of MES, 0.5 μmol of Mn^{2+}), 5 μl of 5 m*M* UDP-[^{14}C]GlcNAc (0.025 μmol), 5 μl of 1.6 *M* GlcNAc (8 μmol), 5 μl of 5% Triton X-100 (0.25 μl), and enzyme to a final volume of 0.040 ml. The GlcNAc is required in crude membrane preparations to inhibit tissue β-*N*-acetylglucosaminidase activity. The latter removes one or both terminal GlcNAc residues from glycopeptide GnGn to form acceptors for GlcNAc-transferases I and II. After incubation at 37° for 1 hr, 10 μl of 2% sodium tetraborate–0.25 *M* disodium EDTA are added to stop the reaction, and the mixture is subjected to high-voltage paper electrophoresis followed

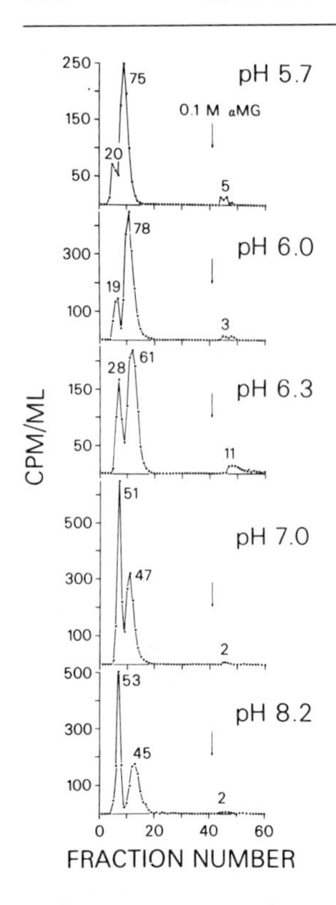

FIG. 2. GlcNAc-transferase assays were carried out under the conditions indicated for GlcNAc-transferase III, at the pH values indicated in the figure, using glycopeptide GnGn as an acceptor. The glycopepetide products were acetylated with nonradioactive acetic anhydride and passed through concanavalin A–Sepharose columns (0.7 × 11 cm) equilibrated with 0.01 *M* Tris-HCl, pH 7.5, containing 0.1 *M* NaCl. The columns were washed with 40 ml of buffer followed by 40 ml of 0.1 *M* methyl α-D-glucopyranoside in buffer (0.1 *M* αMG). The arrows indicate the start of elution with glucopyranoside. The fraction size was 1.0 ml. The percentage distribution of radioactivity among the fractions is indicated in the figure. GlcNAc-transferase III product elutes between fractions 10 and 17, and GlcNAc-transferase IV product elutes between fractions 5 and 8.

by chromatography with 80% ethanol (assay A, as described for GlcNAc-transferase I). If GlcNAc-transferase IV is present, as is indeed the case in crude hen oviduct membranes, the radioactive product remaining at the origin of the electrophoretogram is eluted with water and N-acetylated with nonradioactive acetic anhydride (5 μmol) in 0.1 ml 0.5% sodium bicarbonate[49,50] at room temperature for 1 hr. The sample is desalted on a column (1.7 × 29 cm) of BioGel P-4 equilibrated in water and applied at room temperature to a concanavalin A–Sepharose 4B column (0.7 × 11 cm) equilibrated with 0.01 *M* Tris-HCl, pH 7.5, containing 0.1 *M* NaCl. The column is washed with 40 ml of buffer followed by 40 ml of 0.1 *M* methyl α-D-glucopyranoside in buffer. The GlcNAc-transferase III product passes through the column in a retarded manner and is separated from GlcNAc-transferase IV product, which passes through unretarded, and from any GlcNAc-transferase I or II products that adhere to the column and are eluted by methyl α-D-glucopyranoside (Fig. 2).

The concanavalin A–Sepharose assay is obviously tedious but should be applied at least once to any new enzyme preparation to determine not only the GlcNAc-transferase IV level, but also the β-*N*-acetylglucosaminidase activity that escapes inhibition by GlcNAc. Such activity will be evident by the radioactivity that binds to the lectin column (Fig. 2). If GlcNAc-transferase IV level is low and inhibition of glycosidase by GlcNAc is effective, e.g., at pH 5.7 in crude hen oviduct membranes (Fig. 2), the product can be counted immediately after high-voltage electrophoresis and chromatography with 80% ethanol.

Properties of GlcNAc-Transferase III

Stability. Microsomal preparations from hen oviduct retain activity for several weeks at −20° if suspended in detergent-free buffer.

Kinetic Parameters. The parameters for crude hen oviduct membranes are given in Table I. Although the pH optimum of GlcNAc-transferase III is between 6.0 and 7.0, the standard assay is run at pH 5.7 to minimize interference by GlcNAc-transferase IV (Fig. 2).

Substrate Specificity. The substrate specificity of GlcNAc-transferase III is shown in Table II. The activity of crude hen oviduct membranes toward glycopeptide GnGn is about 85 μunit per milligram of protein under standard assay conditions. GnGn preparations both with and without a Fucα1-6- residue attached to the asparagine-linked GlcNAc give the same activity. Activities with MGn (Fig. 1) and MM [Manα1-6(Manα1-3)Manβ1-4GlcNAc(Fucα1-6)GlcNAc-Asn-X] relative to GnGn are less then 15% and 8%, respectively, at pH 5.7. Activity with glycopeptide GGn (Fig. 1) is about 60% of that with GnGn. However, the enzyme is essentially inactive with the other glycopeptides tested. Thus bisected biantennary complex oligosaccharides (Fig. 1) are formed immediately after both antennae have been initiated by GlcNAc-transferases I and II, respectively.

A bisecting GlcNAc can also be added to glycopeptide GnM_5-R (Fig. 1), but it is not known whether this activity is due to the same enzyme as the one that acts on GnGn. The presence of bisecting GlcNAc prevents the GlcNAc-transferase I-dependent α-mannosidase[11] from removing two Man residues to form the 3-Man core, thereby routing the synthetic pathway into synthesis of bisected hybrid structures (Fig. 1).

UDP-GlcNAc: Glycopeptide GnGn β1-4-GlcNAc-Transferase IV[15]

This enzyme has been described in hen oviduct membranes.[15] It catalyzes the attachment of a GlcNAc residue in β1-4- linkage to the Manα1-3-arm of glycopeptide GnGn, thereby initiating the third antenna or branch (see Fig. 1 for the reaction catalyzed). The enzyme therefore routes the synthetic pathway toward triantennary, bisected triantennary, and tetraantennary complex oligosaccharides. The enzyme has not been purified.

Assay Method

Principle. The assay measures the transfer of radioactive GlcNAc from UDP-GlcNAc to glycopeptide GnGn (structure shown in Fig. 1).

Reagents

MES buffer, 0.5 *M*, pH 7.0, containing 50 m*M* $MnCl_2$

Other reagents as for GlcNAc-transferase III

Assay Procedure The assay procedure is identical to that described for GlcNAc-transferase III except that the pH of the incubation is at 7.0 and the final Triton X-100 concentration is 0.5% (v/v). This pH favors GlcNAc-transferase IV activity, although there is still appreciable GlcNAc-transferase III activity (Fig. 2). GlcNAc-transferase IV activity is further optimized by reducing the Triton X-100 concentration to 0.125% (v/v). However, at that low Triton X-100 concentration and with a crude enzyme preparation, it is difficult to maintain proportionality of enzyme activity to time of incubation and protein concentration. With crude enzyme preparations, therefore, it is advisable to use 0.5% Triton X-100.

Properties of GlcNAc-Transferase IV

Stability. Hen oviduct microsomes suspended in detergent-free buffer retain activity for several weeks at −20°.

Substrate Specificity. The substrate specificity of GlcNAc-transferase IV is shown in Table II. Like GlcNAc-transferase III, the enzyme works on glycopeptide GnGn with or without a Fucα1-6- attached to the asparagine-linked GlcNAc, works on glycopeptides MGn and GGn (Fig. 1), but at a lower rate than on GnGn, and works poorly or not at all on the other glycopeptides tested. An interesting finding is that the presence of a bisecting GlcNAc on glycopeptide GnGn completely inhibits GlcNAc-transferase IV action (Table II). Thus, GlcNAc-transferase IV must act before GlcNAc-transferase III to form bisected triantennary complex oligosaccharides, and this is probably true also for the synthesis of bisected triantennary hybrid oligosaccharides (Fig. 1). This is yet another example of inhibition of enzyme action by the presence of a bisecting GlcNAc. The other enzymes so inhibited are GlcNAc-transferase II, GlcNAc-transferase I-dependent α-mannosidase, and α1-6-fucosyltransferase (discussed below).

UDP-Gal:GlcNAc β1-4-Galactosyltransferase

Enzyme activities capable of transferring Gal from UDP-Gal to GlcNAc, or to oligosaccharides, glycopeptides, or glycoproteins with GlcNAc at their nonreducing termini, have been widely described in many

TABLE II
GLYCOSYLTRANSFERASE SUBSTRATE SPECIFICITIES[a]

	Specific activity (μunits/mg)			
Substrate[b]	Gn-T III pH 5.7	Gn-T III pH 7.0	Gn-T IV pH 7.0	Fuc-T pH 8.0
GnGn (+Fuc)	85	73	83	—[c]
GnGn (−Fuc)	—	75	95	1800
MGn (+Fuc)	≤13	10	18	0
MGn (−Fuc)	—	—	—	3100
MM (+Fuc)	≤ 7	—	≤16	0
MM (−Fuc)	—	—	—	0
GGn (+Fuc)	—	48	30	—
GnG (+Fuc)	—	≤5	≤5	—
GG (−Fuc)	—	≤8	≤8	0[d]
SS (−Fuc)	—	0	0	0
GnGn(Gn) (−Fuc)	—	—	0	0
MGn(Gn) (−Fuc)	—	—	—	0

[a] Hen oviduct membranes were used for studies with GlcNAc-transferase III[13,14] (Gn-T III) at pH 5.7 and 7.0, and GlcNAc-transferase IV[15] (Gn-T IV) at pH 7.0. Golgi-enriched membranes from pig liver were used for the studies on GDP-Fuc:β-*N*-acetylglucosaminide (Fuc to Asn-linked GlcNAc) α1-6-fucosyltransferase[16] (Fuc-T). These membranes are enriched 30-fold in Fuc-T relative to homogenate.

[b] Structures of glycopeptides GnGn, MGn, GGn, GG, and GnGn(Gn) are shown in Fig. 1. Glycopeptide MM is Manα1-6(Manαl-3) Manβ1-4GlcNAcβ1-4(Fucα1-6)GlcNAc-Asn-X. GnG is similar to GGn except that the Gal residue is on the Manα1-3- arm instead of the Manα1-6- arm. SS is fully sialylated biantennary complex glycopeptide with both sialyl residues linked α2-6-. The glycopeptides are named according to the terminal sugar residues with M = Man, Gn = GlcNAc, G = Gal, S = sialyl, and (Gn) = bisecting GlcNAc.

[c] Dashes indicate not done.

[d] Based on work with rat liver Golgi membranes[62] that lack fucosyltransferase activities toward Galβ1-4GlcNAc-terminated acceptors.

species and tissues.[26-28] However, the identification of the product as Galβ1-4GlcNAc has been carried out only with free GlcNAc as acceptor and only with a limited number of enzyme sources. The activity occurs in most tissues in a membrane-bound form localized primarily to the Golgi apparatus and in a soluble form in various tissue fluids. The present discussion will be limited to the bovine milk enzyme, since it has been most thoroughly characterized.[71-75] The reaction catalyzed is as follows:

$$\text{UDP-Gal} + \text{GlcNAc-R} \rightarrow \text{Gal}\beta\text{1-4GlcNAc-R} + \text{UDP}$$

Assay Method

Principle. The assay measures the transfer of radioactive Gal from UDP-Gal to either free GlcNAc or to asialo-,agalacto-transferrin.

Reagents

Sodium cacodylate buffer, 0.25 *M*, pH 7.4
$MnCl_2$, 0.2 *M*
UDP-[^{14}C]galactose (5×10^5 cpm/μmol), 10 m*M*
GlcNAc, 0.2 *M*
Asialo-,agalacto-transferrin (human serum), 2% (w/v) in water
Bovine serum albumin (pretreated at 65° for 15 min), 2% (w/v) in water

Assay Procedure. The standard incubation contains the following in a final volume of 0.1 ml: 20 μl 0.25 *M* sodium cacodylate, pH 7.4 (5 μmol), 20 μl of 0.2 *M* $MnCl_2$ (4 μmol), 10 μl of 10 m*M* UDP-[^{14}C]galactose (0.10 μmol, 50,000 cpm), 10 μl of 0.2 *M* GlcNAc (2 μmol) and enzyme. When highly purified enzyme fractions are assayed , 10 μl of 2% bovine serum albumin (0.2 mg) are added to the incubation. If membrane-bound enzyme is being assayed, it is necessary to make the incubation 0.2% (v/v) in a detergent such as Tween 80 or Triton X-100. After 30 min at 37°, the reaction is stopped by adding 0.5 ml of ice-cold water and passage through a column (0.5 × 5 cm) of AG 1-X8, chloride form, equilibrated in water (assay method C). The column is washed with a further 1.5 ml of water, and the entire effluent is counted. Appropriate controls lacking acceptor are also run. Assays can also be run using asialo,agalacto-transferrin instead of GlcNAc as acceptor. The glycoprotein (0.050 ml of 2% solution, 1 mg), is dried in the assay tube prior to addition of reagents. Incorporation of radioactivity into glycoprotein can be measured by any of the four assay methods described earlier, i.e., high-voltage electrophoresis, acid precipitation, AG 1-X8 columns, or Sephadex G-50 columns.

[71] U. Brodbeck and K. E. Ebner, *J. Biol. Chem.* **241,** 762 (1966).
[72] K. Brew, T. C. Vanaman, and R. L. Hill, *Proc. Natl. Acad. Sci. U.S.A.* **59,** 491 (1968).
[73] H. Babad and W. Z. Hassid, this series, Vol. 8, p. 346.
[74] K. E. Ebner, R. Mawal, D. K. Fitzgerald, and B. Colvin, this series, Vol. 28, p. 500.
[75] K. E. Ebner, R. Mawal, D. K. Fitzgerald, and B. Colvin, this series, Vol. 28, p. 507.

Purification of Bovine Milk UDP-Gal: GlcNAc β1-4-Galactosyltransferase[20]

Brodbeck and Ebner[71] first resolved bovine milk lactose synthetase (UDP-Gal: Glc β1-4-galactosyltransferase) into two proteins, called A and B, both of which are needed for lactose synthetase activity. The B protein was later shown to be α-lactalbumin, a protein totally devoid of any catalytic activity. The A protein was shown[72] to be a β1-4-galactosyltransferase, which in the absence of α-lactalbumin could transfer Gal to GlcNAc to make *N*-acetyllactosamine but had minimal ability to synthesize lactose.

The A protein (UDP-Gal: GlcNAc β1-4-galactosyltransferase) has been purified[19,20,26-28] from bovine and human milk; bovine colostrum; rat, calf, and human serum; human malignant effusions; ovine mammary gland; and swine mesentary lymph nodes, using classical techniques as well as affinity chromatography on α-lactalbumin-agarose, UDP-hexanolamine-agarose, GlcNAc-agarose, *p*-aminophenyl-β-GlcNAc-agarose, and UDP-glucuronyl-6-aminohexyl-agarose. The procedure of Geren *et al.*[20] is described below.

Ammonium Sulfate Fractionation. All operations are carried out at 4°. Skim bovine milk (12 liters) is made 1 m*M* in mercaptoethanol, and a 40 to 60% saturation ammonium sulfate cut is prepared. The precipitate is collected at 10,000 *g* for 45 min and dissolved in 100 ml of 20 m*M* Tris-HCl, pH 7.5, containing 1 m*M* mercaptoethanol and 1.25 *M* ammonium sulfate (buffer A).

Norleucine–Sepharose. Norleucine is coupled to Sepharose 4B by the cyanogen bromide method to give about 10 μmol of amino acid per milliliter of packed gel. Hydrophobic chromatography is carried out by loading the enzyme solution on a column (2.5 × 29 cm) of norleucine–Sepharose and washing the column first with 500 ml of buffer A and then with a decreasing ammonium sulfate gradient formed by 700 ml of buffer A and 700 ml of 20 m*M* Tris-HCl, pH 7.5, containing 1 m*M* mercaptoethanol. The enzyme peak is pooled, protein is precipitated with 65% saturation ammonium sulfate, and the protein is dissolved in about 50 ml of 60 m*M* *N*-ethylmorpholine, pH 8.0, containing 1 m*M* mercaptoethanol, 25 m*M* $MnCl_2$, 25 m*M* GlcNAc, 0.1 *M* ammonium sulfate, and 0.1 *M* 5-amino-*n*-caproic acid (buffer B).

UDP-Hexanolamine–Sepharose. UDP-hexanolamine (Sigma) is coupled to Sepharose by the cyanogen bromide method to give about 4.5 μmol of UDP per milliliter of packed gel. The enzyme is applied to a column (2.2 × 19 cm) of UDP-hexanolamine–Sepharose, and the column is washed with several volumes of buffer B, 0.5 *M* NaCl in buffer B, and buffer B in which the $MnCl_2$ and GlcNAc concentrations are both reduced to 5 m*M*. Galactosyltransferase is eluted with 60 m*M* *N*-ethylmorpholine, pH

8.0, containing 5 m*M* EDTA, 0.1 *M* 6-amino-*n*-caproic acid, 0.1 *M* ammonium sulfate, and 1 m*M* mercaptoethanol. The enzyme is concentrated by ammonium sulfate precipitation and dissolved in 60 m*M* *N*-ethylmorpholine, pH 8.0, containing 0.1 *M* GlcNAc, 1 m*M* $MnCl_2$, and 1 m*M* mercaptoethanol (buffer C).

α-Lactalbumin–Sepharose. The α-lactalbumin is coupled to Sepharose 4B by the cyanogen bromide method to give 1–2 μmol of protein per milliliter of packed gel. The enzyme is applied to a column (2.2 × 19 cm) of α-lactalbumin–Sepharose, and the column is washed with buffer C followed by buffer C in which the concentrations of GlcNAc and $MnCl_2$ are reduced to 50 and 0.5 m*M*, respectively. The enzyme is eluted with 50 m*M* *N*-ethylmorpholine, pH 8.0, containing 5 m*M* EDTA and 1 m*M* mercaptoethanol. The enzyme is concentrated by precipitation with 85% saturated ammonium sulfate and dissolved in a minimum volume of 20 m*M* Tris-HCl, pH 7.6, containing 1 m*M* mercaptoethanol. This solution is stored in small aliquots at −20°. The purified enzyme has a specific activity of 5.6 units per milligram of protein assayed at 23° and about 15 units/mg at 37°. The purification is 6000-fold, and the yield is 65%.

Properties of UDP-Gal: GlcNAc β1-4-Galactosyltransferase

Stability. The highly purified enzyme is labile below pH 6.0 and in dilute solution. Enzyme solution can be stabilized at either 4° or −20° by the presence of ammonium sulfate (as during the above preparation), or by storing at a high protein concentration (achieved either by concentrating the enzyme or by adding bovine serum albumin), or by storing in the presence of 40 m*M* GlcNAc. The enzyme can be lyophilized in the presence of mercaptoethanol and a small amount of salt[74] and stored without loss of activity at −20° for at least a year.

Enzyme Purity. Even the purest bovine milk preparations show at least two bands on SDS–polyacrylamide gel electrophoresis (molecular weights at about 55,000–59,000 and 42,000–44,000). Both forms are catalytically active with similar kinetic constants, are glycoproteins (10–15% carbohydrate), and resemble each other with regard to heat inactivation and inhibition by reagents reactive with sulfhydryl groups. The small form may be a proteolytic product of the larger form. The highly purified preparation of Geren *et al.*[20] is free of protease activity.

Kinetic Parameters. Parameters are listed in Table I. Many kinetic studies have been published, and various enzyme mechanisms have been proposed. There appear to be two Mn^{2+} sites, a high-affinity site I and a low-affinity site II. The mechanism is an ordered addition in which the enzyme first binds Mn^{2+} at site I, followed by a branched pathway in which either Mn^{2+}-UDP-Gal binds to site II or UDP-Gal alone binds to the enzyme. This is followed by binding of GlcNAc and eventual release of products. The K_m for glucose is 1.4 M, but the addition of α-lactalbumin lowers this value to 5 mM. The ordered enzyme mechanism explains the behavior of the enzyme on affinity chromatography. The enzyme adheres strongly to α-lactalbumin–Sepharose only in the presence of acceptor (GlcNAc or Glc), to UDP-hexanolamine–Sepharose only in the presence of Mn^{2+}, and to GlcNAc–Sepharose only in the presence of UDP or UMP and Mn^{2+}.

Substrate Specificity. The purified enzyme transfers Gal to either free GlcNAc or to oligosaccharides, glycopeptides, and glycoproteins with a β-linked GlcNAc at the nonreducing terminus. Glucose, glucosamine, and oligosaccharides with terminal α- or β-linked Glc residues are poor acceptors. The addition of α-lactalbumin lowers the K_m for Glc about 300-fold, lowers transfer to GlcNAc by over 70%, but has only a small inhibitory effect on Gal transfer to larger acceptors. The physiological function of the enzyme in most tissues is the elongation of *N*-glycosyl oligosaccharides (Fig. 1), and it probably also functions in the synthesis of *O*-glycosyl oligosaccharides and glycolipids. In mammary gland during lactation, it combines with α-lactalbumin to make lactose.

UDP-Gal:GlcNAc β1-3-Galactosyltransferase

Many tissues and body fluids are capable of transferring Gal from UDP-Gal to GlcNAc,[26–28] but the product of this reaction has been definitively identified as Galβ1-4GlcNAc (*N*-acetyllactosamine) in only a few studies. N- and O-linked oligosaccharides as well as glycosphingolipids have all been shown to contain both Galβ1-4GlcNAc (sometimes called a type 2 chain in human blood group determinants) and Galβ1-3GlcNAc (type 1 chain). However, the UDP-Gal:GlcNAc β1-4-Gal-transferase (see above) is widely distributed whereas the β1-3-Gal-transferase has more recently been described in pig trachea and rat intestine.[21] It has been purified. The reaction catalyzed is

$$\text{UDP-Gal} + \text{GlcNAc} \rightarrow \text{Gal}\beta\text{1-3GlcNAc} + \text{UDP}$$

Assay Method

Principle. The assay measures the transfer of radioactive Gal from UDP-Gal to GlcNAc. However, the products of the β1-4- and β1-3-Gal-transferases must be distinguished and measured for accurate assays.

Reagents

MES buffer, 0.5 M, pH 6.3

$MnCl_2$, 0.25 M

UDP-[^{14}C]galactose (1×10^6 cpm/μmol), 10 mM

GlcNAc, 0.2 M

Triton X-100, 5.0% (v/v)
7.5% Sodium tetraborate–0.2 *M* EDTA
Phenolphthalein, 1%

Assay Procedure. A 50-μl aliquot of 0.2 *M* GlcNAc (10 μmol) is added to a small test tube and taken to dryness, and the following are added: 10 μl of 0.5 *M* MES, pH 6.3 (5 μmol), 5 μl of 0.25 *M* $MnCl_2$ (1.25 μmol), 5 μl of 10 m*M* UDP-[^{14}C]Gal (0.050 μmol), 5 μl of 5% Triton X-100, enzyme, and water to a final volume of 0.050 ml. Duplicate incubations are carried out at 37° for 40 min, and the reactions are stopped by the addition of 10 μl of 7.5% sodium tetraborate–0.2 *M* EDTA. One incubation is made alkaline by adding 1 μl of 1% phenolphthalein and 35 μl of 0.1 *N* NaOH. The solution is heated at 100° for 5 min to destroy alkali-labile Galβ1-3GlcNAc and cooled on ice; 1.0 *N* HCl is added until the indicator turns colorless. Both solutions are then subjected to high-voltage electrophoresis as described under assay method A (above). The amount of alkali-stable disaccharide product is then subtracted from the total disaccharide product to give the amount of Galβ1-3GlcNAc formed.

Properties of UDP-Gal: GlcNAc β1-3-Galactosyltransferase

Stability. Membrane preparations from pig trachea mucosal lining are stable to freezing. The homogenate must be prepared in the presence of 0.05% (w/v) soybean trypsin inhibitor.

Kinetic Properties. The β1-4-Gal-transferase is inhibited by a GlcNAc concentration of 0.2 *M*, whereas the β1-3-Gal-transferase is not. Comparing Gal transfer to GlcNAc at GlcNAc concentrations of 10 m*M* and 200 m*M* is a useful screening assay for the β1-3-Gal-transferase. Similarly, the β1-4-Gal-transferase is inhibited by α-lactalbumin whereas the β1-3-Gal-transferase is not. A total tissue survey in the rat showed appreciable amounts of the β1-3-Gal-transferase only in rat intestine.

CMP-NeuNAc: Galβ1-4GlcNAc α2-6-Sialyltransferase

Several different sialyltransferases acting on *N*- and *O*-glycosyl oligosaccharides and on glycosphingolipids have been described.[22-28,76] The only enzymes to be considered here are the two sialyltransferases acting on the disaccharide Galβ1-4GlcNAc and on *N*-glycosyl oligosaccharides terminated at their reducing ends by this disaccharide. As in the case of Gal

[76] J. E. Sadler, T. A. Beyer, C. L. Oppenheimer, J. C. Paulson, J.-P. Prieels, J. I. Rearick, and R. L. Hill, this series, Vol. 83, p. 458.

transfer to GlcNAc (above), many tissues are capable of transferring NeuNAc from CMP-NeuNAc to Galβ1-4GlcNAc or to asialoglycoproteins, glycopeptides, or oligosaccharides terminating with this disaccharide, but very few studies have determined the linkage formed between sialic acid and Gal. When in fact such a study was carried out in rat liver,[77] the only linkage formed was α2-6. However, the α2-3-sialyltransferase has also been described[24,25] (below). The α2-6-sialyltransferase has been described in a soluble form in the colostrum of the goat, cow, and human and in human, pig, and rat serum, and bound to Golgi membranes in a variety of tissues, particularly in rat, cow, pig, guinea pig, and human livers.[26-28] It has been purified to homogeneity from bovine colostrum.[22,23] The reaction catalyzed is

CMP-NeuNAc + Galβ1-4GlcNAc-R $\rightarrow$ NeuNAcα2-6Galβ1-4GlcNAc-R + CMP

Assay Method

Principle. The assay measures transfer of radioactive NeuNAc from CMP-NeuNAc to acceptors of the form Galβ1-4GlcNAc-R, where R can be H, or an oligosaccharide, glycopeptide, or glycoprotein.

Reagents

Sodium phosphate buffer, 0.5 *M*, pH 6.8
CMP-[^{14}C]NeuNAc (1 $\times$ 10^7 cpm/μmol), 2 m*M*
Asialotransferrin (human serum), 2% (w/v) in water
Bovine serum albumin (pretreated at 65° for 15 min), 2% (w/v) in water
Phosphotungstic acid, 5% (w/v)–trichloroacetic acid, 15% (w/v)

Assay Procedure. A 50-μl aliquot of 2% asialotransferrin (1 mg or approximately 0.050 μmol terminal Gal residues) is placed in a small test tube and taken to dryness, and the following are added: 10 μl 0.5 *M* phosphate, pH 6.8 (5 μmol), 5 μl of 2 m*M* CMP-[^{14}C]NeuNAc (0.010 μmol, 10^5 cpm), 10 μl of 2% bovine serum albumin (0.2 mg), enzyme, and water to 0.040 ml. If membrane-bound enzyme is being measured, the incubation should be made 0.5% (v/v) in Triton X-100. After incubation at 37° for 15 min (longer times are needed if nucleotide–sugar of lower specific activity is used), 0.5 ml of ice-cold 5% phosphotungstic acid–15% trichloroacetic acid are added, and the precipitate is collected and thoroughly washed by centrifugation or filtration (assay method B). Assay methods A or D can also be used. The assay is not specific for the α2-6-sialyltransferase. If a crude enzyme preparation is being assayed, either the α2-3- (see below) or the

[77] D. H. Van den Eijnden, P. Stoffyn, A. Stoffyn, and W. E. C. M. Schiphorst, *Eur. J. Biochem.* **81,** 1 (1977).

α2-6-sialyltransferase or both may be picked up, and there is no simple assay that is specific for either of these enzymes.

Purification of Bovine Colostrum CMP-NeuNAc: Galβ1-4GlcNAcα2-6-Sialyltransferase[22,23]

Step 1. All procedures are carried out at 4°. Bovine colostrum (22 liters) is defatted by low-speed centrifugation and dialyzed thoroughly against 10 m*M* sodium cacodylate, pH 6.5. An equal volume of 50% (v/v) glycerol is added.

Step 2. CDP–Sepharose I. Either CDP-hexanolamine–Sepharose or CDP-ethanolamine–Sepharose can be used in both steps 2 and 4. Syntheses for both materials are described by Paulson *et al.*[22] However, only CDP-ethanolamine is commercially available (Sigma) and can be readily coupled to Sepharose 4B by the cyanogen bromide method. The procedure for the latter material is therefore presented. Enzyme from 1 liter of bovine colostrum requires a column volume of about 80 ml. The column is equilibrated in 25 m*M* cacodylate, pH 6.5, enzyme is applied, and the column is washed with 15 volumes of 25 m*M* cacodylate, pH 6.5, containing 25% glycerol and 0.1 *M* NaCl followed by one volume of 25 m*M* cacodylate, pH 5.3. Enzyme is eluted with 25 m*M* cacodylate, pH 5.3, containing 0.5 *M* NaCl.

Step 3. Ammonium Sulfate Precipitation. Enzyme is concentrated by precipitation at 80% saturated ammonium sulfate, centrifugation, and dissolution of the precipitate in a minimal volume of 25 m*M* cacodylate, pH 5.3, followed by dialysis overnight against 100 volumes of the same buffer.

Step 4. CDP–Sepharose II. About 1 ml of CDP–Sepharose is used per 10–15 mg of enzyme protein at this stage. The column is equilibrated with 25 m*M* cacodylate (pH 5.3), enzyme is applied, and the column is washed with 50 volumes of 25 m*M* cacodylate (pH 5.3) containing 25% glycerol and 0.1 *M* NaCl. The column is brought to room temperature and washed with 25 m*M* cacodylate (pH 5.3) containing 50 m*M* NaCl. Enzyme is eluted with the same buffer containing 2 m*M* CDP.

Step 5. SP-50-Sephadex. The entire batch of enzyme is diluted fourfold with water, the pH is adjusted to 5.1, and the batch is applied to a column (3 × 0.7 cm) of SP (sulfopropyl)-Sephadex equilibrated with 25 m*M* cacodylate, pH 5.1. The column is washed with the same buffer, and the enzyme is eluted with 25 m*M* cacodylate, pH 5.3, containing 0.5 *M* NaCl. The enzyme is concentrated to about 12 ml by this procedure.

Step 6. Sephadex G-150. The enzyme is passed through a Sephadex G-150 column (2.5 × 90 cm) equilibrated with 50 m*M* cacodylate, pH 5.3, containing 0.2 m*M* CDP to stabilize the enzyme.

Purification during this procedure is 440,000-fold with a yield of 15% and a final specific activity of 27 units/mg. Optimal purification requires careful attention to pH and ionic strength.[22]

Properties of CMP-NeuNAc: Galβ1-4GlcNAc α2-6-Sialyltransferase

Stability. The enzyme is stabilized by high protein concentration or glycerol, low pH (5.2), use of plastic containers, CDP, and low temperature. The pure enzyme is stable for at least 2 months in 50% glycerol containing 12 m*M* cacodylate buffer, pH 5.3, at −20°.

Enzyme Purity. The pure preparation shows only two major bands (molecular weights of 56,000 and 43,000) on SDS–polyacrylamide gel electrophoresis. Both bands are highly active sialyltransferases with the same kinetic properties.

Kinetic Parameters. The kinetic parameters are summarized in Table I. The enzyme appears to have an equilibrium random-order mechanism.[23]

Substrate Specificity. Galβ1-4GlcNAc and glycopeptides and glycoproteins with Galβ1-4GlcNAc termini are the only good acceptors. Galβ1-3GlcNAc, Galβ1-6GlcNAc, Galβ1-4Glc (lactose), and other β-galactosides have much higher K_m values (140–1780 m*M*) than *N*-acetyllactosamine (K_m = 12 m*M*). Product (usually α2-6-linked) can be made with these poor substrates, but only at very high concentrations. Substitution of the terminal Gal with sialic acid or removal of the terminal Gal prevents enzyme action. Galβ1-3GalNAc-terminal glycoproteins, which are excellent acceptors for two (α2-3- to Gal and α2-6- to GalNAc) sialyltransferases acting on mucins, are not acceptors for the bovine colostrum enzyme. It is interesting that the enzyme prefers to attach sialic acid to the Gal on the Manα1-3- arm of the core (Fig. 1).[78] This is compatible with the finding of monosialylated biantennary complex oligosaccharides in IgG[49,50] in which the sialic acid is completely on the Manα1-3- arm. The physiological function of this sialyltransferase is almost certainly the synthesis of the NeuNAcα2-6Galβ1-4GlcNAc- antennae of complex *N*-glycosyl oligosaccharides.

CMP-NeuNAc: Galβ1-4GlcNAc α2-3-Sialyltransferase

The sialylα2-3Gal linkage is found in both *N*- and *O*-glycosyl oligosaccharides, in milk oligosaccharides, and in gangliosides. At least three different and distinct sialyltransferases are now known to be involved in the synthesis of this linkage,[26–28] i.e., enzymes that respectively convert (*a*) lactosyl-ceramide to sialylα2-3Galβ1-4Glc-ceramide (ganglioside G_{M3}); (*b*) Galβ1-3GalNAc-R (where R is H, oligosaccharide, glycoprotein, or ganglioside) to sialylα2-3Galβ1-3GalNAc-R; and (*c*) Galβ1-4GlcNAc-R (where R is a glycoprotein) to sialylα2-3Galβ1-4GlcNAc-R. The synthesis of sialylα2-3Galβ1-4Glc (3′-sialyllactose) has been reported in many tissues (rat mam-

[78] D. H. Van den Eijnden, D. H. Joziasse, L. Dorland, H. Van Halbeek, J. F. G. Vliegenthart, and K. Schmid, *Biochem. Biophys. Res. Commun.* **92,** 839 (1980).

mary gland, liver from several mammalian species, colostrum, rat brain, pig submaxillary gland). It was assumed in earlier work that the synthesis of 3′-sialyllactose was due to CMP-NeuNAc:Galβ1-4GlcNAcα2-3-sialyltransferase, but in fact it is probably catalyzed by CMP-NeuNAc: Galβ1-3GalNAcα2-3-sialyltransferase. The enzyme that transfers sialic acid in α2-3- linkage to Galβ1-4GlcNAc-R has proved to be elusive. Rat and pig liver and rabbit bone marrow have little or none of this enzyme, but it has been described in fetal calf liver, embryonic chicken brain, rat testis, rabbit spleen and kidney, and human placenta.[24,25,77]

Assay Method

There is no simple way to assay separately for the α2-3- and α2-6-sialyltransferases acting on Galβ1-4GlcNAc-R. The α2-3-sialyltransferase was first demonstrated[24,25] by using nonradioactive CMP-NeuNAc as donor and radioactive glycoprotein as acceptor. The terminal Gal residues of the acceptor were made radioactive with the galactose oxidase technique. The product was methylated and hydrolyzed and the hydrolyzate was analyzed for radioactive partially methylated sugars by thin-layer chromatography. The presence of 2,4,6-trimethyl-[^{3}H]galactose and 2,3,4-trimethyl-[^{3}H]galactose indicate, respectively, the action of α2-3- and α2-6-sialyltransferases. Whenever crude membrane preparations are assayed using CMP-NeuNAc as donor and Galβ1-4GlcNAc-R as acceptor, either the α2-3 or the α2-6-sialyltransferases or both may be measured. All tissues so far shown to have the α2-3-sialyltransferase also contain varying levels of the α2-6-sialyltransferase.

GDP-Fuc:β-N-Acetylglucosaminide (Fuc to Asn-linked GlcNAc) α1-6-Fucosyltransferase

Fucose has been reported to occur linked α1-6- or α1-3- to the Asn-linked GlcNAc of *N*-glycosyl oligosaccharides.[26-28] Fucose also occurs in a variety of antigenic determinants (Table III and next section) at the nonreducing termini of milk and urinary oligosaccharides, N- and O-linked protein-bound oligosaccharides, and glycosphingolipids.[26-30] Golgi apparatus-enriched membrane preparations from pig and rat liver and rat and mouse testis, crude extracts from HeLa cells, and human serum, can all transfer Fuc from GDP-Fuc to *N*-glycosyl oligosaccharides with terminal GlcNAc residues. The rat liver enzyme has been shown to transfer Fuc to the Asn-linked GlcNAc residue, but the linkage has not been determined.[79] The pork liver enzyme has been shown to catalyze the following reaction.[16]

[79] J. R. Wilson, D. Williams, and H. Schachter, *Biochem. Biophys. Res. Commun.* **72,** 909 (1976).

GDP-β-L-Fuc + GlcNAcβ1-2Manα1-3(R-Manα1-6)Manβ1-
4GlcNAcβ1-4GlcNAc-Asn-X → GlcNAcβ1-2Manα1-3
(R-Manα1-6)Manβ1-4GlcNAcβ1-4(Fucα1-6)GlcNAc-Asn-X + GDP

where R is defined in the section on substrate specificity (below). No product was detected with Fuc in α1-3 linkage to the Asn-linked GlcNAc.

Assay Method

Principle. The assay determines the incorporation of radioactive Fuc from GDP-Fuc into glycopeptide GnGn prepared from transferrin.

Reagents

Tris-HCl buffer, 0.5 *M*, pH 8.0
$MgCl_2$, 0.4 *M*
GTP, 25 m*M*
GDP-β-L-[^{14}C]fucose (9×10^6 dpm/μmol), 2.5 m*M*
Transferrin glycopeptide GnGn, 5 m*M*
Triton X-100, 10% (v/v)
2% Sodium tetraborate–0.5 *M* disodium EDTA

Assay Procedure. The standard incubation contains the following in a final volume of 0.050 ml: 5 μl of 0.5 *M* Tris-HCl, pH 8.0 (2.5 μmol), 5 μl of 0.4 *M* $MgCl_2$ (2.0 μmol), 5 μl of 25 m*M* GTP (0.125 μmol), 5 μl of 2.5 m*M* GDP-[^{14}C]Fuc (0.0125 μmol, 112,000 dpm), 5 μl of 5 m*M* glycopeptide GnGn (0.025 μmol), 5 μl of 10% Triton X-100 (0.5 μl), and enzyme. One can use asialoagalacto-glycoprotein as acceptor instead of glycopeptide GnGn, but the K_m is rather high with glycoprotein acceptors (17 m*M* in terms of terminal GlcNAc residues[80]), and it is therefore difficult to saturate the enzyme with this acceptor. After 1 hr at 37° the reaction is stopped by addition of 0.020 ml of 2% sodium tetraborate–0.5 *M* EDTA, and formation of product is determined by high-voltage electrophoresis (assay A), AG 1-X8 chromatography (assay C), or gel filtration (assay D).

Properties of α1-6-Fucosyltransferase

Stability. Golgi-rich membranes from pig or rat liver are stable for months at −20° in the absence of detergent.

Kinetic Parameters. See Table I.

Substrate Specificity. The substrate specificity of the α1-6-fucosyltransferase is summarized in Table II. The enzyme will act on substrates with the structure GlcNAcβ1-2Manα1-3(R-Manα1-6)Manβ1-4GlcNAcβ1-4GlcNAc-Asn-X, where R is H (glycopeptide MGn), GlcNAcβ1-2 (glycopeptide GnGn), or Manα1-6 (Manα1-3). The enzyme does not act on

[80] I. Jabbal and H. Schachter, *J. Biol. Chem.* **246,** 5154 (1971).

substrates that lack the GlcNAcβ1-2Manα1-3- sequence or on any substrate carrying a bisecting GlcNAc residue (GlcNAc attached β1-4- to the β-linked Man). Elongation of antennae by addition of Galβ1-4- or sialyl-Galβ1-4- sequences also prevents enzyme action. These reactions are summarized in Fig. 1. These studies predict that high-mannose *N*-glycosyl oligosaccharides should not have a Fuc in the core, and, indeed, this prediction has thus far not been violated.

GDP-Fuc:β-Galactoside α1-2,α1-3 and α1-4-Fucosyltransferases

There are at least three separate fucosyltransferases acting on acceptors with a β-galactoside nonreducing terminus,[17,18,26-30] i.e., (*a*) the human blood group H-dependent GDP-L-Fuc:β-galactoside α1-2-fucosyltransferase; (*b*) the human blood group Lewis-dependent GDP-L-Fuc: Galβ1-3GlcNAc-R (Fuc to GlcNAc) α1-4-fucosyltransferase, and (*c*) GDP-L-Fuc:Galβ1-4GlcNAc-R (Fuc to GlcNAc) α1-3-fucosyltransferase. All three enzymes have an absolute requirement for a terminal Gal residue and are therefore different from the α1-6-fucosyltransferase discussed above. Space allows only a brief discussion of these enzymes.

The blood group H enzyme attaches Fuc in α1-2 linkage to the terminal Gal residue of any β-D-galactoside irrespective of the subterminal sugars and, indeed, the blood group H determinant occurs on both type 1 and type 2 chains (Table III) and on the Galβ1-3GalNAc- sequence. Phenyl β-D-galactoside is a useful specific substrate for the enzyme, since it is inactive with the other known fucosyltransferases. Structures like Galβ1-3(Fucα1-4)GlcNAc-R or Galβ1-4(Fucα1-3)GlcNAc-R are not acceptors, since the second fucose residue inhibits enzyme action. The enzyme occurs in the serum and bone marrow of all humans with the H gene irrespective of secretor status. The enzyme also occurs in human milk, submaxillary glands, and gastric mucosa, but only in individuals who are secretors, i.e., have the Se gene. A similar enzyme occurs in other species, i.e., dog, pig, rat, and cow. Such an enzyme has been purified 124,000-fold to homogeneity from pig submaxillary glands.[28]

The blood group Lewis enzyme attaches Fuc in α1-4 linkage to the subterminal GlcNAc of Galβ1-3GlcNAc-R (type 1) structures (Table III). The enzyme has been found in the submaxillary glands, gastric mucosa, and milk of individuals with the Lewis (Le) gene. The enzyme is not present in human serum irrespective of the Lewis genotype. The best substrates for specific assay of the enzyme are milk oligosaccharides like lacto-*N*-fucopentaose I (Fucα1-2Galβ1-3GlcNAcβ1-3Galβ1-4Glc), which are not acceptors for the other known fucosyltransferases. The enzyme has been purified over

TABLE III
ANTIGENIC DETERMINANTS[29,30,82,83]

Structure of determinant	Name of antigen
Type 1[a]	
Fucα1-2Galβ1-3GlcNAc-	Blood group H, type 1 (Le[d]H)
GalNAcα1-3(Fucα1-2)Galβ1-3GlcNAc-	Blood group A, type 1 (ALe[d])
Galα1-3(Fucα1-2)Galβ1-3GlcNAc-	Blood group B, type 1 (BLe[d])
Galβ1-3(Fucα1-4)GlcNAc-	Blood group Lewis[a] (Le[a])
Fucα1-2Galβ1-3(Fucα1-4)GlcNAc-	Blood group Lewis[b] (Le[b]H)
Type 2[b]	
Fucα1-2Galβ1-4GlcNAc-	Blood group H, type 2
GalNAcα1-3(Fucα1-2)Galβ1-4GlcNAc-	Blood group A, type 2
Galα1-3(Fucα1-2)Galβ1-4GlcNAc-	Blood group B, type 2
Galβ1-4(Fucα1-3)GlcNAc-	Determinant X, stage-specific embryonic antigen-1 (SSEA-1)
Fucα1-2Galβ1-4(Fucα1-3)GlcNAc-	Determinant Y
Polylactosaminoglycan	
Galβ1-4GlcNAcβ1-3Galβ1-4GlcNAcβ1-3-R	Blood group i
Galβ1-4GlcNAcβ1-6\\ Galβ1-4GlcNAc-R / Galβ1-4GlcNAcβ1-3	Blood group I

[a] Determinants containing the sequence Galβ1-3GlcNAc are called type 1.
[b] Determinants containing the sequence Galβ1-4GlcNAc are called type 2.

500,000-fold to homogeneity from human milk.[17,28] The highly purified enzyme carries out all three of the following reactions.

GDP-Fuc + Galβ1-3GlcNAc-R → Galβ1-3(Fucα1-4)GlcNAc-R + GDP
GDP-Fuc + Galβ1-4GlcNAc-R → Galβ1-4(Fucα1-3)GlcNAc-R + GDP
GDP-Fuc + Galβ1-4Glc → Galβ1-4(Fucα1-3)Glc + GDP

The kinetic parameters for these substrates are shown in Table I. The α1-3- and α1-4- activities copurify, are activated to the same extent by various cations, and are inactivated at identical rates by various procedures. Kinetic analysis indicates that a single active site is involved in both α1-3- and α1-4-Fuc incorporation. This is the first example of a single glycosyltransferase catalyzing sugar attachment to two different carbon atoms at appreciable rates and is therefore the first example of an exception to the "one linkage–one glycosyltransferase" rule.[26-28] Enzyme activity is enhanced by the presence of a Fucα1-2- residue on the terminal Gal of a substrate. Enzyme is inactive toward substrates with the structure Galβ1-3GalNAc-R commonly found in *O*-glycosyl oligosaccharides.

It has been demonstrated[18] that saliva from individuals with the Le gene can catalyze the three reactions shown above, in agreement with the finding that the purified Le-dependent fucosyltransferase catalyzes these reactions. Saliva from Le-negative individuals, however, can catalyze only the attachment of Fuc in α1-3 linkage to GlcNAc. There is, therefore, a third enzyme, GDP-L-Fuc:Galβ1-4GlcNAc (Fuc to GlcNAc) α1-3-fucosyltransferase, which is different from both the H- and Le-dependent fucosyltransferases. This enzyme and the Le-dependent enzyme can, in fact, be separated by isoelectric focusing[18] of human saliva from Le-positive individuals. Also, human serum contains the non-Le-dependent α1-3-fucosyltransferase but lacks completely the Le-dependent α1-3/4-fucosyltransferase even in Le-positive individuals. Neither the serum enzyme nor the Le-negative salivary enzyme can add Fuc in α1-4 linkage to Galβ1-3GlcNAc or in α1-3 linkage to Galβ1-4Glc.

Pig liver contains a GDP-L-Fuc:β-galactoside fucosyltransferase which is highly active toward Galβ1-4GlcNAc, but not toward Galβ1-3GlcNAc, Galβ1-6GlcNAc, or Galβ1-4Glc.[80] It thus resembles the human non-Le-dependent α1-3-fucosyltransferase in substrate specificity. The structure Galβ1-4(Fucα1-3)GlcNAc occurs in complex *N*-glycosyl oligosaccharides, e.g., in human serum α_1-acid glycoprotein,[81] and is probably synthesized by the non-Le-dependent α1-3-fucosyltransferase. This enzyme has not as yet been purified.

Table III lists other antigenic determinants that contain fucose.[29,30,82,83] These determinants occur at the nonreducing ends of *N*-glycosyl oligosaccharides of the polylactosaminoglycan type (Table III), *O*-glycosyl oligosaccharides, and glycosphingolipids. Some of these determinants have been attached to the nonreducing ends of complex *N*-glycosyl oligosaccharides by *in vitro* enzymic methods.[28] Several reviews are available on the biosynthesis of these determinants.[26-30,82]

[81] B. Fournet, J. Montreuil, G. Strecker, L. Dorland, J. Haverkamp, J. F. G. Vliegenthart, J. P. Binette, and K. Schmid, *Biochemistry* **17,** 5206 (1978).

[82] S.-I. Hakomori, *Semin. Hematol.* **18,** 39 (1981).

[83] E. F. Hounsell, H. C. Gooi, A. M. Lawson, and T. Feizi, *in* "Glycoconjugates, Proceedings of the Sixth International Symposium on Glycoconjugates" (T. Yamakawa, T. Osawa, and S. Handa, eds.), p. 308. Japan Scientific Societies Press, Tokyo, 1981.

[11] Inhibitors of Glycoprotein Synthesis

By ALAN D. ELBEIN

Many of the glycoproteins of eukaryotic cells contain oligosaccharide chains that are attached to protein via a GlcNAc → asparagine linkage.[1] In these proteins, the oligosaccharide may be of the complex type or of the high-mannose structure as outlined in Fig. 1. Both types of oligosaccharides contain the same pentasaccharide core region composed of a branched trimannose linked to *N,N'*-diacetylchitobiose, which in turn is attached to the amide nitrogen of asparagine. In the high-mannose structures shown in Fig. 1, this pentasaccharide is further substituted by a number of α-linked mannose residues (as many as six), whereas in the complex types the pentasaccharide is elongated by the trisaccharide, sialic acid → galactose → GlcNAc. There may be two, three, or four of these trisaccharides attached to this pentasaccharide, as well as some fucose residues.[2,3]

Both types of oligosaccharides are biosynthesized by a similar series of reactions that involve lipid-linked saccharide intermediates. The postulated series of reactions in this pathway are outlined in Fig. 2. In this pathway, the sugars GlcNAc, mannose, and glucose are transferred from their sugar–nucleotide derivatives to an oligosaccharide–lipid that has been characterized as Glc_3Man_9-$GlcNAc_2$-pyrophosphoryl-dolichol.[4,5] The oligosaccharide is then transferred from its lipid carrier to the protein, apparently while the polypeptide is being synthesized on membrane-bound polysomes.[6-10] After this transfer, the oligosaccharide portion of the glycoprotein undergoes a number of processing or trimming reactions. The initial processing reactions probably begin in the endoplasmic reticulum and continue as the protein is transported through the Golgi apparatus to the membrane.[11,12]

[1] R. Kornfeld and S. Kornfeld, *Annu. Rev. Biochem.* **45,** 217 (1976).

[2] H. Schachter and S. Roseman, *in* "The Biochemistry of Glycoproteins and Proteoglycans" (W. J. Lennarz, ed.), p. 85. Plenum, New York, 1980.

[3] T. A. Beyer, J. E. Sadler, J. I. Rearick, J. C. Paulson, and R. L. Hill, *Adv. Enzymol.,* **52,** 23 (1981).

[4] D. K. Struck and W. J. Lennarz, *in* "The Biochemistry of Glycoproteins and Proteoglycans" (W. J. Lennarz, ed.), p. 35. Plenum, New York, 1980.

[5] A. D. Elbein, *Annu. Rev. Plant Physiol.* **30,** 239 (1979).

[6] M. Kiely, G. S. McKnight, and R. Schimke *J. Biol. Chem.* **251,** 5490 (1976).

[7] V. Czicki and W. J. Lennarz, *J. Biol. Chem.* **252,** 7901 (1977).

[8] R. C. Das and E. C. Heath, *Proc. Natl. Acad. Sci. U.S.A.* **77,** 3811(1980).

[9] E. Rodriquez-Boulan, G. Kriebach, and D. D. Sabatini, *J. Cell Biol.* **78,** 874 (1978).

[10] V. R. Lingappa, J. R. Lingappa, R. Prasad, K. Ebner, and G. Blobel, *Proc. Natl. Acad. Sci. U.S.A.* **75,** 2338 (1978).

[11] L. S. Grinna and P. W. Robbins, *J. Biol. Chem.* **254,** 2255 (1979).

[12] J. J. Elting, W. W. Chen, and W. J. Lennarz, *J. Biol. Chem.* **255,** 2325 (1980).

ISBN 0-12-181998-1

```
(A)
Man   Man   Man
 |     |     |
Man   Man   Man
 |      \   /
Man      Man
   \    /
    Man
     |
   GlcNAc
     |
   GlcNAc
     |
    Asn
    (A)

(B)
NeuNAc NeuNAc  NeuNAc NeuNAc
  |      |       |      |
 Gal    Gal     Gal    Gal
  |      |       |      |
GlcNAc GlcNAc  GlcNAc GlcNAc
    \   /          \   /
     Man            Man
         \        /
            Man
             |
           GlcNAc
             |
           GlcNAc
             |
            Asn
            (B)
```

FIG. 1. General structures of the high-mannose (A) and complex (B) oligosaccharides of the asparagine-linked glycoproteins.

The processing reactions involved in glycoprotein biosynthesis are outlined in Fig. 3. Very soon after transfer of the oligosaccharide to protein, all three of the glucose residues are removed by membrane-bound glucosidase(s)[5] that appear to reside in the endoplasmic reticulum.[13,14] These reactions give rise to a glycoprotein having a $Man_9GlcNAc_2$-Asn structure. This glycoprotein may proceed directly to the high-mannose glycoprotein or

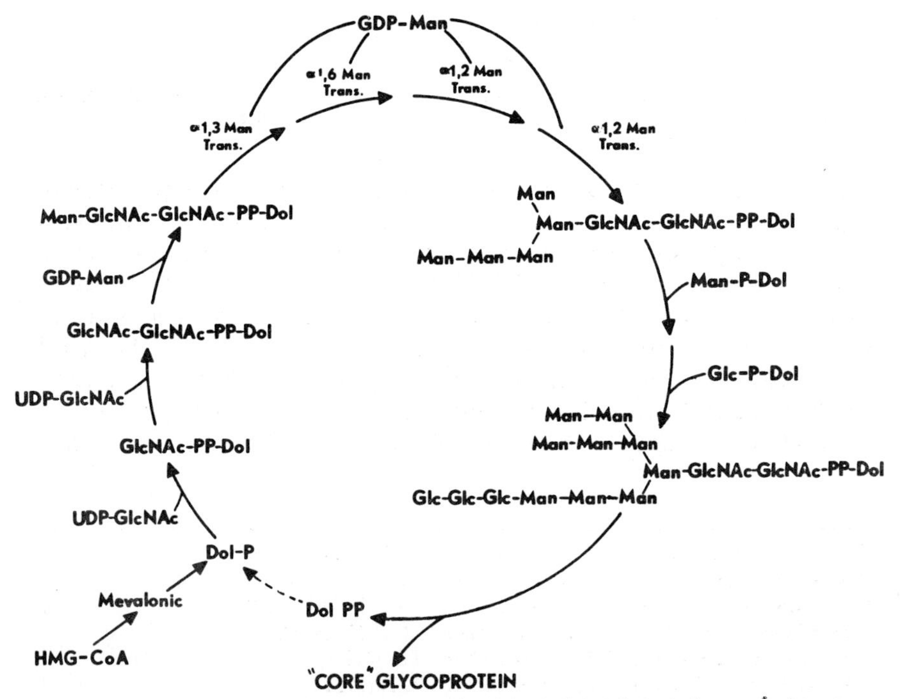

FIG. 2. Proposed reactions for the biosynthesis of the dolichol-linked oligosaccharides that participate in glycosylation of asparagine-linked glycoproteins.

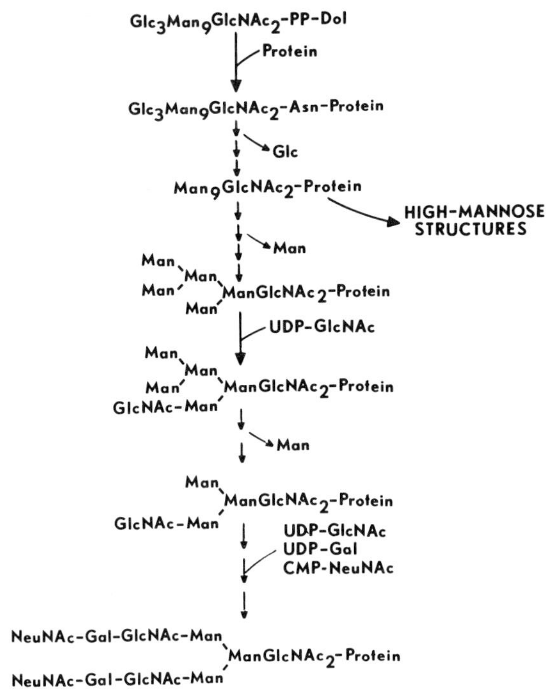

FIG. 3. Proposed pathway of processing for the formation of complex and high-mannose oligosaccharides.

it may be further processed to form the precursor to the complex type of oligosaccharide.[15] These additional processing reactions occur in the Golgi and involve the removal of four mannose residues by membrane-bound mannosidase(s) to give a $Man_5GlcNAc_2$-protein.[16-19] A GlcNAc residue is then added from UDP-GlcNAc to the $Man_5GlcNAc_2$-protein, and this addition apparently signals an α-mannosidase to remove the two branched mannoses.[20,21] Then the remaining sugars, GlcNAc, galactose, and sialic acid (and fucose), are added by sequential transfer catalyzed by individual glycosyltransferases to form the complex chains.[15] These oligosaccharides

[13] S. C. Hubbard and P. W. Robbins, *J. Biol. Chem.* **254,** 4568 (1979).
[14] R. A. Ugalde, R. J. Staneloni, and L. F. Leloir, *FEBS Lett.* **91,** 209 (1978).
[15] S. C. Hubbard and R. J. Ivatt, *Annu. Rev. Biochem.* **50,** 555 (1980).
[16] D. J. Opheim and O. Touster, *J. Biol. Chem.* **253,** 1017 (1978).
[17] I. Tabas and S. Kornfeld, *J. Biol. Chem.* **254,** 11655 (1979).
[18] S. Kornfeld and I. Tabas, *J. Biol. Chem.* **253,** 7770 (1978).
[19] W. J. Forsee and J. Schutzbach, *J. Biol. Chem.* **256,** 6577 (1981).
[20] I. Tabas and S. Kornfeld, *J. Biol. Chem.* **253,** 7779 (1978).
[21] S. Narasimhan, P. Stanley, and H. Schachter, *J. Biol. Chem.* **252,** 3926 (1977).

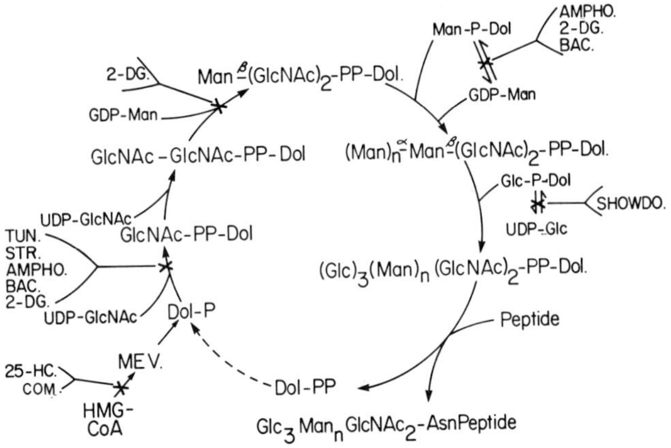

FIG. 4. Site of action of specific inhibitors of the lipid-linked saccharides. Abbreviations of inhibitors are as follows: 25-HC, 25-hydroxycholesterol; COM., Compactin; TUN., tunicamycin; STR., streptovirudin; AMPHO., amphomycin; BAC., bacitracin; 2-DG, 2-deoxyglucose; SHOWDO., showdomycin.

may have two (biantennary), three (triantennary), or four (tetraantennary) chains of the trisaccharide, sialic acid → galactose → GlcNAc.[2,3] However, all the details in the synthesis of these oligosaccharides are not known at this time.

One approach to examining the details of these pathways is the use of inhibitors that block at various steps in the pathway. As shown in Fig. 4, various inhibitors are known that block the formation of lipid-linked saccharides at selected points in the pathway. Some of these inhibitors can also be used to ask questions about the role of the carbohydrate in the function of the glycoprotein. Since glycoproteins play key roles in diverse biological systems, these inhibitors have found a place in cell biology, development biology, physiology, immunology, virology, micrcbiology, biochemistry, and other fields.

Inhibitors of the Lipid-Linked Saccharide Pathway

Tunicamycin and Related Antibiotics

Tunicamycin is a nucleoside antibiotic produced by *Streptomyces lysosuperificus*. This antibiotic was originally isolated by Tamura and associates and shown to be inhibitory toward gram-positive bacteria, yeast, fungi, protozoa, enveloped viruses, and mammalian cells in culture.[22] Its structure was deduced by a number of chemical studies by Tamura and co-workers.[23]

[22] A. Takatsuki, K. Kolino, and G. Tamura, *J. Antibiot.* **24,** 215 (1971).

[23] A. Takatsuki, K. Kawamura, Y. Kodama, I. Teichiro, and G. Tamura, *Agric. Biol. Chem.* **43,** 761 (1979).

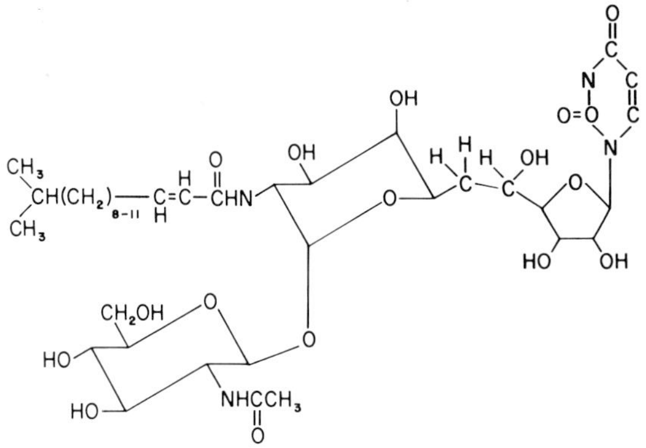

FIG. 5. Structure of tunicamycin. Various tunicamycins (or related antibiotics) differ from each other in the structure of the fatty acid.

As shown in Fig. 5, it contains a uracil to which is attached the unusual 11-carbon amino-dialdose called tunicamine. The tunicamine is substituted at two positions; at the anomeric carbon an *N*-acetylglucosamine (GlcNAc) is attached in glycosidic linkage, whereas at the amino group a long-chain fatty acid is linked in an amide bond. Tunicamycin is produced as a mixture of at least 10 homologous antibiotics that can be separated from each other by high-performance liquid chromatography.[24-26] These homologs differ in the structure of their fatty acid components, which may vary in chain length from C-13 to C-17 and may be either normal or branched, and saturated or unsaturated. The molecular weights of the tunicamycins vary from about 814 to 872.[25,26]

The site of action of tunicamycin was demonstrated in particulate enzyme preparations from mammalian cells and tissues. This antibiotic was shown to inhibit the first reaction in the lipid-linked saccharide pathway (Figs. 2 and 4), i.e., the transfer of GlcNAc-1-P from UDP-GlcNAc to dolichyl-P to form dolichyl-pyrophosphoryl-GlcNAc.[27] This GlcNAc-lipid serves as the acceptor for the remaining sugars to produce the $Glc_3Man_9GlcNAc_2$-pyrophosphoryl-dolichol that is the ultimate donor of oligosaccharide to protein (Fig. 2). In the presence of sufficient amounts of tunicamycin, this lipid-linked oligosaccharide cannot be formed and the protein is not glycosylated. Thus, in mammalian cells incubated in the presence of tunicamycin, at levels of 0.1–5 μg/ml, depending on the cell

[24] W. C. Mahoney and D. Duskin, *J. Biol. Chem.* **254,** 6572 (1979).

[25] T. Ito, A. Takatsuki, K. Kawamura, K. Sato, and G. Tamura, *Agric. Biol. Chem.* **44,** 695 (1980).

[26] R. W. Keenan, R. L. Hamill, J. L. Occolowitz, and A. D. Elbein, *Biochemistry* **20,** 2968 (1981).

[27] J. S. Tkacz and J. O. Lampen, *Biochem. Biophys. Res. Commun.* **65,** 248 (1975).

type, the incorporation of [2-^{3}H]mannose into lipid-linked oligosaccharides and into glycoproteins is inhibited by 90–95%.

The UDP-GlcNAc-dolichyl-P: GlcNAc-1-P transferase that catalyzes the first reaction has been solubilized from several tissues, and the mechanism of tunicamycin inhibition was examined.[28,29] Although tunicamycin bears a structural relationship to the substrate, UDP-GlcNAc, and might be expected to be a competitive inhibitor, it was not possible to reverse the inhibition at high substrate concentrations, probably because tunicamycin has such a high affinity for the enzyme. However, the inhibition by tunicamycin could be prevented by preincubation of the transferase with high concentrations of UDP-GlcNAc.[29] This led the authors to suggest that tunicamycin is a bisubstrate analog mimicking the substrate–product transition state. On the other hand, in the case of streptovirudin (see below), which has a 100-fold lower affinity for the enzyme, the inhibition was reversible at increased UDP-GlcNAc concentrations and the kinetics suggested a competitive type of inhibition.[30]

Several other antibiotics have been described that are closely related in structure to tunicamycin and apparently have the same site of action.[31] These include streptovirudin,[32] mycospocidin,[33] antibiotic 24010,[34] and antibiotic MM 19290.[35] Although these compounds have not been characterized to the same extent as tunicamycin, the method of purification and their chromatographic properties are quite similar to those of tunicamycin. Streptovirudin and antibiotic 24010 were also shown to inhibit the formation of dolichyl-pyrophosphoryl-GlcNAc.[36] Of interest is the fact that this group of antibiotics appears to be quite specific for the GlcNAc-1-P transferase, at least when tested at fairly low concentrations. Thus, tunicamycin did not inhibit the addition of the second GlcNAc to the lipid-linked saccharides (i.e., the formation of dolichyl-pyrophosphoryl-*N*,*N*′-diacetylchitobiose as shown in Fig. 2),[37] nor did it inhibit the addition of terminal GlcNAc residues to the protein.[38] However, high concentrations of these antibiotics also block the formation of dolichyl-phosphoryl-glucose.[36]

[28] A. Heifetz, R. W. Keenan, and A. D. Elbein, *Biochemistry* **18,** 2186 (1979).
[29] R. K. Keller, D. Y. Boon, and F. C. Crum, *Biochemistry* **18,** 3946 (1979).
[30] A. D. Elbein, J. L. Occolowitz, R. L. Hamill, and K. Eckardt, *Biochemistry* **20,** 4210 (1981).
[31] A. D. Elbein, *Trends Biochem. Sci.* **6,** 219 (1981).
[32] K. Eckardt, H. Thrum, G. Bradler, E. Tonew, and M. Tonew, *J. Antibiot.* **28,** 274 (1975).
[33] J. S. Tkacz and A. Wong, *Fed. Proc., Fed. Am. Soc. Exp. Biol.* **37,** 2772 (1976).
[34] M. Mizuno, Y. Shinojima, T. Lugawara, and I. Takeda, *J. Antibiot.* **24,** 896 (1971).
[35] M. Kenig and C. Reading, *J. Antibiot.* **32,** 549 (1979).
[36] A. D. Elbein, J. Gafford, and M. S. Kang, *Arch. Biochem. Biophys.* **196,** 311 (1979).
[37] L. Lehle and W. Tanner, *FEBS Lett.* **71,** 167 (1976).
[38] D.K. Struck and W. J. Lennarz, *J. Biol. Chem.* **252,** 1007 (1977).

Tunicamycin has been utilized in a variety of whole-cell and tissue systems (usually at concentrations of 0.5–5 μg/ml) in order to investigate the role of the carbohydrate portion of the glycoprotein in secretion, membrane function, viral assembly, and so on. The results obtained in these various systems are rather confusing, but taken as a whole they suggest that the function of the carbohydrate depends on the glycoprotein in question. Thus, in terms of secretion, some glycoproteins are still secreted even in their "unglycosylated" form, whereas others are not. For example, yeast invertase and acid phosphatase produced in the presence of tunicamycin were still secreted from protoplasts of yeast although there was a marked reduction in their synthesis.[39] Tunicamycin also did not block the secretion of very-low-density lipoprotein (VLDL) or transferrin by primary cultures of chicken liver or rat liver hepatocytes, although it did prevent glycosylation of these molecules.[40] Secretion of procollagen was likewise unimparied in the presence of this antibiotic.[41]

However, in other systems tunicamycin had a profound effect on the secretion of the glycoprotein or on the amount of glycoprotein present. For example, tunicamycin inhibited the glycosylation and secretion of μ- and α-immunoglobulin heavy chains from myeloma tumor cell lines.[42] In these cells, electron microscopic studies and direct immunofluorescence showed the presence of distended rough endoplasmic reticulum (RER) that contained a dense granular precipitate identified as immunoglobulin. The authors suggested that the absence of carbohydrate caused an alteration in the physical properties of the immunoglobulins and led to decreased solubility and aggregation. In the case of studies on the major cell surface glycoprotein produced by chick embryo fibroblasts, tunicamycin did not change the amount of this protein appearing at the cell surface,[43] although it did reduce the amount present in fibroblast cultures. This decrease in total cell surface glycoprotein was due to an increased degradation in cells treated with tunicamycin. Thus, it appears that, in some cases, unglycosylated glycoproteins are less soluble and/or undergo more rapid proteolytic degradation.

Studies on the formation of various viral glycoproteins in the presence of tunicamycin are in keeping with the above results. Again depending on the specific system under consideration, the virus may be formed and may be infective whereas in other cases, mature virus is not observed. Thus, in baby hamster kidney (BHK) cells infected with vesicular stomatitis virus (VSV), tunicamycin caused a marked decrease in the yield of infective VSV.[44] This

[39] S.-C. Kuo and J. O. Lampen, *Biochem. Biophys. Res. Commun.* **58,** 287 (1974).

[40] D. K. Struck, P. B. Suita, M. D. Lane, and W. J. Lennarz, *J. Biol. Chem.* **253,** 5332 (1978).

[41] D. Duskin and P. Bornstein, *Proc. Natl. Acad. Sci. U.S.A.* **74,** 3433 (1977).

[42] S. Kickman, A. Kilczki, Jr., R. G. Lynch, and S. Kornfeld, *J. Biol. Chem.* **252,** 4402 (1977).

[43] K. Olden, R. M. Pratt, and K.M. Yamada, *Cell* **13,** 461 (1978).

[44] R. Leavitt, S. Schlesinger, and S. Kornfeld, *J. Virol.* **21,** 375 (1977).

decrease was found to result from a block in virion assembly since the nonglycosylated proteins did not migrate to the plasma membrane, which is an obligatory step in the budding process.[45,46] On the other hand, tunicamycin had relatively little effect on the formation of influenza virus in one system, although the virus particle did lack the "spiked" appearance of normal virus particles.[47] But in another case, the carbohydrate-free hemagglutinin of fowl plague virus formed in the presence of tunicamycin was found to undergo rapid proteolysis in chick embryo fibroblasts and could be detected only when a protease inhibitor was added to the medium.[48] Various strains of VSV were used to examine glycoprotein synthesis in a BHK cell line. With one of these VSV strains (San Juan), tunicamycin prevented viral replication more than 90% when the BHK cells were grown at 38° and almost the same when the cells were grown at 30°. In another VSV strain (Orsay), viral replication again was inhibited 85–90% at 38° in tunicamycin, but only 30–55% at 30°. With VSV Orsay in cells grown at 30°, the nonglycosylated G protein could be detected on the external side of the plasma membrane (i.e., in the plasma membrane), but with other VSV strains it was undetectable. Furthermore, differences were observed in the physical properties of the viral proteins formed at the different temperatures in the presence of tunicamycin. Thus the Orsay protein synthesized at 30° could be solubilized by Triton X-100, whereas the proteins from other strains remained insoluble.[47] These data suggested that different proteins may have different requirements for carbohydrate in order to allow the protein to assume the proper conformation.

Glucose and Mannose Analogs

A number of analogs of glucose and mannose, such as 2-deoxyglucose or 2-deoxy-2-fluoro-D-glucose, or D-mannose, are useful inhibitors of protein glycosylation. It is interesting that some of these inhibitors were discovered because they showed antiviral activity against enveloped viruses. Subsequently, it was found that this antiviral activity was due to the fact that these compounds blocked the glycosylation of virus-coded glycoproteins and therefore prevented the formation of the viral envelope.[48,49] However, these sugar analogs are not very specific and may also inhibit protein synthesis and cellular metabolism. These compounds are usually added to cells at concentrations of 0.1–5 m*M*.

[45] J. Lenard and R. W. Compans, *Biochim. Biophys. Acta* **344,** 51 (1974).
[46] K. Nakamura and R. W. Compans, *Virology* **84,** 803 (1978).
[47] R. Gibson, S. Schlesinger, and S. Kornfeld, *J. Biol. Chem.* **254,** 3600 (1979).
[48] R. T. Schwarz, J. M. Rohrschneider, and M. F. G. Schmidt, *J. Virol.* **19,** 782 (1976).
[49] R. T. Schwarz and R. Datema, *Adv. Carbohydr. Chem. Biochem.* (in press).

In cells, deoxyglucose, which is an analog of glucose and mannose, is converted to both UDP-2-deoxyglucose and GDP-2-deoxyglucose as well as to the dolichyl derivative, dolichyl-phosphoryl-2-deoxyglucose.[50] Apparently, the inhibition of protein glycosylation is not the result of depletion of the sugar nucleotide pool, since the levels of UDP-GlcNAc and GDP-mannose are actually increased in the presence of this inhibitor. In fact, GDP-2-deoxyglucose appears to be the major nucleotide responsible for inhibition because this inhibition can be reversed by the addition of mannose, with a resultant reduction in levels of GDP-2-deoxyglucose in the cells.[50]

It seems likely that the actual site of inhibition of protein glycosylation is at the level of lipid-linked oligosaccharides.[51] The 2-deoxyglucose is transferred in cells from GDP-2-deoxyglucose to form dolichyl-pyrophosphoryl-*N,N'*-diacetylchitobiosyl-2-deoxyglucose and dolichyl-pyrophosphoryl-$(GlcNAc)_2$-Man-2-deoxyglucose. These lipid-linked oligosaccharides apparently cannot be further extended by the addition of mannose residues, nor are these oligosaccharides transferred to protein. Since the levels of dolichyl phosphate in the cell are apparently limiting, high levels of 2-deoxyglucose may tie up all of the available dolichyl phosphate and lead to the inhibition of protein glycosylation. And in fact, it was found that in the presence of 2-deoxyglucose, the normally glycosylated K-46 chain produced by myeloma cells was not glycosylated.[52] Nevertheless, the precise mechanism of inhibition of these sugar analogs remains to be established.

Fluoroglucose and fluoromannose are other glucose and mannose analogs that are similar to 2-deoxyglucose. Thus when cells are placed in the presence of these compounds, the lipid-linked oligosaccharides that are found are of intermediate size rather than of the normal $Glc_3Man_9GlcNAc_2$ species. Probably these smaller-sized oligosaccharides, like those found in 2-deoxyglucose-inhibited cells, are not transferred to protein.[53]

Glucosamine and Mannosamine

Glucosamine (when added to cell cultures at millimolar concentrations) was initially reported to inhibit viral multiplication of a variety of enveloped viruses.[54] In the presence of this inhibitor, whole oligosaccharide chains were missing from the proteins, suggesting that this compound might

[50] R. Datema and R. T. Schwarz, *Eur. J. Biochem.* **67,** 239 (1976).
[51] R. Datema and R. T. Schwarz, *Eur. J. Biochem.* **90,** 505 (1978).
[52] P. K. Eagon and E. C. Heath, *J. Biol. Chem.* **252,** 2372 (1977).
[53] R. Datema , R. T. Schwarz, and A. W. Jankowski, *Eur. J. Biochem.* **109,** 331 (1980).
[54] M. F. G. Schmidt, R. T. Schwarz, and H. Ludwig, *J. Virol.* **18,** 819 (1976).

influence the synthesis of lipid-linked saccharides.[55] However, aberrant forms of lipid-linked oligosaccharides were not detected during glucosamine infection in one study.[56] And at inhibiting concentrations of glucosamine, no unusual metabolites were found and no evidence for compounds such as UDP-glucosamine were reported.[57] Thus, glucosamine itself appears to be necessary as an inhibitor and inhibition requires whole cells or tissues.

However, in another study, glucosamine at millimolar concentrations did cause drastic alterations in the nature of the oligosaccharide portion of the lipid-linked oligosaccharides.[58] Normally, the major oligosaccharide in the lipid-linked oligosaccharides of influenza virus-infected Madin–Darby canine kidney (MDCK) cells is the $Glc_3Man_9GlcNAc_2$. However, at low glucosamine concentrations (0.5–1 mM), the $Glc_3Man_9GlcNAc_2$ was replaced by a small oligosaccharide characterized as a $Man_7GlcNAc_2$. And at higher glucosamine concentration (2–10 mM) the $Man_7GlcNAc_2$ disappeared and was replaced by a $Man_3GlcNAc_2$. These data suggest that these oligosaccharides may represent control points in the lipid-linked saccharide pathway or may represent places where there is a change in the nature of the glycosyl donor. In these studies as in those reported above, the effect of glucosamine was reversible, such that washing the cells free of glucosamine restored normal synthesis. This glucosamine effect could not be mimicked by galactosamine or *N*-acetylglucosamine. However, mannosamine also caused alterations in the lipid-linked oligosaccharides. But in the presence of low concentrations of mannosamine, the major oligosaccharide associated with the lipid was a $Man_6GlcNAc_2$,[59] whereas at higher concentrations this oligosaccharide was shifted to a $Man_2GlcNAc_2$. Thus the site of action of mannosamine must be different from that of glucosoamine.

Amphomycin and Tsushimycin

Amphomycin is a lipopeptide antibiotic produced by *Streptomyces canus*.[60] The structure of amphomycin is shown in Fig. 6. This compound was shown to be an undecapeptide containing either 3-isododecenoic or 3-anteisotridecenoic acid attached to the N-terminal aspartic acid via an amide linkage.[61] In gram-positive bacteria, amphomycin inhibited the synthesis of cell wall peptidoglycan by blocking the transfer of phospho-*N*-ace-

[55] R. T. Schwarz, M. F. G. Schmidt, V. Answer, and H. D. Klenk, *J. Virol.* **23,** 217 (1977).
[56] R. Datema and R. T. Schwarz, *Biochem. J.* **184,** 113 (1979).
[57] V. Koch, R. T. Schwarz, and C. Scholtissek, *Eur. J. Biochem.* **94,** 515 (1979).
[58] Y. T. Pan and A. D. Elbein, *J. Biol. Chem.* **257,** 2795 (1982).
[59] Y. T. Pan and A. D. Elbein, unpublished observations.
[60] B. Heinemann, M. A. Kaplan, R. D. Muir, and I. R. Hooper, *Antibiot. Chemother. (Washington, D. C.)* **3,** 1239 (1953).
[61] M. Bodansky, G. F. Syler, and A. Bodansky, *J. Am. Chem. Soc.* **95,** 2352 (1973).

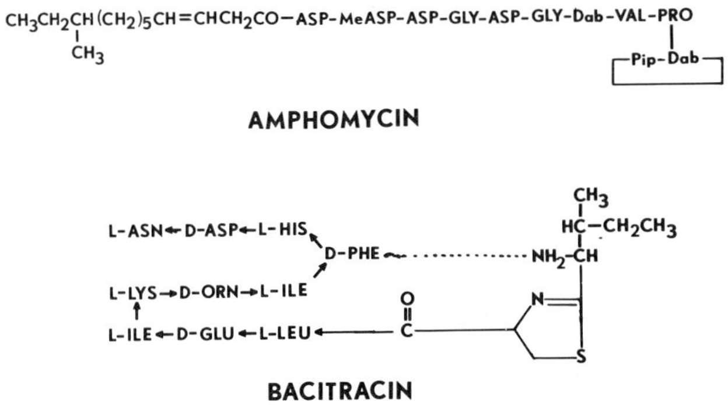

$CH_3CH_2CH(CH_2)_5CH{=}CHCH_2CO$–ASP-MeASP-ASP-GLY-ASP-GLY-Dab-VAL-PRO (branch CH_3; Pip-Dab cyclic)

AMPHOMYCIN

L-ASN←D-ASP←L-HIS → D-PHE ... NH_2-CH($HC(CH_3)$–CH_2CH_3); L-LYS→D-ORN→L-ILE; L-ILE←D-GLU←L-LEU← C(=O)– thiazoline (N, S)

BACITRACIN

FIG. 6. Structures of amphomycin and bacitracin.

tyl-muramoyl-pentapeptide from its UMP derivative to the lipid carrier, undecaprenylphosphate.[62] Tsushimycin is also an acidic acylpeptide that belongs to the amphomycin–glumamycin group of antibiotics, but differs from amphomycin in its amino acid and fatty acid composition.[63] Its complete structure has not been established, but it too appears to be an inhibitor of peptidoglycan synthesis.

Amphomycin was utilized as an inhibitor of the lipid-linked saccharide pathway in a particulate enzyme preparation from pig aorta that has the capacity to transfer mannose from GDP-mannose, GlcNAc from UDP-GlcNAc and glucose from UDP-glucose into the various lipid-linked monosaccharides and lipid-linked oligosaccharides (see Fig. 2). With this enzyme preparation, the transfer of mannose from GDP-mannose into dolichyl-phosphoryl-mannose was much more sensitive to inhibition than was the transfer of mannose into the lipid-linked oligosaccharides. Thus, at amphomycin concentrations of 50–200 μg per incubation mixture that completely blocked the formation of dolichyl-phosphoryl-mannose, mannose was still incorporated into lipid-linked oligosaccharide. Under these conditions, esssentially all of the labeled mannose was found in a single oligosaccharide that was partially characterized as a $Man_5GlcNAc_2$.[64] These data suggest that some, if not all, of the first four α-linked mannose residues in the $Man_5GlcNAc_2$ are derived from GDP-mannose rather than from dolichyl-phosphoryl-mannose. This idea has been confirmed by demonstrating the transfer of mannose directly from GDP-mannose to the $Man_5GlcNAc_2$-lipid by solubilized and partially purified enzyme fractions,[65,66] and also by

[62] H. Tanaka, R. Iowa, S. Matsukara, and S. Omura, *Biochem. Biophys. Res. Commun.* **86,** 902 (1979).

[63] J. Shoji, *Adv. Appl. Microbiol.* **24,** 187 (1978).

[64] M. S. Kang, J. P. Spencer, and A. D. Elbein, *J. Biol. Chem.* **253,** 8860 (1978).

[65] J. P. Spencer and A. D. Elbein, *Proc. Natl. Acad. Sci. U.S.A.* **77,** 2524 (1980).

[66] J. W. Jensen and J. S. Schutzbach, *J. Biol. Chem.* **256,** 12899 (1981).

the isolation of a mutant mammalian cell line that has lost the capacity to synthesize dolichyl-phosphoryl-mannose.[67] This mutant cell synthesizes the $Man_5GlcNAc_2$ oligosaccharide but is not able to elongate it further unless supplemented with exogenous dolichyl-phosphoryl-mannose.

Other studies with amphomycin in brain membrane preparations have confirmed the above results and further demonstrated that this antibiotic also blocks the transfer of glucose from UDP-glucose to dolichyl phosphate but not to particle-bound glucan or to ceramide.[68] Also in these studies, the transfer of GlcNAc-1-phosphate from UDP-GlcNAc to dolichyl phosphate was inhibited by this antibiotic. These workers suggested that amphomycin may form a complex with dolichyl phosphate, thereby obstructing glycosylation reactions. Amphomycin also inhibited similar types of glycosylation reactions in plants.[69]

With the membrane preparation of aorta, tsushimycin was found to be quite similar to amphomycin in inhibiting the formation of dolichyl-phosphoryl-mannose. In the presence of tsushimycin (also at 50–200 μg per incubation mixture) mannose was also incorporated almost exclusively into the $Man_5GlcNAc_2$–lipid.[70] Unfortunately, these antibiotics are apparently not able to penetrate intact eukaryotic cells, and therefore studies in whole cells have been unsuccessful.

Bacitracin

Bacitracin (see Fig. 6 for structure) is also a polypeptide antibiotic that is produced by certain strains of *Bacillus licheniformis* and was shown to inhibit peptidoglycan synthesis in bacteria by preventing the dephosphorylation of undecaprenylpyrophosphate, which is necessary for reutilization of the lipid carrier.[71] This antibiotic has been reported to have multiple effects on the eukaryotic lipid-linked saccharide pathway. For example, in hen oviduct membranes, bacitracin was reported to cause the accumulation of labeled GlcNAc-containing lipids when UDP-[^{14}C]GlcNAc served as substrate.[72] However, the mechanism of inhibition was not elucidated. On the other hand, in calf pancreas microsomes, bacitracin (0.2–1 m*M*) was shown to inhibit the addition of the second GlcNAc to form dolichyl-pyrophosphoryl-*N,N'*-diacetylchitobiose. In this system, the antibiotic had relatively little effect on the formation of dolichyl-phosphoryl-mannose or dolichyl-phosphoryl-glucose from their respective sugar nucleotides.[73]

[67] A. Chapman, K. Fujimoto, and S. Kornfeld, *J. Biol. Chem.* **255,** 4441 (1980).
[68] D. K. Banerjee, M. G. Scher, and C. J. Waechter, *Biochemistry* **20,** 1561 (1981).
[69] M. C. Ericson, J. Gafford, and A. D. Elbein, *Arch. Biochem. Biophys.* **191,** 698 (1978).
[70] A. D. Elbein, *Biochem. J.* **193,** 477 (1981).
[71] K. J. Stone and J. L. Strominger, *Proc. Natl. Acad. Sci. U.S.A.* **68,** 3223 (1971).
[72] W. W. Chen and W. J. Lennarz, *J. Biol. Chem.* **251,** 7802 (1976).
[73] A. Herscovics, B. Bugge, and R. W. Jeanloz, *FEBS Lett.* **82,** 215 (1977).

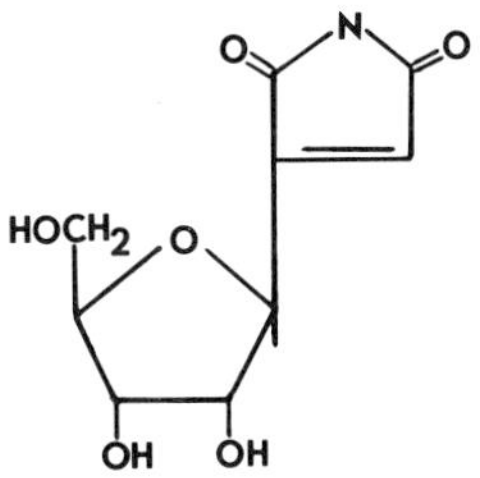

FIG. 7. Structure of showdomycin

It seems likely that bacitracin action in the mammalian systems, as has been demonstrated in the microbial systems,[71] is due to the formation of a complex with the lipid carrier, dolichyl phosphate. Thus, with the aorta particulate enzyme or with the solubilized system, bacitracin at 0.1–0.2 mM inhibited the incorporation of both mannose and GlcNAc from their sugar nucleotides into lipid-linked monosaccharides and lipid-linked oligosaccharides.[74] From these results it appears that bacitracin may block each of the steps of the lipid-linked saccharide pathway.

Showdomycin

Showdomycin is a broad-spectrum nucleoside antibiotic that shows remarkable activity against Ehrlich ascites tumors in mice and against HeLa cells.[75] The structure of this antibiotic is shown in Fig. 7. Showdomycin strongly inhibited bovine liver UDPglucose dehydrogenase, and this inhibition was attributed to its alkylating action on the enzyme.[76] This inhibition could be completely removed by preincubating the showdomycin with cysteine.

With the aorta particulate enzyme, showdomycin effectively inhibited the formation of dolichyl-phosphoryl-glucose, but this inhibition was much more pronounced in the presence of the nonionic detergent Nonidet P-40 (NP-40). At 0.25% NP-40, 50% inhibition required about 10 μg of showdomycin per milliliter. The antibiotic also inhibited transfer of mannose to both dolichyl-phosphoryl-mannose and to lipid-linked oligosaccharides, but this inhibition was evident only in the presence of detergent and required considerably higher concentrations of antibiotic than necessary for glucose inhibition. Relatively little inhibition of GlcNAc transfer was observed either in the presence or the absence of detergent.[77]

[74] J. P. Spencer, M. S. Kang, and A. D. Elbein, *Arch. Biochem. Biophys.* **190,** 829 (1978).

[75] H. Nishimura, M. Mayama, Y. Komatsu, H. Kato, N. Shimaoka, and Y. Tanaka, *J. Antibiot.* **17,** 148 (1964).

[76] S. Roy-Burman, P. Roy-Burman, and D. W. Visser, *Cancer Res.* **28,** 1605 (1968).

[77] M. S. Kang, J. P. Spencer, and A. D. Elbein, *J. Biol. Chem.* **254,** 10037 (1979).

Showdomycin inhibited glucolipid formation in aorta, whereas it greatly stimulated glucose incorporation into lipid in yeast membrane preparations. This glucolipid had chemical and chromatographic properties like those of glucosylceramide. The stimulation appeared to be due to a protection of the substrate UDP-glucose from degradation. That is, it seems likely that showdomycin also inhibits one or more of the enzymes involved in the modification or degradation of UDP-glucose.[77]

Duimycin

Duimycin is a phosphoglycolipid antibiotic produced by *Streptomyces umbrinus*[78] that was shown to inhibit cell wall biosynthesis in *Staphyloccoccus aureus*.[79] With a soluble enzyme fraction from *Acanthamoeba*, duimycin inhibited the transfer of mannose from GDP-mannose and GlcNAc from UDP-GlcNAc into the lipid-linked saccharides.[80] On the other hand, glucose transfer was much less sensitive to inhibition. In yeast membrane preparations, this antibiotic inhibited the different mannosyl transfer reactions. Thus, the drug effectively inhibited the formation of dolichyl-phosphoryl-mannose, 50% inhibition being observed at 10 μg/ml. Mannose transfer from dolichyl-phosphoryl-mannose to protein was also inhibited, and this included transfer to serine and threonine residues as well as to asparagine residues. However, the glycosyltransfer reactions in which mannose is transferred from GDP-mannose directly to protein-bound serine and threonine residues were not sensitive to duimycin.[81]

Other Inhibitors

Several other rather diverse compounds appear to be inhibitors of glycoprotein assembly, but the exact site of action of these compounds is not known. For example, the ionophores monensin and A23187 (at concentrations of 10^{-6} *M*) were found to prevent the secretion of Sindbis virus (SV) and VSV into the medium by BHK and chicken embryo fibroblasts and also blocked the appearance of the viral glycoproteins at the surface membrane.[82] However, the viral glycoproteins were still synthesized at the normal rate, but apparently did not migrate to the cell surface. Probably the ionophores block the movement of the VSV G protein and the SV E2 protein from the Golgi to the cell surface where viral budding occurs.

[78] E. Meyers, D. Smith Slusarchyk, J. L. Bouchard, and F. L. Weisenborn, *J. Antibiot.* **22,** 490 (1969).

[79] E. J. J. Lugtenberg, J. A. Hellings, and G. J. Van de Berg, *Antimicro. Agents Chemother.* **2,** 485 (1972).

[80] C. L. Villemez, and P. L. Carlo, *J. Biol. Chem.* **255,** 8174 (1980).

[81] P. Babczinski, *Eur. J. Biochem.* **112,** 53 (1980).

[82] D. C. Johnson and M. J. Schlesinger, *Virology* **103,** 407 (1980).

It is known that glucose-starved mammalian cells do not form the $Glc_3Man_9GlcNAc_2$–lipid to any extent, but instead produce and accumulate a $Man_5GlcNAc_2$–lipid.[83] *m*-Chlorocarbonyl-cyanide phenylhydrazone (CCCP) is an uncoupler of oxidative phosphorylation and therefore causes energy depletion of cells when added to the culture medium. In the presence of this compound (at about 10 μM concentration), the synthesis of dolichyl-phosphoryl-mannose is inhibited, but not the formation of dolichyl-phosphoryl-glucose or dolichyl-pyrophosphoryl-*N*,*N*′-diacetylchitoboise. When CCCP is added to the cells, the cells produce a $Man_5GlcNAc_2$–lipid and transfer this oligosaccharide to protein. Some of the $Man_5GlcNAc_2$ is apparently glucosylated at the lipid stage and then transferred to protein.[84]

Inhibitors of Dolichol Synthesis

The levels of dolichyl phosphate (and also of dolichol) are apparently limiting in most membranes, and this has been postulated to be one of the control points in glycoprotein biosynthesis.[85,86] Although the turnover rate for dolichyl phosphate in various membranes is not known, one would expect that inhibitors of dolichyl-phosphate synthesis might exert a profound effect on the synthesis of the asparagine-linked glycoproteins. The biosynthesis of dolichyl phosphate proceeds through the same series of reactions as cholesterol from acetate to farnesyl pyrophosphate, that is, via hydroxymethylglutaryl-CoA (HMG-CoA) and mevalonic acid to farnesyl pyrophosphate. At this point, there is a branch in the pathway either to form squalene and then cholesterol or to form 2,3-dihydrodolichyl pyrophosphate and then dolichyl phosphate.[87] The synthesis of cholesterol, and probably also of dolichyl phosphate, is regulated at the level of the HMG-CoA reductase and also probably at other enzymes specific to either branch of the respective pathways. Several inhibitors of HMG-CoA reductase that inhibit cholesterol biosynthesis also have been reported to block the formation of the oligosaccharide portion of asparagine-linked glycoproteins.

When aortic smooth muscle cells were grown in culture in the presence of 25-hydroxycholesterol (10 μg/ml), there was an 80–90% inhibition in the incorporation of acetate into both dolichol and cholesterol, as well as a diminished incorporation of glucose into dolichyl-pyrophosphoryl oligosaccharides and asparagine-linked glycoproteins.[88] However, the incorpora-

[83] S. J. Turco, *Arch. Biochem. Biophys.* **205,** 330 (1980).
[84] R. Datema and R. T. Schwarz, *J. Biol. Chem.* **256,** 11191 (1981).
[85] J. J. Lucas and E. Levin, *J. Biol. Chem.* **252,** 4330 (1977).
[86] J. R. Harford, C. J. Waechter, and F. L. Earl, *Biochem. Biophys. Res. Commun.* **26,** 1036 (1977).
[87] P. K. Grange and W. L. Adair, Jr., *Biochem. Biophys. Res. Commun.* **79,** 734 (1977).
[88] J. T. Mills and A. M. Adamany, *J. Biol. Chem.* **253,** 5270 (1978).

tion of mevalonic acid into cholesterol and dolichol was not altered by growth in 25-hydroxycholesterol; in fact, mevalonic acid restored the normal synthesis of the dolichyl-linked saccharides and glycoproteins. Hydroxycholesterol (at 1–5 μM) also inhibited the incorporation of acetate into cholesterol and dolichol in L-cell cultures, but in this case the relationship between the concentration of 25-hydroxycholesterol and the level of inhibition differed between the two lipids. Thus, large fluctuations in cholesterol synthesis were observed under conditions where dolichol synthesis was only slightly affected. However, under conditions where sterol synthesis was repressed to levels below 25% of controls, further inhibition of cholesterol synthesis was accompanied by a proportionate decline in dolichol synthesis.[89] The authors suggest that a rate-limiting enzyme unique to the dolichol pathway may be saturated at a lower level of intermediates than is required to saturate the cholesterol pathway at the next rate-limiting step beyond the branch point. Therefore, HMG-CoA reductase activity and the rate of cholesterol synthesis may fluctuate with little change in dolichol synthesis as long as the levels of these intermediates are sufficient to saturate the dolichol branch.

Nevertheless these studies do suggest that the rate of synthesis of dolichyl phosphate, and thus the levels of dolichyl phosphate in the cell may play a role in the regulation of glycoprotein biosynthesis. It seems likely that in some tissues the synthesis of dolichyl phosphate is regulated at steps in the pathway unique to the dolichol branch, as well as at the HMG-CoA reductase step. Thus, feeding animals a diet high in cholestyramine, a compound that causes increased activity of HMG-CoA reductase, resulted in a great stimulation in acetate incorporation into cholesterol, but not into dolichyl phosphate, leading these authors to conclude that the rate of dolichyl phosphate synthesis is not regulated by the activity of HMG-CoA reductase, at least in rat liver.[90] They suggested that regulation may be at the level of the dolichyl phosphate synthetase. But animals fed a high-cholesterol diet, which suppresses cholesterol biosynthesis, show increased incorporation of mevalonate into dolichol and dolichyl-pyrophosphoryl oligosaccharides and also increased activities of some of the glycosyltransferases associated with glycoprotein synthesis.[91,92] In this case, inhibiting the cholesterol branch of the pathway may increase the concentrations of various intermediates (farnesyl pyrophosphate, isopentenyl pyrophosphate, etc.) and therefore stimulate biosynthesis of dolichyl phosphate.

[89] M. J. James and A. A. Kandutsch, *J. Biol. Chem.* **254,** 8442 (1979).
[90] R. K. Keller, W. L. Adair, Jr., and G. C. Ness, *J. Biol. Chem.* **254,** 9966 (1979).
[91] I. A. Tavares, T. Coolbear, and F. W. Hemming, *Arch. Biochem. Biophys.* **207,** 427 (1981).
[92] D. A. White, B. Middleton, S. Pawson, J. P. Bradshaw, R. J. Clegg, F. W. Hemming, and G. D. Bell, *Arch. Biochem. Biophys.* **208,** 30 (1981).

A specific nonsteroidal inhibitor of cholesterol biosynthesis called compactin or ML-236B[93] shows great promise for some of these studies on the role of dolichyl phosphate levels on glycoprotein biosynthesis. This compound is produced by several fungal strains such as *Penicillium brevicompactum,*[94] and its structure includes a lactonized ring that resembles the lactone form of mevalonic acid. In Chinese hamster ovary cells in culture, compactin (at a concentration of 1 – 10 μM) blocks the synthesis of cholesterol by inhibiting the HMG-CoA reductase, thus making these cells dependent on external cholesterol.[95] When compactin (1 – 5 μM) is given to developing sea urchin embryos, it induces abnormal gastrulation. This effect is apparently due to an inhibition of dolichyl-phosphate synthesis, since the embryos cultured in compactin show a decreased capacity to synthesize mannose-labeled glycolipids and asparagine-linked glycoproteins. However, the inhibitory effect of compactin on development and glycoprotein biosynthesis could be overcome by supplementing the embryos with exogenous dolichol or with dolichyl phosphate, but not with cholesterol or coenzyme A.[96] Thus, compactin and 25-hydroxycholesterol are examples of inhibitors that can be used to study some of the steps in synthesis of dolichyl-phosphate. However, it would be useful to have inhibitors that block this pathway after the branch point, since these inhibitors would be much more specific for dolichol.

Inhibitors of Glycoprotein Processing

As shown in Fig. 3, following the synthesis of the $Glc_3Man_9GlcNAc_2$-pyrophosphoryl-dolichol and the transfer of this oligosaccharide to protein, the oligosaccharide undergoes a number of processing steps whereby all of the glucose residues and a number of mannose residues are removed by membrane-bound and neutral glycosidases. These processing reactions are apparently necessary to form the precursor for the complex types of glycoproteins.[15] Thus inhibitors of these glycosidases might be expected to produce glycoproteins with altered and aberrant sugar chains and also prevent the formation of the complex glycoproteins.

Swainsonine is an indolizidine alkaloid that was isolated from the plant *Swainsona canescens.*[97] Its structure is shown in Fig. 8. Livestock that ingest the plant show symptoms similar to those of human α-mannosidosis, and

[93] A. Endo, M. Kuroda, and K. Tanzawa, *FEBS Lett.* **72,** 323 (1976).

[94] A. G. Brown T. C. Smale, T. J. King, R. Hasenkamp, and R. H. Thompson, *J. Chem. Soc. Perkins Trans.* **1,** 1165 (1976).

[95] J. L. Goldstein, J. A. S. Helgeson, and M. S. Brown, *J. Biol. Chem.* **254,** 5403 (1979).

[96] D. D. Carson and W. J. Lennarz, *Proc. Natl. Acad. Sci. U.S.A.* **76,** 5709 (1979).

[97] S. M. Colegate, P. R. Dorling, and C. R. Huxtable, *Aust. J. Chem.* **32,** 2257 (1979).

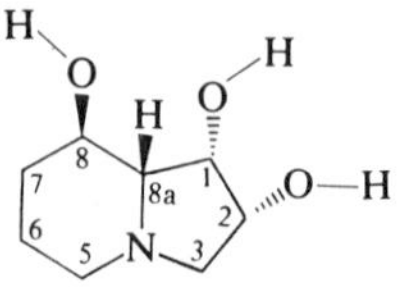

FIG. 8. Structure of swainsonine (8αβ-indolizidine-1α,2α,8β-triol; M_r 173.1).

swainsonine was shown to be a potent inhibitor of lysosomal α-mannosidase.[98] In a rat liver particulate enzyme, this alkaloid, at 10^{-6} to 10^{-7} *M*, also inhibited the neutral α-mannosidase(s) that trims mannose residues from the high-mannose glycopeptides, but it had relatively little effect on the glucosidases that remove glucose residues from these glycoproteins.[99] When confluent monolayers of MDCK cells were incubated for several hours in swainsonine (100–500 ng/ml) and then labeled for several hours with either [2-^{3}H]mannose or [6-^{3}H]glucosamine, these cells showed a significant reduction in the incorporation of these labeled sugars into the complex glycoproteins as compared to controls, and a considerable increase in the label in the high-mannose glycoproteins.[99]

In cells in culture, it is not possible to study glycoprotein synthesis from the start, nor is it possible to examine the synthesis of a single glycoprotein. However, various enveloped viruses produce glycoproteins having asparagine-linked oligosaccharides, and these systems offer excellent models for studies on glycoprotein synthesis. For example, influenza virus hemagglutinin has been reported to have five or six asparagine-linked oligosaccharides of which four or five are of the complex type and one or two are of the high mannose type. In one experiment, MDCK cells were infected with influenza virus, and then swainsonine (1 μg/ml) was added. After several hours of incubation, [^{3}H]mannose or [^{3}H]glucosamine was added to the cell cultures to label the viral glycoproteins, and the cells were incubated to produce mature virus. In normal virus, approximately two-thirds of the mannose was incorporated into complex chains and one-third into high-mannose structures. However, virus raised in swainsonine had over 90% of the mannose incorporated into the high-mannose structure and very little of the labeled complex chains was observed. Similar results were observed with glucosamine labeling. Of interest is the finding that virus raised in swainsonine was equally as infective as control virus and also had equal hemagglutinating activity.[100] Thus, the complex types of chains do not appear to be necessary for biological activity of this virus. A recent study has shown that swainsonine specifically inhibits the mannosidase II that removes the α1,3 and α1,6-linked mannoses from the GlcNAc-Man_5-$GlcNAc_2$-protein, but

[98] P. R. Dorling, C. R. Huxtable, and S. M. Colegate, *Biochem. J.* **191,** 649 (1980).

[99] A. D. Elbein, R. Solf, P. R. Dorling, and K. Vosbeck, *Proc. Natl. Acad. Sci. U.S.A.*, **78,** 7393 (1981).

[100] A. D. Elbein, P. R. Dorling, K. Vosbeck, and M. Horisberger, *J. Biol. Chem.* **257,** 1573 (1982).

did not inhibit mannosidase I, that removes the α1,2-linked mannoses from the $Man_9GlcNAc_2$-protein.[101]

When various mammalian cell lines are grown for 3 or 4 days in the presence of swainsonine, these cells show greatly increased abilities to bind lectins and bacteria (concanavalin A and *Escherichia coli* B-886) that interact with high-mannose glycoproteins. On the other hand, these cells have a diminished ability to interact with lectins that recognize sugars of the complex chains (wheat germ agglutinin and ricin). In addition, cells incubated in swainsonine have a significantly reduced capacity to incorporate glucosamine, fucose, and mannosamine into complex glycoproteins, but the incorporation of mannose, proline, and leucine were not inhibited by swainsonine. Swainsonine was not cytotoxic to cells and did not inhibit the growth of various mammalian cell lines.[102] These results suggest that swainsonine alters the composition of the oligosaccharide chains of glycoproteins and should be valuable for studying the role of the complex glycoproteins.

Inhibitors of Protein Synthesis

One of the major, and still unanswered, questions with regard to biosynthesis of the asparagine-linked glycoproteins is how the formation of the oligosaccharide chain is regulated. That is, is the assembly of the core oligosaccharide linked in any way to the formation of the protein? Some of these questions have been approached by the use of inhibitors of protein synthesis in order to examine the effects of these compounds on the synthesis of the various lipid-linked saccharide intermediates.

In one study, MDCK cells were grown in [2-^{3}H]mannose or [^{3}H]leucine in the presence of cycloheximide (1 – 5 μM) or puromycin, and the effect of these inhibitors on the synthesis of proteins and lipid-linked oligosaccharides was measured.[103] In all cases, the inhibition of protein synthesis resulted in a substantial inhibition in mannose incorporation into the lipid-linked oligosaccharides, although the synthesis of dolichyl-phosphoryl-mannose was only slightly affected. Cycloheximide had no effect on the *in vitro* incorporation of mannose from GDP-[^{14}C]mannose into the lipid-linked saccharides by aorta microsomal preparations. The inhibition of lipid-linked oligosaccharide formation did not appear to be due to a decrease in the amounts of the various glycosyltransferases as a result of inhibition of protein synthesis, nor was it the result of more rapid degradation of lipid-linked oligosaccharides. Although in this study it was not possible completely to rule out limitations in the amount of dolichyl phosphate as one factor, this did not appear to be the major cause of inhibition, suggesting that the formation of lipid-linked oligosaccharides

[101] D. P. R. Tulsiani, T. M. Harris, and O. Touster, *J. Biol. Chem.* **257**, 7936 (1982).
[102] A. D. Elbein, Y. T. Pan, R. Solf, and K. Vosbeck, *J. Cell Physiol.* **115**, 265 (1983).
[103] J. W. Schmitt and A. D. Elbein, *J. Biol. Chem.* **254**, 12291 (1979).

might be regulated, at least in part, by end-product inhibition. In another study, oligosaccharide–lipid synthesis was examined in cells incubated in the presence of actinomycin D (1 μg/ml) to depress levels of mRNA, or in the presence of cycloheximide (100 μg/ml) to abolish protein synthesis.[104] The results indicated that the synthesis of oligosaccharide–lipid was proportional to the rate of protein synthesis. The regulated step appeared to be prior to the formation of the $Man_5GlcNAc_2$–lipid (see Fig. 2) leading these workers to suggest that the most likely control was the availability of dolichyl phosphate. These studies, as well as others described above, suggest that there may be several control points in the pathway of oligosaccharide assembly and that the use of various inhibitors may help to delineate these points.

Conclusions

Various inhibitors have been described that act at specific steps in the assembly of the lipid-linked oligosaccharide ($Glc_3Man_9GlcNAc_2$), or in the processing of the oligosaccharide after its transfer to protein. Several of these inhibitors, such as the glucose analogs and amphomycin, have provided valuable information on the sequence of reactions in lipid-linked oligosaccharide assembly, whereas others, like tunicamycin and swainsonine, have given significant insight into the role(s) of the carbohydrate in glycoprotein function. Thus, the identification and utilization of specific inhibitors of glycoprotein biosynthesis is one fruitful approach to understanding this area of biochemistry. It is hoped that other inhibitors will be found that block other specific sites, allowing each of the steps to be mapped out and each of the intermediates to be characterized.

[104] S. C. Hubbard and P. W. Robbins, *J. Biol. Chem.* **255,** 11782 (1980).

[12] Preparation of Glycopeptide and Oligosaccharide Probes for Receptor Studies

By JACQUES U. BAENZIGER

A number of unique mammalian lectins, including membrane proteins that mediate oligosaccharide-specific endocytosis, plasma proteins, and soluble cell-associated proteins, have been described. The majority of studies directed at examining the specificity of mammalian lectins have relied on monosaccharides and glycoproteins as probes. Monosaccharides yield limited information, since these lectins recognize extended oligosaccharide structures. Glycoproteins yield results that are difficult to interpret because they generally bear multiple oligosaccharides of differing structure. Polyva-

METHODS IN ENZYMOLOGY, VOL. 98

ISBN 0-12-181998-1

lency due to the presence of multiple oligosaccharides on a single polypeptide adds still another complicating feature to specificity studies. It is possible to circumvent these difficulties by utilizing characterized glycopeptides and oligosaccharides to examine the specificity and other properties of these lectins. General methods for the preparation of both glycopeptides and oligosaccharides that can be radiolabeled will be described as well as some of their applications.

Iodinated Glycopeptides. ^{14}C and ^{3}H can be introduced into glycopeptides by acylation with ^{14}C- or ^{3}H-labeled acetic anhydride; however, the products obtained have a relatively low specific activity and require scintillation counting for detection. The introduction of a *p*-hydroxyphenyl moiety onto the amino terminus of glycopeptides makes it possible to prepare ^{125}I-labeled derivatives of much higher specific activity.

Derivatization

Reagents

Borate buffer, 0.1 *M*, pH 8.5

3-(4-Hydroxyphenyl)propionic acid *N*-hydroxysuccinimide ester (ICN Pharmaceuticals, Inc., Cleveland, Ohio) in anhydrous dioxane; prepared immediately prior to use

Ethanolamine, 1.0 *M*, pH 8.5

Sephadex G-10

Solvent A: butanol–ethanol–water, 4 : 1 : 1 (v/v/v)

Procedure. A 10-fold molar excess of the succinimide ester is added to the glycopeptide dissolved in the borate buffer. After 16 hr at room temperature, any remaining ester is consumed by addition of a 10-fold molar excess of ethanolamine. The derivatized glycopeptide is separated from these reactants by passage over Sephadex G-10 in water. The product is dried, applied to Whatman 3MM paper, and subjected to descending paper chromatography with solvent A for 48 hr. The glycopeptide remains at the origin and is eluted with water.

Comments. Any amine-containing compounds, such as small peptides, will compete for the succinimide ester. An estimate of such contaminants is readily obtained using either a ninhydrin assay[1] or Fluram assay.[2] Contaminants must be removed prior to derivatization or the amount of ester increased. The derivatized contaminants in the latter case must still be removed prior to iodination. Both the glycopeptide and ester will remain in solution over a range of 10–50% dioxane in water (v/v); however, at high ester concentrations, 50% dioxane may be necessary to maintain solubility.

[1] C. W. H. Hirs, this series, Vol. 11 [35].

[2] S. Udenfriend, S. Stein, P. Bohlen. W. Dairman, W. Leimgraber, and M. Weigele, *Science* **178,** 871 (1972).

Iodination

Reagents

Tris-HCl. 0.05 *M*, pH 7.5–8.8

1,3,4,6,-Tetrachloro-3*a*,6*a*-diphenylglycouril (Iodo-Gen; Pierce Chemical Co., Rockford, Illinois), 100 μg, coated onto 1.5-ml Eppendorf centrifuge tubes

Sephadex G-10

Procedure. The *p*-hydroxyphenyl-derivatized glycopeptide (5 nmol) in 50–250 μl of 0.05 *M* Tris buffer is mixed with 1 mCi of $Na^{125}I$. The mixture is added to the Iodo-Gen at 4° and allowed to react for 10–15 min. The glycopeptide is separated from reagents by gel filtration over Sephadex G-10 in water or an appropriate buffer.

Comments. Utilizing 5 nmol of homogeneous glycopeptide and 1 mCi of $Na^{125}I$, the specific activity obtained is generally on the order of 5 to 10×10^4 dpm/pmol. This can be increased or decreased by altering the amount of glycopeptide used relative to the iodine in the labeling step. The iodination procedure is essentially that of Fraker and Speck.[3]

[^{125}I]Glycopeptides have been utilized to examine the specificity of plant[4,5] as well as mammalian[6-8] lectins. Additional derivatives can be prepared by sequential enzymic degradation of the [^{125}I]glycopeptides with highly purified exoglycosidases.[4] These derivatives have been particularly useful for examining the specificity and properties of oligosaccharide-specific receptor mediated endocytosis by isolated cells such as hepatocytes[7] and hepatic reticuloendothelial cells.[8] A present shortcoming of the [^{125}I]glycopeptide derivatives, however, is the inability to obtain absolutely homogeneous glycopeptide species in some cases.

High-performance liquid chromatography (HPLC) has provided a means for separating closely related oligosaccharide species on either an analytical or a preparative scale. Oligosaccharides bearing differing numbers or sialic acid, phosphate, or sulfate moieties can be rapidly separated by ion-exchange HPLC.[9] The effectiveness of such separations is illustrated in Fig. 1, where high-mannose-type oligosaccharides with differing numbers of phosphate moieties in mono and diester linkage have been separated. Neutral oligosaccharides can be fractionated with a high degree of resolution by amine adsorption HPLC.[10] Thus, fractionation of fucosylated and non-fucosylated tri- and tetra-branched complex oligosaccharides such as those

[3] P. J. Fraker and J. C. Speck, Jr., *Biochem. Biophys. Res. Commun.* **80,** 849 (1978).

[4] J. U. Baenziger and D. Fiete, *J. Biol. Chem.* **254,** 2400 (1979).

[5] J. U. Baenziger and D. Fiete, *J. Biol. Chem.* **254,** 9795 (1979).

[6] J. U. Baenziger and Y. Maynard, *J. Biol. Chem.* **255,** 4607 (1980).

[7] J. U. Baenziger and D. Fiete, *Cell* **22,** 611 (1980).

[8] Y. Maynard and J. U. Baenziger, *J. Biol. Chem.* **256,** 8063 (1981).

[9] J. U. Baenziger and M. Natowicz, *Anal. Biochem.* **112,** 357 (1981).

[10] S. J. Mellis and J. U. Baenziger, *Anal. Biochem.* **114,** 276 (1981).

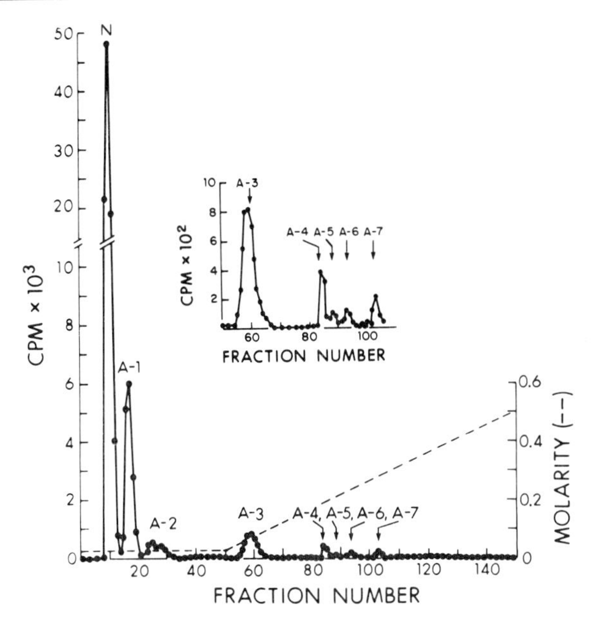

FIG. 1. Preparative fractionation of the endoglycosidase H-released oligosaccharides from human β-glucuronidase on MicroPak AX-10. Endoglycosidase H-released oligosaccharides were labeled by reduction with NaB^3H_4 and applied to a MicroPak AX-10 column in 25 mM KH_2PO_4, pH 4.0. The molarity of the K^+ over the gradient is indicated. All the species eluted consisted of high-mannose-type oligosaccharides with varying numbers of mannose-6-PO_4 moieties in mono or diester linkage. N, neutral; A-1, one phosphate in diester linkage; A-2, one phosphate in monoester linkage; A-3, two phosphates, both in diester linkage; A-4, A-5, and A-6, two phosphates, each with one in monoester and the other in diester linkage; A-7, two phosphates, both in diester linkage. From M. Natowicz, J. U. Baenziger, and W. S. Sly, *J. Biol. Chem.* **257,** 4412 (1982).

present on human orosomucoid is not yet feasible with glycopeptides but can be rapidly achieved utilizing amine adsorption HPLC after removal of the peptide by anhydrous hydrazinolysis as demonstrated in Fig. 2. HPLC fractionation of oligosaccharides has been extended to include anionic as well as neutral oligosaccharide species by what we termed ion-suppression amine-adsorption HPLC utilizing 1% phosphoric acid or 3% acetic acid titrated to pH 5.5 with triethylamine.[11]

Ion Exchange-HPLC of Anionic Oligosaccharide Species

Reagents

Buffer A: 5 mM KH_2PO_4, titrated to pH 4.0 with phosphoric acid
Buffer B: 500 mM KH_2PO_4, titrated to pH 4.0 with phosphoric acid

[11] S. J. Mellis and J. U. Baenziger, *Anal. Biochem.* (in press).

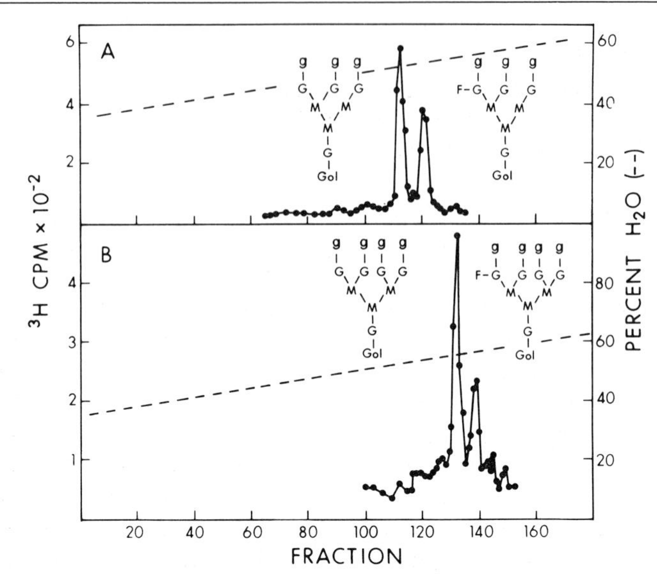

FIG. 2. Separation of fucosylated and nonfucosylated complex tri- and tetra-branched complex oligosaccharides on MicroPak AX-5 The complex oligosaccharides were loaded in acetonitrile–H_2O (65 : 35, v/v). The gradient consisted of increasing water at 0.5% per minute with a flow rate of 1 ml/min. The symbols used are g, F, M, G, and Gol for galactose, fucose, mannose, *N*-acetylglucosamine, and *N*-[^{3}H]acetylglucosaminitol, respectively. From Mellis and Baenziger.[10]

Column

MicroPak AX-10 (Varian Associates), 4 mm × 30 cm column

Conditions

Flow, 1 ml/min

Buffer A for 15 min followed by a linear gradient proceeding from 100% buffer A to 100% buffer B over 30 min.

Amine Adsorption-HPLC of Neutral Oligosaccharides

Reagents

Acetonitrile

Water

Column

MicroPak AX-10 or AX-5 (Varian Associates), 4 mm × 30 cm column

Conditions

Flow, 1 ml/min

Initial conditions: Acetonitrile–H_2O, 65 : 35

Linear gradient starting at the time of injection consisting of increasing H_2O at 0.5% per minute

Comments. Both procedures can be utilized either analytically or preparatively. Samples can be loaded in relatively large volumes (up to 1 ml) with no loss of resolution. Larger oligosaccharides should be dissolved in H_2O prior to addition of acetonitrile to a final 65 : 35 (v/v) mixture to assure solubilization. These columns have long lifetimes with excellent retention of resolution; however, it is essential to wash the column with 0.1% phosphoric acid in H_2O after each analysis done in acetonitrile – H_2O. If this is not done, oligosaccharides will elute at progressively earlier times over a short period of time owing to incomplete elution of contaminants.

Although oligosaccharides prepared by HPLC can be labeled by reduction with NaB^3H_4, the specific activities obtained are rarely sufficient for direct binding or uptake studies. As a result, these oligosacchrides are generally more useful as inhibitors. They can nonetheless be utilized to establish subtle differences between what would appear to be identical lectins when examined with monosaccharides. For example, the hepatic reticuloendothelial cell mannose – *N*-acetylglucosamine-specific receptor is distinguishable from the soluble hepatocyte mannose – *N*-acetylglucosamine-specific lectin by its ability to differentiate hybrid and high-mannose-type oligosaccharides, whereas these different oligosaccharide types are not distinguished by the soluble lectin.[8,12]

Lectins are highly specific in their recognition of extended features of oligosaccharide structures. Recognition of this aspect of their specificity ultimately requires the use of homogeneous glycopeptides and oligosaccharides. The procedures we have described permit the preparation of such well-characterized and homogeneous oligosaccharide probes. They also have applications as substrates for glycosidases and glycosyltransferase.

[12] Y. Maynard and J. U. Baenziger, *J. Biol. Chem.* **257**, 3788 (1982).

[13] Immunological Recognition of Modifications on Functionally Related Proteins

By DAVID A. KNECHT, ROBERT C. MIERENDORF, JR., and RANDALL L. DIMOND

Protein modification is a universal feature of eukaryotic cells, but because of its complexity, the role(s) of modification in cellular biology has been difficult to resolve. It is becoming increasingly clear that for any given gene product, multiple forms may coexist in the cell at any given time as a result of modifications that accompany the synthesis, localization, aging, and secretion of proteins. It would also seem that the same modifications

ISBN 0-12-181998-1

can play different roles on different molecules, as in the case of glycosylations that protect some proteins from intracellular degradations, while being unnecessary for the survival of other proteins.[1] The demonstration of a mannose 6-phosphate modification on lysosomal enzymes and a mannose 6-phosphate receptor in fibroblasts provides an exciting example of the possible roles for modification in targeting of proteins.[2,3] Thus, it is possible that some modifications may be involved in determining the membrane, organellar, or extracellular localization of some proteins. We (and others) have found that it is possible to take an immunological approach to the study of modification that complements the more classical structural approach. For example, when rabbits are immunized with a purified lysosomal enzyme from the cellular slime mold *Dictyostelium discoideum,* the resulting antisera cross-react with all the other known lysosomal enzymes because of similarity in modifications.[4] Although *Dictyostelium* enzymes contain mannose 6-phosphate,[5] the common determinants we have identified are probably different from mannose 6-phosphate and involve some other lysosomal enzyme modifications that are not present on other cellular proteins. The preparation of antisera that recognize unique modifications allows us to study a group of proteins modified in the same way without necessarily identifying all the members of the group. By extending this immunological analysis to include other groups of similarly modified proteins, we will be able to investigate why groups of proteins are modified in similar ways and how modification affects function and targeting.

In analyzing antisera that cross-react with multiple proteins it is very important to characterize the antigenic specificity carefully. When such an antiserum is obtained, several possible sources of cross-reactivity need to be distinguished.

1. The immunized animal is injected with a mixture of antigens, each of which generates an immune response so that the serum contains a mixture of antibody specificities.
2. Multiple proteins, all derivatives of the same gene but processed, modified, or degraded in different ways might appear to be a group of related proteins when actually it is the protein portion itself that is the antigen.
3. The antiserum may recognize a modification common to a number of otherwise structurally unrelated proteins. It is this last type of antiserum that is of interest in these investigations.

[1] C. Sidman, M. J. Potash, and G. Kohler, *J. Biol. Chem.* **256,** 13180 (1981).
[2] E. F. Neufeld, G. N. Sando, J. Gavin, and L. H. Rome, *J. Supramol. Struct.* **6,** 95 (1977).
[3] H. D. Fischer, A. Gonzalez-Noriega, and W. S. Sly, *J. Biol. Chem.* **255,** 5069 (1980).
[4] D. A. Knecht and R. L. Dimond, *J. Biol. Chem.* **256,** 3564 (1981).
[5] H. H. Freeze, A. L. Miller, and A. Kaplan, *J. Biol. Chem.* **255,** 11081 (1980).

Production of Antisera Specific for Modifications

Rabbit Sera. The production of antisera of the desired type in rabbits is aided by the injection of antigens that are as pure as possible so as to avoid multispecific antisera (see No. 1 above). In the case of the *Dictyostelium* lysosomal enzymes, injection of any of the three purified lysosomal enzymes that we have used each led to the production of modification-specific antisera. It appears that the modification present on these proteins is sufficiently immunogenic in rabbits to dominate the protein-specific response. We have seen no evidence for a significant protein-specific response with these enzymes. For proteins that do not elicit this type of response, several options remain. One is to increase the immunogenicity of less dominant domains, such as the modifications. This may be accomplished by simply boiling the antigen in 2% sodium dodecyl sulfate and a reducing agent such as dithiothreitol prior to injection of the antigen. This has the effect of unfolding the polypeptide and may result in a greater contribution of the carbohydrate side chains to the immunogenicity of the molecule. Alternatively, Smith and Ginsburg[6] have demonstrated that a complex carbohydrate group can act as immunogen when attached to a hapten carrier. Thus, the modification of interest could be isolated or synthesized and then conjugated to a protein and used as antigen. It should be pointed out that the simplest approach of injecting pure protein to generate cross-reacting sera is not unique to the *Dictyostelium* lysosomal enzymes. Antisera prepared to T-locus antigens, oncornaviral proteins, blood group antigens, and yeast mannans and invertase all show properties of antigenic specificity residing in the carbohydrate modifications present on protein immunogens.[7-10]

Monoclonal Antibodies. In order to prepare monoclonal antibodies to group-specific determinants, two considerations may be important in addition to the ones mentioned above for rabbit antibodies. We have found a far greater tendency for myeloma × spleen cell fusions to yield group-specific antibodies when mice are immunized with large amounts (>500 μg) of mixtures of partially purified proteins (lysosomal enzymes in our case) rather than individual purified proteins.[11] Thus it is desirable to immunize mice in a number of different ways for any determinant being examined (i.e., pure protein, denatured protein, crude protein) to obtain the widest possible range of specificity. We have been able to screen directly for

[6] D. F. Smith and V. Ginsburg, *J. Biol. Chem.* **255,** 55 (1980).

[7] C. C. Cheng and D. Bennett, *Cell* **19,** 537 (1980).

[8] H. W. Snyder, Jr. and E. Fleissner, *Proc. Natl. Acad. Sci. U.S.A.* **77,** 1622 (1980).

[9] W. M. Watkins, *in* "Glycoproteins" (A. Gottschalk, ed.), p. 830. Elsevier, Amsterdam, 1972.

[10] L. Lehle, R. E. Cohen, and C. E. Ballou, *J. Biol. Chem.* **254,** 12209 (1979).

[11] D. A. Knecht, unpublished observations, 1981.

antibodies that recognize lysosomal enzymes (either crude or pure), and some of these antibodies react with common modifications.[12]

We have described in detail a method for rapid replica screening of tissue-culture fluids of hybrid cells with a simplified system. This type of screening can also be used to detect group-specific determinants of the type described here. Figure 1 outlines the basic approach. A mouse is given an injection of a mixture of proteins thought to contain similar modifications. The hybrid cells resulting from the fusion of spleen cells with NS-1 myeloma cells are screened simultaneously for the production of antibodies that bind to two different purified proteins from the injected mixture. We have used a 96-well adjustable pipetting device to assist in the process of transferring media to and from wells. This device makes it possible easily to replica-sample a large number of tissue culture plates for simultaneous screening against two different proteins. Only clones that appear to react with both proteins are saved for further analysis. The culture fluids can be screened either for antibodies that bind to two different proteins that have been purified from the crude preparation, or antibodies that react with two different enzymic activities present in the crude mixture.

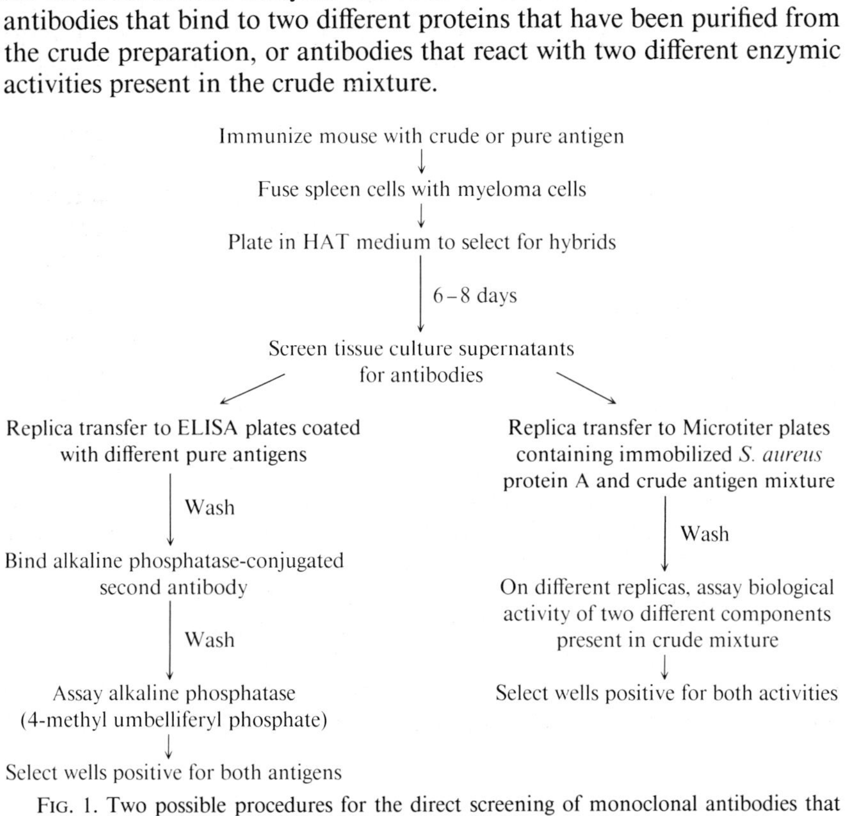

FIG. 1. Two possible procedures for the direct screening of monoclonal antibodies that cross-react with multiple proteins through shared antigenic determinants.

[12] D. A. Knecht, R. C. Mierendorf, Jr., and R. L. Dimond, manuscript in preparation, 1983.

Analysis of Cross-Reacting Antibodies

Once an antiserum has been obtained, there are a number of ways to determine whether it has the desired properties. A cross-reacting antiserum may possess any or all of the following characteristics: immunoprecipitation of multiple proteins from radioactively labeled cell extracts as detected by gel electrophoresis and fluorography, the presence of multiple precipitin lines in Ouchterlony double immunodiffusion or immunoelectrophoresis, reaction with multiple proteins blotted onto nitrocellulose or other support, and immunoprecipitation of multiple enzymic or other biological activities. Caution should be observed particularly in analyzing Ouchterlony double-diffusion or immunoelectrophoresis experiments, because the antibodies to lysosomal enzymes that we have obtained do not form precipitin lines under any conditions that we have tried. They will, however, precipitate both native enzymic activity and radioactively labeled enzyme protein using an indirect second antibody or *Staphylococcus aureus* protein A precipitation technique. If an antiserum exhibits cross-reactivity by one or more of the above methods, then several experiments are required to verify its specificity for shared posttranslational modifications. To use the example of the *Dictyostelium* lysosomal enzymes, we have given rabbits injections of purified *N*-acetylglucosaminidase and obtained antisera that precipitate both *N*-acetylglucosaminidase and β-glucosidase and a large number of proteins from radioactively labeled cell extracts. Several approaches have provided evidence that the different proteins are precipitated by the same antibodies.

1. When the antibody is preadsorbed with the immunogen, the subsequent precipitation of other radioactively labeled proteins and enzyme activities is prevented. However, since the experiment is performed with the immunogen, contaminating proteins that gave rise to multiple antibodies could also compete. Therefore, the possibility that multiple antibody species could be interacting with individual proteins cannot be ruled out.
2. To extend this type of analysis, the possibility of minor contaminating proteins in the purified preparation can be minimized by the further purification of the immunogen by electrophoresis on denaturing polyacrylamide gels followed by excision of the appropriate band and elution of protein from the gel.[13] We have shown that *N*-acetylglucosaminidase and β-glucosidase purified in this way prevent the precipitation of all cross-reacting proteins by our antiserum.[14] If preadsorption of a protein of one molecular weight prevents the precipitation of another protein of a different

[13] D. A. Hager and R. R. Burgess, *Anal. Biochem.* **109**, 76 (1980).

[14] D. A. Knecht and R. L. Dimond, unpublished observations, 1982.

molecular weight, it suggests that cross contamination has not occurred. Thus, the same antibody must be interacting with both proteins.

3. A third approach to the problem is at the level of enzyme activity. We have shown that the antibody made against *N*-acetylglucosaminidase precipitated both *N*-acetylglucosaminidase and β-glucosidase activities. Since these proteins are difficult to purify from each other it is possible that they were cross-contaminated and the minor impurity of β-glucosidase in the *N*-acetylglucosaminidase immunogen was responsible for the precipitating activity. Because a preadsorption experiment should be performed in excess of competitor, this approach would not eliminate the possibility of cross-contamination. Another way to resolve this problem is to use a mixing experiment instead of competition. When a 20-fold excess of β-glucosidase protein was mixed with *N*-acetylglucosaminidase prior to the addition of antibody, the precipitation of *N*-acetylglucosaminidase activity requires 20 times more antibody than in the absence of the added β-glucosidase. In fact, this was the same amount of antibody required to precipitate the β-glucosidase alone (Fig. 2). If two separate antibodies were present and each interacted with one of the enzymes, they would be unaffected by mixing the

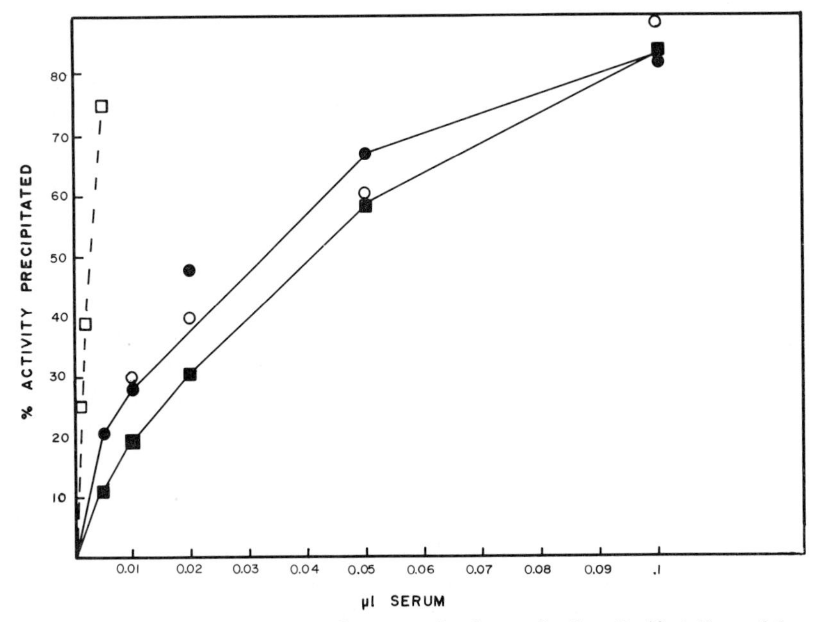

FIG. 2. Mixing of antigens to test for contaminating antibodies. Purified *N*-acetylglucosaminidase (□, ■) and β-glucosidase (○, ●) are precipitated either individually (open symbols) or mixed together (filled symbols) using an antibody prepared against pure *N*-acetylglucosaminidase. Equal amounts of enzymic activity are used, but because of the difference in turnover number there is a 20-fold excess of β-glucosidase protein.

two proteins and each precipitation would follow its unmixed curve. That the *N*-acetylglucosaminidase titration curve is shifted by adding β-glucosidase indicates that the antibody is unable to distinguish between the two proteins and thus one antibody is interacting in a stochastic manner with both proteins in the mixture.

4. It is also important to determine whether the multiple proteins that are precipitated by a given antiserum are recognized because they share similar peptide sequences rather than modifications. One way to investigate this question is to perform proteolytic analysis of the immunoprecipitated proteins (see Section I, this volume). If all are related in protein structure, then they will yield a similar size distribution of peptide fragments when digested with proteases. However, this method cannot exclude the possibility that a small, highly antigenic peptide sequence may be shared by a group of proteins whose overall sequences are unrelated. In these instances, the peptide mapping studies could be complemented by structural analysis where antigens that have undergone various chemical treatments (e.g., complete proteolysis, endoglycosidase H, periodate) would be tested for their ability to cross-react with the intact proteins. In our case, it has been possible to show that oligosaccharides isolated from pure β-glucosidase will inhibit the precipitation of *N*-acetylglucosaminidase when preadsorbed to a cross-reactive antiserum or monoclonal antibody.[15]

5. Monoclonal antibodies, by definition, recognize single antigenic determinants. Thus the preparation of monoclonal antibodies against protein modifications eliminates the possibility of cross-contamination of immunogens as a cause of multiple specificity. We have succeeded in generating mouse monoclonal antibodies that cross-react with determinants found on lysosomal enzymes in a manner similar to the rabbit antisera described above.[14] Even in cases where the results from polyclonal antisera seem clear, we feel that it is important to generate monoclonal antibodies. An antiserum that reacts with modifications common to a group of proteins may in fact contain a mixture of antibodies that recognize different shared modifications. A number of modifications might be shared by a subset of the group, but not by all members of the group. With a collection of monoclonal antibodies it becomes possible to identify these subgroups.

There are at least two ways to study the groups of proteins recognized by modification-specific antibodies. The first is to immunoprecipitate radioactively labeled proteins and analyze them by gel electrophoresis. In our case, all the proteins recognized by the antisera fall in a class of minor cell proteins of extremely acidic p*I* that form a distinct group when subjected to two-dimensional gel electrophoresis. In cell extracts labeled with $^{32}PO_4$ or $^{35}SO_4$ we

[15] R. C. Mierendorf, Jr., H. H. Freeze, R. Wunderlich, and R. L. Dimond, unpublished observations, 1983.

have found a strict correspondence between precipitable proteins and those that are phosphorylated and sulfated. The presence of these acidic groups may account for the low p*I* of these proteins, their antigenicity, and possibly their lysosomal localization. The second method of analysis is the immunological detection of antigenic proteins blotted onto nitrocellulose paper following electrophoresis.[16] In our experience, antibodies appear to vary greatly in their ability to interact with native protein versus protein immobilized on a solid support. Thus, a combination of the two methods may reveal aspects of specificity only demonstrated by one or the other technique.

Potential Advantages of the Immunological Approach

The main advantage of this particular immunological approach is that it ensures the study of protein modifications that are shared by a group of proteins rather than those specific to an individual protein. Thus one can identify groups of proteins that are modified in similar ways, trace the origin of a particular group modification, and determine its role in the function and/or localization of the related proteins. By generating a set of monoclonal reagents that are specific to modifications on lysosomal enzymes and other cell proteins, we hope to be able to follow groups of related proteins in their synthesis, processing, localization, packaging, and secretion. In addition, we will be able to analyze the changes that occur in modification during development and determine what role these changes have in cellular processes. Finally, these antibodies provide new probes that can be used to select cells that are altered in modification. For example, it may be possible to select for cells that no longer possess a particular posttranslational modification pathway. Characterization of the group of proteins that would be affected in such cells may lead to new insights into the function(s) of protein modification.

[16] W. N. Burnette, *Anal. Biochem.* **112,** 195 (1981).

Section III

Exocytosis

[14] Role of Cytoskeleton in Liver: *In Vivo* Effect of Colchicine on Hepatic Protein Secretion

By COLVIN M. REDMAN

The major pharmacological effect of colchicine, an alkaloid, is to bind tubulin and impede its polymerization into microtubules.[1] In most tissues there is a dynamic equilibrium between free tubulin and microtubules, and thus treatment with colchicine leads to a decrease in number, or a disappearance, of microtubules.[2,3] Concentrations of colchicine that impair microtubule polymerization inhibit secretion in a large number of secretory tissues. In the liver, colchicine binds to soluble tubulin,[4] inhibits the secretion of lipoproteins and other plasma proteins, and causes an accumulation of nonsecreted proteins within hepatocytes.[5-7] Doses of colchicine that inhibit hepatic secretion are similar to the concentrations of colchicine that bind to soluble liver tubulin[4] and the reduction in the number of hepatic microtubules has been correlated with inhibition of very-low-density-lipoprotein (VLDL) secretion.[8] Lumicolchicine, an isomer of colchicine, which does not bind to tubulin[9] and thus does not affect the formation of microtubules, does not inhibit hepatic secretion.[7,10] Because of these parallelisms between antimicrotubular action of colchicine and inhibition of secretion, it is thought that microtubules play an important role in hepatic secretion.

In this chapter, the *in vivo* conditions necessary to elicit inhibition of hepatic secretion by colchicine, the changes noted in the secretion of VLDL and other plasma proteins, notably albumin, and the side effects of colchicine are described.

Procedures

Administration of Colchicine. Colchicine is light sensitive and should not be left exposed for long periods of time. Preferably, it should be dissolved in

[1] L. Wilson, J. R. Bamburg, S. B. Mizel, L. M. B. Brisham, and K. M. Creswell, *Fed. Proc., Fed. Am. Soc. Exp. Biol.* **33,** 158 (1974).
[2] S. I. Inoué and H. Sato, *J. Gen. Physiol.* **50,** Suppl., 259 (1967).
[3] J. B. Olmsted and G. G. Borisy, *Annu. Rev. Biochem.* **42,** 507 (1973).
[4] C. Patzelt, A. Singh, Y. LeMarchand, L. Orci, and B. Jeanrenaud, *J. Cell Biol.* **66,** 609 (1975).
[5] Y. LeMarchand, A. Singh, F. Assimacopoulous-Jeannet, L. Orci, C. Rouiller, and B. Jeanrenaud, *J. Biol. Chem.* **248,** 6862 (1973).
[6] O. Stein and Y. Stein, *Biochim. Biophys, Acta* **306,** 142 (1973).
[7] C. Redman, D. Banerjee, K. Howell, and G. Palade, *J. Cell Biol.* **66,** 42 (1975).
[8] E. Reaven and G. Reaven, *J. Cell Biol.* **84,** 28 (1980).
[9] L. Wilson and M. Friedkin, *Biochemistry* **5,** 2463 (1966).
[10] P. R. Dorling, P. S. Quinn, and J. D. Judah, *Biochem. J.* **152,** 341 (1975).

ISBN 0-12-181998-1

0.15 *M* NaCl just prior to injection into the animal. It may be administered intraperitoneally, subcutaneously, or intravenously.

Dose. In studies with perfused liver or with liver slices, the minimal concentration of colchicine needed to inhibit plasma protein secretion was found to be 10^{-6} *M. In vivo,* with mice and rats, intraperitoneal or subcutaneous doses as low as 0.125 to 0.60 μmol/100 g body weight are sufficient to inhibit hepatic secretion.[6,10,11] Larger doses (25 μmol/100 g body weight) may be administered to rats intravenously without, after 3 hr, evidence of any gross morphological damage to the liver.[7] An equal volume of saline is usually given to the control animals, but subcutaneous injection of 0.6 μmol of lumicolchicine per 100 g body weight is an effective control.[10]

Onset and Duration of Action. The time necessary for colchicine to elicit an inhibitory response on hepatic secretion and its duration of action appear to be dose related. However, the methods used to measure inhibition of secretion may be important, since they differ in sensitivity. For example, in measuring the *in vivo* effect of colchicine on lipoprotein secretion, the level of serum triglycerate is often measured, and differences of at least 0.1 m*M* in the serum levels are necessary to measure a decrease accurately, since there is a large pool of circulating triglyceride in the serum. On the other hand, if radioactive precursors such as [^{14}C]palmitic acid are employed to measure VLDL secretion[6] or radioactive amino acids are used to measure albumin secretion,[7] then smaller differences in secretion may be observed.

When 0.125 μmol of colchicine per 100 g body weight is injected intraperitoneally into rats, maximal inhibition of VLDL secretion occurs 4.5 hr after administration, and the effect is reversible after 6–7 hr. With a higher dose (1.2 μmol/100 g body weight), there is no reversibility of the effect at 7 hr.[12] With mice, 0.125 μmol of colchicine per 100 g body weight causes a measurable decrease of triglyceride levels 4 hr after administration, with a maximal effect at 11 hr. At 20–24 hr after administration of colchicine, the triglyceride levels return to normal.[11]

There is evidence that the onset of action of colchicine, when administered intravenously and as a large dose (25 μmol/100 g body weight), may occur within minutes. To measure this early action of colchicine, it is necessary to inject the animal (rat) intravenously with a radioactive amino acid and then, at a time when radioactive plasma proteins (e.g., albumin) are being secreted linearly from the hepatocyte into the blood (20–40 min), colchicine is administered into the femoral vein. At short time intervals following the administration of colchicine, small quantities of blood are

[11] Y. LeMarchand, A. Singh, C. Patzelt, L. Orci, and B. Jeanrenaud, *in* "Microtubules and Microtubule Inhibitors" (M. Borgers and M. de Brabander, eds.), p. 153. North-Holland Publ., Amsterdam, 1975.

[12] O. Stein, L. Sanger, and Y. Stein, *J. Cell Biol.* **62,** 90 (1974).

collected from the tail ven. With this method, it has been determined that within 2 min of the administration of colchicine there is a noticeable reduction in the rate at which radioactive albumin is secreted into the blood.[7]

Hepatocyte Morphology

At 2–4 hr after the *in vivo* administration of colchicine, the general ultrastructure of the hepatocyte is well maintained (Fig. 1). The most prominent effect of colchicine treatment is the accumulation of secretory vesicles, which are marked by their lipoprotein contents and the decrease or absence of VLDL particles in the space of Disse. The VLDL-containing vesicles, besides being concentrated along the sinusoidal front of the hepatocytes, are also found in other parts of the cytoplasm[8,12] In addition, there is usually a marked increase in the number of lysosomes.[7,8] In the hepatocyte of control rats, microtubules are found in the Golgi region and in the peripheral cytoplasm, but their concentration is low. Colchicine causes an apparent reduction in the number of microtubules[8,12] although this is not always noticed.[7]

Intracellular Localization of Nonsecreted Proteins

Lipoprotein particles may be observed morphologically, but in order to determine the intracellular location of other nonsecreted proteins that accumulate in the hepatocyte owing to colchicine treatment, it is necessary to fractionate the liver into the pertinent organelles involved in secretion and to measure at which of these cellular locations there is an accumulation of nonsecreted proteins. Such a study is facilitated by pulse-labeling the animal with radioactive amino acids to label the secretory proteins. It is important in these studies that sufficient time be given, after the administration of the radioactive amino acid, to allow the secretory proteins to travel to, and accumulate at, the subcellular sites. The intracellular transit time, from time of synthesis on the rough endoplasmic reticulum to discharge from the cell, varies for different proteins. Morgan and Peters have calculated that for albumin it is 16 min and for transferrin it is 31 min.[13] Schreiber *et al.* have measured similar transit times: 15 min for albumin and 21 min for transferrin.[14] Thus the liver should not be removed and fractionated after the administration of radioactive amino acids at earlier times. In practice, a

[13] E. H. Morgan and T. Peters, *J. Biol. Chem.* **246,** 3508 (1971).

[14] G. Schreiber, H. Dryburgh, A. Millership. Y. Matsuda, A. Ingles, J. Phillips, K. Edwards, and J. Maggs, *J. Biol. Chem.* **254,** 12013 (1979).

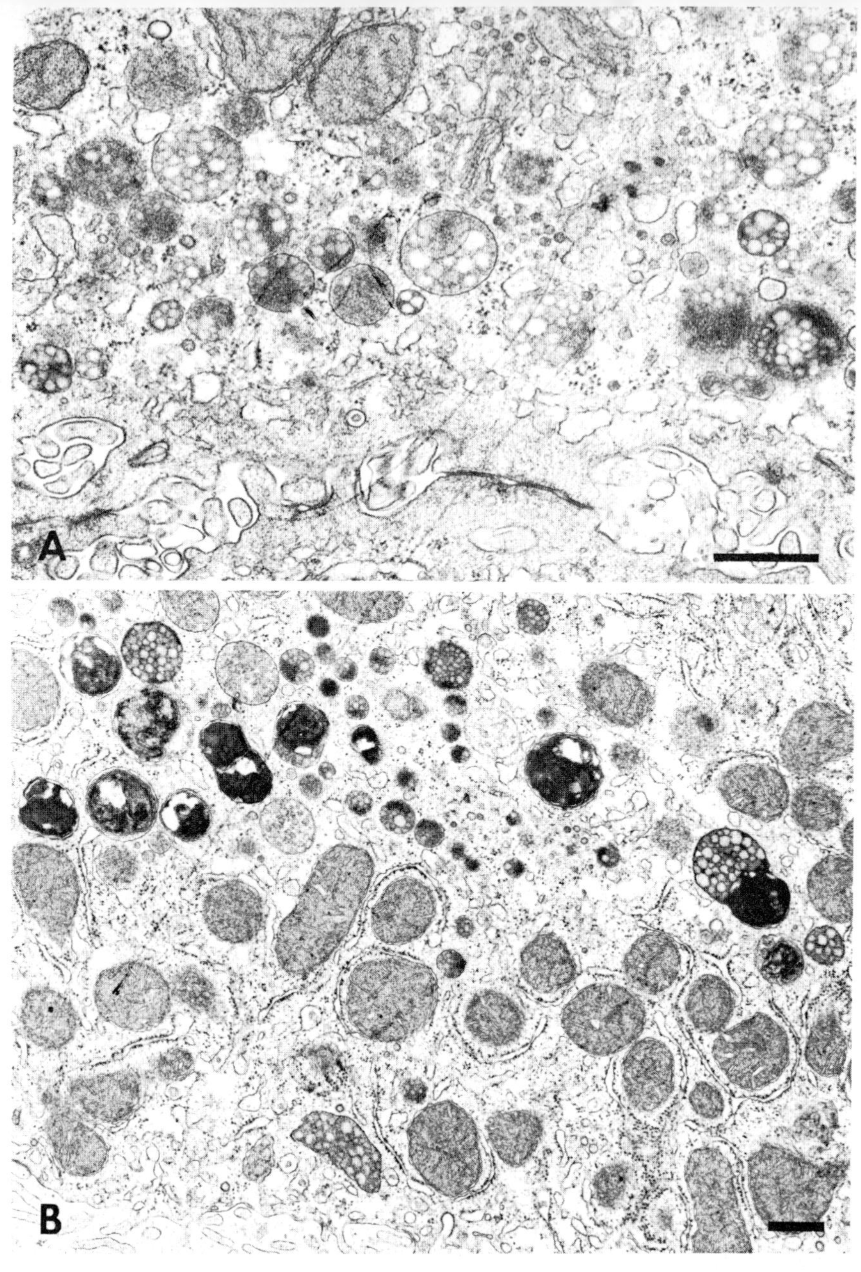

FIG. 1. Portions of the cytoplasm of hepatocytes from colchicine-treated rats. Young male rats were administered colchicine intravenously (25 μmol/100 g body weight), and portions of the liver were removed after 3.5 hr and processed for electron microscopy. (A) Rats were given ethanol by stomach tube (0.6 g of ethanol as a 50% solution per 100 g body weight) 30 min prior to the administration of colchicine. Small pieces of liver were fixed in 2.5% formaldehyde and 2% glutaraldehyde in 0.1 *M* sodium cacodylate buffer, pH 7.4, and postfixed with 1% OSO_4 in 0.1 *M* sodium, cacodylate, pH 7.4. ×28,800; bar = 0.5 μm. (B) The liver was fixed with 3% glutaraldehyde in 0.1 *M* sodium cacodylate, pH 7.4, and postfixed in 1% OSO_4. ×15,840; bar = 0.5 μm. The hepatic cytoplasm of colchicine-treated rats contains numerous secretion vesicles, marked with VLDL particles, which are more concentrated in the Golgi regions and in the periphery of the cell. The numbers of lysosomes and autophagic vacuoles increase after prolonged colchicine treatment. Micrographs by courtesy of Tellervo Huima, New York Blood Center.

longer time is necessary in order to allow sufficient radioactive proteins to accumulate in the pertinent organelle so that they can be easily measured.

In studies on the secretion of albumin by rat liver, 25 μmol of colchicine per 100 g body weight did not affect the movement of L-[U-^{14}C]leucine-labeled albumin from the rough to the smooth endoplasmic reticulum nor from the smooth endoplasmic reticulum to the Golgi apparatus. At 30–40 min after the intravenous injection of radioactive L-leucine, there was a striking accumulation of radioactive proalbumin and albumin in the Golgi apparatus[7,15] (see Fig. 2).

Specificity of Colchicine

The effect of colchicine on the general physiology of the perfused liver has been studied in great detail. Colchicine at concentrations (10 μ*M*) that inhibit VLDL secretion in perfused livers does not affect glucose production, intracellular ATP levels, oxygen consumption, protein or lipid synthesis, or ureogenesis.[3] *In vivo,* larger doses of colchicine (25 μmol/100 g body weight) have been shown not to affect phospholipid metabolism or the intracellular levels of nucleoside phosphates.[7]

The effect of colchicine on secretion, like its effect on tubulin, is specific, since plasma protein secretion by rat liver slices is not affected by lumicolchicine, an inactive isomer of colchicine that does not bind to tubulin; by griseofulvin, an antimitotic agent that does not block the *in vitro* polymerization of tubulin, or by cytocholasin B, which does not bind to tubulin but disorganizes microfilaments.[7] Colchicine preferentially binds to soluble tubulin in liver, but it also shows some affinity for other proteins.[4,16,17]

There are some pharmacological effects of colchicine that are not easily explained by its binding to tubulin, and these should be kept in mind when interpreting results obtained with colchicine. For example, colchicine inhibits nucleotide transport,[18] and it appears to influence membrane structure as shown by its ability to change the shape of erythrocytes.[19]

Side Effects of Colchicine That May Influence the Measurement of Hepatic Protein Secretion

When radiolabeled amino acids are being used to tag nascent secretory proteins, the investigator should be aware of the fact that colchicine influ-

[15] C. M. Redman, D. Banerjee, C. Manning, C. Y. Huang, and K. Green, *J. Cell Biol.* **77,** 400 (1978).

[16] J. Stadler and W. W. Franke, *J. Cell Biol.* **60,** 297 (1974).

[17] E. Reaven, *J. Cell Biol.* **75,** 731 (1977).

[18] S. B. Mizel and L. Wilson, *Biochemistry* **11,** 2573 (1972).

[19] P. Seaman, M. Chau-Wong, and S. Moyen, *Nature (London), New Biol.* **241,** 22 (1973).

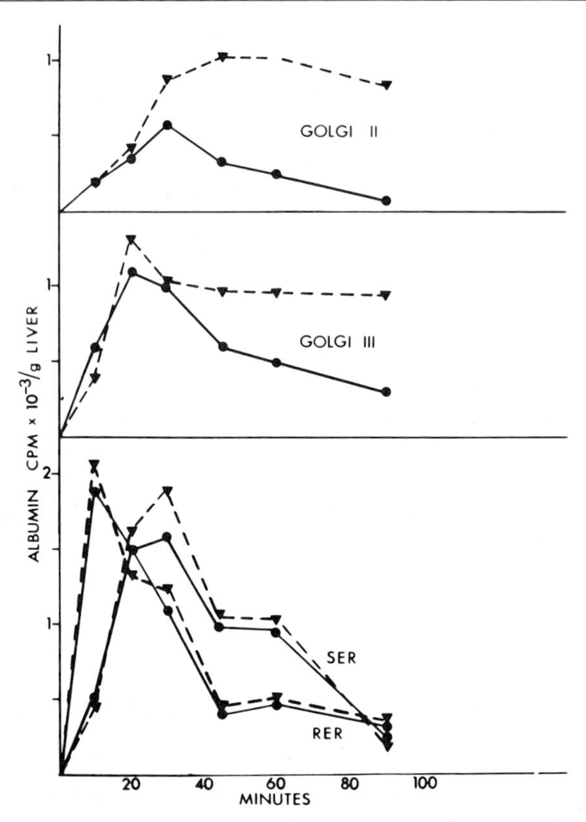

FIG. 2. Effect of colchicine on the intracellular transport of nascent albumin. Rats were injected with 25 μmol of colchicine per 100 g of body weight; control rats received an equal volume of saline. These injections were followed by the intravenous administration of 12.5 μCi of L-[U-^{14}C]leucine per 100 g of body weight. All injections were given in the femoral vein. At the various times indicated, the livers were removed and fractionated into rough endoplasmic reticulum (RER), smooth endoplasmic reticulum (SER), and two Golgi cell fractions (Golgi II and III). Radioactive albumin was isolated from the cell fraction by immunoprecipitation. The dashed lines represent the amounts of radioactive albumin in colchicine-treated rats, and the solid lines represent nascent hepatic albumin in the control rats. From Redman *et al.*,[7] with permission of the *Journal of Cell Biology*.

ences both the uptake of radioactive amino acids by the liver and the intracellular amino acid pools. In rats pretreated for 3 hr with 25 μmol of colchicine per 100 g body weight, the effective hepatic pulse time of injected radioactive leucine is longer than that of saline-treated rats.[15] In addition, Dorling *et al.*[10] have shown that the hepatic amino acid pools vary in colchicine-treated rats (0.5 μmol/100 g body weight). At this dose the intracellular specific radioactivity of L-leucine was raised 150%, but the radioactivities of glutamic acid, glycine, alanine, and histidine were decreased.

In addition to inhibiting protein secretion, colchicine, under certain conditions, inhibits protein synthesis. At a concentration of 0.25 μmol/100 g body weight, colchicine administered three times (0, 90, and 150 min) does not affect protein synthesis, although higher doses (10-fold) have a slight inhibitory effect.[6] At a dose of 0.625 μmol/100 g body weight, colchicine, after 2 hr of administration, is thought to inhibit not only albumin secretion, but also the synthesis of proalbumin and its subsequent conversion to albumin.[10] At higher concentrations (0.5–25 μmol/100 g) colchicine markedly inhibits total protein synthesis after a lag of 1 hr.[15] It has been postulated that the delayed inhibitory effect of colchicine on protein synthesis may be a consequence of a feedback mechanism. This appears not to be the case, however, since it has been determined that the delayed action of colchicine in inhibiting protein synthesis is a general response that affects both secretory and nonsecretory proteins produced by the liver.[15] It is quite possible that the delayed inhibition of protein synthesis by colchicine is due merely to a toxic effect of the drug.

[15] Role of the Cytoskeleton in Secretory Processes: Lacrimal and Salivary Glands

By B. ROSSIGNOL, A. M. CHAMBAUT-GUÉRIN, and P. MULLER

During the past few years an increasing number of reports have appeared on the role of microtubules and microfilaments in protein secretion of various exocrine and endocrine glands. Many of these works are based on the use of compounds that cause either the disorganization or the stabilization of microtubules and of microfilaments.

In various secretory cells, microtubules are scattered in the cytoplasm; they have often been observed to be prominent in the rough endoplasmic reticulum (RER)–Golgi area and sometimes in the region of secretory granules.[1-5a] Microfilaments are distinguishable in the apical region of the acinar cells, close to secretory granules and the plasma membrane; they are

[1] W. J. Malaisse and L. Orci, *Methods Achiev. Exp. Pathol.* **9,** 112 (1979).
[2] C. Patzelt, D. Brown, and B. Jeanrenaud, *J. Cell. Biol.* **73,** 578 (1977).
[3] E. P. Reaven and G. M. Reaven, *J. Cell. Biol.* **84,** 28 (1980).
[4] D. Soifer, ed., "The Biology of Cytoplasmic Microtubules," *Ann. N. Y. Acad. Sci.* **253** (1975).
[5] J. A. Williams and M. Lee, *J. Cell Biol.* **71,** 795 (1976).
[5a] S. C. Nickerson, R. M. Ackers, and B. T. Weinland, *Cell Tissue Res.* **223,** 421 (1982).

ISBN 0-12-181998-1

often described as bundles parallel to the lumen; sometimes actin filaments are observed surrounding secretory granules.[1,6] The question whether microtubules and microfilaments are structural elements that define the spatial organization of the cell (for instance, in the RER–Golgi area) or whether they actually participate in the secretory process (secretory granule transport and discharge) is, up to now, unanswered. The approach in investigating the role of microtubular system in protein secretion mainly consisted of the use of antimitotic compounds, such as colchicine or vinblastine, that disrupt the microtubules: colchicine binds to tubulin and provokes depolymerization of microtubules; vinblastine causes paracrystalline deposits in the acinar cells, with a concomitant disappearance of microtubules.[1,7] Other compounds that stabilize microtubules, such as deuterium oxide and hexylene glycol, have also been used.[1] The function of microfilaments was explored by using drugs, such as cytochalasins or phalloidin, that interact with actin filaments: when gland fragments were incubated in the presence of cytochalasin B, the microfilament network was apparently disrupted; however, it has been shown that cytochalasin B also effects the transport system of small molecules in various tissues, whereas cytochalasin D does not.[8,9] So it appears that cytochalasin D is the best drug for exploring the role of microfilaments. The second (phalloidin), which is a cyclic peptide from *Amanita,* enhances *in vitro* the rate of actin polymerization and stabilizes the F-actin structure. Phalloidin has mainly been used in studies with hepatocytes.[10]

In rat exorbital lacrimal and salivary glands (mainly parotids), the role of microtubules or of microfilaments has been explored by using colchicine or related compounds, vinblastine, and cytochalasin B or D.[2,7,11-21] In these

[6] H. Bauduin, C. Stock, D. Vincent, and J. F. Grenier, *J. Cell Biol.* **66,** 165 (1975).

[7] R. B. Kelly, C. Oliver, and A. R. Hand, *Cell Tissue Res.,* **195,** 227 (1978).

[8] S. J. Atlas and S. Lin, *J. Cell Biol.* **76,** 360 (1978).

[9] E. G. Loten and B. Jeanrenaud, *Biochem. J.* **140,** 185 (1974).

[10] M. Prentki, C. Chaponnier, B. Jeanrenaud, and G. Gabbiani, *J. Cell Biol.* **81,** 592 (1979).

[11] S. Busson-Mabillot, L. Ovtracht, A. M. Chambaut-Guérin, P. Muller, G. Herman, and B. Rossignol, *Eur. J. Cell. Biol.* **22,** 184 (1980).

[12] S. Busson-Mabillot, A. M. Chambaut-Guérin, L. Ovtracht, P. Muller, and B. Rossignol, *J. Cell Biol.* **95,** 105 (1982).

[13] F. R. Butcher and R. H. Goldman, *Biochem. Biophys. Res. Commun.* **48,** 23 (1972).

[14] A. M. Chambaut-Guérin, P. Muller, and B. Rossignol, *Gastroenterol. Clin. Biol.* **2,** 215 (1978).

[15] A. M. Chambaut-Guérin, P. Muller, and B. Rossignol, *J. Biol. Chem.* **253,** 3870 (1978).

[16] A. M. Chambaut-Guérin, P. Muller, and B. Rossignol, *J. Biol. Chem.* **254,** 10734 (1979).

[17] V. Herzog, H. Sies, and F. Miller, *J. Cell. Biol.* **70,** 692 (1976).

[18] G. Keryer and B. Rossignol, *Am. J. Physiol.* 230, 99 (1976).

[19] P. Muller, A. M. Chambaut-Guérin, and B. Rossignol, *Biol. Cell* **45,** 255 (1982).

[20] B. Rossignol, G. Herman, and G. Keryer, *FEBS Lett.* **21,** 189 (1972).

[21] R. Temple, J. A. Williams, J. F. Wilber, and J. Wolff, *Biochem. Biophys. Res. Commun.* **46,** 1454 (1972).

glands, protein secretion can be triggered by cholinergic or adrenergic agonists.[17,22-23c] Colchicine inhibits the discharge of newly synthesized proteins from secretory granules formed and transported in its presence, but does not significantly suppress the release of proteins from secretory granules already stored at the cell apex.[2,11,12] Colchicine does not significantly affect protein biosynthesis, nucleotide level, and [^{32}P]phosphate incorporation into phospholipids, proteins, or nucleotides. Moreover, it does not hinder the effects of muscarinic or adrenergic agonists on various metabolic parameters.[2,15,16] Vinblastine seems to block the discharge of old and new proteins.[18] This drug does not appear to interact with the muscarinic receptors, since it does not inhibit the cholinergic stimulation of ^{45}Ca uptake.[18] When lacrimal or parotid gland fragments are incubated with [^{3}H]colchicine, colchicine binds to a cytosolic protein similar to tubulin.[2,14-16] The high-affinity sites bind 2.5–2.8 nmol of colchicine per gram of gland with a K_d of 2 μM. The colchicine–tubulin complex has been isolated from gland fragments by Sephadex G-100 or BioGel P-300 filtration and characterized by polyacrylamide gel electrophoresis. Although colchicine does not disturb the cellular metabolism, it has been reported that, under some conditions, it induces an increase in the number of lysosomes.[2,11,12] As previously described in the liver, colchicine can produce *in vivo* an accumulation of autophagic vacuoles in parotid glands under a long treatment.[2] In lacrimal glands crinophagy has been observed after 2 hr of *in vitro* treatment.[11,12]

The role of microfilaments has been little investigated in lacrimal or salivary glands; however, it has been shown that cytochalasin B inhibits the release of proteins already stored at the cell apex. Taken together, the available data suggest that colchicine, by binding to tubulin, prevents its polymerization into cytoplasmic microtubules and perturbs the intracellular transport of proteins already in the rough endoplasmic reticulum. It inhibits the discharge of proteins biosynthesized and stored in secretory granules formed in its presence. Besides, cytochalasins hinder the release of proteins from secretory granules stored at the cell apex. Data obtained with cytochalasin D suggest that microfilaments could be also involved in the mechanisms triggered by cholinergic or adrenergic agonists in inducing protein secretion.[19]

Up to now, it has been demonstrated only that the disturbance of microtubule or microfilament systems coexists with the inhibition of protein secretion, but the actual relationship between the two processes remains

[22] G. Herman, S. Busson, L. Ovtracht, C. Maurs, and B. Rossignol, *Biol. Cell* **31,** 255 (1978).
[23] J. W. Putney, C. M. Van de Wall, and B. A. Leslie, *Am. J. Physiol.* **235 C,** 188 (1978).
[23a] Z. Y. Friedman, M. Lowe, and Z. Selinger, *Biochim. Biophys. Acta* **675,** 40 (1981).
[23b] P. Mauduit, G. Herman, and B. Rossignol, *Biol. Cell* **45,** 281 (1982).
[23c] P. Mauduit, G. Herman, and B. Rossignol, *FEBS Lett.* **153,** 21 (1983).

to be fully assessed. Moreover, the role of the cytoskeleton has been investigated only at the level of the microtubule and microfilament functions. The presence and the possible role of other elements of cytoskeleton in acinar cells of salivary or lacrimal glands remains to be established. The presence of intermediate-sized filaments of prekeratin in the myoepithelial cells of salivary glands has been reported.[24]

Localization of the Drug Effect on Secretory Processes in Lacrimal and Parotid Glands

The effect of drugs either on discharge of proteins or on their intracellular transport from the RER toward the cell apex can be explored *in vitro* with gland lobules. The effects are investigated at various drug concentrations, usually 0.01 – 50 μM colchicine, 0.01 – 10 μM vinblastine, 0.001 – 10 μM cytochalasin.

Preparation of Gland Lobules and General Incubation Methods

Male albino rats (5 – 7 weeks old) are stunned by a blow on the head; the parotid or lacrimal glands are quickly removed, and the lobules are isolated as described by Castle *et al.*[25] The lobules are incubated in Krebs-Ringer bicarbonate buffer (KRB) supplemented with 0.55 mM glucose and equilibrated with 95% O_2 + 5% CO_2.

Drug Effect on Protein Discharge

Protein discharge is estimated by measuring either the output of an enzyme (amylase from parotid or peroxidase from lacrimal glands) or the release of labeled proteins.

Enzyme Discharge. Lobules (50 mg) are preincubated in 5 ml of KRB for about 30 min; then they are extensively washed and incubated with or without drug and in the presence or the absence of an inducer of secretion (cholinergic or adrenergic agonist) at the optimal concentration, usually 1 – 10 μM according to the agonist and the gland.[22] Aliquots from the incubation medium are withdrawn at various times and centrifuged in order to remove any particulate matter. At the end of the incubation, gland lobules are blotted, weighed, and homogenized with an Ultra-Turrax homogenizer in 4 ml of phosphate buffer, pH 7.4, with 0.05 M NaCl (for amylase) or of Tris, pH 7.35 (for peroxidase). The homogenate is centrifuged at 37,000 g (30 min at 4°). The supernatant fractions are analyzed for their

[24] W. W. Franke, E. Schmid, C. Freudenstein, B. Appelhans, M. Osborn, K. Weber, and T. W. Keenan, *J. Cell Biol.* **84,** 633 (1980).

[25] J. D. Castle, J. D. Jamieson, and G. E. Palade, *J. Cell Biol.* **53,** 290 (1972).

enzyme activity according to Bernfeld's technique[26] for amylase or to Putney *et al.*[23] for peroxidase.

Labeled Protein Discharge. Lobules (200 mg) are pulse labeled for 3.5 min with 10 μCi of [^{14}C]leucine in 10 ml of KRB. Incorporation is stopped by adding 10 ml of KRB supplemented with 1 mM leucine and washing three times with 10 ml of the same medium. Lobules are then incubated for a chase period of 70 min (a time sufficient for labeled secretory proteins to be transported at the cell apex) in 10 ml of KRB with 1 mM leucine. Subsequently, several washed 50-mg quantities of lobules are submitted to a further incubation for various times, in flasks containing 5 ml of KRB with or without colchicine, vinblastine, or cytochalasin in the presence or the absence of an inducer of secretion. At the end of this period, the gland lobules are blotted, weighed, and then homogenized with an Ultra-Turrax homogenizer in 4 ml of ice-cold water. The homogenate is centrifuged at 37,000 g (30 min at 4°); the incubation medium too is centrifuged. The proteins of both supernatants are precipitated by 10% trichloroacetic acid, and the radioactivity is counted.[15,16,22]

Secretion of enzyme or of labeled proteins is expressed as the amount of enzyme activity or labeled protein present in the incubation medium as a percentage of the total enzyme activity or labeled proteins in tissue and medium. Secretion of peroxidase is often expressed as the amount of peroxidase activity (units per gram of gland) released in the incubation medium.

Drug Effect on Intracellular Transport

Since colchicine does not affect exocytosis, it can be used to investigate the role of microtubules in the intracellular transport of proteins by measuring their discharge according to the method described above. In contrast, cytochalasin, which seems to inhibit the last step of the secretory process induced (or not) by cholinergic or adrenergic agonists, cannot be used. Lobules are incubated and pulse labeled as described above, except that colchicine (1–10 μM) is added at various times during the intracellular migration of labeled proteins (i.e., before, at the onset, or after pulse labeling). At the end of the 70-min chase period, the lobules are washed several times with a colchicine-free KRB. Then sets of 50 mg of lobules are incubated for various times in 5 ml of KRB in the presence or the absence of an inducer of secretion and in the absence of colchicine. Protein secretion is measured as described above.

The effect of drugs on the intracellular transport can also been investigated by electron microscope autoradiography. Lobule samples are col-

[26] P. Bernfeld, this series, Vol. 1, p. 149.

lected at various times during the incubation period after the pulse labeling with [^{3}H]leucine and fixed as usual for electron microscopy. Thin sections are prepared according to the conventional dipping method (see summary[27]) and measurements of radioactivity are made as described by Droz.[28]

Specificity of Drugs

The specificity of the drug effect on protein discharge or on intracellular transport has to be checked by investigating its effects on other cellular parameters (for instance, labeled amino acid or sugar incorporation into proteins; ATP level; ^{32}P incorporation into phospholipids, stimulated or not by cholinergic or adrenergic agonists), under strictly identical conditions.[2,15]

Evaluation of the Drug Effect on the Microtubule or Microfilament Systems

It is of great interest to connect the inhibitory effect of a drug on protein secretion and the stage of microtubule and microfilament systems. The degree of polymerization of microtubules or microfilaments after drug treatment can be estimated either by electron microscopy or by biochemical methods.

Electron Microscopy

The effects of colchicine and of cytochalasin at the ultrastructural level have been described for various secretory tissues. A method of quantitative evaluation of microtubules has been used by Reaven *et al.*[3] for hepatocytes. Up to now, few ultrastructural observations have been reported for lacrimal or salivary glands to explore the effects of these drugs. Only the qualitative data concerning microtubules in parotids reported by Patzelt *et al.*[2] are available.

Biochemical Methods

The amount of free tubulin and microtubules can be determined by the method used by Patzelt *et al.* in parotid gland.[2] This method is based on the properties of microtubules to be depolymerized by cold treatment and to be stabilized by glycerol. The amount of tubulin present as microtubules is determined by the difference between the total colchicine-binding sites, measured in the supernatants from cold- and glycerol-treated gland homogenates, respectively.

The relationship between the drug effect on secretion and the degree of

[27] J. Boyenval and J. Fischer, *J. Microsc. Biol. Cell.* **27**, 115 (1976).

[28] B. Droz, *J. Microsc. Biol. Cell* **27**, 191 (1976).

disorganization of microtubules or microfilaments can be based on the comparison between the dose-dependent effect of the drug on protein secretion and the dose-dependent binding of the drug on a high-affinity site. This method has been used to explore the effect of colchicine by measuring its binding to tubulin in the lacrimal gland[15,16]; no data on cytochalasin binding in these glands have been reported.

Determination of Colchicine Binding Sites in Gland Lobules.[15,16] Lobules are incubated under conditions strictly identical to those under which the effect of colchicine on protein secretion is investigated. After a 70-min incubation period in the presence of various concentrations of [^{3}H]colchicine, the lobules are homogenized with an Ultra-Turrax homogenizer in MES [2-(*N*-morpholino)ethanesulfonic acid]–$MgCl_2$–sucrose buffer (0.01 *M*–0.005 *M*–0.24 *M*), pH 6.8 (160 mg of gland per milliliter), then centrifuged at 37,000 *g* for 15 min at 4°. The amount of bound [^{3}H]colchicine is determined by submitting the supernatant fraction either to gel filtration [Sephadex G-100 column 0.9 × 30 cm, equilibrated with MES–$MgCl_2$ (0.01 *M*–0.005 *M*), buffer, pH 6.8], or to filtration through DEAE-cellulose paper disks according to Borisy.[29]

Determination of Total Binding Sites.[15,16] One hundred milligrams of lobules are homogenized in 2 ml of ice-cold MES–$MgCl_2$–sucrose (0.01 *M*–0.005 *M*–0.24 *M*) buffer, pH 6.8, with an Ultra-Turrax homogenizer. The homogenate is kept for 30 min at 4° (in order to depolymerize the microtubules[4]) then centrifuged at 37,000 *g* for 15 min. Colchicine binding is measured by incubating aliquots of the supernatant fraction at 37° for 70 min with various concentrations of colchicine (0.3–50 μM). The reaction is stopped by cooling the mixture. The [^{3}H]colchicine-bound fraction is measured by the filter disk assay method.[29] The K_d value and the number of sites are determined from a Scatchard plot. The K_d can be compared to the K_d of colchicine for purified brain tubulin taken as a reference.[15]

The number of high-affinity sites for colchicine is expressed as the number of nanomoles of colchicine bound per gram of gland. It is also the number of free tubulin residues in the tissue.

Characterization of Colchicine-Binding Protein Complex. The identification of tubulin as the binding protein complexed with colchicine can be made by comparing its properties to those of the colchicine–brain tubulin complex taken as a reference, by various techniques, such as gel filtration on Sephadex G-100 or Biogel P-300, ion-exchange chromatography on DEAE-Sephadex A-50,[15,16,30] and polyacrylamide gel electrophoresis under nondissociating and nondenaturating conditions.[16] Gel electrophoresis can be a suitable method for the identification of the colchicine–tubulin complex

[29] G. G. Borisy, *Anal. Biochem.* **50,** 573 (1972).

[30] G. G. Borisy and E. W. Taylor, *J. Cell Biol.* **34,** 525 (1967).

formed in a tissue during incubations in the presence of [^{3}H]colchicine and also for the simultaneous quantitative assessment of the complex, and therefore of tubulin.[16] Furthermore, the complex isolated from incubated lacrimal gland lobules inhibits the polymerization of brain tubulin, like the colchicine–brain tubulin complex.[16]

[16] A Subcellular Fractionation Approach for Studying Insulin Release Mechanisms and Calcium Metabolism in Islets of Langerhans[1]

By MICHAEL L. MCDANIEL, JERRY R. COLCA, NIRMALA KOTAGAL, and PAUL E. LACY

D-Glucose is the primary physiological stimulus that initiates the release of insulin from the islets of Langerhans. At present, it is not clear whether the initiation of insulin release by D-glucose involves interactions with a membrane receptor and/or the transport and subsequent metabolism of this hexose. There is little doubt, however, that cellular Ca^{2+} is central to the coupling of the stimulus to the secretory event. Insulin secretion is absolutely dependent on extracellular Ca^{2+}, and all known insulinogenic stimuli affect Ca^{2+} fluxes in intact islets. The precise means by which Ca^{2+} concentrations in the β cell are regulated during stimulus–secretion coupling and the manner in which Ca^{2+} mediates its effects on insulin secretion may be approached on a molecular level by subcellular fractionation of islet cells.[2] Limitations of the quantity of islets available for fractionation make isolation of highly purified cell fractions difficult. However, the utilization of techniques for the isolation of large quantities of islets and microassay procedures makes it feasible to obtain organelle-enriched fractions for characterizing these cellular processes.

This chapter describes techniques for the mass isolation of islets of Langerhans and procedures for obtaining characterized and purified subcellular fractions. Techniques are described for the evaluation of the functions of the plasma membrane and endoplasmic reticulum in the regulation of cellular Ca^{2+} levels. Techniques are also presented to study the effects of Ca^{2+} and calmodulin on protein phosphorylation in subcellular fractions

[1] This work was supported in part by Grants AM06181 and AM03373 from the National Institutes of Health.

[2] The islets of Langerhans represent a heterogeneous population of cell types; however, adult rat islets are composed of approximately 80% β cells [B. Hellman, *Acta Endocrinol.* **32,** 92 (1959)].

ISBN 0-12-181998-1

obtained from islet cells and the correlation of these events with the secretion of insulin.

Isolation of Islets

Islets are isolated by a modification of the technique of Lacy and Kostianovsky.[3]

Animals. Male Wistar strain rats (175–220 g) are obtained from Charles River Laboratories (Wilmington, Massachusetts) and housed individually for at least 3 days prior to use. During this time the animals are allowed access to feed (Purina Rat Chow) and water *ad libitum.*

Reagents

All glassware is siliconized (1% Siliclad; Clay–Adams, Parsippany, New Jersey) and sterilized prior to use. Hank's balanced salt solution 10× concentration (GIBCO Laboratories, Grand Island, New York), buffered to pH 7.4 with $NaHCO_3$. The solution is supplemented with 1% penicillin and 1% streptomycin.

CMRL medium 1066 (GIBCO Laboratories) supplemented with 10% heat-inactivated fetal bovine serum, 1% L-glutamine, 1% penicillin, 1% streptomycin, and 100 or 150 mg of D-glucose per deciliter. All media are prepared or diluted with sterile water.

Ficoll, type 400, M_r 400,000 (Sigma Chemical Co.), is dialyzed (Spectrapor, M_r cutoff 12,000–14,000, Spectrum Medical Industries, Inc. Los Angeles, California) five times against a 20-fold excess of distilled-deionized water. The dialyzed Ficoll is lyophilized and diluted to 27% (w/w) with Hank's solution supplemented with 25 mM HEPES, pH 7.3. The 27% Ficoll is gravity filtered through Whatman No. 1 filter paper and autoclaved prior to dilution to 23, 20.5, and 11% Ficoll with Hank's solution containing 25 mM HEPES, pH 7.3.

Procedure. Rats are anesthetized by an intraperitoneal injection of Nembutal (10–20 mg per animal), and the pancreas is visualized by a central abdominal incision. The bile duct is clamped at the tip of the duodenum and cannulated (Intramedic polyethylene tubing, PE 50; Clay–Adams, Parsippany, New Jersey) at a point sufficiently proximal to the liver (1 cm) to allow inflation of the head and tail of the pancreas with Hank's solution at 24°. Hank's solution (15–20 ml) is introduced by syringe slowly (1 min) into the bile duct until the pancreas is clearly distended. The inflated pancreas is gently removed by blunt dissection. Pancreases are collected in covered petri dishes and kept moist with Hank's solution.

The pancreases are freed of fat and lymph nodes using forceps and scissors with the visual aid of a dissecting microscope. The cleaned pan-

[3] P. E. Lacy and M. Kostianovsky, *Diabetes* **16,** 35 (1967).

creases are then collected in a 50-ml beaker, where they are minced into pieces (1 mm in diameter) with fine scissors. The minced pancreas is distributed among calibrated 13 × 100 mm glass tubes with screw-cap lids. Five milligrams of collagenase (Worthington, CLS IV-Millipore Corp., Freehold, New Jersey) are added for each milliliter of gravitated tissue. Usually 20–25 mg of collagenase are dissolved into each glass tube containing 4–5 ml of tissue.

Tubes are fastened to a wrist-action shaker (Burrell Model 75; Burrell Corp., Pittsburgh, Pennsylvania), submerged in a 37° water bath, and vigorously agitated for 10–12 min. Digestion is terminated when translucent gelatinous spots are located along the wall of the tube. The digested tissue is diluted with Hank's solution at 24° and rapidly pelleted at 500 *g* for 1 min (Model HN-S centrifuge; International Equipment Co., Needham Heights, Massachusetts). The pellet is resuspended in 8–10 ml of Hank's solution and resedimented two additional times to remove collagenase. Three milliliters of 27% Ficoll is added to each tube and vortexed vigorously. Coalesced connective tissue is removed with a Pasteur pipette, and the remaining suspension is transferred to 40-ml centrifuge tubes (Pyrex No. 8400) with a 3-ml wash of 27% Ficoll. Typically, 5–6 tubes of tissue suspension are transferred to 4 centrifuge tubes (40 ml) containing a total of 7.5 ml of 27% Ficoll per tube. The Ficoll suspension is gently mixed, and any remaining connective tissue is removed. Ficoll dilutions, 4 ml each of 23, 20.5, and 11%, are sequentially layered over the 27% Ficoll–tissue mixture. The Ficoll gradients are centrifuged at 800 *g* for 10 min at 24°. Islets rise to the interface of 11% and 20.5% Ficoll and are harvested with a Pasteur pipette.[4] To remove residual Ficoll, islets are diluted with Hank's solution and pelleted at 500 *g* for 1 min. The pellet is resuspended in CMRL 1066 medium containing 100 mg of glucose per deciliter and transferred to a small plastic petri dish 60 × 15 mm (Falcon, Oxnard, California). Islets are swirled to the center of the petri dish and passed through a series (2–3) of these dishes containing CMRL 1066 medium until the islets are free of debris. Clean islets are transferred to a final petri dish, washed twice with CMRL medium containing 100 mg of glucose per deciliter, and suspended in 5 ml of CMRL 1066 medium with 150 mg of glucose per deciliter. The culture plates are placed in a Plexiglas chamber (Belco Glass; Vineland, New Jersey), which is gassed with CO_2–air (5 : 95%) and cultured at 24° (Lab-line Ambi Hi-Lo Chamber; Lab-line Instruments Inc., Melrose Park, Illinois) for 24–48 hr prior to fractionation.

Routinely, islets from 15 rats are harvested in 3–3.5 hr when 4 technicians are employed (two for surgical removal of the pancreas and two for the

[4] A. Shibata, C. W. Ludvigsen, S. P. Naber, M. L. McDaniel, and P. E. Lacy, *Diabetes* **25**, 667 (1976).

remainder of the isolation technique). The average yield is 200–400 islets per rat,[5] and an adequate quantity of islets for subcellular fractionation is achieved by repeating this isolation technique.

Subcellular Fractionation of Islets

Isolated pancreatic islets are fractionated by a modification of the technique of Naber *et al.*,[6] which allows for simultaneous preparation of fractions enriched in plasma membrane and endoplasmic reticulum.

Reagents

MES (2[*N*-morpholino]ethanesulfonic acid), 50 m*M*, containing 250 m*M* sucrose and 1 m*M* EDTA, pH 7.2, at 4° (fractionation buffer)

MES, 10 m*M*, containing 1 m*M* EDTA, pH 6.0, at 4°

Sucrose solutions prepared in 10 m*M* MES, pH 6.0, at densities 1.14, 1.16, 1.18, 1.20. The densities are adjusted using a refractometer.

Procedure. The islets (10,000–12,000) are pooled (Fig. 1) and transferred with a Pasteur pipette to a Teflon–glass, Potter–Elvehjem tissue grinder (size 0019; Kontes Biomedical Products, Vineland, New Jersey) and washed with 50 m*M* MES, pH 7.2. The islets are homogenized in 800 μl of fractionation buffer using 14 strokes with a motor drive for the pestle at 1170 rpm (Polyscience RZ R10 at setting 3). The homogenate is transferred to a 10 $\times$ 75 mm Pyrex tube with 200 μl of fractionation buffer. The homogenate is centrifuged at 600 g_{max} (2250 rpm) for 5 min in a Sorvall RC 5B centrifuge using a SS-34 rotor. The supernatant (S_1) is transferred to a second 10 $\times$ 75 mm Pyrex tube, and the pellet is resuspended in 1 ml of fractionation buffer, vortexed vigorously, and again pelleted at 600 g_{max} (2250 rpm) for 5 min. The resulting pellet (P_1) contains nuclei and cellular debris. The combined supernatants (S_1) are centrifuged as above at 20,000 g_{max} (13,000 rpm) for 20 min. The resulting supernatant (S_2) is essentially free of mitochondria and secretory granules and is centrifuged at 150,000 g for 90 min to yield an endoplasmic reticulum-enriched pellet (P_4) and cytosol (S_4). The pellet (P_4) is resuspended in 400–800 μl of fractionation buffer (minus EDTA).

The 20,000 g pellet (P_2), which contains mitochondria, secretory granules, and heavy membrane components, is rehomogenized (14 strokes as above) in 500 μl of 10 m*M* MES, pH 6.0, and layered over a discontinuous sucrose gradient (1 ml each with sucrose densities of 1.14, 1.16, 1.18, and 1.20) in a Beckman cellulose nitrate centrifuge tube ($\frac{1}{2} \times 2$ in.). The sucrose gradient containing the rehomogenized P_2 is centrifuged at 150,000 g_{max} (40,000 rpm) in a Beckman L5-50 ultracentrifuge using a SW 50.1 rotor for

[5] Yields vary depending on the lot of collagenase.

[6] S. P. Naber, J. M. McDonald, L. Jarett, M. L. McDaniel, C. W. Ludvigsen, and P. E. Lacy, *Diabetologia* **19,** 439 (1980).

FIG. 1. Pooled islets (10,000–12,000) obtained from 30 rats prior to subcellular fractionation.

90 min. Membrane bands are removed from the top of the gradient interfaces to yield bands 1 and 2, which contain plasma membranes, and bands 3 and 4, which contain a mixed fraction of mitochondria, secretory granules, and other heavy membranes. The membrane bands are individually resuspended in 4 ml of 10 m*M* MES, pH 6.0, and pelleted at 150,000 g_{max} (40,000 rpm) for 60 min to remove sucrose. The membrane pellets are visualized with the aid of a dissection microscope and individually resuspended in 200–500 μl of 10 m*M* MES, pH 6.0. This fractionation scheme is summarized in Fig. 2, and Fig. 3 illustrates the discontinuous sucrose gradient for isolation of plasma membrane-enriched fractions (bands 1 and 2).

Biochemical and Morphological Characterization of Islet-Cell Subcellular Fractions

Limitations of the quantity of isolated islets of Langerhans make extensive purification of islet-cell fractions difficult. The limited yields of protein from this technique require the use of microassay procedures and of constriction pipettes calibrated to microliter quantities.

Protein Determination. Protein is measured by a fluorometric procedure utilizing fluorescamine.[7] The sensitivity of this micromethod minimizes the amount of sample required to determine total protein.

Reagents

Borate buffer, 0.2 *M* adjusted to pH 9.0 with NaOH

Fluorescamine (Fluram; Roche) 0.5 mg per milliliter of acetone prepared fresh every third day

Bovine serum albumin, 20–100 μg/ml (Armour Pharmaceutical Co., Kankakee, Illinois)

Procedure. The assay is performed in disposable 10 × 75 mm borosilicate culture tubes and utilizes 2–10 μl of sample or protein standard. The protein is diluted with 200 μl of 0.2 *M* borate buffer, pH 9.0. Assay tubes are vortexed immediately after addition of 10 μl of fluorescamine reagent and diluted by addition of 1 ml of water prior to measurement of fluorescence on a modified Farrand Model A fluorometer (primary filter CS-7-37 and secondary filters CS 4-72 and 3-72; Corning Glassware, Corning, New York). The fluorometric scale is adjusted with a quinine standard,[8] and as little as 0.2 μg of protein can be used with reliable results.

Plasma Membrane Fraction. Enrichment of plasma membrane within

[7] P. Bohlen, S. Stein, W. Dairman, and S. Udenfriend, *Arch. Biochem. Biophys.* **155,** 213 (1973).

[8] O. H. Lowry and J. V. Passonneau, "A Flexible System of Enzymatic Analysis." Academic Press, New York, 1972.

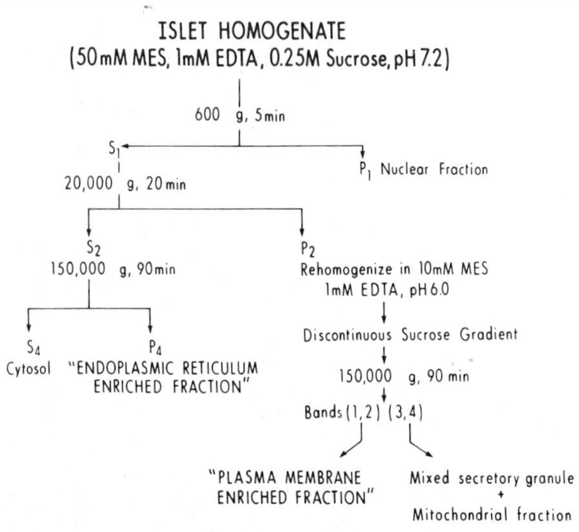

FIG. 2. Flow sheet for subcellular fractionation of islets.

the islet-cell fractions is assessed by the binding of ^{125}I-wheat germ agglutinin and by the activity of 5′-nucleotidase. In the former case, islets are incubated with iodinated wheat germ agglutinin, and its binding is determined in the different subcellular fractions.[6] 5′-Nucleotidase activity is measured by the method of Avruch and Wallach[9] with modifications described previously.[6] This procedure measures the liberation of tritiated adenosine after separation of [^{3}H]AMP by $ZnSO_4-Ba(OH)_2$ precipitation and requires 0.4–10 μg of protein per assay tube. The table demonstrates the enrichment of plasma membrane marker (5′-nucleotidase) in bands 1 and 2, which are obtained from the 20,000 *g* pellet after separation on sucrose density gradients. Figure 4A illustrates the morphological appearance of band 1.

Endoplasmic Reticulum. Enrichment of endoplasmic reticulum in the subcellular fractions is assessed by measurement of glucose-6-phosphatase and NADH-cytochrome *c* reductase activity in the various fractions. Glucose-6-phosphatase is measured at pH 6.5 by the isotopic procedure of Kitcher *et al.*,[10] utilizing uniformly labeled [^{14}C]glucose-6-phosphate as substrate. The assay volume is 25 μl and requires 2–10 μg of protein per sample. NADH-cytochrome *c* reductase is measured by the technique of Dallner *et al.*,[11] utilizing 2–20 μg of sample in a 1-ml volume with a Gilford 250 recording spectrophotometer. Fraction P_4 (Fig. 2 and the table) contains a two- to threefold enrichment of NADH-cytochrome *c* reductase and

[9] J. Avruch and D. F. H. Wallach, *Biochim. Biophys. Acta* **233,** 334 (1971).
[10] S. A. Kitcher, K. Siddle, and J. P. Luzio, *Anal. Biochem.* **88,** 29 (1978).
[11] G. Dallner, P. Siekovitz, and G. E. Palade, *J. Cell Biol.* **30,** 97 (1966).

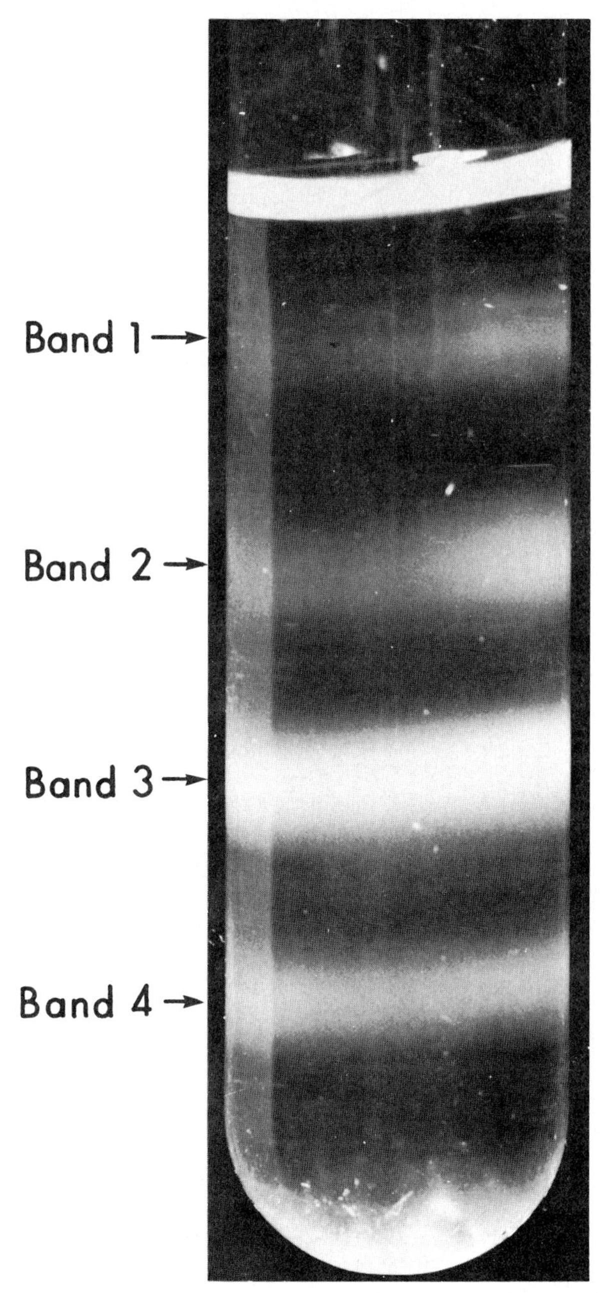

FIG. 3. Discontinuous sucrose gradient for isolation of plasma membrane-enriched fractions (bands 1 and 2).

BIOCHEMICAL CHARACTERIZATION OF ISLET-CELL FRACTIONS[a]

Fraction[b]	Homogenate	P_1	B_1	B_2	B_3	B_4	P_4	S_4
Total protein (μg)	4277 ± 679	575 ± 250	73 ± 15	124 ± 18	34.6 ± 27	22 ± 8	181 ± 34	1667 ± 154
% Recovery	(100)	12.4 ± 3.3	1.8 ± 3.3	3.0 ± 0.4	8.5 ± 0.8	0.5 ± 0.2	4.4 ± 0.9	40.7 ± 4.4
5′-Nucleotidase[c] (nmol/μg/60 min)	1.3 ± 0.1	3.5 ± 0.5	7.4 ± 0.3	4.5 ± 0.2	1.8 ± 0.1	2.6 ± 0.5	2.4 ± 0.3	0.3 ± 0.1
Fold enrichment	(1.0)	2.7 ± 0.5	5.7 ± 0.5	3.4 ± 0.3	1.4 ± 0.2	2.1 ± 0.5	1.8 ± 0.4	0.2 ± 0.1
% Recovery	(100)	28.8 ± 2.1	9.6 ± 0.6	10.0 ± 0.9	11.9 ± 1.3	1.2 ± 0.5	7.4 ± 0.6	8.9 ± 2.5
NADH-cytochrome *c* reductase (nmol/mg/min)	22.0 ± 7.5	42.0 ± 17.2	34.2 ± 6.4	54.5 ± 12.8	41.1 ± 7.8	21.4 ± 4.5	43.6 ± 7.3[d]	1.5 ± 0.5
Fold enrichment	(1.0)	1.7 ± 0.3	1.4 ± 0.2	2.2 ± 0.2	1.8 ± 0.5	1.2 ± 0.1	2.4 ± 0.4	0.1 ± 0.1
% Recovery	(100)	18.8 ± 1.5	2.2 ± 0.1	6.4 ± 0.6	15.3 ± 3.0	0.9 ± 0.3	7.8 ± 0.6	3.4 ± 1.3
Succinate-cytochrome *c* reductase (nmol/mg/min)	8.8 ± 0.8	15.0 ± 3.6	1.3 ± 0.8	28.2 ± 3.3	52.6 ± 7.7	14.7 ± 9.2	ND[e]	ND
Fold enrichment	(1.0)	1.7 ± 0.5	0.2 ± 0.1	3.2 ± 0.1	5.9 ± 0.4	1.5 ± 0.9	ND	ND
% Recovery	(100)	16.3 ± 5.8	0.2 ± 0.1	9.6 ± 1.6	52.8 ± 1.4	1.1 ± 0.6	ND	ND
Extractable insulin (mU/μg)	1.6 ± 0.3	1.5 ± 0.2	0.4 ± 0.1	1.4 ± 0.2	2.5 ± 0.1	3.3 ± 0.3	0.3 ± 0.1	1.0 ± 0.1
Fold enrichment	(1.0)	1.0 ± 0.2	0.2 ± 0.1	1.0 ± 0.2	1.7 ± 0.3	2.0 ± 0.2	0.2 ± 0.1	0.9 ± 0.1
% Recovery	(100)	11.4 ± 2.7	0.5 ± 0.2	3.1 ± 1.1	15.1 ± 3.7	1.2 ± 0.4	0.7 ± 0.1	29.1 ± 7.7

[a] This table presents the distribution of total protein and selective biochemical markers among the islet-cell subcellular fractions. Data from three to four fractionations are shown. Mean values ± SEM are shown as an index of variability in the technique.

[b] Fractions: P_1 = nuclear pellet; B_1 and B_2 = plasma membrane-enriched fractions; B_3 and B_4 = mixed secretory granule and mitochondrial fractions; P_4 = endoplasmic reticulum-enriched fraction; S_4 = cytosol.

[c] Similar results are obtained when the binding of ^{125}I-wheat germ agglutinin is used as a marker for plasma membrane.

[d] Similar results are obtained when glucose-6-phosphatase is used as a marker for endoplasmic reticulum. For example, in four tissue preparations the specific activity of glucose-6-phosphatase in P_4 was 0.42 pmol/μg in 60 min, which represented a 2.2 ± 0.3-fold increase over the homogenate.

[e] ND, not detectable.

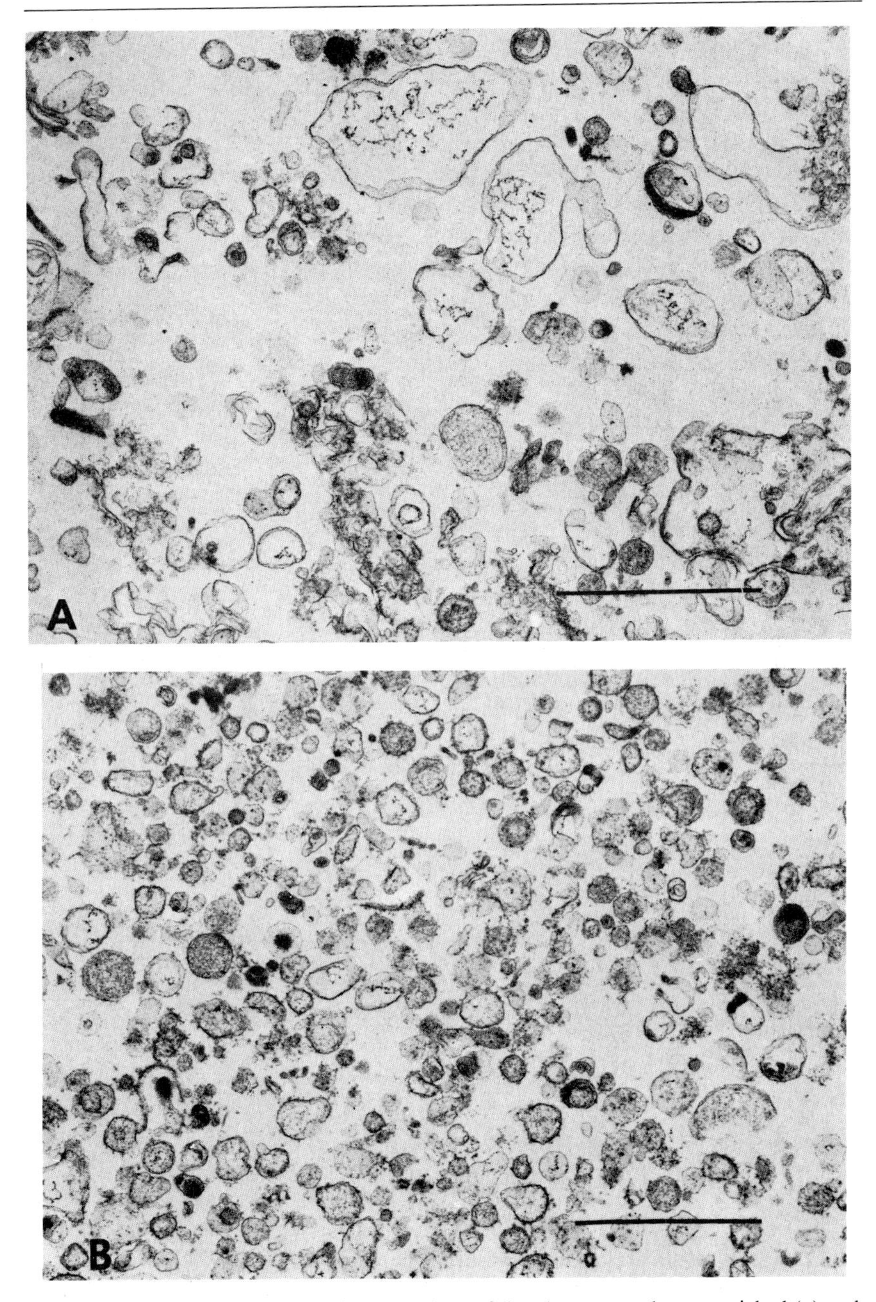

FIG. 4. Electron micrographs of thin sections of the plasma membrane-enriched (*a*) and endoplasmic reticulum-enriched (*b*) islet-cell fractions. Bar = 1 μm ($\times$ 25,000).

glucose-6-phosphatase activity with virtually no contamination with mitochondria or secretory granules. Figure 4B shows the morphological appearance of this fraction.

Mitochondria and Secretory Granules. Succinate cytochrome *c* reductase is used as an index of mitochondrial contamination and is measured as described by Fleischer and Fleischer[12] in a total volume of 200 μl at 30° utilizing 2–20 μg of protein. Extractable immunoreactive insulin is used as a marker for secretory granules. The sample (20 μl) is extracted with 40 μl of acid alcohol (1.5% H_2SO_4 in 75% ethanol) overnight at 4°. Immunoreactive insulin[13] is determined in a 100-μl aliquot of a 200- to 2000-fold dilution of the acid alcohol extract. Mitochondria and secretory granules are primarily confined to the lower membrane bands obtained after fractionation of the 20,000 *g* pellet on a discontinuous sucrose gradient (Fig. 2 and the table). At present, no further attempts have been made to resolve these mixed fractions.

Plasma Membrane Ca^{2+}, Mg^{2+}-ATPase

Intracellular Ca^{2+} concentration plays an essential role in insulin secretion from islets of Langerhans. A Ca^{2+} extrusion pump associated with the plasma membrane may regulate these Ca^{2+} levels. A well-characterized Ca^{2+}-stimulated and Mg^{2+}-dependent ATPase (Ca^{2+},Mg^{2+}-ATPase) activity localized in the erythrocyte plasma membrane is considered to represent the enzymic basis for the Ca^{2+} extrusion pump.[14] A similar Ca^{2+},Mg^{2+}-ATPase activity has been identified in islet-cell plasma membranes,[15] which displays two apparent affinities for Ca^{2+}, a high affinity ($K_m = 0.085\ \mu M$) and a low affinity with a requirement for Ca^{2+} greater than 10 μM (Fig. 5). The isotopic method of Seals *et al.* for measuring Ca^{2+},Mg^{2+}-ATPase activity has been modified in order to utilize microgram quantities, or less, of protein.[16] The assay measures Ca^{2+}-dependent liberation of $^{32}P_i$ from [γ-$^{32}P_i$]ATP.

Reagents

PIPES [piperazine-*N*,*N'*-bis(2-ethanesulfonic acid)], 500 m*M*, pH 7.5, at 37°; pH adjusted with Tris base

EGTA, 4 m*M*; adjusted with Tris base to pH 7.5 at 37°

ATP, 20 m*M*; vanadate free and concentration standardized by the procedure of Lowry and Passonneau[8]

NaN_3, 1 *M*

[12] S. Fleischer and B. Fleischer, this series, Vol. 10, p. 406.

[13] P. H. Wright, P. R. Makulu, D. Vichick, and K. E. Sussman, *Diabetes* **20,** 33 (1971).

[14] H. J. Schatzmann, *Experientia* **22,** 364 (1966).

[15] H. A. Pershadsingh, M. L. McDaniel, M. Landt, C. G. Bry, P. E. Lacy, and J. M. McDonald, *Nature (London)* **288,** 492 (1980).

[16] J. R. Seals, J. M. McDonald, D. E. Bruns, and L. Jarett, *Anal. Biochem.* **90,** 785 (1978).

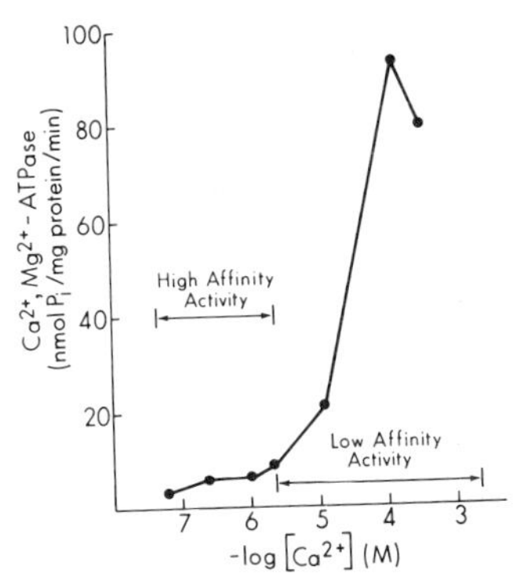

FIG. 5. Dependence of plasma membrane Ca^{2+},Mg^{2+}-ATPase on the concentration of ionized Ca^{2+}.

$CaCl_2$, 100 m*M*, prepared in 50 m*M* Tris-PIPES, pH 7.5, and standardized by atomic absorption spectrophotometry
SDS (sodium dodecyl sulfate), 3%
[γ-$^{32}P_i$]ATP, 3000 Ci/mmol
Phosphate extraction reagent is prepared fresh from the following three solutions in the ratio 2 : 2 : 1, respectively: 10 *N* H_2SO_4, 10% ammonium molybdate, 0.1 *M* silicotungstic acid (Fisher Co.)
Organic solvent: 65% xylene–35% isobutanol

Procedure. The incubations are performed in triplicate in 13 × 100 mm disposable borosilicate culture tubes. The tubes are maintained on ice prior to incubation, and all additions are made on ice. The final assay mixture (100 μl) consists of 50 m*M* Tris–PIPES, pH 7.5, 20 m*M* NaN_3, 0.25 m*M* ATP (1 μCi per tube [γ-$^{32}P_i$]ATP), and 0.2 m*M* EGTA. Ca^{2+} when present is 0.16 to 0.50 m*M*, resulting in a free Ca^{2+} range of 0.06 to 300 μM calculated using a stability constant of $10^{7.83}$ for Ca^{2+}-EGTA at pH 7.5 and verified using a Ca^{2+} selective electrode.[17] In the assay protocol, a set of tubes without tissue is used to determine the amount of spontaneous hydrolysis of [γ-$^{32}P_i$]ATP under the specific experimental conditions. The total amount of radioactivity introduced into the assay is determined in a 10-μl aliquot of the assay mixture.

The assay tubes are preincubated at 37° for 5 min, and the reaction is initiated by the addition of 0.5–1.0 μg of protein (bands 1 or 2) in a volume of 10 μl. The reaction mixture is incubated at 37° for 30 min in a Dubnoff water bath with gentle shaking and terminated by the addition of 50 μl of 3% SDS followed by vigorous vortexing.

The amount of $^{32}P_i$ liberated is extracted with the addition of 75 μl of

[17] N. Kotagal, J. R. Colca, and M. L. McDaniel, *J. Biol. Chem.* **258,** 4808 (1983).

phosphate extraction reagent followed by the addition of 1.5 ml of the organic solvent. This mixture is vortexed vigorously for 20 sec and centrifuged to separate the aqueous and organic phases. An aliquot (1 ml) of the organic phase is removed, and $^{32}P_i$ liberated is quantitated by liquid scintillation counting. Ca^{2+}-stimulated ATPase activity is determined by subtracting values obtained with 0.2 m*M* EGTA alone, i.e., tissue blank, from those obtained in the presence of chelator and Ca^{2+}.

Comments. The isotopic assay for determining the release of $^{32}P_i$ from $[\gamma\text{-}^{32}P_i]$ATP is very sensitive and allows the use of tissue in the range of 1 μg or less. The spontaneous hydrolysis of ATP under assay conditions is negligible (<0.2%), whereas hydrolysis greater than 1% becomes limiting to the sensitivity of the assay. The assay volume has been reduced from 1 ml described in the original procedure,[16] and the use of disposable assay tubes aids in avoiding contamination. Sodium azide is present in the assay medium to inhibit any possible mitochondrial contamination. Although there is a Mg^{2+} requirement for the enzyme activity, no exogenous Mg^{2+} is added. The concentration of endogenous Mg^{2+} (8.5–9.0 μM) is sufficient for the expression of the enzyme activity. Current studies in our laboratory[17] and others[18] indicate that the high-affinity Ca^{2+},Mg^{2+}-ATPase may be significantly modulated by $[Ca\text{-}EGTA]^{2-}$ concentration in the medium as well as by the presence of calmodulin.

Ca^{2+} Uptake by Islet-Cell Endoplasmic Reticulum

The function of the sarcoplasmic reticulum in the release and storage of Ca^{2+} ions is known to play a critical role in the contraction–relaxation cycle of contractile cells. A similar function with regard to the regulation of intracellular Ca^{2+} has been ascribed to the reticulum of noncontractile cells.[19] Given the essential role of Ca^{2+} ions in the insulin release mechanism, it seems most probable that elements of the β-cell endoplasmic reticulum are centrally involved in the control of insulin secretion. The properties of Ca^{2+} handling by the islet-cell endoplasmic reticulum are studied using oxalate-sustained transport with tracer $^{45}Ca^{2+}$ followed by rapid filtration. These investigations utilize the endoplasmic reticulum-enriched fraction (fraction P_4), which has been described above.[20]

[18] A. Al-Jobore and B. D. Roufogalis, *Biochim. Biophys. Acta* **645,** (1981).

[19] D. E. Bruns, J. M. McDonald, and L. Jarett, *J. Biol. Chem.* **251,** 7191 (1976).

[20] The Ca^{2+} transport properties of the islet-cell endoplasmic reticulum and plasma membrane have been shown to be fundamentally distinct.[23] The contribution of contaminating organelles to the measured quantity of Ca^{2+} uptake in this fraction was assessed by simultaneous measurement of Ca^{2+} uptake in other fractions or by the addition of metabolic inhibitors. Although the plasma membrane-enriched fraction (i.e., band 1) contains a much greater enhancement of plasma membrane marker than does P_4 (see the table), the rate of Ca^{2+}

Reagents

Tris buffer, 100 mM, pH 6.8 at 37°
KCl, 1 M
$MgCl_2$, 0.5 M, standardized by atomic absorption spectrophotometry
$CaCl_2$, 100 mM, standardized as above
Tris oxalate, 250 mM
Tris ATP, 20 mM (vanadate free)
EGTA, 40 mM, buffered with Tris base to pH 7.0
Wash solution: 250 mM sucrose, 40 mM NaCl

Materials

Millipore filters 0.22 μm pore size, type GSWP (Millipore Corp., Bedford, Massachusetts), presoaked at least 15 min in 250 mM KCl
Gelman filter apparatus (Gelman, Ann Arbor, Michigan)
Vacuum pump (Gast, Benton Harbor, Michigan)

Tissue. Fraction P_4 (endoplasmic reticulum-enriched fraction) is resuspended in 400–800 μl of 50 mM MES, pH 7.2, containing 250 mM sucrose (minus EDTA). The suspension is rehomogenized with a Teflon–glass tissue grinder (0019; Kontes Biomedical Products, Vineland, New Jersey) using a motor drive for the pestle at 1170 rpm. The tissue is used immediately or stored at −80° with minimal loss of activity.

Procedure. The Ca^{2+} uptake assay is performed in 1.5 ml of polypropylene microfuge tubes in a total volume of 100 μl. The final assay contains 50 mM Tris, pH 6.8, at 37°, 100 mM KCl, 5 mM $MgCl_2$, 10 mM oxalate,[21] 10–80 μM $CaCl_2$ (0.1–0.2 μCi of $^{45}Ca^{2+}$ per tube), ± 1.25 mM ATP. The reaction medium is temperature equilibrated at 37°, and Ca^{2+} uptake is initiated by the addition of 2–5 μg of protein. After incubation (1–30 min) an 80-μl aliquot is removed and collected on a Millipore filter (0.22 μm) supported on a single-unit Gelman filter apparatus and rinsed twice with 5 ml of wash solution under vacuum (22 in. Hg); the filtration time is 25 sec per sample. The filters are placed in scintillation counting vials, air dried, and then dissolved in 0.5 ml of ethylene glycol monomethyl ether. $^{45}Ca^{2+}$ is quantitated by liquid scintillation counting in 10 ml of Aquasol II (New England Nuclear, Boston, Massachusetts). Appropriate blanks consist of the complete assay medium minus tissue and are quantitated by the same filtration procedure. The rate of Ca^{2+} uptake is expressed as nanomoles of Ca^{2+} per milligram of protein per minute.

uptake for band 1 was only 11% the rate of uptake in P_4 when assessed under the same conditions.[22] Similarly, consistent with the biochemical and morphological characterization of the endoplasmic reticulum-enriched fraction, the mitochondrial inhibitors azide and ruthenium red did not affect Ca^{2+} uptake by this fraction.[22]

[21] Rapid addition of cold oxalate to the buffer may precipitate Ca^{2+} oxalate. The reaction buffer and oxalate stock are both equilibrated to 37°, and the oxalate is added drop by drop with stirring.

Comments. The rate of active or ATP-dependent Ca^{2+} uptake under these conditions (i.e., in the presence of 10 mM oxalate and 1.25 mM ATP) remains linear for at least 30 min, whereas passive binding (i.e., Ca^{2+} accumulation in the absence of ATP) is rapidly saturated within 1 min. This allows for the measurement of the initial rate of ATP-dependent Ca^{2+} uptake with 10–30 min incubations.

An important consideration in these experiments is the actual concentration of free Ca^{2+} in the medium that is available for transport. In the absence of added ligands, the Ca^{2+} levels are buffered by the presence of oxalate and ATP^{4-}. The influence of these chelators can be estimated by consideration of the stability constants for Ca^{2+} and Mg^{2+}-oxalate and Ca^{2+} and $MgATP^{2-}$. Under these experimental conditions approximately 36% of the total Ca^{2+} remains ionized,[22] However, it is possible that experimental manipulations of the Ca^{2+} uptake system may artifactually change the rate of Ca^{2+} uptake by altering the concentration of ionized Ca^{2+}. It is therefore advisable to carry out the assay at higher concentrations of total Ca^{2+} with free Ca^{2+} buffered to a physiological level. This is accomplished by using a Ca^{2+}–EGTA buffer system where free Ca^{2+} concentrations are determined from the stability constant for Ca^{2+}-EGTA.[24] In these experiments, 200 μM EGTA is added to the uptake medium and total Ca^{2+} is varied from 100 to 195 μM with free Ca^{2+} buffered at 0.7–7 μM.[22] Under these conditions it is less likely that an experimental treatment may artifactually alter the concentration of free Ca^{2+}. However, the reduction of specific activity of $^{45}Ca^{2+}$ necessitates the use of two to three times more protein per assay. Figure 6 shows a representative experiment depicting the dependence of the rate of Ca^{2+} uptake on the concentration of ionized Ca^{2+} in the medium.

Ca^{2+}-Stimulated Protein Phosphorylation

One mechanism by which Ca^{2+} is known to mediate cellular functions is by alteration of the phosphorylation of regulatory proteins. Many of the cellular effects of Ca^{2+}, including those involving protein phosphorylation, are mediated by the Ca^{2+} binding protein calmodulin.[25] Pancreatic islets possess an endogenous protein kinase activity that is modulated by Ca^{2+} and calmodulin.[26] The importance of Ca^{2+} in the regulation of protein phosphorylation is investigated by measurement of protein kinase activity in

[22] J. R. Colca, J. M. McDonald, N. Kotagal, C. Patke, C. J. Fink, M. H. Greider, P. E. Lacy, and M. L. McDaniel, *J. Biol. Chem.* **257,** 7223 (1982).

[23] J. R. Colca, N. Kotagal, and M. L. McDaniel, *Biochim. Biophys. Acta* **729,** 176 (1983).

[24] H. J. Schatzmann, *J. Physiol.* (*London*) **235,** 551 (1973).

[25] W. Y. Cheung, *Science* **207,** 19 (1980).

[26] M. Landt, M. L. McDaniel, C. G. Bry, N. Kotagal, J. R. Colca, P. E. Lacy, and J. M. McDonald, *Arch. Biochem. Biophys.* **213,** 148 (1982).

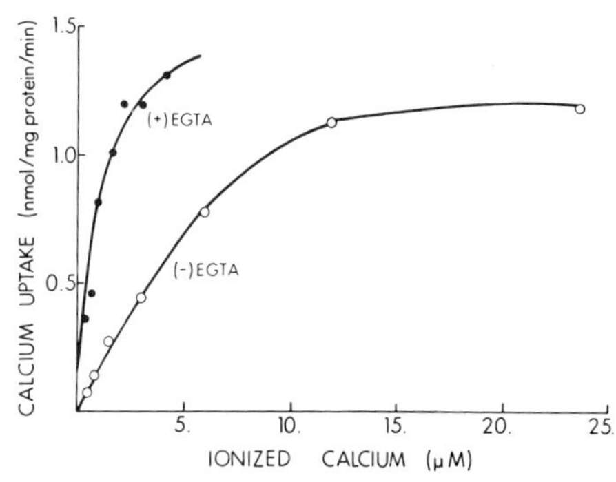

FIG. 6. Dependence of Ca^{2+} uptake by the islet-cell endoplasmic reticulum-enriched fraction on the concentration of ionized Ca^{2+}. Ca^{2+} may be buffered by ATP and oxalate alone (O-O) or by the addition of 200 μM EGTA (●-●), which allows the use of higher total Ca^{2+} concentrations.

islet-cell homogenates and subcellular membrane fractions and by studying protein phosphorylation in intact islets.[27] *In vitro* assays allow controlled manipulation of endogenous protein kinase activity, including direct measurements of Ca^{2+} requirements, whereas studies with intact islets allow direct correlation of protein phosphorylation with the secretory state of the islets in response to physiological stimuli.

Measurement of Endogenous Protein Kinase Activity

Reagents

Reaction buffer: 250 mM PIPES, pH 7.0, 50 mM $MgCl_2$, 1 mM EGTA, 0.5 mM dithiothreitol

$CaCl_2$, 20 mM

Calmodulin, 300–400 μg/ml; prepared from rat brain[26]

ATP, 25 μM: [γ-$^{32}P_i$]ATP

Stop solution: 9% SDS, 6% 2-mercaptoethanol, 15% glycerol

Protosol (New England Nuclear)

Materials

10% Polyacrylamide slab gels

Film (Kodak XAR-5)

Wolf cassette

Procedure. Protein kinase activity is assayed in 12 × 75 mm conical

[27] J. R. Colca, C. L. Brooks, M. Landt, and M. L. McDaniel, *Biochem. J.* **212,** 819 (1983).

polypropylene tubes in a final volume of 100 μl. The final assay medium contains 50 mM PIPES–NaOH (pH 7.0), 10 mM $MgCl_2$, 0.2 mM EGTA, 0.1 mM dithiothreitol, 40–160 μg of islet protein (homogenate or membrane fraction), and 5 μM ATP (5–10 μCi [γ-$^{32}P_i$]ATP per tube). When present, the final concentrations of $CaCl_2$ and calmodulin are 1 mM and 1 μM (1.8 μg/tube), respectively. The assay is initiated by the addition of [γ-$^{32}P_i$]ATP and terminated after a 5-sec incubation[28] at 37° with 50 μl of stop solution. The samples are vortexed and placed in a boiling water bath for 2 min; 50-μl aliquots are subjected to SDS–polyacrylamide gel electrophoresis on 10% slab gels. After electrophoresis, the gels are stained for protein (0.5% Coomassie Blue in 10% acetic acid, 10% isopropanol), destained (five changes of 10% acetic acid, 10% isopropanol), and autoradiographed for 6–24 hr. Using the exposed film as a guide, the protein bands of interest are cut from the gel, placed in scintillation vials in 0.5 ml of water, and heated with 2 ml of Protosol (3 hr at 50°) to elute the radioactively labeled protein. The samples are then quantitated by liquid scintillation counting. Figure 7 illustrates an autoradiogram indicating a Ca^{2+}- and calmodulin-sensitive 57,000 M_r protein band.

Protein Phosphorylation Determined in Intact Islets

Reagents

Krebs–Ringer buffer (KRB): 115 mM NaCl, 24 mM $NaHCO_3$, 5.0 mM KCl, 1 mM $MgCl_2$, 2.5 mM $CaCl_2$, 0.1% bovine plasma albumin (Armour Pharmaceutical Co.), 5 mM HEPES (pH 7.4) containing 100–500 mg of D-glucose per deciliter

KRB stock (above), 2× concentrated

[$^{32}P_i$]Orthophosphoric acid, 5 mCi/ml (New England Nuclear, Bedford, Massachusetts)

Stop solution: 9% SDS, 6% 2-mercaptoethanol, 15% glycerol

Procedure. Freshly isolated islets obtained from 10–11 rats are washed with KRB and preincubated at 37° for 60 min in 3 ml of KRB. The islets are collected by centrifugation (500 g for 30 sec), washed with KRB, and transferred to a siliconized 13 × 100 mm culture tube, and all residual medium is removed. The islets are loaded with $^{32}P_i$ for 20 min at 37° in 200 μl of 2× KRB and 200 μl of $^{32}P_i$ (400–700 μCi). The islets are sedimented (500 g for 30 sec), the radioactive medium is removed, and the islets are washed with KRB at 4° and distributed into two siliconized petri dishes in an ice bath. One hundred islets are selected for each sample with the aid of a

[28] Incubation times are varied depending on the intent of the experiment. Shorter time periods (e.g., 5 sec) allow measurement of kinase activity per se, whereas at longer incubations the endogenous substrate becomes rate limiting.

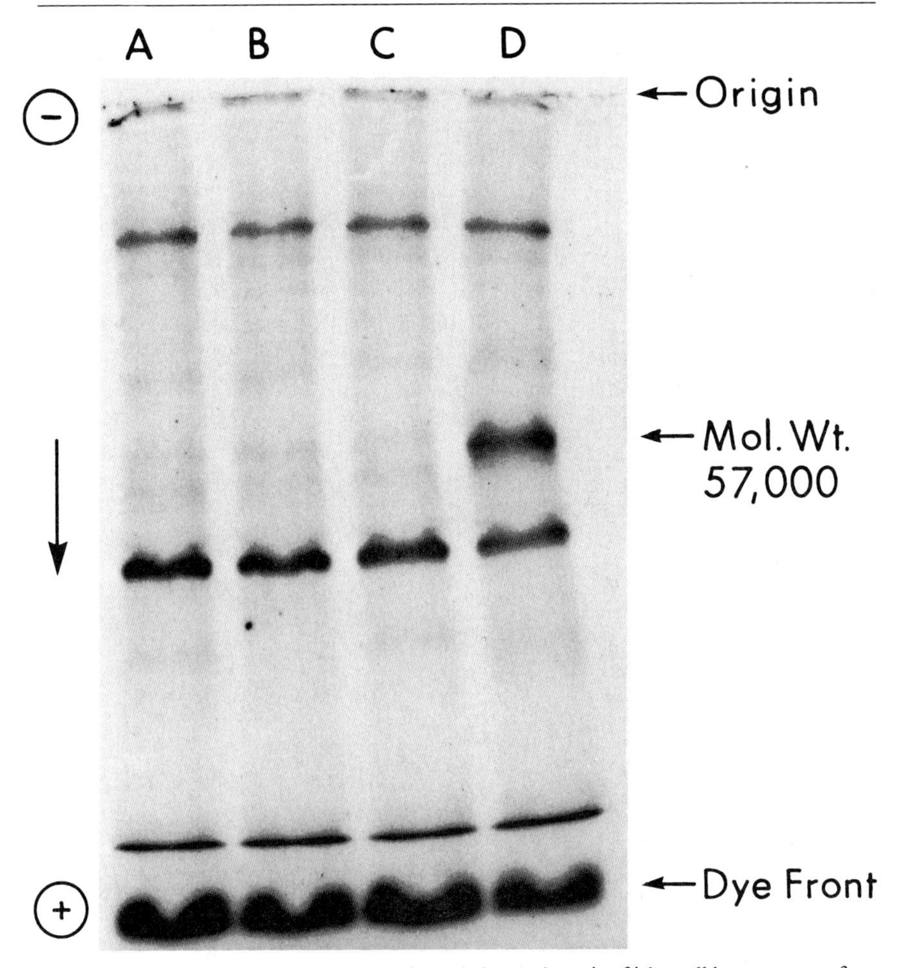

FIG. 7. Autoradiogram of polyacrylamide gel electrophoresis of islet-cell homogenate after assay of protein kinase activity using [γ-$^{32}P_i$]ATP. Lane A, (−) Ca^{2+}, (−) calmodulin; lane B, (+) Ca^{2+} (200 μM),(−) calmodulin; lane C, (−) Ca^{2+}, (+) calmodulin (7 μg/ml); lane D, (+) Ca^{2+} (200 μM) and (+) calmodulin (7 μg/ml). Arrow shows 57,000 M_r protein band. From Landt *et al.*[26]

dissection microscope. Fifty islets are randomly chosen by each of two technicians and are delivered to randomized 10 × 75 mm borosilicate culture tubes at 4°. The samples containing 100 islets in 200 μl of KRB are preincubated for 5 min at 37°. This medium is removed, and the experimental period is initiated by the addition of 50 μl of KRB containing either 100 or 500 mg of D-glucose per deciliter (37°). The incubation is terminated by the addition of 25 μl of stop solution. In cases where insulin secretion is also determined, the incubation medium is 250 μl. The incubation is terminated by removal of the medium and addition of 1:3 dilution of stop solution directly to the islets. Insulin is determined by the method of Wright *et al.*[13] After addition of the stop solution, the sample are placed in a boiling water bath for 2 min, and 5-μl aliquots are removed for determination of total $^{32}P_i$ incorporated. One hundred islets contain 0.2–0.4% of the total counts added during the loading period. A 50-μl aliquot is subjected to SDS–gel electrophoresis and autoradiography as described above.

Comments. The quantitative pattern of protein phosphorylation obtained from experiments performed with intact islets varies somewhat from the results obtained under protein kinase conditions. Electrophoresis of phosphorylated proteins obtained under both conditions allows comparison of specific protein bands separated on the same slab gels. Thus it is possible to determine whether protein bands that are sensitive to Ca^{2+} *in vitro* are also affected by conditions that alter insulin secretion and Ca^{2+} metabolism *in vivo.*[27]

Section IV

Internalization of Plasma Membrane Components during Adsorptive Endocytosis

A. Receptor-Mediated Uptake
Articles 18 through 26

B. Coated Vesicles
Articles 27 through 32

[17] Use of Electron-Opaque Tracers for Studies on Endocytosis and Membrane Recycling[1]

By VOLKER HERZOG and MARILYN GIST FARQUHAR

Electron-opaque tracers have proved to be valuable tools for determining the pathways of endocytosis. Indeed much of what we know about endocytic traffic stems from the use of such probes. In addition to their use as general tracers, electron-dense markers may be covalently coupled to various proteins (ligands, antibodies) and used to follow the fate of specific molecules. Largely through the use of such markers (either as general tracers or specific tags), it has become clear that the pathways of endocytosis vary among different cell types (see overview, this volume [1]). Our ability to detect specific pathways in a given cell depends on the nature of the tracer and on its interaction with components of the cell surface. Tracers may bind to sites on the cell membrane, enter the cell by adsorptive endocytosis, i.e., bound to the endocytic vesicle membrane, and accordingly act as *membrane markers,* or they may fail to bind to the membrane but nevertheless be taken up as part of the fluid contents of the vesicle (*fluid-phase* or *content markers*).[1a] We have also learned that the critical use of tracers for studies on membrane recycling requires the careful selection of markers that least perturb cell functions.

It is the purpose of this overview to point out some principles in the use of tracers for studies on endocytosis, giving examples in which membrane retrieval has been traced successfully.

Tracers for Studies on Endocytosis and Membrane Traffic

There are two major classes of tracers—so-called *mass tracers* best exemplified by hemoproteins, such as horseradish peroxidase (Table I), and *particulate tracers* best exemplified by native ferritin and colloidal gold (Table II). At the electron microscope level, mass tracers appear to be nonparticulate and are visible either by virtue of their inherent electron density (e.g., hemoglobin[2] or lysozyme[3]) or become visible after a cytochemical reaction usually based on their peroxidatic activity (in the case of hemoproteins). Particulate tracers, as their name implies, appear as discrete

[1] Supported by the Deutsche Forschungsgemeinschaft (to V.H.) and USPHS Grant AM 17780 (to M.G.F.).

[1a] S. C. Silverstein, R. M. Steinman, and Z. A. Cohn, *Annu. Rev. Biochem.* **46,** 669 (1977).

[2] M. G. Farquhar and G. E. Palade, *J. Cell Biol.* **17,** 375 (1963).

[3] J. P. Caulfield and M. G. Farquhar, *Proc. Natl. Acad. Sci. U.S.A.* **73,** 1646 (1976).

ISBN 0-12-181998-1

TABLE I
HEMOPROTEIN TRACERS[a]

Tracers	Molecular weight	Molecular radius (nm)	Isoelectric point	Optimal pH for visualization with DAB as hydrogen donor	Commercial source or preparation procedure
Microperoxidases					
Heme-octapeptide[b]	1,550	1.0	5.4	12[c]	
Heme-nonapeptide[d]	1,630	1.0	4.95	12.5[c]	Sigma Chemical Co., St. Louis, Missouri
Heme-undecapeptide[e]	1,880	1.0	4.85	13.5[c]	
Cytochrome c[f]	12,800	1.5	10.65	2.5[c]	Sigma; Boehringer, Mannheim (Germany)
Myoglobin[g]	16,900	1.8	7.0	5.0[c]	
Horseradish peroxidase (HRP)	44,000	3.0	Several isoenzymes ranging from 4.8 to 9.1	4.3[c] 7.6	Sigma type II
Acid isoenzyme	—	—	4.3 to 5.8	4.5[c]	Types VII and VIII
Basic isoenzyme	—	—	8.7 to 9.2	3.5[c]	Type X
Chemically modified HRP					
Succinylated	—	—	<4	4.3[c]	Modified according to Rennke *et al.*[44]
Cationized	—	—	8.4 to 9.2	4.3	

Hemoglobin[h]	65,000	3.4	6.8	7	Sigma; Boehringer Mannheim (Germany)
Lactoperoxidase[i]	82,000	3.6	8.0	7	Sigma
Myeloperoxidase[j]	160,000	4.4	10.0	7.5	Prepared after Odajima and Yamazaki[k]
Catalase[l]	240,000	5.2	5.7	10.5[c]	Sigma

[a] Data compiled from M. G. Farquhar [*Kidney Int.* **8,** 197 (1975)], Simionescu,[30] and R. Tiggemann, H. Plattner, I. Rasched, P. Bäuerle, and E. Wachter *J. Histochem. Cytochem.* **29,** 1387 (1981).

[b] J. P. Kraehenbuhl, R. E. Galardy, and J. D. Jamieson, *J. Exp. Med.* **139,** 208 (1974).

[c] Optimal pH was determined with the colorimetric DAB assay.[11] For cytochemical visualization, other (neutral) pH values may give satisfactory results. However, it is advantageous to use the pH values as indicated (e.g., pH 10.5 for catalase[14] or pH 4.3 for HRP[12]) because endogenous peroxidases are usually much less active at acidic or alkaline pH.

[d] H. Plattner, E. Wachter, and P. Gröbner, *Histochemistry* **53,** 223 (1977).

[e] N. Feder, *J. Cell Biol.* **51,** 339 (1971).

[f] M. J. Karnovsky and D. F. Rice, *J. Histochem. Cytochem.* **17,** 751 (1969).

[g] W. A. Anderson, *J. Histochem. Cytochem.* **20,** 672 (1972).

[h] J. E. L. Ericson, *Nephron* **5,** 7 (1968).

[i] R. C. Graham and R. W. Kellermeyer, *J. Histochem. Cytochem.* **16,** 275 (1968).

[j] R. C. Graham and M. J. Karnovsky, *J. Exp. Med.* **124,** 1123 (1966).

[k] T. Odajima and I. Yamazaki, *Biochim. Biophys. Acta* **284,** 360 (1972).

[l] M. A. Venkatachalam and H. D. Fahimi, *J. Cell Biol.* **42,** 480 (1969).

TABLE II
PARTICULATE ELECTRON-DENSE TRACERS[a]

Tracer	Molecular weight	Particle size	Charge	Commercial source or preparation procedure
Dextrans (D)				Dextran preparations of varying size (up to $M_r = 2,000,000$) and charge; commercially available from Pharmacia Fine Chemicals, Uppsala, Sweden
Neutral D	10,000	—	Uncharged	Visualized by the fixation procedure of Simionescu *et al.*[4]
	44,000	—	Uncharged	
	70,000	—	Uncharged	
DEAE-D	500,000	—	Positive	
D Sulfate	500,000	—	Negative	
Glycogen				
Shellfish	—	20 nm	Uncharged	Commercially available from Schwarz/Mann. Visualized by the procedure of Simionescu *et al.*[4]
Rabbit liver	—	30 nm	Uncharged	
Hemin-dextran	Variable according to the size of dextran used		—	Prepared after Aronson *et al.*[19] and visualized by the DAB technique of Graham and Karnovsky[10]

Iron-dextrans				Available from Fisons Ltd., Pharmaceutical Division, Loughborough, Leics., England
Imferon	73,000	11 × 17 nm (iron core: 3 × 3 nm)[b]	Neutral	Particles are visible by virtue of the electron density of their iron core. Does not require special staining procedures. Best contrast in tissue is achieved with light lead staining of sections.
Imposil	200,000	21 × 12 nm (iron core: 11 × 3 nm)	Neutral	
Gleptoferron	2,000,000	43 × 15 nm (iron core: 31 × 17 nm)	Neutral	
Ferritins				Cadmium-free preparations are available from Miles Laboratories, Elkhart, Indiana
Native / Cationized	480,000	11 nm (iron core: 5.5 nm)	Negative (p*I* 4.7); Positive (p*I* > 8.5)	Visible by the inherent electron density of their iron core. Staining of the apoferritin with bismuth enhances the size and the electron density of the particles.
Colloidal gold	Variable	5–20 nm	Negative	Preparation after Horisberger and Rosset[50]
Colloidal silver	Variable	20–130 nm	Negative	Available as Neosilvol from Park-Davis and Co., Detroit, Michigan
Colloidal thorium	Variable	3–10 nm	—	Available as Thorotrast from Fellows Testagar, Detroit, Michigan, or prepared after Schlüter[c]
Colloidal carbon	Variable	250 × 300 nm	Neutral	Available from Günter Wagner, Hannover, Germany as India ink No. C 11/143A. Contains phenol, which must be removed by dialysis.

(continued)

TABLE II (*continued*)

Tracer	Molecular weight	Particle size	Charge	Commercial source or preparation procedure
Polyvinylpyrrolidone	36,000 to 160,000	—	Neutral	Available from Matheson, Coleman, and Bell, Los Angeles, California. Visualized by the reduced osmium technique[d]
Latex particles, polystyrene beads (PSB)		Variable, useful sizes between 0.05 and 1.0 μm		Available from Polysciences, Warrington, Pennsylvania. Latex particles are used for endocytosis either uncoated or coated with ligands (proteins) for specific (adsorptive) endocytosis. Because they are solubilized by propylene oxide, dehydration in ethanol and embedding in Spurr's low-viscosity medium[e] is recommended.
Native PSB			Neutral	
Derivatized PSB				
Carboxy,			Negative	
Hydroxy,			Negative	
Amino, functionality			Positive	

[a] Data compiled from M. G. Farquhar [*Kidney Int.* **8,** 197 (1975)] and Herzog and Miller.[18]
[b] P. R. Marshall and D. Rutherford *J. Colloid Interface Sci.* **37,** 390 (1971).
[c] G. Schlüter, *Microsc. Acta* **7666,** 428 (1975).
[d] S. K. Ainsworth, *J. Histochem. Cytochem.* **25,** 1254 (1977).
[e] A. R. Spurr, *J. Ultrastruct. Res.* **26,** 31 (1969).

particles and are recognizable either by their inherent electron density (ferritin, iron dextran, and colloidal gold, silver, or carbon) or after special staining procedures (dextrans[4]). Either type of tracer can be used as a direct tag for other molecules (e.g., ferritin- or peroxidase-labeled lectins or antibodies). We concentrate here on some special problems in the use of tracers as general membrane and content markers.

It should be emphasized at the outset that the same tracers may function as a fluid-phase marker under some conditions and as a membrane marker under other conditions. Hence, it is necessary to consider the possible interactions between a potential marker and the cell surface.

Fluid-Phase (Content) Markers

The most frequently used fluid-phase markers are horseradish peroxidase (Table I), native (anionic) ferritin (Table II), and neutral dextrans (Table II). Since they are internalized in proportion to their concentration in the intercellular space or surrounding medium, they are used at comparatively high concentrations (1 – 50 mg/ml).

Hemoproteins

Horseradish peroxidase (HRP) has been used to quantitate fluid-phase pinocytosis and has been successfully detected cytochemically and colorimetrically in mouse peritoneal macrophages[5] and fibroblasts[6] when applied at concentrations ranging from 0.1 to 1.0 mg/ml. At this high concentration, HRP is internalized predominantly by fluid phase uptake in most cells. In some cell types, however, e.g., acinar cells of the rat exocrine pancreas (Fig. 1)[7] and rat alveolar macrophages,[8] HRP binds to the plasma membrane and is also taken up by adsorptive endocytosis. In the latter case this binding is specifically inhibited by 10^{-3} *M* mannose or 10^{-3} *M* mannan and is assumed to be due to binding of HRP (a polymannose glycoprotein) to the mannose receptor.[8] In fact, HRP has been used to detect the mannose receptor.[8,9] Therefore, the use of HRP as a fluid-phase marker is valid only at high concentrations and does not apply to all cell types.

Horseradish peroxidase is by far the most useful hemoprotein tracer because of its high enzyme activity. A number of other hemoproteins that

[4] N. Simionescu, M. Simionescu, and G. E. Palade, *J. Cell Biol.* **53,** 365 (1972).
[5] R. M. Steinman and Z. A. Cohn, *J. Cell Biol.* **55,** 186 (1972).
[6] R. M. Steinman and Z. A. Cohn, *J. Cell Biol.* **63,** 949 (1974).
[7] V. Herzog and H. Reggio, *Eur. J. Cell Biol.* **21,** 141 (1980).
[8] W. S. Sly and P. Stahl, *in* "Transport of Macromolecules in Cellular Systems" (S. C. Silverstein, ed.), p. 229 (Dahlem-Konferenzen). Berlin, 1978.
[9] W. Straus, *Histochemistry* **73,** 39 (1981).

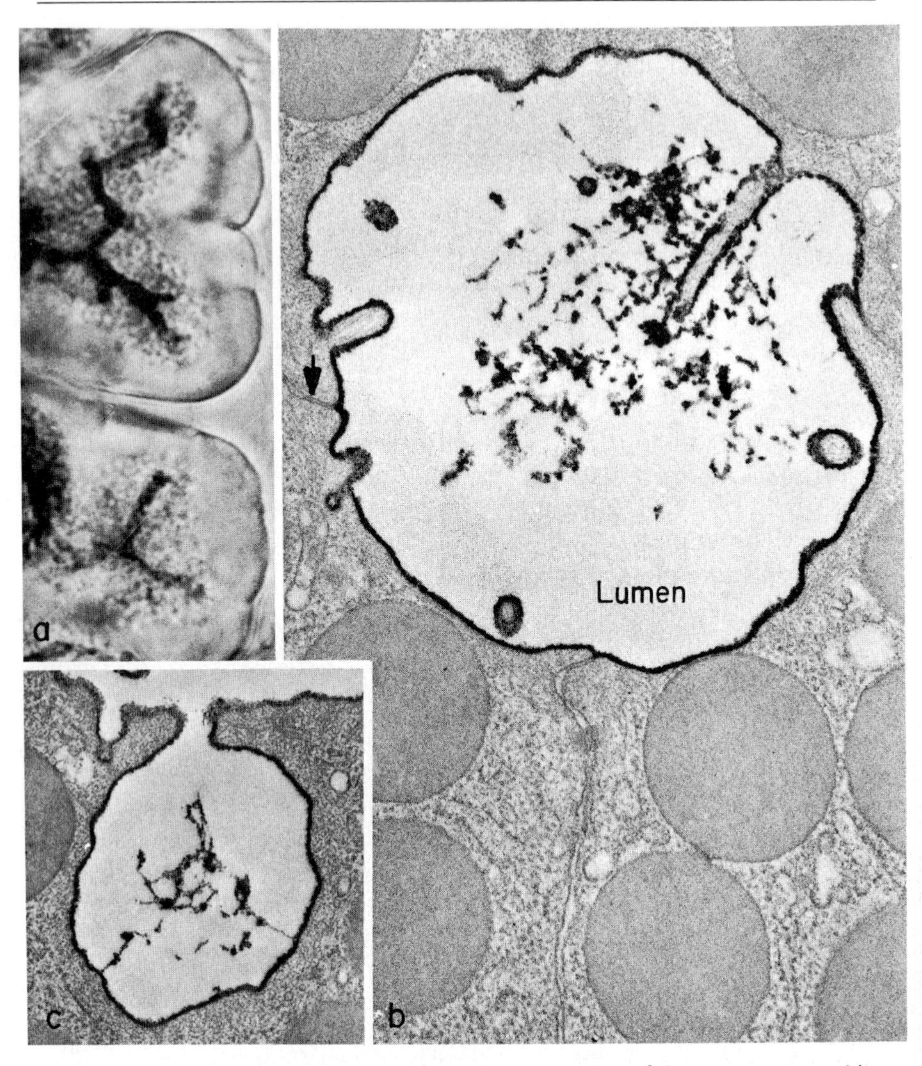

FIG. 1. Horseradish peroxidase was infused through the duct of the rat pancreas, and its distribution (seen here in the ductal lumina) was monitored under a stereomicroscope (a). By electron microscopy the tracer can be seen adhering to the apical plasma membrane facing the lumen (b). Note that tight junctions remain intact (arrow), and that the tracer also binds to the membrane of the exocytotic lacunae (c).

differ from HRP in their molecular weight and in their ability to interact ionically with the plasma membrane (because of differences in their isoelectric point or charge density) are also available (Table I). They differ also in their sensitivity to glutaraldehyde and in their pH optimum (Table I).

Usually HRP and other hemoproteins are visualized cytochemically using 3,3′-diaminobenzidine (DAB) as the hydrogen donor.[10] Owing to the toxicity of DAB (classified as a carcinogen), other donors have been introduced to replace DAB.[11] However, the DAB procedure is still the most widely used for electron microscopy because of its high sensitivity. In addition, it has the advantage that a colorimetric assay is available for determining the optimal conditions of fixation and incubation for the cytochemical visualization of HRP and other hemoproteins.[12]

Fixation of Tissue. Tissues are fixed for periods of 30 min to 2 hr in glutaraldehyde (1–2% in 0.1 *M* cacodylate buffer) or in a combined formaldehyde–glutaraldehyde fixative, buffered at pH 7.4, for periods of 30 min to $1\frac{1}{2}$ hr. It is highly advantageous that fixation with glutaraldehyde has little effect on the activity of HRP and most other hemoproteins, but the activity of various endogenous peroxidases is markedly inhibited by glutaraldehyde.[13] This allows one to detect the exogenous tracer and to discriminate it from endogenous peroxidases. Catalase constitutes an unusual exception: its peroxidatic activity is greatly enhanced upon fixation with glutaraldehyde while its catalytic activity is inhibited.[14]

Cytochemical Reaction for Peroxidatic Activity. After fixation, the tracer is detected by incubation of tissue in a DAB medium (originally introduced by Graham and Karnovsky[10]) for the localization of HRP. This procedure results in the deposition of an electron-dense reaction product (oxidized and polymerized DAB) at the sites of peroxidatic activity. Optimal pHs for the various hemoproteins (listed in Table I) have been determined using a colorimetric DAB assay.[12] Increasing concentrations of DAB give rise to denser deposits of oxidized and polymerized DAB reaction product. Maximum concentration is 3 mg of DAB per milliliter, but 2 mg/ml also gives satisfactory results. To achieve this concentration, the DAB is first dissolved in H_2O before the buffer (0.1 *M* Tris-HCl) is added and the pH is adjusted.

DAB medium for the visualization of peroxidases: DAB (40 mg) is dissolved in 5 ml of H_2O (under the hood to avoid the fumes that develop upon addition of H_2O). 10 ml of 0.2 *M* Tris-HCl buffer are added. Then the pH is adjusted to the pH optimum of the particular peroxidase (see Table I), and H_2O is added to give a final volume of 20 ml (2 mg of DAB/ml).

The optimal concentrations of H_2O_2 vary among peroxidases, catalase requiring the highest (0.15% H_2O_2), lacrimal gland peroxidase the lowest (0.003% H_2O_2), and HRP an intermediate (0.01%) concentration.[10,13]

[10] R. C. Graham and M. J. Karnovsky, *J. Histochem. Cytochem.* **14,** 291 (1966).
[11] J. S. Hanker, P. E. Yates, C. B. Metz, and A. Rustoni, *Histochem. J.* **9,** 789 (1977).
[12] V. Herzog and H. D. Fahimi, *Anal. Biochem.* **55,** 554 (1973).
[13] V. Herzog and H. D. Fahimi, *Histochemistry* **46,** 273 (1976).
[14] V. Herzog and H. D. Fahimi, *J. Cell Biol.* **60,** 303 (1974).
[15] M. J. Karnovsky, *Annu. Meet. Am. Soc. Cell Biol.* Abstr. 284 (1971).

Postincubation Processing. After the cytochemical reaction, the tissue is processed for electron microscopy by routine procedures (postfixation in OsO_4, dehydration in graded alcohols, and embedding in Epon). Often OsO_4 reduced by addition of 5 – 15 mg of ferrocyanide[15] per milliliter is used to increase the electron density of the oxidized and polymerized DAB reaction product and to enhance membrane staining.

Dextrans

Neutral Dextrans. These are uncharged polymers of D-glucopyranose that are available commercially as preparations of distinct size (i.e., molecular weight range) (see Table II). Charged derivatives are polyanionic dextran-sulfate and the polycationic diethylaminoethyl dextran (DEAE-dextran). DEAE-dextran binds strongly to cell membranes (Fig. 2b), whereas uncharged dextrans (Fig. 2a) and dextran sulfate exhibit no or only weak interaction with the plasmalemma and have been used as fluid-phase markers for studies on capillary[4] and glomerular[16] permeability and on membrane retrieval in secretory cells.[7,17,18] Dextrans are prepared as 5 – 10% (w/v) solutions that are heated to disperse aggregates and subsequently filtered through a Millipore filter (pore size = 0.2 μm) to remove aggregates.[17] A special fixation and staining procedure is required for their ultrastructural visualization, as well as for glycogen, which has been also used for studies on capillary permeability.[4]

Special Fixative for the Visualization of Dextrans and Glycogen[4]:

Solution A, 6 ml: 1 g of paraformaldehyde in 8.3 ml of H_2O is heated to 80 – 90° and solubilized by the addition of 3 or 4 drops of 1 N NaOH. After cooling, 15.6 ml of 8% glutaraldehyde (EM Sciences, Worthington, Pennsylvania) and 5 ml of 0.15 M arsenate buffer, pH 7.4, are added.

Solution B, 4 ml: 2% OsO_4 in 0.075 M arsenate buffer, pH 7.4

Solution C, 2 ml: saturated solution of lead citrate in 0.075 M arsenate buffer, pH 7.4

All solutions must be kept at 4° (to retard OsO_4 reduction). Freshly prepared fixative is colorless, and it must be replaced when it darkens. Tissue or cells are usually fixed for 2 hr at 4°.

During fixation, dextran molecules form aggregates of variable size (Fig. 2a). Sometimes fine lead precipitates occur in the tissue. This is a disturbing artifact that can be avoided by the use of clean glassware, "aged" (several

[16] J. P. Caulfield and M. G. Farquhar, *J. Cell Biol.* **63,** 883 (1974).

[17] V. Herzog and M. G. Farquhar, *Proc. Natl. Acad. Sci. U.S.A.* **74,** 5073 (1977).

[18] V. Herzog and F. Miller, *In* "Secretory Mechanisms" (C. R. Hopkins, and C. J. Duncan, eds.), p. 101. Cambridge Univ. Press, London and New York, 1979.

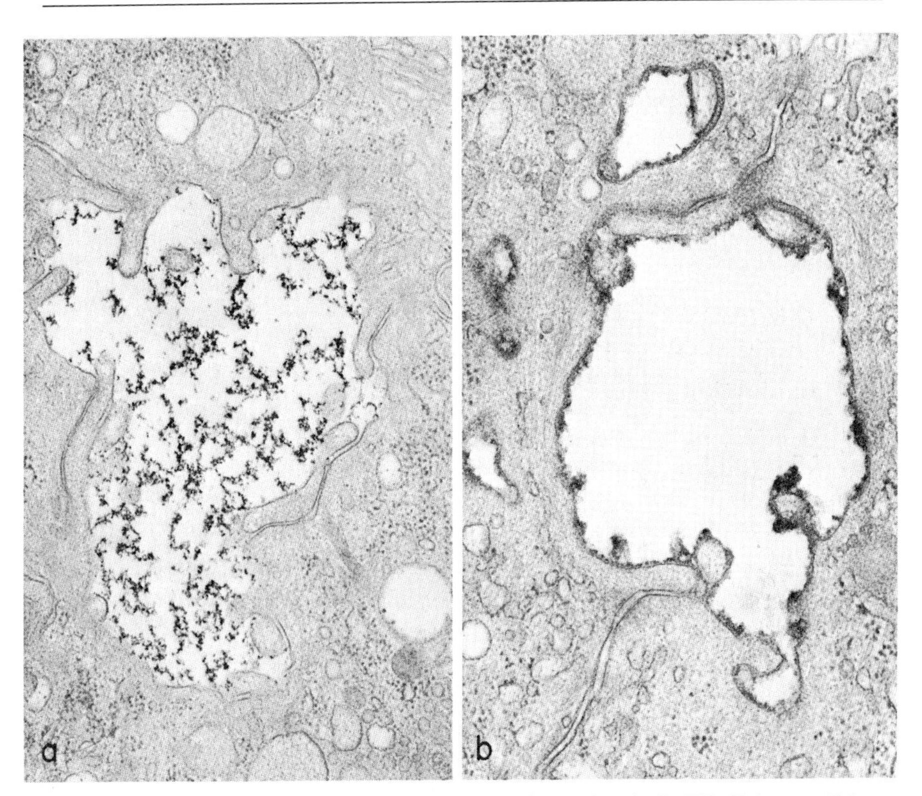

FIG. 2. Neutral (uncharged) dextran ($M_r \sim 44,000$) and cationic DEAE-dextran ($M_r \sim 500,000$) were infused *in vivo* into the duct of the rat parotid. Uncharged dextran aggregates fill the acinar lumen (a), whereas cationic DEAE-dextran adheres to the apical cell surface (b). Note that the tracers do not penetrate through the tight junctions.

weeks old) lead solutions, and fixative mixtures prepared immediately prior to use.

Complexed Dextrans. Hemin-dextran[19] (which becomes detectable after the DAB reaction) or iron-dextrans (Table II) (synthetic complexes of iron oxide coated with dextran) may be used instead of higher molecular weight dextrans ($M_r > 70,000$). Iron-dextran molecules, which are directly visible by virtue of their electron-dense iron core, do not require special fixation or staining for their visualization. They have a characteristic rodlike structure (Fig. 3) and have been used for double-labeling experiments[20] because they can be clearly distinguished from the more spherical ferritin molecules. Iron-dextran has been used for labeling of antibodies[20] and for studies on

[19] J. F. Aronson, G. G. Pietra, and A. P. Fishman, *J. Histochem. Cytochem.* **21,** 1047 (1973).
[20] A. H. Dutton, K. T. Tokuyasu, and S. J. Singer, *Proc. Natl. Acad. Sci. U.S.A.* **76,** 3392 (1979).

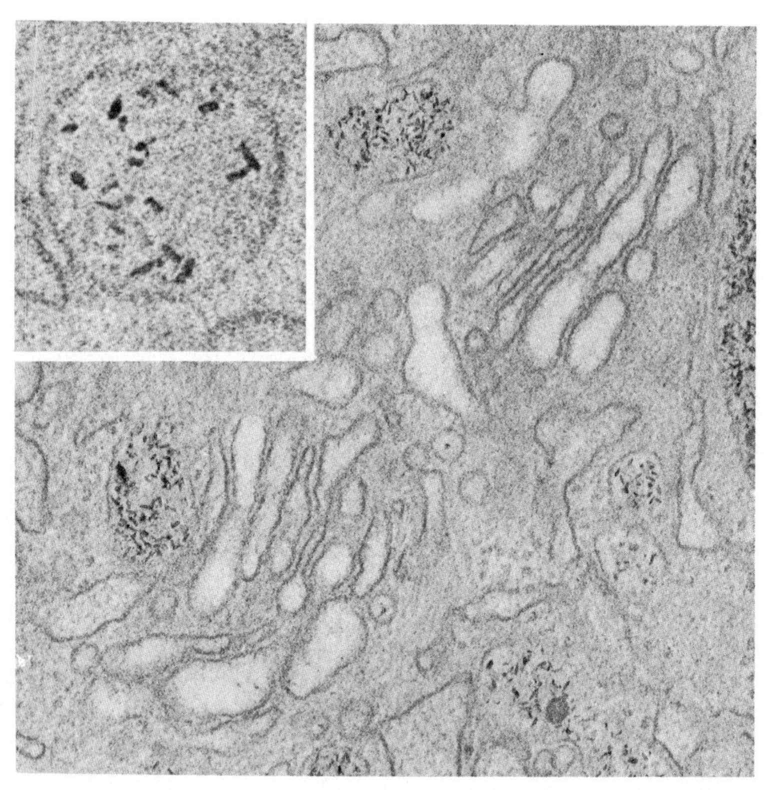

FIG. 3. Golgi area from a thyroid follicle cell after exposure to iron-dextran (Imposil). The tracer particles (which are visible by virtue of the inherent electron density of their iron core) can be distinguished by their rodlike appearance (inset).

fluid-phase uptake in thyroid follicle cells.[21,22] Commercially available iron-dextrans contain 0.5% phenol as a preservative, which must be removed by dialysis before use.

Ferritin

Native ferritin, which is anionic (p*I* 4.7) (Table II) and, accordingly, does not adhere electrostatically to cell surfaces, is the oldest fluid-phase marker.[23–25] It is visualized by virtue of its ~ 5.5 nm iron core surrounded by an apoferritin shell (total diameter ~ 11 nm) that is not usually visible

[21] V. Herzog and F. Miller, *Eur. J. Cell Biol.* **24,** 74 (1981).
[22] V. Herzog, *Philos. Trans. R. Soc. London Ser. B* **296,** 67 (1981).
[23] M. G. Farquhar, S. L. Wissig, and G. E. Palade, *J. Exp. Med* **113,** 47 (1961).
[24] M. G. Farquhar and G. E. Palade, *J. Exp. Med.* **114,** 699 (1961).
[25] H. Ryser, J. B. Caulfield, and J. C. Aub, *J. Cell Biol.* **14,** 255 (1962).

but can be seen after staining of sections with bismuth.[26] Such staining considerably facilitates the detectability of the molecule by enhancing its opacity (metallic bismuth binds to and stains the apoferritin shell and thereby increases the effective diameter of the visible molecule). It is normally used at relatively high concentrations (up to 1 mg/ml).

Pitfalls and Side Effects in the Use of Fluid-Phase Markers

As already mentioned, relatively high concentrations of fluid-phase markers must usually be applied in order to detect the internalized tracer. At such high concentrations, impurities and contaminants may be raised to toxic levels. For example, when impure preparations of native ferritin are used, trace amounts of cadmium present may be toxic.[24] At present, ferritin preparations with a very low cadmium content are commercially available (Miles Laboratories, Inc., Elkhart, Indiana).

Horseradish peroxidase (or contaminants present therein) may also have toxic effects on cells. Modifications in the distribution of synaptic vesicles have been observed after application of HRP to frog neuromuscular junctions.[27] Moreover, HRP may induce vascular side effects probably mediated by the release of histamine, causing leakage of tracer along intercellular junctions; induction of leakage from endothelial junctions along postcapillary venules has been reported for HRP.[28,29] Similar effects have been reported for myoglobin[30] and for dextrans.[31] Histamine release varies among different species and strains; dextran injections produced marked elevation of serum histamine in Lewis rats, but not in those of the Wistar–Furth strain.[31] Therefore, strains such as Wistar–Furth rats, known to be genetically resistant to histamine-releasing agents, should be used. Dextran may also induce swelling of organelles containing the tracer,[17] presumably as an osmotic side effect. Such an effect was observed[7,17] in Golgi cisternae of exocrine glandular cells when dextran was used as a tracer for endocytosis (Fig. 4).

Membrane Markers

Since membrane markers bind to the plasma membrane and are internalized by adsorptive endocytosis, they are effectively concentrated during uptake. Accordingly, they can be used at much lower concentrations (1–50 μg/ml) than fluid-phase markers. Two classes of membrane markers

[26] S. K. Ainsworth and M. J. Karnovsky, *J. Histochem. Cytochem.* **25,** 1254 (1977).
[27] D. T. Rutherford and J. F. Gennaro, *J. Cell Biol.* **83,** 438a (1979).
[28] R. S. Cotran and M. J. Karnovsky, *Proc. Soc. Exp. Biol. Med.* **126,** 557 (1967).
[29] T. Vegge and R. Haye, *Histochemistry* **53,** 217 (1977).
[30] N. Simionescu, *J. Histochem. Cytochem.* **27,** 1120 (1979).
[31] G. B. West, *Int. Arch. Allergy Appl. Immunol.* **47,** 296 (1974).

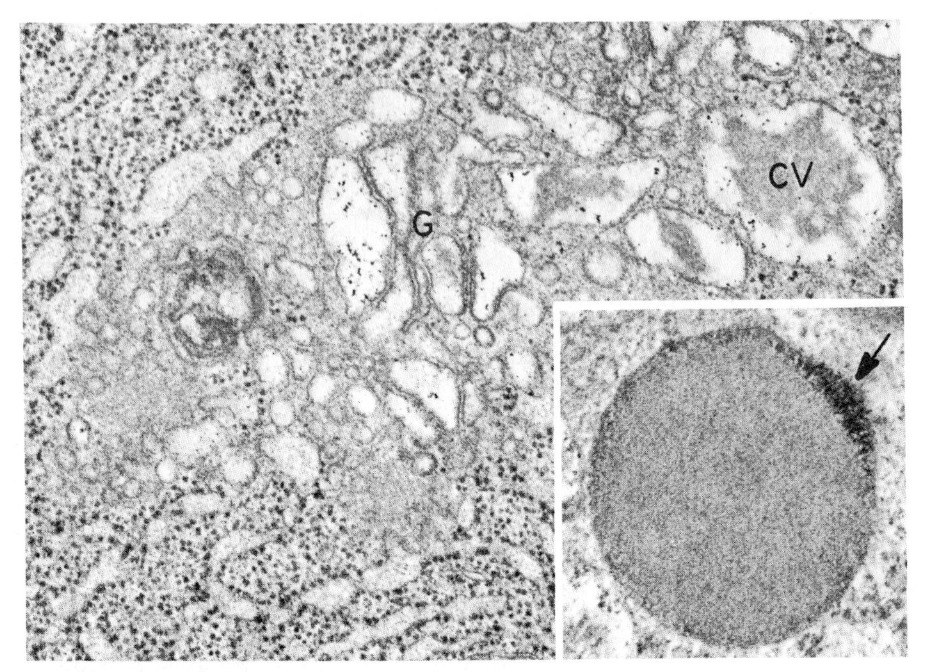

FIG. 4. Golgi complex from a rat exocrine pancreatic cell fixed 2 min after infusion of dextran (M_r 44,000) through the pancreatic duct. Dextran particles are present in Golgi cisternae (G) and condensing vacuoles (CV), which appear distended. At 15 min, dextran is also seen concentrated around mature secretion granules (arrow, inset).

can be distinguished: general or *nonspecific markers* and *specific markers.* The nonspecific markers, of which the most commonly used is cationic ferritin, bind to the cell surface by ionic interaction and are visible by their inherent electron density. Specific membrane markers can be made by selecting a ligand that binds to a specific receptor on the cell surface and labeling it with any one of a number of electron-dense tracers (HRP, ferritin, gold).

Nonspecific Membrane Markers

The general markers used so far are all cationic probes that bind (by electrostatic interaction) to cell membranes. The most frequently used nonspecific membrane marker is cationized ferritin[32–37] (CF) (Figs. 5 and 6),

[32] M. G. Farquhar, *J. Cell Biol.* **78,** R35 (1978).
[33] V. Herzog and F. Miller, *Eur. J. Cell Biol.* **19,** 203 (1979).
[34] P. Ottosen, P. Courtoy, and M. G. Farquhar, *J. Exp. Med.* **152,** 1 (1980).
[35] M. G. Farquhar, *Methods Cell Biol.* **23,** 43 (1981).
[36] M. G. Farquhar, *Membr. Recycling Ciba Found. Symp.* **92,** 157 (1982).
[37] N. Simionescu, M. Simionescu, and G. E. Palade, *J. Cell Biol.* **90,** 605 (1981).

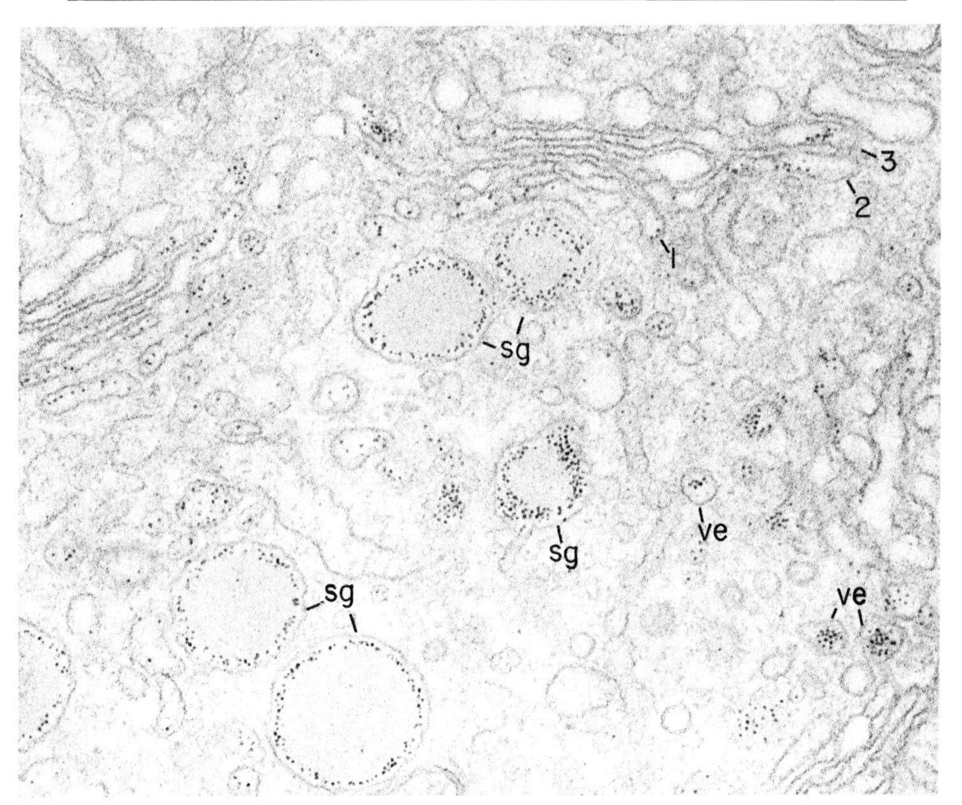

FIG. 5. Golgi region of an anterior pituitary cell incubated for 60 min with cationized ferritin (50 μg/ml). Cationized ferritin molecules are seen in several stacked Golgi cisternae (1–3), in numerous small vesicles (ve), and around all of the secretory granules (sg) present.

which was introduced by Danon and co-workers[38,39] to study the distribution of sialic acid and other negatively charged groups on the cell surface; CF with a high isoelectric point (pI > 8.5) is available commercially (Table II). The affinity of CF binding to anionic sites on the cell surface can be modulated by modifying the charge densities of either the tracer or of the cell-surface components. Cationized ferritins of varying isoelectric points (pI 5–>9) can be made using modifications[40,41] of the original cationization procedure. Additional information can be obtained on the nature of the anionic sites to which CF binds by pretreatment of the cells with appropriate

[38] D. Danon, L. Goldstein, Y. Marikowsky, and E. Skutelsky, *J. Ultrastruct. Res.* **38,** 500 (1972).

[39] E. Skutelsky and D. Danon, *J. Cell Biol.* **71,** 232 (1976).

[40] H. G. Rennke, R. S. Cotran, and M. A. Venkatachalam, *J. Cell Biol.* **67,** 638 (1975).

[41] Y. S. Kanwar and M. G. Farquhar, *J. Cell Biol.* **81,** 137 (1979).

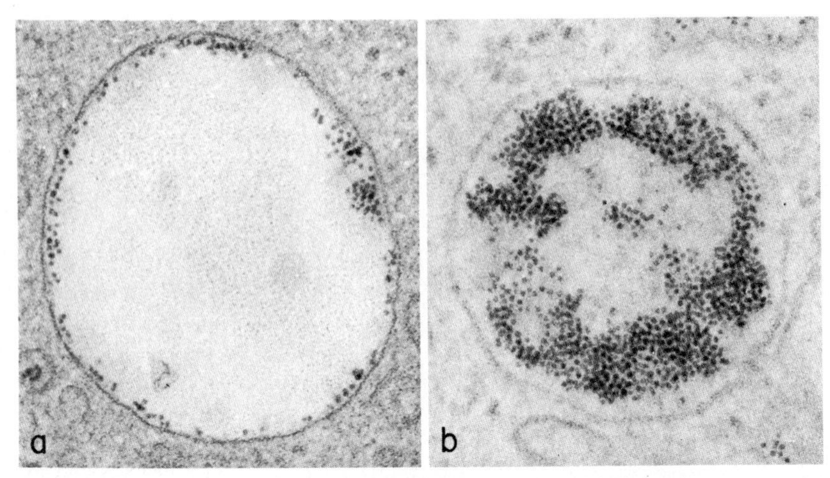

FIG. 6. After endocytosis of cationized ferritin (CF) by thyroid follicle cells, tracer particles appear first in endosomes (a), where they are bound to the internal surface of the membrane. When these structures become converted into lysosomes (i.e., when acid phosphatase activity becomes demonstrable), CF particles dissociate from the membrane and accumulate in the lysosomal matrix (b).

enzymes. For example, neuraminidase treatment results in reduced binding of CF to endothelial and other cells,[38,39,42] thereby indicating the presence of sialic acid along the corresponding cell surfaces. Such an approach was also used to determine the nature of proteoglycan-containing anionic sites in the glomerular basement membrane[43] and on the endothelium.[42]

Other cationic, nonspecific membrane markers that have been successfully used in different cell types are DEAE-dextran,[18] the cationic isozyme of HRP (Table I), and cationized HRP.[44] DEAE-dextran was reported to produce profound alterations of the rough ER in myeloma cells,[45] but such effects were not observed in thyroid follicular cells.[7]

Special precautions are required to achieve satisfactory binding of membrane markers to the cell surface. In particular, cationic proteins may form aggregates with anionic components (serum proteins) in the medium, so binding is usually done in PBS or simple, serum-free media (e.g., MEM). Furthermore, some cells release anionic secretory proteins (e.g., parotid secretory proteins, thyroglobulin) that will form aggregates with cationic tracers and prevent their binding to the cell surface. Thus, the cells have to be washed extensively with protein-free culture medium shortly before

[42] M. Simionescu, N. Simionescu, J. E. Silbert, and G. E. Palade, *J. Cell Biol.* **90,** 614 (1981).

[43] Y. S. Kanwar and M. G. Farquhar, *Proc. Natl. Acad. U.S.A.* **76,** 1303 (1979).

[44] H. G. Rennke and M. A. Venkatachalam, *J. Histochem. Cytochem.* **27,** 1352 (1979).

[45] A. Tartakoff, P. Vassalli, and R. Montesano, *Eur. J. Cell Biol.* **26,** 188 (1981).

exposure to CF; CF should be added to the culture medium at low concentrations.

Specific Membrane Markers

Molecules that bind specifically to receptors include lectins, toxins, peptide hormones, and antibodies directed against membrane constituents. Such molecules are not visible unless bound to an appropriate electron-dense detector molecule. Successful covalent labeling of ligands has been obtained with ferritin,[46,47] HRP,[48,49] or iron-dextran.[20] Noncovalent conjugates of ligands with gold particles[50–52] have also been prepared. Both types of conjugates have been very useful for the localization of various receptors on cell surfaces and for studies on the fate of various ligands upon internalization. However, it has to be kept in mind that the conjugation of a ligand with a tracer of considerable size may alter its behavior. Appropriate tests should be done to establish the specificity of the binding.[47,53] The mildest labeling procedures are probably those that involve radiolabeling (iodination, reductive methylation) of the ligand, which can then be detected by autoradiography. The latter approach also has the advantage that the kinetic data can be obtained on the uptake process. However, the spatial resolution achieved in autoradiograms (900 – 1500 Å) is much poorer than that achieved with the electron-dense tracers mentioned.

Pitfalls and Limitations in the Use of Membrane Markers

The major pitfall in the use of any membrane marker is the strong interaction with the cell surface and the possibility of introducing structural and functional perturbations, particularly when markers are used at high concentrations. In the case of polycationic tracers, it has been reported that they may induce shedding,[39] cause disturbances in the architecture of epithelial cells,[54,55] induce capping in lymphocytes,[56] and stimulate endocy-

[46] S. J. Singer, *Nature* (*London*) **183,** 1523 (1959).
[47] J. A. McKanna, H. T. Haigler, and S. Cohen, *Proc. Natl. Acad. Sci. U.S.A.* **76,** 5689 (1979).
[48] N. K. Gonatas, S. U. Kim, A. Stieber, and S. Avrameas, *J. Cell Biol.* **73,** 1 (1977).
[49] M. E. Schwab, K. Suda, and H. Thoenen, *J. Cell Biol.* **82,** 798 (1979).
[50] M. Horisberger and J. Rosset, *J. Histochem. Cytochem.* **25,** 295 (1977).
[51] D. A. Handley, C. M. Arbeeny, L. D. Witte, and S. Chien, *Proc. Natl. Acad. Sci. U.S.A.* **78,** 368 (1981).
[52] I. H. Pastan and M. C. Willingham, *Science* (*Washington, D. C*) **214,** 504 (1981).
[53] D. A. Wall, G. Wilson, and A. L. Hubbard, *Cell* **21,** 79 (1980).
[54] P. M. Quinton and C. W. Philpott, *J. Cell Biol.* **56,** 787 (1973).
[55] M. W. Seiler, M. A. Venkatachalam, and R. S. Cotran, *Science* (*Washington D. C.*) **189,** 390 (1975).
[56] B. Larsen, *Mol. Cell Biochem.* **15,** 117 (1977).

tosis.[32,57,58] Therefore, it is important to use as low a concentration of such tracers as possible. Cationized ferritin was found *not* to affect secretion in myeloma cells[34] or thyroid cells[21] when used at low concentrations (50 – 100 μg/ml).

Tracers that initially bind to the cell surface membrane may detach upon internalization and therefore lose their usefulness as a membrane marker (Table III). Dissociation of ligands from the membrane has been observed in secretion granules of anterior pituitary cells (Fig. 5) that are known to have a low pH and to contain acidic macromolecules (sulfated glycoproteins and proteoglycans), which can be assumed to compete with sites on the membrane for binding to internalized CF.[32] Detachment of CF from the membrane was also observed in thyroid follicle cells when endosomes are transformed into lysosomes[33] (Fig. 6). In the latter case, detachment of CF from the membrane could be due either to the low pH environment encountered in the lysosome or endosome[59] or to intralysosomal digestion of the protein ligand (apoferritin).

It should be kept in mind that in all likelihood no presently available method for labeling—including radioiodination—can be considered to be entirely innocuous. All such procedures lead to some modification in the charge or conformation of molecules exposed on the cell surface. Procedures such as radioiodination and periodate oxidation followed by borohydride reduction are also perturbing to living cells. Our experience[60] as well as that of others indicates that these radiolabeling methods can cause blebbing and reduced viability to cultured cells. Nonperturbing membrane markers of high affinity and resolution are sorely needed but are not yet available. However, until they are, useful information can be obtained with available procedures as long as the conditions of labeling are carefully controlled and their effects on cells are properly monitored.

Tests for Evaluation of Side Effects

Membrane markers or tracers for fluid-phase uptake may affect the specific cell function under investigation. Therefore, it is important to assess the level of pinocytosis in the presence and in the absence of tracers (e.g., as was done in macrophages[1,5]). In secretory cells, endocytosis is coupled to exocytosis. It is therefore essential to study possible side effects of the tracer on synthesis, intracellular transport, and exocytosis of secretory proteins. In myeloma cells[34] and in acinar cells of rat parotid or lacrimal

[57] H. J. P. Ryser, *Nature* (*London*) **215,** 934 (1967).

[58] R. Duncan, M. K. Pratten, and J. B. Lloyd, *Biochim. Biophys. Acta* **587,** 463 (1979).

[59] B. Tycko and F. R. Maxfield, *Cell* **28,** 643 (1982).

[60] P. Wilson, D. Sharkey, N. Haynes, P. Courtoy, and M. G. Farquhar, *J. Cell Biol.* **91,** 417a (1981).

TABLE III
DETACHMENT OF LIGANDS FROM MEMBRANE BINDING SITES

Tracer	Cell type	Compartment	Potential cause of detachment	Authors
Cationized ferritin	Anterior pituitary cells	Secretion granules	Anionic contents (glycoproteins and proteoglycans) that compete with original membrane binding sites	Farquhar[32]
Cationized ferritin	Thyroid follicle cells	Lysosomes	Low pH and digestion by lysosomal enzymes	Herzog and Miller[33]
	Choroid epithelium			Van Deurs *et al.*[a]
Ferritin-labeled EGF	Human carcinoma cells	Endosomes or lysosomes	Low pH or lysosomal digestion	McKanna *et al.*[47]

[a] B. van Deurs, F. von Bülow, and M. Moller, *J. Cell Biol.* **89,** 131 (1981).

gland,[17] no changes in the rate of protein synthesis or in the levels of secretion were detected when cells were exposed to CF or dextran, respectively.

Selective Application of Tracers to Distinct Domains of the Cell Surface

In polarized epithelial cells, appropriate experimental steps are required to expose the different macrodomains (apical or basolateral regions of the plasmalemma) to electron dense tracers. In addition, most cell surfaces contain small regions or microdomains of different composition that can also be labeled selectively.

Detection of Macrodomains in Vivo. Under appropriate conditions, the apical or the basolateral surfaces of epithelial cells can be selectively labeled. The apical plasma membrane of exocrine gland cells can be reached *in situ* by infusion of the tracer through the duct against the flow of secretion.[7,17] The infusion pressure usually ranges between 15 and 25 mm Hg, which is sufficient to overcome the intraductal pressure and low enough to avoid the rupture of tight junctions. The pH of the tracer solution must be adjusted to the pH of the secretion fluid to avoid the precipitation of secretory proteins. Such precipitates may obstruct the duct and cause the rupture of junctional complexes and leakage into the intercellular spaces.

The following is an example of conditions whereby the successful infusion of tracer into the rat exocrine pancreas[7] can be achieved. The duodenal portion of the exocrine pancreas is surgically exposed, and a 5-cm polyeythylene catheter (inner diameter = 0.28 mm) is inserted into the duct to a depth of ~ 1 cm. When filled with secretion fluid, the catheter is connected to a 30-gauge needle attached to a syringe containing the tracer solution: dextran, $M_r \sim 44{,}000$ (200 mg/ml) and dextran blue as a visible cotracer (10 mg/ml) in 150 mM Tris-HCl, pH 8.3. The pressure (15–25 mm) is precisely controlled by a manometer, and the appearance of the tracer fluid (recognized by the blue stain of the cotracer in the acinar lumina of the tail area of the pancreas) is monitored under a stereomicroscope, usually at intervals ranging from 2 to 60 min, and fixed and processed for visualization of dextran.[4] Other tracers, e.g., HRP (Fig. la), may be similarly infused. Acinar lumina from the rat parotid gland infused with dextran and DEAE-dextran are shown in Fig. 2.

The basolateral plasma membrane of epithelial cells *in situ* is usually reached by vascular infusion or interstitial injection of the tracer.

Detection of Macrodomains in Vitro. When epithelial cell lines (e.g., Madin–Darby canine kidney, MDCK) or epithelial cells derived from enzyme-dissociated tissues are cultivated *in vitro,* the apical plasma membrane of epithelial cells is easily reached if cells are kept in monolayer culture, because the apical plasma membrane is directed toward the medium and is separated from the basolateral cell surface by tight junctions. However, labeling of the basolateral domain is more difficult because the cells must be detached from the dish by procedures (physical lifting or enzymic digestion), which may perturb the cells.

Detection of Microdomains. Microdomains are small, structurally distinct areas on the plasma membrane that carry specific binding sites and therefore can be visualized electron microscopically after decoration with appropriate ligands.[37,42] The best-characterized microdomains on the cell surface are clathrin-coated pits. The composition of their membrane differs from that of the remainder of the plasmalemma in that they contain a higher concentration of cell surface receptors for many bound ligands [low-density lipoproteins (LDL), α_2-macroglobulin, peptide hormones[61-63]] and lack (or selectively exclude) other cell surface receptors.[64] In several systems they have been shown to contain higher concentrations of anionic sites (as detected with cationic ferritin[34,35]) than the remainder of the cell surface.

[61] J. L. Goldstein, G. W. Anderson, and M. S. Brown, *Nature* (*London*) **179,** 679 (1979).
[62] M. C. Willingham, F. R. Maxfield, and I. H. Pastan, *J. Cell Biol.* **82,** 614 (1979).
[63] J. Roth, M. Ravazzola, M. Bendayan, and L. Orci, *Endocrinology* **108,** 247 (1981).
[64] M. S. Bretscher, J. N. Thomson, and B. M. F. Pearse, *Proc. Natl. Acad. Sci. U.S.A.* **77,** 4156 (1980).

Besides coated pits, other microdomains (fenestrae, transport vesicles) of differing composition have been detected in fenestrated capillaries using labeled lectins and CF coupled with specitic enzymic digestion.[42]

In macrophages the available evidence (based on radioiodination[65]) indicates that pinocytic vesicles involved in fluid-phase uptake do not represent specialized domains differing from that of the remainder of the plasma membrane since the plasmalemma and the pinocytic vesicle membranes have the same distribution of labeled species.

Results on Membrane Recycling Using Electron-Dense Tracers in Secretory Cells

Evidence for membrane recovery and reutilization in secretory cells was obtained from rat lacrimal and parotid glands[17] and the exocrine pancreas[7] using dextran as a fluid-phase marker. After infusion of dextran up the parotid or pancreatic duct *in vivo,* within 2–5 min dextran was seen to be taken up by endocytosis and carried from the apical plasma membrane to Golgi cisternae and condensing vacuoles as well as to lysosomes (Fig. 4), indicating that the incoming vesicles had fused with all these compartments. Later (after 15 min), the tracer was seen in condensed form at the periphery of secretion granules (inset, Fig. 4) indicating that internalized tracer was copackaged together with newly synthesized secretory proteins. Using cationized ferritin as a nonspecific membrane marker, internalized membrane was traced to stacked Golgi cisternae and to secretion granules or vacuoles in anterior pituitary cells[32] (Fig. 5) and myeloma cells,[34] respectively. From these findings, it is assumed that some of the retrieved plasma membrane represents granule membrane that is recycled—i.e., it is recovered and reutilized in packaging of secretory products.[7,17,35,36]

In some secretory cells—e.g., in thyroid follicle cells, where the secretory product is processed in lysosomes—the main pathway traced for retrieved surface membrane is to lysosomes. After labeling of the apical plasma membrane with CF, tracer particles are found within 5 min in lysosomes and within 30 min in Golgi cisternae.[21,33] Similarly, if CF-coated latex beads (inset, Fig. 7) are internalized, the CF particles dissociate from the latex beads after entering the lysosome. CF particles remain closely bound to the lysosomal membrane, and 30 min later are also found in Golgi cisternae (Fig. 7). The latex beads remain within the lysosomes. The observations suggest that part of the lysosomal membrane may be recovered and be carried on to Golgi cisternae.

[65] I. S. Mellman, R. M. Steinman, J. C. Unkeless, and Z. A. Cohn, *J. Cell Biol.* **86,** 712 (1980).

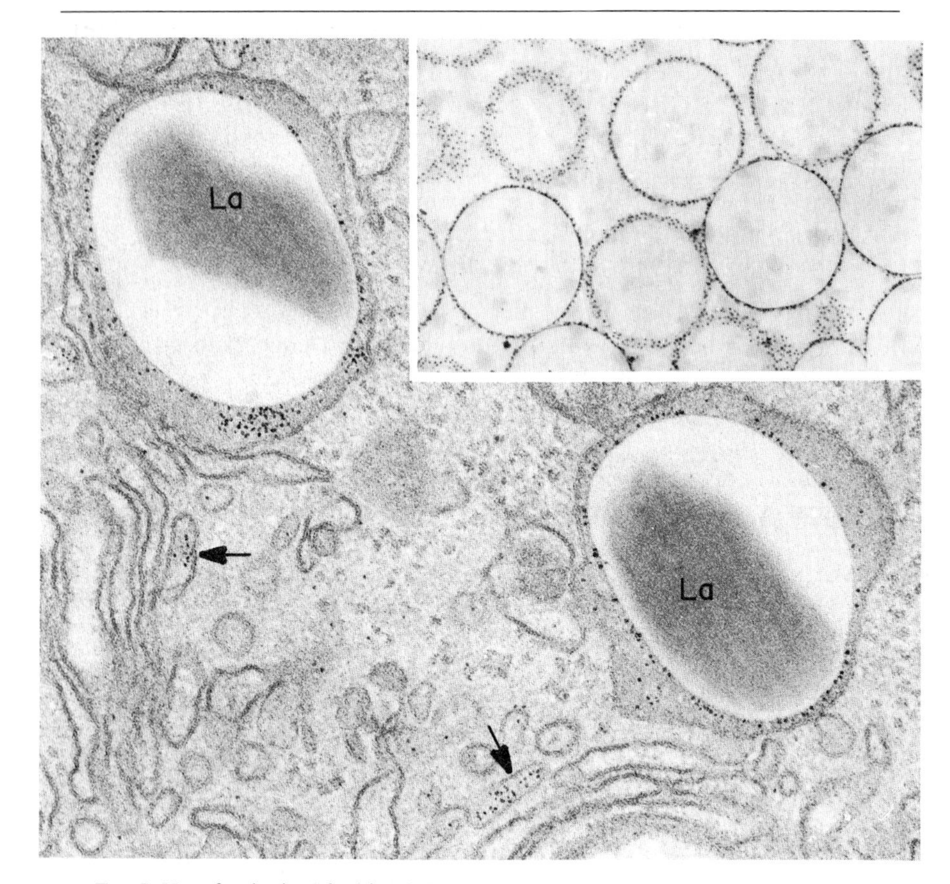

FIG. 7. Use of cationized ferritin (CF)-coated latex particles (inset) as a tracer for endocytosis in thyroid epithelial cells. The particles are taken up and reach lysosomes within 5 min, where the CF dissociates from the latex surface; by 30 min, free CF particles reach the Golgi cisternae. Here two latex particles (La), which are partially extracted (during the dehydration and embedding procedures for electron microscopy), are seen within two large lysosomes, and free CF particles are seen in several of the stacked Golgi cisternae (arrows).

Concluding Comments

It should be mentioned that tracers detect primarily the predominant traffic in a given cell type; e.g., plasmalemma to lysosomal traffic in macrophages and fibroblasts in culture[5,6] and plasmalemma to Golgi traffic in secretory cells.[17,32-36] However, variations in traffic can be seen with the same tracer in different systems. For example, in some secretory cells, e.g., anterior pituitary cells[66] and β cell of the exocrine pancreas,[67] HRP was

[66] G. Pelletier, *J. Ultrastruct. Res.* **43,** 445 (1973).

[67] L. Orci, A Perrelet, and P. Gorden, *Recent Prog. Horm. Res.* **34,** 95 (1978).

demonstrated to reach the Golgi cisternae. However, in other secretory cells, e.g., exocrine pancreatic cells[7] and myeloma cells,[34] HRP was carried exclusively to lysosomes. The different results obtained may depend not only on the different cell types studied, but also on the HRP preparations used; these, depending on the mixture of isozymes, would vary in their ability to interact with the cell and/or pinocytic vesicle membranes.

In several systems, double-labeling experiments have been done and it has been shown that membrane markers and content markers have different destinations. In myeloma cells a membrane marker (CF) was traced to both lysosomes and Golgi cisternae, whereas a content marker (HRP) was traced only to lysosomes.[34] Similarly, in the intestinal epithelial cells of the suckling rat, a content marker (HRP) and a membrane marker (ferritin-conjugated IgG) were taken up in the same vesicles, but the content marker later appeared only in lysosomes and the membrane marker (IgG) was transported across the cell and released at the basolateral cell surface.[68] From these experiments it was concluded that sorting of vesicle membrane and contents must occur intracellularly.

These examples serve to illustrate that, in order to obtain maximum information on membrane traffic, more than one tracer should be used to detect the pathways of endocytosis in a given cell type.

[68] D. R. Abrahamson, and R. Rodewald, *J. Cell Biol.* **91,** 270 (1981).

[18] Use of Isolated, Perfused Liver in Studies of Receptor-Mediated Endocytosis

By WILLIAM A. DUNN, DORIS, A. WALL, and ANN L. HUBBARD

In recent years the process of receptor-mediated endocytosis has become the focus of studies by us and many other workers. Our emphasis has been on the asialoglycoprotein (ASGP) recognition system in rat liver, which mediates the clearance of glycoproteins having galactose- or *N*-acetylgalactosamine-terminating carbohydrate side chains. We have studied the dynamics of the ligand pathway *in vivo,*[1,2] in isolated hepatocytes,[3] and in the

[1] A. L. Hubbard and H. Stukenbrok, *J. Cell Biol.* **83,** 65 (1979).
[2] D. A. Wall and A. L. Hubbard, *Cell* **21,** 79 (1980).
[3] P. L. Zeitlin and A. L. Hubbard, *J. Cell Biol.* **92,** 634 (1982).

METHODS IN ENZYMOLOGY, VOL. 98

ISBN 0-12-181998-1

isolated perfused liver.[4-6] The latter system has been used in a variety of forms by many investigators for the study of a number of liver functions.[7-10]

The isolated perfused liver offers numerous advantages over the intact animal: (*a*) reagents such as high-specific-activity radioisotopes, enzyme-impermeant chemical labels, or metabolic inhibitors can be introduced into the liver circulation without dilution by total blood volume and without flow to other organs as occur in the intact animal; (*b*) these reagents can be recirculated, then removed from the perfusion medium under defined conditions (to establish a "pulse") and replaced with fresh medium lacking the reagent (the "chase"); (*c*) the temperature of the perfusion medium and, thus, that of the liver can be varied over a wide range (4–37°) and the effect on various parameters easily assessed; (*d*) temperature can be rapidly switched from one level to another, allowing the use of a temperature-shift protocol to synchronize ligand entry; and yet (*e*) existing subcellular fractionation procedures for intact liver can still be used. The perfusion system that we use is a simple one,[8] but it is fully adequate for maintaining liver viability up to 4 hr at 37°, or for much longer times (> 10 hr at 16°) at lower temperatures. All the equipment is commercially available and can be set up in a small space.

In the next section we describe the components and assembly of the perfusion system in detail. In the three following main sections we present a step-by-step procedure for isolating a rat liver; procedures for performing temperature-shift experiments; and an outline of the method we have used to label selectively the contents of endocytic compartments, using the asialoglycoprotein system as a model. Finally, in the last section we evaluate selected functions of the perfused liver that relate to endocytosis.

The Isolated Liver Perfusion System

General Description

The liver perfusion system as diagrammed in Fig. 1 comprises three primary components: (*a*) a temperature-regulation system; (*b*) a pump; and (*c*) gas-permeable fibers mounted in a beaker assembly (biofiber reservoir). The temperature of the perfusate is maintained by pumping the medium

[4] D. A. Wall and A. L. Hubbard, *J. Cell Biol.* **90,** 687 (1981).
[5] W. A. Dunn, J. H. LaBadie, and N. N. Aronson, Jr., *J. Biol. Chem.* **254,** 4191 (1979).
[6] W. A. Dunn, A. L. Hubbard, and N. N. Aronson, Jr., *J. Biol. Chem.* **255,** 5971 (1980).
[7] N. N. Aronson, Jr., *Biochem. J.* **203,** 141 (1982).
[8] L. G. deGaldeano, R. Bressler, and K. Brendel, *J. Biol. Chem.* **248,** 2514 (1973).
[9] W. A. Dunn and S. Englard, *J. Biol. Chem.* **256,** 12437 (1981).
[10] G. E. Mortimore and W. F. Ward, *Front. Biol.* **45,** 157 (1976) ("Lysosomes in Biology and Pathology," Vol. 5).

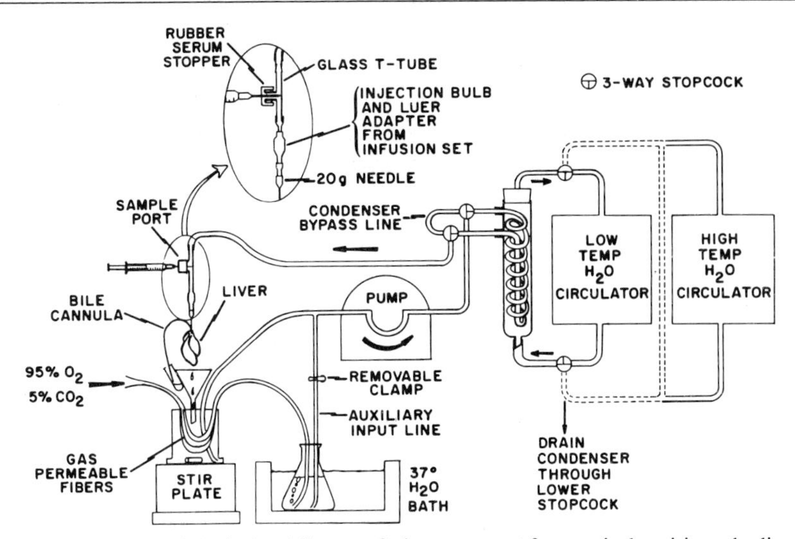

FIG. 1. Diagram of the isolated liver perfusion system. After surgical excision, the liver is suspended over the collecting funnel of reservoir containing perfusate. The medium in the reservoir is oxygenated via gas-permeable fibers, and the effluent gas is bubbled through prewarmed Krebs–Hensleit medium kept in a 37° water bath. The medium that drains from the liver is mixed with the contents of the reservoir with a small stirring bar. Under normal operation, the perfusate is pumped out of the reservoir at 40 ml/min, enters a condenser, and then is carried back to the liver. A sample port through which aliquots of the perfusion medium can be taken (see enlarged inset) is located just above the liver. During temperature shifts, the condenser is bypassed via a short piece of tubing attached by three-way stopcocks, and perfusion medium at the elevated temperature enters the system via the auxiliary input line. The temperature of the condenser is maintained at either a high or a low level by attachment to either warming or cooling circulating water baths. Three-way stopcocks in the tubing from the water circulators allow rapid switching from one temperature to the other and allow the condenser to be drained before switching via tubing attached to the lower stopcock.

through a microcondenser, which, in turn, is connected to one of two recirculating water baths. Stopcocks make it possible to use either the low- or high-temperature circulator and, when necessary (i.e., in the temperature-shift experiments described below), to divert the perfusing medium around the microcondenser. To avoid mixing the solutions in the low- and high-temperature circulating baths (50% ethanol and water, respectively), the micrcondenser is drained from the bottom stopcock prior to switching the temperature circulators. The liver temperature (as measured by the temperature of the perfusate leaving the liver) can be maintained between 4 and 37° with ~ 1° variation. The perfusate is circulated through the liver at 40 ml/min, a high flow rate that compensates for the lack of hemoglobin, red blood cells, or oxygen carriers in the medium and allows adequate oxygenation of the liver.[8,11] The gas-permeable fibers are connected to a tank of 95% O_2–5% CO_2 that both oxygenates and buffers the bicarbonate-based perfusate medium surrounding the fibers. Constant stirring of the medium within the biofiber reservoir ensures rapid equilibration of the perfusate with oxygen and carbon dioxide and with any additions (i.e., ligands, drugs) introduced into the system through the funnel atop the reservoir. When nonvolatile buffers are used, 100% oxygen is passed through the fibers.

The equipment necessary to assemble the perfusion system in Fig. 1 is listed below along with its source. The perfusion system is assembled as seen in Fig. 1. By placing as few connections as possible between the pump and the reservoir, the chance of drawing air into the system through a leak is minimized. It becomes a good habit to check *all* tubing connections for leaks before each experiment. Tubing lengths are not critical, although shorter lengths favor more efficient cooling during temperature shifts. The volume of the tubing in our system is 20–25 ml and that of the reservoir is 100–150 ml.

Periodically the perfusion tubing is cleaned of bacteria with detergent (Axion) and Clorox bleach. The reservoir is cleaned with detergent after each experiment and stored in detergent solution or in 1% formaldehyde.

Equipment

- Pump assembly
  - Masterflex pump (No. 7553-10) (A. Daigger & Co., Chicago, Illinois)
  - Pump head (No. 7016-30)
  - Tubing (thick-walled silicon; i.d. 0.125 in; o.d. 0.251 in.)
- Biofiber reservoir
  - Fleaker HF concentration system (Spectrum Medical Industries, Queens, New York)
  - Hollow fiber bundle: Special order—shorten to 12-cm fiber length so that stir bar is not in contact
  - Collecting funnel, 65-mm short stem (American Scientific Products)
- Recirculating water baths
  - 0–20° (low): Refrigerated circulator bath RTE-5 (NESLAB Instruments, Silver Spring, Maryland)
  - 20–37° (high): VWR circulator bath (No. 1130) (VWR)
- Water bath: Tekbath (American Scientific Products)
- Cannula setup: Saftiset i.v. set (Cutter Laboratories, Inc. Berkeley, California)
- Sample port
  - Glass T-tube, for tubing; i.d. $\frac{3}{16}$ in. (American Scientific Products)
  - Rubber serum stopper; size 14 (American Scientific Products)

[11] D. L. Schmucker and J. C. Curtis, *Lab Invest.* **30,** 201 (1974).

Tubing
  For recirculating water bath (rubber or Tygon, i.d. 3/8 in.) (Fisher Scientific)
  For auxiliary input line (polyethylene): PE-190, i.d. 0.047 in.
3-Way stopcocks
  For condenser bypass: 3-way Nalgene Pharmaseal; i.d. 1/8 in. (American Scientific Products)
  For water bath lines: Nalgene polypropylene; i.d. 4 mm (American Scientific Products)
Other accessories
  Magnestir (VWR)
  Microcondenser, (No. 457000); 175-mm length, 24/40 adaptor (Kontes Glass Co., Vineland, New Jersey)
  O_2–CO_2 gas cylinder at a ratio of 95% O_2, 5% CO_2

Excision and Perfusion of a Rat Liver

The Perfusion Medium[12]

Component	Concentration	
	g/liter	m*M*
NaCl	6.9	120
KCl	0.36	4.8
KH_2PO_4	0.13	1.0
$MgSO_4$ (anhydrous)	0.14	1.2
$CaCl_2$ (anhydrous)	0.24	2.2
Glucose	0.9	5.0
$NaHCO_3$	2.0	23.8
(or HEPES)	(4.74)	(20.0)
PVP-40	20.0	—
(or bovine serum albumin)	(20–30)	—
Penicillin	0.06	—
Phenol red	0.01	—

Prior to use, the perfusing medium is filtered through a Millipore type GS (0.22 μm) filter. Buffers such as PIPES and sodium acetate have been substituted for $NaHCO_3$ or HEPES to obtain pH values other than 7.4.

Initial Preparations

The recirculating water baths are precooled or prewarmed to the appropriate temperatures. Water baths maintained at 39°, 12°, and −2° will maintain the liver at approximately 37°, 16°, and 4°, respectively.

[12] R. Dawson, *in* "Data for Biochemical Research" (R. Dawson, D. Elliot, W. Elliot, and K. Jones, eds.), p. 507. Oxford University Press, New York, 1969.

Before the actual surgery is started, the perfusion system and surgical platform are set up as described below:

1. *All* the perfusion lines and the reservoir are filled with perfusate, with care taken that there are no trapped air bubbles that may interrupt hepatic circulation.

2. Both the perfusate and the extra medium (~ 100 ml) needed to replenish that lost during liver removal are preoxygenated and prewarmed to 37°.

3. Four lengths of suture (4-0 silk) and PE-10 tubing (for bile cannulation) are precut to 6 and 4–5 in., respectively.

4. The 20-gauge needle (1 in.), which will be used to cannulate the hepatic portal vein, is etched and dulled with a file to ensure a firm suture tie and to minimize the possibility of puncturing the vein during the operation.

The Surgery

Cannulation of the bile duct and the hepatic portal vein and subsequent removal of the liver are done as described in detail below.

1. Male rats are anesthetized with an intraperitoneal injection of sodium phenobarbitol (8–10 mg/100 g body weight).

2. The internal organs of the abdominal cavity are exposed by cutting through the body wall and retracting the skin and muscle layers with hemostats. Care is taken not to cut into the thoracic cavity.

3. To allow easy access to the bile duct and the hepatic portal vein, the intestines are wrapped in saline-moistened gauze and displaced outside the cavity. In some cases it is necessary to gently pull aside the left lateral lobe of the liver and hold it in place with moistened gauze.

4. Four separate sutures of 4-0 silk are then tied in a loose knot around (*a*) the inferior vena cava just above the kidney on the right side; (*b*) the bile duct 1–2 cm below the liver; (*c*) the hepatic portal vein 4–6 cm before the vessel enters the liver; and (*d*) the hepatic portal vein 1–2 cm from the liver. The placement of suture *d*, which eventually will hold in place a 20-gauge cannula, is critical and must be located close enough to the liver that branch vessels (i.e., gastric vein) that enter the portal vein are not between the suture tie and the liver.

5. A part of the bile duct near where the cannula will be inserted is cleaned of pancreatic tissue and cannulated by first positioning the duct with forceps and carefully "nicking" the bile duct with curved iris scissors. A 4- to 5-in. length of PE-10 tubing is then inserted and tied in place with suture *b* (see above), with care taken not to interrupt bile flow by overtightening the knot.

6. The hepatic portal vein is cannulated by inserting the "dulled"

20-gauge needle (not yet connected to the perfusion system) into the vein somewhere between sutures *c* and *d*. Either forceps or suture *c* is used to lift and steady the vein during insertion of the needle. Suture *d* is immediately tied about the cannula. After the needle is completely filled with blood, the perfusion system is connected to the inserted cannula and the perfusing medium is pumped into the liver at a rate of ~30 ml/min. The time between interruption of *in situ* hepatic blood circulation due to cannulation and the onset of perfusing the liver should be no longer than 30 sec. A longer delay can result in anoxia and subsequent cell death. Also, it is important to cut the vena cava *soon* after connecting the cannula to the pump, or back pressure will damage the liver.

7. To prevent excess blood from entering the cavity during surgical removal of the cannulated liver, sutures *a* and *c* can be tied around the inferior vena cava and portal vein, respectively.

8. The liver is carefully removed by cutting away the connective tissue starting at the diaphragm and rolling the cannulated liver out to the left until it is completely free. Care must be taken when freeing the lobes of the liver that are closely associated with the stomach and when cutting around the bile and liver cannulas. The freed cannulated liver is suspended by its cannula above a collecting funnel (see Fig. 1), and the recirculating perfusate flow is adjusted to 40 ml/min. The perfusate volume is adjusted to 100–125 ml, and the bile produced by the liver is collected in a small test tube.

9. Prior to any additions or temperature shifts, the cannulated liver is allowed to equilibrate at 37° for at least 10 min.

Temperature-Shift Protocol

In the following descriptions of temperature-shift protocols, the liver is maintained either in the normal recirculating perfusion mode or in a single-pass mode. The latter is used for changes of solution and for rinsing and fixation steps. In the recirculating mode (see Fig. 1) the liver drains into a funnel attached to the biofiber reservoir. To switch over to the single-pass mode, the entire stir plate and reservoir are moved enough so that an empty beaker can be placed under the liver. Alternatively, the liver can be held by a clamp attached to a horizontal crossbar held by a ring stand. In this case, by moving the crossbar in an arc, one can readily swing the liver itself to a position that allows it to drain into the empty beaker.

Procedure for Switching from High to Low Temperature, or Vice Versa

1. Approximately 200 ml of fresh perfusate are held ready at the temperature to be maintained after the shift. If the temperature of this medium is

greater than ~10°, the medium is preoxygenated to avoid transient anoxia of the liver during the switchover. One of the recirculating water baths is precooled (or prewarmed) to a suitable temperature.

2. The reservoir is drained of the old medium by switching to the single-pass perfusion mode until the level of the medium in the reservoir is near the bottom. Care should be taken not to introduce air bubbles into the perfusion line at this point, since bubbles could block flow through the liver sinusoids.

3. While the used medium is draining, the H_2O recirculator in use is shut off and the condenser is drained through an auxiliary line attached to the bottom three-way stopcock (see Fig. 1). The second recirculator is attached to the condenser, and the flow is started.

4. When the level of the old medium in the reservoir is low, 20–30 ml of the new perfusate medium are added through the top funnel of the apparatus and allowed to drain; then the reservoir is filled with fresh medium.

5. The liver is switched back to the recirculating perfusion mode by moving it so that it will drain into the funnel of the reservoir.

If the timing of the temperature shift is not critical, it may not be necessary to drain the condenser before attaching the new H_2O recirculator, since the system will equilibrate almost as rapidly to the new temperature when this step is omitted.

Selective Labeling of Endocytic Compartments

The isolated perfused liver system allows an investigator to synchronize ligand entry into cells during receptor-mediated endocytosis, since the ligand-binding step can be separated from the internalization step by temperature manipulation. Our protocol involves binding of ligand to the cell surface at 2–4°, a temperature at which no endocytosis occurs, in order to obtain the desired degree of receptor occupancy. Excess ligand is then removed, and the liver is warmed to a higher temperature for varying times to allow ligand to enter the membrane compartments of the endocytic pathway. After this step, the liver is cooled back to 2–4° and can either be fixed for microscopic examination or used for biochemical studies after it is perfused through the desired homogenization medium. To avoid the complication of plasma membrane labeling, any remaining surface-bound ligand can be removed (in the ASGP case, by EGTA addition) prior to homogenization or processing for morphology.

A detailed protocol for this type of experiment follows.

1. The liver is shifted from perfusion at 37° to 2–4° as described above.
2. Ligand, at trace or saturating levels, is introduced into the perfusate

and allowed to bind to cell-surface receptors. Clearance is monitored by periodically sampling the perfusate.

3. Unbound ligand is rinsed out:
   a. The liver is switched to the single-pass perfusion mode, and as much perfusate as possible is drained from the reservoir.
   b. The reservoir is rinsed several times with fresh cold perfusion medium (~ 200 ml total rinse volume), then refilled to the original level with more cold perfusate.
   c. The liver is returned to the recirculating mode.
4. The liver is switched to higher temperature:
   a. The auxiliary input line is placed into a flask containing preoxygenated 37° perfusion medium (see Fig. 1).
   b. The clamp is moved from the auxiliary input line to the line draining the biofiber reservoir.
   c. The pump is stopped long enough to switch the small three-way stopcocks such that the condenser is bypassed; then the pump is started again.
   d. The warming phase is timed from this point.
   e. The liver is switched to the single-pass perfusion mode.
   f. The cooling H_2O recirculator is turned off so that perfusate does not freeze in the coils of the condenser.

The exact order of steps during the warming step can be modified, so long as the same procedure is used every time.

5. The liver is shifted back to low temperature:
   a. When the desired time of the warm period has elapsed, the auxiliary input line is clamped and the reservoir drain line is opened.
   b. The pump is turned off and the three-way stopcocks are switched back to allow the perfusate to enter the condenser.
   c. The cooling H_2O recirculator is turned back on.
   d. After ~ 20 ml of the perfusate that is still in the lines has passed through the liver, the organ is switched back into the recirculating perfusion mode and fresh cold perfusate is added to the reservoir.
6. The remaining surface-bound ligand is removed (if desired):
   a. For ASGP dissociation, 1 ml of 0.5 *M* EGTA, pH 7.5 (4 m*M* final concentration) is added together with sufficient 2 *N* NaOH (~ 2 or 3 drops) to prevent the pH from dropping.
   b. EGTA is circulated for 10 min, and aliquots of the perfusate are removed to monitor dissociation of surface-bound ligand.

7. The liver is given a final rinse in the homogenization medium (if the liver is to be used for subcellular fractionation):
   a. Cold homogenization medium (~ 50 ml) is introduced via the auxiliary input line, with the liver in the single-pass perfusion mode.

b. The homogenization medium is passed through the cold condenser before it enters the liver to ensure that no warming occurs at this time.

c. The liver is removed from the cannula, weighed, and processed.

8. The whole liver is fixed by perfusion (if the liver is to be examined by electron microscopy):

a. Cold 2% glutaraldehyde (100 ml) in 0.1 *M* sodium cacodylate (pH 7.4) is introduced via the auxiliary input line, with the liver in the single-pass perfusion mode.

b. After the fixative has passed through the liver, the organ is removed from the cannula and selected regions of the hardened tissue are cut into 1-mm cubes. The tissue cubes are adequately fixed at this point and can be processed by standard procedures.[4]

9. A small portion of the liver is fixed by immersion (this procedure allows one to examine the morphology of the same liver as that used for subcellular fractionation):

a. After step 5 or step 7a, a piece of liver tissue is removed with fine iris scissors and is immersed in cold half-strength Karnovsky's fixative (2.5% glutaraldehyde, 1.5% formaldehyde, 2.5 m*M* $CaCl_2$, 0.1 *M* sodium cacodylate, pH 7.4) or in cold 3% glutaraldehyde in 0.1 *M* sodium cacodylate, pH 7.4.

b. The tissue is cut into pieces approximately $2 \times 2 \times 3$ mm and is fixed on ice for 2 hr.

c. The tissue is minced into 1-mm cubes and fixed for a further 30 min; it is then rinsed in 0.1*M* sodium cacodylate and processed by standard procedures.

An alternative protocol, which can be used to label all prelysosomal membrane compartments (endosomes) involved in ligand transport, is perfusion of the liver at 16–18°. At this temperature, fusion of endosomes with lysosomes is prevented, and ligands accumulate within the hepatocyte without being degraded. Continuous perfusion with saturating doses of the ligand at 16–18° therefore allows one to load the endosome compartments to a degree not possible with the temperature-shift or 37° formats.

Assessment of Liver Function

Many studies have dealt with the morphological, functional, and physiological properties of the isolated liver perfused at 37° under a variety of conditions. There is not universal agreement regarding the most reliable index of liver viability, and the reader is directed to several studies for relevant discussions.[11,13] We have used the perfused liver principally in

[13] L. L. Miller, *Isol. Liver Perfusion, Its Appl.* p. 11 (1973).

studies of receptor-mediated endocytosis by hepatocytes. Consequently, we have confined our analysis of liver function to those parameters that are directly relevant to this process. We have examined liver morphology, uniformity of perfusion among the lobes at different temperatures, receptor-mediated endocytosis of ASGPs, and the selectivity of endosome labeling as described above. Our goal has been to determine the extent to which the endocytic process in this isolated perfused liver system resembles that of the liver *in vivo.*

Morphology

Figures 2a and b illustrate sections of a liver perfused at 37° for 0 or 8 hr prior to fixation by immersion. The liver architecture is quite similar to that seen after fixation *in situ,* even at 8 hr, although we do not routinely perfuse livers for more than 4 hr. The major difference from normal morphology at the light microscopic (LM) level is an increased dilation of the sinusoids over time (Figs. 2a and b), although the absence of blood elements at 0 hr indicates that the sinusoids are well perfused. When livers perfused at 37° are fixed by perfusion, rather than immersion, the sinusoids appear much more dilated, even at the earliest times examined (Fig. 2c).

At the electron microscopic (EM) level, the cellular morphology is very similar to that seen *in situ.* That is, the endoplasmic reticulum is in parallel arrays, mitochondria are in the orthodox state, and Golgi complexes are intact. We have noted an increase in autophagic vacuoles over time at 37°. Occasionally, large, electron-lucent vacuoles are evident at the LM level at the sinusoidal periphery of hepatocytes. At the EM level, the vacuoles are membrane limited and similar in appearance to those present in hypoxic cells.[11] However, in contrast to hypoxic livers, all other organelles appear normal, which suggests that oxygen deprivation may not be the cause of vacuolation. Furthermore, these vacuoles are not concentrated in centrilobular regions as would be expected if hypoxia were the cause for their appearance. We do not know the origin of the vacuoles at present, but their presence at low levels does not appear to affect the kinetics or pathway of receptor-mediated endocytosis.

Livers perfused at temperatures below 37° (e.g., 16°) are morphologically indistinguishable from those maintained at 37°.

Receptor-Mediated Endocytosis of Asialoglycoproteins

We have compared the clearance rate of saturating levels of ^{125}I-labeled ASGPs *in vivo* to that in the perfused liver and in isolated hepatocytes. Regoeczi *et al.*[14] have reported a rate of ~ 5 μg of asialofetuin per minute per

[14] E. Regoeczi, M. T. Debanne, M. W. C. Hatton, and A. Koj, *Biochim. Biophys. Acta* **541,** 372 (1978).

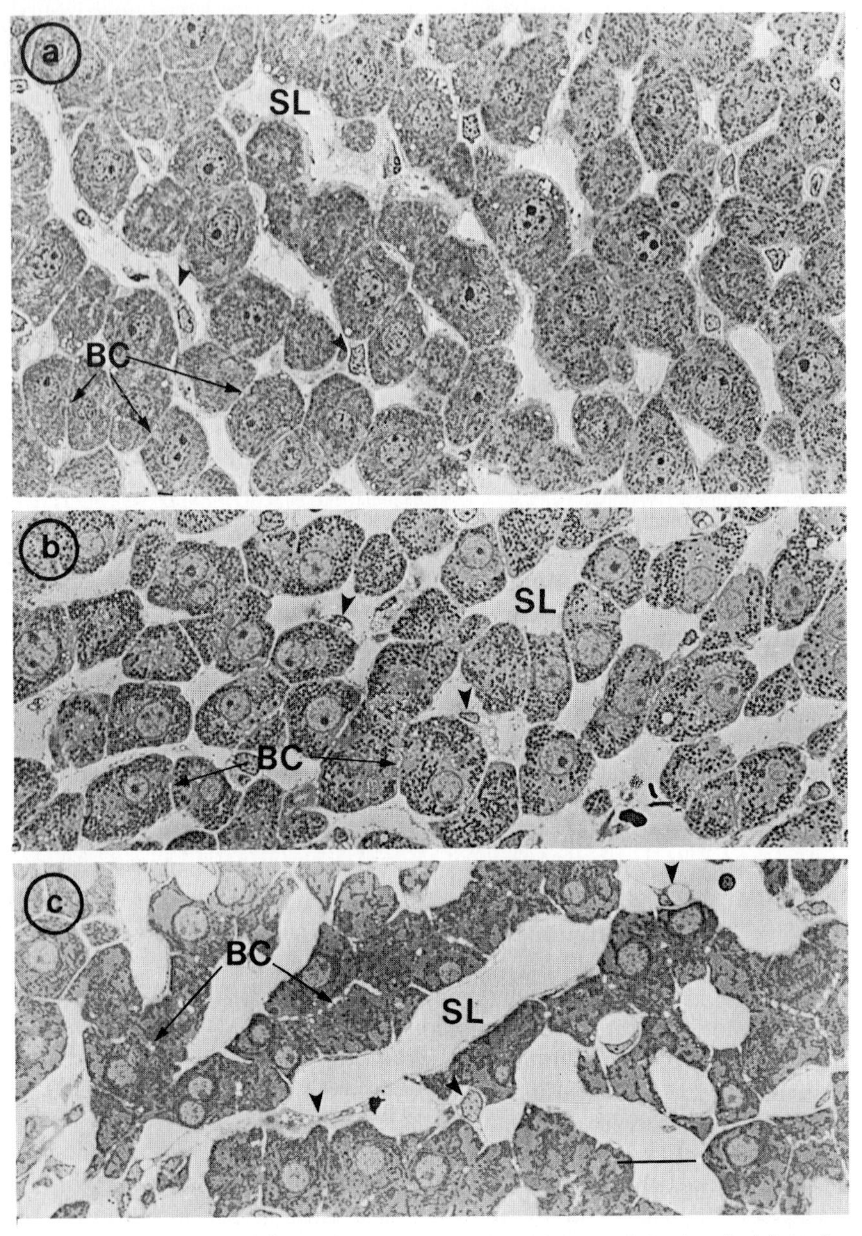

FIG. 2. Light micrographs of livers perfused at 37°. (a) Tissue (~ 0.25 g) excised from the median lobe of a liver perfused at 37° for ~ 5 min was fixed by immersion, processed, embedded, and sectioned (0.5 μm) as previously described.[3,4] (b) Tissue from the same liver as in (a) perfused at 37° for 8 hr was fixed and treated as in (a). The bile canaliculi (BC) are discernible as small clear areas between adjacent cells. The sinusoidal lumen (SL) appears to be more dilated with increased perfusion. Cells lining the sinusoids (arrowheads) are evident. (c) Section from a liver fixed by perfusion, then treated as in (a). The sinusoidal lumen is much more dilated, and the bile canaliculi are more prominent. × 560; bar = 20 μm.

gram of liver *in vivo.* We have measured a very similar rate of 4.5 μg/g per minute in the perfused liver with the same ligand. In contrast, rat hepatocytes isolated by us[3] and by others[15,16] internalize ^{125}I-labeled asialoorosomucoid (ASOR) at much slower rates, from 0.1 to 0.6 μg/g per minute. Clearly, the rate of ASGP endocytosis by cells in the perfused liver is more comparable to that seen *in vivo* than is the endocytic rate by isolated hepatocytes.

Effect of Temperature Shifts on Morphology and Endocytosis

The methods we have described in the earlier sections involve rapid cycling of the liver between low and high temperatures. We were concerned that these temperature shifts might alter tissue or cellular morphology and thus affect receptor-mediated endocytic pathways. This has not proved to be the case. That is, morphologically, both the tissue and the cells appear normal at the LM and EM levels. Furthermore, the kinetics of ^{125}I-labeled ASOR internalization and degradation in livers kept in 4° for up to 2 hr before warming fall within the range of those measured in livers maintained at 37°.

Uniformity of Perfusion Among the Lobes

We have measured the extracellular fluid volume within isolated livers perfused at three different temperatures: 4°, 16°, and 37°. We were concerned that circulation to and through the lobes might change at the different temperatures used in one experiment, resulting in unequal delivery of ligand to all cells. Since the cell-surface receptor number is measured at 4°[4] and internalization rates are measured at 16° or 37°, uneven perfusion would make comparison of these measurements questionable. We addressed this issue in the following way.

Livers were isolated and perfused at 37° for 15 min; the tissue was then cooled to 4° and [^{3}H]inulin (50 μCi) was added. (Inulin is a linear polymer of fructose used to measure extracellular volume.) After 30 min of cyclic perfusion at 4°, liver biopsies (10–40 mg net weight each) were taken using a small biopsy needle (10–12 gauge). The same liver was then warmed to 16° for 30 min, and additional tissue samples were taken. The last of the liver biopsies were taken after a final temperature shift to 37° for 30 min. The sequence of temperatures was also reversed (37° $\rightarrow$ 16° $\rightarrow$ 4°), and in one series the temperature was maintained at 4° throughout in order to assess the effect of time on perfusion volume. Tissue samples were hydrolyzed, and the radioactivity was determined.

[15] C. Steer and G. Ashwell, *J. Biol. Chem.* **255,** 2008 (1980).

[16] P. H. Weigel and J. A. Oka, *J. Biol. Chem.* **256,** 2615 (1981).

EFFECT OF TEMPERATURE ON EXTRACELLULAR FLUID VOLUME AMONG THE VARIOUS LOBES OF PERFUSED LIVER[a]

Liver lobes	[³H]Inulin content (per gram) at 0 hr		1 hr		2 hr
Experiment A (3)	4°	→	16°	→	37°
1. Caudate and right lateral	1 (57,746 ± 7922)[b]		1.13 ± 0.33		1.47 ± 0.38
2. Median	1 (63,136 ± 9281)		1.06 ± 0.20		1.50 ± 0.36
3. Left lateral	1 (64,900 ± 13,240)		1.03 ± 0.15		1.26 ± 0.43
Experiment B (2)	37°	→	16°	→	4°
1. Caudate and right lateral	1 (50,239 ± 2098)		1.08 ± 0.14		1.43 ± 0.21
2. Median	1 (48,366 ± 3628)		1.15 ± 0.24		1.29 ± 0.28
3. Left lateral	1 (53,797 ± 6554)		1.22 ± 0.22		1.15 ± 0.36
Experiment C (2)	4°	→	4°	→	4°
1. Caudate and right lateral	1 (46,124 ± 6123)		1.22 ± 0.10		1.52 ± 0.36
2. Median	1 (61,319 ± 3039)		1.15 ± 0.19		1.32 ± 0.29
3. Left lateral	1 (58,556 ± 9243)		1.22 ± 0.23		1.44 ± 0.26

[a] Three separate livers were used in A, 2 in B, and 2 in C. In each experiment, three biopsies from each lobe (three total from category 1) were taken at each temperature (nine total per lobe). The [³H]inulin content (cpm/g) at the initial temperature (4° in A and C and 37° in B) was normalized to 1, and the other two measurements were expressed as a fraction. Averages of the experiments ± SD are presented here.

[b] Representative values (cpm/g) from a single experiment.

The table summarizes the results of seven experiments performed as described above. The general trend is similar for all lobes and appears to be independent of temperature or the direction of temperature shift. That is, the extracellular fluid volumes in all lobes of the liver were not significantly different at the first two temperatures tested (4°, then 16° or 37°, then 16°) but increased at the final temperature. In fact, it appears that the duration of the perfusion is the important factor, since livers maintained at 4° throughout also showed the same increase during the last time interval. However, the fact that the liver binds similar amounts of ^{125}I-labeled ASOR at 1 hr and at 4 hr of perfusion argues against the possibility that additional lobules will

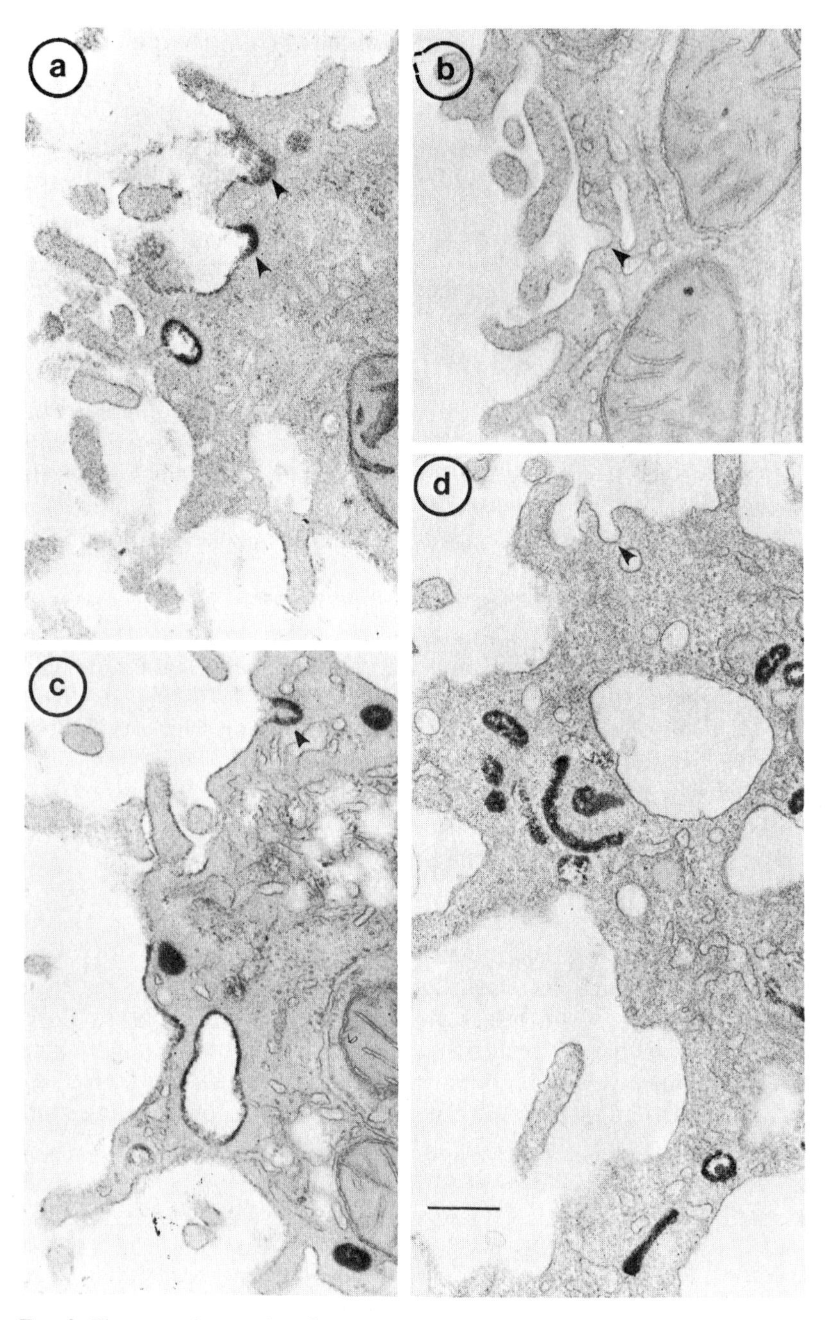

FIG. 3. Electron micrographs of peripheral cytoplasm of hepatocytes fixed by perfusion after incubation with the EM conjugate ASOR–horseradish peroxidase (ASOR–HRP). (a, b) Livers were exposed to 2.2 nmol of ASOR–HRP for 60 min at 4°; excess ligand was rinsed out, and the liver was fixed and processed for HRP cytochemistry either without exposure to EGTA (a) or after a 10-min perfusion with 10 m*M* EGTA (b). EGTA removed all the surface-bound

become perfused with time. We conclude that the temperature-shifts do not dramatically change the extracellular fluid volume, a result that allows us to compare various kinetic and numerical parameters of receptor-mediated endocytosis in the perfused liver at different temperatures. However, the duration of a perfusion experiment may influence measurements, although it is possible that the multiple biopsies taken (27 total per liver) affected the result).

Selectivity of Endosome Labeling

We have examined several aspects of the temperature-shift protocol used selectively to label, or "mark," the contents of various endocytic compartments of the ASGP pathway: (*a*) the uniformity of 4° cell surface labeling in different regions of the liver; (*b*) the extent to which EGTA removes only surface-bound ligand; and (*c*) the behavior of labeled endosomes after subcellular fractionation.

Surface Labeling at 4°. The uniformity of surface labeling was assessed by perfusion of saturating amounts of ^{125}I-labeled ASOR at 4° for 60 min, removal of excess unbound ligand, and measurement of the radioactivity present in regions of each lobe. We found the ratios of ^{125}I-labeled ASOR per gram of liver in the different lobes to be quite similar, which indicates that the perfusion at 4° was uniform.

Effect of EGTA. Since our endosome labeling protocol includes a step in which residual surface-bound ligand is dissociated by EGTA, we were concerned that this reagent quantitatively remove ligand at the surface without affecting internalized ligand. Biochemically, we obtain >90% removal of the ^{125}I-labeled ASOR bound at 4° by the addition of 4–10 m*M* EGTA.[4] Morphological examination reveals that virtually all of a specific EM ligand bound at 4° is localized to the sinusoidal surface of hepatocytes (Fig. 3a) and that EGTA removes this ligand (Fig. 3b). If both endosome contents and the plasma membrane are labeled by a brief temperature shift after 4° binding, EGTA treatment removes only the surface-bound ligand, not that sequestered within the cell (Fig. 3c, d). Thus, our labeling protocol selectively marks only endocytic vesicles of the hepatocyte.

ligand, demonstrating complete accessibility of ASOR–HRP bound at 4°. (c, d) Livers were exposed to 2.2 nmol of ASOR–HRP for 60 min at 4°, excess unbound ligand was rinsed out, and the livers were warmed to 31–32° for 1.5 (c) or 2 min (d). The liver shown in (d) was then exposed to 10 m*M* EGTA for 10 min at 4° to remove remaining surface-bound ligand before fixation. The liver shown in (c) was fixed without EGTA exposure. EGTA treatment eliminated ASOR–HRP from the cell surface, but ligand sequestered within the cell was not accessible to EGTA dissociation. Arrowheads indicate membrane invaginations 100 nm in diameter at the base of microvillar projections. All at × 38,000; bar = 0.25 μm.

Subcellular Fractionation of Tissue Containing Labeled Endosomes. Approximately 85% of the ^{125}I-labeled ASOR sequestered within endosomes by our temperature-shift protocol can be recovered in a microsomal pellet.[17] After resuspension, >90% of the ligand in this fraction is still sedimentable. These results indicate that after homogenization, the endocytic vesicles containing sequestered ligand are small and reasonably stable to handling. Furthermore, the ligand is contained within the vesicles, since it can be released only by the addition of detergents (e.g., digitonin) at concentrations that permeabilize but do not solubilize the membrane.[18] Therefore, our labeling and subcellular fractionation procedures yield intact endosome vesicles in substantial quantity for further analysis.

[17] J. H. Ehrenreich, J. J. M. Bergeron, P. Siekevitz, and G. E. Palade, *J. Cell Biol.* **59,** 45 (1973).
[18] D. A. Wall and A. L. Hubbard, manuscript in preparation.

[19] Receptor-Mediated Endocytosis of Low-Density Lipoprotein in Cultured Cells

By JOSEPH L. GOLDSTEIN, SANDIP K. BASU, and MICHAEL S. BROWN

Studies of the cell surface binding, internalization, and metabolism of low-density lipoprotein (LDL) in cultured cells have provided useful information regarding the general aspects of receptor-mediated endocytosis.[1,2] LDL is found in the plasma of humans and all other vertebrate species. The lipoprotein consists of a hydrophobic core composed of approximately 1500 molecules of cholesteryl ester, surrounded by a polar phospholipid coat that contains free cholesterol and a poorly characterized protein called apoprotein B.[3]

Animal cells in tissue culture require cholesterol for synthesis of plasma membranes. Although the cells can synthesize cholesterol from acetyl-CoA, they do so only at a low rate. Rather, they derive their cholesterol from the LDL that is present in the serum of culture medium. To obtain cholesterol from LDL, the cells bind and internalize the lipoprotein in a series of steps outlined in Fig. 1. Uptake of LDL has been studied most extensively in human fibroblasts, but the LDL receptor pathway is also known to occur in other cultured cells, such as Chinese hamster fibroblasts, mouse L cells, and bovine adrenocortical cells.

[1] J. L. Goldstein, R. G. W. Anderson, and M. S. Brown, *Nature* (*London*) **279,** 679 (1979).
[2] M. S. Brown, R. G. W. Anderson, S. K. Basu and J. L. Goldstein, *Cold Spring Harbor Symp. Quant. Biol.* **46,** 713 (1982).
[3] R. J. Havel, J. L. Goldstein, and M. S. Brown, *in* "Metabolic Control and Disease" (P. K. Bondy and L. E. Rosenberg, eds.), 8th ed., Chapter 7, p. 393. Saunders, New York, 1980.

METHODS IN ENZYMOLOGY, VOL. 98

ISBN 0-12-181998-1

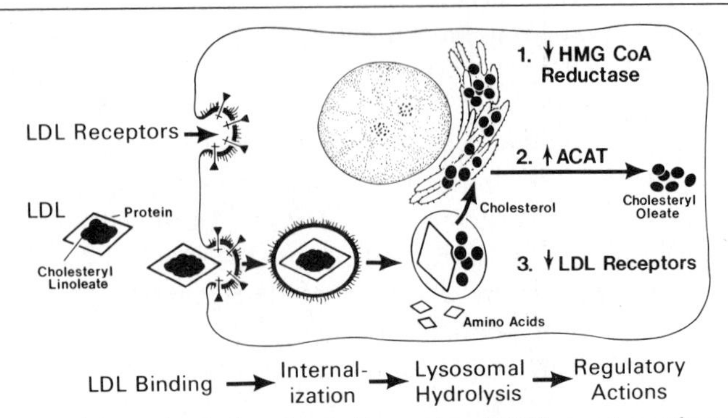

FIG. 1. Sequential steps in the low-density lipoprotein (LDL) receptor pathway in cultured mammalian cells. HMG-CoA reductase denotes 3-hydroxy-3-methylglutaryl-CoA reductase; ACAT denotes acyl-CoA: cholesterol *O*-acyltransferase.

Steps in the LDL Receptor Pathway

When human fibroblasts require cholesterol for plasma membrane formation, they synthesize relatively large numbers of LDL receptors (up to about 70,000 receptors per cell). The receptors migrate laterally in the plane of the plasma membrane until they reach coated pits, which are regions of the plasma membrane that are indented and coated on the cytoplasmic surface by a protein, clathrin.[1,4] Although coated pits comprise only 2% of the surface of human fibroblasts, they contain about 50–80% of the LDL receptors. The coated pits continually invaginate into the cell and pinch off to form coated endocytic vesicles. Any LDL that is bound to the receptor is trapped in the coated vesicle and carried into the cell. The vesicle rapidly loses its coat and appears to undergo a series of fusions with other vesicles beneath the plasma membrane. Eventually the LDL is delivered to lysosomes. The protein component of LDL is rapidly degraded to amino acids, and the cholesteryl esters are hydrolyzed by an acid lipase. The liberated cholesterol leaves the lysosome and is used by the cell for membrane synthesis.[5]

The cholesterol derived from the catabolism of LDL regulates three events in intracellular cholesterol metabolism. First, it suppresses 3-hydroxy-3-methylglutaryl-CoA reductase (HMG-CoA reductase), the rate-controlling enzyme in cholesterol biosynthesis.[6] This turns off cholesterol production by the cell. Second, the incoming cholesterol activates an acyl-CoA: cholesterol *O*-acyltransferase (cholesterol acyltransferase; ACAT), which reesterifies excess cholesterol, allowing it to be stored in the cytoplasm

[4] B. M. F. Pearse and M. S. Bretscher, *Annu. Rev. Biochem.* **50,** 85 (1981).

[5] J. L. Goldstein and M. S. Brown, *Annu. Rev. Biochem.* **46,** 897 (1977).

[6] M. S. Brown, S. E. Dana, and J. L. Goldstein, *J. Biol. Chem.* **249,** 789 (1974).

as cholesteryl ester droplets.[7] And third, the LDL-derived cholesterol suppresses the synthesis of new LDL receptors, thereby preventing an overaccumulation of cholesterol within the cell.[8]

By virtue of the above regulatory mechanisms, cultured cells maintain a relatively constant level of free cholesterol despite widely varying conditions of growth and availability of exogenous cholesterol.[9] When the cells are grown in serum containing LDL, they adjust the number of receptors to supply just sufficient cholesterol for growth. If LDL is removed from the culture medium, the cells continue to grow because HMG-CoA reductase activity rises by approximately 50-fold. The number of LDL receptors also increases. When LDL is added back to the cells, it binds to the induced receptors and enters the cell, where the liberated cholesterol suppresses HMG-CoA reductase activity. The initial burst of LDL entry also stimulates the ACAT enzyme, so that excess cholesterol is stored as cholesteryl oleate. At the same time, the incoming cholesterol derived from LDL suppresses the synthesis of LDL receptors. Within 24–28 hr a new steady state is reached in which HMG-CoA reductase remains suppressed, the number of receptors has been reduced by 75% of the maximal number, and the ACAT enzyme is no longer active, since excess LDL-cholesterol is no longer entering the cell.[5,9]

LDL Receptor Mutations and Their Detection

The study of the LDL receptor has been facilitated by analysis of mutant fibroblasts obtained from human subjects with disorders of cholesterol metabolism. The most informative cells, obtained from patients with familial hypercholesterolemia (FH), have defects in the gene encoding the LDL receptor.[10] The existence of three classes of mutant alleles at the LDL receptor locus has been deduced on the basis of genetic and kinetic data. One of these alleles specifies a receptor that is unable to bind LDL (receptor-negative phenotype). The second type of allele specifies a receptor that can bind small amounts of LDL (receptor-defective phenotype); and the third type of allele specifies a receptor that can bind LDL, but cannot be incorporated into coated pits and hence cannot carry the LDL into the cell (internalization-defective phenotype). The first two alleles are common among FH patients, whereas the third allele is extremely rare.

[7] J. L. Goldstein, S. E. Dana, and M. S. Brown, *Proc. Natl. Acad Sci. U.S.A.* **71,** 4288 (1974).

[8] M. S. Brown and J. L. Goldstein, *Cell* **6,** 307 (1975).

[9] M. S. Brown and J. L. Goldstein, *in* "Disturbances in Lipid and Lipoprotein Metabolism" (J. M. Dietschy, A. M. Gotto, and J. Ontko, eds.), p. 173. American Physiological Society, Bethesda, Maryland, 1978.

[10] J. L. Goldstein and M. S. Brown, *Annu. Rev. Genet.* **13,** 259 (1979).

Heterozygotes for the receptor-negative allele and the receptor defective allele constitute approximately 1 in 500 persons in the population. These individuals produce only half the normal number of LDL receptors, and therefore they degrade LDL with reduced efficiency. From the time of birth these heterozygotes have an LDL-cholesterol level that is about 2.5-fold above normal, and they develop heart attacks at about age 45. About 1 in 1,000,000 persons in the population inherits two mutant alleles at the LDL receptor locus and is an FH homozygote. These individuals have a severely impaired ability to degrade LDL, and as a result they maintain 6- to 10-fold elevations in plasma LDL levels and develop heart attacks in childhood.[11]

The analysis of LDL receptor activity and diagnosis of the various FH mutations is based upon quantitative measurements of the amount of binding, internalization, and degradation of ^{125}I-labeled low-density lipoprotein (^{125}I-LDL) in cultured fibroblasts.[10,11] In determining whether a given fibroblast strain comes from a receptor-negative FH homozygote (<2% of normal LDL receptor activity), a receptor-defective homozygote (2–25% of normal LDL receptor activity), or an FH heterozygote (30–70% of normal receptor activity), the cells must be examined under conditions in which the number of receptors is maximally induced. In normal cells the activity of the LDL receptor varies 10-fold according to the degree of cholesterol deprivation[8] and the rate of cell growth.[12,13] It is therefore imperative that growth and assay conditions be rigidly standardized in order to induce receptors maximally so as to allow meaningful comparisons of LDL receptor activity between cell strains from different patients.

Factors Affecting Expression of LDL Receptor Activity

The expression of maximal receptor activity requires that the cells be actively growing, since the number of receptors declines markedly when fibroblasts reach confluence and no longer require net amounts of cholesterol for new membrane synthesis.[12,13] Maximal induction also requires that the cells be deprived of all usable exogenous sources of cholesterol.[8,14] Rapid growth is accomplished by a protocol in which the cells are seeded initially in medium containing 10% (v/v) fetal calf serum. After a few days of growth the cells are switched to medium containing human serum from which the lipoproteins have been removed by ultracentrifugation (i.e., lipoprotein-de-

[11] J. L. Goldstein and M. S. Brown, *in* "The Metabolic Basis of Inherited Disease" (J. B. Stanbury, J. B. Wyngaarden, D. S. Fredrickson, J. L. Goldstein, and M. S. Brown, eds.), 5th ed., pp. 672–712. McGraw-Hill, New York, 1983.

[12] J. L. Goldstein and M. S. Brown, *J. Biol. Chem.* **249,** 5153 (1974).

[13] M. S. Brown and J. L. Goldstein, *Cell* **9,** 663 (1976).

[14] J. L. Goldstein, M. K. Sobhani, J. R. Faust, and M. S. Brown, *Cell* **9,** 195 (1976).

ficient serum) to deprive them of exogenous cholesterol. Although human fibroblasts will not grow indefinitely in lipoprotein-deficient serum, they will divide for 48 hr, and a maximal induction of LDL receptor activity occurs during this period.

Even when such precautions are taken, LDL receptor activity varies by two- to threefold in the same cell strain studied on different days and with different batches of lipoprotein-deficient serum.[15] Therefore, if two cell strains are to be compared, it is necessary to grow the cells simultaneously and to compare them in assays performed with the same batch of lipoprotein-deficient serum and on the same day. Moreover, it is necessary that the cells have relatively similar growth rates and that the cells be at the same passage number, i.e., before the 20th population doubling. The table shows representative values for the amounts of ^{125}I-labeled LDL binding, uptake, and degradation in normal fibroblasts when they are grown according to a standardized protocol.

In addition to direct measurements of the binding, uptake, and the degradation of LDL, it is also possible to assess LDL receptor activity by measuring the consequences of receptor-mediated LDL uptake—namely, suppression of HMG-CoA reductase activity and stimulation of cholesteryl ester synthesis by the ACAT enzyme. In order to assess the suppression of HMG-CoA reductase, the cells must be actively growing in the presence of lipoprotein-deficient serum, so that they will express a maximal enzyme activity (see the table for a typical value). The degree to which HMG-CoA reductase is suppressed by LDL in growing cells is not precisely proportional to the number of LDL receptors. The degree of suppression depends on the cumulative uptake of cholesterol and hence on the duration of exposure to LDL as well as on the number of receptors. Even when cells have only 5% of normal receptor number (receptor-defective FH homozygotes), they show at least a 50% suppression of HMG-CoA reductase after 6–8 hr of incubation with saturating levels of LDL.[16] (Normal cells show a 90–95% suppression after 6–8 hr.) Cells that have no detectable receptor activity (receptor-negative FH homozygotes) show no suppression after incubation with saturating levels of LDL for 6–8 hr and less than 20% suppression after 24 hr.[15,16]

Whereas the suppression of HMG-CoA reductase after short-term incubation wth LDL is a sensitive index of receptor activity, the stimulation of cholesteryl ester synthesis is a less sensitive index. To develop enhanced cholesteryl ester synthesis, the cells must take up LDL-cholesterol in amounts that exceed their ability to incorporate the sterol into newly

[15] J. L. Goldstein, M. S. Brown, and N. J. Stone, *Cell* **12,** 629 (1977).

[16] J. L. Goldstein, S. E. Dana, G. Y. Brunschede, and M. S. Brown, *Proc. Natl. Acad. Sci. U.S.A.* **72,** 1092 (1975).

TYPICAL VALUES FOR LOW-DENSITY LIPOPROTEIN RECEPTOR ACTIVITY IN GROWING HUMAN FIBROBLASTS CULTURED FROM NORMAL SUBJECTS[a]

Parameter measured at 37°	Concentration of LDL in medium (μg protein/ml)	Duration of incubation with LDL (hr)	Typical value
Cell surface binding of ^{125}I-LDL	10	5	100 ng/mg protein
Internalized ^{125}I-LDL	10	5	600 ng/mg protein
Rate of degradation of ^{125}I-LDL	10	5	3000 ng/mg protein
HMG-CoA reductase activity			
Induced	0	—	250 pmol min^{-1} mg protein^{-1}
Suppressed	10	7	25 pmol min^{-1} mg protein^{-1}
Cholesteryl [^{14}C]oleate formation (ACAT)			
Basal	0	—	<0.05 nmol hr^{-1} mg protein^{-1}
Induced	200	7	3.5 nmol hr^{-1} mg protein^{-1}

[a] Fibroblasts are grown in monolayer, used between the fifth and twentieth population doubling, and maintained in a humidified 5% CO_2 incubator at 37° in stock flasks (75 cm^2) containing 10 ml of growth medium consisting of Eagle's minimum essential medium supplemented with 24 m*M* $NaHCO_3$, 1% (v/v) nonessential amino acids, penicillin (100 units/ml), streptomycin (100 μg/ml), and 10% (v/v) fetal calf serum. For experiments, confluent monolayers of cells from stock flasks are dissociated with 0.05% trypsin–0.02% EDTA solution in physiological saline. On day 0, a total of $\sim 6 \times 10^4$ fibroblasts are seeded into each 60-mm petri dish containing 3 ml of growth medium with 10% fetal calf serum. On day 3, the medium is replaced with 3 ml of fresh medium containing 10% fetal calf serum. On day 5, when the cells are in late logarithmic growth, each monolayer is washed with 3 ml of Dulbecco's phosphate-buffered saline, after which is added 2 ml of fresh medium containing 10% (v/v) human lipoprotein-deficient serum (final protein concentration, 5 mg/ml). Experiments are carried out on day 7, after the cells have been incubated with lipoprotein-deficient serum for 48 hr. The cellular content of protein on day 7 is typically 180–230 μg per dish.

synthesized membranes, and this requires the expression of a reasonable number of receptors. Cells with only 5% of receptor activity may show less than 5% of the normal amount of stimulation of ACAT activity when incubated with LDL for 6–8 hr.[16]

In measuring the stimulation of ACAT activity or the suppression of HMG-CoA reductase, it is helpful to perform a control experiment in which the cells are incubated with a mixture of 25-hydroxycholesterol plus cholesterol added to the medium in ethanol. This combination of sterols will enter cells and suppress HMG-CoA reductase and activate ACAT independently of the LDL receptor.[5] The activity of LDL can then be expressed relative to the activity of the sterol mixture in any given cell strain, and this helps to circumvent problems arising from the different responses of cells to LDL-cholesterol after it enters the cell.[16]

The expression of LDL receptors and the level of HMG-CoA reductase is influenced by the presence in lipoprotein-deficient serum of agents that act to remove cholesterol from the cell and hence increase the cell's cholesterol requirement. For this reason and also because lipoprotein-deficient serum is only a marginal supporter of cell growth, different batches of lipoprotein-deficient serum differ in their ability to induce these activities. Moreover, induction of HMG-CoA reductase[6] and of LDL receptors[17] is stimulated by insulin, which may vary in amount between various batches of serum. In general, lipid-depleted serum, which is prepared by solvent extraction, does not give values for LDL receptor activity and HMG-CoA reductase activity that are as high as those obtained with lipoprotein-deficient serum, which is prepared by ultracentrifugation (see below).

Radioiodination of LDL

Principle. The procedure of choice for preparing ^{125}I-labeled LDL for tissue culture studies is based on Bilheimer's modification[18] of the iodine monochloride method of MacFarlane.[19] The method described below produces biologically active ^{125}I-LDL that has the following properties: (*a*) <1 atom of iodine incorporated into each 100,000 daltons of LDL; (*b*) a final specific radioactivity of 200–600 cpm per nanogram of protein; (*c*) $>98\%$ of the ^{125}I radioactivity is precipitable by incubation with 10% (v/v) trichloroacetic acid; (*d*) $>95\%$ of the ^{125}I radioactivity is precipitable by a monospecific antiserum directed against human LDL; and (*e*) $<5\%$ of the ^{125}I radioactivity is extractable into chloroform–methanol (2 : 1).

Reagents

Glycine–NaOH buffer, 1 *M*, pH 10

Iodine monochloride (stock solution, 33 m*M*; diluted solution, 2.64 m*M*). Mix 150 mg of sodium iodide dissolved in 8 ml of 6 *N* HCl together with 99 mg of sodium iodate (anhydrous) dissolved in 2 ml of water. The iodate solution is forcibly injected into the iodide–HCl solution in order to avoid precipitation of iodine. Dilute the mixture with water to a final volume of 40 ml and shake in a glass-stoppered cylinder with 5 ml of carbon tetrachloride. The upper organic phase

[17] A. Chait, E. L. Bierman, and J. J. Albers, *Biochem. Biophys. Acta* **529,** 292 (1978).

[18] D. W. Bilheimer, S. Eisenberg and R. I. Levy, *Biochim. Biophys. Acta* **260,** 212 (1972).

[19] A. S. McFarlane, *Nature (London)* **182,** 53 (1958).

is discarded. If the organic phase has a faint red color owing to the presence of free iodine, the extraction with carbon tetrachloride should be repeated. Residual carbon tetrachloride is removed by aerating the aqueous phase with moist air for 1 hr, after which the volume is adjusted with water to 45 ml. This solution is 33 m*M* with respect to iodine monochloride and approximately 1 *N* with respect to HCl; it can be stored as a stock solution at 4° in an amber bottle for several months. Immediately prior to iodination, dilute the stock iodine monochloride solution with 12.5 volumes of 2 *M* NaCl.

$Na^{125}I$ (carrier free), 10 mCi in 0.1 ml of 0.1 *N* NaOH. This is purchased from either New England Nuclear or Amersham/Searle.

Buffer A: 150 m*M* NaCl and 0.24 m*M* disodium EDTA at pH 7.4 The final pH is adjusted with NaOH.

Plasma LDL at a protein concentration of 20–40 mg/ml in buffer A.

Sephadex G-25 column (Pharmacia PD-10; bed volume, 9.1 ml) equilibrated with 20 ml of buffer A

Red stopper 3-ml Vacutainer glass tubes (Becton-Dickinson)

Preparation of Plasma LDL. Plasma LDL (density 1.019 to 1.063 g/ml) is prepared from the venous blood of individual healthy human subjects who are fasted for 12 hr prior to venipuncture. The blood (500 ml) is collected in a 600-ml blood bag (Fenwal Laboratories, Deerfield, Illinois) containing 7.5 ml of a sterile solution of 0.25 *M* disodium EDTA, pH 7.4. The plasma fraction is obtained by low-speed centrifugation of the whole blood. The plasma is kept at 4° and used within 24 hr for preparation of LDL. Lipoproteins are isolated by sequential flotation in a preparative ultracentrifuge according to the method of Havel *et al.*[20] The plasma (*initial* density = 1.006 g/ml) is adjusted to a *final* density of 1.019 g/ml by adding solid potassium bromide according to the Radding–Steinberg formula,[21] as indicated below.

$$X = [V(d_f - d_i)]/[1 - (0.312 \times d_f)]$$

where X = grams of potassium bromide, d_i = initial density, d_f = final or desired density, V = volume of plasma in milliliters, and 0.312 = partial specific volume of potassium bromide. After density adjustment to 1.019 g/ml, the plasma is centrifuged in a Beckman 60 Ti rotor at 59,000 rpm for 16–20 hr at 4°. After centrifugation, each 36-ml polyallomer centrifuge tube is sliced with a tube slicer (obtained from either Nuclear Supply Co., Washington D. C., or Beckman Instruments, Palo Alto, California) so as to obtain two fractions: a top fraction (26 ml) that contains a mixture of very low-density lipoproteins (VLDL) and intermediate-density lipopro-

[20] R. J. Havel, H. A. Eder, and J. H. Bragdon, *J. Clin. Invest.* **34,** 1345 (1955).

[21] C. M. Radding and D. Steinberg, *J. Clin. Invest.* **39,** 1560 (1960).

teins (IDL); and a bottom fraction (10 ml) that contains a mixture of LDL, high-density lipoproteins (HDL), and the lipoprotein-deficient fraction of plasma. The top fraction is discarded, and a pellet of red blood cells and debris at the bottom of the tube is also discarded. The bottom fraction (whose initial density is assumed to be 1.019 g/ml) is then adjusted to a final density of 1.063 g/ml with solid potassium bromide according to the above formula and centrifuged for 24 hr at 4° in a 60 Ti Beckman rotor at 59,000 rpm.

After centrifugation each 36-ml polyallomer centrifuge tube is sliced with a tube slicer so as to obtain two fractions: a top fraction (9 ml) that contains LDL (bright orange color), and a bottom fraction (27 ml) that contains HDL and the lipoprotein-deficient fraction of plasma. To determine the density of the top fraction, a blank tube containing potassium bromide at a density of 1.063 g/ml is subjected to centrifugation in parallel with the sample-containing tube. The 9-ml top fraction from the blank tube is obtained, and a 1-ml aliquot is weighed in a tared pipette. The value obtained is used to calculate the density of the top fraction from the sample-containing tube. The density is typically 1.040 g/ml. The density of the sample is then adjusted to 1.063 g/ml with solid potassium bromide as described above. The fraction is centrifuged for 18 hr in a 50 Ti Beckman rotor at 49,000 rpm at 4°. This latter centrifugation serves two functions: (*a*) to clear the LDL of contaminating proteins such as albumin and HDL; and (*b*) to concentrate the LDL. Each 12-ml polyallomer centrifuge tube is sliced with a tube slicer to obtain a top fraction (2.5 ml) of bright orange color, which contains the LDL. The bottom fraction (9.5 ml) is discarded. If the LDL is in the form of a soft gel, it should be dissolved by gentle pipetting through a Pasteur pipette. The LDL solution is then dialyzed against a total of 12 liters of buffer A for 36–48 hr at 4° (two changes of 6 liters each). Care is taken to avoid overly vigorous stirring, which can cause the LDL to precipitate in the dialysis bag. After dialysis, the LDL solution is centrifuged at 10,000 rpm for 30 min in a Sorvall SS-34 rotor at 4°, and the supernatant is sterilized by passage through a Millipore filter (Millex-HA, 0.45 μm). The protein content of the solution is determined by the Lowry method,[22] and the LDL solution is adjusted with buffer A to a protein concentration of 20–40 mg/ml, stored in a sterile plastic tube at 4°, and used within 2 weeks.

Procedure. All steps are performed at 0–4°. The iodination reaction is conducted behind lead bricks in a fume hood. Gloves are worn. Remove the red stopper of a 3-ml Vacutainer tube and add to the tube 0.5 ml of 1 *M* glycine–NaOH buffer and 16 mg of protein of LDL in 0.8 ml of buffer A. To a V vial containing 10 mCi of $Na^{125}I$ in 0.1 ml of 0.1 *N* NaOH, add 0.4 ml of 1 *M* glycine–NaOH buffer. Withdraw the contents of this vial and add them

[22] O. H. Lowry, N. J. Rosebrough, A. L. Farr, and R. J. Randall, *J. Biol. Chem.* **193,** 265 (1951).

to the Vacutainer tube containing the LDL solution. Replace the stopper on the Vacutainer tube. With the use of a tuberculin syringe, quickly and forcefully inject 0.42 ml of diluted (2.64 m*M*) iodine monochloride solution through the stopper of the Vacutainer tube. At the same time that the injection is being made, the Vacutainer tube is vortexed. The duration of vortexing should not exceed 2 sec. Incubate the mixture on ice for 5 min. Add 0.28 ml of buffer A to the iodination mixture to bring the final volume to 2.5 ml. Apply the whole mixture (2.5 ml) to a Sephadex G-25 column (bed volume, 9.1 ml) and discard the initial 2.5 ml that flows through the column. Apply 3.5 ml of buffer A to the column, and collect the eluate. Dialyze the eluate against 10 liters of buffer A (two changes of 2 liters each followed by one change of 6 liters) until the final dialyzate has less than 3000 cpm/ml (usually 18–24 hr).

After dialysis, the ^{125}I-LDL solution is centrifuged at 10,000 rpm for 30 min at 4° in a Sorvall SS-34 rotor, and the supernatant is sterilized by passage through a Millipore filter (Millex-HA, 0.45 μm). The ^{125}I-LDL solution is stored in a plastic tube at 4° and used within 2 weeks.

The extent of radiolabeling of lipid in the ^{125}I-LDL solution is measured by extracting a small aliquot with chloroform–methanol (2 : 1)[23] and subjecting the organic phase to scintillation counting after evaporating the solvents under nitrogen. Protein-bound radioactivity is measured by precipitating an aliquot of ^{125}I-LDL with 10% (v/v) trichloroacetic acid. The ^{125}I-LDL is not used for studies unless it meets the following criteria: (*a*) protein concentration of 3–5 mg/ml with a specific radioactivity of 200–600 cpm/ng (LDL with specific radioactivity above this level often undergoes denaturation); (*b*) $>$98% of the ^{125}I radioactivity is precipitable by trichloroacetic acid; and (*c*) $<$5% of the ^{125}I radioactivity is extractable into chloroform–methanol.

Preparation of Lipoprotein-Deficient Serum

Principle. In order to induce a maximal number of LDL receptors, cells must grow for some time in the absence of LDL or other source of cholesterol. Cells grown in medium containing 10% (v/v) whole serum (which contains LDL) express a level of LDL receptors that is 10–20% of that found when the cells are cultured for 36–48 hr in medium containing 10% (v/v) lipoprotein-deficient serum.

Procedure. Lipoprotein-deficient serum is prepared from human plasma by ultracentrifugation, using the same techniques as described above for the preparation of LDL. Whole plasma (*initial* density, 1.006 g/ml) is adjusted to a *final* density of 1.215 g/ml with solid potassium bromide and centrifuged for 36 hr at 4° at 59,000 rpm in a 60 Ti Beckman rotor. Each 36-ml

[23] J. Folch, I. Ascoli, M. Lees, J. A. Meath, and F. N. LeBaron, *J. Biol. Chem.* **191,** 833 (1951).

polyallomer tube is sliced with a tube slicer to obtain two fractions: a top fraction (18 ml) that contains lipoproteins; and a bottom fraction (18 ml) that contains the lipoprotein-deficient fraction of plasma. (A salt pellet at the bottom of the tube is discarded.) The bottom fraction is dialyzed extensively at 4° against a total of 30 liters of 150 mM NaCl for 48–72 hr (five changes of 6 liters each). After dialysis, the lipoprotein-deficient plasma is converted to lipoprotein-deficient serum by incubating it at 4° for 24 hr with USP thrombin (Thrombostat, Parke-Davis, Morris Plains, New Jersey) at a final concentration of 10 US (NIH) units per milliliter. The resulting clot is removed by centrifugation at 18,000 rpm for 2 hr at 4° in a Sorvall SS-34 rotor. The lipoprotein-deficient serum is sterilized by passage through a Millipore filter (Millex-HA, 0.45 μm), adjusted to a protein concentration of 50 mg/ml by dilution with 150 mM NaCl, and kept frozen at −70° until use. The total cholesterol content of the lipoprotein-deficient serum should be <5% of the level in whole serum.

Assays of Surface Binding and Internalization of ^{125}I-LDL in Monolayers of Cultured Cells

Principle. ^{125}I-labeled LDL binds with high affinity to intact human fibroblasts and other cell types grown in monolayer culture in the presence of lipoprotein-deficient serum. When the binding is performed at 4°, 90% of the receptor-bound ^{125}I-LDL remains on the cell surface. When the cells are incubated with ^{125}I-LDL at 37°, the receptor-bound ^{125}I-LDL is internalized by the cells and degraded. The internalized ^{125}I-LDL is replaced at the receptor site by a new molecule of ^{125}I-LDL from the medium. After about 2 hr, a dynamic steady state is reached in which the amounts of ^{125}I-LDL bound to the receptor and contained within the cell are constant, but the cells are internalizing ^{125}I-LDL and excreting the degradation products at a steady rate.[2,12]

The binding of ^{125}I-LDL to the receptor is competitively inhibited by plasma lipoproteins containing apoprotein B (LDL), apoprotein E (canine apo-E-HDL$_c$), or both apoproteins B and E (VLDL and β-migrating VLDL).[24-26] Lipoproteins devoid of apoproteins B and E, such as typical HDL, do not compete for binding.[24,26] Binding of ^{125}I-LDL to the receptor absolutely requires a divalent cation, either Ca^{2+} or Mn^{2+}, and is inhibited by excess EDTA.[27] Receptor binding of LDL can also be prevented by

[24] M. S. Brown and J. L. Goldstein, *Proc. Natl. Acad. Sci. U.S.A.* **71,** 788 (1974).

[25] T. P. Bersot, R. W. Mahley, M. S. Brown, and J. L. Goldstein, *J. Biol. Chem.* **251,** 2395 (1976).

[26] T. L. Innerarity and R. W. Mahley, *Biochemistry* **15,** 1440 (1978).

[27] J. L. Goldstein, M. S. Brown, and R. G. W. Anderson, *in* "International Cell Biology 1976–1977" (B. R. Binkley, and K. R. Porter, eds.), p. 639. Rockefeller Univ. Press, New York, 1977.

chemically modifying the lipoprotein in one of two ways: by acetylation[28,29] or methylation[29] of the lysine residues; or by treatment of the arginine residues with cyclohexanedione.[30]

The rate of dissociation of receptor-bound ^{125}I-LDL is very slow at 0–4°. This dissociation can be enhanced by incubating the cells in the presence of polyanionic macromolecules such as heparin or dextran sulfate.[31] After incubation of cells with ^{125}I-LDL and subsequent washing, the amount of ^{125}I-LDL that can be released by treatment with heparin or dextran sulfate corresponds to the amount of ^{125}I-LDL bound to the high-affinity cell-surface receptors. The amount of ^{125}I radioactivity resistant to release by heparin or dextran sulfate corresponds to the amount of ^{125}I-LDL that is contained within the cells.[31]

The affinity of the LDL receptor for ^{125}I-LDL at 37° is lower than the affinity at 4°; half-maximal binding at 37° requires 10–15 μg of protein per milliliter of LDL as opposed to 1–1.5 μg of protein per milliliter at 4°. On the other hand, the maximal amount of ^{125}I-LDL bound to the surface receptors is two- to threefold higher at 37° than at 4°.[31]

To determine the amount of nonspecific binding, cells are incubated with ^{125}I-LDL (usually 10 μg of protein per milliliter) in the absence and in the presence of a level of unlabeled LDL that is at least 10-fold above saturation (usually 500 μg of protein per milliliter). The specific binding is calculated by subtracting the radioactivity bound in the presence of LDL from that in its absence. At an ^{125}I-LDL concentration of 10 μg of protein per milliliter, nonspecific binding in normal cells is generally less than 5–10% of the total binding.

Reagents

Medium 1 and medium 2 for incubation of cell monolayers with ^{125}I-LDL. Medium 1 consists of Eagle's minimal essential medium without bicarbonate supplemented with 10 mM N-2-hydroxyethylpiperazine-N'-2-ethanesulfonic acid (HEPES) and 10% (v/v) human lipoprotein-deficient serum at pH 7.4. Medium 2 consists of Eagle's minimum essential medium supplemented with 24 mM bicarbonate, 1% (v/v) nonessential amino acids, and 10% human lipoprotein-deficient serum at pH 7.4.

Buffer B and buffer C for washing the cell monolayers. Buffer B consists of 150 mM NaCl, 50 mM Tris-HCl, and 2 mg/ml of bovine serum

[28] S. K. Basu, J. L. Goldstein, R. G. W. Anderson, and M. S. Brown, *Proc. Natl. Acad. Sci. U.S.A.* **73,** 3178(1976).

[29] K. H. Weisbraber, T. L. Innerarity, and R. W. Mahley, *J. Biol. Chem.* **253,** 9053 (1978).

[30] R. W. Mahley, T. L. Innerarity, R. E. Pitas, K. H. Weisgraber, J. H. Brown, and E. Gross, *J. Biol. Chem.* **252,** 7279 (1977).

[31] J. L. Goldstein, S. K. Basu, G. Y. Brunschede, and M. S. Brown, *Cell* **7,** 85 (1976).

albumin at pH 7.4. Buffer C is the same as buffer B except that bovine serum albumin is omitted.

Buffer D for heparin or dextran sulfate release of receptor-bound ^{125}I-LDL. Buffer D consists of 50 m*M* NaCl and 10 m*M* HEPES at pH 7.4 to which either 10 mg/ml of heparin (Sigma Chemical Co., H-3125) or 4 mg/ml of dextran sulfate (Pharmacia, M_r 500,000) is added.

NaOH, 0.1 *N*

Procedure. Cells are grown in monolayer in 60-mm plastic petri dishes containing 10% fetal calf serum. They are used before confluency is reached. For the final 48 hr of cell growth, they are cultured in medium containing lipoprotein-deficient serum (see footnote legend to the table). Prior to a 4° binding experiment, the cells are placed for 30 min in a 4° cold room. The growth medium is then replaced with 2 ml of ice-cold medium 1 containing various concentrations of ^{125}I-LDL in the presence or the absence of a 50-fold excess of unlabeled LDL. Surface binding of ^{125}I-LDL reaches equilibrium within 1–2 hr at 4°. After equilibrium is reached, the cell monolayers are washed in a cold room three times rapidly with 3 ml of ice-cold buffer B. Each monolayer is then incubated twice for 10 min with 3 ml of ice-cold buffer B followed by one rapid wash with 3 ml of ice-cold buffer C. Each monolayer then receives 2 ml of buffer D (containing either heparin or dextran sulfate) and is placed on a rotary shaker (60 rotations per minute) for 60 min at 4°. Buffer D is then collected, and an aliquot (1 ml) is counted for determination of the amount of ^{125}I-LDL released from the cell surface (i.e., heparin-releasable ^{125}I-LDL). The cells are dissolved by incubation at room temperature for at least 15 min in 1 ml of 0.1 *N* NaOH. The cell suspension is removed quantitatively from the dish by trituration with a Pasteur pipette. One aliquot of the cell suspension (0.5 ml) is counted to determine the amount of ^{125}I-LDL that has been internalized by the cells (i.e., heparin-resistant ^{125}I-LDL), and another aliquot (50 μl) is used to determine the content of cellular protein by the Lowry procedure.[22] In experiments performed at 4°, at steady state 80–90% of the total cell-associated ^{125}I radioactivity is found in the heparin-releasable fraction. Nonspecific binding should be less than 5–10% of the total binding.

For measurement of the binding and uptake of ^{125}I-LDL at 37°, the cells are grown as indicated above. At the start of the experiment, the growth medium is removed and each dish receives 2 ml of warm (37°) medium 2 containing various concentrations of ^{125}I-LDL in the absence or the presence of excess unlabeled LDL. The cells are then placed in a CO_2 incubator and kept at 37° for at least 2 hr or until a steady-state cellular content of ^{125}I-LDL is reached. If degradation of ^{125}I-LDL is also to be measured, the incubations are prolonged to 5–6 hr to allow sufficient ^{125}I-labeled degradation products to accumulate in the medium (see below). The cells are then transferred to a 4° cold room, the medium is immediately removed and replaced with 3 ml of ice-cold buffer B. When desired, the medium is saved for measurement of ^{125}I-labeled degradation products (see below). The cells are then washed, incubated with heparin or dextran sulfate, and harvested exactly as described above for a 4° experiment. After a steady state is reached at 37°, about 15% of the total cell-bound radioactivity should be on the surface (heparin-releasable) and 85% should be within the cell (heparin-resistant). Nonspecific binding and internalization should be less than 5–10% of the total.

Assay of Proteolytic Degradation of ^{125}I-LDL in Monolayers of Cultured Cells

Principle. After the binding to high-affinity cell surface receptors at 37°, LDL is internalized in endocytic vesicles and transported to lysosomes where its protein and cholesteryl ester components are hydrolyzed.[5] The hydrolysis of the protein component can be quantified by following the fate of ^{125}I-labeled LDL incubated with intact fibroblasts at 37°.[12] The hydrolysis of the cholesteryl ester component of LDL can be quantified by following the fate of reconstituted LDL in which the cholesteryl ester core has been removed and replaced with [^{3}H]cholesteryl linoleate.[32,33] The assay for proteolytic hydrolysis of ^{125}I-LDL is quite simple and is described below; the assay for hydrolysis of [^{3}H]cholesteryl linoleate-labeled LDL is more involved and is described in detail elsewhere.[33]

Reagents

Trichloroacetic acid, 50% (w/v)
Potassium iodide, 40% (w/v)
Hydrogen peroxide, 30%

Procedure. Cells are incubated at 37° with ^{125}I-LDL as described in the preceding section. Afterward, the medium (2 ml) from each cell monolayer is removed and added to a glass tube containing 0.5 ml of 50% trichloroacetic acid to precipitate undegraded ^{125}I-LDL. After incubation at 4° for at least 30 min, the precipitable material is removed by low-speed centrifugation. An aliquot (1 ml) of the trichloroacetic acid soluble supernatant is mixed with 10 μl of 40% potassium iodide as carrier, followed by the addition of 40 μl of 30% hydrogen peroxide. The hydrogen peroxide converts any [^{125}I]iodide ions to [^{125}I]iodine. After thorough mixing the tubes are kept at room temperature for 5–10 min. Then 2 ml of chloroform is added and the free [^{125}I]iodine is extracted into the chloroform layer with

[32] J. L. Goldstein, S. E. Dana, J. R. Faust, A. L. Beaudet, and M. S. Brown, *J. Biol. Chem.* **250,** 8487 (1975).

[33] M. Krieger, M. S. Brown, J. R. Faust, and J. L. Goldstein, *J. Biol. Chem.* **253,** 4093 (1978).

Vortex agitation.[34] After 15 min at room temperature, an aliquot of 0.5 ml is removed from the upper aqueous layer and its content of ^{125}I radioactivity is determined. This aqueous material consists almost exclusively of [^{125}I]monoiodotyrosine obtained from the degradation of ^{125}I-LDL in cellular lysosomes.[12] The oxidation–extraction step eliminates any ^{125}I that contaminates the ^{125}I-LDL preparation and thus lowers the blank for the assay. However, if cells form ^{125}I during the incubation, this will be removed during the extraction.[12] Thus, the assay measures the formation of [^{125}I]monoiodotyrosine, but does not measure deiodination. As a result, it gives a slight underestimate of the total degradation of ^{125}I-LDL by the cells.[12]

The degradation of ^{125}I-LDL can be blocked if the intact cells are incubated with an inhibitor of lysosomal function, such as chloroquine at concentrations of 50–100 μM.[35] It is important that all cell incubations be accompanied by a "no cell" blank in which the ^{125}I-LDL containing medium is incubated at 37° in a petri dish containing no cells. The trichloroacetic acid-soluble radioactivity obtained from the "no cell" blank is subtracted from that obtained in parallel incubations with cell monolayers. The value for the "no cell" blank should represent <0.02% of the initially added ^{125}I radioactivity. All incubations are conducted in the absence and in the presence of excess unlabeled LDL, and the degradation mediated by the receptor uptake process is calculated by subtracting the value obtained in the presence of unlabeled LDL from that obtained in its absence. Nonspecific degradation should be less than 5–10% of the total.

Assay of HMG-CoA Reductase Activity in Cultured Cells

Principle. Cells are lysed in a detergent,[36] and the supernatant is incubated with ^{14}C-labeled HMG-CoA in the presence of NADPH. The [^{14}C]mevalonate formed is isolated by thin-layer chromatography using added [^{3}H]mevalonate as an internal standard for recovery.[37]

Reagents

Buffer E: 50 mM Tris-HCl and 150 mM NaCl at pH 7.4

Buffer F: 50 mM potassium phosphate, 5 mM dithiothreitol, 5 mM disodium EDTA, 200 mM KCl, and 0.25% (v/v) Kyro EOB at pH 7.4. Kyro EOB, obtained from Miami Valley Research Laboratories of Proctor and Gamble Co., is a synthetic nonionic detergent that solubilizes plasma membranes, but not endoplasmic reticulum, of

[34] E. L. Bierman, O. Stein, and Y. Stein, *Circ. Res.* **35,** 136 (1974).

[35] J. L. Goldstein, G. Y. Brunschede, and M. S. Brown, *J. Biol. Chem.* **250,** 7854 (1975).

[36] M. S. Brown, S. E. Dana, and J. L. Goldstein, *Proc. Natl. Acad. Sci. U.S.A.* **70,** 2162 (1973).

[37] D. J. Shapiro and V. W. Rodwell, *J. Biol. Chem.* **246,** 3210 (1971).

cultured cells.[36,38] A commercially available detergent, Brij 96 (Sigma Chemical Co.), has also been used for reductase assays in place of Kyro EOB.[39]

Buffer G: 200 m*M* potassium phosphate, 12 m*M* dithiothreitol, 40 m*M* D-glucose 6-phosphate (monosodium salt), and 4 m*M* NADPH at pH 7.4.

DL-3-hydroxy-3-methyl[3-^{14}C]glutaryl-CoA (624 μM; specific radioactivity ~20,000 dpm/nmol). This solution is prepared by diluting DL-[^{14}C]HMG-CoA (40–60 mCi/mmol; New England Nuclear) with unlabeled DL-HMG-CoA in 10 m*M* potassium phosphate at pH 5.5. The radioactive solution is stored in small aliquots at −20°.

DL-[2-^{3}H]Mevalonic acid lactone (0.1*M*; 5000 dpm/μl). This solution is prepared by diluting [^{3}H]mevalonic acid lactone (Amersham/Searle, 100–500 mCi/mmol) with unlabeled mevalonolactone in water (Fluka Chemical Corp., Hauppauge, New York).

HCl, concentrated

Diethyl ether (anhydrous)

Sodium sulfate (granular)

Silica gel G (without gypsum) thin-layer chromatography sheets with plastic backs (Brinkmann Instruments)

Benzene

Acetone

Aquasol scintillation fluid (New England Nuclear)

Preparation of Cell-Free Extracts. After incubation of cell monolayers with or without LDL, the medium from each petri dish is discarded and the cells are washed twice with 2 ml of ice-cold buffer E. The cells are scraped with a rubber policeman into 1 ml of ice-cold buffer E and placed in a glass test tube. Each dish is rinsed briefly with 1 ml of ice-cold buffer E, and this is added to the suspension of scraped cells. The cell suspension is centrifuged at low speed (900 *g*, 5 min, 4°). The resulting supernatant is discarded, and the cell pellet is frozen in a liquid nitrogen freeze and kept at −190° until the time of assay.

Cell-free extracts are prepared by dissolving the thawed pellet in 0.1 ml of buffer F at room temperature. The resulting detergent-solubilized suspension is centrifuged for 1 min at 12,000 rpm at room temperature in a Beckman microfuge, and aliquots of the supernatant are assayed both for HMG-CoA reductase activity (see below) and for protein content by the Lowry procedure,[22] after precipitation with 10% trichloroacetic acid.

Enzyme Assay Procedure. Aliquots of the cell-free extract (20–100 μg of protein in 50 μl of buffer F) are mixed with 100 μl of buffer G and 40 μl of

[38] P. J. Birckbichler and I. F. Pryme, *Eur. J. Biochem.* **33,** 368 (1973).

[39] I. Kaneko Y. Hazama-Shimada, and A. Endo, *Eur. J. Biochem.* **87,** 313 (1978).

water. Ten microliters of DL-[^{14}C]HMG-CoA are added, and the mixture (final volume, 200 μl) is incubated for 30–120 min at 37°. The final concentration of DL-[^{14}C]HMG-CoA in the assay is 31 μM at a specific activity of ~ 20,000 dpm/nmol. The K_m of the human fibroblast enzyme is approximately 5 μM for DL-HMG-CoA and 50 μM for NADPH under these assay conditions. The reaction is stopped by addition of 20 μl of 5 N HCl; 20 μl of the [2-^{3}H]mevalonolactone (2 μmol; ~ 100,000 dpm) are added as an internal standard, followed by addition of 50 μl of a solution containing 5 mg of unlabeled mevalonolactone, which acts as a marker for visualization on the thin-layer sheets. The mixture is incubated at 37° for 30 min to allow lactonization of the [^{14}C]mevalonic acid formed during the assay. The solution is then transferred to large glass tubes (20 × 150 mm) that contain ~ 1 g of granular sodium sulfate. The slurry is extracted twice with 10 ml of diethyl ether, which is added with a syringe. The ether extracts are poured into 50-ml conical tubes and evaporated to dryness under air at 37–40°. The residue is taken up in 60 μl of chloroform–methanol (2 : 1), spotted on plastic-backed silica gel thin-layer sheets, and placed in a tank containing benzene–acetone (1 : 1). The upper limit of the mevalonolactone band is visible on the dry chromatogram without staining. A section of each lane extending 0.5 in. below the top of the mevalonolactone band is cut out with scissors, placed in a vial containing 10 ml of Aquasol, and subjected to double-label scintillation counting. The R_f of mevalonolactone is about 0.7. The recovery of added [^{3}H]mevalonolactone averages 50%.

Several types of blank reactions are conducted: (*a*) "zero time blank"—addition of 5 N HCl to the reaction mixture prior to addition of enzyme; (*b*) "no extract blank"—omission of extract from the reaction mixture; and (*c*) "no substrate blank"—omission of NAPDH from the reaction mixture (in some cases prior dialysis of the extracts is required to remove endogenous NADPH). In all cases the assay blank for each tube should be $<$60 dpm of ^{14}C radioactivity.

Assay of Cholesteryl Ester Formation in Cultured Cells

Principle. Cells grown in lipoprotein-deficient serum synthesize sufficient cholesterol for growth, but do not synthesize excess cholesterol. When these cells are incubated with [^{14}C]oleate, there is little , if any, incorporation of [^{14}C]oleate into cholesteryl oleate.[7] However, when the cells have taken up LDL by the receptor route, cholesterol liberated from LDL activates ACAT.[40] When such cells are incubated with [^{14}C]oleate, there is a markedly enhanced incorporation of the [^{14}C]oleate into cholesteryl [^{14}C]oleate (see the table). The ability of LDL to stimulate this reaction is dependent on its

[40] M. S. Brown, S. E. Dana, and J. L. Goldstein, *J. Biol. Chem.* **250,** 4025 (1975).

ability to enter cells via receptor-mediated endocytosis. Hence, the degree of stimulation of cholesteryl [^{14}C]oleate synthesis is an index of the number of receptors in a given cell line. The ACAT enzyme is also stimulated when cells are incubated with oxygenated sterols, such as 25-hydroxycholesterol, added to the medium in ethanol.[40] This stimulation does not require the LDL receptor, and hence it serves as a check on the ability of the cells to develop enhanced ACAT activity once cholesterol is liberated in the cell. A standard screening protocol can be established in which replicate dishes of fibroblasts are grown in 10% lipoprotein-deficient serum and the individual dishes are incubated with LDL or with a mixture of 5 μg of 25-hydroxycholesterol plus 10 μg of cholesterol per milliliter (the cholesterol is added to ensure the presence of excess substrate for the ACAT reaction). After a suitable interval (usually 5 hr), the cells are pulse-labeled for 2 hr with [^{14}C]oleate and the incorporation into cholesteryl [^{14}C]oleate is measured. The ratio of cholesteryl [^{14}C]oleate synthesis in the presence of LDL divided by the synthesis in the presence of the sterol mixture serves as an index of the number of LDL receptors.

Reagents

Sodium [1-^{14}C]oleate–albumin complex (stock solution, 10 m*M* sodium [1-^{14}C]oleate at a specific radioactivity of ~ 10,000 dpm/nmol and an albumin concentration of 120 mg/ml). [1-^{14}C]Oleic acid in heptane (55–60 mCi/mmol) is obtained from New England Nuclear. An aliquot of [^{14}C]oleic acid (500 μCi) is added to a 25-ml glass Erlenmeyer flask and evaporated to dryness with a stream of nitrogen. The [^{14}C]oleate is resuspended in 8.7 ml of a solution containing 12.7 m*M* nonradioactive sodium oleate in complex with 12% (w/v) bovine albumin in 150 m*M* sodium chloride at pH 7.4 (see below), followed by the addition of 3.2 ml of 12% (w/v) bovine albumin in 150 m*M* NaCl at pH 7.4 (see below). The solution is stirred gently with a magnetic stirrer for 4–6 hr at room temperature, distributed in multiple 1-ml aliquots in glass tubes, and kept frozen at −20° until use. In calculating specific radioactivity, no correction is made for the small amount of endogenous oleate bound to the bovine albumin.

Solution of 12.7 m*M* sodium oleate in complex with 12% (w/v) albumin in 150 m*M* NaCl. This solution is prepared by a modification of the method of Van Harken *et al.*[41] Ninety milligrams of oleic acid (Applied Science) are transferred to a 50-ml glass beaker containing 2 ml of ethanol; 100 μl of 5 *N* NaOH are added to the beaker, and the contents are thoroughly mixed. The ethanol is removed by evaporation under a stream of nitrogen, after which 10 ml of 150 m*M* NaCl solution are added to the dried sodium oleate. The clear solution is

[41] D. R. Van Harken, C. W. Dixon, and M. Heimberg, *J. Biol. Chem.* **244,** 2278 (1969).

heated for 3–5 min at 60°, after which the beaker is placed on a stirring plate at room temperature and stirred with a magnetic stirrer. While the fatty acid solution is still warm, 12.5 ml of ice-cold 24% (w/v) bovine albumin in 150 m*M* NaCl (see below) are added rapidly, and the clear solution is stirred for 10 min. The final volume is adjusted to 25 ml with 150 m*M* NaCl and kept frozen at −20° until use. The 24% (w/v) bovine albumin solution is prepared by adding 12 g of bovine albumin (fraction V, Reheis Chemical Co., Armour Pharmaceutical Co., Phoenix, Arizona) to 35 ml of 150 m*M* NaCl. The albumin is added in 6 aliquots (2 g per aliquot) over a 5-hr period while the mixture is being stirred at room temperature. The solution is adjusted to a final pH of 7.4 with 5 *N* NaOH and the final volume is adjusted to 50 ml with 150 m*M* NaCl. The solution is kept at −20° until use.

Buffers B and C for washing cell monolayers. Buffer B consists of 150 m*M* NaCl, 50 m*M* Tris-HCl, and 2 mg/ml of bovine serum albumin at pH 7.4. Buffer C is the same as buffer B except that bovine serum albumin is omitted.

[1,2-^{3}H]Cholesteryl oleate (50 μg; ~25,000 dpm). [^{3}H]Cholesteryl oleate is synthesized from [1,2-^{3}H]cholesterol (Amersham/Searle, 43 Ci/mmol) and oleoyl chloride in dry pyridine as described by Goodman.[42] The resulting [^{3}H]cholesteryl oleate is isolated as described[43] and diluted with unlabeled cholesteryl oleate to give the indicated specific radioactivity.

Silica gel G (without gypsum) thin-layer chromatography sheets with plastic backs (Brinkmann)

Hexane

Isopropanol

Ethyl ether (anhydrous)

Glacial acetic acid

Heptane

Aquasol scintillation fluid (New England Nuclear)

NaOH, 0.1 *N*

Assay Procedure for Cholesteryl [^{14}C]Oleate Formation. Each cell monolayer is incubated at 37° in a 5% CO_2 incubator in a 60-mm petri dish containing 2 ml of culture medium with 5% (v/v) human lipoprotein-deficient serum and varying concentrations of LDL (see the table) or 25-hydroxycholesterol plus cholesterol. After the indicated time of preincubation (usually 5 hr), each monolayer receives 20 μl of sodium [^{14}C]oleate–albumin complex (10 m*M* [^{14}C]oleate–1.2 mg/ml albumin, ~10,000 dpm/

[42] DeW. S. Goodman, this series, Vol. 15, p. 522.

[43] J. R. Faust, J. L. Goldstein, and M. S. Brown, *J. Biol. Chem.* **252,** 4861 (1977).

nmol of [^{14}C]oleate). After a 2-hr incubation at 37°, each monolayer is washed twice with 2 ml of buffer B, followed by one wash with buffer C. Two milliliters of hexane–isopropanol (3 : 2) is added to each dish and incubated for 30 min at room temperature.[44] The organic solvent is removed, each monolayer is rinsed briefly with 1 ml of solvent of the same composition, and the two organic solvent extracts are combined in a glass tube. An internal standard containing [^{3}H]cholesteryl oleate (50 μg; ~25,000 dpm), unlabeled triolein (50 μg), and unlabeled oleic acid (50 μg) in chloroform–methanol (2 : 1) is added directly as a single 10-μl addition to each tube. Each tube is agitated on a Vortex and allowed to stand at room temperature for at least 30 min. The solvent is then evaporated to dryness under air, and the lipids in each tube are resuspended in 60 μl of hexane and spotted on a silica gel G thin-layer chromatogram. The chromatogram is developed in heptane–ethyl ether–acetic acid (90 : 30 : 1), and the cholesteryl ester spot is identified with iodine vapor ($R_f = 0.9$), cut from the chromatogram, and counted in 10 ml of Aquasol.

After the lipids are extracted *in situ* from the petri dish, the cells in each monolayer are dissolved in 1 ml of 0.1 *N* NaOH and aliquots are removed for protein determination by the Lowry procedure.[22]

Acknowledgment

This work was supported by Research Grant HL-20948 from the National Institutes of Health.

[44] M. S. Brown, Y. K. Ho, and J. L. Goldstein, *J. Biol. Chem.* **255**, 9344 (1980).

[20] Binding, Endocytosis, and Degradation of Enveloped Animal Viruses

By MARK MARSH, ARI HELENIUS, KARL MATLIN, and KAI SIMONS

The assays described in this chapter are designed to monitor the interaction of viruses with monolayer tissue culture cells. The techniques used in studies with Semliki Forest virus (SFV; togaviridae) and baby hamster kidney (BHK21) cells are described in detail. The adapted procedures for vesicular stomatitis virus (VSV; rhabdoviridae) and fowl plague virus (FPV; orthomyxoviridae) interaction with Madin–Darby canine kidney (MDCK) cells are also described. Metabolically radiolabeled viruses are used: [^{35}S]methionine, [^{3}H]uridine, [^{3}H]mannose, or [^{32}P] is incorporated into the

ISBN 0-12-181998-1

viral protein, glycoprotein, RNA, and phospholipid,[1-4] allowing the fate of these components to be followed individually. The cell-surface viruses are distinguished from internalized viruses biochemically using hydrolytic enzymes, and the degradation of viral proteins is determined by acid precipitation and polyacrylamide electrophoresis. We found it useful to validate biochemical data with morphological observation of the cells. Viral particles are readily recognized in thin sections by electron microscopy, and specific antibodies for immunofluorescence microscopy can be easily obtained.

The following viruses and cells have been used: SFV, a prototype strain (School of Hygiene and Tropical Medicine, London); FPV, Rostock strain; VSV, Indiana serotype; BHK21 cells, and MDCK cells (cloned and selected for growth rate and ability to form blisters) obtained from EMBL, Heidelberg. Growth conditions and preparation methods for viruses have been described.[1,3-5]

Binding of SFV to BHK21 Cells

To study virus binding on live cells, low temperatures (0–4°) must be used to prevent endocytosis.[5,6] We have used two types of binding medium (BM): RPMI-1640, without bicarbonate (GIBCO), buffered to pH 6.8 with 10 m*M* 4-(2-hydroxyethyl)-1-piperazineethanesulfonic acid (HEPES) and HCl; and G-MEM (without bicarbonate and similarly buffered). Virus binding to cells is highly pH sensitive,[7] and careful buffering is essential: the absence of bicarbonate enables use of the media in normal atmosphere without changes in pH. Many enveloped animal viruses have potent low pH-induced membrane fusion activities[8,9]; pH values above those at which fusion occurs should be used. Serum reduces binding to cells and is replaced with 0.2% bovine serum albumin (BSA) to help prevent nonspecific interactions.

The cells, on 35-mm plastic tissue culture dishes (3×10^6 to 4×10^6 cells per dish), are placed on ice, the growth medium is aspirated, and the

[1] L. Kääriäinen, K. Simons, and C.-H. von Bonsdorff, *Ann. Med. Exp. Biol. Fenn.* **47,** 235 (1969).

[2] A. Helenius and H. Söderlund, *Biochim. Biophys. Acta* **307,** 287 (1973).

[3] K. S. Matlin, H. Reggio, A. Helenius, and K. Simons, *J. Mol. Biol.* **156,** 609 (1982).

[4] K. S. Matlin, H. Reggio, A. Helenius, and K. Simons, *J. Cell Biol.* **91,** 601 (1981).

[5] A. Helenius, J. Kartenbeck, K. Simons, and E. Fries, *J. Cell Biol.* **84,** 404 (1980).

[6] M. Marsh and A. Helenius, *J. Mol. Biol.* **142,** 439 (1980).

[7] E. Fries and A. Helenius, *Eur. J. Biochem.* **97,** 213 (1979).

[8] J. White and A. Helenius, *Proc. Natl. Acad. Sci. U.S.A.* **77,** 3273 (1980).

[9] J. White, K. Matlin, and A. Helenius, *J. Cell Biol.* **89,** 674 (1981).

cells are washed twice with 1.0 ml of Hank's balanced salt solution or BM (0°). Viruses in a minimum volume of cold BM (0.5 ml for a 35-mm plate) are added, and the plates are placed on ice and rocked, in a cold room, for 1 hr. The half-time of association of free SFV with the cell surface is about 45 min under these conditions.[7]

The unbound viruses are removed, and the monolayer is washed twice with 1.0 ml of BM at 0°. Washing with other media, especially phosphate-buffered saline (PBS), may cause dissociation of bound viruses. The cells are scraped from the dish in a total of 10 ml of BM at 0° using a silicon rubber scraper, collected in glass tubes, centrifuged at 1500 rpm for 5 min, resuspended in BM, and recentrifuged. The resulting cell pellets are dissolved directly in a scintillation fluid designed for aqueous samples. The radioactivity associated with the cell pellet is a direct measure of the number of viruses bound. Some viruses bind nonspecifically to tissue culture plastic; therefore we have found it more reliable to measure the cell-associated viruses rather than the viruses remaining in the medium after binding. Similarly, if cells are removed from the dishes by detergent, or NaOH, exaggerated values are obtained due to elution of viruses bound to plastic and inclusion of viruses that may have been trapped in the spaces between cells.

Immunofluorescence and electron microscopy of vertical thin sections are valuable indicators of the distribution of viruses in the cell monolayer after the initial washes. Established procedures have been used to prepare samples for electron microscopy.[3-5] Direct (rhodamine- and fluorescein-labeled SFV) and indirect (affinity-purified rabbit antibodies to virus spike glycoproteins, and rhodamine or fluorescein conjugated goat anti-rabbit IgG) techniques have been used for immunofluorescence.[3-5] Note that for biochemical assays both high and low multiplicities of viruses to cells can be used; however, in order to observe the interaction of viruses and cells by microscopy, high multiplicities of viruses should be used.

Endocytosis of SFV in BHK21 Cells

Internalization of SFV into BHK21 cells occurs when the cells are warmed to 37°.[6] Rapid warming of the cells is achieved by placing the plates at 37° and adding 37° medium (1 ml per plate). Under these conditions a half-time for virus particles on the cell surface of 8–10 min is measured.[6] Alternatively, cold plates can be placed in a 37° incubator, which gives synchronous, but slow, warming of the cells; the half-time of a virus particle on the cell surface under these conditions is 20–25 min.[1] Sodium bicarbonate (2.2 g liter^{-1}) should be added to BM if a CO_2 incubator is to be used. Several plates are always kept at 0° as time zero points.

Proteinase K (Boehringer Mannheim GmbH) digestion is used to differentiate between internalized viruses and those remaining on the cell surface.

This enzyme releases SFV by removing the SFV receptors on BHK21 cells; it works efficiently at 0° (90–100% of the bound virues are removed) and is inhibited by phenylmethanesulfonyl fluoride (PMSF) or aprotinin (both Sigma Chemical Co., St. Louis, Missouri). It also releases BHK21 cells from the substrate, facilitating quantitative cell recovery and washing. Trypsin can also be used to release SFV from BHK21s cell but is less efficient.

The procedure is as follows: duplicate plates of cells are cooled to 0°, the medium is removed and kept on ice, and the cells are washed once with 0° BM. The cells from one plate are scraped as above, without protease digestion, to determine total cell-associated virus. To the other plate 1.0 ml of 0.5 mg ml^{-1} proteinase K in PBS with Ca^{2+}, Mg^{2+} (0°) is added, and the plate is placed on a platform shaker (essential). After 45 min the cells, released from the plate by the enzyme, are collected by pipette into glass tubes, and 1 ml of 2 m*M* PMSF in PBS with Ca^{2+}, Mg^{2+} (newly diluted from a fresh 40 m*M* stock in absolute ethanol) is added. The tubes are made up to 10 ml with PBS with Ca^{2+}, Mg^{2+}, containing 0.2% BSA, centrifuged as above, resuspended in PBS–BSA, and centrifuged once more; the cell pellets are dissolved in scintillation fluid. The activity associated with the pellets corresponds to the internalized portion of the cell-associated viruses. The method can equally well be used in experiments in which virus is added to cells at 37° rather than prebound in the cold; however, endocytosis must be stopped by cooling the cells to 0° prior to protease digestion. Thin-section electron microscopy and immunofluorescence microscopy can be used to confirm that proteinase K does remove the cell-surface viruses and that viruses are internalized at 37°.

Degradation of SFV in BHK21 Cells

To determine how much of the intracellular [^{35}S]methionine radioactivity represents degraded viruses, the cell pellets after proteinase K treatment are dissolved in 0.2% sodium dodecyl sulfate (SDS). The lysates are made 10% with 0° trichloroacetic acid (TCA) and left on ice for 1 hr. Subsequently, the precipitates are sedimented (5 min in a table-top centrifuge), and an aliquot of the supernatant is counted. The activity in the supernatant corresponds to degraded viral proteins. The integrity of labeled proteins in the TCA precipitates may be analyzed by SDS–polyacrylamide gel electrophoresis. With SFV, very little intracellular TCA-soluble activity is found at any time.[6]

The medium, removed from cells after incubation at 37°, is similarly analyzed with TCA. With [^{35}S]methionine-labeled virus most of the degraded radioactive protein is found in this fraction. With SFV, the acid-precipitable fraction of the medium radioactivity is usually low and corresponds mainly to viruses eluted from the cell surface. To be degraded the viruses have to be internalized; therefore, the TCA-soluble radioactivity in the medium should be added to the intracellular (proteinase K resistant) and total cell-associated radioactivity to obtain a true value for the total internalized activity.

Uncoating of the SFV Nucleocapsid

The nucleocapsids of SFV are ribonuclease (RNase) sensitive.[8] This property can be exploited to measure the penetration of the viral nucleocapsids to the cytosol and the release of the nucleocapsids from the protective viral envelope.[10,11] [^{3}H]Uridine-labeled SFV is bound to cells at 0° as above, the unbound viruses are washed away, and the cells are warmed to 37°. To measure uncoating, the plates are returned to 0° by replacing the 37° medium with 0° medium, and the cells are scraped and collected in glass tubes. The cells are washed once with 10 ml of BM, then transferred to 1.5-ml Eppendorf tubes in 1.0 ml of 0.25 *M* sucrose containing 0.01 *M* triethanolamine and 0.01 *M* acetic acid, pH 7.5, at 0°. They are pelleted in a table-top centrifuge, resuspended in 0.5 ml of the same buffer containing 1 m*M* EDTA, and lysed by 30 passes through a 24-gauge needle.[11,12] More than 99% lysis is obtained, and 70–90% of the lysosomal hydrolases remain latent. Nucleases [0.5 mg, 37 Kunitz units, RNase A and 1 unit of micrococcal nuclease (both Sigma), in 0.05 ml of buffer] are added to each lysate for 30 min at 37°, and the lysate is subsequently precipitated for 1 hr by addition of an equal volume of 20% 0° TCA. The precipitates are removed by centrifugation, and the radioactivity is determined in aliquots of the supernatants. The TCA-soluble radioactivity reflects the viral RNA that has been uncoated and released into the cytoplasmic compartment. Several controls are included: (*a*) lysates not treated with nucleases; (*b*) a sample of viruses in the absence of cells; (*c*) lysates of cells not incubated at 37°, (*d*) cells warmed to 37° in the presence of penetration inhibitors such as ammonium chloride[11]; and (*e*) detergent-treated virus. Freeze–thawing has been used as an alternative method of opening cells.[10] In our hands, this has not proved to be as reliable as the methods described.

FPV and VSV Interaction with MDCK Cells

The MDCK cells are of epithelial origin, and in culture they form differentiated polarized monolayers. With confluent monolayers on tissue

[10] W. C. Koff and V. Knight, *J. Virol.* **31,** 261 (1979).

[11] A. Helenius, M. Marsh, and J. White, *J. Gen. Virol.* **58,** 47 (1982).

[12] E. Harms, H. Kern, and J. A. Schneider, *Proc. Natl. Acad. Sci. U.S.A.* **77,** 6139 (1980).

culture plastic (1×10^6 to 2×10^6 cells per 35-mm dish) only the apical microvillar surfaces are available for binding viruses. Cells do not overgrow each other, and the intercellular spaces are isolated from the apical surface by junctional complexes between the cells. The junctions are resistant to proteases applied apically,[3,13] and, since the cells are not removed from the dish by proteases, they must be scraped after enzyme treatment.

The recommended binding medium in studies with FPV is Earle's minimum essential medium (MEM) (growth media for MDCK cells) without bicarbonate, containing 0.2% (w/v) BSA and buffered to pH 7.4 with 10 m*M* 3-(*N*-morpholino)propanesulfonic acid (MOPS), 10 m*M* *N*-Tris(hydroxymethyl)methyl-2-aminoethanesulfonic acid (TES), 15 m*M* HEPES, and 2 m*M* NaH_2PO_4.[4] Cell surface-bound FPV is removed using 5 mg ml^{-1} neuraminidase for 90 min at 0° (type V, purified from *Clostridium perfringens*, 0.86 unit mg^{-1} using NAN-lactose; Sigma). Neuraminidase removes the sialic acid from glycoproteins and glycolipids and thereby removes the receptors for influenza virus.[14] The FPV carries its own neuraminidase, which is ineffective at low temperature, but on warming, a large fraction of bound FPV is released into the medium through the action of the viral neuraminidase.[15] This phenomenon is equivalent to the classical reversal of hemagglutination observed for influenza virus.[15]

As binding media in VSV studies we have used Earle's MEM, without bicarbonate, containing 0.2% BSA and buffered to pH 6.3 with 10 m*M* bis(2-hydroxyethyl)amino–tris(hydroxymethyl)methane (BIS–TRIS), 10 m*M* 1,4-piperazine-diethanesulfonic acid (PIPES), 10 m*M* MOPS, and 10 m*M* NaH_2PO_4. It should be noted that the binding of VSV to MDCK[3] and BHK cells[16] is low and variable and requires long incubation times (>2 hr) to reach plateau levels. In addition, the binding of VSV to MDCK cells[3] shows a strong dependence on pH; a 10-fold increase in binding is observed at the optimal pH 6.5 compared to pH 7.4. However, even at optimal pH the binding remains low and irreproducible. Proteinase K has been very effective in removing bound VSV from MDCK cells over a broad concentration range. Trypsin also removes bound viruses, but again the backgrounds are higher. As with SFV, proteases probably degrade the receptors. A potential nondestructive method of removing VSV from MDCK cells is treatment with 20 m*M* EDTA in 0° PBS (without Ca^{2+} and Mg^{2+}); 70% of the bound viruses are removed in 5 min.[17]

[13] A. Martinez-Palomo, I. Menza, G. Beaty, and M. Cerejido, *J. Cell Biol.* **87**, 736 (1980).

[14] I. T. Schulze, *in* "The Influenza Viruses and Influenza" (E. D. Kilbourne, ed.), p. 53. Academic Press, New York, 1975.

[15] B. D. Davis, R. Dulbecco, H. N. Eisen, H. S. Ginsberg, and W. B. Wood, "Microbiology," 2nd ed. Harper & Row, New York, 1973.

[16] D. K. Miller and J. Lenard, *J. Cell Biol.* **84**, 430 (1980).

[17] M. Pesonen, K. S. Matlin, and K. Simons, unpublished results.

Acknowledgments

A. Helenius was supported by Grant AI-18582 from the National Institutes of Health and M. Marsh by a fellowship from the European Molecular Biology Organization. K. Matlin was a European Molecular Biology Laboratory fellow.

[21] Image Intensification Techniques for Detection of Proteins in Cultured Cells

By MARK C. WILLINGHAM and IRA H. PASTAN

The principles and applications of some image intensification techniques currently available to record weak images from fluorescence microscopes are described here. The major application has been in the study of the endocytosis of various fluorescently labeled ligands by cultured cells. However, the principles governing these experiments and the technical problems encountered are relevant for other applications. These techniques allow the recording of images generated by the presence of extremely small amounts of fluorescently labeled material on or in living cells; at the same time, they obtain the highest possible resolution and the lowest possible damage to both the cells and the fluorochromes used for labeling. These techniques have allowed direct visualization of the movement and fate of proteins inside living cells over relatively long periods of time. The combination of these intensification techniques and direct microinjection of fluorescently labeled proteins[1] have allowed direct experiments on living cells not possible previously. This review is intended as a practical guide to the use of these techniques.

General Principles

Image Recording—Sensitivity

The human eye is a detection system of extremely high resolution and sensitivity. It responds to a certain number of photons per unit time in a resolvable small part of the total image field and accurately registers that light intensity in correct spatial relationship to the rest of the field. For images of limited intensity, the retina reaches a lower limit at some point where it fails to respond faithfully in that small resolution element to the number of photons per unit time per unit area, and the image begins to lose recognition and detail, and finally is not be recorded at all. The eye,

[1] D. L. Taylor, Y.-L. Wang, and J. M. Heiple, *J. Cell Biol.* **86,** 590 (1980).

ISBN 0-12-181998-1

therefore, is a real-time sensing device that cannot store the accumulated image for any long period of time. This property of requiring the registration of an image within a short time frame is similar to many man-made devices, such as video cameras, which either scan or in some way temporally relate the amount of light in an image field.

A different form of image recording is typified by photographic film. While it too has resolution and sensitivity limits, the image can be accumulated over a long period of time. This property of accumulation makes it possible for the astronomer to record the image of a distant star by accumulating the same number of photons on the image field over a period of hours rather than a fraction of a second, unlike the human eye or the video camera. In practical terms, however, many images cannot be accumulated over many hours because the object being photographed moves or for some other reason disappears during the observation period. Thus, image recording can be either temporally limited or cumulative depending on the detection system.

Image Recording—Resolution

The retina has a very high degree of resolution compared to most types of photographic film. An extremely sensitive detection system does little good if its resolution cannot create a clear image of the object of interest. The extreme example, of course, is the single photomultiplier tube, which has essentially only a single picture element. Expand those picture elements by millions,[2] however, and the image recording quality satisfies almost any application. Man-made detection systems all have limits to their resolution. Often, as with photographic films, the higher the sensitivity, the worse the resolution (the grainier the image). With scanning systems, such as video cameras, the resolution is to some extent a function of the scanning rate; the slower the scan rate, the better the resolution. Practical considerations make resolution a constant factor in image intensifier systems, and for many microscopic images the limit of usefulness of a particular system is almost always a combination of sensitivity and resolution together. Inherent in the problem of resolution is the question of intrinsic noise in the system; the signal-to-noise ratio of any intensifier system generally decreases at higher gain or intensification levels. Noise can be due to electronic causes, cosmic rays, or spontaneous silver reduction, depending on the detection system.

One method of increasing sensitivity is to project the same image on a smaller area of the detector, thus increasing the number of photons per area on the target. Such an approach is quite useful, for example, when observing a weak image by eye in the microscope directly. By changing eyepieces from

[2] M. Lampton, *Sci. Am.* **245,** 62 (1981).

15× to 10×, or from 10× to 5×, the same number of photons per unit time is projected on a smaller area of the retina, greatly increasing the brightness of the image. The balance is that the spatial resolution is now much smaller. If the image requires very little resolution, this creates no problems; but if the resolution required to see the details of interest is high, then the image becomes a bright blur.

In general, a practical balance must be created between sensitivity, time, and resolution to achieve a useful system.

Fluorescence Microscope Optics

Fluorescence Methods

Fluorescence microscopy is a common, accepted technique for visualizing small numbers of molecules selectively in a biological specimen. By coupling fluorochromes to antibodies, affinity labels, or purified cell components, the parts of tissue and cells that contain them can be identified with high sensitivity. The two most common fluorochrome dyes used for this purpose are fluorescein and rhodamine. Fluorescein can be excited with narrow-band blue-violet light (450 nm) and fluoresces in the yellow–green range (520 nm). Rhodamine can be excited with narrow-band green light (546 nm) and fluoresces in the red range (590 nm). These two dyes can be used simultaneously on the same specimen, and both images can be selectively recorded in double-label experiments.

For experiments with low amounts of fluorescence, rhodamine has certain advantages over fluorescein. Living cells are damaged less by green excitation than by blue. Cells have less autofluorescent material in them in the rhodamine spectrum than in the fluorescein spectrum. Rhodamine bleaches at a much slower rate than fluorescein. The overall total light output over its lifetime is much higher for rhodamine than fluorescein, in spite of the initially higher light output of fluorescein. Many of the sensitive detection systems used, such as SIT video cameras (see below), are as sensitive or even more sensitive to red light. Derivatives of both rhodamine and fluorescein that covalently bind to proteins are commercially available. Thus, for use with image-intensification systems, rhodamine is the preferred fluorochrome.

Commercially available optical microscopes are quite advanced, and a number of manufacturers produce high-quality instruments. The requirements of image-intensifier systems, however, demand the highest possible light gathering and resolution in optics. Furthermore, the efficiency of fluorescence illumination systems should be as high as possible. For this reason, microscopes equipped with epifluorescence illumination are preferred, since the high-resolution objective acts as both condenser and objective. In addition, total specimen thickness is of no importance, and separate condenser systems are not required, such as in dark-field fluorescence. These epifluorescence systems are usually available as a cube insert composed of exciter filter, dichroic reflector, and barrier filter in one unit. In Zeiss standard microscopes, for example, a double cube that allows two independent channels to be rapidly switched back and forth on the same specimen with one movement is standard. This is a considerable advantage in double-label experiments. These systems can be used with the same ease in either upright or inverted microscopes. Also, a large number of selective, special filter combinations are available from Zeiss, which allow narrow-band excitation or emission comparisons using small interference filters. These filter-reflector systems produce bright, clear fluorescence images with a minimum of effort. They also allow simultaneous observations by phase-contrast (transmitted light) from a second light source and condenser.

Microscope Optics

Since the amount of light being emitted from fluorescent specimens is very small, it is important to use optics with the highest light-gathering ability possible. This generally requires the use of high-numerical-aperture objectives, usually Planapochromats. With Zeiss microscopes, the most useful objectives are the 63×, N.A. (numerical aperture) 1.4, oil Planapochromat (usually purchased with a phase ring) (No. 461341); the 40×, N.A. 1.0, oil Planapochromat (with phase ring) (No. 461747); or the 25×, N.A. 0.8, oil Plan-neofluar (with phase ring) (No. 461626). Although all of these objectives are expensive, their superior light-gathering and resolution abilities are extremely valuable. In many instances with images inside single cells requiring image intensification, only the 63×, N.A. 1.4 objective produces a detectable image. One of the disadvantages of such high-numerical-aperture objectives is their very short working distance. This requires the use of thin coverslip interfaces with the cells under study (see below). On upright microscopes, it is occasionally possible to use direct-immersion objectives in the culture medium, and the Plan-neofluar (oil–water–glycerin immersion) objectives with correction collars are quite useful.

Since we generally wish to have the ability to record phase-contrast images of cells at the same time as fluorescence images, we always use objectives with phase rings. While this theoretically should reduce the sensitivity for use with fluorescence, we have never found any practical difference. Phase contrast has proved to be the most useful transmitted light imaging technique because of its convenience in combination with fluorescence. Differential interference contrast methods could be used, but they require other objects in the light path beyond the objective, which must be removed for maximum fluorescence sensitivity and cannot be used with plastic culture chambers or dishes, which are very convenient to use. Since most high-magnification objectives require a large (phase 3) condenser phase ring with a short specimen to condenser working distance, the most practical condensers are the intermediate (0.63 N.A.) working-distance condensers (e.g., IV Z/7), which can accommodate the phase 3 condenser ring. The working distance is then adequate for use of a standard 35-mm culture dish with cover as the culture chamber. With short working distance objectives, the 1-mm thickness of the dish is too great to focus on cells grown on the bottom surface. Substituting a thin (No. $1-1\frac{1}{2}$) coverslip for the bottom of the dish makes this an ideal chamber (see below).

Since the image from some fluorescence samples is quite low, it is important to remember that many of these can be visualized by eye if the optics are properly arranged. For example, while in the standard microscope with 10× eyepieces the images used with intensifier systems may not be visible by eye, the same image may be made visible by reducing the field size to place it on a smaller part of the retina by exchanging the 10× eyepieces with 5× or 3.2× eyepieces. The resolution on the retina is lower and the eyepiece image is generally not parfocal with the other optics of cameras in the microscope, but the image is visible. In practical terms, this can be quite useful.

Intensifier-Microscope Connections

Intensifier systems generally have a flat plane target onto which the microscope image should be focused. The goal of this interface is to present the part of the microscope output field for which one wants to produce an image in a flat focal plane precisely on the target. For video cameras the target lies right in front of the vidicon, generally a circular target ~ 1 in. in diameter. For the EMI intensifier described below, the target is a 2-in.-diameter photocathode located at the front end of the intensifier tube buried deep within the magnetic focusing coils. In any case, the microscope image has to be projected in a flat plane on the target, and this is most easily done by using an eyepiece and placing the target a certain distance from the eyepiece depending on the target size. It might be important not to put the entire microscope output on the target, so mechanical positioning and use of eyepieces of differing focal lengths can be selected. The goal in this interface is to produce a stable mechanical lighttight junction between microscope and intensifier. Zeiss has many ways in which to handle the microscope image, but the best arrangement is one that allows 100% of the output to be focused on the target, generally with a way of putting the image back onto an exterior eyepiece to allow preliminary alignment of the image. Photo-changers or image prisms made for regular camera outputs are quite useful for this purpose. For video cameras, Zeiss has an adapter (C mount) also compatible with cine cameras which fits the same dovetail mounts as the basic camera body mounts (Zeiss part No. 477921) and can be mounted beyond the eyepiece output of most of their microscopes. With less forgiving intensifier systems, it becomes crucial not to expose them to sudden high light levels during operation. As a result, it is important to have output control devices such as photochangers or shutters to ensure that too much light does not expose the intensifier target inadvertently. For the EMI intensifier shown below, this is accomplished by previewing the microscope output by a SIT camera through a photochanger (Zeiss No. 473051). As this implies, the SIT intensifier cameras are quite forgiving and have internal control protection circuits that minimize damage to the tube due to accidental light exposure (see below).

It is also of value to be able to control the illumination of the specimen externally. Because of the complexity of the entire microscope system, it is convenient to have remote shutter controls on all light sources in a central location so that the operator can selectively control light input and, therefore, output while focusing or observing the image. With the large physical size of some microscopes this is difficult to do without looking away from the field or producing vibrations in the microscope itself. Since image-intensification systems can have three or four separate image outputs to be concerned with, the more centralized and simpler the controls of the whole system are, the better.

Culture Chambers

High-numerical-aperture objectives have short working distances to the specimen, often less than 0.1 mm. This requires the use of culture systems that have thin coverslips between the cells and the end of the objective, generally with immersion oil between the objective and the coverslip. One simple and convenient solution to this problem with inverted microscopes is to modify a standard 35-mm plastic culture dish by cutting a circular 22-mm-diameter hole in its bottom surface using a high-speed grinder (Dremel) and a cutting burr. This hole is then covered by gluing a 25-mm circular (No. 1) coverslip over the hole using silicon adhesive (Dow-Corning Silastic). After this adhesive dries (24 hr), the coverslip dish can then be sterilized by UV light and handled as a normal culture dish. For observations, the dish is then viewed using an inverted microscope from below through the thin coverslip. When the experiment is finished the dish can be cleaned, resterilized, and used again. This system has the advantages of a thin substrate for short-working-distance objectives, but without the complexities of some culture chambers.

For upright microscopes or for general use, a number of high-quality culture chambers have been designed,[3,4] some of which are commercially available (Zeiss Dvorak-Stotler chamber,[4] Carl Zeiss, Inc.; Bionique-Gabridge chamber, Bionique, Inc.). These provide the short working distance needed at the coverslip interface, and some can be used with higher-numerical-aperture, short-working-distance condensors.

For observations of living cells, the microscope should be equipped with a method of maintaining an appropriate temperature (e.g., 37°) as well as the proper CO_2–air mixture if vented chambers are to be used with bicarbonate-buffered media. Our solution to this has been to enclose the stage, both above and below, in an accessible plastic box and to recirculate the atmosphere within the box through a heater–recirculator device.[5] The temperature is controlled by a proportional temperature controller (Yellow-Springs Model No. 72), and a small fan constantly recirculates the atmosphere. In addition, a CO_2–air mixing device (Napco Model 3634) constantly adds 4–8 liters/min of the proper atmosphere (e.g., 10% CO_2–90% air). This system maintains the temperature within $\pm 0.1°$ and allows rapid changes in temperature when desired for temperature-shift experiments (between 23° and 42°). Some of these components are shown in Figs. 1 and 2.

Image Intensifiers

Video Systems

A standard plumbicon video camera has a target sensitivity for a useful image of ~ 1 fc. More sensitive targets (Newvicon, Ultricon) are available with a sensitivity of ~ 0.1 fc. Beyond this, intensification devices are available that amplify the incoming light through one, two, or three stages of electron acceleration intensification before hitting the target. One of these cameras, the silicon intensifier target (SIT) video camera has a useful sensitivity of ~ 10^{-5} fc. While the resolution of the image decreases somewhat, the immense gain in sensitivity makes this a very useful device.[5] A number of manufacturers now produce these intensifier-video cameras.[5,6] Our experience has been with those made by RCA (Models TC-1030/H and TC-1040/H). The three-stage TC-1040/H has such low resolution that we have not been able to use it for most purposes, in spite of its greater sensitivity. The TC-1030/H SIT camera has been the most useful, and we

[3] R. O. Poyton and D. Branton, *Exp. Cell Res.* **60,** 109 (1970).

[4] J. A. Dvorak and W. Stotler, *Exp. Cell Res.* **68,** 144 (1971).

[5] M. C. Willingham and I. H. Pastan, *Cell* **13,** 501 (1978).

[6] D. J. O'Kane and B. A. Palevitz, *J. Cell Biol.* **83,** 304a (1979).

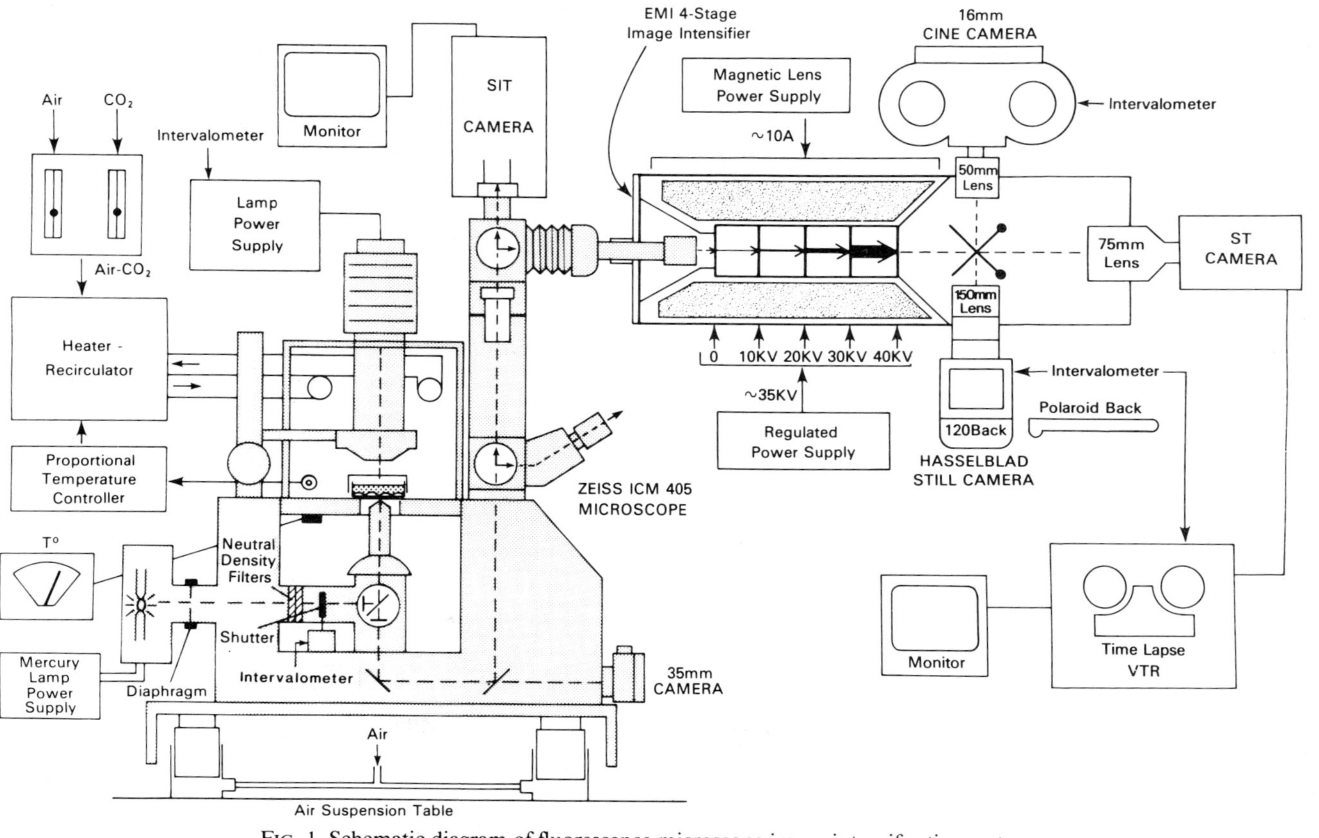

FIG. 1. Schematic diagram of fluorescence microscope image-intensification system.

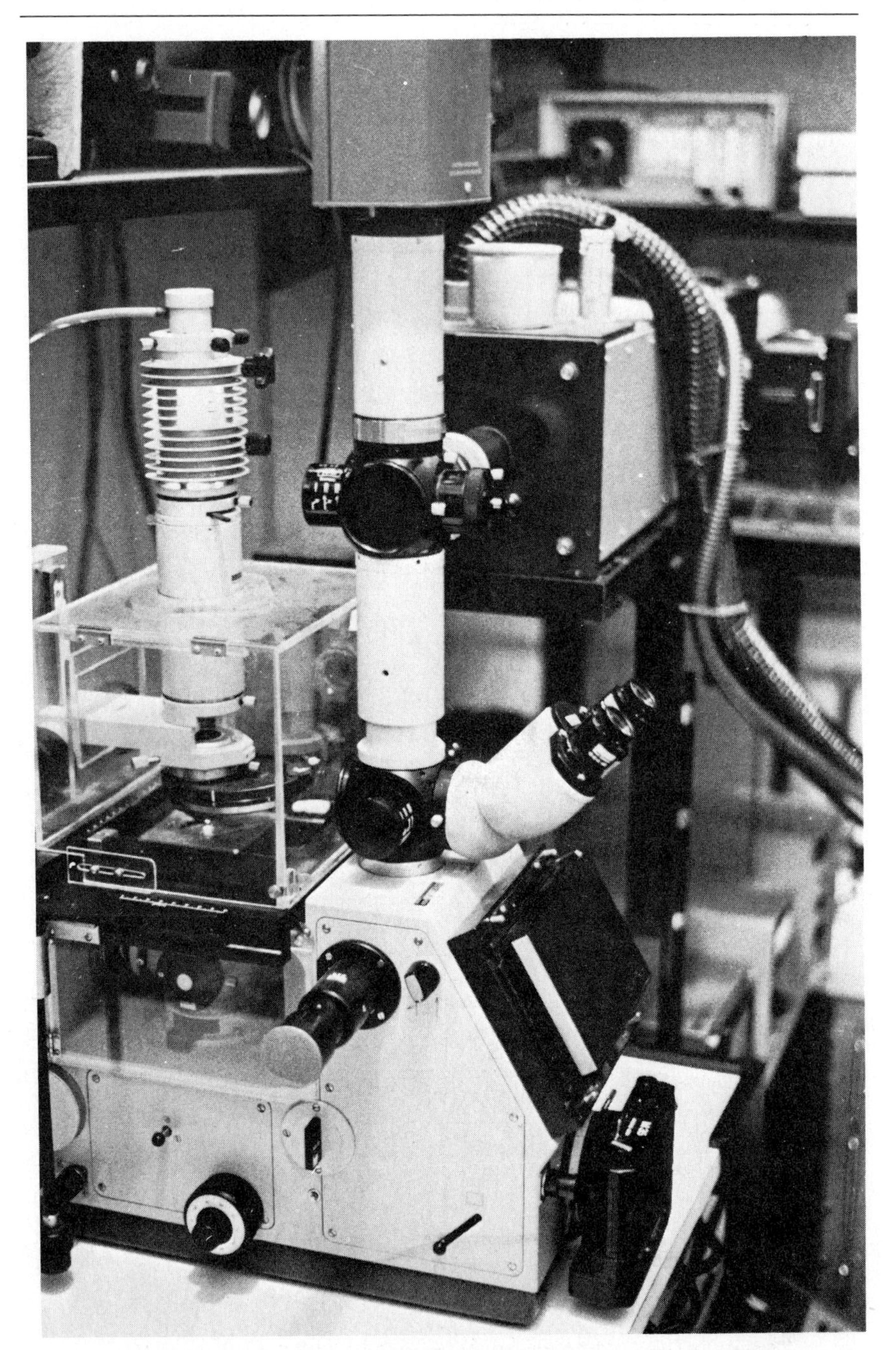

FIG. 2. Photograph of fluorescence microscope image-intensification system.

have over 15 camera-years of experience with this device. It allows one simply to mount it beyond the output of the eyepiece of a fluorescence microscope and continuously record low-light-level images. Combined with a video tape recorder (our preference for time-lapse and real time recording is the Panasonic Model NV-8030 $\frac{1}{2}$-in. recorder) and monitor (e.g., Hitachi VM-904 AU), one can immediately produce continuous or time-lapse tape records of microscope images.[5,7,8]

The video image can be photographed directly from the monitor using 35-mm cameras or with Polaroid film (e.g., Polaroid CU-5 fixed-focus camera). The exposure times should exceed $\frac{1}{8}$ sec to ensure that more than one trace of the video raster will expose the film (video images are formed by two sequential alternate-line traces of $\frac{1}{60}$ sec to complete a full image every $\frac{1}{30}$ sec, referred to as 2:1 interlace). For rapidly bleached images it is useful to record the camera output continuously on tape and then replay the brightest early part of the recording to photograph the monitor image. Continuous-playback photographs also increase the resolution of the image by averaging many frames of the image together. Time-lapse or continuous recordings can be converted to cine films using special cine camera equipment to put the retrace bar in proper register (such a service for $\frac{1}{2}$-in. video tape to 16-mm cine film is available from Windsor Total Video, New York, New York). Tape recorders capable of single-frame playback allow sequential single-frame images to be held for photography while contrast and brightness adjustments on the monitor are optimized.

Video systems are quite practical and relatively inexpensive, but they do have certain inherent drawbacks. They are scanning systems that prevent the accumulation of images for greater sensitivity and resolution. Also, these systems generally increase their noise level as the gain increases, so that at the lowest levels of light the signal-to-noise ratio and resolution are relatively poor. For these reasons, video systems have specific limitations in their application, such as when very high resolution or extremely low light levels over long periods must be used. They have the advantage, however, of registering images quickly ($\frac{1}{30}$ sec) so that fast events can be faithfully recorded. A relative comparison of the SIT camera and photographic film for image recording is shown in Fig. 3. Polaroid 107 (3000 ASA) film is compared on similar objects to 35-mm high-speed films (Tri-X and XP-1). The 35-mm films are more sensitive mainly because of their smaller camera factor of enlargement compared to the $3\frac{1}{4} \times 4$ in. Polaroid film. An image requiring 5 sec of exposure on these fast 35-mm films ($\sim$ 1000 ASA) (Fig. 3B, C) is faithfully recorded by the SIT camera in $\frac{1}{30}$ sec (Fig. 3D, E). The real comparison comes in the recording of much lower light levels in Fig. 3 (F, G,

[7] M. C. Willingham and S. S. Yamada, *J. Cell Biol.* **78,** 480 (1978).

[8] M. C. Willingham, S. S. Spicer, and R. A. Vincent, *Exp. Cell Res.* **136,** 157 (1981).

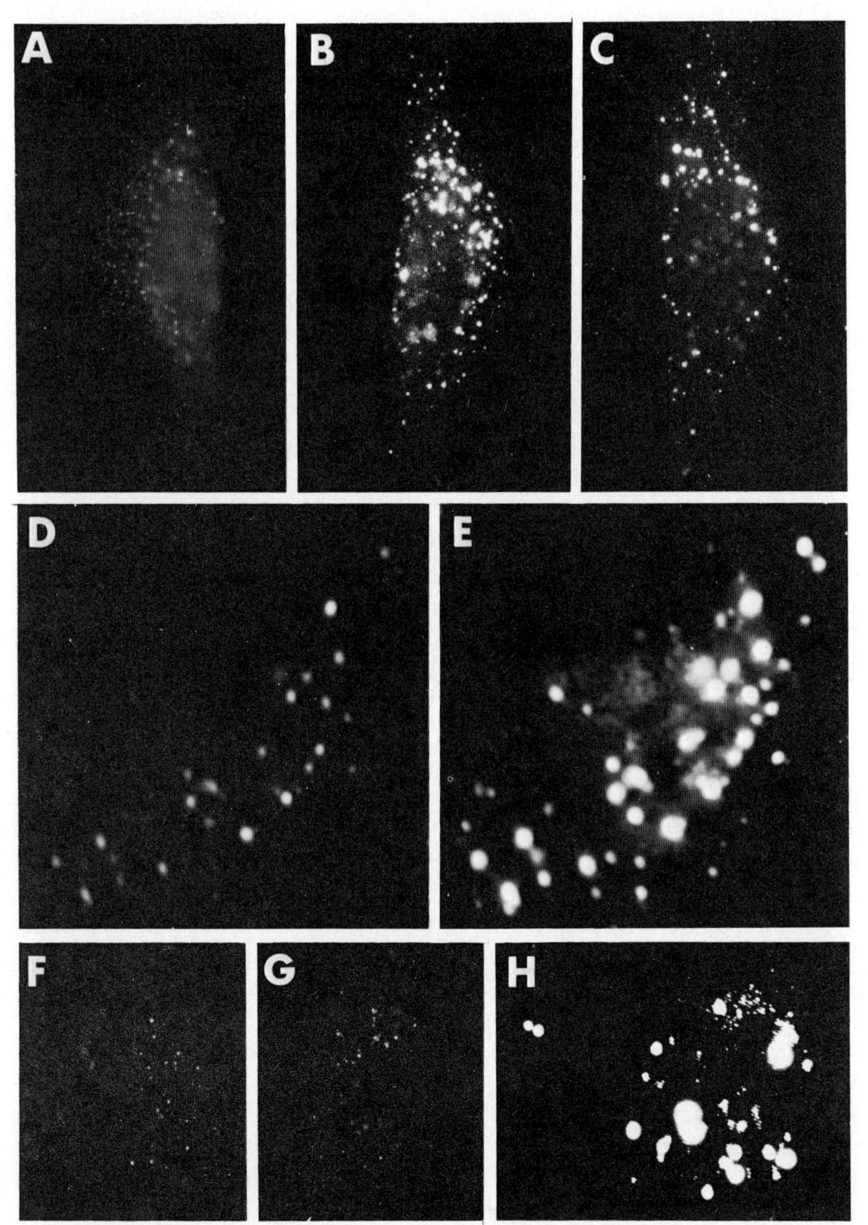

FIG. 3. Fluorescence images recorded by various detection systems. Swiss 3T3 mouse fibroblasts were incubated for 10 min at 37° with 150 μg of rhodamine-labeled α_2-macroglobulin (α_2M) per milliliter as a model endocytic fluorescent ligand. At this point, α_2M is concentrated in receptosomes, the intracellular endocytic vesicles derived from coated pits on the cell surface. At 100% illumination (100-W Hg source) the image of a cell containing these labeled receptosomes can be recorded directly on photographic film. Examples of this are shown using Polaroid 107 film (45-sec exposure, A), Kodak Tri-X 35 mm film (5-sec exposure, B), or Ilford XP-1 35-mm film (5-sec exposure, C). This same image was also recorded using the SIT video

H), in which the high-speed 35-mm films require 2 min to record an image that the SIT camera records in $\frac{1}{30}$ sec. This time-sensitivity ratio (120 sec : $\frac{1}{30}$ sec) shows that the SIT system is >3000 times more sensitive than photographic film, even film rated at 1600 ASA (XP-1). Thus, for a rapidly bleaching fluorescence image, the SIT camera could record something in $\frac{1}{30}$ sec that might be totally bleached in 5 sec and not be recorded by any photographic film. As a result, the closer the bleach time of the image comes to the scan time of the SIT camera ($\frac{1}{30}$ sec), the greater the advantage the SIT camera will have over cumulative detection methods (e.g., photographic film). This high sensitivity in short time scans is one of the advantages used when recording the motion of intracellular organelles containing labeled ligands.[5]

Electron Acceleration Devices

The principle of image intensifiers relies on the increase in energy induced in electrons in a vacuum accelerated over a voltage differential. The electrons are initially produced at a photocathode, which produces them after photons strike the special photocathode material. After acceleration across a potential difference (e.g., 1000–10,000 V), the electrons then strike a phosphor surface that releases photons. Because of the gain in energy induced by the voltage, more photons are produced at the phosphor end than originally hit the photocathode, producing a net gain in light. The problem comes in keeping the photons at the phosphor end in the same part of the image as those that entered the photocathode end. By putting the two surfaces very close together the image elements can be kept in place, but this severely limits the amount of voltage possible across the photocathode to phosphor space. A specialized approach to this has come in the development of multichannel plate intensifiers,[2] which are not yet commercially available. Presently available devices often use magnetic lenses to keep the electron–photon interactions in their proper place in the image. To increase the amplification, multiple stages of intensification can be put together in series to produce an overall higher gain. Such a device is the four-stage EMI intensifier shown in Figs. 1 and 2. This device (EMI Model T2001/4, tube type 9912, cathode S20B) has an overall gain of $>10^6$ with an acceleration voltage of 36 kV.

The output of this intensifier appears on a phosphor screen, which can

camera ($\frac{1}{30}$-sec exposure) and photographed on Polaroid 611 film from the video monitor using the SIT camera with low (D) or high (E) gain. When the illumination was reduced to 1.5% of the total light available from the 100-W Hg light source, the image could be recorded on Tri-X (F) or XP-1 (G) film after only a 2-min exposure. The same image could still easily be recorded at $\frac{1}{30}$-sec exposure using the SIT camera at full gain (H). A, ×500; B, C, F, G, ×750; D, E, H, ×1500.

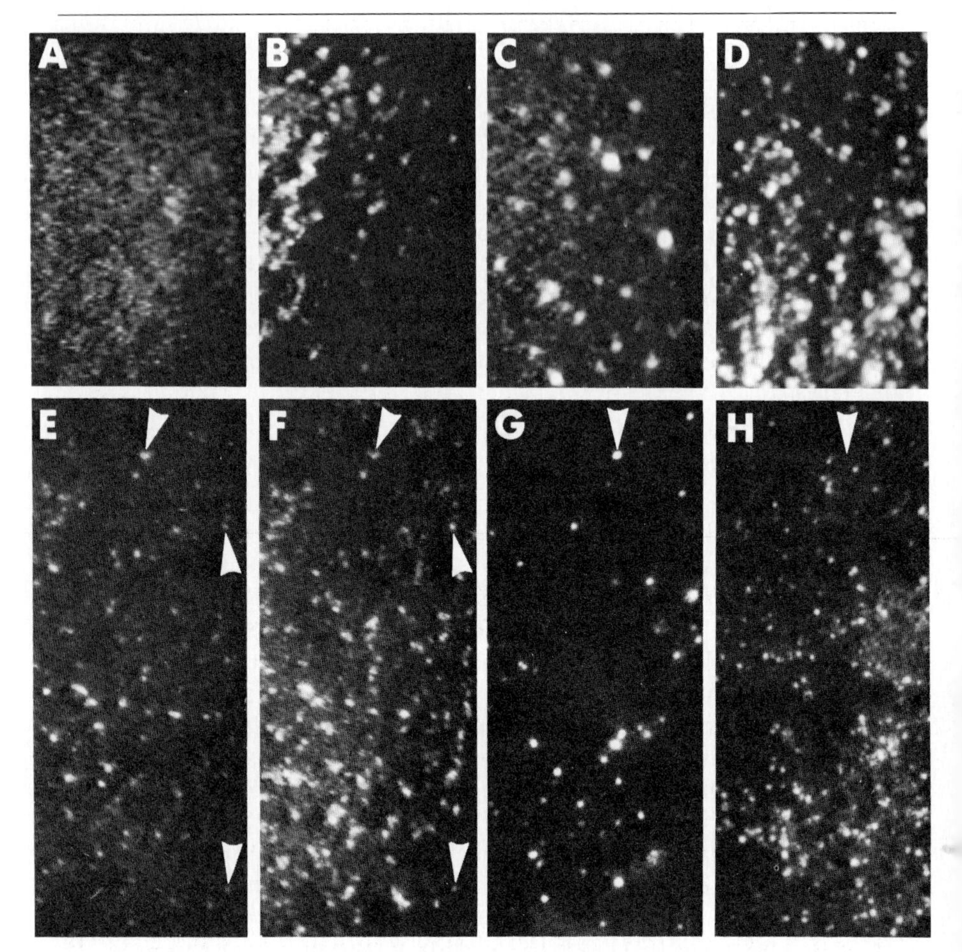

FIG. 4. Fluorescence images recorded by SIT or EMI 4-stage intensifier systems at maximum resolution. To produce a test system with very small amounts of fluorescence requiring a high level of resolution, Swiss 3T3 cells were incubated at 4° with a small amount (50 μg/ml) of rhodamine-α_2M (Rh-α_2M) and either fixed directly at 4° (A, B, E, F) or warmed to 37° for 5 min (C, D, G, H). Under these conditions, α_2M will mainly be confined to clathrin-coated pits at 4° and present in receptosomes after warming to 37°. Rh-α_2M shows less fluorescence when clustered in coated pits than when present in receptosomes. The light source was reduced to 12% of the total available light. The samples were further processed after fixation by permeabilization with Triton X-100 and labeling with fluorescein-labeled anti-clathrin antibody. A, C, E, and G represent images on the rhodamine channel (α_2M location), and B, D, F, and H represent images on the fluorescein channel (clathrin location). A–D are images taken from the SIT camera output; E–H are images recorded on Tri-X 120 film from the EMI 4-stage intensifier output. In A, the presence of Rh-α_2M is not detected by the SIT camera in the coated pits on the cell surface. Clathrin localization is barely detected and resolved by the SIT camera (B). This same image is easily detected at high resolution by the EMI intensifier for Rh-α_2M in E and fluorescein anti-clathrin in F. Note the correspondence of the clusters seen in E at the

then be recorded by video or photographic film. This four-stage device could be coupled with a SIT camera at its output to give an extremely high-gain video output.[9] We have chosen to use a nonintensified silicon target camera in the output because of its higher resolution, which we use for focusing the output image. Actual recording is done using direct photographic film, either cine or still images, since the main advantage we have found with this device is its very high resolution, which can be produced by accumulating the image on film. The noise is random and not determined by video raster lines, and the light output from the device is essentially proportional to its input. This device, unlike the SIT camera, can record and resolve images that cannot be seen by the dark-adapted eye, even when optimized by lower magnification eyepieces. The video output camera must be kept more than 12 in. from the end of the lens housing because of magnetic interference. Because the final phosphor screen is recessed deeply in the end of the magnetic lens case, long-working-distance lens and mirror devices are necessary to transfer and focus the output image in a lighttight box. Since none of these processing components are commercially available, the investigator must manufacture these items, a time-consuming task. This device also must be handled carefully, since even slight exposure of the photocathode to stray light while in operation could permanently damage the tube. The magnetic lens case is very heavy, the power supplies take a fair amount of space, and this equipment is rather expensive. Even so, the images possible with such a device cannot be produced with any other presently available system. For example, Fig. 4 demonstrates a test sample that is below both the resolutions and sensitivity limits of the SIT camera. However, the EMI system produces a high-resolution, clear image of this specimen. The major reason for this is that the EMI output can be accumulated on photographic film and the noise is randomized. Because of this improvement in resolution and the greater sensitivity of such a system, this device allows experiments not possible with other systems. Similar devices have been used previously in some biological experiments.[9,10]

[9] D. L. Taylor, J. R. Blinks, and G. Reynolds, *J. Cell Biol.* **86**, 599 (1980).

[10] G. T. Reynolds, *Q. Rev. Biophys.* **5**, 295 (1972).

same position as some of the coated pits seen in F. Examples of this correspondence are pointed out by arrowheads. In C, receptosomes containing Rh-α_2M can be barely detected by the SIT camera, as well as the clathrin pattern on the same part of the cell in D. This is, thus, the lower limit of resolution and detection by the SIT system. A similar area, resolved at higher resolution by the EMI intensifier, is shown in G for Rh-α_2M and in H for anti-clathrin. Note the lack of correspondence of the rhodamine-labeled receptosmes in G (example shown by arrowhead) and the clathrin pattern shown in H. A–D, $\cong \times 2000$; E–H $\cong \times 1000$.

Experimental Procedures

Fixed Samples

Hormones and other ligands labeled with fluorochromes bind to cultured cells with appropriate receptors. Many of these ligands are internalized into these cells by a specialized pathway involving clustering of receptor–ligand complexes into clathrin-coated pits on the cell surface followed by endocytosis into specialized uncoated endocytic vesicles called receptosomes.[11] We have directly observed a number of these ligands in living cells using both the SIT and EMI intensifier systems. Employing these techniques, the characteristic jumping pattern of saltatory motion could be seen for these receptosomes, indicating that they were isolated vesicles moving by this form of directed specific intracellular motion. At higher levels of excitation light, the motion of these fluorescently labeled vesicles was stopped, indicating interference due to photochemical damage. Thus, without the image intensification systems, this motion of the fluorescently labeled vesicles would have been very difficult or impossible to observe.

The procedures for this type of experiment are fairly straightforward: Cells can be incubated in appropriate amounts of fluorescently labeled ligand in a medium that allows binding to receptors at 4° where endocytosis is inhibited, or at higher temperatures (e.g., 37°) where endocytosis proceeds. The 4° experiment allows one to follow and analyze the entry of a synchronous wave of ligand when the cells are later warmed to 37°. The continuous 37° incubation allows other processes that might not occur at 4°, such as transport of some drugs across membranes or binding of some ligand–receptor systems. Since the internalization of many ligands can occur in only 1–2 min at 37°, and further intracellular processing proceeds during the observation period, one can fix cells at various times and look at their fluorescence later. A typical procedure might be to incubate a cell in the rhodamine-labeled ligand for 10 min at 37°, wash the excess ligand away at 23°, and fix the cells in 2–4% formaldehyde in buffered saline. Such a routine experiment could be done directly in a 35-mm culture dish in which the cells had been planted two days before. After fixing for 5–10 min, the fixative is washed away with saline, and the cells are mounted under a circular 25-mm diameter No. 1 coverslip in buffered glycerol (Difco FA mounting fluid). The dish can then be observed later by placing immersion oil on the coverslip and placing the dish upright on an upright microscope or upside down on an inverted microscope (taped to the specimen stage holder). After focusing the image of the cells using phase contrast, the fluorescence light source is opened and the image is transferred to an intensifier, where it is recorded.

[11] I. H. Pastan and M. C. Willingham, *Science* (*Washington, D. C.*) **214,** 504 (1981).

Alternatively, cells could be incubated with labeled ligand at 4° for 30 min to 12 hr, depending on binding characteristics, washed free of ligand at 4°, then warmed to 37° for various times before fixing in formaldehyde. Other fixatives that are useful in some experiments include diimidoesters, carbodiimides, acetic acid, trichloracetic acid, acetone, and methanol. For those ligand systems in which a multiple-step label must be used, such as labeled antibody instead of directly labeled ligand, those incubations are carried out on the cell surface at 4°, prior to warming and internalization of the entire labeled complex. In each case, an appropriate control, usually competition with excess unlabeled ligand if possible, must be carried out to ensure that the fluorescence image seen really corresponds to the binding of the ligand, not to some nonspecific interaction or binding of labeled contaminants.

Living Cells

When cells are to be labeled and then observed while still alive at 37°, special chambers are necessary to allow short-working-distance objectives to be used. This requires coverslip dishes, the Dvorak–Stotler chamber,[4] or other special coverslip chambers that allow the cells to be maintained in medium without evaporation or contamination. Inverted microscopes (as in Figs. 1 and 2) are ideal for these experiments, and with the appropriate chamber and heater devices, cells can be kept under observation for days without damage. Fluorochromes have limited life-spans under intense illumination, so that long-term observations require that images be recorded only at intervals, not continuously. Intensifiers allow the use of lower and lower levels of excitation light for brightly fluorescent samples. As a result, a typical sample using the EMI system may be *continuously* observed for up to 1 hr without significant bleaching of the rhodamine label or damage to the cells. This is one of the major uses for intensification in cultured cell experiments: not simply to produce a single image, but to produce many sequential images without cell damage or loss of fluorescence. This makes it possible to observe continuously movement of intracellular organelles and perhaps detect new phenomena that would not be evident from fixed-time images.

Problems to Avoid

1. When cultured cells flatten against their underlying substrate, they often become only 1–2 μm thick, producing very flat cells with large, thin areas of cytoplasm. Such flat cells are ideal for following fluorescent ligands because individual organelles are more easily resolved. Sometimes the cell type of interest is not flattened, making visualization of intracellular details impossible. One must try either to pick an appropriate flat cell type or to

flatten the cells by increasing their adhesiveness to substrates by coating the dish with agents like polylysine. Otherwise, labeling experiments can be very difficult to interpret.

2. Autofluorescence of cytoplasm and organelles can sometimes be a problem. Rhodamine is attractive as a label because cellular autofluorescence is often not in the rhodamine spectrum. Sometimes, fixation can induce autofluorescence that did not exist in the living state. An appropriate choice of fixative can often minimize this problem. Some cell types, however, are almost impossible to work with because of autofluorescence; a preliminary examination of the cell type to be studied is advisable.

3. On occasion, some labeled ligands will show large amounts of nonspecific binding to extracellular matrix components, the substrate, or the cell membrane. Usually these ligand systems have been examined beforehand by radioactive binding studies where specificity of binding has been demonstrated. However, the fluorescent derivatives can often show binding characteristics different from those of the radioactive or native ligands, and this possibility must be carefully controlled by appropriate specificity (competition) controls. Free fluorescent dye is also a common problem that can occur on storage of labeled derivatives. Characteristically, free rhodamine will enter cells readily and is concentrated in the nucleus. Results showing fluorescence in the nucleus should be examined carefully for this artifact.

4. Light damage can produce strange results in living cell experiments, often related specifically to the area of the cell containing fluorescent material. Alterations in fluorescent organelles even at low light exposure have been seen for internalized ligand-containing vesicles, as well as for structures after microinjection of fluorescently labeled proteins or antibodies. When mitochondria in living cells are labeled with vital dyes such as rhodamine-123 and exposed to light, often they fragment. Another example is the explosion of lysosomes that commonly occurs after labeling with acridine orange and exposure to light.

5. Immersion oils vary in their viscosity and autofluorescence. Most oils, even though labeled as containing no solvents, do etch plastic and painted surfaces with time, a process made much more rapid by heating in a 37° chamber. For inverted microscopes, higher-viscosity oils are recommended (Cargille type B), but autofluorescence of this type of oil, particularly in the fluorescein spectrum, is a problem. Zeiss sells a nonfluorescent oil of low viscosity that is excellent for fluorescein, but it contains more solvents than many other oils. For long-working-distance oil objectives for use with culture dishes and flasks in a 37° chamber (e.g., Zeiss objective achromat 40×, N.A. 0.85, 1.5 mm working distance, *phase No. 2 special order,* No. 461709-811), no immersion oil has been found to be totally satisfactory. Our solution has been to layer a glass coverslip onto the plastic flask with paraffin oil, and then place the immersion oil between the

coverslip and the objective. This has succeeded in producing a system stable for more than 1 week. When using an inverted microscope with oil immersion objectives, it is important to prevent excess oil from running down the sides of the objective into the microscope. The simplest solution is to wrap pipe cleaners around the objective at intervals and monitor their contamination with oil.

Summary

We describe here the components and uses of two image intensifier systems. The SIT camera system is convenient, relatively inexpensive, readily adaptable to most microscopes, and reliable. It suffers from lack of resolution at high gain levels and the inability to extend its sensitivity by accumulating an image over time. The EMI system is expensive and bulky and requires special adaptation to the microscope and special image recording devices in its output. It has extraordinary sensitivity and resolution, however, and allows experiments to be carried out that are otherwise not possible. Other systems similar to those described here are also commercially available and, in general, have similar advantages and disadvantages. The choice of the proper type of system varies with the particular application. These systems amplify the amount of light in an available image within constraints of gain and resolution and produce a publishable record of what otherwise might not be able to be recorded. These systems cannot improve images that contain too much background nor improve the resolution inherent in the microscopic method employed.

See Addendum, page 635.

Acknowledgment

The authors wish to thank Drs. Frederick R. Maxfield and Jürgen Wehland for helpful scientific discussions, and S. Sorota, I. Tice, R. L. Steinberg, W. Poppe, R. E. W. Trevor, and R. Ellis for helpful technical advice.

[22] Receptor-Mediated Endocytosis of Epidermal Growth Factor

By HARRY T. HAIGLER

A number of polypeptide hormones have been shown to enter cells by the process of receptor-mediated endocytosis, and the list is steadily growing. The internalization of these ligands by cultured cells provides a well-defined system for investigating the basic cellular processes involved in endo-

ISBN 0-12-181998-1

cytosis, the cellular fate of plasma membrane glycoproteins that serve as receptors, and the role endocytosis plays in the biological activity of the hormones.

Since Carpenter and Cohen presented the first evidence that the mitogenic polypeptide hormone epidermal growth factor (EGF) is internalized by cultured cells,[1] it has been used extensively as a model system for studying hormone internalization. The EGF is a very stable, well-characterized polypeptide[2] that can be isolated in relatively large amounts from the mouse submaxillary gland[3] and, therefore, lends itself to chemical modification. A protein (human EGF or urogastrone) nearly identical in biological activity and very similar in structure to mouse EGF can be isolated in small amounts from human urine.[2]

Radioactive, fluorescent, and ferritin conjugates of EGF can be prepared that bind to cellular receptors in a highly specific manner and permit the study of the interactions of EGF with its target cells at different levels of resolution.[4] With these derivatives, it has been shown that EGF binds to specific receptors on the plasma membrane that are initially diffusely distributed. The hormone-receptor complex then clusters in clathrin-coated pits and is internalized into endocytic vesicles that eventually deliver their contents to lysosomes. The methods involved in these studies are described below with emphasis on the preparation of EGF conjugates.

Fluorescent EGF

Fluorescein-conjugated EGF (Fl-EGF) can be prepared by allowing EGF to react with fluorescein isothiocyanate (FITC) as previously described.[5] The EGF (1 mg in 300 μl of 0.2 M $NaHCO_3$, pH 8.75) was allowed to react at 4° with FITC (Research Organics, Inc., Cleveland, Ohio) by slowly adding 32 μl of a 30 mg/ml solution of FITC in dimethyl sulfoxide (DMSO) over a 1-hr period with stirring. The molar ratio of FITC to EGF was 15 : 1. The entire reaction was applied to a Sephadex G-25 column (0.9 × 9 cm) equilibrated with water, and the excluded peak was collected and adjusted to pH 7.1 by the addition of 0.17 volume of 0.1 M Tris-HCl (pH 7.1). The solution was applied to a 0.9 × 5-cm column of DEAE-Sephacel (Pharmacia Chem, Uppsala, Sweden) equilibrated with 20 mM Tris-HCl (pH 7.1) at

[1] G. Carpenter and S. Cohen, *J. Cell Biol.* **71,** 159 (1976).
[2] G. Carpenter and S. Cohen, *Annu. Rev. Biochem.* **48,** 193 (1979).
[3] S. Cohen and C. R. Savage, Jr., this series, Vol. 37, p. 424.
[4] H. T. Haigler, *in* "Growth and Maturation Factors" (G. Guroff, ed.), Vol. 1, p. 117. Wiley, New York, 1983.
[5] H. T. Haigler, J. F. Ash, S. J. Singer, and S. Cohen, *Proc. Natl. Acad. Sci. U.S.A.* **75,** 3317 (1978).

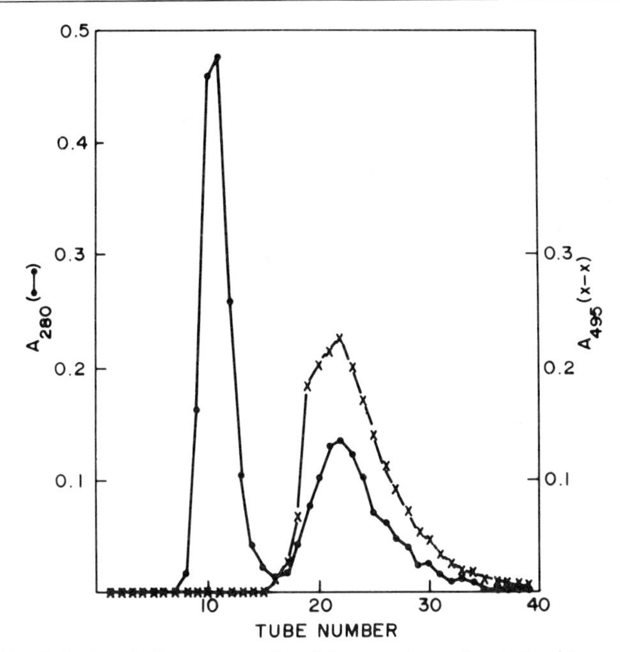

FIG. 1. DEAE-Sephacel chromatography of fluorescein-conjugated epidermal growth factor (Fl-EGF). The excluded peak from the Sephadex G-25 column (containing EGF and Fl-EGF) was applied to a DEAE-Sephacel column. The adsorbed protein was eluted with a NaCl gradient, and 2.1-ml fractions were collected.

4°. The absorbed protein was washed with 6 ml of equilibration buffer, then eluted with NaCl gradient prepared by allowing 1.0 *M* NaCl to flow into a constant-volume mixing chamber containing 50 ml of 20 m*M* Tris-HCl buffer (pH 7.1). The elution of native EGF was monitored by OD_{280} and of Fl-EGF by OD_{495}. Fl-EGF is more negatively charged than native EGF and is eluted later in the profile (Fig. 1). The fractions containing OD_{495} were pooled and concentrated to an OD_{280} of approximately 0.15 by ultrafiltration using a UM-10 membrane (Amicon Corp., Oosterhout, Holland). The concentration of EGF was determined by the method of Lowry *et al.*[6]

Fl-EGF contained EGF and fluorescein in a 1:1 molar ratio as was expected, since the only free amino group in EGF is the α-amino of the NH_2-terminal amino acid—mouse EGF contains no lysyl residues. The yield of Fl-EGF was low (approximately 10%) and could not be improved by increasing the reaction time, temperature, or FITC concentration. Fl-EGF retained 45% of the binding capacity of the native hormone and was as mitogenic as the unmodified hormone. Fl-EGF was stable when stored at either 4° or −20° for 1 year.

[6] O. H. Lowry, N. J. Rosebrough, A. L. Farr, and R. J. Randall, *J. Biol. Chem.* **193,** 265 (1951).

Epidermal growth factor has also been allowed to react with tetramethylrhodamine isothiocyanate (TRITC; Research Organics, Inc., Cleveland, Ohio)[7] under conditions identical to the reaction with FITC. After unreacted TRITC was removed by gel filtration, tetramethylrhodamine-labeled EGF (R-EGF) was separated from free EGF by ion-exchange chromatography as described for Fl-EGF, except the DEAE column was equilibrated with Tris-HCl (20 mM, pH 7.6) and was eluted with a NaCl gradient prepared by allowing 0.5 M NaCl to flow into a 100-ml constant-volume mixing chamber containing Tris-HCl (20 mM, pH 7.6). The yield of R-EGF was also low (less than 5%). However R-EGF is more useful for fluorescent microscopy than Fl-EGF because tetramethylrhodamine is a much brighter fluorochrome and resists photobleaching.

Although the NH_2-terminal amino group of EGF does not react well with isothiocyanate, it will react nearly quantitatively with *N*-hydroxysuccinimidyl esters. This reactivity has been exploited to couple EGF to biotin (B-EGF).[7] Biotin-labeled proteins are useful derivatives for an indirect fluorescent staining method that exploits the extraordinarily high affinity of avidin for biotin.[8] The biotin-labeled ligand is bound to its receptor, then fluorescein-labeled avidin is added. If more intense staining is required, fluorescein-labeled antibodies that are specific for avidin then can be added.

B-EGF was prepared as follows: A solution of EGF (1 mg in 0.5 ml) in 0.2 M sodium bicarbonate (pH 8.75) was allowed to react with gentle stirring for 1 hr at 20° with a fivefold molar excess of *N*-hydroxysuccinimidobiotin (Pierce Chemical Company, 11.3 μl of a 25-mg/ml solution in dimethylformamide). The conjugate was separated from unreacted *N*-hydroxysuccinimidobiotin by gel filtration on a Sephadex G-25 column (0.9×25 cm) equilibrated with phosphate-buffered saline. Since B-EGF is formed in greater than 95% yield, it is not generally necessary to separate free EGF from B-EGF by ion-exchange chromatography prior to use in indirect staining procedures.

Visualization of the cellular binding of fluorescent derivatives of EGF has been greatly facilitated by using the A-431 cell line isolated by Fabricant *et al.* that contains an extraordinarily high binding capacity[9]—approximately 2.5×10^6 EGF receptors per cell.[5] It is possible to visualize directly the binding of Fl-EGF to A-431 cells. However, to visualize directly binding to cells containing fewer EGF receptors (e.g., NRK-VB_4 that contains approximately 2×10^5 receptors), a brighter derivative, such as R-EGF, must be used. Fluorescent derivatives of EGF can also be studied in culture cells containing approximately 100,000 receptors per cell (e.g., 3T3 fibro-

[7] H. T. Haigler, unpublished results.

[8] M. H. Heggeness and J. F. Ash, *J. Cell Biol.* **73**, 783 (1977).

[9] R. N. Fabricant, J. E. DeLarco, and G. J. Todaro, *Proc. Natl. Acad. Sci. U.S.A.* **74**, 565 (1977).

blasts) by observing the cell-bound derivatives with a silicon intensifier target camera system.[10] This system allows detection of very low levels of emitted light. By using low amounts of excitation light, the cellular damage caused by illumination could be reduced to the point that it is possible to visualize fluorescent probes in living cells.[11] This system is described in this volume.[12]

Ferritin-Conjugated EGF

By exploiting the fact that EGF contains only one free amino group, it was possible to design a synthetic procedure for preparing EGF conjugated to ferritin in a 1 : 1 molar ratio. This procedure has been described in detail previously[13] and is outlined below. The EGF was allowed to react with a 1000-fold excess of glutaraldehyde, and the unreacted glutaraldehyde was removed by gel filtration. The monoactivated EGF was then made to react with ferritin—a 10-fold molar excess of ferritin was used to minimize the attachment of more than one EGF molecule per ferritin particle. After purification by gel filtration and affinity chromatography, a biologically active ferritin–EGF (F–EGF) conjugate was isolated. Cellular binding of F–EGF could be quantitated by counting ferritin particles in electron micrographs of thin sections prepared from cells incubated with the conjugate. Binding of F-EGF was very specific—the amount of binding was reduced more than 99% when an excess of native EGF was added simultaneously with the conjugate. Ferritin conjugates of other polypeptide ligands often have been plagued with unacceptable levels of nonspecific binding. The above method is probably successful because the ferritin is not exposed to the free chemical cross-linking agent—it is only exposed to activated EGF.

Owing to the low reactivity of the N-terminal amino group of EGF with glutaraldehyde and the multistep nature of the procedure, F–EGF was isolated in low yield (less than 5%). It should be possible considerably to increase the yields of the above procedure by substituting a bifunctional *N*-hydroxysuccinimide ester [e.g., disuccinimidyl suberate or dithiobis (succinimidyl propionate); Pierce Chemical company] for glutaraldehyde as the cross-linking agent. The heterobifunctional cross-linking agent *N*-succinimidyl-3(2-pyridyldithio) propionate (Pharmacia) has been used to couple EGF to diphtheria toxin and ricin in high yields.[14]

[10] J. Schlessinger, Y. Schechter, M. C. Willingham, and I. Pastan, *Proc. Natl. Acad. Sci. U.S.A.* **75,** 2659 (1978).

[11] M. C. Willingham and I. Pastan, *Cell* **13,** 501 (1978).

[12] M. C. Wilingham and I. Pastan, this volume [21].

[13] H. T. Haigler, J. A. McKanna, and S. Cohen, *J. Cell Biol.* **81,** 383 (1979).

[14] D. B. Cawley, H. R. Herschman, D. G. Gilliland, and R. J. Collier, *Cell* **22,** 563 (1980).

^{125}I-Labeled EGF

^{125}I-labeled EGF (^{125}I-EGF) of high specific activity can be prepared by techniques that use either lactoperoxidase,[15] chloramine-T,[1] or 1,3,4,5,6-tetrachloro-3α,6α-diphenylglycoluril (Iodo-Gen, Pierce Chemical Company).[16] These derivatives retain the same affinity for cellular receptors as the native hormone and bind with high specificity. The chloramine-T procedure of Hunter and Greenwood[17] as modified by Carpenter and Cohen has been used extensively and will be described in detail. The reaction was run in the Combi-V-Vial, in which New England Nuclear (Boston, Massachusetts) packages its "high concentration" Na^{125}I. The EGF (5 μg in 40 μl of 0.5 *M* sodium phosphate, pH 7.5) was added to 1.0 mCi of carrier-free Na^{125}I. Under dim lighting, chloramine-T (20 μg in 10 μl of phosphate buffer, dissolved immediately before use) was added; 25 sec later, the reaction was stopped by adding sodium metabisulfite (40 μg in 10 μl of phosphate buffer). The labeled EGF was separated from unreacted Na^{125}I by gel filtration through a column (0.9 $\times$ 8 cm) containing Sephadex G-25 equilibrated with 0.05 *M* sodium phosphate (pH 7.5). The columm was previously washed with phosphate buffer containing 1.0 mg of bovine serum albumin per milliliter, and then extensively with phosphate buffer. Labeled EGF eluted from the column was frozen at $-70°$ in phosphate buffer containing 1 mg of bovine serum albumin per milliliter at a concentration of about 1 μg/ml. Specific activities of approximately 150,000 cpm/ng were obtained. Nonspecific binding increases with time of storage so that it reaches an unacceptable level after about 1 to 2 months.

^{25}I-EGF can be used to determine the binding affinity and number of EGF binding sites in isolated membranes using the method of Scatchard. However, since Scatchard anaysis requires that the system be at equilibrium, it cannot be used to determine these parameters in intact cells at physiological temperature. In the latter system, cell-bound ^{125}I-EGF is rapidly internalized by receptor-mediated endocytosis and then degraded in the lysosomes with the release of [^{125}I]tyrosine into the medium. Under physiological conditions, the binding of EGF to living cells cannot be viewed solely as the interaction of EGF with its receptor; the cellular processes involved must be considered.

A method has been developed that permits the study of EGF internalization using a simple biochemical method that allows a clear discrimination between cell-surface-bound and internalized hormone.[18] Monolayer cul-

[15] G. Carpenter, J. J. Lembach, M. M. Morrison, and S. Cohen, *J. Biol. Chem.* **250,** 4297 (1975).

[16] H. S. Wiley and D. D. Cunningham, *J. Biol. Chem.* **257,** 4222 (1982).

[17] W. M. Hunter and F. C. Greenwood, *Nature* (*London*) **194,** 495 (1962).

[18] H. T. Haigler, F. R. Maxfield, M. C. Willingham, and I. Pastan, *J. Biol. Chem.* **255,** 1239 (1980).

tures of cells in 35-mm dishes were incubated with ^{125}I-EGF for the desired period of time at physiological temperature, then washed five times at 4° with PBS containing 1 mg of BSA per milliliter to remove unbound hormone. Cultures were then treated with 0.7 ml of acetic acid (0.2 *M*, pH 2.5) containing 0.5 *M* NaCl for 6 min at 4°. The dishes were then rinsed with 0.4 ml of the same solution, and the remaining cell-associated radioactivity was removed by incubating for 1 hr at 37° with 1 *N* NaOH. The radioactivity removed by acid, and NaOH was determined in a gamma spectrometer. It was demonstrated that more than 90% of the cell-surface-bound ^{125}I-EGF was removed with the acid washes while removing about 5% of the internalized hormone. Thus, although the acetic acid kills the cells and removes the cell surface hormone, the plasma membrane remains impermeable to the internalized hormone during the low pH wash. This rapid assay permits the study of EGF internalization on a large number of samples and can be used to study the internalization of other ^{125}I-labeled polypeptide ligands including insulin[7] and human choriogonadotropin (hCG).[19] Ascoli found that, if the 6-min acetic acid wash was replaced by a 2-min wash with a glycine buffer (50 m*M*, pH 3.0 containing 100 m*M* NaCl), ^{125}I-labeled hCG could be selectively removed from the cell surface of cultured Leydig tumor cells, and that when the cells were returned to growth medium, they remained viable.[19] The glycine treatment had little or no effect on the ability of the cells to rebind hCG and to produce steroids in response to the hormone. This method was used to evaluate the role of receptor-mediated endocytosis of hCG in the process of hormone stimulated steroidogenesis.

A significant advance in the quantitative investigation of the cellular fate of polypeptide ligands was made by Wiley and Cunningham. They reasoned that, under physiological conditions, the interaction of polypeptide ligands with cells can be more accurately modeled on steady-state assumptions than on equilibrium assumptions. They derived steady-state equations for analyzing the binding, internalization, and degradation of ligands.[20] The equations include four new rate constants that correspond to the rate of (*a*) insertion of new receptors into the cell surface; (*b*) endocytosis of occupied receptors; (*c*) turnover of unoccupied receptors and (*d*) hydrolysis of internalized ligands. Using the interaction of ^{125}I-EGF with human fibroblasts as a model system, they developed procedures for the experimental determination of these rate constants.[16,20] Computer simulations, based on the derived steady-state equations, showed a close correspondence to experimental data and indicated that the equations provide an accurate model of the cellular processes involved in the interaction of EGF and cells under physiological conditions. Using these methods, Wiley and Cunningham determined that occupied receptors are internalized more than 14 times faster than unoccu-

[19] M. Ascoli, *J. Biol. Chem.* **257,** 13306 (1982).
[20] H. S. Wiley and D. D. Cunningham, *Cell* **25,** 433 (1981).

pied receptors, thereby explaining the phenomenon of "down regulation." On the basis of steady-state equations, a method was developed that permitted a rapid and sensitive method for determining the lag time between ^{125}I-EGF internalization and the initiation of degradation. The time course of cellular processing of ^{125}I-EGF determined by these methods is consistent with previous morphological studies of ferritin-labeled EGF. The approach developed by Wiley and Cunningham should be applicable to the study of a wide variety of polypeptide ligands and may permit a quantitative description of the cellular processes that control the turnover of cell-surface hormone receptors.

[23] Determinants in the Uptake of Lysosomal Enzymes by Cultured Fibroblasts

By KIM E. CREEK, H. DAVID FISCHER, and WILLIAM S. SLY

Many lysosomal acid hydrolases are subject to adsorptive pinocytosis by human fibroblasts in culture via phosphomannosyl recognition markers on the enzymes for which there are high-affinity enzyme receptors on the cell surface.[1-6] Lysosomotropic amines greatly enhance secretion of lysosomal enzymes by fibroblasts, and the acid hydrolases secreted by fibroblasts in the presence of amines are enriched for high-uptake enzyme forms bearing the phosphomannosyl recognition marker.[7,8] Here we describe methods for the partial purification of β-hexosaminidase B from secretions of Tay–Sachs disease fibroblasts, produced in medium containing ammonium chloride, and describe its use as a ligand to study adsorptive pinocytosis by fibroblasts. In addition, we present methods for the preparation of [2-^{3}H]mannose-labeled glycopeptides, for separation of phosphorylated oligosaccharides released from the glycopeptides by endoglycosidase H, and describe the use of these ligands to study the phosphomannosyl uptake system in human fibroblasts.

[1] E. F. Neufeld, T. W. Lim, and L. J. Shapiro, *Annu. Rev. Biochem.* **44,** 357 (1975).
[2] A. Kaplan, D. T. Achord, and W. S. Sly, *Proc. Natl. Acad. Sci. U.S.A.* **74,** 2026 (1977).
[3] G. N. Sando and E. F. Neufeld, *Cell* **12,** 619 (1977).
[4] A. Kaplan, D. Fischer, D. Achord, and W. Sly, *J. Clin. Invest.* **60,** 1088 (1977).
[5] K. Ullrich, G. Mersmann, E. Weber, and K. von Figura, *Biochem. J.* **170,** 643 (1978).
[6] K. von Figura and U. Klein, *Eur. J. Biochem.* **94,** 347 (1979).
[7] A. Gonzalez-Noriega, J. H. Grubb, V. Talkad, and W. S. Sly, *J. Cell Biol.* **85,** 839 (1980).
[8] A. Hasilik, U. Klein, A. Waheed, G. Strecker, and K. von Figura, *Proc. Natl. Acad Sci. U.S.A.* **77,** 7074 (1980).

ISBN 0-12-181998-1

Collection of β-Hexosaminidase B from Tay–Sachs Disease Fibroblast Secretions

Tay–Sachs disease fibroblasts (GM-502) are obtained from the Human Mutant Cell Repository, Camden, New Jersey. Since Tay–Sachs disease fibroblasts are deficient for β-hexosaminidase A and secrete predominantly β-hexosaminidase B, which is heat stable, they are a useful cell type to use for the collection of high-uptake forms of β-hexosaminidase B. Tay–Sachs disease fibroblasts, split 1 : 5, are grown to confluence in 490 cm^2 roller bottles (Corning). These cells grow well when maintained at 37° in minimal essential medium (MEM Earle's medium, GIBCO), supplemented with 15% heat-inactivated (60° for 30 min) fetal calf serum (GIBCO), 1 m*M* sodium pyruvate, 100 U of penicillin per milliliter and 100 μg of streptomycin sulfate per milliliter; HEPES (15 m*M*) is added to maintain the pH at 7.6. The cells reach confluence 7–10 days after transplantation, at which time they are washed with 50 ml of 0.9% NaCl solution. Collection of secretions begins by adding 50 ml per roller bottle of serum-free Waymouth medium (KC Biological) containing 1 mg of human serum albumin per milliliter and 10 m*M* NH_4Cl. The cells are maintained at 37° for 24 hr, at which time the medium is changed; the enzyme-containing medium is stored at −20°. When assayed using the synthetic fluorometric substrate 4-methylumbelliferyl-2-acetamido-2-deoxy-β-D-glucopyranoside, approximately 15–40 mU of β-hexosaminidase per milliliter of secretion medium is obtained. One unit is defined as the amount of enzyme that catalyzes the release of 1 μmol of 4-methylumbelliferone per minute.[9] This represents approximately a fivefold enhancement of the amount of β-hexosaminidase accumulated in serum-free medium is the absence of 10 m*M* NH_4Cl. Furthermore, as judged by its susceptibility to pinocytosis by hexosaminidase-deficient fibroblasts, the much larger amount of enzyme secreted in the presence of NH_4Cl is also enriched for high-uptake forms. Collection of β-hexosaminidase under these serum-free conditions has been carried out for as long as twenty-one 24-hr collections.

Partial Purification of β-Hexosaminidase B from Tay–Sachs Disease Fibroblast Secretions

The secretion medium is thawed, pooled, and filtered first by vacuum on a Büchner funnel (Whatman No. 40 paper) and then through a membrane filter (Gelman, 0.45 μm) under pressure. The medium is then concentrated by ultrafiltration to 5% of original volume using an XM-50 membrane filter

[9] H. D. Fischer, A. Gonzalez-Noriega, W. S. Sly, and D. J. Morré, *J. Biol. Chem.* **255,** 9608 (1980).

(Amicon) and dialyzed extensively against 0.01 *M* sodium phosphate, pH 6.0. The dialyzed enzyme is applied to a DEAE-Sephadex A-25 column (2.8 × 8.0 cm) equilibrated with 0.01 *M* sodium phosphate, pH 6.0. Under these condtions, β-hexosaminidase B does not adsorb to the column, but many other proteins are retained. The β-hexosaminidase B that passes through this column is dialyzed overnight against 0.01 *M* Tris-HCl, 0.01 *M* sodium phosphate, pH 7.2, and then divided into aliquots and stored frozen at −20°. The specific activity is 5000 mU per milligram of protein, which indicates a 150- to 200-fold purification from the initial albumin-containing secretion medium. When stored frozen, the enzyme is stable for at least 6 months with no loss of enzymic activity or susceptibility to pinocytosis by fibroblasts.

Pinocytosis Measurements of β-Hexosaminidase B by Sandhoff Disease Fibroblasts

Measurements of β-hexosaminidase pinocytosis by β-hexosaminidase-deficient Sandhoff disease fibroblasts (available from the Repository for Mutant Human Cell Strains, Montreal Children's Hospital, Montreal, Canada, cell strain code WG98) is carried out as follows. Sandhoff disease fibroblasts are grown at 37° in 5% CO_2 in mimimum essential medium (MEM Earle's medium, GIBCO), supplemented with 15% heat-inactivated (60° for 30 min) fetal calf serum (GIBCO), 1 m*M* sodium pyruvate, 100 U of penicillin per milliliter, and 100 μg of streptomycin sulfate per milliliter in 35-mm petri dishes (Corning) to confluence (approximately 0.2–0.3 mg of protein per plate, ~2.5 × 10^5 cells). Uptake is conveniently measured in duplicate 35-mm petri dishes with 15–45 mU of partially purified secretion β-hexosaminidase B in 1.0 ml of the medium described above. Uptake is usually conducted for a period of 2 hr at 37° in a 5% CO_2 atmosphere. The uptake is terminated by chilling the cells on ice, removing the uptake medium, and carefully washing the cells six times with 3-ml portions of ice-cold phosphate-buffered saline. The cells are then disrupted by the addition of 1.0 ml of distilled water followed by freezing for 20 min at −20° and then thawing. β-Hexosaminidase internalized is measured with 25–50 μl of disrupted cells using the fluorometric substrate 4-methylumbelliferyl-2-acetamido-2-deoxy-β-D-glucopyranoside.[9] To measure the small amount of uptake of β-hexosaminidase that occurs by means other than phosphomannosyl recognition, duplicate plates containing β-hexosaminidase and 2 m*M* mannose 6-phosphate are also included.

The pinocytosis of the high-uptake β-hexosaminidase partially purified from secretions is inhibited 95% by 2 m*M* mannose 6-phosphate, which indicates that the enzyme is pinocytosed by the phosphomannosyl recognition system present on human fibroblasts. The very low levels of endoge-

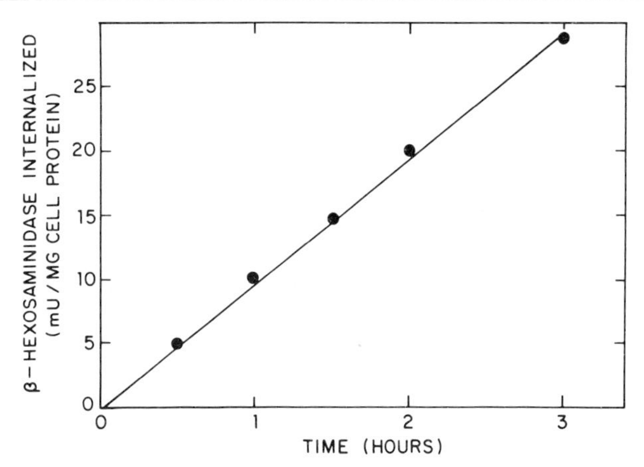

FIG. 1. Pinocytosis of partially purified, secretion β-hexosaminidase B by Sandhoff disease fibroblasts as a function of time. Fibroblast secretion β-hexosaminidase B, 42 mU/plate, was added in 1.0 ml of medium to duplicate 35-mm petri dishes containing β-hexosaminidase-deficient fibroblasts. Pinocytosis of β-hexosaminidase was measured at the indicated times as described.

nous β-hexosaminidase present in Sandhoff disease fibroblasts should also be measured and subtracted from pinocytosis measurements. The rate of enzyme internalization is expressed as the net amount of enzymic activity that becomes cell associated per unit of time (units per milligram of cell protein per hour). After 2 hr, less than 5% of the cell-associated enzyme is enzyme bound to cell-surface receptors, but not internalized. Therefore, no correction is generally necessary for bound but not internalized enzyme in pinocytosis measurements. Under these conditions the uptake of β-hexosaminidase is linear for at least 3 hr (Fig. 1). The pinocytosis process is saturable and inhibited by mannose 6-phosphate (Fig. 2). The K uptake, calculated from a double reciprocal plot of the data in Fig. 2 is 0.8×10^{-9} *M*, which compares with K uptakes of 1 to 6×10^{-9} *M* previously reported for other high-uptake enzymes.[2,3,10] The β-hexosaminidase B prepared under these conditions is a suitable ligand for the study of the phosphomannosyl–enzyme receptors that mediate enzyme pinocytosis in fibroblasts.

Preparation of [2-^{3}H]Mannose-Labeled Glycopeptides from Tay–Sachs Disease Fibroblast Secretions

Tay–Sachs disease fibroblasts (GM-221) are obtained from the Human Mutant Cell Repository, Camden, New Jersey, and grown to confluence in

[10] K. von Figura and H. Kresse, *J. Clin. Invest.* **53,** 85 (1974).

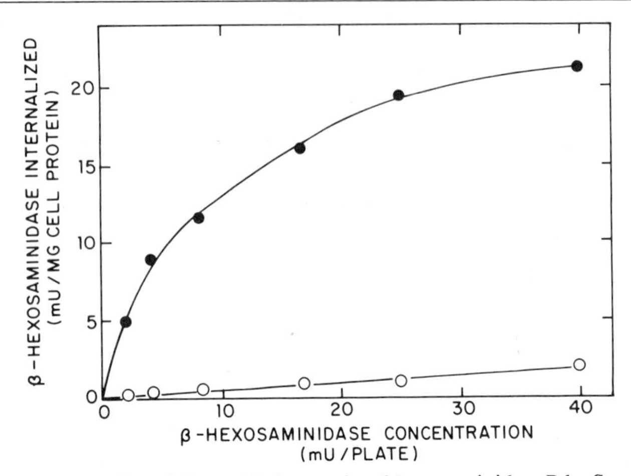

FIG. 2. Pinocytosis of partially purified, secretion β-hexosaminidase B by Sandhoff disease fibroblasts as a function of the added β-hexosaminidase B concentration. Fibroblast secretion β-hexosaminidase B in the absence (●) or in the presence (○) of 2 m*M* mannose 6-phosphate was incubated at the indicated concentrations for 2 hr in duplicate 35-mm petri dishes containing β-hexosaminidase-deficient fibroblasts. Pinocytosis of β-hexosaminidase was measured as described.

490 cm^2 roller bottles (Corning) as described above for Gm-502. The cells are then radiolabeled for 24 hr at 37° in 50 ml, per roller bottle, of low-glucose (0.25 m*M*), serum-free Waymouth medium (KC Biological) containing human serum albumin (1 mg/ml), glutamine (2 m*M*), pyruvate (0.1 mg/ml), NH_4Cl (10 m*M*), mannose 6-phosphate (2 m*M*), sodium phosphate (2 m*M*), HEPES (15 m*M*), and [2-^{3}H]mannose (20 μCi/ml; 15.8 Ci/mmol). After labeling, the medium is collected and may be stored frozen at −20° until used. The secretions are then thawed at 5° and applied at a flow rate of 25 ml/hr to a column (5.0 ml of packed beads) of concanavalin A (Con A)–Sepharose (Pharmacia) equilibrated with 10 m*M* Tris, 500 m*M* NaCl, 50 m*M* sodium phosphate, pH 7.5 (buffer A), at 5°. The column is then washed with 50 ml of buffer A (flow rate of 25 ml/hr), warmed to room temperature, and eluted at 14 ml/hr with 50 ml of buffer A containing 750 m*M* α-methylmannoside and 0.02% sodium azide. Fractions of 2.0 ml are collected, and aliquots are monitored for radioactivity (10 μl) and β-hexosaminidase activity (25 μl) using 4-methylumbelliferyl-2-acetamido-2-deoxy-β-D-glucopyranoside as the substrate.[9]

Peak fractions are pooled (from four roller bottles we obtain approximately 21×10^6 cpm of [2-^{3}H]mannose-labeled glycoproteins and 3620 mU of β-hexosaminidase), 1.0 mg of human serum albumin per milliliter are added, and the mixture is extensively dialyzed against 10 m*M* Tris, pH 8.0, at 5°. After dialysis, the recovery of β-hexosaminidase activity is quantitative, and approximately 10% of the [2-^{3}H]mannose counts are found in the dialysis buffer. The dialyzed [2-^{3}H]mannose-labeled glycoproteins are now frozen and lyophilized. The lyophilized secretions are solubilized in 3.0 ml of buffer containing 200 m*M* glucose 6-phosphate (to inhibit phosphatase activity present in Pronase), 50 m*M* *N*-acetylglucosamine, and 20 m*M* $CaCl_2$. The pH is then carefully adjusted to 8.0 with 1.0 *M* Tris, giving a final Tris concentration of approximately 0.25 *M*. To this suspension is added 0.1 ml of a 150 mg/ml Pronase (Boehringer Mannheim) solution that had been preincubated without substrate for 2 hr at 37°, in 100 m*M* Tris, 20 m*M* $CaCl_2$, pH 8.0. The Pronase digest is incubated at 37° for 6 hr, at which time an additional 0.1 ml of Pronase solution is added, and the incubation is continued for an additional 18 hr. The digestion is terminated by boiling for 5 min, and the digest is added undiluted to a column (0.7 × 4.0 cm) of Con A–Sepharose, equilibrated with buffer A at room temperature. The column is then washed with 10.0 ml of buffer A and eluted twice with 5.0 ml of buffer A heated to 60° and containing 500 m*M* α-methylmannoside.

Approximately 35–40% of the [2-^{3}H]mannose-labeled glycopeptides applied to the Con A–Sepharose column adhere and are subsequently eluted with 500 m*M* α-methylmannoside. These represent a population of glycopeptides containing the high mannose-type oligosaccharide core some (20–25%) which contain phosphomannose. We obtain 5 to 7 × 10^6 cpm of [2-^{3}H]mannose-labeled glycopeptides from the secretions collected from 4 roller bottles of Tay–Sachs disease fibroblasts. The glycopeptides eluted from the Con A–Sepharose column are desalted at 5° in 5.0-ml aliquots on a column (1.4 × 52 cm) of BioGel P-2 (100–200 mesh) (Bio-Rad Laboratories). The material that is voided from the desalting column is pooled, frozen, and lyophilized. The lyophilized [2-^{3}H]mannose-labeled glycopeptides are solubilized in 1.0 ml of 5 m*M* sodium phosphate buffer, pH 7.0, and are now ready for use as ligands for pinocytosis measurements in I-cell disease fibroblasts.

Pinocytosis of [2-^{3}H]Mannose-Labeled Glycopeptides by I-Cell Disease Fibroblasts

Measurements of pinocytosis of [2-^{3}H]mannose-labeled glycopeptides are carried out in I-cell disease fibroblasts (available from the Human Mutant Cell Repository, Camden, New Jersey). I-cell disease fibroblasts are grown to confluence in 35-mm petri dishes as described in a preceding section for Sandhoff disease fibroblasts. Uptake is measured in duplicate 35-mm petri dishes containing 15,000–30,000 cpm of the [2-^{3}H]mannose-labeled glycopeptide ligand in 1.0 ml of medium. Uptake is usually conducted for a period of 12 hr at 37° in a 5% CO_2 atmosphere. The uptake is

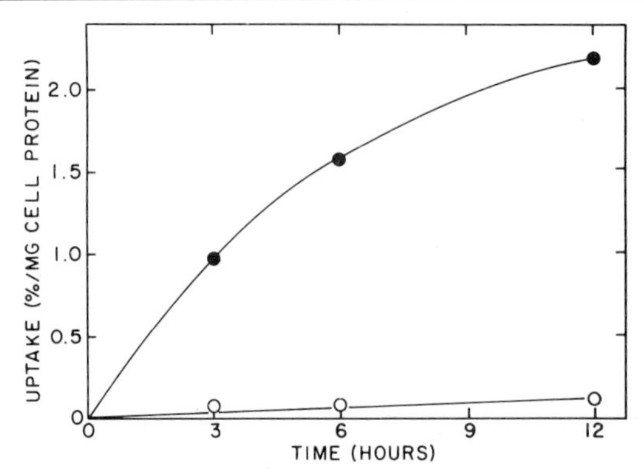

FIG. 3. Pinocytosis of [2-^{3}H]mannose-labeled glycopeptides by I-cell disease fibroblasts as a function of time. [2-^{3}H]mannose-labeled glycopeptides (17,000 cpm/plate) were added to I-cell disease fibroblasts in the absence (●) or the presence (○) of 2 m*M* mannose 6-phosphate. Pinocytosis of the labeled ligand was measured in duplicate 35-mm petri plates at the times indicated as described.

terminated by chilling the cells on ice, removing the uptake medium, and carefully washing the cells six times with 3-ml portions of ice-cold phosphate-buffered saline. The cells are then lysed with 0.5 ml of 1% sodium deoxycholate. Radioactivity is determined in 0.4 ml of the lysate in 4.0 ml of Scinti Verse liquid scintillation cocktail (Fisher Scientific Co.). Protein can then be determined in the remaining lysate by the method of Lowry *et al.*[11] A typical time course for the uptake of [2-^{3}H]mannose-labeled glycopeptides into I-cell disease fibroblasts is presented in Fig. 3. The uptake is inhibited 95% by 2 m*M* mannose 6-phosphate, which suggests that the uptake being measured is occurring via the phosphomannosyl recognition system present on fibroblasts. We have also measured the uptake of this ligand into Sandhoff disease fibroblasts, but the rate of uptake is generally three to six times lower than that observed for I-cell disease fibroblasts.

Preparation and Pinocytosis of [2-^{3}H]Mannose-Labeled Oligosaccharides by I-Cell Disease Fibroblasts

[2-^{3}H]Mannose-labeled glycopeptides are prepared as described in the preceding section. These glycopeptides are then incubated with 50 mU of endo-*β*-*N*-acetylglucosaminidase H from *Streptomyces griseus* (Miles Laboratories) in 0.2 ml of 0.1 *M* citrate–phosphate buffer, pH 5.5, at 37° under a toluene atmosphere for 17 hr. The digest is terminated by dilution with 20 volumes of H_2O and applied to a column (0.5 × 5.0 cm) of AG 50W-X2

[11] O. H. Lowry, N. J. Rosebrough, A. L. Farr, and R. J. Randall, *J. Biol. Chem.* **193,** 265 (1951).

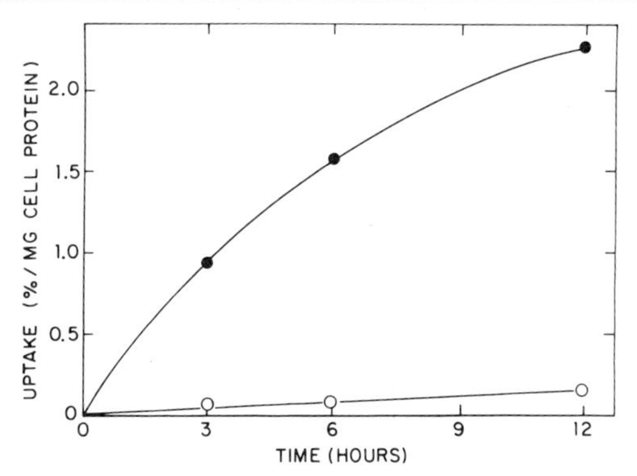

FIG. 4. Pinocytosis of [2-^{3}H]mannose-labeled oligosaccharides by I-cell disease fibroblasts as a function of time. [2-^{3}H]Mannose-labeled oligosaccharides (30,000 cpm/plate) were added to I-cell disease fibroblasts in the absence (●) or in the presence (○) of 2 m*M* mannose 6-phosphate. Pinocytosis of the labeled ligand was measured in duplicate 35-mm petri plates at the times indicated as described.

(200–400 mesh, Bio-Rad Laboratories) equilibrated with H_2O. Under these conditions most of the peptide moieties bind to the resin, but neutral and anionic oligosaccharides flow through. The flow through and three 1.0 ml H_2O washes of the column, containing the [2-^{3}H]mannose-labeled oligosaccharides, are frozen and lyophilized. The lyophilized [2-^{3}H]mannose-labeled oligosaccharides are solubilized in 1.0 ml of 5 m*M* sodium phosphate, pH 7.0, and are ready for use as ligands for pinocytosis measurements in I-cell disease fibroblasts. Uptake of the [2-^{3}H]mannose-labeled oligosaccharides is measured in I-cell disease fibroblasts exactly as described in the preceding section for [2-^{3}H]mannose-labeled glycopeptides. As shown in Fig. 4, the uptake of the oligosaccharides into I-cell disease fibroblasts is linear for approximately 6 hr and the uptake is inhibited 95% by 2 m*M* mannose 6-phosphate. Properties of the uptake of the [2-^{3}H]mannose-labeled oligosaccharides are similar in every respect to those observed for the uptake of [2-^{3}H]mannose-labeled glycopeptides. (Compare Figs. 3 and 4.)

Separation and Pinocytosis of Individual Phosphorylated [2-^{3}H]Mannose-Labeled Oligosaccharide Species Isolated from Tay–Sachs Disease Fibroblast Secretions

[2-^{3}H]Mannose-labeled oligosaccharides are prepared from Tay–Sachs disease fibroblast secretions as described in the preceding section. Prior to separation, they are further purified on a column (0.7 × 4.0 cm) of Con A–Sepharose equilibrated at room temperature with 150 m*M* NaCl,

2.6 m*M* KCl, 10 m*M* sodium phosphate, 1 m*M* $MgCl_2$, pH 7.2 (buffer B). After the [2-^{3}H]mannose-labeled oligosaccharides are added to the Con A-Sepharose column, it is washed with 10 1.0-ml volumes of buffer B and then with 30 1.0-ml volumes of buffer B containing 10 m*M* α-methylglucoside. The α-methylglucoside wash elutes oligosaccharides that have a low affinity for Con A–Sepharose (approximately 15% of the added [2-^{3}H]mannose-labeled oligosaccharides) and selects for high mannose-type oligosaccharides, which remain bound to the column under these conditions. The high mannose-type oligosaccharides are eluted with 5.0 ml of buffer B containing 500 m*M* α-methylmannoside and preheated to 60°. These labeled oligosaccharides are desalted at 5° on a column (1.4 × 52 cm) of BioGel P-2 (100–200 mesh) (Bio-Rad Laboratories). The material that is voided from the desalting column is pooled, frozen, and lyophilized. The lyophilized [2-^{3}H]mannose-labeled oligosaccharides are solubilized in 2.0 ml of 2 m*M* pyridinium acetate, pH 5.3.

Phosphorylaed oligosaccharides with one or two phosphomonoesters (uncovered phosphates) or phosphodiesters (covered phosphates) may be separated on a column of quaternary aminoethyl (QAE)-Sephadex (Q-25-100, Sigma Chemical Co.) by the method of Goldberg and Kornfeld.[12] Oligosaccharide samples in 2.0 ml of 2 m*M* pyridinium acetate, pH 5.3. are applied to a column (0.7 × 15.0 cm) of QAE-Sephadex equilibrated with 2 m*M* pyridinium acetate, pH 5.3. The column, at 5°, is washed with 2 m*M* pyridinium acetate, pH 5.3, and fractions of 2.0 ml are collected. At fraction 15 a linear gradient (200 ml) of pyridinium acetate, pH 5.3, from 2 to 400 m*M* is begun. The flow rate is 25 ml/hr, and a total of 100 fractions are collected. Aliquots (10 μl) of each fraction are monitored for radioactivity. Since we found significant uncovering of the oligosaccharides if they were lyophilized in the pyridinium acetate buffer, peak fractions are immediately pooled and applied to columns (0.5 × 2.0 cm) of Con A–Sepharose equilibrated with buffer B at room temperature. The columns are washed with 5.0 ml of buffer B and eluted with 3.0 ml of buffer B containing 500 m*M* α-methylmannoside and preheated to 60°. Samples are then desalted on columns of BioGel P-2 (100–200 mesh) at 5°. The [2-^{3}H]mannose-labeled oligosaccharides that are voided from the desalting columns are frozen, lyophilized, and solubilized in distilled H_2O or 5 m*M* sodium phosphate, pH 7.0. As shown in Fig. 5, the [2-^{3}H]mannose-labeled oligosaccharides are separated into a large neutral peak (N), 79% of the total, and five anionic species. The sensitivity of each peak to alkaline phosphatase before and after mild acid hydrolysis determines whether the negative charge on each oligosaccharide species is imparted by a phosphate moiety and also whether the phosphate is in monoester linkage (phosphatase sensitive) or

[12] D. E. Goldberg and S. Kornfeld, *J. Biol. Chem.* **256,** 13060 (1981).

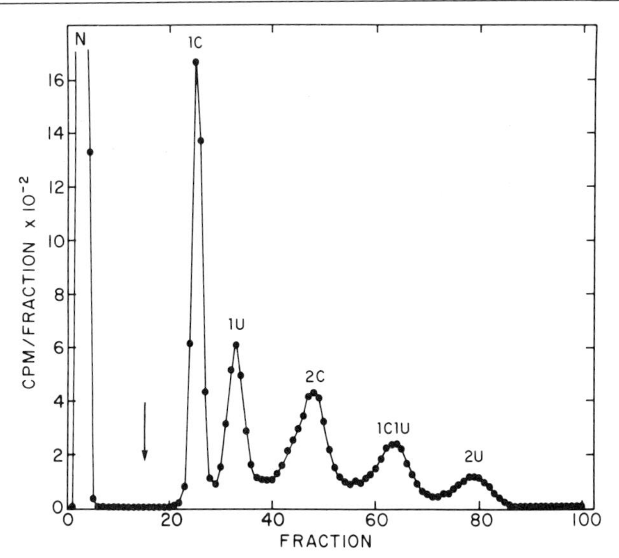

FIG. 5. Separation of [2-^{3}H]mannose-labeled oligosacchrides by gradient elution from QAE-Sephadex. Oligosaccharides (75,000 cpm) were applied to a column of QAE-Sephadex and separated as described in text. A 2 to 400 m*M* linear gradient of pyridinium acetate, pH 5.3, was started at fraction 15 (arrow). N, neutral oligosaccharides; 1C, oligosaccharides containing one covered phosphate; 1U, oligosaccharides containing one uncovered phosphate; 2C, oligosaccharides containing two covered phosphates; 1C1U, oligosaccharides containing one covered and one uncovered phosphate; 2U, oligosaccharides containing two uncovered phosphates.

diester linkage (sensitive to phosphatase only after mild acid hydrolysis). See Varki and Kornfeld[13] for further details of this procedure.

Using this analysis, we identified each of the five anionic peaks as [2-^{3}H]mannose-labeled oligosaccharides with one covered phosphate (1C), one uncovered phosphate (1U), two covered phosphates (2C), one covered phosphate and one uncovered phosphate (1C1U), and two uncovered phosphates (2U) (Fig. 5; see Creek and Sly[14]). Individual oligosaccharide species can be tested for uptake into I-cell disease fibroblasts exactly as discussed in a previous section for [2-^{3}H]mannose-labeled glycopeptides. As shown in the table, uptake of neutral oligosaccharides (N) and 1C oligosaccharides could not be detected. Uptake of 1U, 2C, and 1C1U oligosaccharides was very low. However, the uptake of 2U oligosaccharides was remarkable and showed an uptake rate about 30-fold greater than the other oligosaccharide species. Therefore oligosaccharides with two exposed phosphates are much better ligands for the phosphomannosyl receptor than the

[13] A. Varki and S. Kornfeld, *J. Biol. Chem.* **255,** 10847 (1980).

[14] K. E. Creek and W. S. Sly, *J. Biol. Chem.* **257,** 9931 (1982).

PINOCYTOSIS BY I-CELL DISEASE FIBROBLASTS OF THE DIFFERENT PHOSPHORYLATED OLIGOSACCHARIDE SPECIES ISOLATED BY GRADIENT ELUTION FROM QAE-SEPHADEX[a]

Oligosaccharide	Uptake (%/mg cell protein/12 hr)
N	ND[b]
1C	ND
1U	2.0
2C	1.7
1C1U	3.8
2U	58.7

[a] Pinocytosis of the different oligosaccharide species (1000–10,000 cpm/plate), separated as in Fig. 5, was measured in I-cell disease fibroblasts as described in the text. Inhibition by 2 m*M* mannose 6-phosphate was 96%.

[b] ND, none detected.

other phosphorylated oligosaccharide species present on Tay–Sachs disease fibroblast secretions.[14,15]

These results point to two important criteria for producing an optimal high mannose-type oligosaccharide ligand for uptake into fibroblasts: (*a*) the oligosaccharide must have uncovered phosphate; (*b*) the oligosaccharide must have more than one uncovered phosphate. The fact that the phosphorylated oligosaccharides themselves are subject to pinocytosis indicates that they are the principal determinants of the uptake of lysosomal enzymes (i.e., the recognition marker) and that the polypeptide portions of acid hydrolases are not essential for recognition by the phosphomannosyl–enzyme receptors on fibroblasts. See Fischer *et al.*[15] and Varki and Kornfeld[16] for details of the structural requirements for binding of phosphorylated oligosaccharides to immobilized phosphomannosyl receptors purified from bovine liver.

[15] H. D. Fischer, K. E. Creek, and W. S. Sly, *J. Biol. Chem.* **257,** 9938 (1982).

[16] A. Varki and S. Kornfeld, *J. Biol. Chem.* **258,** 2808 (1983).

[24] Uptake of Glycoproteins and Glycoconjugates by Macrophages

By M. KONISH, V. SHEPHERD, G. HOLT, and P. STAHL

Macrophages express a cell surface pinocytosis receptor specific for glycoproteins and glycoconjugates that terminate in mannose of L-fucose. After binding, the mannose receptor mediates the rapid transfer of ligands from the cell surface to the lysosomal system. Binding and uptake require Ca^{2+} and proceed optimally at pH 7.0–7.4.[1]

Cells

Alveolar Macrophages. Alveolar macrophages are obtained from rats or rabbits by pulmonary lavage.[2,3] After anesthesia with Nembutal (30–50 mg/kg, ip) and exsanguination, the lungs are washed *in situ* three or four times with saline. A convenient way to proceed is to hang a standard saline intravenous drip unit about 2 ft above the animal. The trachea and lungs are exposed surgically, and polyethylene tubing with a three-way stopcock is inserted into the trachea, connecting the saline reservoir with the lungs. The lungs are allowed to fill with saline until they are fully distended. The saline is then removed via the three-way stopcock. This is repeated three or four times. The cells are collected by centrifugation and resuspended in assay medium (see below) to a concentration of 10^7 cells/ml. Cell yields are usually 2 to 5×10^6 cells per rat, 2 to 5×10^7 cells per normal rabbit, and $>5 \times 10^8$ cells per bacillus Calmette-Guérin-stimulated rabbit.

Bone Marrow-Derived Macrophages. Bone marrow cells will differentiate in culture to mature macrophages in the presence of colony-stimulating factor (CSF).[4] The CSF is secreted by mouse L cells and can be used in the form of spent medium from L cells in culture (L-cell conditioned medium, or LCM). LCM can be prepared as follows: 10^6 L cells are seeded into a T-150 flask in minimum essential medium (MEM) with 10% fetal calf serum. On day 4, the medium is replaced with serum-free MEM. After 3 more days, the spent medium (LCM) is removed, filtered, and frozen.

Bone marrow-derived macrophages are prepared by a modification of

[1] P. Stahl, J. Rodman, J. Miller, and P. Schlesinger, *Proc. Natl. Acad. Sci. U.S.A.* **75,** 1399 (1978).

[2] J. Brain and R. Frank, *J. Appl. Physiol.* **34,** 75 (1973).

[3] E. S. Leake and Q. Myrvik, *Res. J. Reticuloendothel. Soc.* **5,** 33 (1968).

[4] H. S. Lin and S. Gordon, *J. Exp. Med.* **150,** 231 (1979).

ISBN 0-12-181998-1

the method of Lin and Gordon.[4] Mouse or rat femurs are removed from the animal, and the marrow is flushed out with sterile Hank's balanced salt solution. The cells are washed, counted and resuspended in MEM containing 10% fetal calf serum; penicillin, streptomycin, and kanamycin; and 10% LCM. After 4–6 days in culture, most of the cells are attached and have begun to spread and take on a macrophage-like appearance. These cells display characteristics of mononuclear phagocytes, such as esterase activity and phagocytic activity.

Glycoprotein Ligands

β-Glucuronidase. β-Glucuronidase is prepared from rat preputial glands as follows: 15 g of rat preputial glands, freed of excess fat, are homogenized in a Polytron (Brinkmann Instruments) in 120 ml of 0.1 *M* Tris-acetate buffer, pH 7.8, at 4°, using 6 to 10-sec bursts at a high setting. The homogenate is centrifuged at 100,000 *g* for 30 min, and the supernatant is carefully decanted. Solid ammonium sulfate (44 g/100 ml) is added slowly to the supernatant at 4° (final concentration of ammonium sulfate is 70%). The mixture is centrifuged at 15,000 rpm for 20 min. The pellet is dissolved in 75 ml of 0.02 *M* Tris–acetate buffer, pH 7.8, and resuspended with a Dounce homogenizer. Freezer-chilled ethanol (0.4 ml/ml of enzyme preparation) is added slowly, and after the mixture is stirred for 10 min at 4°, it is centrifuged at 15,000 rpm for 20 min. Cold ethanol (0.8 ml/ml of solution) is added slowly to the supernatant. The mixture is stirred for 10 min, then centrifuged at 15,000 rpm for 20 min. The pellet is resuspended in 10–20 ml of 0.02 *M* Tris–acetate buffer, pH 7.8, using a Dounce homogenizer, and clarified by centrifugation if necessary. The resulting enzyme-containing supernatant is fractionated on a Sephadex G-200 column (2.5 × 85 cm) buffered with 5 m*M* $NaPO_4$, pH 7.5, containing 0.15 *M* NaCl. Fractions are assayed for β-glucuronidase activity using phenolphthalein glucuronide as described by Stahl and Touster.[5] Active fractions are pooled and concentrated using an Amicon ultrafilter to a protein concentration of 1–2 mg/ml. Homogeneous enzyme can be obtained by this method, with a specific activity of 2000 units per milligram of protein (1 unit = 1 μmol of product released per minute at 37°. Enzyme is stable at 4° for 1–2 months, or can be stored at −20° in 50% glycerol for >6 months.

Neoglycoproteins. Synthetic glycoconjugates (i.e., neoglycoproteins) have proved to be potent ligands for uptake and binding studies in macrophages. The best ligands are mannose, fucose, and *N*-acetylglucosamine covalently linked to bovine serum albumin (BSA). Neoglycoproteins can be prepared according to the method of Lee *et al.*[6] Active precursors of many of

[5] P. Stahl and O. Touster, *J. Biol. Chem.* **246,** 5398 (1971).
[6] Y. C. Lee, C. P. Stowell, and M. J. Krantz, *Biochemistry* **15,** 3956 (1976).

the monosaccharides are available commercially as the cyanomethyl-1-thioglycosides. The procedure for preparation of mannose–BSA, starting with the above precursor, is as follows: To 54 mg of sodium methylate dissolved in 20 ml of absolute methanol is added 0.4 g of precursor. The reaction mixture is stirred for 24 hr at room temperature. The solvent is then removed on a rotary evaporator at 30°. To the resulting white powder is added 6 mg of bovine serum albumin dissolved in 6.6 ml of 0.1 *M* sodium borate, pH 8.5. The reaction is allowed to proceed for 0.5–24 hr, depending on the degree of mannose substitution desired and the activity of the precursor. The reaction is terminated by the addition of 0.660 ml of 1 *M* acetic acid. After dialysis versus distilled water, the mannose–BSA product is lyophilized.

Iodination of Ligands. β-Glucuronidase is iodinated by the following procedure. One hundred micrograms of β-glucuronidase is mixed with 1 mCi of $Na^{125}I$ and 10 μg of chloramine-T in 0.1 *M* $NaPO_4$, pH 7.6. The reaction is allowed to proceed for 5 min on ice and is terminated by addition of 332 μl of a 1 : 10,000 dilution of 2-mercaptoethanol. Free iodine is removed by gel filtration or dialysis.

Mannose–BSA[7] (100 μg) is mixed with 1 mCi of $Na^{125}I$ and 300 μg of chloramine-T in 0.1 *M* $NaPO_4$ buffer, pH 7.6, in a total volume of 80 μl. The reaction is allowed to proceed for 10 min on ice and is terminated by addition of 190 μl of sodium metabisulfite (2.4 mg/ml) and 190 μl of potassium iodide (10 mg/ml). Free iodine is removed by dialysis.

Uptake and Binding Assays

Alveolar Cells (in Suspension). Assays are conducted at 4° (binding) or 37° (uptake) in MEM or Hank's balanced salt solution buffered with non-volatile buffers [10 m*M* HEPES and 10 m*M* *N*-tris(hydroxymethyl)methyl-2-aminoethanesulfonic acid (TES)] at pH 7.4 containing 10% calf serum or 10 mg of bovine serum albumin per milliliter. The medium containing ligand ± inhibitor and cells (5×10^5) in a volume of 0.1 ml is placed in a 0.4-ml microfuge tube over 0.15 ml of oil [4 parts silicon oil (No. 660 Accumetric, Elizabethtown, Kentucky) : 1 part mineral oil]. After incubation, the tubes are spun for 10–30 sec in a microfuge, and the tips (containing only cells) are cut off for counting. Assays are run in the presence or the absence of yeast mannan (1–2 mg/ml). Uptake or binding in the presence of yeast mannan is considered nonspecific.

Cultured Bone Marrow Cells. Cells for assay are grown in 24-well plates, with an initial seeding density of 10^6 cells per well per 1.0 ml of medium. For uptake or binding assays, the medium is removed and replaced with Hank's balanced salt solution, buffered with 10 m*M* HEPES and 10 m*M* TES to pH

[7] P. Stahl, P. Schlesinger, E. Sigardson, J. Rodman, and Y. C. Lee, *Cell* **19**, 207 (1980).

7.4 and containing 0.1% glucose and 10 mg of BSA per milliliter. Radiolabeled ligand is added in the presence or the absence of yeast mannan (1 mg/ml) in a total volume of 0.4 ml. After incubation, the cell layer is washed two or three times with Hank's balanced salt solution and solubilized in 0.5 ml of 0.1% Triton X-100 in H_2O.

Internalization of Prebound Mannose-BSA

Cells (5×10^6/ml) are incubated in Hank's buffered salt solution, pH 7.4, containing 10 m*M* HEPES and TES, 0.1% glucose, and 10 mg of BSA per milliliter, with ^{125}I-labeled mannose–BSA (5 μg/ml) for 60 min at 4°. The cells are then washed in Hank's balanced salt solution and resuspended in medium containing BSA. The cells are warmed to 37° in the absence of added ligand for 0–60 min. After warming, the cells are cooled and separated from the medium by centrifugation. The cells are then resuspended in 0.1% trypsin containing 10 m*M* EDTA, and 100-μl aliquots are placed over oil in 0.4-ml microfuge tubes. After 15 min of incubation on ice, the cells are spun through the oil. Trypsin-releasable (cell surface) and trypsin-resistant (internalized) ligand are estimated from the medium and cell pellets, respectively.

[25] Isolation of a Phosphomannosyl Receptor from Bovine Liver Membranes[1]

By DIANE C. MITCHELL, G. GARY SAHAGIAN, JACK J. DISTLER, RENATE M. WAGNER, and GEORGE W. JOURDIAN

The uptake of exogenous lysosomal enzymes by cultured cells and the intracellular translocation of newly synthesized lysosomal enzymes are mediated by the binding of a recognition marker associated with the enzymes to a membrane-associated receptor. The receptor recognizes residues of mannose 6-phosphate that comprise a portion of the oligosaccharide side chains on lysosomal enzymes.[2] Phosphomannosyl receptors with the same apparent molecular weight and similar binding specificity have been iso-

[1] This work was supported in part by Grant Am-10531 from the National Institute of Arthritis, Metabolism, and Digestive Diseases, The National Foundation—March of Dimes, and the Arthritis Foundation, Michigan Chapter.

[2] G. W. Jourdian, G. G. Sahagian, and J. J. Distler, *Biochem. Soc. Trans.* **9**, 510 (1982).

ISBN 0-12-181998-1

lated from human, bovine, hamster, and rat liver, Chinese hamster ovary cells, human skin fibroblasts,[3,4] and rat chondrocytes.[5]

This chapter describes a modification of a previously described procedure[6] for the isolation of phosphomannosyl receptor from bovine liver. Zwittergent is used as the extracting detergent in place of Triton X-100. The modified procedure offers the advantages of higher yields, little interference with protein determinations, and ready removal of detergent by dialysis.

Assay Method

Principle. A radioimmunoassay based on the ability of unlabeled phosphomannosyl receptor (PMR) to inhibit the binding of [^{125}I]PMR to anti-PMR is described. In this procedure [^{125}I]PMR is adsorbed to a limited amount of anti-PMR, and the complex is precipitated with formalin-fixed *Staphylococcus aureus* cells containing protein A. Addition of unlabeled PMR decreases [^{125}I]PMR in the precipitated complex.

Reagents

Suspending medium: One volume of 0.1 *M* Tris-HCl, pH 7.0, is mixed with 1 volume of physiological saline containing 0.04% Zwittergent 3-14 (Calbiochem–Behring Corp., La Jolla, California), 0.04% NaN_3, and 0.1% bovine serum albumin (BSA).

IgGsorb suspension: 10% lyophilized, formalin-fixed *S. aureus* cells, Cowan's strain A (The Enzyme Center Inc., Boston, Massachusetts)

Anti-PMR is produced in New Zealand white rabbits (3–4 kg). Each animal is injected intramuscularly with 1 ml of an emulsion composed of 1 volume of complete Freund's adjuvant and 1 volume of physiological saline containing 250 μg of purified PMR. Initially, injections are administered every 2 weeks for 6 weeks. Thereafter, injections are made at 6-week intervals using incomplete adjuvant with an equivalent concentration of PMR. Antibody is detected approximately 1 month after the injections are initiated. The antisera are diluted with saline to a titer sufficient to bind 25–50% of the [^{125}I]PMR in the assay described below (in the absence of unlabeled PMR).

[^{125}I]PMR (1–3 μCi/100 μg): PMR (100 μg) is iodinated with Bio-Rad Enzymobeads radioiodination reagent (Bio-Rad, Richmond, California). After iodination, the protein is subjected to gel-permeation

[3] G. G. Sahagian, J. J. Distler, and G. W. Jourdian, unpublished results.

[4] G. G. Sahagian, J. J. Distler, and G. W. Jourdian, *Proc. Natl. Acad. Sci. U.S.A.* **78,** 4289 (1981).

[5] A. W. Steiner and L. H. Rome, *Arch. Biochem. Biophys.* **214,** 681 (1982).

[6] G. G. Sahagian, J. J. Distler, and G. W. Jourdian, this series, Vol. 83 [34].

chromatography on Sephadex G-100 and is further purified by affinity chromatography on phosphomannan–Sepharose 4B as described below. The [^{125}I]PMR is diluted with suspending medium without BSA to the concentration required.

Unlabeled PMR, 2–35 μg/ml, purified as described below.

Procedure. Each assay mixture is prepared in duplicate. The mixtures contain (in a final volume of 225 μl): 50 μl of [^{125}I]PMR (approximately 5×10^4 cpm), 100 μl of unlabeled PMR (0.2–3.5 μg), 50 μl of diluted anti-PMR, and 25 μl of IgGsorb. Each assay is incubated at 25° for 2 hr, 3 ml of suspending medium is added, and the mixture is centrifuged for 2 min at 2200 *g*. The supernatant is removed by aspiration, and the ^{125}I content of the pellet is measured in a gamma spectrometer. The PMR content of each sample is determined from a standard curve (Fig. 1).

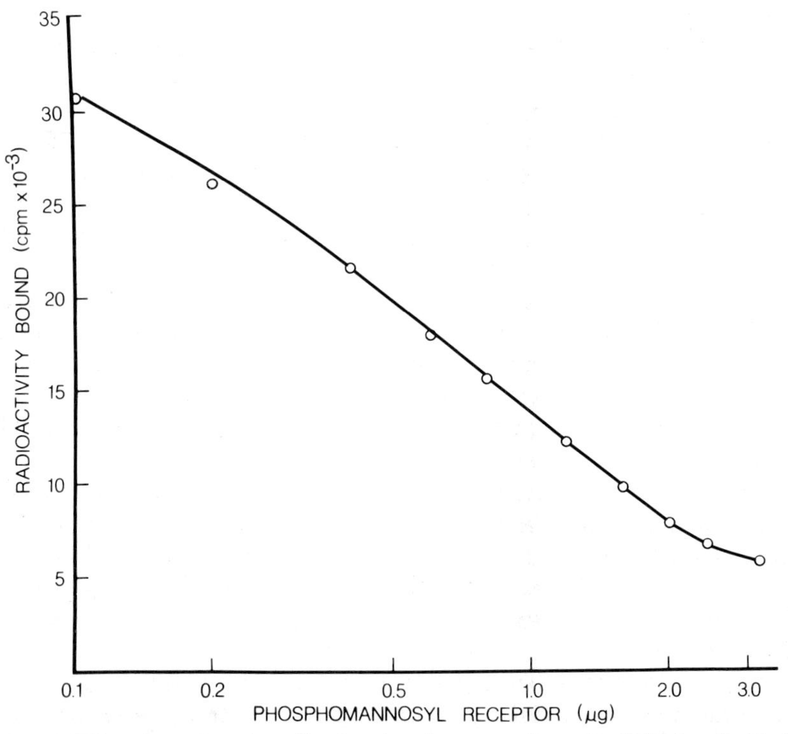

FIG. 1. Effect of concentration of bovine phosphomannosyl receptor (PMR) on the binding of [^{125}I]PMR to antiserum against PMR. Incubations contained 50 μl of a 600-fold diluted rabbit antiserum, 5×10^4 cpm [^{125}I]PMR, the indicated amount of PMR, and *Staphylococcus aureus* cells to immobilize the immunocomplex. The precipitate was washed as described in the text, and the precipitated radioactivity was determined.

Purification Procedure

All operations are conducted at 0–4° unless otherwise stated. Protein content is estimated by the procedure of Bradford.[7]

Step 1. Preparation of Liver Membranes. Fresh bovine liver is freed of connective tissue and processed as soon as possible. The tissue can be stored at −20° for up to 2 months; however, storage for longer periods results in a large decrease in yield of receptor. The tissue (500 g) is diced into 1-cm cubes and homogenized for 1 min in a Waring blender in 300 ml of a solution composed of 0.5 m*M* $CaCl_2$, 1 m*M* $NaHCO_3$, and the following protease inhibitors: 10 m*M* Na_2EDTA, 100 m*M* 6-aminohexanoic acid, 10 m*M* *N*-ethylmaleimide, 1 m*M* benzamidine–HCl, and 1 m*M* phenylmethylsulfonyl fluoride. The homogenate is adjusted to pH 5.0 by dropwise addition of 4 *N* acetic acid and centrifuged for 15 min at 10,000 *g*. The pellet is suspended in 1.2 liters of the same solution, the pH is adjusted again to 5.0, and the suspension is again centrifuged at 10,000 *g*.

Step 2. Acetone Powder Preparation. The washed pellet from step 1 is homogenized for 1 min in a Waring blender in 3 liters of acetone cooled to −20°. The suspension is rapidly filtered through Whatman 3MM filter paper on a Büchner funnel, and the residue is washed two additional times with acetone and once with 2 liters of diethyl ether (−70°). Residual ether is removed under vacuum, and the resulting powder is stored at −70°. No loss in binding activity has been detected when the acetone powder has been stored at this temperature for 3 months.

Step 3. Extraction with Zwittergent. Acetone powder (50 g) is suspended in 1.5 liters of a solution containing 0.2 *M* NaCl, 0.1 *M* sodium acetate (pH 6), and the protease inhibitors (at the same concentrations used in the homogenization buffer described in step 1). The suspension is stirred for 30 min and then centrifuged for 15 min at 10,000 *g*. The supernatant is discarded, the pellet is resuspended in 1.4 liters of distilled water containing the protease inhibitors, and the suspension is recentrifuged. The washed pellet containing receptor is then suspended in 3.0 liters of 0.4 *M* KCl, 1% Zwittergent, 50 m*M* imidazole-HCl (pH 7.0), and protease inhibitors. The suspension is stirred for 1 hr and then centrifuged for 30 min at 10,000 *g*. The clear, red-brown supernatant contains approximately 150 μg of soluble PMR for each gram of extracted acetone powder.

Step 4. Affinity Chromatography. The conditions for the preparation of the affinity matrix have been described previously.[6] Briefly summarized, an extracellular phosphomannan protein complex is isolated from the growth medium of cultures of *Hansenula hostii*. The complex is subjected to mild

[7] M. M. Bradford, *Anal. Biochem.* **72**, 248 (1976).

acid hydrolysis to yield a fragment containing residues of mannose 6-phosphate. This fragment is covalently attached to cyanogen bromide-activated Sepharose 4B. The resulting affinity matrix is washed and packed into a column; 2 ml of phosphomannan–Sepharose are required for each gram of acetone powder extracted. The column is equilibrated with an elution buffer consisting of the suspending medium (described under Assay Method) without bovine serum albumin, and the supernatant from step 3 is applied to the column at a rate of 200 ml/hr. The affinity column is washed with 50 column volumes of the elution buffer. The receptor is eluted from the affinity column with 5–7 column volumes of the elution buffer containing 5 m*M* mannose 6-phosphate.

Fractions containing receptor as determined by radioimmunoassay are pooled and concentrated by pressure filtration over an Amicon YM 10 filter to a final protein concentration of 1 mg/ml. The yield of receptor (150 μg of PMR per gram of acetone powder extracted) is approximately 1.5 times greater than that obtained when receptor is extracted with Triton X-100. Zwittergent can be removed from the preparation by dialysis below its critical micelle concentration for 24 hr against 1000 volumes of elution buffer containing 20% ethyl alcohol. Ethyl alcohol is removed by dialysis against the same solution lacking ethyl alcohol.

Preliminary results suggest that the extraction and purification of PMR also from human, rat, and hamster liver may be performed by the above procedure. Receptors obtained from these species may be measured by radioimmunoassay using bovine [^{125}I]PMR and rabbit antisera against bovine PMR. Nonradioactive PMR from the corresponding species must be used as standard.

Properties of PMR

Stability. Purified PMR retains its biological activity in the presence of 0.05% Triton X-100, 5 m*M* mannose 6-phosphate, and 0.02% NaN_3 for at least 6 months when stored at 4° or at −20°.

Specificity. In this laboratory the binding properties of PMR have been examined using the lysosomal enzyme β-galactosidase.[4,6] Binding of β-galactosidase to PMR is saturable with a K_d for β-galactosidase of 25 n*M*. The binding of enzyme to receptor is inhibited by mannose 6-phosphate (K_i = 0.064 m*M*) and fructose 1-phosphate (K_i = 0.24 m*M*) but not by other monosaccharide phosphates. Treatment of β-galactosidase with alkaline phosphatase or endo-β-*N*-acetylglucosaminidase H (EC 3.2.1.96) destroys the binding of enzyme to receptor without altering the catalytic activity of the enzyme. Other lysosomal enzymes known to contain the phosphomannosyl recognition marker inhibit the binding of β-galactosidase to PMR; these include α-*N*-acetyl-D-glucosaminidase (K_i = 42 n*M*), β-D-glucuronidase (K_i = 8.0 n*M*), and α-L-iduronidase (K_i = 1.8 n*M*).

pH Optimum and Activators. Binding of β-galactosidase to PMR is optimal between pH values of 6 and 8 and drops sharply below pH 5.8; below pH 5.6 less than 10% of the binding activity remains. No requirements for divalent cations could be demonstrated.

Physical Properties. The 40,000-fold purified receptor is a glycoprotein containing approximately 5% carbohydrate. When reduced and subjected to polyacrylamide gel electrophoresis in the presence of sodium dodecyl sulfate, the receptor exhibits a single protein-staining band with an apparent molecular weight of 215,000.[4]

[26] Purification of Rat Liver Fucose Binding Protein

By MARK A. LEHRMAN and ROBERT L. HILL

Other chapters in this volume describe the purification of liver lectins (carbohydrate binding proteins) with a high affinity for ligands containing galactose, mannose/*N*-acetylglucosamine, and mannose 6-phosphate at their nonreducing termini. This chapter describes the preparation of a fourth type of liver lectin that has a high affinity for fucose-containing ligands. The starting material for preparation of the fucose binding protein (F-BP) is a Triton X-100 extract of rat liver, which contains F-BP[1] in addition to the galactose (G-BP)[2] and mannose/*N*-acetylglucosamine (M-BP)[3] binding proteins. Thus, the methods described below permit resolution of the different lectins from one another and further purification of each of the different lectins. Moreover, in developing the following methods, a form of the M-BP with 10–20 times the specific activity of most of the M-BP in liver extracts was found to be a major contaminant of the F-BP; methods for preparation of this form of M-BP, designated HM-BP, are also presented.

Affinity chromatography on columns of the neoglycoprotein L-fucosyl-bovine serum albumin (Fuc-BSA)[4] is the principal method for purifying the rat liver lectins. F-BP, M-BP, and HM-BP are adsorbed to Fuc-BSA-agarose whereas G-BP is not adsorbed. The bulk of the M-BP is subsequently removed from F-BP by rechromatography on Fuc-BSA-agarose in the

[1] M. A. Lehrman and R. L. Hill, *Fed. Proc., Fed. Am. Soc. Exp. Biol.* **41,** 1030 (1982).
[2] T. Tanabe, W. E. Pricer, Jr., and G. Ashwell, *J. Biol. Chem.* **254,** 1038, (1979).
[3] Y. Mizuno, Y. Kozutsumi, T. Kawaski, and I. Yamashina, *J. Biol. Chem.* **256,** 4247 (1981).
[4] C. P. Stowell and Y. C. Lee, *Biochemistry* **19,** 4899 (1980).

ISBN 0-12-181998-1

presence of *N*-acetylglucosamine, which inhibits adsorption of M-BP, but not of F-BP or HM-BP. The HM-BP is removed by adsorption on anti-rat-M-BP rabbit antibodies bound to agarose.

Reagents

L-Fucosyl-β-bovine serum albumin (Fuc-BSA) is employed in the assay of F-BP and is also used to prepare the affinity adsorbent Fuc-BSA-agarose. Fuc-BSA is prepared by coupling 2-imino-2-methoxyethyl-1-thio-β-L-fucopyranoside (IME-Fuc) to BSA. The following sections describe the synthesis of IME-Fuc from free fucose as well as preparation of Fuc-BSA and Fuc-BSA-agarose. The methods for synthesis of IME-Fuc are essentially similar to those reported earlier for synthesis of the same derivative of galactose.[5-7]

Step 1. 1,2,3,4-Tetra-O-acetyl-β-L-fucopyranoside (*Ac-Fuc*).[5] Dry acetic anhydride (80 ml), stored over sodium acetate and molecular sieves, is added to dry sodium acetate (20 g) and brought to a gentle boil on a hot plate. L-Fucose (10 g, Aldrich) and sufficient L-[1-^{14}C]fucose (New England Nuclear) to give 300–400 cpm/μmol are slowly added with stirring. The heat source is turned off, and the mixture is allowed to stand for 10 min. Sufficient water (about 50 ml) is added to dissolve the remaining sodium acetate, and the resulting solution is cooled on ice, transferred to a 1-liter beaker, and 120 ml of cold, saturated sodium bicarbonate solution is added. Solid sodium bicarbonate (about 50 g) is then added slowly, with stirring, until the product precipitates, making stirring difficult. The product is extracted with 300 ml of chloroform, and the extract is washed three times with 400 ml of water. The chloroform extract is dried over calcium chloride, and the clear solution is evaporated under vacuum. The product (Ac-Fuc) is a syrup obtained in essentially 100% yield and is stable at room temperature for weeks; it is used directly in the next step and can be handled more easily if traces of chloroform remain to prevent hardening of the syrup.

Step 2. 1-Bromo-2,3,4-tri-O-acetyl-α-L-fucopyranoside (*Br-Fuc*).[5] The Ac-Fuc obtained in the preceding step is dissolved with 50 ml of 32% hydrobromic acid in glacial acetic acid (Eastman), and molecular sieves are added to the solution. After the syrup dissolves, the mixture is allowed to stand in a hood at 23° for 90 min. Chloroform (200 ml) is added, and the solution is then washed successively with ice-cold water (twice), saturated sodium bicarbonate (once), and water (twice). The chloroform layer is dried over calcium chloride, and a syrupy product is obtained by evaporation under vacuum. Since Br-Fuc is very unstable, it must be used immediately.

[5] H. M. Flowers, A. Levy, and N. Sharon, *Carbohydr. Res.* **4,** 189 (1967).
[6] S. Chipowsky and Y. C. Lee, *Carbohydr. Res.* **31,** 339 (1973).
[7] Y. C. Lee, C. P. Stowell, and M. J. Krantz, *Biochemistry* **15,** 3956 (1976).

Step 3. 2-S-(2,3,4-Tri-O-acetyl-β-L-fucopyranosyl)2-thiopseudourea Hydrobromide (T-Fuc).[6] The Br-Fuc obtained in the preceding step is dissolved in acetone (40 ml; dried over molecular sieves), thiourea (22 g) is added, and the solution is filtered while hot. The solution of T-Fuc is then used immediately to prepare the next product.

Step 4. Cyanomethyl-2,3,4-tri-O-acetyl-1-thio-β-L-fucopyranoside (CNM-Fuc).[7] The T-Fuc in 40 ml of acetone as obtained in the preceding step is mixed with water (50 ml), acetone (10 ml), chloroacetonitrile (10 ml), potassium bicarbonate · $1.5H_2O$ (10 g) and sodium bisulfite (9 g), and the mixture is stirred vigorously for 30 min at room temperature. Ice water (350 ml) is then added, and the mixture is stirred on ice for 3 hr. The resulting yellow precipitate[8] is collected by filtration; it is recrystallized from methanol to give CNM-Fuc in a yield of 25%. It is important to obtain white crystals by repeated recrystallization, since the colored contaminants that may be present will react with bovine serum albumin. The product (mp = 109 – 100°) is stable when stored at −20° with desiccant.

Step 5. 2-Imino-2-methoxyethyl-1-thio-β-L-fucopyranoside (IME-Fuc).[7] [1-^{14}C]CNM-Fuc (300 – 400 cpm/μmol; 1.0 g) is dissolved in methanol (40 ml, stored over molecular sieves). A portion (100 μl) is removed and counted to obtain the exact specific radioactivity of the CNM-Fuc (M_r = 345). Sodium methoxide (108 mg) is then added with a few molecular sieves, and the reaction mixture is incubated for 10 – 15 hr at 23°. Deacetylation proceeds to completion, and conversion of CNM-Fuc to IME-Fuc is monitored by a trinitrobenzenesulfonic acid assay.[7]

Step 6. Fucosyl-BSA.[7] The IME-Fuc solution obtained in the preceding step is filtered to remove the molecular sieves and then flash-evaporated in two 20-ml aliquots to give a syrup or a white solid. To one aliquot is added 10 ml of BSA (Boehringer-Mannheim; 10 mg/ml) in 0.2 *M* sodium borate, pH 8.5. The residual sodium methoxide is sufficient to raise the pH to 9.4 After 30 min at 23°, the solution is transferred to the second aliquot of IME-Fuc, and the mixture should be about pH 9.7; the mixture is allowed to stand for 2 hr at 23°, and the pH maintained between pH 9.4 and 9.7 if necessary. The Fuc-BSA is obtained by dialyzing the mixture exhaustively against 0.1 *M* ammonium bicarbonate.

The product should give a single band on gel electrophoresis in sodium dodecyl sulfate[9] with an M_r of 70,000 and migrating slightly slower than BSA. Since amidination does not cause gross changes in protein structure,[10] protein concentration is monitored by absorbance at 280 nm with BSA as a

[8] If a yellow precipitate does not form after 3 hr at 0°, the product may be recovered by chloroform extraction of the aqueous mixture.[4]

[9] U. K. Laemmle, *Nature (London)* **227,** 680 (1970).

[10] M. J. Hunter and M. I. Ludwig, this series, Vol. 25, p. 585.

standard. The amount of L-fucose coupled to Fuc-BSA can be quantitated by counting L-[^{14}C] fucose or by the anthrone or Diesche–Shettles[11] reactions. We routinely obtained 50 ± 2 mol of fucose per mole of BSA. Fuc-BSA is stable when stored lyophilized with desiccant at −20°.

Rabbit-(Anti-Rat-Mannose Binding Protein)-IgG (Anti-M-BP-IgG)

Mannose binding protein (M-BP) is obtained from the unabsorbed fraction in step 4 (Fuc-BSA–Sepharose chromatography with 0.5 *M* D-GlcNAc) below and repurified on mannan–Sepharose (see step 8). Sufficient M-BP (600 μg) to produce antibodies can be obtained from 100 g of liver.

Aliquots of M-BP (100 μg) are precipitated for 30 min with 80% ethanol at 0°. The pellet obtained by centrifugation is dissolved in 1 ml of 0.01 *M* sodium phosphate buffer, pH 7.5, containing 0.15 *M* sodium chloride and emulsified with 1 ml of Freund's complete adjuvant (Difco) with a VirTis-45 mixer for 1 min on a medium setting. A white male New Zealand rabbit is given 7 or 8 subcutaneous/intramuscular injections on each hindquarter. This is repeated 8 and 14 days later, except that incomplete adjuvant is used. Immune serum is taken on day 21. Subsequent immune sera are obtained after further immunization with 100 μg of M-BP in incomplete adjuvant, followed by bleeding within 30 days. Immune sera are routinely tested by Ouchterlony double-diffusion analysis in 1% agarose containing 0.1% Triton X-100, 0.01 *M* sodium phosphate, pH 7.4, 0.15 *M* sodium chloride, 1 m*M* EDTA, and 0.02% sodium azide,[12] using 10 μl of serum and 1 μg of M-BP. Immunoglobulin G (IgG) is purified from the antiserum by protein A–Sepharose chromatography.[13]

Affinity Adsorbents

Four different affinity chromatographic adsorbents are used in the following procedures: Fuc-BSA–Sepharose (1 mg/ml), Fuc-BSA–Sepharose (10 mg/ml), mannan–Sepharose (1 mg/ml), and anti-mannose binding protein IgG–Sepharose (5 mg/ml). Each is prepared as follows.

CNBr Activation of Sepharose 4B.[14] CNBr (1 g) is dissolved in 3 ml of dimethylformamide and slowly added to Sepharose 4B (10 ml) suspended in 20 ml of 2.5 *M* potassium phosphate, pH 12.0, on ice, with constant mixing. After 10 min, the adsorbent is washed on a sintered-glass funnel with excess cold water, and then the buffer is used for coupling the ligands to the agarose (see below).

[11] R. G. Spiro, this series, Vol. 8, p. 3.

[12] A. J. Crowle, "Immunodiffusion," 2nd ed., p. 179. Academic Press, New York, 1973.

[13] J. W. Goding, *J. Immunol. Methods* **13,** 215 (1976).

[14] J. Porath, this series, Vol. 34, p. 13.

Neoglycoprotein and Mannan Adsorbents. Neoglycoproteins are dissolved in 0.1 *M* sodium pyrophosphate, pH 8.5 (coupling buffer), in a volume equal to the volume of the CNBr-activated Sepharose 4B to be used, and added to the activated Sepharose; the mixture then is agitated for 48 hr at 4°, resulting in coupling of more than 90% of the ligand to the agarose. The resulting adsorbent is washed successively at 23° with 0.2 *M* sodium chloride, 1.0 *M* ethanolamine–HCl (pH 8.0), 0.2 *M* sodium chloride, 0.05 *M* glycine–HCl (pH 4.0), 0.2 *M* sodium chloride, 0.05 *M* glycine–NaOH (pH 10.0), 0.02 *M* Tris-HCl (pH 7.8), 0.2 *M* sodium chloride and stored in 0.2 *M* sodium chloride + 0.02% NaN_3 until used. Mannan–Sepharose is prepared similarly, but the coupling efficiency is 50%.

IgG Adsorbents. These are prepared as described for the neoglycoprotein adsorbents except that the coupling reaction is allowed to proceed for 16 hr at 23° in 0.2 *M* sodium citrate, pH 6.5.

[^{125}I]-Labeled Fuc-BSA. Fuc-BSA is labeled with [^{125}I]NaI (Amersham) with lactoperoxidase–glucose oxidase beads (Enzymobeads, Bio-Rad) by the method recommended by the manufacturer.[15] Specific activities of 20–40 μCi/μg are obtained. [^{125}I]Fuc-BSA does not undergo rapid autoradiolysis and remains active for 2 months, at which time it is discarded.

Peroxide-Free Triton X-100

Residual peroxides in commercially available Triton X-100 chemically modify proteins, including oxidation of sulfhydryl groups,[16] and are reduced by modification[17] of a published method.[18] Triton X-100 (100 g) is dissolved in absolute ethanol (475 ml); sodium borohydride (2.5 g) is added, and the mixture is stirred at 23°, with concomitant H_2(g) release. After 3 hr additional sodium borohydride (2.5 g) is added, and the reaction mixture is stirred overnight. Excess borohydride is decomposed by dropwise addition of concentrated HCl until the evolution of gaseous hydrogen ceases. The insoluble sodium chloride is removed by filtration through fritted glass, and the filtrate is flash-evaporated to remove the ethanol. The residual Triton X-100 is then dissolved in 500 ml of nitrogen (g) saturated water, and the solution is passed through a column containing 200 ml of (water-washed) Dowex AG 501-X8(D) to attain a final conductivity of $<$0.1 mmho. The Triton X-100 concentration is estimated spectrophotometrically at 275 nm (1 mg/ml gives an adsorbance of 2.25), and the final concentration is adjusted to 20% (w/v). Sufficient butylated hydroxytoluene (Sigma) and EDTA are added to give final concentrations of 0.0025% and 0.1 m*M*,

[15] Technical Bulletin 1071, Bio-Rad Laboratories, Richmond, California, 1979.
[16] G. J. Giotta, *Biochem. Biophys. Res. Commun.* **71,** 776 (1976).
[17] K. R. Westcott, M. A. Lehrman, and R. L. Hill, unpublished observation.
[18] H. W. Chang and E. Bock, *Anal. Biochem.* **104,** 112 (1980).

respectively. The solution is stored in the dark at 4° and is added to all detergent-containing solutions just before use.

Binding Assays

The binding assays depend upon the removal of receptor–ligand complexes from unbound ligand by filtration on glass fiber filters. Complexes of M-BP or HM-BP with ligand are trapped with the aid of a 5% polyethylene glycol precipitation step (assay A). In contrast, F-BP–ligand complexes do not require a precipitation step (assay B), although addition of 5% polyethylene glycol does not affect binding (assay A). M-BP can also be assayed with an ammonium sulfate precipitation assay.[3]

Assay A: 5% Polyethylene Glycol 6000 (PEG) Precipitation (F-BP, M-BP, HM-BP). Standard assays contain 0–100 ng of receptor, 10 ng of [^{125}I]Fuc-BSA, 1 mg of BSA per milliliter, 0.2 *M* sodium chloride, 0.02 *M* Tris-HCl (pH 7.8), 0.02 *M* calcium chloride, 0.5% sodium azide, and 0.5% (w/v) Triton X-100 in a total volume of 0.25 ml. The mixture is incubated in polystyrene tubes for 60 min at 23°. Human fibrinogen (65 μl of 5 mg/ml), which aids precipitation and 2.5 ml of PEG solution [5% (w/v) PEG 6000, containing 0.01 *M* calcium chloride, 0.02 *M* Tris-HCl (pH 7.8), 0.05% sodium azide, and 0.2 *M* sodium chloride] are then added. The light precipitate that is visible after 10 min at 23° is collected by vacuum filtration on Whatman GF/C filter disks presoaked in a solution containing 0.5 mg of BSA per milliliter 0.01 *M* calcium chloride, 0.02 *M* Tris-HCl (pH 7.8), 0.05% sodium azide, and 0.2 *M* sodium chloride. The assay tubes and filters are rinsed with 5 ml of PEG solution, and the filters are counted for ^{125}I in a gamma counter. Nonspecific binding, determined by omission of lectin or addition of 10 μg of unlabeled Fuc-BSA, was $<5\%$ of the total counts. F-BP, M-BP, and HM-BP are detected by this assay.

Assay B (for FBP). Standard assays contain the same components as assay A in a final volume of 1.0 ml and are incubated for 30 min at 23°. Fibrinogen (65 μl, 5 mg/ml), which decreases the nonspecific binding without effecting specific binding, is then added. After 10 min at 23°, the reaction mixture is filtered as in assay A, except that tubes and filters are rinsed with 10 ml of 0.01 *M* sodium glycine (pH 10.0) containing 0.2 *M* sodium chloride, 0.01 *M* calcium chloride, and 0.02% sodium azide. Nonspecific binding is usually $<0.5\%$ of the counts. Whereas F-BP is detected equally in assays A and B, the lower backgrounds in assay B make this assay preferable. M-BP and HM-BP, however, are best detected with assay A.

One unit is defined as an amount of activity that binds 1% of the counts in the above assays. F-BP has a specific activity of 10–20 U/μg in assay A or B; M-BP has a specific activity of 10–20 U/μg in assay A, and HM-BP has a specific activity of 400–600 U/μg in assay A.

Purification of Rat Liver Fucose Binding Protein (F-BP)

Figure 1 summarizes the purification procedure.[19] Unless otherwise indicated, all steps are performed at 0–4°. Since liver extracts contain other lectins that bind either galactosides[2] or mannosides/*N*-acetylglucosaminides,[3] it is important to achieve separation of these lectins from the desired fucose lectin.

Step 1. Crude Homogenate. Rat livers (300–400 g), stored at −80°, are suspended in deionized water (1 liter) and agitated to aid defrosting and removal of blood. When soft but still semisolid, the livers are rinsed quickly with water, minced with a cleaver, and then homogenized in a Waring blender (4-liter) in 2 liters of an aqueous solution containing 50 KU of Trasylol per milliliter (Mobay), 0.25 m*M* EDTA, and 0.2% sodium azide, adjusted to pH 7.7 with solid sodium bicarbonate. Homogenization is obtained by adjusting the motor to the highest setting and blending three times (30 sec) with intervals of 30 sec between each blending. Triton X-100 (200 ml, 20% solution, peroxide free) is added, and the solution is adjusted to pH 7.7 with 1 *M* Tris.[20] The homogenate is stirred for 1 hr, then filtered through four layers of cheesecloth and centrifuged at 7100 *g* for 20 min. The clarified supernatant is used for the next step.

Step 2. 6% Polyethlene Glycol 6000 (PEG) Precipitation. Solid PEG 6000 (Scientific Products) is slowly added with stirring to the supernatant to give a final PEG concentration of 6%. Stirring is then continued for 15 min to allow dissolution of the PEG, and the suspension is allowed to stand for an additional 30 min. The mixture is centrifuged at 7100 *g* for 90 min; the clear red-brown supernatant is discarded, and the tan residue is resuspended with vigorous stirring for 30 min in 400 ml of 0.02 *M* Tris-HCl buffer, pH 7.8, containing 0.2 *M* sodium chloride, 0.05% sodium azide, 0.03 *M* calcium chloride, and 0.5% (w/v) Triton X-100. The resulting suspension is centrifuged at 16,300 *g* for 20 min; the pellet is discarded, and the clear brown supernatant is used for the next step.

Step 3. Fuc-BSA–Sepharose 4B Chromatography. The supernatant from step 2 is applied to a column of Fuc-BSA–Sepharose 4B (35 ml, 2 mg/ml) equilibrated with buffer A [0.02 *M* Tris-HCl, pH 7.8, containing 0.2 *M* sodium chloride, 0.05% sodium azide, 0.01 *M* calcium chloride, and 0.5% (w/v) Triton X-100]. The top of the column is stirred as necessary to disperse a gummy precipitate that tends to collect as the column develops and thereby reduce the flow rate. After all the supernatant is applied, the column is washed with 350 ml of buffer A and the adsorbed lectins are eluted with 100 ml of buffer B (0.02 *M* Tris-HCl, pH 7.8, containing 0.2 *M*

[19] M. A. Lehrman and R. L. Hill, in preparation.

[20] The glass electrode must be cleaned frequently of solids in the homogenate to obtain correct pH measurements.

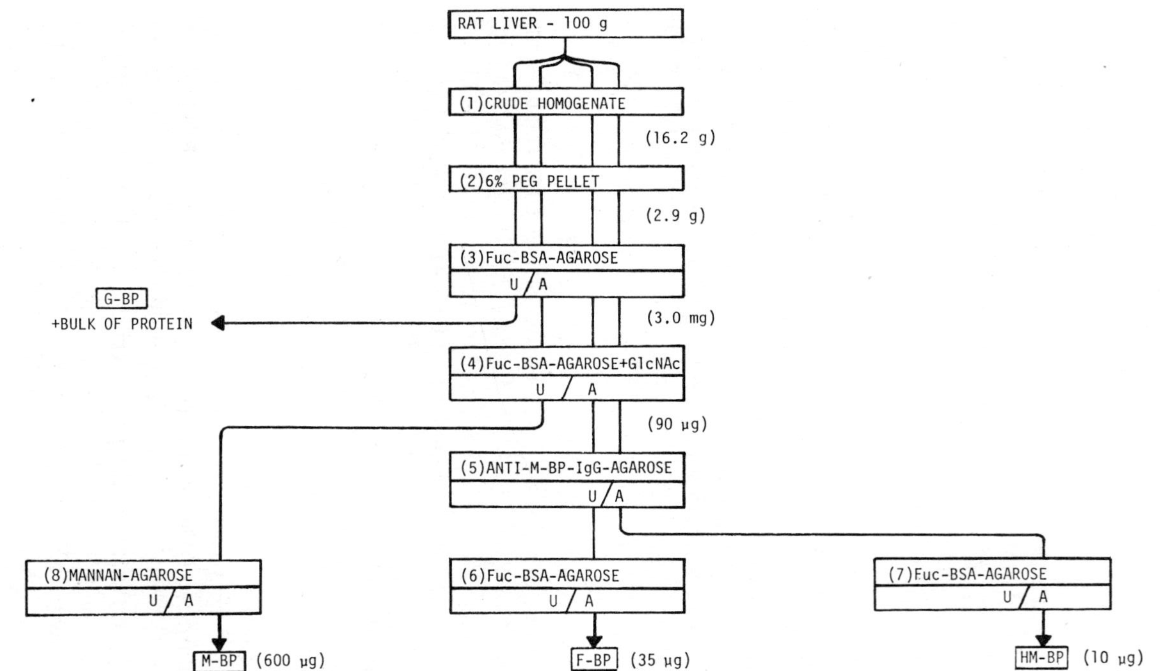

FIG. 1. Summary of the procedures used to purify rat liver lectins. Lectin abbreviations are as follows: F-BP, fucose binding protein; M-BP, mannose binding protein; HM-BP, hyperactive mannose binding protein; G-BP, galactose binding protein. F-BP, M-BP, and HM-BP are purified to homogeneity as described in the text. G-BP may be recovered from the unadsorbed fraction of step 3 by affinity chromatography on asialoorosomucoid-agarose,[2] but the yield is lower than with other purification procedures.[2] The protein content is given in parentheses. U = unadsorbed fraction; A = adsorbed fraction.

sodium chloride, 0.05% sodium azide, 1 mM Na_3 EDTA, and 0.5% (w/v) Triton X-100).[21]

Step 4. Fuc-BSA–Sepharose 4B Chromatography with 0.05 M N-Acetyglucosamine. Solid N-acetyl-D-glucosamine (GlcNAc; Pfanstiehl) is added to 100 ml of eluate from step 3 to give a final concentration of 0.05 M. A solution of calcium chloride (1.0 M) is also adsorbed to give a concentration of 0.01 M. The eluate is then applied to a column of Fuc-BSA–Sepharose 4B (3 ml, 10 mg/ml) and equilibrated with buffer A plus 0.05 M GlcNAc. Then the column is adjusted to a flow rate of 1 ml/min. The flow-through is retained to prepare M-BP (step 8). The column is washed with 10 ml of buffer A containing 0.05 M GlcNAc, followed by 10 ml of buffer A, and then eluted with buffer B. The fractions containing Fuc-BP in the buffer B eluate are located (assay A) and pooled for further purification in step 5.

Step 5. Anti-M-BP-IgG–Sepharose 4B Adsorption. Rabbit (anti-rat M-BP) IgG-Sepharose 4B (5 ml; 5 mg/ml) is washed with 50 ml of buffer B containing 1 mg of BSA per milliliter and then with 100 ml of buffer B. The pooled fractions from step 4 are added to the adsorbent, and the suspension is mixed by rotation at 60 rpm for 16 hr. The adsorbent is packed into a column, which is washed with buffer B (10 ml). The F-BP is present in the unadsorbed fractions and buffer B.

Step 6. Fuc-BSA–Sepharose 4B Chromatography. The fractions from step 5 containing Fuc-BP are made 0.01 M in $CaCl_2$ and purified as in step 4 except that N-acetylglucosamine is not present during the adsorption or washing steps.

Step 7. Recovery of HM-BP. The anti-M-BP-IgG Sepharose after step 5 is suspended in 0.2 M glycine-HCl, pH 2.2, for 60 min at 23°, which results in elution of HM-BP. The eluate is then neutralized to pH 7.8 with 50% NaOH and 1 M Tris base. Pure HM-BP is obtained by making this solution 0.01 M with calcium chloride and repeating step 4, but omitting GlcNAc from the buffers. The anti-M-BP-IgG–Sepharose is regenerated by washing with 0.01 M Tris-HCl, pH 7.8, containing 0.01 M sodium chloride and 0.2% sodium azide, and then with 0.2 M sodium chloride containing 0.2% sodium azide and stored at 4°.

Step 8. Recovery of M-BP. The flow-through material from step 4 is dialyzed exhaustively against 0.2 M sodium chloride containing 0.02 M Tris-HCl, pH 7.8, and 0.002% sodium azide to remove GlcNAc. The dialyzed protein is made 0.01 M with calcium chloride and 0.5% (w/v) with Triton X-100 and applied to a column of mannan–Sepharose 4B (10 ml, 1 mg/ml) equilibrated with buffer A. After washing with buffer A, pure M-BP

[21] The G-BP in the unadsorbed fraction can be brought to apparent homogeneity by affinity chromatography on asialoorosomucoid-agarose,[2] as judged by one experiment. The yield, however, is lower than that reported by other methods.[2]

is obtained by eluting with buffer B, and the active fractions (assay A) are pooled.

Properties of Fucose Binding Protein

Purity. On gel electrophoresis in sodium dodecyl sulfate (SDS) under reducing conditions, the F-BP obtained after step 6 is visibly free of the other rat liver lectins.[19] The F-BP preparations at this stage migrate as two proteins of similar size (M_r = 73,000 and 67,000), with somewhat variable amounts of one or the other present in different preparations. Since the two bands give very similar peptide maps on gel electrophoresis in SDS after digestion with cyanogen bromide or proteolytic enzymes, the smallest species is assumed to be a proteolytic degradation product of the larger species. This assumption is supported by the observation that the binding of Fuc-BSA to rat liver membranes cannot be detected unless the protease inhibitor Trasylol (bovine pancreatic trypsin inhibitor) is present during preparation of the membranes.[22]

The different rat liver lectins do not cross-react with anti-lectin antibodies as judged by Ouchterlony double diffusion.[12] Thus, rabbit anti-rat F-BP IgG[19] gives a precipitin reaction only with rat F-BP and shows no cross-reaction with rat G-BP or M-BP. Similarly, rabbit anti-rat-M-BP IgG cross-reacts with M-BP and HM-BP, but not with F-BP or G-BP. In addition, goat anti-rat G-BP[23] cross-reacts only with G-BP. Also noteworthy is the fact that anti-rat M-BP IgG completely inhibits the binding of ligands to M-BP and HM-BP in assay A but is without effect on the binding of Fuc-BSA to F-BP.

The presence of G-BP, M-BP, or HM-BP in the different fractions obtained during purification of F-BP (Fig. 1) is most easily determined by gel electrophoresis in SDS. The mobilities of each lectin (see the table) are sufficiently different in both the absence and the presence of reducing agents to be diagnostic. Analysis on reduced gels, however, is preferred, since nonreduced gels reveal higher oligomers of each lectin that may be present in variable amounts from preparation to preparation. Rabbit anti-M-BP IgG is particularly useful in detecting trace amounts of HM-BP that may contaminate F-BP after step 5. This antibody preparation completely inhibits the binding activity of M-BP and HM-BP, but not of F-BP. Thus, when added to an assay mixture for F-BP (assay A with Fuc-BSA), any diminution of binding activity likely reflects contamination with HM-BP. Since HM-BP has 10–20 times the specific binding activity of M-BP, its

[22] M. A. Lehrman, M. J. Imber, S. V. Pizzo, and R. L. Hill, *Fed. Proc., Fed. Am. Soc. Exp. Biol.* **39,** 1968 (1980).

[23] Gift of Dr. G. Ashwell.

PROPERTIES OF RAT LIVER LECTINS IN SODIUM DODECYL SULFATE–POLYACRYLAMIDE GEL ELECTROPHORESIS[a]

	Major electrophoretic species (9% acrylamide)		
	+Mercaptoethanol		
Lectin	R_f	Apparent molecular weight	−Mercaptoethanol[b] (R_f)
Fucose binding protein	0.30	73,000	0.03
	0.34	67,000	0.15
Mannose binding protein, hyperactive mannose binding protein	0.79	32,000	0.03[c]
			0.29
			0.55
Galactose binding protein	0.32	86,000	ND
	0.63	43,000	

[a] From Laemmle.[9]

[b] Electrophoretic species present in the absence of mercaptoethanol are assumed to be higher-molecular-weight oligomers of those present with mercaptoethanol. ND, not determined.

[c] The 0.03 R_f species is present in all hyperactive mannose binding protein preparations tested, but varies among mannose binding protein preparations.

presence in preparations of F-BP gives incorrectly high binding activity for Fuc-BSA.

Stability and Storage. F-BP is stable for at least 2 months with little loss of binding activity when stored at a protein concentration of 20 μg/ml in buffer B at −20°. Repeated freezing and thawing of such solutions, however, results in a 0% to 25% loss of binding activity, the exact extent of inactivation varying from preparation to preparation. Thus, storage without extensive loss in activity is best accomplished by freezing F-BP in buffer B at −20° in 500-μl samples that will be consumed in experimental studies after being frozen and thawed only 2 or 3 times.

M-BP and HM-BP are very stable on storage in buffer B at −20°. Unlike F-BP, both forms may be precipitated from solution by 50% ethanol or acetone and remain active when reconstituted in buffer B; M-BP and HM-BP are also stable when stored in buffer B at 4° for at least 2–3 weeks.

Binding Specificity. Like other liver lectins, F-BP displays a spectrum of binding activities for monosaccharides that are found at the nonreducing termini of glycoproteins and glycolipids.[19] Thus, for inhibition of the binding to Fuc-BSA by neoglycoproteins (assay B), the relative order is as follows: Fucβ-BSA = 100; Galβ-BSA = 0.89, Galβ-1,4[Fucα-1,3]-GlcNAcβ-BSA = 0.74; Manα-BSA < 0.58; GlcNAcβ-BSA < 0.53. Only upper limits of the binding affinities of the latter two neoglycoproteins could be determined, and the actual affinities may be much less. Monosaccharides inhibit binding of Fuc-BSA (assay B) by F-BP; the concentrations of monosaccharides for 50% inhibition of Fuc-BSA is D-fucose (3 m*M*), D-*N*-acetylgalactosamine (3 m*M*), D-galactose (5 m*M*), D-mannose (6 m*M*), L-fucose (12 m*M*), L-galactose (18 m*M*), D-glucose (26 m*M*), D-mannose 6-phosphate (54 m*M*), and D-*N*-acetylglucosamine (300 m*M*).

Acknowledgment

This work was supported by the National Institutes of Health, Institute of General Medical Sciences, Grant GM-2747.

[27] Isolation of Coated Vesicles

By B. M. F. PEARSE

Coated vesicles occur in most eukaryotic cells. They mediate the selective transfer of certain membrane components and their ligands from one membrane-bound compartment to another inside cells. They appear to exclude other membrane proteins characteristic of the membrane from which they bud, thus acting as molecular filters and preventing intermixing of the essential components of the donor and receptor compartments.[1,2]

For example, coated vesicles are formed as the result of absorptive endocytosis. In this process,[3,4] coated pits on the plasma membrane of the cell bind specific membrane receptors and their ligands and bud into the cytoplasm to form coated vesicles, entrapping those receptors and ligands in their interior. The coats then dissociate and recycle to form new coated pits, and the vesicles fuse with another membrane-bound compartment, thus delivering their contents to that compartment. However, coated vesicles also bud from internal organelles such as the Golgi apparatus.[e.g.,5] The biochemical functions of this intracellular traffic are, at present, less clearly defined, but it is evident that coated vesicles are involved in extensive transfer routes from one cell compartment to another where, in each cycle, the coat is involved in the formation of a new vesicle. Thus, coated vesicles

[1] M. S. Bretscher, J. N. Thomson, and B. M. F. Pearse, *Proc. Natl. Acad. Sci. U.S.A.* **77,** 4156 (1980).

[2] B. M. F. Pearse and M. S. Bretscher, *Annu. Rev. Biochem.* **50,** 85 (1981).

[3] T. F. Roth and K. R. Porter, *J. Cell Biol.* **20,** 313 (1964).

[4] J. L. Goldstein, R. G. W. Anderson, and M. S. Brown, *Nature* (*London*) **279,** 679 (1979).

[5] D. S. Friend and M. G. Farquhar, *J. Cell. Biol.* **35,** 357 (1967).

ISBN 0-12-181998-1

derived from any tissue or cell type are likely to be heterogeneous in terms of their contents. Nevertheless, the bulk of the coated vesicles from many sources are sufficiently similar in mean size and density that they can be purified from cell extracts by differential centrifugation in density gradients.

Assay

Principle. The progress of the purification is monitored in two ways: (*a*) by electron microscopy to check that enrichment for coated vesicles is occurring; and (*b*) by sodium dodecyl sulfate (SDS)–polyacrylamide gel electrophoresis to locate the fractions containing the highest concentrations of clathrin. Clathrin is the major coat protein of coated vesicles and has a polypeptide chain of molecular weight 180,000.[6]

Electron Microscopy. Samples are negatively stained in unbuffered 1% uranyl acetate[7] on carbon-coated specimen grids and examined in a Philips EM301 microscope operating at 80 kV at a magnification of 25,000. Particles may be photographed on the carbon film or, preferably, suspended in stain over holes in the film. An electron micrograph of coated vesicles purified from human placenta is shown in Fig. 1.

SDS–Polyacrylamide Gel Electrophoresis. Electrophoresis of coated vesicle proteins is carried out on 7.5% polyacrylamide gels according to the procedure of Laemmli.[8] Minislab gels (8 × 4 cm) without stacking gels are used to monitor samples during the purification. Samples of 5–20 μl are applied to the gel, and electrophoresis is carried out at 100 V per slab for about 30 min when the dye band has moved about 4 cm. The gels are stained briefly with Coomassie Blue R-250. On destaining, clathrin usually becomes visible fairly quickly, thus identifying the chief coated vesicle fractions.

Figure 2 shows the major structural proteins of coated vesicles analyzed by SDS–polyacrylamide gel electrophoresis.

Purification of Coated Vesicles and Coated Particles[9]

Isolation Buffer (*HEPES Buffer*)

N-2-Hydroxyethylpiperazine-*N′*-2-ethanesulfonic acid (HEPES), 10 m*M*, pH 7.2, with NaOH

NaCl, 0.15 *M*

Ethylene glycol-bis(*β*-aminoethyl ether)-*N,N′*-tetraacetic acid (EGTA), 1 m*M*

[6] B. M. F. Pearse, *J. Mol. Biol.* **97,** 93 (1975).

[7] H. E. Huxley and G. Zubay, *J. Mol. Biol.* **2,** 10 (1960).

[8] U. K. Laemmli, *Nature* (*London*) **227,** 680 (1970).

[9] B. M. F. Pearse, *Proc. Natl. Acad. Sci. U.S.A.* **79,** 451 (1982).

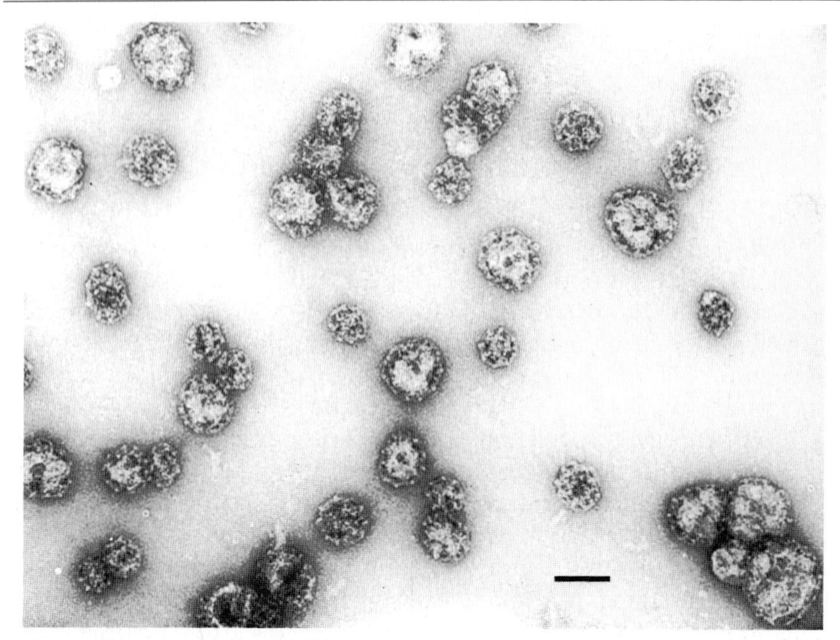

FIG. 1. Coated vesicles, purified from human placenta by the route I procedure, negatively stained in 1% uranyl acetate at a magnification of 75,000. Bar = 1000 Å.

$MgCl_2$, 0.5 m*M*
NaN_3, 0.02%
Phenylmethylsulfonyl fluoride (PMSF) 0.2 m*M*

Isolation from Human Placenta

The essential features of the purification scheme are outlined in Fig. 3.

Step 1. Human placentas that have been delivered normally are processed within 2 hr. The superficial membranes, blood vessels, and cord are removed, and the tissue is cut up, teased out, and rinsed in ice-cold Tris-HCl buffered saline, pH 7.2. Subsequent procedures are carried out at 0–4°. The placental fragments are suspended in three volumes of HEPES buffer, homogenized for 1 min at top speed in a Waring blender, and centrifuged at 1000 *g* for 30 min.

Step 2. The resulting extract is incubated at room temperature for 30 min with 10 units of pancreatic RNase (Worthington) per milliliter and recentrifuged at 55,000 *g* for 1 hr to give pellets.

Step 3. These initial pellets (from one placenta) are resuspended in the HEPES buffer, layered onto 3 × 50-ml 10–90% D_2O gradients (containing HEPES buffer throughout), and centrifuged at 45,000 *g* for 30 min. The

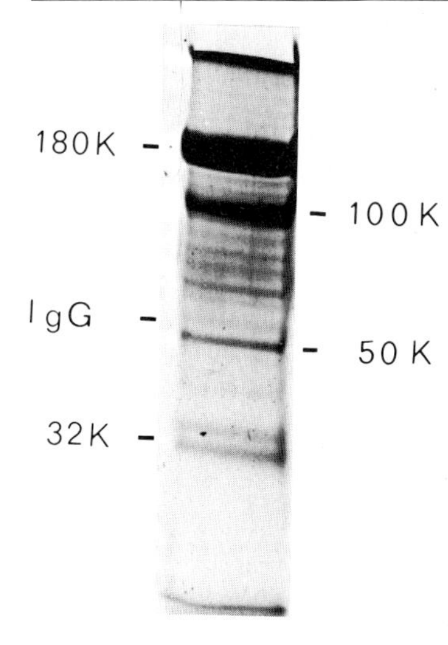

FIG. 2. Sodium dodecyl sulfate–polyacrylamide gel electrophoresis of a sample of human placental coated vesicle proteins prepared by the route II procedure. The major structural proteins are labeled: the M_r 180,000 and 32,000 polypeptides of the clathrin trimers, and the M_r 100,000 and 50,000 polypeptides of the Triton-extracted coated vesicle (TCV) cores. K = 1000.

supernatants, apart from any very turbid fractions above the solid residues, are pooled and diluted fourfold. The crude coated vesicles, contaminated mainly with smooth vesicles, are then pelleted at 100,000 *g* for 1 hr.

This crude coated vesicle fraction can be handled in different ways to yield useful coated vesicle preparations: route I, by centrifugation on an isotonic density gradient; route II, by extraction in 1% Triton X-100. A third alternative is to remove microvillous membranes by precipitation with wheat germ agglutinin.[10]

Route I. The crude coated vesicle pellets are resuspended in the minimum volume of the HEPES buffer and layered onto a gradient from 9% D_2O–2% Ficoll to 90% D_2O–20% Ficoll containing the HEPES buffer throughout. The Ficoll should be extensively dialyzed against water and freeze dried before use. The sample is centrifuged at 80,000 *g* for 16 hr. The coated vesicles are collected from the lowest region of the gradient and are washed and concentrated by centrifugation in the HEPES buffer.

Route II, Step 4. The crude coated vesicle pellets from step 3 are resuspended in the HEPES buffer containing 1% Triton X-100 to the same volume from which they were previously centrifuged and are left standing at room temperature for 30 min. The extracts are cleared at 10,000 *g* for 10 min and then centrifuged at 100,000 *g* for 1 hr.

Route II, Step 5. The pellets from step 4 are resuspended in a small volume of the HEPES buffer containing 1% Triton X-100, layered onto a 10 to 90% D_2O gradient (12 ml) containing the HEPES buffer plus 1% Triton X-100 throughout, and centrifuged at 45,000 *g* for 25 min. The pellet is

[10] A. G. Booth and M. J. Wilson, *Biochem. J.* **196,** 355 (1981).

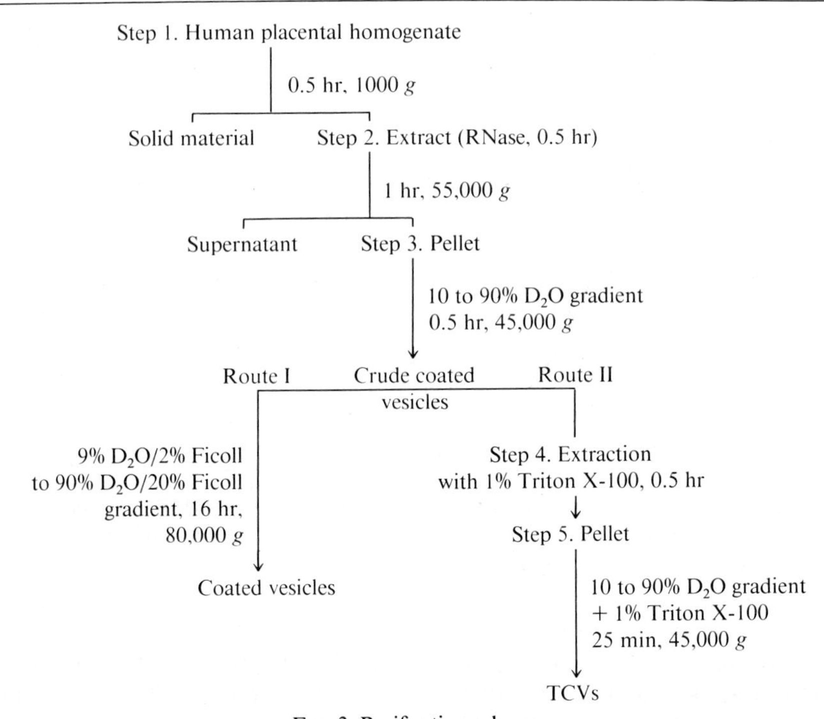

FIG. 3. Purification scheme.

discarded, and the fractions richest in clathrin are pooled, diluted fourfold in the HEPES buffer, and centrifuged at 100,000 *g* for 1 hr. The purified Triton X-100-extracted coated vesicles (TCVs) are resuspended in the HEPES buffer and stored at 0–4°.

Generally the yield of coated particles, after Triton extraction, is between 7 and 10 mg of protein from one placenta.

Isolation of Coated Vesicles from Other Sources

The methods described here are general and have been applied successfully to other tissues and cells. When cells are used, the initial homogenization is performed in the HEPES buffer using a Sorvall Omni-mixer at full speed for 1 min. In many cases (e.g., brain and H6/31 hybridoma cells[11]) clathrin is of the order of 1% of the extracted protein after removal of mitochondrial fractions. Generally, by the end of step 3 of the purification, the 180,000-dalton clathrin band is fairly prominent in SDS–gel analyses of coated vesicle-containing fractions. The final yields of coated vesicle protein

[11] T. Pearson, G. Galfré, A. Ziegler, and C. Milstein, *Eur. J. Immunol.* 7, 684 (1977).

obtained from a number of sources are in the region of 3–5 mg per 100 g wet weight of tissue.

Components of Coated Vesicles

Placental coated vesicles consist of about 75–80% protein, 20–25% lipid, and 1–2% carbohydrate.[9]

The outer cages of coated vesicles, characterized by their lattices of hexagons and pentagons, are composed of clathrin trimers. These trimers (or triskelions) consist of three polypeptides of M_r 180,000 associated with three heterogeneous light chains.[12–14] Clathrin trimers can be extracted from coated vesicles in a number of different ways. The structures can be dissolved in 2.5% cholate,[12] or, without disrupting the lipid bilayers of the vesicles, the coats can be dissociated using 2 *M* urea,[15,16] 0.5 *M* Tris-HCl,[17] or buffers of low ionic strength.[16,18] The trimers can then be separated by gel filtration. Good yields of human placental clathrin trimers can be isolated[19] from the initial pellets obtained in step 2 of the purification scheme for coated vesicles described above.

Purified clathrin trimers can be repolymerized to form closed polyhedral cages in 0.1 *M* 2-(*N*-morpholino)ethanesulfonic acid (MES) buffer, pH 6.5, containing 1 m*M* $MgCl_2$, 0.2 m*M* EGTA, and 0.02% NaN_3.[13] Assembly takes place in a few seconds and appears to be a condensation process[20] where cages exist in equilibrium with a constant low concentration of monomer triskelions.[19]

In coated vesicles, clathrin is apparently bound to a "core" of particles containing the auxiliary structural proteins of M_r 100,000 and 50,000.[9] The relationship of these core particles to the membranes of the vesicles is not yet understood, although a large proportion of their mass appears to occur on the cytoplasmic surfaces of the vesicles.[18] If TCVs, from preparative route II (Fig. 3), are further disrupted using 2 *M* urea, soluble core particles retaining some bound clathrin are released that may be separated by gel filtration. These core particles will reassociate in clathrin reassembly conditions to give particles reminiscent of small TCVs in size and shape but presumably

[12] B. M. F. Pearse, *J. Mol. Biol.* **126,** 803 (1978).
[13] E. Ungewickell and D. Branton, *Nature* (*London*) **239,** 420 (1981).
[14] T. Kirchhausen and S. C. Harrison, *Cell* **23,** 755 (1981).
[15] A. L. Blitz, R. E. Fine, and P. A. Toselli, *J. Cell. Biol.* **75,** 135 (1977).
[16] W. Schook, S. Puszkin, W. Bloom, C. Ores, and S. Kochwa, *Proc. Natl. Acad. Sci. U.S.A.* **76,** 116 (1979); W. Schook and S. Puszkin, this volume [30].
[17] J. H. Keen, M. C. Willingham, and I. H. Pastan, *Cell* **16,** 303 (1979).
[18] E. R. Unanue, E. Ungewickell, and D. Branton, *Cell* **26,** 439 (1981).
[19] R. A. Crowther and B. M. F. Pearse, *J. Cell. Biol.* **91,** 790 (1981).
[20] F. Oosawa and M. Kasai, *J. Mol. Biol.* **4,** 10 (1962).

depleted in clathrin and lacking the fine definition of the complete clathrin lattice.[9]

Coated vesicles and TCVs prepared by the methods described here contain receptors and content molecules depending on the cells from which they originated. For example, placental coated vesicles contain transferrin receptor and transferrin plus many other specific contents including ferritin and immunoglobulin G. These content molecules are trapped inside the coat structure and are released only when both the coats and the membranes are disrupted.[9]

Each one of these specific receptors or contents forms only a small proportion of the total coated vesicle protein. Ferritin, as judged by electron microscopy, is statistically distributed between the coated vesicles. Those that contain ferritin have, on average, 1.4 molecules of ferritin per coat, the maximum observed being six ferritins per coat. Transferrin and its receptor each form about 0.25% of the total coated vesicle protein. This would mean there might be an average of one to two molecules of transferrin per vesicle bound to its receptor, which exists as a dimer. Immunoglobulin G accounts for about 2% of the total protein, suggesting an average of four of these molecules per vesicle.

Thus coated vesicles are quite complex cellular organelles where a number of proteins are involved in mediating the transient association of the clathrin cages with the large variety of ligands that they selectively entrap for transfer steps from one cell compartment to another.

Acknowledgment

This work was supported by an advanced fellowship from the Science Research Council.

[28] Isolation and Characterization of Coated Vesicles from Rat Liver

By F. ANTHONY SIMION, DAVID WINEK, ENRIQUE BRANDAN, BECCA FLEISCHER, and SIDNEY FLEISCHER

Coated vesicles have been implicated in the selective, receptor-mediated endocytosis of a numer of different protein ligands in a variety of eukaryotic cells. After endocytosis, the coated vesicles transport the protein ligand from its receptor site on the plasma membrane to selected locations within the cell. Examples of transport processes mediated by coated vesicles include the

ISBN 0-12-181998-1

transport of yolk proteins to specialized storage vesicles in oocytes[1] and the uptake of low-density lipoprotein and epidermal growth factor in fibroblasts.[2,3] It is probable that coated vesicles also play a role in the recycling of synaptosomal plasma membranes.[4] During excitation, endocytosis by way of coated vesicles from the plasma membrane would balance the increase in plasma membrane caused by the fusion of synaptic vesicles with the cell membrane. It has also been suggested that coated vesicles serve a role in protein secretion, although the evidence for this is circumstantial. Rothman *et al.* have shown that coated vesicles are involved in the transport of maturing G protein of vesicular stomatis virus (VSV) within infected CHO cells.[5]

We describe the isolation of coated vesicles from rat liver using anion-exchange chromatography. This method is a modification of that described by Kanaseki and Kadota for coated vesicles from brain.[6] The coated vesicles derived from liver by anion-exchange chromatography have been biochemically characterized. For comparison, we have also prepared coated vesicles from rat brain.

Alternative methods for isolating coated vesicles that utilize density gradient centrifugation are described in this volume.[7,8]

Isolation of Coated Vesicles

Reagents

Sucrose, 0.32 *M*, pH 7.1 (Enzyme grade, Schwarz/Mann, Orangeburg, New York)
KCl solution, 2 *M*
Tris–malate buffer, 400 m*M*, pH 6.5
Tris–malate buffer, 20 m*M*, pH 6.5, containing 10 m*M* KCl
Tris–malate buffer, 20 m*M*, pH 6.5, containing 200 m*M* KCl
Tris–malate buffer, 20 m*M*, pH 6.5, containing 300 m*M* KCl
Tris–malate buffer, 20 m*M*, pH 6.5, containing 400 m*M* KCl
Tris–malate buffer, 20 m*M*, pH 6.5, containing 500 m*M* KCl
DEAE-Dephadex A-50 [Pharmacia Fine Chemicals (Piscataway, New Jersey)] equilibrated with 20 m*M* Tris–malate buffer, pH 6.5, containing 10 m*M* KCl

[1] T. F. Roth, J. A. Cutting, and S. B. Atlas, *J. Supramol. Struct.* **4,** 527 (1972).
[2] J. L. Goldstein and M. S. Brown, *Annu. Rev. Biochem.* **46,** 897 (1977).
[3] J.-L. Carpentier, P. Gorden, R. G. W. Anderson, J. L. Goldstein, M. S. Brown, S. Cohen, and L. Orci, *J. Cell. Biol,* **95,** 73 (1982).
[4] J. E. Heuser and T. S. Reese, *J. Cell Biol.* **57,** 315 (1973).
[5] J. E. Rothman, H. Burstzyn-Pettegrew, and R. E. Fine, *J. Cell Biol.* **86,** 162 (1980).
[6] T. Kanaseki and K. Kadota, *J. Cell Biol.* **42,** 202 (1969).
[7] B. M. F. Pearse, this volume [27].
[8] J. L. Daiss and T. F. Roth, this volume [29].

The pH of the solutions was adjusted at room temperature. Deionized water was used to prepare all solutions.

Equipment. Potter–Elvehjem-type glass homogenizers with Teflon pestles were used. The Teflon pestles supplied with 55-ml volume homogenizers (type C; Arthur H. Thomas Co., Philadelphia, Pennsylvania) were milled to produce clearances of either 0.012 in. or 0.026 in. between the inner diameter of the glass homogenizer (1.000 in.) and the Teflon pestle (cf. footnote 14 in Fleischer and Kervina[9]). To homogenize smaller volumes, a 10-ml Potter–Elvehjem type homogenizer with a Teflon pestle (type A; Arthur H. Thomas Co., Philadelphia, Pennsylvania) or a 1 ml Duall glass tissue grinder (Kontes Co., Vineland, New Jersey) with a Teflon pestle was used. The Teflon pestles for the 1-ml or 10-ml homogenizers were used as supplied.

Centrifugation was carried out in Beckman Spinco (La Jolla, California) ultracentrifuges when 12,000 g_{av} force or greater was required. A Beckman J-21B centrifuge was used for lesser g_{av} forces. Fixed-angle rotors were used, the type of rotor being selected according to the volume to be centrifuged. The given time of centrifugation is the setting for the time control. It therefore includes the acceleration time and time at set speed, but does not include deceleration time.

All operations beginning with the homogenization of the tissue were carried out in the cold, approximately at 4°.

Preparation from Rat Liver

Male Holtzman rats, 200–250 g fed ad libitum, are used. The rats are decapitated and exsanguinated prior to removal of the livers. Three livers (approximately 15 g each) are diced and homogenized in a Potter–Elvehjem type of homogenizer in 5 volumes of cold 0.32 *M* sucrose, pH 7.1 (based on tissue wet weight), using three up and down passes of a Teflon pestle with a clearance of 0.026 in., motor driven at 1200 rpm.[9] Each pass takes about 6 sec. The homogenization is repeated, but using a tighter pestle with a clearance of 0.012 in. The homogenate is then filtered through a No. 110 nylon mesh (Tetko, Elmsford, New York). The homogenate is centrifuged at 4500 rpm (2000 g_{av}) for 12 min (cf. Fig. 1). The supernatant is decanted and centrifuged at 12,500 rpm (10,000 g_{av}) for 30 min. The supernatant of the second centrifugation is recentrifuged at 29,000 rpm (65,000 g_{av}) for 30 min to sediment the heavy microsomal fraction. The supernatant is decanted and discarded. The heavy microsomal pellet is mechanically transferred into a Potter–Elvehjem homogenizer. Any remaining pellet is transferred by rinsing with cold deionized water. The volume of deionized water

[9] S. Fleischer and M. Kevina, this series, Vol. 31, p. 6.

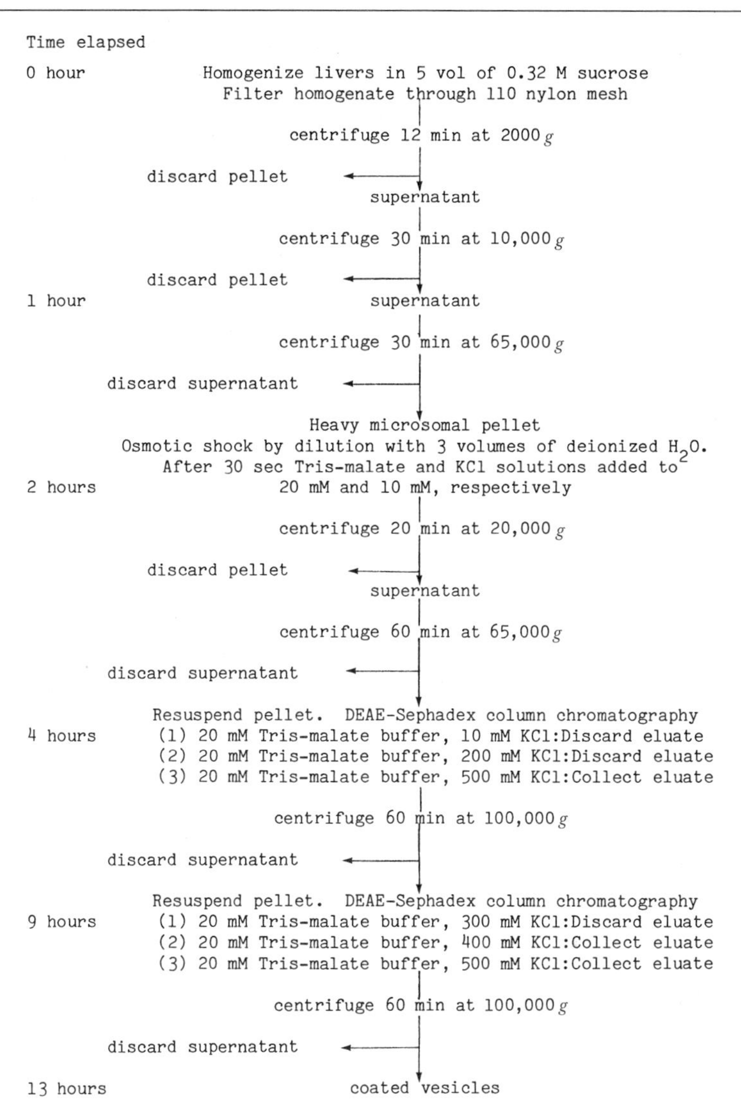

FIG. 1. Preparation of coated vesicles from rat liver.

added to the pellet is increased to three volumes of the original wet weight of liver used. The pellets are suspended in the water by homogenization at 1200 rpm as described above, using the pestle with a clearance of 0.012 in. After 30 sec, Tris–malate buffer and KCl solution are added to give final concentrations of 20 m*M* Tris–malate and 10 m*M* KCl, respectively. The suspension is mixed by homogenization and then centrifuged at 16,000 rpm

(20,000 *g*) for 20 min. The resulting supernatant is recentrifuged at 29,000 rpm (65,000 g_{av}) for 60 min. The final pellet is resuspended in a minimal volume of 40 m*M* Tris–malate buffer containing 20 m*M* KCl and then loaded onto a column containing 12 ml of DEAE-Sephadex A-50 (1.15 cm inner diameter) that has been pre-equilibrated with 20 m*M* Tris–malate, pH 6.5, containing 10 m*M* KCl. The column is then washed sequentially with 25 ml of (*a*) 20 m*M* Tris–malate containing 10 m*M* KCl; (*b*) 20 m*M* Tris–malate, pH 6.5, containing 200 m*M* KCl and (*c*) 20 m*M* Tris–malate, pH 6.5, containing 500 m*M* KCl. The flow rate of the elution buffer was not restricted and varied from 3 drops per minute with the buffer containing 10 m*M* KCl to 20 drops per minute with the buffer containing 500 m*M* KCl. The coated vesicles are in the third eluate and are recovered by centrifugation at 38,000 rpm (100,000 g_{av}) for 60 min. The pellet is resuspended in 20 m*M* Tris–malate, pH 6.5, containing 300 m*M* KCl and loaded onto a second columm (1.65 cm inner diameter) containing 12 ml of DEAE-Sephadex A-50 previously equilibrated with 20 m*M* Tris–malate, pH 6.5, containing 10 m*M* KCl. The column is washed sequentially with 25ml each of (*a*) 20 m*M* Tris–malate, pH 6.5, containing 300 m*M* KCl; (*b*) 20 m*M* Tris–malate, pH 6.5, containing 400 m*M* KCl; and (*c*) 20 m*M* Tris–malate, pH 6.5, containing 500 m*M* KCl. The eluates containing the 400 m*M* KCl and 500 m*M* KCl are combined and centrifuged for 1 hr at 38,000 rpm (100,000 g_{av}).

The pellets containing the coated vecisles are clear and colorless. They are suspended in 0.32 *M* sucrose using a 1-ml Potter–Elvehjem homogenizer and stored in liquid nitrogen. The yield of coated vesicles from liver using this procedure is 2.2 ± 0.7 (SD) μg of protein per gram wet weight of liver. The isolation procedure takes 13 hr to complete.

Purification from Rat Brain

The purification of coated vesicles from rat brain is essentially that of Kanaseki and Kadota[6] for the purification of coated vesicles from guinea pig brain (cf. Fig. 2). Briefly, the crude synaptosomal pellet is osmotically shocked to release its contents. The larger membrane fragments are removed by centrifugation, and then the coated vesicles are purified by DEAE-Sephadex column chromatography. The fraction eluted by 20 m*M* Tris–malate, 500 m*M* KCl contains the coated vesicle fraction.

Characterization of Coated Vesicles

Coated vesicles are characterized by their distinctive morphology (see Fig. 3)[10] and the presence of clathrin as the predominant protein observed by

[10] A. Saito, C. -T. Wang, and S. Fleischer, *J. Cell Biol.* **79**, 601 (1978).

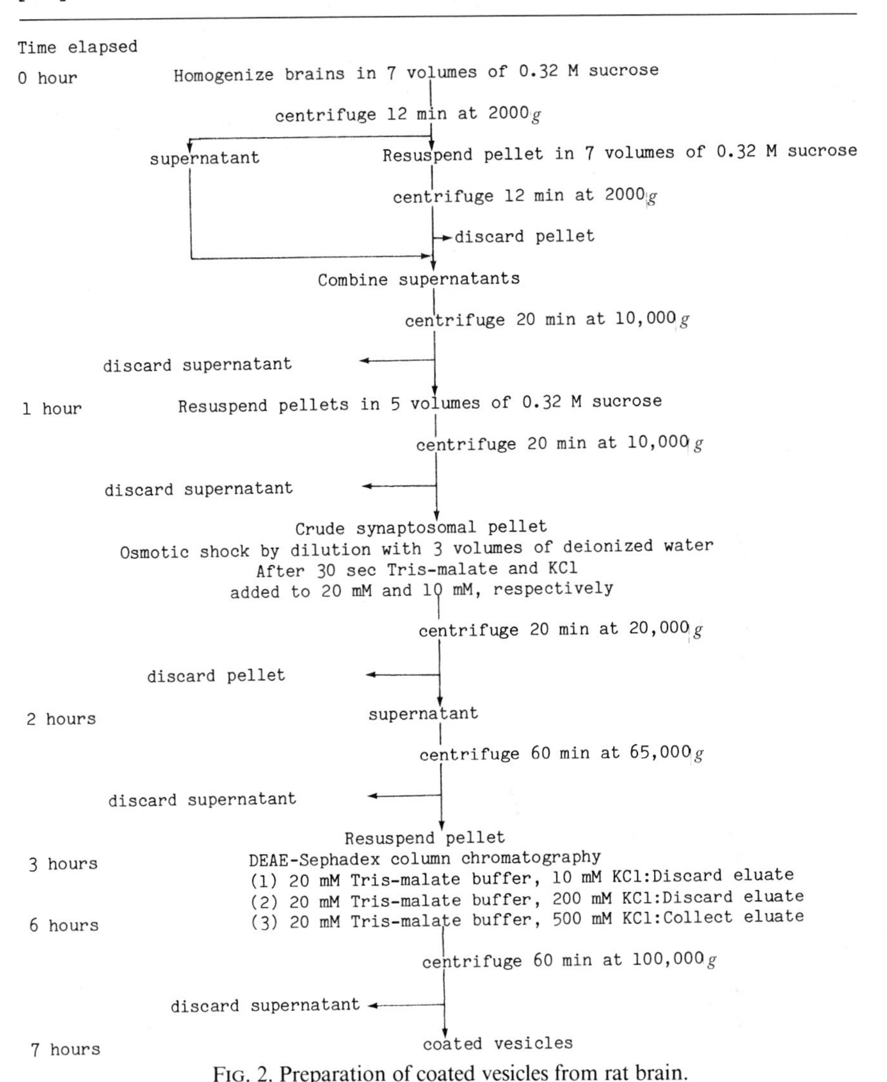

FIG. 2. Preparation of coated vesicles from rat brain.

SDS–polyacrylamide gel electrophoresis (see Fig. 4).[11-13] Coated vesicles prepared from rat liver fulfill these two criteria.

The SDS–polyacrylamide gel electrophoresis pattern for rat brain coated vesicles purified by anion-exchange chromatography (see Fig. 4) appears similar to that seen in coated vesicles prepared from either bovine

[11] U. K. Laemmli, *Nature* (*London*) **227,** 680 (1970).

[12] B. M. F. Pearse, *Proc. Natl. Acad. Sci. U.S.A.* **73,** 1255 (1976).

[13] M. P. Woodward and T. F. Roth, *Proc. Natl. Acad. Sci. U.S.A.* **75,** 4394 (1978).

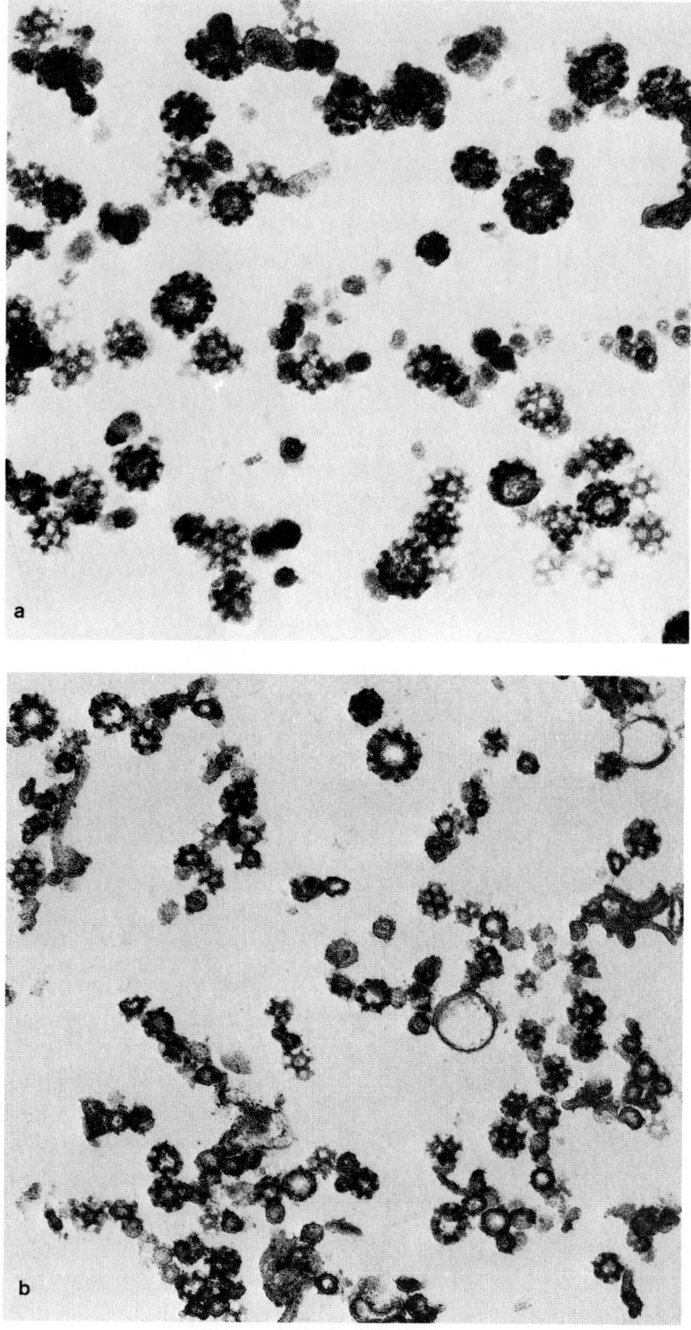

FIG. 3. Representative fields of coated vesicles from rat liver and brain. The coated vesicles fraction was fixed overnight at 4° in 0.1 *M* cacodylate buffer, pH 7.3, solution containing 1% tannic acid and 2.5% glutaraldehyde.[10] The sample was prepared for sectioning as described by Saito *et al.*[10] (a) Rat liver coated vesicles. (× 70,000). (b) Rat brain coated vesicles. (× 70,000).

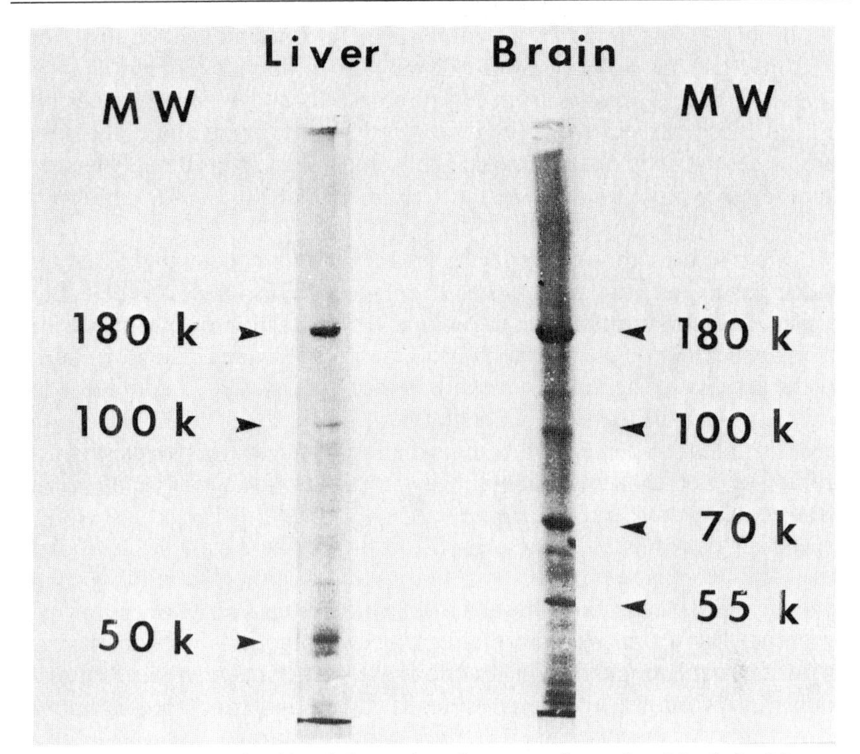

FIG. 4. SDS–polyacrylamide gel electrophoresis pattern of coated vesicles derived from rat liver and rat brain. The slab gel was prepared according to Laemmli[11] and contained 7.5% acrylamide with a stacking gel of 3% acrylamide. The molecular weights of the coated vesicle proteins were determined by comparison to proteins of known molecular weights. The molecular weight standards obtained from Bio-Rad (Richmond, California) were myosin (M_r 200,000) β-galactosidase (M_r 116,250), phosphorylase *b* (M_r 92,500), bovine serum albumin (M_r 66,200), and ovalbumin (M_r 45,000). Electrophoresis was carried out at constant current (20 mA) for 5 hr. The gel was washed in 50% methanol containing 12% acetic acid to fix the proteins, then stained for 2 hr with 2% Coomassie blue in 50% methanol containing 12% acetic acid and destained in a solution of 7% acetic acid in 15% methanol. k = thousand.

brain or pig brain by sucrose gradient sedimentation methods.[13,14] The clathrin band migrates with Stokes' radius equivalent to a molecular weight of 180,000 and by densitometry consists of about 50% of the total protein present when stained with Coomassie blue. Other protein bands can be observed on the gel at apparent molecular weights of 100,000, 70,000, 55,000, and 33,000. The most abundant minor protein band (M_r 70,000) contributes less than 19% to the total protein. SDS–polyacrylamide gels of rat liver coated vesicles also show clathrin (M_r 180,000) to be the predomi-

[14] B. M. F. Pearse, *J. Mol. Biol.* **126**, 803 (1978).

nant protein band. However, the minor protein bands appear to differ from those in coated vesicles of brain. Pearse also observed differences when comparing coated vesicles from lymphoma cells and bovine adrenal medulla to those derived from bovine brain.[12] In our experience, the silver stain[15] does not stain clathrin appreciably, and does so considerably less well than the other protein components. This may be due to the low amount of readily oxidizable amino acid residues in clathrin.[12]

Little is known about the enzymic characteristics of coated vesicles, and no diagnostic enzymic activity has been specially localized in the coated vesicles. The contamination of the coated vesicle fraction by other subcellular organelles can be estimated by using marker enzymes. The ratio of the specific activity of the marker enzyme in the coated vesicle fraction divided by that in the purified subcellular fraction, multiplied by 100%, gives the upper limit for the percentage contamination of the coated vesicles by that subfraction, as a small amount of the diagnostic enzyme activity may belong to the coated vesicle fraction.

Marker enzymes are characteristic for a specific organelle in a given tissue only, so that estimates of contamination cannot be reliably made using the specific activities obtained from a different tissue. Marker enzymes sometimes vary from one mammalian species to another. Marker enzymes for purified rat liver subcellular fractions are well characterized.[9] Brain is a much more complex and heterogeneous organ, and much less is known about diagnostic enzymes in evaluating contamination.

The enzymic characterization of rat liver coated vesicles is summarized in the table.[9,16-19] The activity in the coated vesicles of the diagnostic enzymes for mitochondria, Golgi, plasma membrane, and endoplasmic reticulum (cholate: CoA ligase) are low and compatible with contamination by these organelles. The specific activity of rotenone-insensitive NADH–cytochrome *c* redustase in coated vesicles from liver is almost two-thirds that of endoplasmic reticulum and cannot be attributed to contamination by that organelle. Other highly purified, subcellular membrane fractions, such as the Golgi apparatus and mitochondrial outer membrane, have appreciable rotenone-insensitive NADH–cytochrome *c* reductase activity[20,21] (cf. the table).

The amount of cholesterol in coated vesicles from rat liver (96 μg of

[15] C. R. Merril, D. Goldman, and M. L. Van Keuren, this series, Vol. 96 [18].

[16] B. Fleischer and M. Smigel, *J. Biol. Chem.* **252**, 1632 (1978).

[17] F. A. Simion, B. Fleischer, and S. Fleischer, *Biochemistry* (1983) in press.

[18] B. Fleischer, *Arch. Biochem. Biophys.* **212**, 602 (1981).

[19] O. H. Lowry, N. J. Rosebrough, A. L. Farr, and R. J. Randall, *J. Biol. Chem. 193,* 265 (1951).

[20] G. L. Sottocasa, B. Kuylenstierna, L. Ernster and A. Bergstrand, this series, Vol. 10, p. 448.

[21] A. Ito and G. E. Palade, *J. Cell Biol.* **79**, 590 (1978).

ENZYMIC CHARACTERIZATION AND CHOLESTEROL CONTENT OF COATED VESICLE FRACTION OF RAT LIVER[a]

Fraction	5′-AMPase	Galactosyl transferase	Succinate–cytochrome c reductase	Cholate: CoA ligase	Rotenone-insensitive NADH–cytochrome c reductase	Cholesterol (μ/mg protein)
	Organelle enzyme marker					
	Plasma membrane	Golgi apparatus	Mitochondria	Endoplasmic reticulum		
Homogenate	0.016	12	0.13	0.78	0.39	—
Coated vesicles	0.096	56	0.011	0.66	1.00	96
(percentage maximal concentration)	(12%)	(6%)	(7%)	(6%)		
Rough endoplasmic reticulum	0.046	8	0.003	11.7	1.66	14
Golgi apparatus	0.10	952	0.008	1.70	0.38	71
Plasma membrane	0.80	18	0.026	1.48	0.33	123
Mitochondria	0.006	0	0.44	0.34	0.21	3
Nuclei	0.017	1	0.004	0.64	0.13	—

[a] The activities of 5′-AMPase and the reductases are expressed as micromoles of product formed per minute per milligram of protein. Galactosyltransferase activity is expressed in nanomoles of galactose transferred per hour per milligram of protein, and the ligase activity is expressed in nanomoles of cholyl-CoA formed per minute per milligram of protein. Galactosyltransferase[16] and ligase[17] assays were carried out at 37°, whereas 5′-AMPase activity[18] and the two reductase[9] assays were carried out at 32°. Cholesterol was measured spectrophotometrically using a kit obtained from Sigma Chemical Co. (St. Louis, Missouri). Protein determination is by the method of Lowry *et al.*[19] using bovine serum albumin as a standard.

cholesterol per milligram of protein) is intermediate between that of the Golgi apparatus and the plasma membrane. Coated vesicles from rat brain also have an appreciable cholesterol content (219 μg of cholesterol per milligram of protein), which is similar to that in the synaptosomal plasma membrane (213 μg of cholesterol per milligram of protein).[22]

Comments

A simple and rapid method for isolating coated vesicles from rat liver and rat brain using anion-exchange chromatography is described. The coated vesicles have a negative surface charge and are eluted only at a fairly high salt concentration (0.4 *M* KCl). The negative surface charge of the coated vesicles is a characteristic that may be utilized in the isolation of coated vesicles as an alternative to the sedimentation properties of the coated vesicles. Rubenstein *et al.* used agarose gel electrophoresis to further purify a coated vesicle fraction from bovine brain obtained by sucrose density centrifugation.[23] Like anion-exchange chromatography, electrophoresis relies on differences in surface charge to separate different components.

Using the criteria of electron microscopy and SDS–polyacrylamide gel electrophoresis, the coated vesicles isolated appear to be highly purified. No enzymic activity has as yet been reported to be localized in the coated vesicle fraction, and the presence of various marker enzymes was used to assess the contamination of the rat liver coated vesicles by other subcellular organelles. Rotenone-insensitive NADH–cytochrome *c* reductase is the only marker enzyme that is high in rat liver coated vesicles and cannot be accounted for by contamination with other subcellular organelles. This reductase in rat brain coated vesicles has a 25-fold lower specific activity.

Both rat liver and rat brain coated vesicles have a cholesterol content that is the range of that of the plasma membrane of the tissue from which they are isolated. The coated vesicles from liver are intermediate in cholesterol content per milligram of protein between the plasmalemma and Golgi and may reflect the cycling between the Golgi and plasma membrane. It should be noted that the enzyme markers from both the plasma membrane and Golgi apparatus are excluded from the coated vesicles.

Acknowledgments

This work was supported in part by NIH Grants AM 17223 and AM 14632. We are grateful to Akitsugu Saito for skilled electron microscopy and Kathryn Dewey for capable technical assistance.

[22] P. E. North, Ph. D. Dissertation, Vanderbilt University, Nashville, Tennessee, 1982.

[23] J. L. R. Rubenstein, R. E. Fine, B. D. Luskey, and J. E. Rothman, *J. Cell Biol.* **89**, 357 (1981).

[29] Isolation of Coated Vesicles: Comparative Studies

By JOHN L. DAISS and THOMAS F. ROTH

Background

General Features of Coated Vesicles. Coated vesicles (CV) are transient. intracellular organelles that mediate the selective transport of certain macromolecular components among membrane-bound, topologically extracellular compartments. Ubiquitous among the eukaryotes, CV have been implicated in the selective internalization of numerous polypeptide ligands from their extracellular milieu,[1,2] the retrieval of excess plasma membrane after extensive exocytosis,[3] the transport of newly synthesized lysosomal enzymes from the endoplasmic reticulum (ER) to the lysosomes,[4] and the transport of newly synthesized membrane proteins from the ER to the Golgi and from the Golgi to the cell surface.[5] Each of these processes is thought to be accomplished by the endocytosis of a small area of membrane enriched for the transported element. This endocytosis results in the separation of discrete areas of membrane away from both the neighboring membrane and the surrounding aqueous environment. Selective endocytosis is mediated by the assembly of a dense protein coat on the cytoplasmic face of the membrane to form a coated pit. Through internal rearrangements,[6] the coated pit invaginates pulling the attendant membrane into a small closed sphere, the coated vesicle.

The vesicles produced by this process are among the smallest observed in biological specimens (diameters range from 300 to 2000 Å), although there is some variation in size depending on the tissue and intracellular location.[4,7] Because of their small diameters, CV have very high surface-to-volume ratio, thus maximizing the amount of surface transferred and minimizing irrelevant trapped volume. This property is vital to their function as agents of selective transport. In addition to their small size, CV are characteristically rather dense ($\rho = 1.20-1.25$ mg/ml) owing to their proteinaceous coats. It is their density that is typically exploited to achieve purification.

Tissue Sources. Coated vesicles (CV) have been prepared from a rather small number of tissues that include brains from rat, rabbit, bull, and

[1] T. F. Roth and K. R. Porter, *J. Cell Biol.* **20,** 313 (1964).
[2] J. L. Goldstein, R. G. W. Anderson, and M. S. Brown, *Nature* (*London*) **279,** 679 (1979).
[3] J. E. Heuser and T. S. Reese, *J. Cell Biol.* **57,** 315 (1973).
[4] D. S. Friend and M. G. Farquhar, *J. Cell Biol.* **35,** 357 (1966).
[5] J. E. Rothman and R. E. Fine, *Proc. Natl. Acad. Sci. U.S.A.* **77,** 780 (1980).
[6] J. E. Heuser, *J. Cell Biol.* **84,** 560 (1980).
[7] J. W. Woods, M. P. Woodward, and T. F. Roth, *J. Cell Sci.* **30,** 87 (1978).

ISBN 0-12-181998-1

hog,[7-10] bovine adrenal medulla,[11] bovine adrenal cortex,[12] chicken liver, chicken oocytes,[7] and human placenta.[13,14] In addition, CV have been isolated from three cell lines: CHO cells,[5] the lymphoma line E.L.4,[11] and the hybridoma cell line H6/31.[13]

Because CV are not particularly abundant ($\sim$ 0.1 mg per gram of tissue, wet weight[5,8]), tissues of choice are those readily available in fairly large quantities ($\sim$ 100 g). Although frozen tissue is sufficient, fresh tissue gives a better product, in terms of both yield and behavior in assembly and disassembly.

Brain has been the most frequently used tissue, in part for historical reasons and in part because axonal endings are very actively endocytic.[3] However, other tissues (adrenal medulla, liver, placenta) are becoming more widely used because of the better characterized interaction of particular ligands, receptors, and CV.

Isolation Buffers. The buffer originally employed by Pearse[8] was essentially that used in the preparation of brain microtubules [0.1 m*M* 2-(*N*-morpholino)ethanesulfonic acid (MES), pH 6.5, 1 m*M* $MgCl_2$, 0.5 m*M* EGTA, 0.02% NaN_3]. The stability of the coat is pH dependent,[15] presumably reflecting shifts in the relative rates of assembly and disassembly.[16] Thus, neutral or mildly acidic pH is desirable, and pH $>$ 7.5 is not recommended. Further, certain buffers are better than others: MES and HEPES are widely used, but Tris is a potent dissociating agent and should be used only for that purpose.[10]

Although there is some evidence that coat assembly is promoted by divalent cations,[16,17] there is little evidence for a need for divalent cations in CV isolation buffers. Nonetheless, the practice of including 1 m*M* $MgCl_2$ and 0.5 m*M* EGTA in the isolation buffer is conventional. One report claims that inclusion of 1 m*M* $CaCl_2$ promotes the copurification of calmodulin with CV.[18]

Finally, protease inhibitors are frequently added, although the particular

[8] B. M. F. Pearse, *J. Mol. Biol.* **97,** 93 (1975).

[9] A. L. Blitz, R. E. Fine, and P. A. Toselli, *in* "Contractile Systems in Non-muscle Tissues (S. V. Perry *et al.,* eds.). Elsevier/North-Holland Biomedical Press, Amsterdam, 1976.

[10] J. H. Keen, M. C. Willingham, and I. H. Pastan, *Cell* **16,** 303 (1979).

[11] B. M. F. Pearse, *Proc. Natl. Acad. Sci. U.S.A.* **73,** 1255 (1976).

[12] R. J. Mello, M. S. Brown, J. L. Goldstein, and R. G. W. Anderson, *Cell* **20,** 829 (1980).

[13] B. M. F. Pearse, *Proc. Natl. Acad. Sci. U.S.A.* **79,** 451 (1982).

[14] C. D. Ockleford, A. Whyte, and D. E. Bowyer, *Cell Biol. Int. Rep.* **1,** 137 (1977).

[15] M. P. Woodward and T. F. Roth, *Proc. Natl. Acad. Sci. U.S.A.* **75,** 4394 (1978).

[16] P. P. Van Jaarsveld, P. K. Nandi, R. E. Lippeldt, H. Saroff, and H. Edelhoch, *Biochemistry* **20,** 4129 (1981).

[17] M. P. Woodward and T. F. Roth, *J. Supramol. Struct.* **11,** 237 (1979).

[18] C. D. Linden, J. R. Dedman, J. G. Chafouleas, A. R. Means, and T. F. Roth, *Proc. Natl. Acad. Sci. U.S.A.* **78,** 308 (1981).

inhibitors used vary. For preparation from brain, proteolysis is not a major problem, although many investigators add 1 mM phenylmethylsulfonyl fluoride (PMSF) to their buffers.

Criteria for Purification. Because CV have no unequivocally established enzyme markers, quantitation during purification is difficult. Nonetheless, purification can be monitored by two criteria. Pearse[11] discovered that CV are composed principally of a single polypeptide component, which she named clathrin because of the cagelike assemblies it forms. Clathrin is a polypeptide of M_{app} = 180K on sodium dodecyl sulfate–polyacrylamide gel electrophoresis (SDS–PAGE) (with or without reduction of disulfide bonds), is the principal structural element of the coat, and is apparently unique to CV (Fig. 1). Other proteins copurify with CV, notably a group of polypeptides with M_{app} = 100–125K, three polypeptides of M_{app} = 50–55K (two of which are α- and β-tubulin[19] and two polypeptides having

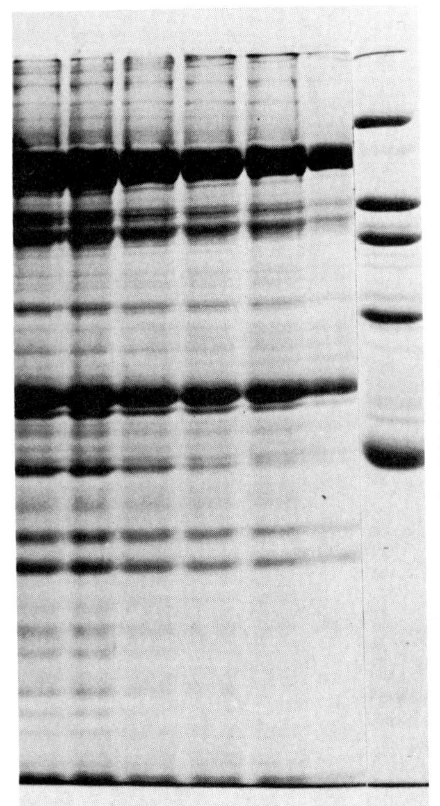

FIG. 1. Polypeptide profiles of coated vesicle proteins on SDS–polyacrylamide gels. Several preparations are shown. The left lane has molecular weight markers: myosin, 200K; β-galactosidase, 116K; phosphorylase b, 92.5K; bovine serum albumin, 67K; and ovalbumin, 45K.

[19] S. R. Pfeffer, D. G. Drubin, and R. B. Kelly, *J. Cell. Biol.* (in press); W. G. Kelly, A. Passiniti, J. W. Woods, J. L. Daiss, and T. F. Roth, *J. Cell Biol.* (in press).

$M_{app} = 30-36K$. These other polypeptides are not as well characterized as clathrin, and their relative stoichiometries and structural roles are not as well understood. Clathrin is at present the only polypeptide marker diagnostic for CV on SDS–PAGE, and it is the most sensitive marker, since it comprises 30–70% of the protein mass in purified CV.[8,15]

Electron microscopic examination of negatively stained samples is an alternative method for monitoring purification. Typically, one drop of sample is placed on a carbon-coated copper grid and then stained with 1% uranyl acetate. It is our observation that purity can be overestimated by this method because protein-coated vesicles are selectively retained on the grid during specimen preparation. Specimens prepared by fixing, embedding, and sectioning give a more faithful account of quality of the preparation. Nonetheless, the speed and ease with which negative staining can be done makes it an easy method for monitoring purification.

Heterogeneity of CV. The purification procedures detailed below yield a suspension of discrete small particles consisting of various proportions of smooth-membrane vesicles, CV, and empty coats. Coated vesicles vary in size owing, in part, to the pleiomorphic character of the coat[16,20] and to the variety of sites at which they function. The CV in the ER–Golgi region tend to be smaller than those forming from the plasma membrane.[4] Further, most isolated CV in the small end of the distribution of sizes of CV are seen in thin sections of intact cells (diameter less than 1000 Å[7]). Whether this reflects a selective enrichment for smaller CV, greater stability, and/or number of CV in the ER–Golgi region relative to those from the plasma membrane, or damage and loss of the initially larger CV[21] is uncertain.

Similarly, the origin of empty coats is not known. Preexisting pools of unassembled clathrin may polymerize during homogenization of the tissue, or the empty coats may represent damaged CV, perhaps ruptured during isolation due to osmotic shock in the dense sucrose solutions used. One must, therefore, be circumspect about how faithfully isolated CV represent CV *in vivo*.

Storage and Handling. Coated vesicles purified from brain have been stored in the sodium azide-containing isolation buffer at 4° for long periods (>6 months) without gross decomposition or significant alteration of their polypeptide composition on SDS–PAGE. There is, however, a tendency for the CV to aggregate upon long-term storage (~ days to weeks). This aggregation can be retarded by storing the CV in isolation buffer containing 10% sucrose.

[20] R. A. Crowther, J. T. Finch, and B. M. F. Pearse, *J. Mol. Biol.* **103,** 785 (1976).
[21] B. M. F. Pearse, *Trends Biochem. Sci.* **5,** 131 (1980).

Purification of CV Using Density Gradients

Early preparative procedures for CV employed sucrose density gradients[22] and differential centrifugation,[23] but the first generally useful procedure, which combined these methods and yielded a consistently satisfactory preparation, was that of Pearse.[8] This method involves homogenization of the tissue, differential centrifugation to obtain a microsomal pellet, and then separation of CV from other microsomal vesicles on a series of sucrose density gradients. It is our experience that this method is particularly well suited to purifying CV from brain, but it has been less successful for chicken liver and chicken oocytes. The whole procedure can be performed in 20–30 hr, and the yields are on the order of 50%.

Below is a step-by-step account of this method as it is practiced routinely in our laboratory for bovine brain and chicken liver. Figure 2 illustrates the same information in a flow chart.

Preparation of Tissue. Because CV are small and spherical, they are relatively insensitive to high shear forces. Thus, conventional high shear methods for homogenization are acceptable. Typically, fresh tissue (or tissue frozen immediately after slaughter) is homogenized in two volumes of ice-cold isolation buffer. We put 75 g of tissue plus 150 ml of buffer into a Waring blender and homogenize this mixture in four 15-sec bursts. Conventionally, the homogenization and all subsequent steps are performed at 0–4° or on ice.

Differential Centrifugation. The resulting homogenate is transferred to centrifuge bottles and centrifuged at 20,000 *g* for 30 min (GSA rotor, 10,000 rpm). The resulting pink, turbid supernatant, which constitutes about 70% of the total volume, is decanted through glass wool or cheesecloth to remove any congealed lipid. At this point, we add PMSF, freshly dissolved in absolute ethanol (2 ml of 20 mg of PMSF per milliliter of ethanol stock per liter of supernatant). Also, some investigators include at this point a 30-min incubation at room temperature with RNase.[21]

The supernatant is then centrifuged for 6×10^6 *g*-min, and a microsomal pellet is collected. At this point the volume will be ~2 ml per gram of tissue wet weight, so the total volume will be ~600 ml per bovine brain. High-capacity rotors like the Beckman type 19 can handle large volumes (1200 ml) but require 4 hr at top speed to pellet microsomes. For smaller volumes, high-speed rotors are more suitable (e.g., Beckman type 60 Ti run at 45,000 rpm for 1 hr; 1.2×10^7 *g*-min).

[22] O. A. Schjeide, R. I.-San Lin, E. A. Grellart, F. R. Galey, and J. F. Mead, *Physiol. Chem. Phys.* **1,** 141 (1969).

[23] T. Kanaseki and K. Kadota, *J. Cell Biol.* **42,** 202 (1969).

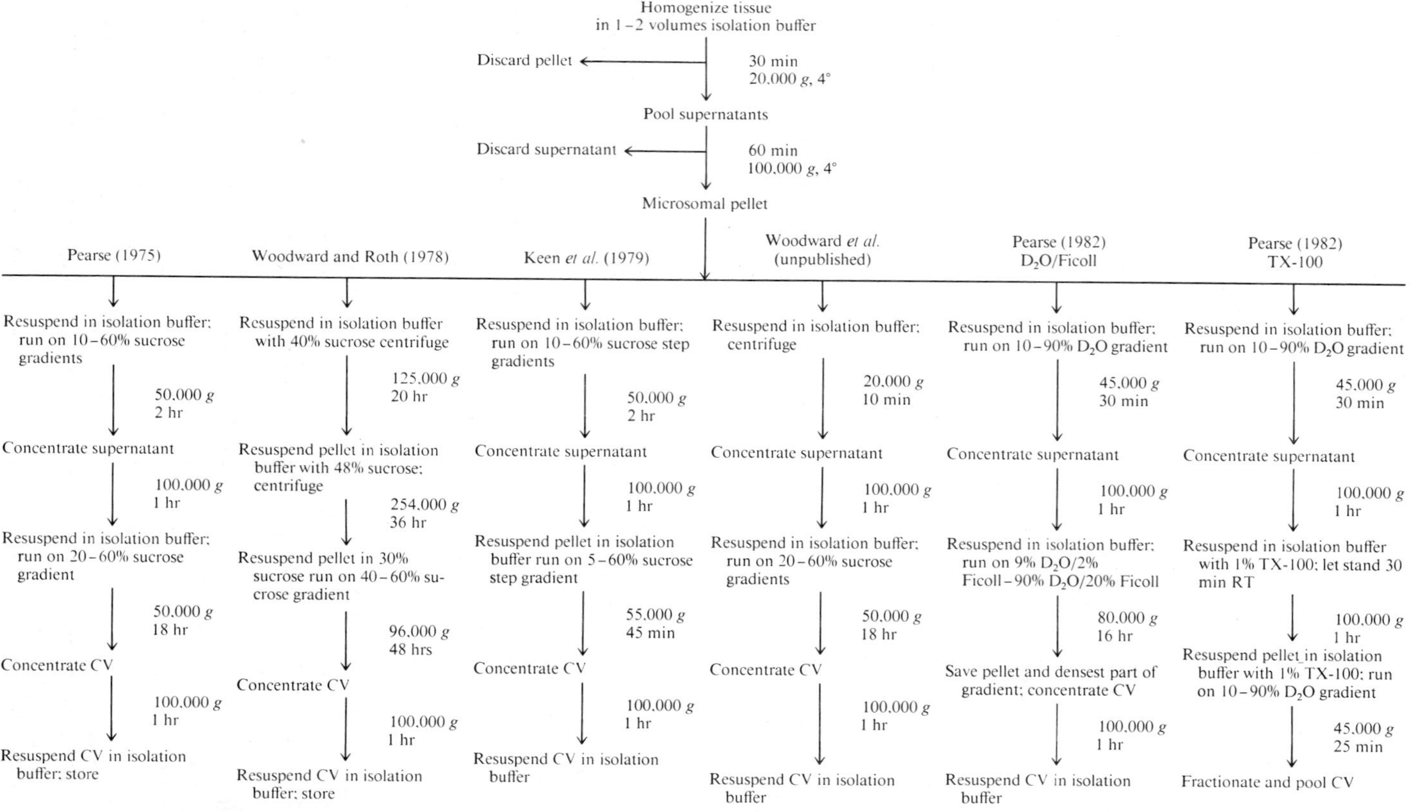

FIG. 2. Flow chart showing the various methods for purifying coated vesicles. TX-100 = Triton X-100; RT, room temperature.

The collected microsomes are conspicuously heterogeneous, and the pellet has a layered appearance. A button at the bottom of the pellet appears dark and hard; it is overlaid by a soft-looking pale-pink or white layer. The button, which contains no CV, is hard to resuspend, but the upper layer is readily broken by trituration with a Pasteur pipette or spatula. The upper layer can then be more thoroughly dispersed by homogenization in Dounce, Polytron, or Potter–Elvehjem homogenizers or by mild sonication. It is our experience that this dispersion does not yield a suspension of discrete small vesicles, but rather that much of the material is aggregated, as judged by the fact that it can be sedimented in less than 2×10^5 *g*-min. It is our practice first to sediment the aggregated material at 2×10^5 *g*-min (10,000 rpm, 10 min in an SS34 rotor) prior to further purification. The aggregated material can be reextracted several times to improve the final yield of CV. The final volume of the pooled resuspended microsomes should be 5–10 ml per 50 g of tissue wet weight.

This preparation of resuspended microsomal vesicles is the starting point for all the methods for purification of CV outlined in Fig. 2. Each method employs essentially two steps of further purification, each of which exploits the high density of CV or their relatively low sedimentation rate.

Velocity Sedimentation in Sucrose Gradients. Typically, 10 ml of resuspended microsomes are loaded on top of linear 10 to 60% sucrose gradients. We use a Beckman SW27 rotor with six $1 \times 3\frac{1}{2}$-in. polyallomer tubes, each with a 28-ml gradient. The gradients can be formed by layering equal volumes of 60, 50, 40, 30, 20, and 10% sucrose, which will diffuse during centrifugation to yield a continuous gradient, or continuous gradients can be poured by the method of Lakshmanan and Lieberman[24] using a peristaltic pump and very aggressive stirring of the mixing chamber because of the high viscosity of the 60% sucrose.

Sedimentation for 2 hr at 22,000 rpm (1.1×10^7 *g*-min) yields a clear, pink or orange supernatant where the sample started a broad turbid zone centered at about 20% sucrose, which is rich in vesicles including CV, and a dense band of material at 30–40% sucrose, which is mostly aggregated material (Fig. 3). The middle zone is collected and diluted with 2–3 volumes of isolation buffer, and vesicles are pelleted by centrifugation at 45,000 rpm for 1 hr in a type 60 rotor (1.2×10^7 *g*-min). The resulting pellet of crude CV is then resuspended in a small volume (10–40 ml) of isolation buffer.

Equilibrium Sedimentation in Sucrose Gradients. Ten milliliters of the concentrated and resuspended crude CV are then layered onto 28-ml linear 20 to 60% (w/w) sucrose gradients in $1 \times 3\frac{1}{2}$-in. polyallomer tubes and centrifuged in an SW27 rotor at 22,000 rpm for 18 hr until the particles reach their isopycnic points in the sucrose gradient. Alternatively, identical

[24] T. K. Lakshmanan and S. Lieberman, *Arch. Biochem. Biophys.* **53,** 258 (1954).

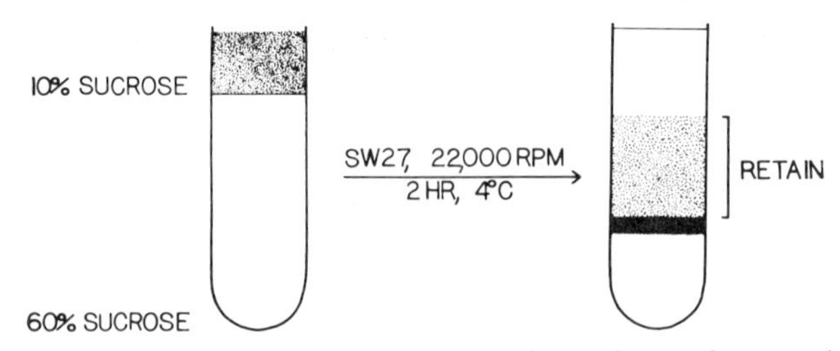

FIG. 3. Typical separation achieved by the velocity sedimentation step in the Pearse procedure.[8]

gradients can be run in vertical tube rotors that shorten the centrifugation time to less than 8 hr (e.g., VTi50, 8 hr, 4°, 45,000 rpm).

The material floating at or above 40% sucrose is depleted of CV and can be discarded. Coated vesicles typically band at 50–55% sucrose, but a minority population is present in the zone between the CV band and the material banding at 40% sucrose (Fig. 4). This heterogeneity in density might be anticipated since CV vary in both diameter and content. Within the CV-containing zones, there is no apparent enrichment for large or small CV, nor are empty baskets resolved from CV.

Coated vesicles can be collected from the top or the bottom of the gradient by standard methods. We collect them from the side of the tube by aspiration with a 20-gauge needle on a 10-ml syringe, thus minimizing the risk of contaminating the CV with vesicles from other portions of the gradient.

The CV are then diluted with five volumes of isolation buffer and pelleted as before. The CV pellet is resuspended in a small volume of isolation buffer supplemented with 10% sucrose and stored at 4°. Contaminating proteins can be removed by sedimenting the CV into a 5 to 30% sucrose gradient on a 60% cushion (1 hr, 90,000 *g*). The CV are harvested from a broad band centered around 15% sucrose and stored as described above.

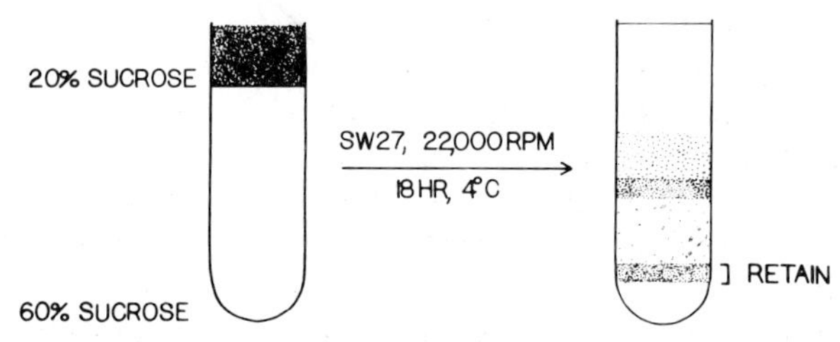

FIG. 4. Typical separation achieved by the equilibrium sedimentation step in the Pearse procedure.[8]

FEATURES OF METHODS FOR ISOLATION OF COATED VESICLES (CV)

Method	Total time[a] (hr)	Hands-on time[b] (hr)	Capacity (g)	Quality of CV[d] (%)
Pearse[8]	20–38	10–12	300	90
Woodward and Roth[15]	60–120	4–12	1000	80–90
Keen *et al.*[10]	10–16	8–12	300	80
Woodward *et al.*[c]	20–30	12–14	1000	90
Pearse[13]				
D_2O/Triton X-100	8–12	8–12	300	90
D_2O/Ficoll	20–30	10–12	300	90

[a] Total time = time from beginning to end of preparation.
[b] Hands-on time = time the investigator is physically working on the preparation.
[c] Capacity = wet weight of tissue conveniently handled, assuming one SW27 or other rotor that limits the capacity.
[d] Quality = percentage of CV-like particles in negatively stained samples.
[e] M. P. Woodward and T. F. Roth, unpublished results.

Variations. Several variations of this method have been used, and their relative advantages are summarized in the table.

Woodward and Roth[15] substitute sedimenting CV out of high-density sucrose solutions in place of the sucrose gradients in the original Pearse procedure. Microsomal vesicles are resuspended in 40% (w/w) sucrose and pelleted (1.5×10^8 *g*-min; 1.25×10^5 *g*, 20 hr). The pelleted crude CV are subsequently resuspended in 48% (w/w) sucrose and centrifuged (5×10^8 *g*-min; 2.54×10^5 *g*, 36 hr). This method enriches for CV solely on the basis of density and is particularly suitable for preparing CV from brain. It has the advantage of being able to accommodate large quantities of tissue (~ 1 kg), and, although the total time is long, it requires little hands-on time. Keen *et al.*[10] employ the Pearse method through the first sucrose gradient, but they then substitute a short (45 min) sedimentation through a sucrose gradient like that depicted in Fig. 5. This procedure attempts to exploit the relatively small size of CV, hence their relatively low sedimentation rate. Unfortunately, this method produces a relatively impure preparation of CV, but it has the advantage of producing an enriched preparation in a single day.

A third variation[25] substitutes three cycles of pelleting (2×10^5 *g*-min) and reextracting the microsomes in place of the first sucrose gradient in the Pearse procedure. Aggregates are removed from the resuspended microsomes at each step by a low-speed centrifugation (10,000 rpm SS34, 10 min). After each of three successive low-speed sedimentations, the CV-containing supernatants are saved and the pellets are resuspended in isolation buffer.

[25] M. P. Woodward, C. C. Cobbs, and T. F. Roth, unpublished observations.

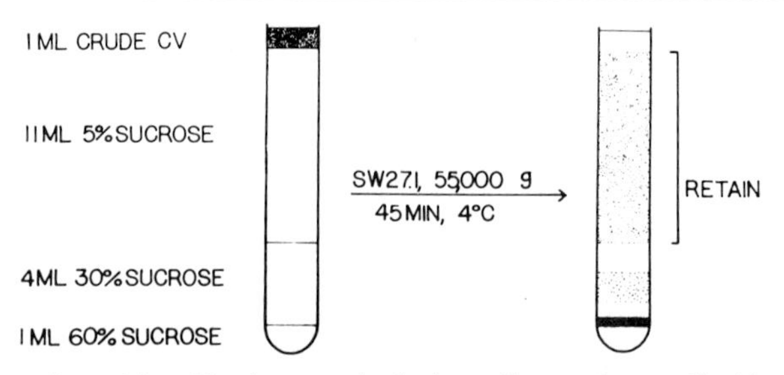

FIG. 5. Separation achieved by the second velocity sedimentation gradient in the method of Keen *et al.*[10]

The last pellet is discarded, and the pooled supernatants are then centrifuged (1 hr, 100,000 *g*). The pelleted vesicles are subjected to three more cycles of resuspension and sedimentation. The pooled supernatants are then pelleted, resuspended in isolation buffer, and fractionated on a 20 to 60% (w/w) sucrose gradient as in the original Pearse method. This method takes as long as the Pearse method and yields a similar product, but it substitutes the tedium of multiple, short centrifugations for the tedium of making and fractionating sucrose gradients. Its principal advantage is that it can handle a larger volume of tissue.

Density Gradients Employing D_2O and Ficoll. Pearse[13] introduced the use of gradients of D_2O and Ficoll for the purification of CV. While the range of densities one can spin is more limited ($1.00 < \rho < 1.2$), the CV are not subjected to the osmotic stress that they experience in high concentrations of sucrose. Coated vesicles isolated by this procedure are larger and, therefore, more like those observed in thin sections of cells than those purified in sucrose gradients. Further, CV isolated by this method appear to retain their contents more faithfully.

Resuspended microsomes are prepared as above and layered onto linear 10 to 90% D_2O gradients ($1.01 < \rho < 1.10$). We make such gradients using 99.8 atom % D_2O (Sigma or Merck) and a 10-fold concentrated isolation buffer (100 m*M* HEPES, pH 7.2, 1.5 m*M* NaCl, 10 m*M* EGTA, 5 m*M* $MgCl_2$, 0.2% NaN_3). These gradients are then centrifuged for 30 min at 45,000 *g* (16,000 rpm in SW27), yielding a microsomal fraction essentially the same as that in the first sucrose gradient in the original Pearse procedure (Fig. 6). The gradients have three fractions: an upper avesicular fraction that can be discarded, a dense pellet at the bottom that can also be discarded, and a broad, turbid middle zone that retains the CV. This zone is collected, diluted threefold, pelleted (100,000 *g*, 1 hr), and resuspended in isolation buffer for fractionation on a second D_2O density gradient.

The second gradient is a density equilibrium gradient that exploits the rather high density of CV, and it is linear from 9% D_2O–2% Ficoll at the top

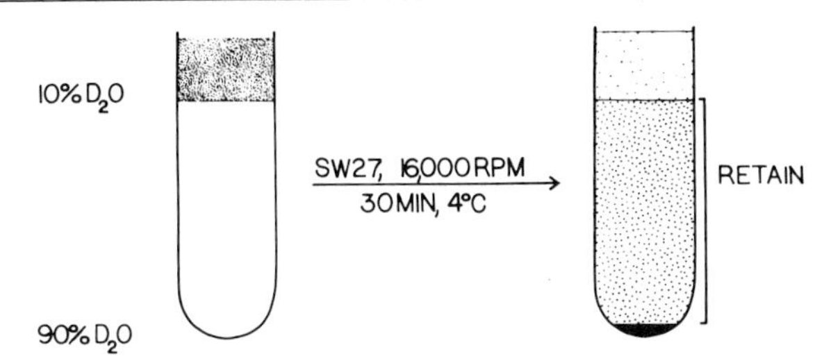

FIG. 6. Separation achieved by the velocity sedimentation step in a gradient of D_2O according to Pearse.[13]

to 90% D_2O–20% Ficoll at the bottom. Ficoll should be dialyzed exhaustively against deionized water and stored lyophilized; alternatively, dialyzed and lyophilized Ficoll can be purchased (Sigma).

The crude CV are layered onto these gradients and centrifuged for 16 hr at 80,000 *g* (21,000 rpm in an SW27). The CV sediment to the bottom of the gradient, well resolved from lighter components banding in the middle (Fig. 7). The collected CV can then be washed and stored in isolation buffer.

The advantages of this method, other than the isoosmolarity of the gradients, are that it can be performed in a single, long day (if vertical rotors are used), and it has produced the purest CV from chicken liver that we have yet obtained. This suggests that this may be a useful method for preparation of CV from many sources. Its only drawback is the expense of D_2O ($0.35–$1.00/ml and 100–200 ml per preparation).

Another variation on this method utilizes the observation that the cagelike structure, physicochemical properties, and polypeptide composition of CV are little altered by treatment with 1% Triton X-100,[15] whereas contaminating membrane vesicles are dissolved. Thus, partially purified CV from the first D_2O gradient described above can be resuspended in isolation buffer containing 1% Triton X-100 and sedimented through a second 10 to 90% D_2O, 1% Triton X-100 gradient to remove the dissolved components of contaminating vesicles.

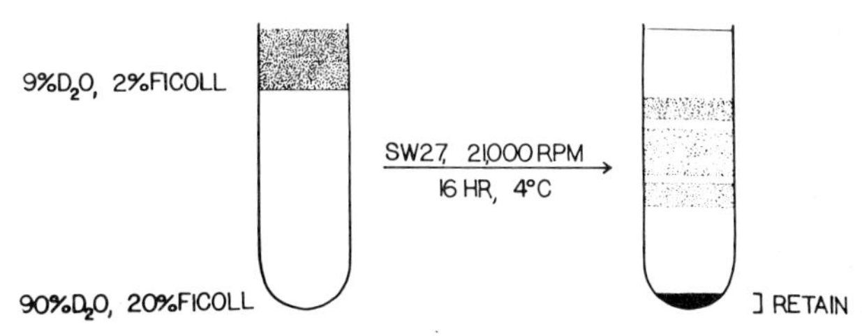

FIG. 7. Typical resolution of coated vesicles by equilibrium sedimentation on D_2O–Ficoll gradients.

Purification of CV Exploiting Other Properties

Three alternative methods of CV purification have been introduced. In general, these techniques are not very useful for purification of CV from large quantities of tissue; but they should all be valuable as analytical tools and as procedures supplemental to those described above under Purification of CV Using Density Gradients.

Permeation Chromatography. Because most CV isolated on sucrose gradients have a fairly narrow size distribution (500–1000 Å in diameter[7]), the use of permeation chromatography is theoretically possible. However, chromatography media of sufficient rigidity and pore size have only recently become available. Pfeffer and Kelly[26] have introduced the use of controlled pore-size glass beads to remove large membrane vesicles that contaminate CV preparations. Controlled-pore glass beads (Electro-Nucleonics, Inc., Fairfield, New Jersey) can be obtained with mean pore size over the range of 1000–3000 Å, large enough to include CV. Before the column is packed, the glass beads must be pretreated with 1% polyethylene glycol (PEG, M_r 20,000; prefiltered through 0.45-μm Millipore filters). After several cycles of swirling in the PEG solution and settling, the beads are washed exhaustively with deionized water. These PEG-treated beads are then resuspended in the column buffer (Pfeffer and Kelly use 0.2 *M* sucrose, 0.3 m*M* NaCl, 10 m*M* HEPES, 10 m*M* EGTA, 0.02% NaN_3, pH 7.0) and exhaustively degassed. Packing the glass beads into a tall, thin column (height : diameter $\geq$ 50) requires constant, aggressive vibration to facilitate settling of the beads. This can be done by sealing the bottom of the column and packing the bead bed while agitating the bottom of the column on a vortex mixer.

The column should then be run at fairly slow flow rate ($\sim$ 2 ml/cm^2 hr^{-1}) to ensure inclusion of the CV in the bead matrix. The sucrose in the column buffer seems to improve the recovery of CV from the column, presumably because it reduces aggregation.

The controlled-pore glass beads are expensive ($1 per packed milliliter), but other recently introduced media are not as useful, e.g., Sephacryl S-1000 (Pharmacia Fine Chemicals, Piscataway, New Jersey).

Immunoprecipitation. Even though clathrin is a rather highly conserved protein,[11] sufficient interspecific heterogeneity exists to make it antigenic, and a number of laboratories have made antisera of varying quality using either clathrin extracted from purified CV[27–29] or clathrin excised from SDS–PAGE gels.[30] Merisko *et al.*[30] have employed rabbit immunoglobulin G (IgG) anti-clathrin antibodies to purify CV from porcine brain micro-

[26] S. R. Pfeffer and R. B. Kelly, *J. Cell Biol.* **91,** 385 (1981).

[27] R. E. Fine, A. L. Blitz, and D. H. Sack, *FEBS Lett.* **94,** 59 (1978).

[28] J. H. Keen, M. C. Willingham, and I. Pastan, *J. Biochem.* (*Tokyo*) **256,** 2538 (1981).

[29] J. Kartenbeck, E. Schmid, H. Muller, and W. W. Franke, *Exp. Cell Res.* **133,** 191 (1981).

[30] E. M. Merisko, M. G. Farquhar, and G. E. Palade, *J. Cell Biol.* **92,** 846 (1982).

somes. Using anti-clathrin IgG adsorbed to formalin-fixed *Staphylococcus aureus* bacteria as a solid phase immunoadsorbent, they were able to purify CV or clathrin from disassembled CV. Procedures for preparation and handling of *S. aureus* can be found in other volumes in this series.[31,32]

Agarose Gel Electrophoresis. Coated vesicles bear a net negative charge at pH 6.5, and consequently they can be electrophoresed in very porous anticonvective matrices. Rubenstein *et al.*[33] have used this property of CV to resolve CV from certain contaminants. They use flat beds of 0.2% agarose in isolation buffer as their anticonvective matrix, and they apply an electric field of 0.75 V/cm. Coated vesicles migrate toward the anode at a rate of 0.08 cm/hr; thus long times are necessary for significant separation (>24 hr). This type of electrophoresis can also be done in narrow glass tubes (like those used in SDS–PAGE) with dialysis tubing sealing the bottom (J. L. Daiss and T. F. Roth, unpublished).

The principal drawback of this procedure is the difficulty in recovering the purified CV from the agarose gel. Alternative anticonvective media, which are microgranular rather than gels of long filaments, may help to circumvent this problem. Percoll, which is colloidal silica particles coated with poly(vinylpyrrolidone) (Pharmacia), has been used (R. Kelly, personal communication).

Conclusion

Methods for the purification of CV have improved as the importance of CV in numerous biological phenomena has come to be appreciated. However, one is still necessarily concerned about how faithfully purified CV represent CV *in vivo.* The isoosmotic density gradient procedures[13] are likely to be a substantial step forward. Alternative methods for purification (permeation chromatography, affinity chromatography, electrophoresis) should enhance our ability to define and resolve CV from minor contaminants.

Other analytical tools await development and reduction to practice. In particular, there is no facile method available for determining the amount of clathrin in various subcellular fractions, nor is there a way to measure absolute concentrations of clathrin in order to estimate yield and purity in the preparative procedures described herein. Further, we have no means of quantitating the amount of clathrin *in vivo* that is assembled or disassembled. Each of these problems may, at least in part, be addressed in an immunoassay utilizing specific anti-clathrin antibodies either from conventional antisera[30] or from hybridoma cell lines (J. L. Daiss and T. F. Roth, unpublished).

[31] S. W. Kessler, this series, Vol. 73, p. 442.

[32] J. M. McSween and S. L. Eastwood, this series, Vol. 73, p. 459.

[33] J. L. R. Rubenstein, R. E. Fine, B. D. Luskey, and J. E. Rothman, *J. Cell Biol.* **89,** 357 (1981).

[30] Dissociation and Reassociation of Clathrin

By W. SCHOOK and S. PUSZKIN

One of the most basic processes for cell survival is that of endocytosis. In order to interact with the external environment, cells must have a mechanism for sensing important changes, especially in the level of appropriate signal molecules, for example, hormones. The cellular mechanism for processing this information has been under study for some time, but only recently has the requisite fusion of biochemical and morphological techniques occurred, with the resultant explosion in information as evidenced by the variety of review articles and texts available[1-3] and by chapters of this volume.

Coated vesicles have been investigated as the probable mediator of endocytotic events since their identification in the mid-1960s.[4] After the isolation of coated vesicles by Pearse,[5] a surge in studies on their formation and function began, leading to the isolation and partial characterization of the coat protein, clathrin, by various investigators.[6-8] Clathrin was shown to exist as a trimer of 180,000 M_r polypeptides that copurifies with a doublet of polypeptides of approximately 30,000 M_r.[9,10] The likely number of these copurifying polypeptides is one 36,000 and two 33,000 per clathrin trimer, based on dye-binding intensity of the protein bands on dodecyl sulfate–polyacrylamide gels.[11] They are known as CAPS (clathrin-associated proteins) or light chains. The clathrin trimer is capable of assembling into the geodesic dome structure observed on coated vesicles by simply decreasing the pH of clathrin solutions from 7.5 to some value between 6 and 6.8, as long as chaotropic anions are absent and the ionic strength is not too

[1] C. D. Ockleford and A. Whyte, eds., "Coated Vesicles." Cambridge Univ. Press, London and New York, 1980.

[2] J. L. Goldstein, R. G. W. Anderson, and M. S. Brown, *Nature (London)* **279,** 679 (1979).

[3] B. M. F. Pearse and M. S. Bretscher, *Annu. Rev. Biochem.* **50,** 85 (1981).

[4] T. F. Roth and K. R. Porter, *J. Cell Biol* **20,** 313 (1964).

[5] B. M. F. Pearse, *J. Mol. Biol.* **97,** 93 (1975).

[6] W. Schook, S. Puszkin, W. S. Bloom, C. Ores, and S. Kochwa, *Proc. Natl. Acad. Sci. U.S.A.* **76,** 116 (1979).

[7] J. H. Keen, M. C. Willingham, and I. H. Pastan, *Cell* **16,** 303 (1979).

[8] M. P. Woodward and T. F. Roth, *Proc. Natl. Acad. Sci. U.S.A.* **75,** 4394 (1978).

[9] H. T. Pretorius, P. K. Nandi, R. E. Lippoldt, M. L. Johnson, J. H. Keen, I. Pastan, and H. Edelhoch, *Biochemistry* **20,** 2777 (1981).

[10] E. Ungewickell and D. Branton, *Nature (London)* **289,** 420 (1981).

[11] J. Kirchhausen and S. C. Harrison, *Cell* **23,** 755 (1981).

ISBN 0-12-181998-1

high.[6-8,12] The 30,000-M_r doublet, which we call CAPs, is required for the proper formation of the basket structure[13]; since brief treatment with chymotrypsin, which partially degrades the CAPs, without apparent effect upon clathrin, destroys the ability of clathrin to assemble correctly. These observations have been supported by others[11] using elastase, which apparently cleaves the CAPs into smaller fragments than chymotrypsin. The importance of understanding the mechanism of clathrin recruitment, *in vivo* assembly, and interaction with the cellular cytoskeleton is self-evident, and our approach to this problem basically has been a biochemical one starting with the isolation of clathrin.

The available procedures for clathrin purification depend upon the relative enrichment of this protein by virtue of its tendency to form or remain assembled as macromolecular aggregates under conditions of slightly acidic pH. In addition, cosedimentation of both coated vesicles and clathrin baskets in sucrose gradients allows a further enrichment from the starting homogenate. A second property utilized is clathrin's sensitivity to conditions of basic pH. When assemblies of clathrin are incubated in buffers of low ionic strengths and basic pH (7.5 or somewhat higher), they disassemble into trimers (triskelions), which can be separated from the bulk of the vesicular membrane and other proteins by sedimentation and column chromatography. There are two principal approaches to clathrin purification described in the literature: one utilizes 2 *M* urea following low-ionic-strength extraction of a crude coated vesicle fraction; the other utilizes a high concentration of Tris buffer and a more highly purified coated vesicle preparation. Our procedure will be described in detail, and the method of Keen and co-workers will be described briefly, since it is basically quite similar.

Purification Procedure

The purification scheme presented below [6,14] results in a preparation which is 90–95% pure; the major copurifying proteins ae the CAPs that are bound to clathrin.

Step 1. Fresh bovine brains, packed in ice, are obtained from a local slaughterhouse. The usual preparation requires three to five brains, and the exact volume of solutions for four brains is given. Meninges are removed,

[12] P. P. Van Jaarsveld, P. K. Nandi, R. E. Lippoldt, H. Saroff, and H. Edelhoch, *Biochemistry* **20,** 4129 (1981).

[13] M. P. Lisanti, W. Schook, N. Moskowitz, K. Beckenstein, W. S. Bloom, C. Ores, and S. Puszkin, *Eur. J. Biochem.* **121,** 617 (1982).

[14] W. S. Bloom, W. Schook, E. Feageson, C. Ores, and S. Puszkin, *Biochim. Biophys. Acta* **598,** 447 (1980).

along with the cerebellum and brainstem. The gray matter is separated from the white by suction through a suction flask kept on ice. The tissue is suspended into 0.1 2-(*N*-morpholino)ethanesulfonic acid *M* (MES) buffer, pH 6.5, containing 1 m*M* EGTA, 0.5 m*M* $MgCl_2$, 7 m*M* 2-mercaptoethanol, and 0.02% sodium azide. Approximately half the tissue suspension is transferred to a Waring blender and homogenized for 10 sec at top speed followed by a 1-min rest period. This procedure is repeated twice, and the homogenate is transferred to a 2-liter beaker. The second half is treated similarly, and all the homogenates are combined. With this number of brains, 1 liter of buffer is used and the resulting homogenate volume is approximately 2.0 liters. The homogenate is centrifuged at 30,000 *g* for 15 min in a Sorvall SS-34 rotor, and the supernatant is saved for the next step.

Step 2. The low-speed supernatants obtained above (approximately 900 ml) are combined and loaded into a Beckman 30 rotor and centrifuged at 80,000 *g* for 1 hr. The pellets obtained are suspended in 500 ml of 20 m*M* Tris, pH 7.5, containing 1 m*M* EDTA and 2 m*M* 2-mercaptoethanol and extracted overnight with stirring at 4°. The extract is centrifuged for 1 hr at 100,000 *g* to remove membranous debris and fractionated with solid ammonium sulfate. The salt is added to 30% saturation with stirring, and the solution is left on ice for 30 min and finally centrifuged for 10 min at 30,000 *g* to pellet the protein. Protein pellets are suspended in a minimal volume, approximately 20 ml of 20 m*M* Tris, pH 7.5 buffer, containing 2 *M* urea and 7 m*M* 2-mercaptoethanol (buffer A) and dialyzed against 20 volumes of the same buffer overnight.

Step 3. The dialyzate is centrifuged at 30,000 *g* for 10 min to remove any denatured protein and loaded onto a 5 × 100-cm column of Sepharose 4B equilibrated with buffer A. A flow rate of 15–20 ml/hr is maintained with a peristaltic pump, and fractions of 8 ml are collected. Those containing proteins are detected by UV absorbance at 280 nm. A peak eluting in the void volume containing residual membranes and other aggregates is discarded; the second peak observed contains clathrin, and these fractions are pooled, adjusted to 50% saturation by the addition of solid ammonium sulfate, and sedimented at 30,000 *g* for 10 min after a 30-min incubation on ice. The pellets obtained are resuspended in a minimal volume of buffer A (approximately 10 ml) and dialyzed against 20 volumes of 20 m*M* Tris, pH 7.6, containing 0.5 m*M* $MgCl_2$, 7 m*M* 2-mercaptoethanol, and 0.02% sodium azide (buffer B), followed by a buffer change and overnight dialysis.

Step 4. Dialyzed protein from two large column runs (120–140 mg) is centrifuged at 100,000 *g* for 1 hr to remove aggregates and loaded on another Sepharose 4B column (2.5 × 100 cm) equilibrated with buffer B; 5-ml fractions are collected. Material eluting in the void volume is composed mainly of aggregated clathrin and much of the copurifying proteins. The second peak observed by UV absorbance at 280 nm is composed of 90% pure clathrin, and major copurifying polypeptides consisting of the CAPs that are associated with clathrin in the triskelion. The clathrin peak is concentrated again by ammonium-sulfate fractionation (50% using the solid salt), and the protein pellets are resuspended in 10–15 ml of buffer A, dialyzed overnight against at least 20 volumes of buffer A, followed by two changes of buffer B. The dialyzed protein can be stored frozen in 10% sucrose for 6 months with no loss in its ability to reassemble into baskets. Figure 1 demonstrates the purity obtainable from a typical clathrin preparation.

Keen and co-workers[7] have used 0.5 *M* Tris buffer to effect solubilization of clathrin from a more purified starting material. This buffer results in the solubilization of more protein from the vesicle membrane, and this results in a third clearly discernible peak following chromatography. Starting with a more purified coated vesicle fraction is more important in this procedure, as higher levels of contamination are likely to be observed in clathrin preparations because the buffer extracts a greater amount of protein from the membrane. Usually quite pure clathrin can be obtained with only one pass through a 1 × 90 cm column, using the coated vesicles obtained from one bovine brain; however, the yield is quite low (approximately 3–5 mg). The assembly properties of clathrin prepared by this procedure are also somewhat different from preparations using 2 *M* urea, and this will be discussed fully later.

Biophysical Properties of Clathrin

Irrespective of the procedure used to purify clathrin, it is always associated with the CAPs of molecular weights, as reported by various investigators, ranging from 30,000 to 36,000.[10–13] Preparations of clathrin contain approximately 2 mol of faster-moving polypeptide to one of the slower, as judged by densitometry of Coomassie Blue-stained bands in polyacrylamide gels. When stored in solutions that favor depolymerization, clathrin has been shown by several groups to have a sedimentation constant of 8.6 $s^0_{20,w}$, which, when combined with other biophysical data, suggests that clathrin exists as a trimer or triskelion in solution. This proposal is supported by electron microscopic evidence using rotary shadowing[10] and traditional[15] negative staining techniques that clearly show three bent legs radiating from a central point. If such triskelions are assembled into baskets and briefly treated with chymotrypsin,[13] the CAPs are partially proteolyzed, whereas the clathrin is apparently unaffected. When such preparations are examined in the analytical ultracentrifuge, an $s^0_{20,w}$ value of 8.2 is obtained, which clearly demonstrates that removal of the susceptible segment of the CAPs polypeptide chain does not result in disassembly of the triskelion. The partial removal of the CAPs alters the ability of the modified triskelions to

[15] R. A. Crowther and B. M. F. Pearse, *J. Cell Biol.* **91,** 790 (1981).

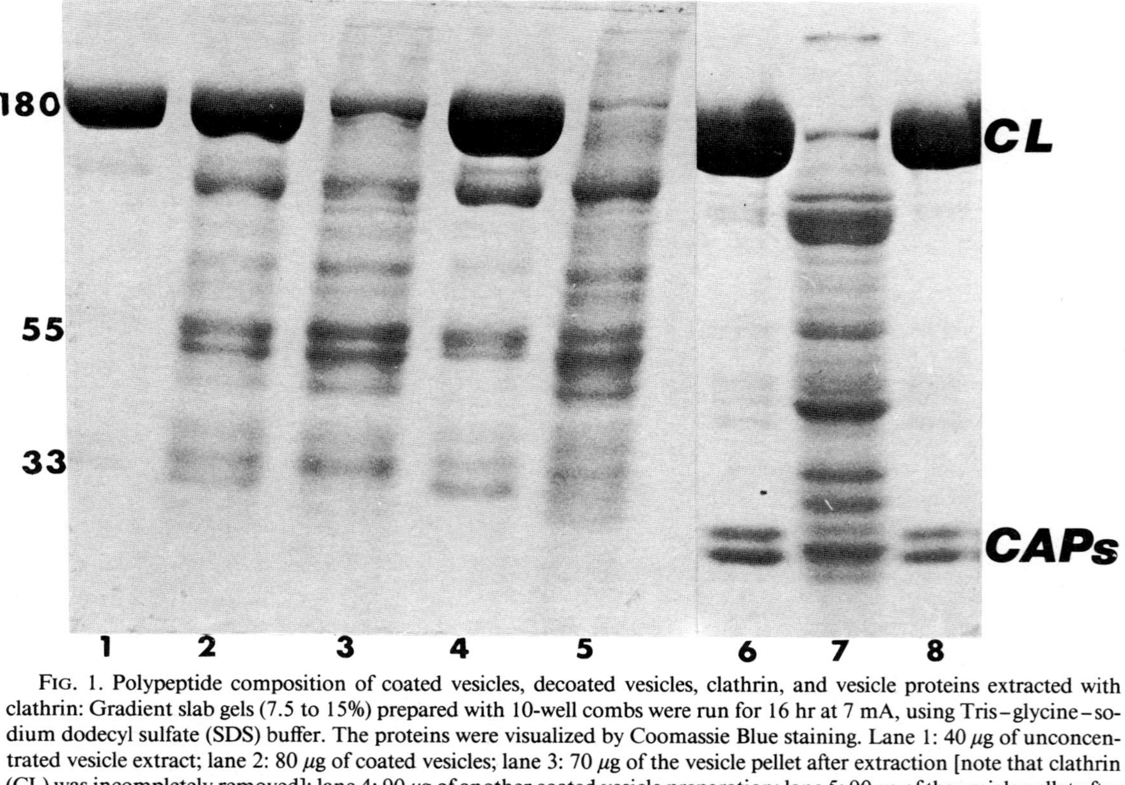

FIG. 1. Polypeptide composition of coated vesicles, decoated vesicles, clathrin, and vesicle proteins extracted with clathrin: Gradient slab gels (7.5 to 15%) prepared with 10-well combs were run for 16 hr at 7 mA, using Tris–glycine–sodium dodecyl sulfate (SDS) buffer. The proteins were visualized by Coomassie Blue staining. Lane 1: 40 μg of unconcentrated vesicle extract; lane 2: 80 μg of coated vesicles; lane 3: 70 μg of the vesicle pellet after extraction [note that clathrin (CL) was incompletely removed]; lane 4: 90 μg of another coated vesicle preparation; lane 5: 90 μg of the vesicle pellet after extraction (note that more complete removal of clathrin is evident in this case); lane 6: 100 μg of clathrin after chromatography on Sepharose 4B; lane 7: 100 μg of the proteins recovered from the Sepharose column after the elution of clathrin (note the presence of proteins migrating in the 100,000 M_r region of the gel, one of which is probably the attachment point for clathrin on the vesicle surface); lane 8: 80 μg of same clathrin preparation (note that major copurifying proteins are the CAPs). Numbers to the left of the figure indicate molecular weight in thousands.

assemble into baskets.[11,13] There is a marked shift in the equilibrium from assembled structures to that of trimers under conditions usually conducive to cage assembly. This change in assembly properties points to a pivotal role for the CAPs in guiding the way triskelions interact with each other during the formation of cages. On the basis of the proposed model for clathrin assembly by Crowther and Pearse[15] and on data obtained with trypsin-treated clathrin by Schmid *et al.*,[16] it seems likely that the CAPs are found near the vertex of the triskelion and interact with the triskelion's legs to determine their orientation and therefore guide the formation of points of association during clathrin assembly.

Utilizing chymotrypsin bound to the surfaces of polystyrene beads, we have shown that the CAPs are oriented to the outside of the basket structure in coated vesicles and clathrin baskets[17] and that the fragmentation of CAPs by this protease results in the retention of a portion of the CAP molecule by clathrin.

We have devised a procedure for the purification of CAPs from triskelions that reveals two interesting properties of these polypeptides—their heat stability and binding affinity for calmodulin.[18]

Step 1. Clathrin preparations or even crude coated vesicles (2–5 mg/ml) are placed in buffer B and heated for 5 min in a boiling water bath and centrifuged at 100,000 *g* for 60 min to remove denatured material. The supernatant is concentrated to approximately 1 mg/ml with dry Sephadex G-200 or Ficoll. An estimate of the amount of CAPs expected from such treatment can be obtained by assuming that CAPs constitute 5–10% of the starting clathrin concentration.

Step 2. A calmodulin affinity column is prepared as previously described[19] using CNBr-activated Sepharose and calmodulin. The CAPs preparation is dialyzed overnight against 20 m*M* Tris, pH 6.5, buffer containing 1 m*M* Ca^{2+}, 2 m*M* 2-mercaptoethanol, and 0.02% sodium azide and allowed to bind to this column for 30–60 min at room temperature. The column then is washed with more buffer to remove unbound protein, washed with a similar buffer containing 0.5 *M* NaCl to remove nonspecifically bound protein, and finally with a similar buffer containing 2 m*M* EGTA in place of Ca^{2+}. The CAPs are eluted in the EGTA buffer and are dialyzed against 50 volumes of buffer B overnight. They are concentrated again and stored in 10% sucrose. The protein composition at each step in this procedure is illustrated in Fig. 2.

[16] S. L. Schmid, A. K. Matsumoto, and J. E. Rothman, *Proc. Natl. Acad. Sci. U.S.A.* **79,** 91 (1982).

[17] M. P. Lisanti, W. Schook, N. Moskowitz, C. Ores, and S. Puszkin, *Biochem. J.* **201,** 297 (1982).

[18] M. P. Lisanti, L. Shapiro, N. Moskowitz, E. Hua, S. Puszkin, and W. Schook, *Eur. J. Biochem.* **125,** 463 (1982).

[19] D. M. Watterson and T. C. Vanaman, *Biochem. Biophys. Res. Commun.* **73,** 40 (1976).

Step 2 is necessary when crude preparations of starting material are used rather than purified clathrin. This purification procedure will allow preparation of CAPs adequate for investigation of the chemical and biophysical properties of these regulatory proteins that attach to clathrin. The properties already observed, i.e., heat stability, requirement for efficient cage assembly, ability to bind calmodulin, and their presence in all coated vesicle or clathrin preparations, suggest that these polypeptides are important for clathrin function and have not been formed as a result of proteolytic degradation, as suggested by immunological cross-reactivity with antibodies purified using a

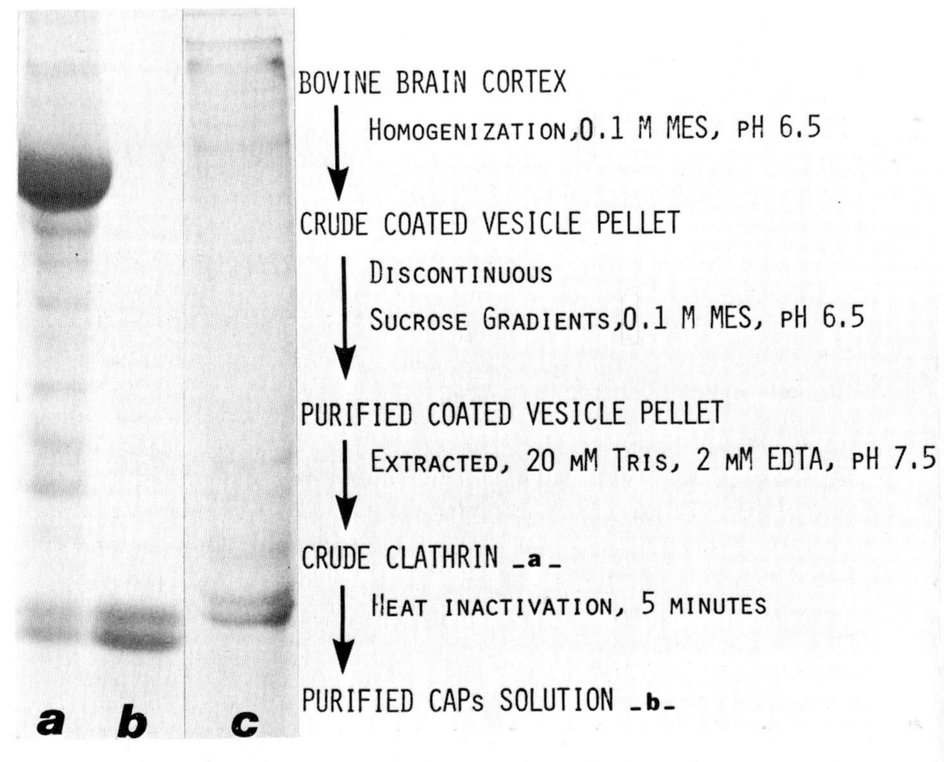

FIG. 2. Schematic summary of the purification steps leading to a homogeneous solution of clathrin-associated proteins (CAPs): The protein composition of selected steps is illustrated by the Coomassie Blue-stained bands on SDS–gradient slab gels. Lane a: Extract of purified coated vesicles containing clathrin prior to heat treatment (100 μg); lane b: concentrated supernatant after heat treatment of clathrin (30 μg); lane c: supernatant of an extract of crude coated vesicles after heat treatment. Note that the preparation shown in lane c is considerably less pure than that shown in lane b. The purity of this CAP preparation can be improved to that illustrated in lane b by use of affinity chromatography using immobilized calmodulin as described in the text (30 μg).

denatured clathrin affinity column.[20] Furthermore, recent results from our laboratory have demonstrated that the CAPs colocalize with clathrin at both the light and electron microscopic levels,[18,21,22] and, although these studies failed to detect a soluble pool of CAPs, the possibility that these proteins may exist independently from clathrin has been raised by the conditions of their interaction with calmodulin. This possibility has received support from a recent report that CAP-like polypeptides present in a cytosol fraction derived from adrenal chromaffin cells are bound to the chromaffin granule membrane in a Ca^{2+}- and calmodulin-dependent manner.[23]

Conditions Conducive to Basket Assembly

Since the discovery that clathrin was capable of *in vitro* assembly into cages, a number of groups have been investigating the environmental changes that can alter their rate and/or final state of assembly. The principal conclusions from these studies can be summarized as follows. Divalent cations can increase the rate of basket formation at a constant pH. Control of pH is important, since the rate of assembly is dependent on this variable.[2] The order of effectiveness for divalent cations is as follows: Mn^{2+} greater than Ca^{2+} greater than Mg^{2+} [21]; that for anions was found to follow basically the Hofmeister ranking, although sulfate seemed to exhibit more specific inhibitory properties.[12] Several drugs were shown to have stimulatory effects upon clathrin assembly[13] and to share a common factor, hydrophobic structure and inherent basicity. They appear to stimulate the rate of assembly and possibly act by mimicking the effect of protons, making the environment appear more acidic than it actually is. Using a variety of polyamine bases, Nandi *et al.*[21] have shown that a minimum of four basic groups appear to be required for significant binding unless a nonpolar region is present as in dansyl cadavarine. This group confirmed our observations of chlorpromazine effects upon basket assembly and found that lysozyme was capable of stimulating assembly, whereas other basic proteins tested had no effect. These studies all point to a region of the clathrin molecule capable of interacting with a variety of agents that can affect the conformation of the triskelion and stimulate the assembly process. These observations suggest that the assembly process depends upon both ionic and hydrophobic interactions. The type of buffer used also modulates the pH at which basket

[20] J. H. Keen, M. C. Willingham, and I. Pastan, *J. Biol. Chem.* **256,** 2538 (1981).

[21] S. Puszkin, C. Ores, A. Andres, M. P. Lisanti, and W. J. Schook, *Cell Tissue Res.* **231,** 495 (1983).

[22] M. P. Lisanti, A. Andres, C. Ores, A. C. Puszkin, W. J. Schook, and S. Puszkin, *Cell Tissue Res.* **231,** 507 (1983).

[23] M. J. Geisow and R. D. Burgoyne, *Nature (London)* **301,** 432 (1983).

assembly is normally observed, cacodylate appearing to enhance basket formation under more alkaline conditions.[6]

Some differences in assembly properties have been reported by various laboratories for clathrin preparations obtained by different procedures.[22,24,25] It appears that clathrin prepared by 0.5 *M* Tris extraction is more sensitive to substances present in the assembly buffer and requires Ca^{2+} as well as additional protein cofactors. Urea-purified clathrin assembles readily without such cofactors or similar requirements. An additional protein component of approximately 110,000 M_r[26] seems to be required for clathrin assembly onto membranes. This protein is probably responsible for the stimulatory effect of the peak III protein fraction observed by Keen *et al.*[7] We have compared the assembly capacity of clathrin obtained by both procedures and have found that Tris-treated clathrin yields quatitatively less assembly when tested under identical conditions, but that urea treatment will partially reverse the effects of the high Tris concentration used in the isolation procedure. These observations suggest that the effects of these agents upon clathrin's ability to assemble are complex and may involve changes in the properties of accessory factors as well as in clathrin itself. An especially promising recent development was the discovery of a cytosolic factor capable of effecting the disassembly of clathrin from isolated coated vesicles under conditions of neutral pH. This factor appears to be a protein that requires ATP and results in a modification of the clathrin triskelion which makes it refractory to *in vitro* reassembly under usually effective conditions.[27] The identification of a membrane-bound protein as the probable site of attachment of clathrin triskelions to the membrane as well as the purification of the clathrin-associated proteins (the CAPs) and the discovery of their interaction with calmodulin suggest that studies on the interrelationship of these and the possibly other yet-to-be-described effectors will continue and that descriptions of the biochemical and biophysical properties of clathrin and its associated proteins will remain an important area for investigation.

[24] P. K. Nandi, P. P. Van Jaarsveld, R. E. Lippoldt, and H. Edelhoch, *Biochemistry* **20,** 6706 (1981).

[25] M. P. Woodward and T. F. Roth, *J. Supramol. Struct.* **11,** 237 (1979).

[26] E. R. Unanue, E. Ungewickell, and D. Branton, *Cell* **26,** 439 (1981).

[27] E. J. Patzer, D. M. Schlossman, and J. E. Rothman, *J. Cell Biol.* **93,** 230 (1982).

[31] Preparation of Antibodies to Clathrin and Use in Cytochemical Localization

By JAMES H. KEEN

Clathrin, the structural component of the coat structure, is a ubiquitous protein that is intimately involved in many different intracellular processes, including endocytosis, secretion, and membrane recycling. Antibodies to clathrin are useful for quantitating amounts of clathrin, determining its subcellular localization, and as a probe of its function in cells and tissues.

A. Preparation of Clathrin as an Antigen

Principle. Clathrin is isolated from bovine brain by first preparing partially purified coated vesicles (see this volume [27]–[29]). The clathrin is dissociated from the vesicles, concentrated, and purified by gel filtration.[1] This material is cross-linked and/or peroxidized and is then injected into rabbits. Homogeneous clathrin is prepared separately and used for the affinity purification of the immune serum.[2]

Procedure. Clathrin-coated vesicles (see this volume [27]–[29]) containing 40 mg of protein are sedimented at 100,000 *g* for 60 min, and the pellet is resuspended with a Dounce homogenizer in buffer A [0.5 *M* Tris-HCl, 0.05 *M* sodium 2(*N*-morpholino)ethanesulfonate (MES), 0.5 m*M* EGTA, 0.25 m*M* calcium chloride, 0.01% sodium azide, pH 7.0] to yield a final protein concentration of 1 mg/ml. After incubation for 15 min at room temperature, the suspension is recentrifuged as before and the supernatant is withdrawn. Protein is concentrated by addition of solid ammonium sulfate to 50% saturation, centrifuged for 30 min at 10,000 *g*, and resuspended in a minimal volume of buffer A at room temperature. Resuspension and the subsequent gel filtration step should be performed at room temperature, since chromatography at 4° yields incomplete fractionation. If excessively turbid, the solution should be briefly spun in a table-top centrifuge. The sample (up to 7 ml) is applied to a column (1.5×90 cm) containing Sepharose CL-4B preequilibrated with buffer A. It is eluted at 10 ml/hr, and 3.0-ml fractions are collected. A typical absorbance profile is shown in Fig. 1. The first peak eluted (fractions 22–26) contains proteins excluded from the resin; the second (generally the major) peak, containing approximately 10 mg of protein, consists predominantly of highly purified clathrin (frac-

[1] J. H. Keen, M. C. Willingham, and I. H. Pastan, *Cell* **16,** 303 (1979).
[2] J. H. Keen, M. C. Willingham, and I. H. Pastan, *J. Biol. Chem.* **256,** 2538 (1981).

ISBN 0-12-181998-1

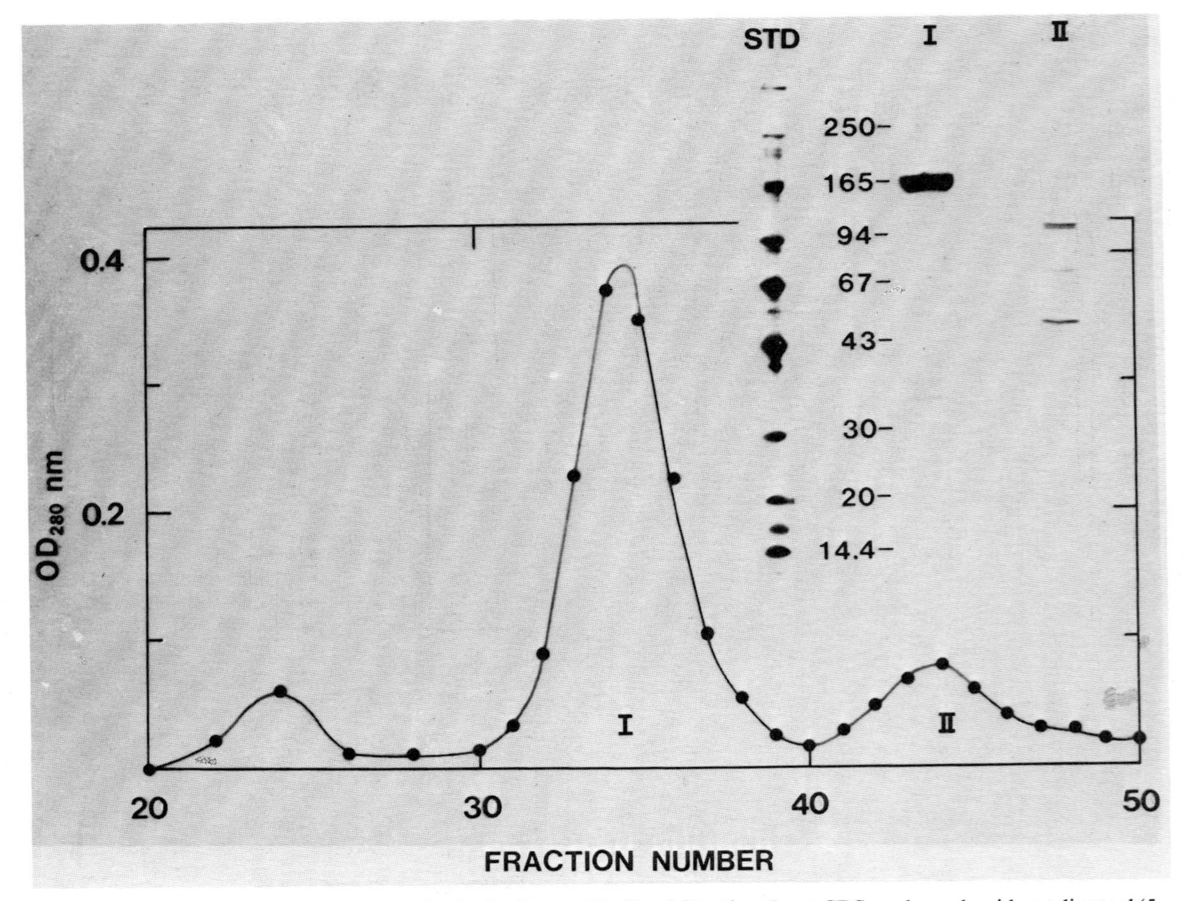

FIG. 1. Analysis of extracted coat proteins by Sepharose CL-4B gel filtration. Inset: SDS–polyacrylamide gradient gel (5 to 15%) electrophoresis of samples from peaks I and II. From Keen *et al.*[1]

tions 32–37). Polypeptides of lower molecular weight (approximately 33,000 and 38,000) are also present, each constituting less than 5% of the total protein. These polypeptides are generally found tightly associated with clathrin when the latter is isolated in native form (i.e., capable of reassembly).[3,4] This material is suitable for use as an antigen.

B. Immunization Procedure

Principle. Clathrin has been found to be a very weak antigen, presumably because it is highly conserved,[5] and it is therefore desirable to derivatize the protein to stimulate immunogenicity. Glutaraldehyde cross-linking has been used for this purpose, and we have found peroxidation to enhance this effect. A procedure for cross-linking is described below; a peroxidation method is described elsewhere.[6]

Procedure. Clathrin purified by gel filtration, 0.5–1.0 mg/ml, is dialyzed extensively against Dulbecco's phosphate-buffered saline (D-PBS) (8.1 m*M* Na_2HPO_4, 1.5 m*M* KH_2PO_4, 137 m*M* sodium chloride, 2.7 m*M* potassium chloride) to remove all Tris-HCl. Cross-linking is performed by adding fresh 50% glutaraldehyde (TEM grade, Tousimis, Rockville, Maryland) to the dialyzate, to a final concentration of 1%, and allowing dialysis to continue overnight at 4°. The sample is then again dialyzed extensively against D-PBS. Rabbits to be injected (we have used white New Zealand males) are bled to obtain preimmune sera, and each is then injected intradermally in multiple sites along the back with 100–500 μg of protein (in 0.1–1.0 ml) mixed with an equal volume of complete Freund's adjuvant for the initial injection, or incomplete adjuvant for subsequent injections, at 3-week intervals.[7] Blood is withdrawn from the ear vein at weekly intervals and allowed to clot overnight at 4°, and the serum is isolated by centrifugation. The antibody titer can be quantitatively assessed directly in the serum by immunoprecipitation of isotopically labeled clathrin.[8] A partially purified immunoglobulin fraction prepared by precipitation of the serum with 35% saturated ammonium sulfate can also be used semiquantitatively to titer the bleeds by immunofluorescence (see below).

Comments. Using this procedure, we have obtained anti-clathrin antisera in approximately 8 of 16 rabbits. The immune sera are generally of only moderate titer, containing approximately 20–50 μg of anti-clathrin anti-

[3] B. M. F. Pearse, *J. Mol. Biol.* **126,** 803 (1978).

[4] T. Kirchhausen and S. C. Harrison, *Cell* **23,** 755 (1981).

[5] B. M. F. Pearse, *Proc. Natl. Acad. Sci. U.S.A.* **73,** 1255 (1976).

[6] C. H. W. Hirs, this series, Vol. 11, p. 197.

[7] J. L. Vaitukaitus, this series, Vol. 73, p. 46.

[8] This series, Vol. 96 [8].

body per milliliter of serum. Similar results, in terms of titer and specificity, were obtained whether rabbits were injected with clathrin in dissociated form or with clathrin reassembled into coat structures.

C. Affinity Purification of Immune Sera

Principle. For most experiments involving cytochemistry, it is desirable to purify the antisera, either to increase the anti-clathrin titer or to remove an unwanted antibody contaminant. To accomplish this, a column containing covalently attached homogeneous clathrin is prepared. The immune serum is passed over this resin, and the anti-clathrin antibodies specifically retained are eluted in a greatly purified form.

Materials

Buffer B: 20 m*M* sodium phosphate, 6 *M* guanidine hydrochloride, 5 m*M* glutathione, pH 6.25

Buffer C: buffer B + 10 m*M* dithiothreitol

Buffer D: 0.1 *M* sodium borate, 0.5 *M* sodium chloride, pH 8.0

CnBr–Sepharose (Pharmacia)

Collodion dialysis tube, UH 100/25, Schleicher & Schuell (Keene, New Hampshire)

Preparation of Clathrin–Sepharose. Approximately 5 mg of gel filtration-purified clathrin (see Section A) is precipitated with 50% saturated ammonium sulfate and resuspended in a minimal volume of freshly prepared buffer C. This sample is applied to a column (1.5 × 68 cm) of Sephacryl S-300 equilibrated with freshly prepared buffer B. The column is eluted at 5 ml/hr, and 1.7-ml fractions are collected. The major peak (fractions 13–16) containing homogeneous clathrin (Fig. 2) is pooled and dialyzed against 0.1 *M* sodium borate, 0.5 *M* sodium chloride, pH 9.0. This preparation, containing 1–3 mg of protein, is coupled to CNBr–Sepharose (1 g dry weight) as described elsewhere.[9]

Affinity Purification. All steps are performed at 0–4°. Immune serum (10–20 ml) is twice precipitated with ammonium sulfate (35% saturation), resuspended in one-half the original volume, dialyzed against buffer D, and applied over 2–3 hr to the clathrin–Sepharose column preequilibrated with buffer D. The column is washed with buffer D and then with 0.1 *M* sodium acetate, 0.5 *M* sodium chloride, pH 4.8, until little or no protein is detected in the effluent. Specifically bound protein is then released by eluting with 0.2 *M* acetic acid, 0.5 *M* sodium chloride, pH 2.6; 1.0-ml fractions are collected into tubes containing 0.6 ml of 1 *M* Tris-HCl, pH 8, to immediately neutralize the acid. Protein-containing fractions (detected by absorbance at 280 nm) are concentrated by precipitation with ammonium sulfate

[9] S. Kaufman, this series, Vol. 22, p. 233.

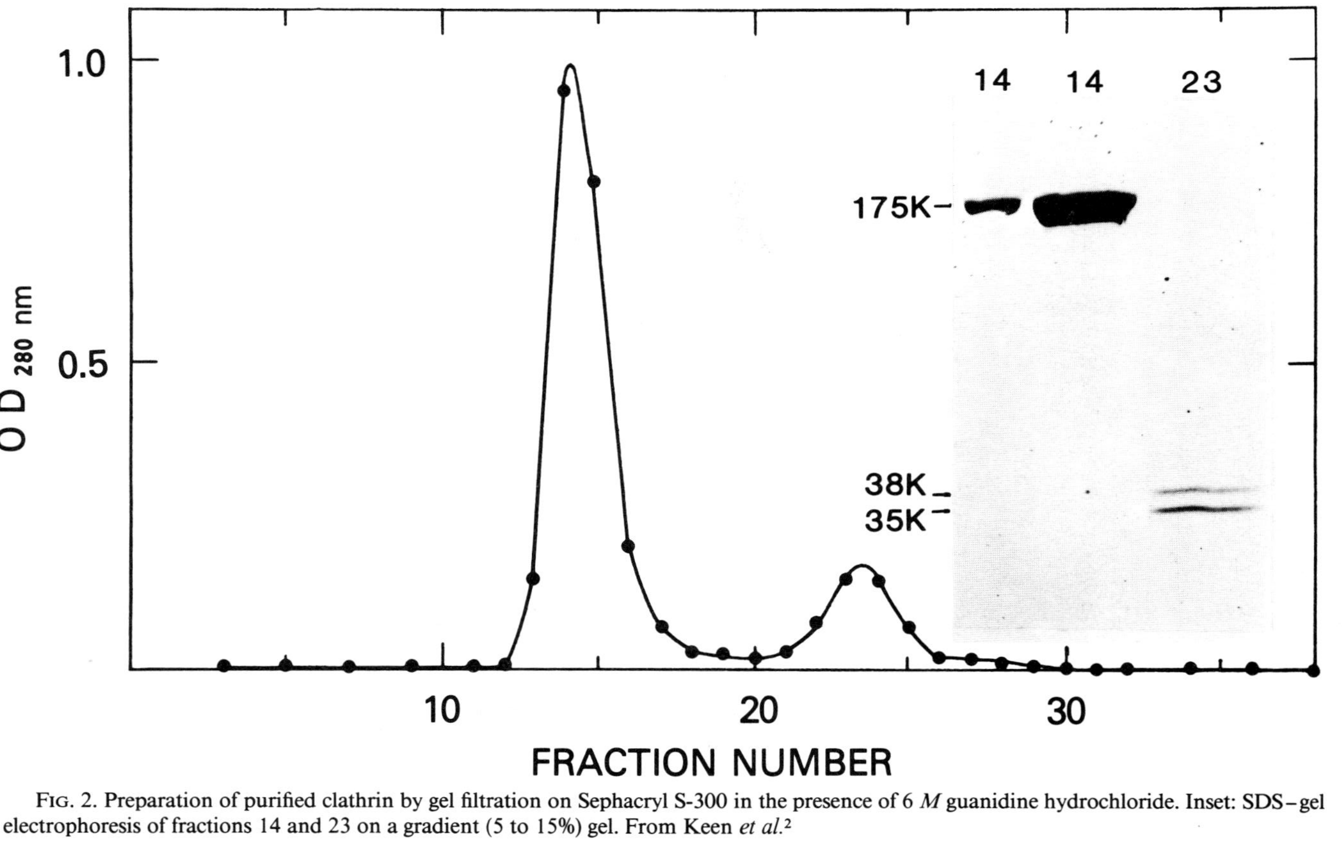

FIG. 2. Preparation of purified clathrin by gel filtration on Sephacryl S-300 in the presence of 6 *M* guanidine hydrochloride. Inset: SDS–gel electrophoresis of fractions 14 and 23 on a gradient (5 to 15%) gel. From Keen *et al.*[2]

(40% saturation), resuspended in a minimal volume and dialyzed in a collodion bag against D-PBS. Samples may be kept at 4° for up to 1 month or, if the protein concentration is 1 mg/ml or greater, they may be stored at −20°.

The specificity of the purified antibodies may be checked by immunoprecipitation of labeled cell proteins, preferably with the cell type to be used in subsequent experiments.[8]

Comments. The Sephacryl S-300 column fractions may be checked for purity by sodium dodecyl sulfate–polyacrylamide gel electrophoresis (SDS–PAGE) if they are first dialyzed against buffer B containing 6 *M* urea instead of guanidine HCl (see Fig. 2, inset). With the sera we have obtained, yields of antibody are improved if the sample is repeatedly recycled over the clathrin–Sepharose column for 2–3 hr rather than subjected to a single pass. The acidic elution procedure described here has also provided greater yields than elution with high concentrations of $MgCl_2$ or thiocyanate.

D. Cytochemical Localization

Principle. Once monospecific antibody has been obtained, it can be used to probe for clathrin localization in cells and tissues. Two examples of clathrin localization, at the light and electron microscopic levels, using immunofluorescence and ferritin labeling techniques, will be presented.

Immunofluorescence

Adherent cells grown in plastic tissue-culture dishes can be studied using the technique of indirect immunofluorescence. The washed and fixed cells are permeabilized by treatment with low concentrations of nonionic detergent and are sequentially treated with rabbit anti-clathrin antibodies and fluorescently labeled anti-rabbit immunoglobulin antibodies.

Materials

Formaldehyde solution, 37% (w/w)
Dulbecco's phosphate-buffered saline (D-PBS)
Normal goat immunoglobulin (NGG; Cappel Laboratories, Cochranville, Pennsylvania)
Rhodamine-conjugated goat anti-rabbit immunoglobulins (Cappel Laboratories)

Procedure. Subconfluent dishes of cells (e.g., mouse 3T3 fibroblasts) are washed three times in D-PBS and fixed with freshly prepared 3% formaldehyde in D-PBS for 10 min at room temperature. They are then permeabilized with 0.05% Triton X-100 in D-PBS for 10 min at room temperature. All subsequent steps are performed at 37° in the continued presence of

0.05% Triton X-100 and 1 mg/ml of NGG to minimize nonspecific adsorption. Rabbit antibodies (10–30 μg/ml) are added and incubated for 30 min followed by washout for 30 min. Rhodamine-labeled goat anti-rabbit immunoglobulin is then incubated with the cells, at 1:20 to 1:200 dilution in D-PBS, for 30 min; the unbound protein is removed by a 30-min incubation. The cells are then washed with D-PBS, mounted, and examined by epifluorescence microscopy using appropriate filters.

Comments. In adherent fibroblastic cells (e.g., the Swiss mouse 3T3 cell) a characteristic punctate pattern of fluorescence is observed extending to the periphery of the cell with increased density in the perinuclear region (Fig. 3, A and B), a pattern that correlates with the distribution of coated membranes in these cells.[10] Controls using equivalent amounts of preimmune serum should be run (Fig. 3C). Anti-bovine brain clathrin antibodies have been found to yield immunofluorescent patterns similar to those shown here for mouse cells when either rat or chicken fibroblasts have been tested.[2] When neuronal cell cultures have been tested, substantial diffuse immunofluorescence has been observed.[11,12]

Electron Microscopy

Localization of clathrin at the ultrastructural level has been performed using peroxidase and ferritin as electron-dense markers.[10,11,13] A ferritin-bridge procedure, which employs a series of antibodies to link ferritin to the cellular antigen, is presented here.

Procedure. Fixed cells are washed in 0.02% saponin in D-PBS for 30 min at 25°. The dishes are subsequently incubated, in the continuous presence of 0.02% saponin in D-PBS, with (*a*) affinity-purified rabbit anti-clathrin antibody (50 μg/ml) or normal rabbit globulin (50 μg/ml); (*b*) goat anti-rabbit immunoglobulin ($\frac{1}{50}$ dilution); (*c*) affinity-purified rabbit anti-horse spleen ferritin (50 μg/ml); (*d*) horse spleen ferritin (100 μg/ml). Between incubations the dishes are washed in 0.02% saponin in D-PBS for 30 min at 25°. The samples are subsequently postfixed with 3% glutaraldehyde in D-PBS for 30 min at 25° and prepared for electron microscopy and ferritin localization.[10,14]

Comments. Thin sections prepared without uranyl salt counterstaining

[10] M. C. Willingham, J. H. Keen, and I. H. Pastan, *Exp. Cell Res.* **132,** 329 (1981).

[11] T. P.-O. Cheng, F. I. Byrd, J. N. Whitaker, and J. G. Wood, *J. Cell Biol.* **86,** 624 (1980).

[12] W. S. Bloom, K. L. Fields, S. H. Yen, K. Haver, W. Schook, and S. Puszkin, *Proc. Natl. Acad. Sci. U.S.A.* **77,** 5520 (1980).

[13] R. G. W. Anderson, E. Vasile, R. J. Mello, M. S. Brown, and J. L. Goldstein, *Cell* **15,** 919 (1978).

[14] M. C. Willingham, *J. Histochem.* **12,** 419 (1980).

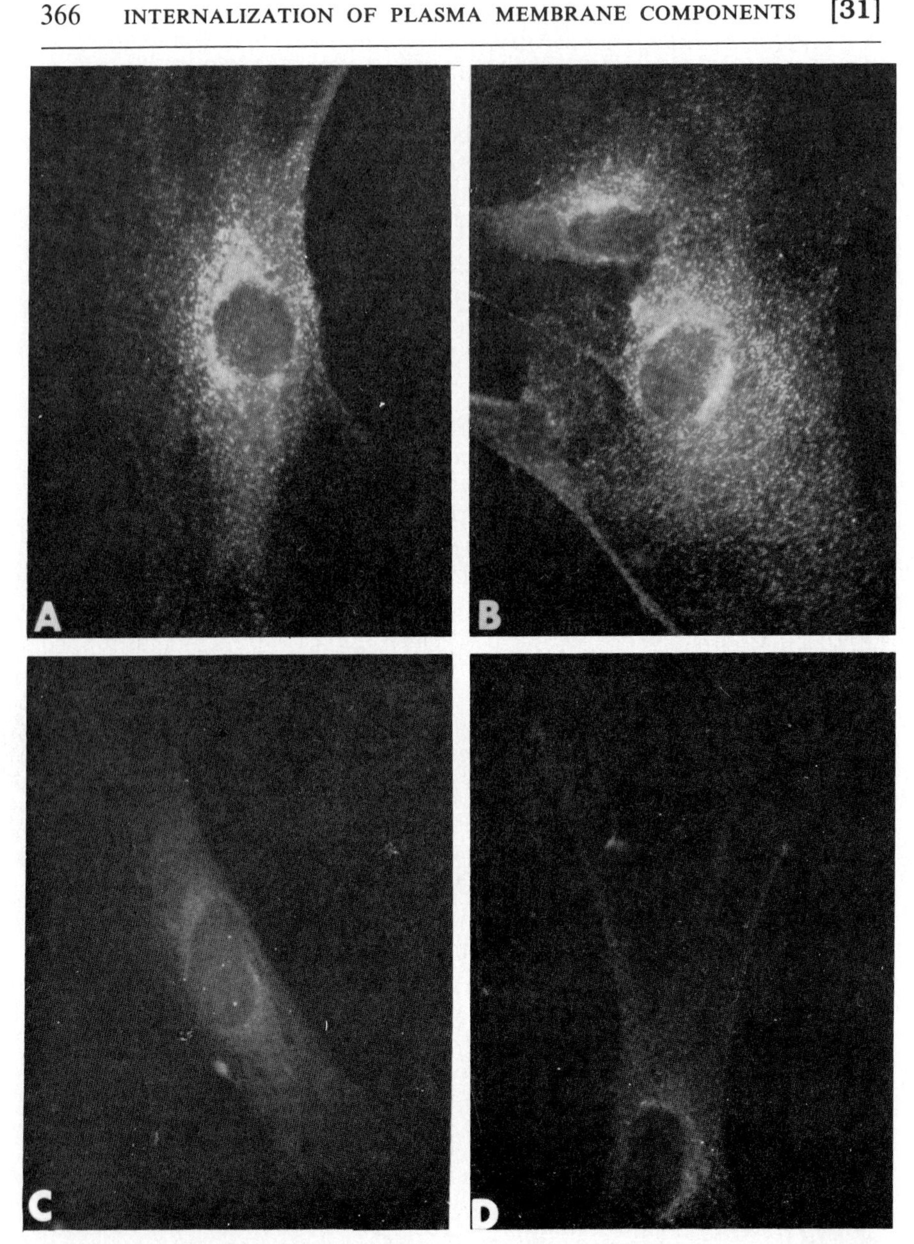

FIG. 3. Immunofluorescent localization of anti-clathrin antibodies in Swiss mouse 3T3 fibroblasts. (A, B) Affinity-purified anti-clathrin; (C) preimmune serum; (D) preadsorption of affinity-purified anti-clathrin antibodies with excess clathrin. From Keen *et al.*[2]

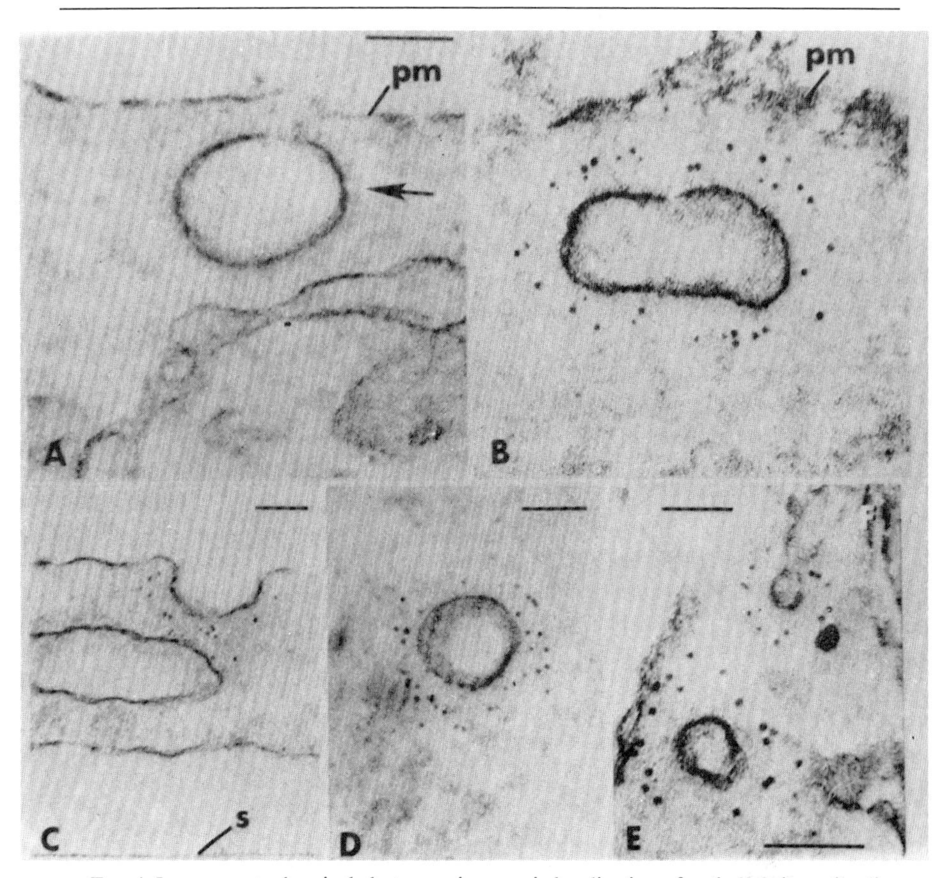

FIG. 4. Immunocytochemical electron microscopic localization of anti-clathrin antibodies in Swiss mouse 3T3 fibroblasts using preimmune (A) or affinity-purified anti-clathrin antibodies (B–E). (B–D) Sections demonstrating localization of label to coated pit structures in the plasma membrane. (E) Localization of anti-clathrin to small coated structures in the Golgi–GERL region. pm, Plasma membrane. Bar = 0.1 μm. From Willingham *et al.*[10]

facilitate the identification of ferritin cores. These sections do not reveal the characteristic bristle-coat structure, but the protein density of the clathrin coat can generally be detected. In mouse 3T3 fibroblasts, ferritin localization surrounds coated membranes in and near the plasma membrane (Fig. 4, A–D). The ferritin cores are often present as closely spaced doublets and are at some distance from the membrane, consistent with the thickness of the clathrin-coat structure and the accumulated length of the antibody bridge molecules involved. Localization is also seen in the Golgi–GERL region, where smaller membrane segments are labeled (Fig. 4E). In this cell type very little label is seen freely distributed in the cytosol or associated with other organelle membranes using these antibodies. Studies with rodent cerebellar tissue[11] demonstrated dense (peroxidase reaction product) labeling of coated membrane in the presynaptic terminal. In addition, considerable diffuse labeling of the intervesicular cytoplasmic space in the terminal was observed. The significance of this localization and its functional implications for the role of clathrin in neuronal cells remains to be elucidated.

[32] Structural Investigations on the Role of Microfilaments in Ligand Translocation

By JEFFREY L. SALISBURY and GILBERT A. KELLER

Receptor-mediated endocytosis along the coated-vesicle pathway is of fundamental importance to cells in dealing with a diverse group of molecular ligands. This process involves several discrete and separable steps that utilize distinct aspects of cellular machinery.[1,2] Ligand concentration occurs through binding to specific cell-surface receptor molecules that cluster or patch to further concentrate ligand in the plane of the membrane. Receptor clusters subsequently associate with specialized regions of the plasma membrane, termed clathrin-coated pits. Coated pits become internalized by a membrane vesiculation process resulting in the formation of ligand-laden coated vesicles within the cytoplasm. Translocation of ligand–receptor clusters at the cell surface and of ligand-laden vesicles within the cell occurs in a rapid and directed fashion. Evidence implicating a role for actin-based cortical microfilaments in ligand translocation is accumulating.[1-5] Nonmuscle cells contain actin and myosin and these proteins form functional hybrid complexes with their muscle counterparts that are capable of tension development and contraction *in vitro*.[6-8] These properties of the nonmuscle cytoskeleton suggest that the essential functional domains of contractile proteins such as actin and myosin are highly conserved and can be utilized

[1] J. L. Salisbury, J. S. Condeelis, and P. Satir, *J. Cell Biol.* **87,** 132 (1980).

[2] J. L. Salisbury, J. S. Condeelis, and P. Satir, *Cold Spring Harbor Symp. Quant. Biol.* **46,** pt. 2 (1982).

[3] J. Flanagan and G. Koch, *Nature* (*London*) **273,** 278 (1978).

[4] G. Koch and M. Smith, *Nature* (*London*) **273,** 274 (1978).

[5] B. E. Batten, J. J. Aalberg, and E. Anderson, *Cell* **21,** 885 (1980).

[6] J. A. Spudich, *J. Biol. Chem.* **249,** 6013 (1974).

[7] T. Pollard, *J. Cell Biol.* **67,** 93 (1975).

[8] J. Condeelis, and D. L. Taylor, *J. Cell Biol.* **74,** 901 (1977).

ISBN 0-12-181998-1

by nonmuscle cells for movement. In addition, a wide variety of actin-binding proteins occur in muscle and nonmuscle cells that nucleate, cross-link, or cap actin filaments and thereby play a fundamental role in regulating microfilament polarity, length, and specific association with other structural elements of the cytoplasm.[9-11]

Receptor clustering, or patching, is not sensitive to metabolic poisons or to the actin-directed cytochalasins, but is inhibited by cold. Thus, receptor clustering appears to be a diffusion mediated process not dependent on the integrity of cortical cytoplasmic actin filaments. In contrast, the translocation of ligand–receptor complexes over long distances at the cell surface, or within vesicular compartments inside the cytoplasm is inhibitable by both metabolic poisons and the cytochalasins (B and D and dihydrocytochalasin B).[1,2,12,13] Cytochalasins do not, in general, affect internalization of ligand by clathrin-coated vesicles.[1,2] Thus, it appears that, although the internalization of ligand by coated vesicles is independent of cytoplasmic actin, translocation of ligand is dependent on functional actin-based microfilaments in the cell cortex. Direct biochemical evidence for an association of actin with the cell surface[6,14] and with antigen receptors on B lymphoblastoid cells[3,4] has lead a number of laboratories to investigate the structural organization of the cortical actin cytoskeleton and its relationship with receptor complexes and the machinery of receptor-mediated endocytosis.[1,2,5] These studies provide strong structural evidence for a direct association between actin microfilaments and cell-surface receptor molecules, coated pits, and coated vesicles, and they support the earlier suggestion[1,6] that the actin-based nonmuscle contractile apparatus is involved in active translocation of these organelles (see Figs. 1 and 2).

We detail here methods for identification and visualization of actin microfilaments associated with cell surface receptors, coated pits, and coated vesicles in glycerol-extracted cell models by decoration with the myosin subfragments HMM or S1. The cultured human lymphoblastoid cell line WIL2, which displays antigen–receptor immunoglobulin M on its cell surface, has been a system of choice for these studies.[15] In addition, we have studied other cultured cell lines, including adherent 3T3-L1 cells, and believe our observations will prove to represent a general phenomenon.

[9] H. Yin, K. Zaner, and T. Stossel, *J. Biol. Chem.* **255,** 9494 (1980).

[10] K. Burridge and J. Feramisco, *Nature (London)* **294,** 565 (1981).

[11] J. Condeelis, *in* "International Cell Biology" (H. G. Schweiger, ed.), p. 306. Springer-Verlag, Berlin, 1981.

[12] R. B. Taylor, W. Duffus, M. Raff, and S. De Petris, *Nature (London)* **1233,** 225 (1971).

[13] J. L. Salisbury, J. S. Condeelis, N. J. Maihle, and P. Satir, *Nature (London)* **294,** 163 (1981).

[14] J. Condeelis, *J. Cell Biol.* **80,** 751 (1979).

[15] J. L. Salisbury, *Proc. Annu. Meet. Electron Microsc. Soc. Am.* **39,** 528 (1981).

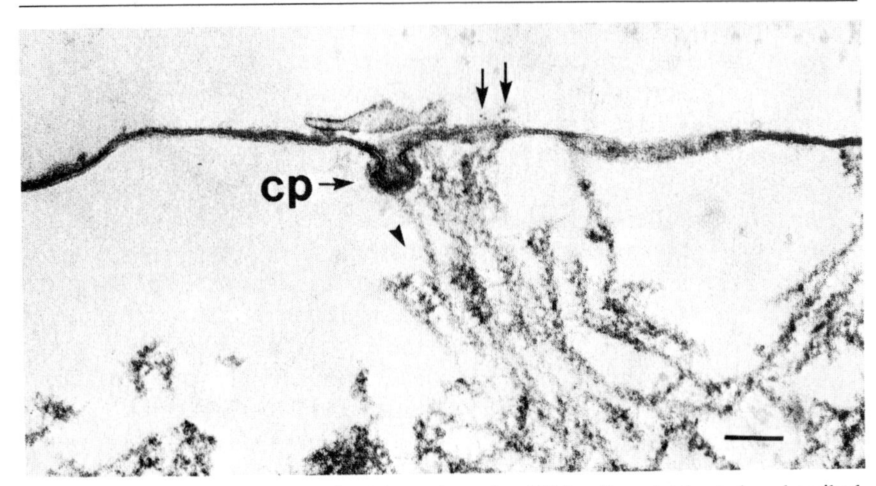

FIG. 1. Electron micrograph of a thin section of a WiL2 cell model treated as described under Materials and Methods. Microfilaments are observed in association with ligand–receptor complexes (small arrows) and coated membranes (cp). In all instances observed, the polarity of arrowhead decoration points away from the coated membrane, into the cytoplasm (arrowhead). Bar = 0.1 μm. From Salisbury *et al.*[1] with permission.

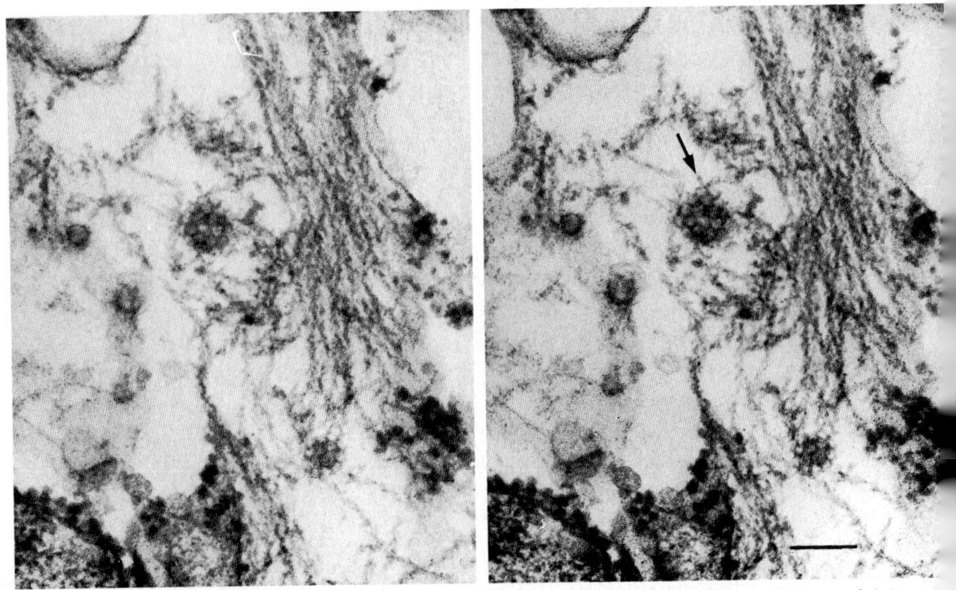

FIG. 2. Stereomicrographs of a thick sectin of a 3T3 L1 cell model. A coated pit (lower right) has several actin filaments associated with it. Another coated organelle can be seen (arrow) in this section. Bar = 0.1 μm.

Materials and Methods

Solutions

Extraction solution: 0.5 *M* KCl, 0.1 *M* phosphate buffer, pH 7.0
Standard salt solution: 0.1 *M* KCl, 5 m*M* $MgCl_2$, and 6 m*M* phosphate buffer, pH 7.0
Glassware wash solution: 0.5 *M* KCl
KCl, 2 *M*
Acetic acid, 1.0 *M*
Water: 10–15 liters, cold glass-double-distilled or deionized
Saturated ammonium sulfate: Add 780 g of Schwarz Mann Ultrapure ammonium sulfate in 1 liter of water and heat until dissolved. Add EDTA to make the solution 10 m*M* and cool. Adjust the pH to an apparent value of 8.2 (will read approximately pH 7.0 when the stock solution is diluted 1 to 10 with water). Caution: ammonium sulfate will poison certain pH electrodes.
Dialysis buffer: 0.5 *M* KCl, 10 m*M* PIPES, pH 7.0, 1 m*M* EDTA, and 5 m*M* 2-mercaptoethanol
Iodoacetic acid, 0.1 *M*, pH 6.5

Preparation of Myosin

Detailed methods for the preparation of myosin subfragments HMM or Sl by proteolytic digestion of myosin are outlined in a number of works.[16–23] We have found the following slight modification of these techniques[16,22] to produce S1 in good yield, which results in excellent actin-decorating morphology and good storage for extended periods of time.

Food is withdrawn from a small (5–7 kg) New Zealand white rabbit one day prior to the extraction. All buffers should be made with double-distilled or deionized water. All steps should be carried out at 4° unless otherwise noted. Exsanguinate the rabbit by cutting the jugular vein or kill by the method of choice. Quickly skin and dissect out back and hindleg muscles, rinse in phosphate-buffered saline (PBS) and place immediately on ice. Grind the muscles twice in a chilled standard meat grinder and weigh.

Extract the minced muscle for exactly 10 min using 3 volumes of

[16] W. W. Kelley and L. B. Bradley, *J. Biol. Chem.* **218,** 653 (1956).
[17] D. M. Young, S. Himmelfarb, and W. F. Harrington, *J. Biol. Chem.* **240,** 2428 (1965).
[18] S. Lowey, H. S. Slayter, A. G. Weeds, and H. Baker, *J. Mol. Biol.* **42,** 1 (1969).
[19] T. D. Pollard, E. Shelton, R. R. Weihing, and E. D. Korn, *J. Mol. Biol.* **50,** 91 (1970).
[20] R. Cooke, *Biochem. Biophys. Res. Commun.* **49,** 1021 (1971).
[21] J. A. Spudich and S. Watt, *J. Biol. Chem.* **246,** 4866 (1971).
[22] S. S. Margossian and S. Lowey, *J. Mol. Biol.* **74,** 313 (1973).
[23] A. Forer and O. Behnke, *J. Cell Sci.* **11,** 491 (1972).

extraction solution (0.5 *M* KCl, 0.1 *M* phosphate buffer, pH 7.0) with constant stirring at 4°. This is a high-salt extraction, intended to extract myosin, but actin extraction increases with time. Spin down the muscle residue in a GSA fixed-angle rotor (Sorval) at 12,000 rpm for 10 min. The supernate contains the myosin. The pellet can be stored frozen and used later for actin extraction.

Measure the volume of the myosin-containing supernate and adjust the pH to 6.6 with 1.0 *M* acetic acid, while stirring. Wash the pH meter electrode and any contaminated glassware with 0.5 *M* KCl to remove any adsorbed protein. It is important to perform the following steps with great care. Dilute the supernate *slowly,* with 10 volumes of cold glass double-distilled or deionized water while stirring to precipitate the myosin. Check the pH during this procedure and readjust to 6.6 if necessary. Stir for 10 min. Allow the myosin to settle for 1–2 hr; this time will vary depending on how slowly the water was added. Aspirate the supernate and centrifuge the sediment in a GSA rotor at 12,000 rpm for 10 min.

Resuspend the pellet in a small known (10–15 ml) volume of 2 *M* KCl. Measure the volume of the suspension and determine the volume of the pellet. Add more KCl to bring the solution to a concentration of 0.5 *M* KCl. (The pellet is 0.045 *M* KCl, so the volume of 2 *M* KCl to be added will be equal to 0.303 times the volume of the pellet.) Check the pH of the solution, and adjust to 6.7–6.8 with 1 *M* $NaHCO_3$. Carefully dilute the sample to 0.28 *M* KCl with cold glass-distilled water and constant stirring. Stir for 5 min. This will precipitate actomyosin. Centrifuge the precipitate in an SS 34 fixed-angle rotor (Sorval) at 19,000 rpm for 10 min and discard the pellet.

Measure the volume of the supernate and dilute it slowly with 6 volumes of cold glass-distilled water (to a final concentration of 0.04 *M* KCl) to precipitate the myosin. Centrifuge the precipitate in a GSA rotor at 12,000 rpm for 15 min, and discard the supernate. Dissolve the pellet in a small volume of 2 *M* KCl (10–15 ml), then measure the solution volume and determine the volume of the pellet. Add enough 2 *M* KCl to bring the final concentration to 0.5 *M* KCl as before.

Add saturated ammonium sulfate to the myosin preparation with constant stirring to bring the solution to 40% saturation. The volume of ammonium sulfate added will be two-thirds the volume of the myosin solution. Stir for 5 min and allow to stand for 10 min. Centrifuge out the precipitate in an SS 34 rotor at 13,000 rpm for 10 min and discard the pellet. Add saturated ammonium sulfate to bring the solution to 50% saturation with constant stirring. The amount of ammonium sulfate to add will be one-fifth the volume of the solution. Stir for 5 min. The myosin can be stored at 4° for several months in 50% ammonium sulfate.

To prepare the myosin for proteolytic digestion, centrifuge down the precipitate in an SS 34 rotor at 19,000 rpm for 10 min and resuspend the pellet in 2.5 pellet volumes of 0.5 *M* KCl. Dialyze the myosin against several changes of 0.5 *M* KCl, 10 m*M* PIPES, pH 7.0, 1 m*M* EDTA, and 5 m*M* 2-mercaptoethanol over a 24-hr period. Clarify the myosin solution by centrifugation in an SS 34 rotor at 15,000 rpm for 30 min. One should expect to obtain at this point approximately 10 ml of myosin, 25 mg/ml, if one started with about 300 g of muscle (myosin $E_{280}^{1\%} = 5.4$).[17] Adjust the myosin concentration to 20 mg/ml, and dialyze the clarified solution against 0.2 *M* ammonium acetate for 24 hr to precipitate the myosin in the form of thick filaments.

Preparation of Myosin Subfragment S1

Myosin subfragment S1 is prepared according to the procedure outlined by Margossian and Lowey.[22] Warm an aliquot of the myosin to room temperature and add papain (Worthington Biochemicals) made up in 30 m*M* cysteine, pH 7.5, to a final concentration of 0.05 mg/ml. Stir slowly for 8 min at room temperature. Add 0.01 ml of 0.1 *M* iodoacetic acid, pH 6.5, per milliliter of protein solution to stop the digestion, and return the solution to 4°. Centrifuge out the insoluble material, leaving myosin subfragment S1 in the supernatant.

The concentration of S1 can be determined spectrophotometrically[17] on the basis of an extinction coefficient $E_{280}^{1\%}$ of 7.7.

At this stage, myosin subfragment S1 can be filtered through a sterile 0.2-μm Millipore filter and stored for several months to one year in one part glycerol and one part standard salt solution containing 0.1 *M* KCl, 5 m*M* $MgCl_2$, and 6 m*M* phosphate buffer, pH 6.8, at −20°.[18,22,23]

Cell Culture and Ligand Challenge

WiL2 cultures[24] are grown in RPMI medium supplemented with 1 m*M* glutamine and 10% fetal calf serum (Flow Laboratories, Inc., Rockville, Maryland) at 37° in 5% CO_2 in air to 1.2×10^6 cells per milliliter. Cells are washed by centrifugation in Hanks' balanced salts (North American Biologicals, Inc., Miami, Florida) containing 5% fetal calf serum and 10 m*M* HEPES, pH 7.4 (0–4°), and incubated for 10 min with antibody (10 μl of stock per milliliter, Cappel Laboratories, Cochranville, Pennsylvania) against the cell surface antigen–receptor, integral membrane IgM. Cells are washed as above, labeled with a one-to-five dilution of ferritin-conjugated secondary antibody (Cappel Laboratories), washed again, and incubated at 37° for 15 min to allow them to begin processing the ligand along the coated vesicle pathway.

[24] J. A. Levy, M. Virolainen, and U. Defenoli, *Cancer* (*Philadelphia*) **22**, 517 (1968).

Cell Models and Actin Decoration

Cell models are made by extraction at 0–4° with 50% glycerol in a standard salt solution containing 0.1 *M* KCl, 5 m*M* $MgCl_2$, and 6 m*M* phosphate buffer, pH 7.0, and a protease inhibitor cocktail including 1 m*M* phenylmethylsulfonyl fluoride (Sigma), 20 μg of chymostatin (Sigma) and 1 mg of soybean trypsin inhibitor (Sigma) per milliliter for 4 hr as outlined by Ishikawa and co-workers.[25] The glycerol concentration is reduced to 5% by dilution with ice-cold standard salt solution over a period of 4–6 hr. Care must be taken to avoid disruption of the cell models when changing the glycerol concentration and throughout the actin-labeling procedure outlined below. This is achieved by using dilution and sedimentation at 1 *g* rather than centrifugation and pelleting of cells. Cells grown on coverslips can be treated identically.

Cell models are incubated with myosin subfragment S1 (2–4 mg/ml) in the standard salt solution for 12 hr at 4° and washed twice in the same solution without added S1. Control cells are incubated with S1 and 5 m*M* pyrophospate or treated with 2 m*M* ATP prior to and during the wash steps.

Electron Microscopy

Fix the cell models for 1 hr at room temperature in 2% glutaraldehyde in the standard salt solution containing 0.1% tannic acid[26]; wash in standard salt solution (buffer B); postfix for 20 min at 4° with 1% aqueous OsO_4 in phosphate buffer, pH 6.0; and, finally, wash in deionized water. Dehydrate the material through an ethanol series; stain en bloc with 1% uranyl acetate in the 25% ethanol step; and infiltrate with Epon 812. Thin (75 nm thick) sections can be made with a diamond knife and thick sections (250 nm) with a glass knife on a standard ultramicrotome. Poststain sections with uranyl actetate and lead citrate. The micrographs illustrated in this study were obtained with a JEOL 100 CX electron microscope operated at an accelerating voltage of 80 kV. Stereomicrographs are taken of thick-sectioned material in successive exposures with the specimen holder tilted through various tilt angles from the horizontal axis (usually between 6° and 15°). The micrograph pairs can be viewed through a stereoviewer to obtain a three-dimensional image.

Acknowledgment

The authors thank Dr. N. J. Maihle for careful reading of the manuscript and Drs. A. B. Novikoff, P. Satir, and J. Condeelis for their help with experimental protocols and support.

[25] H. Ishikawa, R. Bischoff, and H. Holtzer, *J. Cell Biol.* **43,** 312 (1969).

[26] D. A. Begg, R. Rodewald, and L. Rebhun, *J. Cell Biol.* **79,** 846 (1979).

This work was supported in part by U.S. Public Health Service Grants GM-29321 (to J. L. S.), CA-06576 (to A. B. N.), and GM-25813. J. L. S. is a Junior Faculty Research Award recipient from the American Cancer Society and G. A. K. is a Swiss National Science Foundation postdoctoral fellow.

Section V

Recycling of Plasma Membrane Proteins

[33] Use of Immunocytochemical Techniques in Studying the Biogenesis of Cell Surfaces in Polarized Epithelia

By HUBERT REGGIO, PAUL WEBSTER, and DANIEL LOUVARD

Several cultured cell lines retain *in vitro* many features of a differentiated epithelium.[1] In these polarized cells the plasma membrane is divided into two distinct domains that possess characteristic sets of proteins.[2-5] The two domains are separated from each other by the junctional complex, which prevents the passage of macromolecules into the intercellular space[6,7] and the diffusion of membrane proteins from one domain to the other.[8] Confluent monolayers display an electrical resistance that reflects the tightness of the epithelium formed by the cell type under consideration.[7] In certain cases the cultured epithelium can transport solutes from one face to the other.[9] In such a polarized culture only the apical surface is accessible for labeling with large probes such as antibodies. The leading edge of nonconfluent monolayers is in close contact with the substratum and also prevents the diffusion of macromolecules underneath the monolayer. This allows one to study the mechanism by which the apical membrane is established and maintained.

Endocytosis and/or recycling of apical membrane protein markers can be used to follow the intracellular pathway of internalized surface proteins. Their recycling provides useful information on the way membrane proteins are inserted into the apical plasma membrane and can be studied by light and electron microscopy using fluorescent antibodies.[4,5] In this chapter we describe the use of immunochemical techniques to monitor the antibody-induced endocytosis and recycling of apical membrane proteins. The methods used for immunofluorescence studies and the use of double labeling to distinguish between the probes that are internalized from those that

[1] M. J. Rindler, L. M. Chuman, L. Shaffer, and M. H. Saier, *J. Cell Biol.* **81,** 635 (1979).
[2] E. Rodriguez-Boulan and D. D. Sabatini, *Proc. Natl. Acad. Sci. U.S.A.* **75,** 5071 (1978).
[3] J. C. W. Richardson and N. L. Simmons, *FEBS Lett.* **105,** 201 (1979).
[4] D. Louvard, *Proc. Natl. Acad. Sci. U.S.A.* **77,** 4132 (1980).
[5] H. Reggio, E. Coudrier, and D. Louvard, *in* "Membrane Growth and Development" (J. F. Hoffman, G. H. Giebisch, and L. Bolis, eds.), p. 89. Liss, New York, 1982.
[6] M. Cereijido, I. Ehrenfeld, A. Meza, and A. J. Martinez-Palomo, *J. Membr. Biol.* **52,** 147 (1980).
[7] M. Cereijido, E. S. Robbins, W. J. Dolan, C. A. Rotunno, and D. D. Sabatini, *J. Cell Biol.* **77,** 853 (1978).
[8] M. Pisam and P. Ripoche, *J. Cell Biol.* **71,** 907 (1978).
[9] D. S. Misfeldt, S. T. Hamamoto, and D. R. Pitelka, *Proc. Natl. Acad. Sci. U.S.A.* **73,** 1212 (1976).

ISBN 0-12-181998-1

remain on the surface are described. The last section deals with electron microscopic techniques used for similar studies as well as for general immunochemical labeling of intracellular membranes or subcellular fractions. The methods were designed mainly to study polarity in Madin-Darby canine kidney cells (MDCK), but they should be applicable to a variety of cellular systems.

General Considerations for Immunolabeling Techniques

A vast literature exists on immunolabeling techniques.[10-21] Immunofluorescent methods have been reviewed by Wang *et al.*[21] The reader is referred to this excellent paper, since the same basic principles are followed here. Indirect labeling methods were used both for light and electron microscopy. The antibodies were always affinity-purified on antigens, coupled to activated Ultrogel (LKB) according to Ternynck and Avrameas.[22] Their specificity was tested by immunoprecipitation or Western blotting.[23] The quality of antibodies was routinely tested by immunofluorescence staining of thin frozen sections of a tissue in which the localization of the antigens is known.[24-28] This electron microscopic technique is described in this series.[28] It can be used extensively at the light level for preliminary characterization of antigens before electron microscopic localization.[5,29-32] Second antibod-

[10] U. Groschel-Stewart, *Int. Rev. Cytol.* **65,** 193 (1980).

[11] B. R. Brinkley, S. H. Fistel, J. M. Marcum, and R. L. Pardue, *Int. Rev. Cytol.* **63,** 59 (1980).

[12] D. L. Taylor and Y. L. Wang, *Nature (London)* **284,** 405 (1980).

[13] F. A. Pepe, *Int. Rev. Cytol.* **24,** 193 (1968).

[14] E. Lazarides and K. Weber, *Proc. Natl. Acad. Sci. U.S.A.* **71,** 2268 (1974).

[15] M. Goldman, "Fluorescent Antibody Methods." Academic Press, New York, 1968.

[16] R. C. Nairn, "Fluorescent Protein Tracing." Churchill-Livingstone, Edinburgh and London, 1976.

[17] J. H. Peters and A. H. Coon, *Methods Immunol. Immunochem.* **5,** 424 (1976).

[18] W. Hijmans and M. Schaeffer, eds., *Ann. N. Y. Acad. Sci.* **254** (1975).

[19] R. D. Goldman, T. D. Pollard, and R. Rosenbaum, eds., "Cell Motility," *Cold Spring Harbor Conf. Cell Proliferation* **3** (Book A-C) (1976).

[20] A. A. Thaer and M. Sernetz, "Fluorescence Techniques in Cell Biology." Dekker, New York, 1971.

[21] K. Wang, J. R. Feramisco, and J. F. Ash, this series, Vol. 85, p. 514.

[22] T. Ternynck and S. Avrameas, *Scand. J. Immunol. Suppl.* **3,** 29 (1976).

[23] W. N. Burnette, *Anal. Biochem.* **112,** 195 (1981).

[24] K. T. Tokuyasu, *J. Cell Biol.* **57,** 551 (1973).

[25] K. T. Tokuyasu, *J. Ultrastruct. Res.* **63,** 287 (1978).

[26] K. T. Tokuyasu and S. J. Singer, *J. Cell Biol.* **71,** 894 (1976).

[27] K. T. Tokuyasu, *Histochem. J.* **12,** 381 (1980).

[28] G. Griffiths, K. Simons, G. Warren, and K. T. Tokuyasu, this series, Vol. 96 [37].

[29] E. Coudrier, H. Reggio, and D. Louvard, *J. Mol. Biol.* **152,** 49 (1981).

[30] E. Coudrier, H. Reggio, and D. Louvard, *Cold Spring Harbor Symp. Quant. Biol.* **46,** 881 (1982).

[31] D. Louvard, H. Reggio, and G. Warren, *J. Cell Biol.* **92,** 92 (1982).

[32] H. Reggio, E. Coudrier, and D. Louvard, *Proc. Natl. Acad. Sci. U.S.A.* (in press).

ies were prepared and affinity-purified in our laboratory. They were conjugated to rhodamine,[33] fluorescein,[34] ferritin,[35] or Imposil.[36] In our hands they were superior to commercial conjugates, giving lower background and higher resolution. Peroxidase coupled to immunoglobulin G (IgG) was obtained from Institut Pasteur Production, Paris.

Use of Immunofluorescence Techniques to Monitor the Distribution of Proteins and Their Internalization in Polarized Epithelia in Culture

Materials and Reagents

MDCK cells: The cells are grown on coverslips until they are about 60% confluent. They are then reasonably flat, giving better resolution at the light microscopic level.

Formaldehyde fixative: A 3% w/v solution of paraformaldehyde in phosphate-buffered saline (PBS) is heated at 60° until the solution clears, at which point it is depolymerized. $CaCl_2$ and $MgCl_2$ are then added to a final concentration of 0.1 m*M*, and the solution is adjusted to pH 7.4. Frozen aliquots can be stored at −20° for up to 1 month.

PBS–gelatin: Dissolve 0.2% w/v gelatin in PBS without Ca^{2+} and Mg^{2+}. This solution is used to saturate the nonspecific binding sites on the cell surface before any exposure to antibodies. A rather crude gelatin preparation is used, since this saturates even more nonspecific sites (Sigma; from calfskin grade III). The solution can be stored at 4° if sterile.

Triton X-100, 0.1% (w/v) in PBS

Antibodies: Dilute to the required concentration in PBS–gelatin (usually 10–30 μg/ml). Centrifuge just before use at 10,000 *g* for 2 min, and discard any pellet (Beckman microfuge).

Hank's balanced salt solution

Incubation medium for endocytosis: Use minimum essential medium (MEM) buffered with 10 m*M* HEPES to pH 7.2. It is essential to use medium without serum, since this contains components that can interfere with the kinetics of endocytosis.

Mounting medium[37]: Six grams of analytical grade glycerol is placed in a 50-ml disposable plastic conical centrifuge tube; 2.4 g of Moviol 4-88 (Hoechst) is added and stirred thoroughly. Distilled water (6 ml) is then added, and the solution is left for 2 hr at room temperature. Then 12 ml

[33] P. Brandtzaeg, *Scand. J. Immunol.* **2,** 273 (1973).

[34] J. W. Goding, *J. Immunol. Methods* **13,** 215 (1976).

[35] Y. Kishida, B. R. Olsen, R. A. Berg, and D. J. Popskrop, *J. Cell Biol.* **64,** 331 (1975).

[36] A. H. Dutton, K. T. Tokuyasu, and S. J. Singer, *Proc. Natl. Acad. Sci. U.S.A.* **76,** 3392 (1979).

[37] G. V. Heimer and C. E. Taylor, *J. Clin. Pathol.* **27,** 254 (1974).

of 0.2 *M* Tris buffer, pH 8.5, is added and the solution is incubated in a water bath at 50° for 10 min with occasional stirring to dissolve the Moviol. The mixture is centrifuged at 5000 *g* for 15 min, aliquoted, and stored at −20°.

Microscope slides, equipped with spacers as described by Wang *et al.*[21] or as described later for electron microscopy

Distribution of Membrane Proteins

When MDCK cells are grown on glass coverslips, the apical membrane faces the medium and can be easily labeled with antibodies to proteins found specifically in this membrane (for instance, aminopeptidase). In contrast, the basolateral face is not accessible. It can be labeled, however, by permeabilization of the cells with a mild Triton treatment. This also reveals other intracellular antigens, such as those associated with the cytoskeleton[21,38] or with internal membranes.[4,39] By using a variety of conditions, one can label either the whole cell (inside and outside) or the apical surface exclusively. Internal membrane antigens can be distinguished from those of the apical surface by the use of double-labeling techniques.

Permeabilizing the Cells to Antibodies. All procedures are carried out at room temperature in the dishes where the cells were grown. The changes of solutions should be fast to avoid drying, which irreversibly damages the cells and produces high nonspecific binding. Pouring the solutions directly on the coverslips should be avoided, as this may detach the cells. Labeling is also done at room temperature, but at this step the coverslips are transferred onto spacers as described later.

1. Wash the cells three times for 5 min with PBS.
2. Replace by formaldehyde fixative for 20 min.
3. Wash twice for 5 min with PBS.
4. Quench the remaining aldehyde groups by incubating with PBS containing 50 m*M* NH_4Cl for 10 min.
5. Wash once for 5 min with PBS.
6. Replace by 0.1% Triton X-100 in PBS for 4 min.
7. Wash three times with PBS containing 0.2% gelatin.
8. Place 100 μl of the first antibody solution between the spacers of the microscope slide, which is placed on a flat support.
9. Invert the coverslips onto the antibody solution (i.e., cell side down) and incubate for 20 min.
10. Return coverslips to the dish in PBS (cell side up). Wash three times for 5 min.

[38] M. H. Heggeness, K. Wang, and S. J. Singer, *Proc. Natl. Acad. Sci. U.S.A.* **74,** 3883 (1977).
[39] J. F. Ash, D. Louvard, and S. J. Singer, *Proc. Natl. Acad. Sci. U.S.A.* **74,** 5584 (1977).

11. Repeat the labeling with the second antibody (go back to step 8).

12. Finally wash three times with PBS.

13. Wash the glass side of the coverslip with H_2O to eliminate salt precipitation that would perturb the observation, as it is made through the coverslip. To do this, hold the coverslip with tweezers and carefully rub the glass side with a wet Q-tip.

14. Mount the coverslip by inverting it onto one drop of mounting medium placed on a microscope slide. The slide can be looked at immediately. The medium becomes hard in less than 24 hr at room temperature. The slide can be stored for up to 1 year in a lightproof box at 4°.

If the edges of the coverslip are damaged during this procedure, only the center may be used for observation.

Labeling of the Plasma Membranes. The labeling is exclusively restricted to the apical membrane when the cells are not permeabilized. This is achieved by omitting steps 5 and 6 in the preceding protocol.

Simultaneous labeling of both the apical and basolateral plasma membranes can be performed by treatment of the monolayer with 2 m*M* EGTA in PBS without Ca^{2+} and Mg^{2+} for 5 min at 37° before fixation.[40,41] In MDCK cells this opens the junctional complex and allows the diffusion of antibodies into the intercellular space. The treatment is mild enough to prevent significant diffusion of membrane proteins within the plan of the membrane from one face to another. This possibility should be carefully controlled and adapted in each experimental conditions.

Differential Labeling of the Intracellular Membranes and of the Apical Plasma Membrane. The use of a double-labeling technique[21,38,39] makes it possible to visualize separately the same antigen at the apical plasma membrane and/or intracellularly. For this purpose one needs two independent sets of primary antibodies raised in different species. The apical plasma membrane is first incubated with the first antibody (for instance, guinea pig anti-aminopeptidase). The antibody should be used in excess to saturate all antigens on the apical surface. This surface labeling is achieved by using the preceding protocol, steps 1–4 and 7–10. The cells are then permeabilized with Triton and allowed to react with the second antibody (steps 6–9) (for instance, rabbit anti-aminopeptidase). The antibodies do not recognize the antigens at the apical surface, since these are already covered by the first set, but they recognize antigens eventually present in intracellular or basolateral membranes. These two sets of antibodies can now be labeled by fluorescent antibodies, each specific for one of these primary antibodies. These second

[40] E. J. Rodriguez-Boulan and M. Pendergast, *Cell* **20**, 45 (1980).

[41] K. Matlin, D. F. Bainton, M. Pesonen, D. Louvard, N. Genty, and K. Simons, *J. Cell Biol.* (in press).

antibodies are coupled to different probes (for instance, rhodamine goat anti-rabbit and fluorescein goat anti-guinea pig). Labeling with these second antibodies is carried out using the preceding protocol beginning at step 8. In the example here, observation with the rhodamine filter would show the apical distribution of aminopeptidase, and the fluorescein filter would reveal its intracellular distribution.

To prevent the effects of undesirable cross-reactivity, it is necessary to absorb each second antibody against purified IgG to which the other second antibody is directed. This is easily done by coupling the IgG to Ultrogel beads.[22] The diluted antibody is mixed with an appropriate amount of IgG-coupled beads and incubated for 10 min at room temperature; the immunocomplexes formed are sedimented. The supernatant can be used directly, and the two second antibodies can then be mixed together.

Dynamic Aspects: Endocytosis and Membrane Recycling

Rationale. The cells grown on glass coverslips are cooled to 0° to inhibit membrane traffic. They are labeled with fluorescent antibodies specific to apical membrane antigens. In this example, the cells are incubated at 4° with rabbit antibodies to aminopeptidase (R × AP) and then with rhodamine-conjugated goat anti-rabbit (Rh-G × R). The cells are then transferred to culture medium prewarmed to 37°, at which temperature the antibodies induce clustering of the immunocomplexes formed at 4°. These are rapidly endocytosed together with the fluorescent probes. At fixed times thereafter the cells are fixed with formaldehyde. The problem that arises then is to distinguish the markers that have been endocytosed from those still present at the surface of the cells. This can be solved by using double-labeling methods as described in Fig. 1. The immunocomplexes remaining on the apical surface can be detected with a third antibody raised against the second antibody and conjugated to another fluorescent probe (here fluorescein-R × G). During observation, what is labeled with both fluorescent probes is outside the cell; what is labeled only with the first probe (Rh-G × R) must have been endocytosed (Fig. 1b and c).

Labeling Living Cells to Study Endocytosis. Living cells are incubated on ice with antibodies against surface proteins. Great care should be taken to prevent any warming, as this could initiate antibody-induced clustering of the antigens. All washing solutions are kept on ice.

1. Place the cells in their incubation medium onto ice and let them cool slowly.
2. Wash cells twice for 5 min with cold PBS.
3. Replace with cold PBS–gelatin and leave for 5 min.
4. Place a microscope slide (equipped with spacers) on a stainless steel plate that is lying on ice.

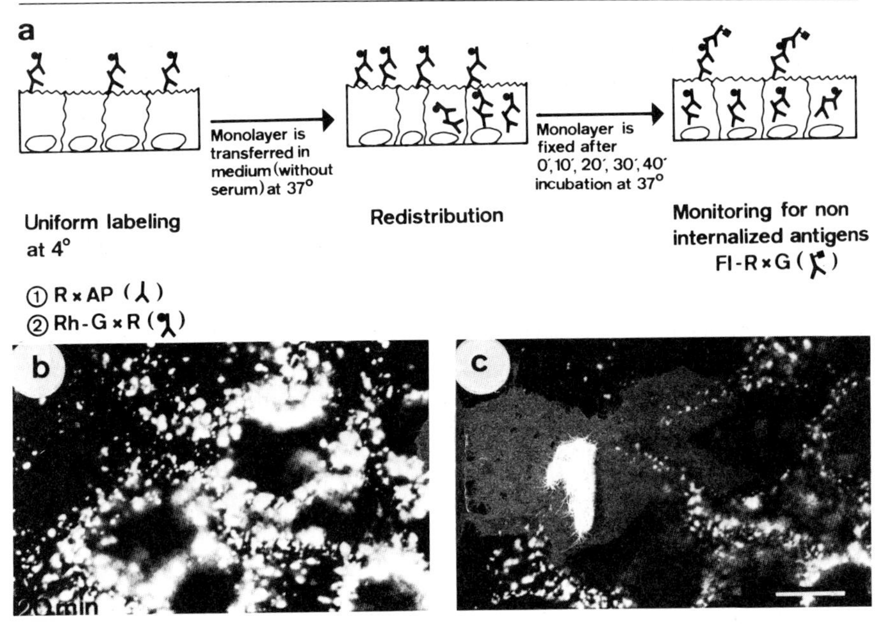

FIG. 1. Endocytosis induced by antibodies to aminopeptidase in MDCK cells. (a) Representation of the labeling steps of the apical plasma membrane before and after endocytosis. (b, c) Cells labeled as described in (a). After 20 min of endocytosis at 37° the two probes show striking differences in the labeling pattern; (b) observation with the rhodamine filter (surface + intracellular); (c) observation with the fluorescein filter (surface). ×900; bar = 10 μm.

5. Put 100 μl of the primary antibody mixture between spacers (e.g., R × Ap).

6. Invert the coverslips onto the solution (i.e., cell side down) and incubate for 20 min on ice.

7. Return coverslips to cold PBS and wash three times for 5 min with cold PBS.

8. Repeat the labeling with the second antibody (e.g., Rh-G × R). Go back to step 5.

9. Finally wash three times for 5 min with cold PBS.

10. Replace with cold Hank's solution for at least 5 min to restore metabolite levels in the cells. The cells are kept in cold Hank's solution until endocytosis is initiated.

Endocytosis

1. Transfer coverslips into serum-free medium prewarmed at 37° in a CO_2 incubator.

2. At required time points, transfer coverslips directly into formalde-

hyde fixative solution at room temperature for 20 min. All subsequent operations can be done now at room temperature.

3. Wash twice for 5 min with PBS.

4. Quench the free aldehyde groups using PBS containing 50 m*M* NH_4Cl for 10 min.

5. Wash for 5 min with PBS.

6. Stain with the third antibody (e.g., Fl-R × G) as previously described.

7. Wash three times for 5 min with PBS.

8. Mount the coverslips by inverting them (cells face down) onto one drop of mounting medium placed on a regular microscope slide.

Use of Immunochemical Techniques on Cultured Cell Monolayers for Ultrastructural Studies

The same basic principles described previously for immunofluorescence apply to electron microscopy, but specific problems are encountered leading to some adaptation of the techniques. After labeling, processing, and embedding, a relatively small number of cells are observed on a section. All precautions should be taken to ensure that these cells have been exposed to the antibodies under optimal conditions. This is done with the help of the slides described below. We have also designed a processing protocol that allows one to have many cell sheets on one electron microscopic section.

The staining of basolateral or internal membranes is also more difficult. Permeabilization of the cells with Triton is usually not satisfactory for ultrastructural studies, although it has been used for studies on cytoskeleton.[42] Considerable damage occurs to the membrane during this treatment. These are compatible with immunofluorescence studies but not with electron microscopic ones. In fact, major problems seem to arise during dehydration of the Triton-containing membranes with solvents. We describe here a preembedding technique which uses a milder detergent, saponin. The method is extremely sensitive and the morphology can be adequate.

Other methods have been described in the literature but will not be discussed here. Immunostaining of thin frozen sections is very powerful.[24-28] They are also recommended at the light microscopic level (immunofluorescence) to determine the optimal staining conditions for electron microscopy. Postembedding techniques are also useful in specific cases.[43-46]

[42] J. De Mey, M. Moeremans, G. Geuens, R. Nuydens, and M. de Brabander, *Cell Biol. Int. Rep.* **5,** 889 (1982).

[43] J. P. Kraehenbuhl, L. Racine, and J. D. Jamieson, *J. Cell Biol.* **72,** 406 (1977).

[44] J. P. Kraehenbuhl, L. Racine, and G. W. Griffiths, *Histochem. J.* **12,** 317 (1980).

[45] M. Ravazzola, A. Perrelet, J. Roth, and L. Orci, *Proc. Natl. Acad. Sci. U.S.A.* **78,** 566 (1981).

[46] J. Roth, M. Bendayan, and L. Orci, *J. Histochem. Cytochem.* **26,** 1074 (1978).

Materials and Reagents

Slides for electron microscopic immunolabeling. All the labeling procedures described below use simple, modified microscope slides similar to the one described by Wang *et al.*[21] Microscope slides, of tissue culture quality glass, have two 24 × 24 mm square glass coverslips attached to them with either silicon rubber glue or epoxy resin (Fig. 2). The space between the coverslips is approximately 15 mm. The cells are grown on these slides face up. Only the center well is used for the labeling. This minimizes the amount of antibody necessary and precisely delineates the area of the monolayer exposed to antibodies that will be collected. For this reason it is necessary that the spacer coverslips have the same width as the microscope slide.

Fixative for electron microscopy. In the author's hands the best preservation for MDCK cells is obtained using a modified Karnovsky[47] fixative, 2% (w/v) formaldehyde, 4.5% glutaraldehyde (Ladd, Burlington, Vermont), in 0.1 *M* cacodylate buffer, pH 7.4. The cells are fixed for 1 hr at room temperature in a Coplin jar. It is essential to prevent rapid temperature variation during the first few minutes of the fixation, since this can result in considerable damage to the monolayer and junctional complexes. If for some reason the cells are in the cold they should be immersed in cold fixative, which is then left to warm up slowly.

Immunolabeling of Cell-Surface Antigens

Labeling. The same steps described earlier for immunofluorescence labeling of the apical cell surface are used, but a second fixation including glutaraldehyde is necessary before processing the cells for electron microscopy. During the immunolabeling the cells should be handled with great care, since they are just lightly fixed with formaldehyde. This is relatively

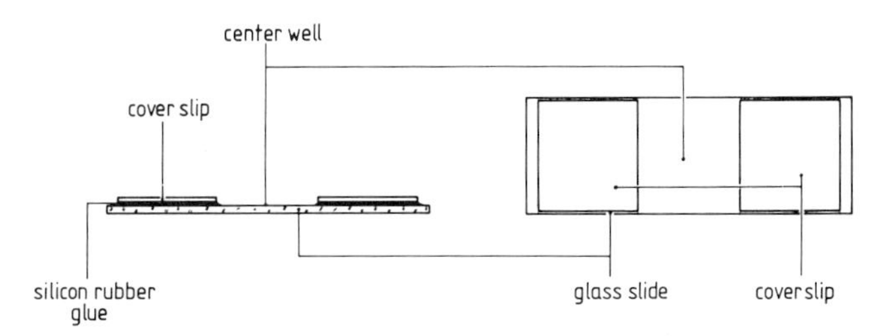

FIG. 2. Modified glass microscope slides used to grow the cells for immunolabeling in electron microscopic studies.

[47] M. J. Karnovsky, *J. Cell Biol.* **27**, 137A (1965).

easy, since all washes and fixations are done by transferring the slides from one Coplin jar to another.

The labeling is done on the slides themselves as follows. Rapidly draw off excess buffer from the slide by touching the edge of the slide with filter paper and place it in a wet chamber. Add approximately 100 μl of the antibody solution to the center well and cover it with a coverslip to prevent drying. To mix the antibody solution with the layer of buffer covering the cells, quickly, but carefully, raise and replace the coverslip. It is essential that the entire center well surface be well covered with the antibody solution and bubble-free. A colored solution provides a control so that one can visualize the mixing process.

Postfixation and Processing. The cells are postfixed with modified Karnovsky fixative or with 4.5% glutaraldehyde to prevent damage during further processing for electron microscopy. They are then washed three times for 5 min in cacodylate buffer and detached from the slide with a Teflon policeman made by cutting a beveled edge on a small piece of Teflon held in an hemostatic clamp (Fig. 3). To do this, rapidly drain the slide and place it on a stable horizontal support (e.g., a 35-mm Petri dish). Holding the Teflon policeman almost perpendicular to the slide, gently push the cells

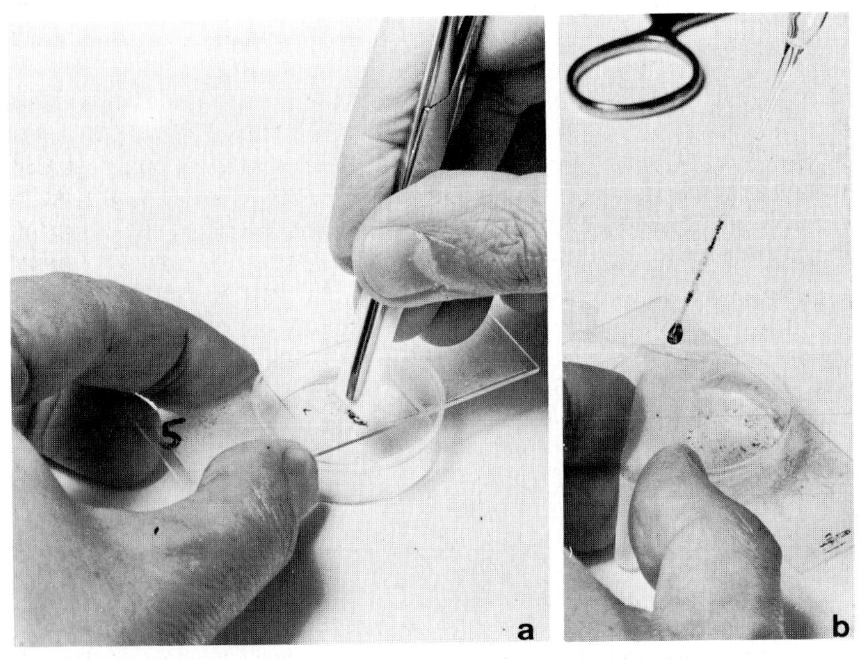

FIG. 3. Methods used to scrape the cell from the glass slide. (a) This shows the position of the Teflon policeman during scraping; the pressure should be vertical, and the policeman should not bend. (b) transfer to the Eppendorf tube. In this particular case the cells are clearly visible, since they were allowed to react with DAB for immunoperoxidase labeling.

toward the center of the well. Vertical pressure works better than trying to scrape the cells from underneath (Fig. 3a). With a little practice it is possible to detach the monolayer without damaging the basal membrane. Scrape only the center well, without drying the pile of cells. With a Pasteur pipette, transfer the cells to an Eppendorf centrifuge tube containing about 200 μl of cacodylate buffer. Avoid drawing the cells up into the Pasteur pipette, as occasionally they stick to the glass, but retain them in a drop at the tip of the pipette (Fig. 3b). Immerse the cells directly into the buffer trying not to touch the side of the tube, as they may stick here as well. The cells are then centrifuged horizontally at about 10,000 *g* for 5 min in a Beckman microfuge.

The buffer is replaced by osmium fixative in cacodylate buffer, and the pellet is left on ice for at least 30 min. It is then carefully detached with a needle to allow osmium penetration underneath and left in suspension for another 30 min or more according to its size. After osmium treatment the pellet is harder and can be processed for electron microscopy according to conventional techniques.

Staining en bloc[48] can be performed by soaking the pellet overnight in 0.5% Mg, uranyl acetate in H_2O. Care should be taken to wash the pellet extensively with H_2O before or after the treatment. In our hands this improves membrane morphology and does not alter the ferritin, in contrast to what is usually believed.

During dehydration, while in 95% ethanol, the pellet is cut in two with a scalpel and into small strips perpendicular to the first cut.[49] It may fall apart a little at this stage, but enough pieces will be recovered. Infiltration should be rather extensive, otherwise the blocks are difficult to cut; 50% Epon in propylene oxide is used at room temperature for 12 hr on a spinning wheel, then replaced by 100% Epon. The strips are left to spin for another 4–6 hr leaving off the caps. Embedding is carried out in a flat mold.

The advantage of this procedure is that the cell monolayers pile up during the centrifugation and form a series of parallel sheets. The orientation is very easy to recognize under a dissecting microscope during or after embedding. One can cut either parallel or perpendicular to the monolayers. In this case it is possible to obtain many cell layers on a section and thus facilitate the sampling and observation (Fig. 4).

Immunolabeling of Basolateral or Intracellular Membranes Using Saponin to Permeabilize the Cells

The same two-step fixation can be used to stain intracellular membranes. Saponin is a mild detergent that opens the cells with a reasonable

[48] M. G. Farquhar and G. E. Palade, *J. Cell Biol.* **26,** 263 (1965).
[49] A. Tartakoff and J. D. Jamieson, this series, Vol. 31, p. 41.

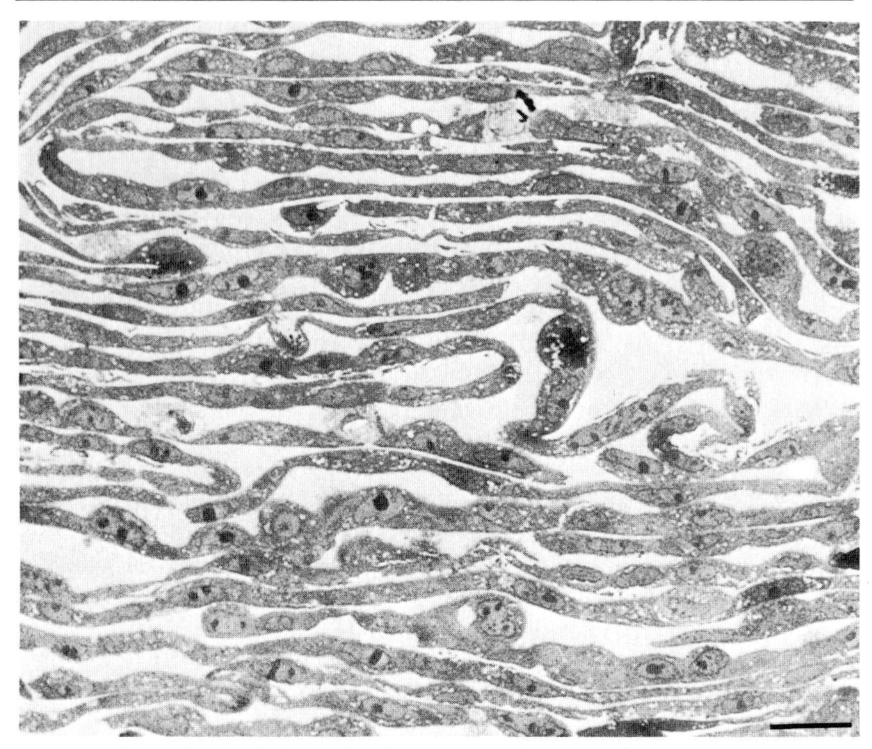

FIG. 4. Light micrograph of the pellet formed by MDCK cells during processing for electron microscopy. Many layers from different parts of the monolayer are seen in cross section. ×370; bar = 30 μm.

preservation of the morphology if the first fixation is carefully controlled.[50] In this case a small amount of glutaraldehyde is necessary to preserve the morphology, since the labeling conditions are more drastic. In fact, the glutaraldehyde fixation is rather critical in this technique, since a small extraction of cytosolic proteins is necessary to allow the penetration of the antibodies into the cells. Too much glutaraldehyde causes too much cross-linking of the cytosol and a poor penetration of the antibodies. Too little glutaraldehyde produces a poor morphology. Other fixation conditions have been used with saponin,[51] but in our hands they were not as satisfactory. These techniques have been discussed by Tougard *et al.*,[52] who combined them with immunoperoxidase staining. This greatly improved the results, since Fab-coupled peroxidase diffused more easily than electron-dense

[50] I. Ohtsuki, R. M. Manzi, G. E. Palade, and J. D. Jamieson *Biol. Cell.* **31,** 119 (1978).
[51] M. C. Willingham and S. S. Yamada, *J. Histochem. Cytochem.* **27,** 947 (1979).
[52] C. Tougard, R. Picart, and A. Tixier-Vidal, *Am. J. Anat.* **158,** 471 (1980).

probes, such as ferritin or gold particles. The method as described here uses the same glass slides as described above.

Reagents for Peroxidase Reaction[53]

Solution A: Dissolve 0.2 g of diaminobenzidine tetrahydrochloride (DAB) in 80 ml of distilled water (use hydrochloride salt, otherwise it is very difficult to dissolve). Add 100 ml of 0.1 *M* unbuffered Trizma base. Adjust pH to 7.6 with 10 *N* HCl. Make up to 200 ml with distilled water. Filter the solution. The red precipitate is very fine, and it usually needs filtering off first with a paper filter before using a Millipore filter (0.45 μm). The final solution should be straw colored, not red. Divide this solution into two parts. Take 100 ml and keep in a dark place until needed. This is solution A.

Solution B: Take 1.5 ml of 30% hydrogen peroxide (H_2O_2), and make up to 50 ml with distilled water. Add 1 ml of this solution to 100 ml of the remaining DAB solution.

Permeabilization Procedure

1. Wash cells in PBS and fix for 1 hr at room temperature with 2% formaldehyde, 0.05% glutaraldehyde in 100 m*M* phosphate buffer, pH 7.4.
2. Quench aldehydes by washing three times for 5 min in PBS containing 50 m*M* NH_4Cl.
3. Soak the cells in PBS containing 0.2% gelatin and 0.05% saponin for 30 min.
4. Treat the cells with the first antibody, diluted with the PBS–gelatin–saponin solution for 90 min.
5. Remove the first antibody and wash three times for 5 min in the PBS–gelatin–saponin solution.
6. Apply the second antibody. This is an Fab-peroxidase fragment directed against the first antibody. Leave for 90 min.
7. Wash three times for 5 min with PBS–gelatin–saponin.
8. Wash three times for 5 min with 0.1 *M* sodium cacodylate buffer, pH 7.4 to remove saponin.
9. Postfix for 1 hr in 4.5% glutaraldehyde in the same buffer.
10. Wash three times for 5 min with sodium cacodylate buffer and with 0.1 *M* Tris-HCl buffer, pH 7.6.
11. Peroxidase reaction[53]: Use DAB solution A for 30 min in the dark. Follow by DAB solution B containing H_2O_2 for 30 min in the dark.

[53] R. C. Graham and M. J. Karnowsky, *J. Histochem. Cytochem.* **14,** 291 (1966).

12. Wash three times with Tris-HCl, pH 7.6, then with 0.1 *M* sodium cacodylate buffer, pH 7; scrape and pellet the cells, then osmicate, dehydrate, and embed as previously.

The cells are observed without staining or after a short exposure to lead citrate. The use of reduced osmium[54,55] improves the visualization of the membranes. When the cell pellet is infused with osmium (as described previously) it is transferred to a 1% OsO_4 solution containing 5 mg of potassium ferrocyanide per milliliter for 15 min. It is then washed in pure 1% osmium for at least 2 min before continuing conventional processing.[55]

Remarks

1. These conditions were optimal for labeling different cell lines with antibodies to organelles[31] or clathrin (Fig. 5). They have to be adapted to

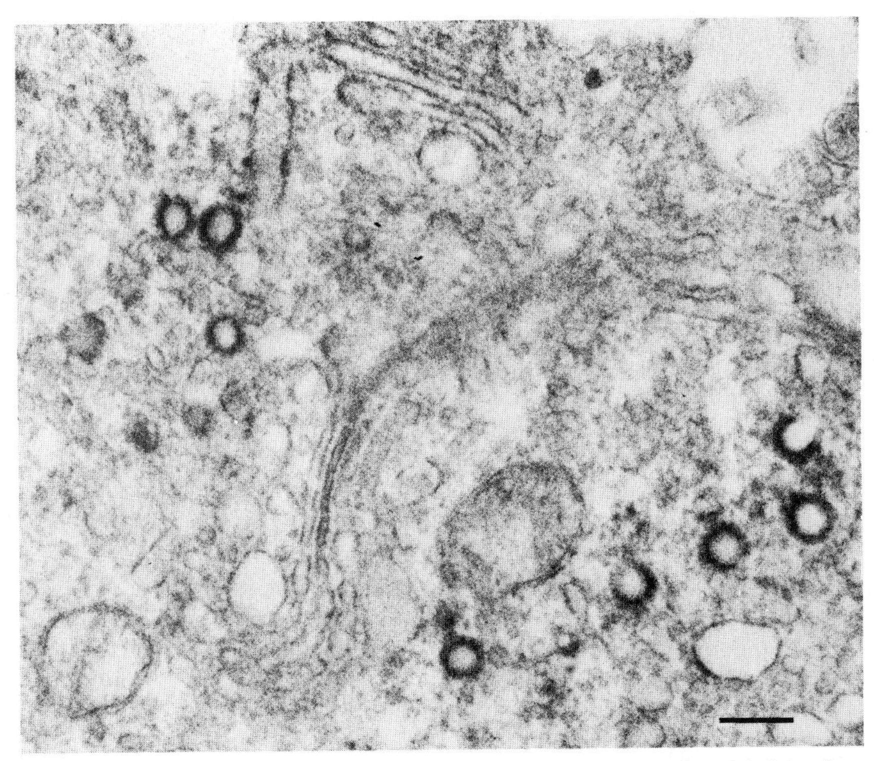

FIG. 5. Immunoperoxidase labeling of the Golgi complex using antibodies to clathrin after permeabilizing the cells with saponin. ×5100; bar = 0.2 μm. From D. Louvard *et al., EMBO J.* (in press).

[54] M. J. Karnowsky, *Meet. Am. Soc. Cell Biol.,* Abstract 284 (1971).
[55] V. Herzog and F. Miller, *Eur. J. Cell Biol.* **19,** 203 (1979).

each cell type and to the subcellular structure of interest. Since saponin presumably binds to cholesterol, the different intracellular membranes are permeabilized to different extents. With mild conditions lysosomes are opened first; Golgi elements need more drastic conditions; rough endoplasmic reticulum is even more difficult to label. Variables are glutaraldehyde concentration, saponin concentration, time of labeling, osmolarity during fixation [hypoosmolarity during fixation facilitates penetration].[50] These variables are rapidly determined at the light microscopic level by comparing the labeling with fluorescent antibodies of Triton opened cells to that of cells treated with saponin in various conditions.

2. A major problem of immunoperoxidase techniques is the production of diffusion artifacts due to the diffusion either of the antigens or of the DAB reaction products. These must be recognized as described later.

3. There is great variability in the technique, and only a fraction of the cells is successfully labeled. We recommend using several fixation conditions in parallel: e.g., 0.05%, 0.03%, 0.01%, and even 0% glutaraldehyde (w/v). This allows one to assess the differential permeability of different intracellular organelles and the diffusion artifacts.

Dynamic Aspects: Endocytosis

Endocytosis and membrane recycling may be followed at the electron microscopic level using electron-dense probes coupled to second antibodies. The method follows exactly the same steps described for immunofluorescence. The labeling is carried out on the modified microscope slides. The final fixation is made in Karnovsky's modified fixative, taking care to avoid temperature shock. The processing is done as described above for surface staining.

It is easy to see whether the antigens detected are inside or outside the cells. However, depending on the plane of the section, electron-dense probes located in invaginations of the plasma membrane may appear to be in a vesicle. These two possibilities can be distinguished by using electron-dense tracer on fixed cells.[56]

Another problem arises because of the size of the electron-dense probe used. The cross-linking of the antibodies on the cell surface induces the formation of patches that are endocytosed. If these patches also contain many large particles, such as ferritin molecules, it may seriously perturb this endocytosis. In our experience, this induces two phenomena: (*a*) more of the immunocomplexes are shed to the lysosomes and escape recycling; (*b*) some

[56] D. A. Wall, G. Wilson, and A. Hubbard, *Cell* **21**, 79 (1980).

immunocomplexes are too big to be internalized by the cells. This may be demonstrated by experiments using double labeling. The cells are first labeled with a primary antibody and subsequently with a second antibody coupled to ferritin. The monolayer is then transferred to serum-free medium at 37° just as described previously. After 30 min of endocytosis, they are cooled again to 0° and labeled with an antibody anti-ferritin coupled to Imposil molecules. After another transfer to prewarmed incubation medium for different time periods, they are fixed and processed for electron microscopy. It is then observed that the Imposil molecules, associated with the ferritin, do not enter the cells. This problem can be overcome by reducing the antibody concentration to the lower limit compatible with specific observations. This diminishes the extent of cross-linking possible, and thus the size of the patches. Another possibility is to use ferritin–avidin with a biotinylated primary antibody. This also reduced the size of the aggregates.[57]

Technique for Mounting Subcellular Fractions on Glass Slides Prior to Antibody Staining for Electron Microscopy

The same modified microscope slides and the same methods are also very useful in labeling subcellular or membrane fractions. The center well of the slides is covered with a mixture of agarose and gelatin, and the subcellular fraction is adsorbed to this layer. The labeling is performed as discussed above, since the adsorbed material is now comparable to a monolayer. The technique is much easier to use than the centrifugation–resuspension methods, which often induce the formation of large aggregates and clumps that are difficult to wash and are damaged during resuspension. In the present case the labeling is carried out on a thin layer of the subcellular fraction that is easily washed by immersing the slides in Coplin jars. This gives a better preservation of morphology and a lower background. Another advantage of the technique is again the small amounts of antibody and of the fraction needed.

Gelatin (0.1 g) and agarose (0.5 g) are dissolved in 100 ml of H_2O by heating to 90°. A small layer of the hot agarose–gelatin mix is spread over the center well part of the slide, then left, covered, to cool and dry on a perfectly level surface. When dry the agarose–gelatin sheet is seen as a cloudy covering on the slide. Thicker sheets are easily formed by spreading too much mixture on the slide, but these are less useful, as the increased thickness of the sheet in the final block means less tissue per section. When dry, the slides are immersed in a solution of 0.01% polylysine for at least 1 hr, then removed, quickly rinsed with distilled H_2O, and left to air dry.

[57] H. Reggio and D. Louvard, unpublished observations.

To load the slides, the sheets are first rewetted with the appropriate buffer (i.e., the one containing the cell fractions) and overlaid with about 100 μl of the fraction suspension, covered with a coverslip to prevent drying, and left to settle for 30 min to 1 hr (at room temperature in a wet chamber, or in the cold if the preparation requires this). Since adsorption is never complete, the concentration of the fraction must be rather high to produce a sufficient covering of the agarose–gelatin sheet. A 1 : 1 dilution of a loose pellet is a good starting concentration, but this has to be adjusted experimentally. Then the slide is gently flooded with fixative, left for 5 min, and immersed in a larger amount of fixative for 25 min. Occasionally one sees a part of the fraction floating on the fixative. A suitable fixative is 2% formaldehyde with 0.05% glutaraldehyde in 0.1 *M* phosphate buffer. The slides are ready for staining after they have had three washes in PBS containing 50 m*M* NH_4Cl (to quench the fixative). Staining and processing are performed as described above.

Acknowledgments

We would like to thank B. d'Arcy, A. McDowall, and S. Robinson for excellent technical assistance; E. Coudrier and K. Soderberg for helpful advice; D. Meyer, J. Siegler, and G. Warren, for reading the manuscript; and A. Steiner for typing it.

[34] Use of Antibody to 5′-Nucleotidase as a Marker to Study Membrane Flow During Pinocytosis

By CHRISTOPHER C. WIDNELL

There is now considerable evidence that plasma membrane that enters the cell as a result of fluid-phase pinocytosis, phagocytosis, or receptor-mediated endocytosis is not degraded, but, instead, reutilized.[1-5] In addition, the rate of recycling of receptors seems to be relatively rapid, especially when compared to the rate of membrane flow in fluid-phase pinocytosis,[1,3] suggesting that different plasma membrane proteins may be internalized and return to the surface at different rates. Since this could complicate the interpretation of results obtained by studying a large number of membrane

[1] R. M. Steinman, S. E. Brodie, and Z. A. Cohn, *J. Cell Biol.* **68,** 665 (1976).
[2] S. C. Silverman, R. M. Steinman, and Z. A. Cohn, *Annu. Rev. Biochem.* **46,** 699 (1977).
[3] J. L. Goldstein, R. G. Anderson, and M. A. Brown, *Nature* (*London*) **279,** 679 (1979).
[4] Y.-J. Schneider, P. Tulkens, C. de Duve, and A. Trouet, *J. Cell Biol.* **82,** 466 (1979).
[5] W. A. Muller, R. M. Steinman, and Z. A. Cohn, *J. Cell Biol.* **86,** 304 (1980).

ISBN 0-12-181998-1

proteins,[4,5] there are obvious advantages in analyzing the internalization of a single membrane protein. The potential disadvantage is that the analysis of a protein that is not a receptor must be correlated with a specific aspect of endocytosis.

The enzyme 5′-nucleotidase is localized in the plasma membrane of most cells,[6] is antigenically similar in different tissues,[7] and is, apparently, an integral membrane protein.[8] The kinetics of internalization and return to the cell surface correlate reasonably well with the rate of fluid-phase pinocytosis,[9,10] and the binding of antibody to the enzyme on the surface of cells does not seem to provoke its internalization.[9] These findings suggest that the enzyme and its antibody represent a useful marker for the membrane events associated with plasma membrane flow in fluid-phase pinocytosis.

Purification of 5′-Nucleotidase and Preparation of Antisera

The procedure for the purification of 5′-nucleotidase as a lipoprotein complex has been described in detail.[11] More recent techniques, which involve affinity chromatography, have resulted in the successful isolation of the enzyme in a homogeneous and soluble form.[12-14] Unfortunately, the soluble form of the enzyme has been a less consistent and potent antigen than the lipoprotein complex, so that the latter preparation remains the antigen of choice. Even though most purified fractions contain traces of impurities, the antisera obtained exhibit a precise correlation between antibody binding to intact cells and inhibition of 5′-nucleotidase activity,[9] and affinity techniques (described below) permit the isolation of antibody that is at least as pure as the antigen.

For immunization, purified 5′-nucleotidase (step 8 in Widnell[11]) is sedimented by centrifugation at 100,000 g for 1 hr; the enzyme is resuspended in 0.15 M NaCl so that the protein concentration is 1 mg/ml, and 0.5 ml is emulsified with 0.5 ml of complete Freund's adjuvant. The antigen is administered to rabbits: 0.3 ml to each Achilles tendon and 0.1 ml subcutaneously at four different sites along the back. Antibody production is determined, initially, by comparing the inhibition of 5′-nucleotidase activ-

[6] W. H. Evans, "Laboratory Techniques in Biochemistry and Molecular Biology," Vol. 7, Part 1. Elsevier/North-Holland, Amsterdam, 1978.
[7] B. L. Riemer and C. C. Widnell, *Arch. Biochem. Biophys.* **171,** 343 (1975).
[8] E. M. Merisko, G. K. Ojakian, and C. C. Widnell, *J. Biol. Chem.* **256,** 1983 (1981).
[9] C. C. Widnell, Y.-J. Schneider, B. Pierre, P. Baudhuin, and A. Trouet, *Cell* **28,** 61 (1982).
[10] D. K. Wilcox, R. P. Kitson, and C. C. Widnell, *J. Cell Biol.* **92,** 859 (1982).
[11] C. C. Widnell, this series, Vol. 32, p. 368.
[12] J. Dornand, J.-C. Bonnafous, and J.-C. Mani, *Eur. J. Biochem.* **87,** 459 (1978).
[13] Y. Naito and J. M. Lowenstein, *Biochemistry* **20,** 5188 (1981).
[14] G. K. Duker and C. C. Widnell, *Fed. Proc., Fed. Am. Soc. Exp. Biol.* **41,** 917 (1982).

ity by serum obtained before and after immunization as described below. Inhibitory activity is evident after 2–3 weeks and declines after 6–8 weeks, at which stage booster injections (100 μg of enzyme emulsified with incomplete Freund's adjuvant, administered intramuscularly into each thigh) are given three times at intervals of 3 weeks. Animals are bled 1 week after the final injection and at regular intervals thereafter until the antibody titer falls.

Two procedures are employed for the assay of antibody. Since every antiserum obtained has inhibited 5′-nucleotidase activity, routine estimations are performed by this technique. Solubilized 5′-nucleotidase (extract; see step 2 in Widnell[11]) is mixed with antiserum (20 μg of extract protein and 1–5 μl of antiserum, diluted as necessary) in a final volume of 0.1 ml in 0.05 *M* Tris-HCl, pH 7.5. After 30 min at 25°, 5′-nucleotidase activity is determined at 37°.[11] Antiserum concentrations are adjusted so that the inhibition ranges from 10 to 60% and the titer is determined as inhibitory units per milliliter of serum (1 unit inhibits 1 unit of 5′-nucleotidase activity).

To determine precisely the antibody concentration in the antiserum, antibody is adsorbed to purified 5′-nucleotidase. The purified enzyme (step 8 in Widnell[11]) is added to antiserum in an amount (determined as above) such that 50% of the activity is inhibited; the solution is mixed by rotation overnight at 2°, and the antigen–antibody complex is sedimented by centrifugation at 100,000 *g* for 30 min. The pellet is washed at room temperature by resuspension and resedimentation in 0.5 *M* NaCl until the A_{280} of the supernatant is less than 0.05. The antibody is then extracted at room temperature by suspending the pellet in 0.5 *M* NaCl, 0.25 *M* acetic acid (0.5 ml per milliliter of antiserum), the enzyme is removed by centrifugation at 100,000 *g* for 15 min, and the extract is neutralized by the addition of solid $NaHCO_3$. The extract is dialyzed against 0.15 *M* NaCl for 4 hr at room temperature, and the antibody concentration is determined from the A_{280}. Analysis of the protein by polyacrylamide gel electrophoresis in sodium dodecyl sulfate (SDS)[15] generally indicates that the antibody is at least 90% immunoglobulin G (IgG). A typical antiserum analyzed in this manner is found to contain 0.5–1.5 mg of anti-5′-nucleotidase IgG per milliliter. It should be noted that conventional immunodiffusion techniques cannot be applied to the analysis of the antiserum, since the antigen is insoluble.

In some experiments a second antibody is desirable for tracing the fate of anti-5′-nucleotidase bound to cells. Since the Fc region of IgG is particularly susceptible to the action of lysosomal proteases,[16] the antibody of choice is prepared in sheep against $F(ab')_2$ fragments of rabbit IgG. Although these

[15] K. Weber and M. Osborn, *J. Biol. Chem.* **244**, 4406 (1969).

[16] Y.-J. Schneider and A. Trouet, *Eur. J. Biochem.* **118**, 33 (1981).

antibodies may be obtained commercially, it is more economical to prepare them in the laboratory if large amounts are required. Consistently satisfactory results have been obtained with the procedure of Kraehenbuhl *et al.*[17]

Preparation of Labeled Antibodies

It is essential to use iodination as the labeling procedure when antibodies with very high specific activity are required. However, this technique yields preparations with a limited half-life, which are also susceptible to radiochemical decomposition. In many experiments, in particular those described below, such high specific activities are not required and a more stable reagent is desirable. Antibodies labeled with [^{3}H]acetic anhydride are thus very satisfactory alternatives.

Before labeling, antibodies are affinity-purified; the procedure for anti-5′-nucleotidase is described above, and the procedure for sheep anti-rabbit F(ab′)^2 is as described in Kraehenbuhl *et al.*[17] except that the rabbit F(ab′)$_2$ is coupled to Sepharose using a water-soluble carbodiimide,[18] and the antibody is eluted with 0.5 *M* NaCl, 0.25 *M* acetic acid. In experiments where F(ab′)$_2$ fragments are to be labeled, the IgG is isolated by the affinity technique and then converted to F(ab′)$_2$.[17]

The labeling is carried out by a modification of the procedure of Fraenkel-Conrat.[19] The antibody (10 mg) is concentrated to 0.5 ml by ultrafiltration, dialyzed against 0.1 *M* sodium phosphate, pH 7.3, overnight, mixed with 0.5 ml of saturated sodium acetate, and stirred magnetically. A solution of [^{3}H]acetic anhydride ($\sim$ 5 Ci/mmol) in benzene ($\sim$ 20 mCi in 50 μl) is added in aliquots of 5 μl at 10-min intervals, and stirring is continued for 30 min after the last addition. The rate of stirring is adjusted so that the benzene forms minute droplets in the aqueous phase and the temperature is maintained at 8 – 10°. The procedure is carried out in a hood with appropriate precautions for radiation safety, and preparations involving 100 mCi have been carried out safely. The aqueous phase is dialyzed against two changes of 0.1 *M* sodium phosphate, pH 7.3, to remove most of the unconjugated radioactivity, and the antibody is purified by chromatography on Sephadex G-150 to separate aggregates that form during the conjugation. It is also possible to repurify the antibody by the affinity techniques at this stage; it is necessary, however, only for occasional preparations that exhibit unusually high nonspecific binding. This procedure results in the incorporation of $\sim$ 2 mol of [^{3}H]acetate per mole of antibody, with a typical specific activity of 2×10^5 dpm/μg. It should be noted that the use of toluene instead

[17] J. P. Kraehenbuhl, L. Racine, and J. D. Jamieson, *J. Cell Biol.* **72,** 406 (1977).

[18] A. Ito and G. E. Palade, *J. Cell Biol.* **79,** 590 (1978).

[19] H. Fraenkel-Conrat, this series, Vol. 4, p. 247.

of benzene to dissolve the acetic anhydride has usually yielded antibodies with a lower binding affinity.

Assay of Surface and Total 5'-Nucleotidase Activity

The procedure, which has been described in outline,[9] determines first the release of P_i from 5'-AMP by intact cells on the culture dish, and then the release of P_i by solubilized cell extracts. The procedure is applicable in principle to any cell in culture and has been applied in this laboratory to fibroblasts and primary cultures of mouse peritoneal macrophages and rat hepatocytes. The procedure has been developed so that cells may be assayed for surface 5'-nucleotidase activity and then returned to culture without any ill effects. Even though the assay is performed at a pH below that of the optimum fc · activity,[11] results obtained by this procedure are directly proportional to V_{max}. The assay is performed at room temperature to avoid complications resulting from culture dishes floating in water baths; the activity at 20–25° is directly proportional to that determined at 37°.[9] The method described is for use with 35-mm culture dishes, but may be applied to dishes of any size by appropriate changes in the volumes used; it is obviously essential that the cells be completely covered with medium at all stages during the assay.

The cells are removed from the incubator, and the medium is aspirated immediately. The cells are then rinsed three times with HEPES–saline (10 m*M* HEPES, pH 7.5, 0.15 *M* NaCl, 2.5 m*M* KCl), and care is taken to wipe any medium (which contains P_i and therefore interferes with P_i assays) from the lid of the dish. The assay medium (1.2 ml of HEPES–saline containing 1 m*M* 5'-AMP and 1 m*M* $MgCl_2$) is added, and the dishes are swirled at intervals of ~ 1 min during the assay to ensure complete mixing of the medium. Samples (0.3 ml) are withdrawn immediately and after appropriate periods of incubation (usually 5 min and 10 min) added directly to 0.7 ml of ascorbate–molybdate[11] and stored on ice until all the samples are processed for color development. Provided that no more than 30% of the substrate is hydrolyzed, P_i release is linear with time for at least 30 min and is linear also with respect to the number of cells on the culture dish.

At the end of the assay for surface activity, the dishes are rinsed twice with HEPES–saline. The cells are scraped from the dish in the presence of 0.2 ml of HEPES–saline containing either 0.2% Triton X-100 or 0.1% sodium deoxycholate, using a plastic transfer pipette; the suspension is transferred to a 10 × 75 mm tube, and the dish is rinsed with 0.3 ml of the same medium, which is then added to the cell suspension. The combined extracts are cooled on ice and sonicated for 10 sec at 50 W using the microprobe of a sonicator, with care taken that the temperature does not exceed 25°; 0.2 ml of the cell extract is then added to 0.8 ml of HEPES–saline containing 1.25 m*M* 5′-AMP and 1.25 m*M* $MgCl_2$, and samples are incubated again at room temperature. At appropriate time intervals (usually 0, 10, and 20 min) samples of 0.3 ml are withdrawn and added to 0.1 ml of 40% trichloroacetic acid. After centrifugation to remove precipitated protein and detergent, P_i is determined in 0.3 ml of the supernatant.

To calculate the specific activity of 5′-nucleotidase, one determines the P_i release during the first and second time intervals, remembering that for the assay of surface activity all the cells remain on the culture dish and that the volume of medium decreases during the assay. These values are usually in good agreement and are averaged to provide the rate of P_i release (nanomoles of P_i per minute per culture). To determine the protein content of the cultures, 0.1 ml of the cell extract is mixed with 0.1 ml of 2% sodium dodecyl sulfate; the sample is boiled for 2 min, and protein is determined[20] using samples of 50 and 100 μl, with bovine serum albumin (treated in the same way) as a standard. The small contribution of HEPES to the color that develops is determined with appropriate blanks.

The specificity of the assay for 5′-AMP is determined by running assays in parallel with 2′-AMP or β-glycerophosphate as substrate, since neither of these is hydrolyzed by 5′-nucleotidase,[11] and also with no substrate. Release of P_i in these circumstances is generally less than 5% of that with 5′-AMP. The only problems that can occur are that some commercial samples of Triton X-100 and sodium deoxycholate may either interfere with the P_i assay or inhibit 5′-nucleotidase activity. The first is monitored easily by incubating cell extracts without substrate and then adding P_i standards at the end of the assay. The second is monitored by preparing cell extracts in the absence of detergent and then adding detergent to the assay for 5′-nucleotidase. In general, the activity of cell extracts is 1.5–2 times that of the intact cells, and the activity of cell extracts in the presence of detergent is 2–3 times that of intact cells.[9]

Antibody Binding Assays

The techniques described here have been developed for the assay of the binding of anti-5′-nucleotidase and for the binding of sheep anti-rabbit antibodies to anti-5′-nucleotidase. More recently,[21] they have been employed for the analysis of other cell surface antigens, with equally satisfactory results. They seem, therefore, to be generally applicable to the analysis of cell surface constituents.

[20] O. H. Lowry, N. J. Rosebrough, A. L. Farr, and R. J. Randall, *J. Biol. Chem.* **193,** 265 (1951).

[21] D. K. Wilcox, P. Whitaker-Dowling, J. S. Youngner, and C. C. Widnell, *J. Cell Biol.* (in press).

Since different preparations of antibody can exhibit slightly different characteristics, it is essential to characterize the time course of binding, and to perform a Scatchard[22] analysis of the binding, with each new preparation. It must be emphasized that binding data obtained without such a firm kinetic basis can be very difficult to interpret. In general, when concentrations of antibody are employed at which nonspecific binding is insignificant, the time required to achieve saturation is impractically long. The conditions must therefore be selected to provide a working compromise between these two extreme solutions. The antibodies described below have generally worked well at concentrations in the range 5–25 μg/ml.

The initial characterization of the binding of anti-5′-nucleotidase is carried out by analyzing the inhibition of surface 5′-nucleotidase activity. Cells are washed twice with HEPES–saline containing 10% newborn calf serum (1 ml for a 35-mm dish) at room temperature and then cooled to 2° in 1 ml of HEPES–saline containing 7.5% newborn calf serum and 2.5% rabbit serum. The rabbit serum is obtained commercially, checked to ensure that it contains no inhibitor of 5′-nucleotidase, and added to minimize nonspecific binding of the anti-5′-nucleotidase. When the cells are at 2°, usually after 30 min on a bench in a cold room, anti-5′-nucleotidase is added at concentrations ranging from 1 to 50 μg/ml either as antiserum or as affinity-purified IgG. After 30, 60, or 120 min the cells are washed three times with HEPES–saline containing 10% newborn calf serum and then removed to room temperature in the same medium. After 15 min the cells are washed twice with HEPES–saline, and surface and total 5′-nucleotidase activity is determined as described above. In a typical experiment, the amount of anti-5′-nucleotidase required to inhibit 85% of the surface activity after 1 hr at 2° is approximately five times that needed to inhibit the total activity of the cells as judged by the inhibition assay described above. There should be no inhibition of the internal activity (total activity minus surface activity).

To analyze the binding of [^{3}H]anti-5′-nucleotidase to cells, the procedure for the incubation with antibody is the same as that for the characterization of the inhibition of enzyme activity. At the end of the incubation, cells are washed, also at 2°, three times with HEPES–saline containing 10% newborn calf serum and then three times with HEPES–saline. The final HEPES–saline wash should contain less than 5% of the radioactivity that remains associated with the cells. The cells are solubilized in 1% sodium dodecyl sulfate (0.7 ml for a 35-mm dish) and boiled for 5 min; radioactivity is determined by scintillation counting using a sample of 0.4 ml and an appropriate aqueous scintillation cocktail, and protein is determined[20] using samples of 50 and 100 μl. In general, the rapid phase of binding is complete between 1 and 2 hr, but continues for at least 12 hr, so that the amount

[22] G. Scatchard, *Ann. N.Y. Acad. Sci.* **51,** 660 (1949).

bound increases almost twofold between 2 and 12 hr. However, cell viability can decrease appreciably after 8 hr, so that binding is usually determined at the end of the initial rapid phase.

At low antibody concentrations (1 – 20 μg/ml) there is usually an excellent correlation between binding of [^{3}H]anti-5′-nucleotidase and inhibition of surface 5′-nucleotidase activity. However, at high antibody concentrations, antibody binding continues after the maximum inhibition of 5′-nucleotidase has been achieved. This is assumed to be a consequence of the antibody's being polyclonal, so that it can bind to more than one site on the 5′-nucleotidase molecule.

Two procedures are employed to monitor nonspecific binding. The simplest is to determine the binding of ^{3}H-labeled nonimmune rabbit IgG, which is prepared at the same time as the [^{3}H]anti-5′-nucleotidase. The second is to saturate 5′-nucleotidase on the cell surface with nonradioactive antibody and then incubate cells with the ^{3}H-labeled ligand.

To determine the amount of antibody required to saturate the surface, the number of sites is determined by a Scatchard[22] analysis of the binding of [^{3}H]anti-5′-nucleotidase. The cells are then incubated with anti-5′-nucleotidase at a concentration (usually more than 200 μg/ml) and for a time 2 – 4 hr) that results in at least 90% saturation. It should be emphasized that complete saturation cannot be achieved in practice and that, after binding of nonradioactive antibody to 90% of the sites, the remaining 10% still bind ^{3}H-labeled antibody. Nonspecific binding can be calculated, however, from the binding under these conditions and should agree closely with results obtained by the first procedure.

It is also desirable to determine that, for a fixed concentration of anti-5′-nucleotidase in the medium, binding of anti-5′-nucleotidase is directly proportional to the number of sites available on the cell surface. Cells are first preincubated with nonradioactive antibody, at concentrations such that the number of sites saturated ranges from 20 to 80%. The specific binding of [^{3}H]anti-5′-nucleotidase is then determined, using a single concentration of antibody in the medium, and the binding is compared to the number of sites available. This relationship is usually linear for the concentration range 5 – 20 μg/ml but must be confirmed if the antibody is to be used to quantitate surface 5′-nucleotidase without performing a Scatchard[22] analysis.

To analyze the binding of ^{3}H-labeled sheep-anti-rabbit antibody to anti-5′-nucleotidase on the cell surface, the assay procedure is essentially the same as that for the binding of [^{3}H]anti-5′-nucleotidase. The sheep-anti-rabbit antibodies prepared in this laboratory have not reacted with newborn calf serum, as judged by immunodiffusion analysis, so that the cells are cooled to 2° and the binding is carried out in HEPES – saline containing 10% newborn calf serum; the washing procedure following binding is the same.

Some additional protein is required to minimize nonspecific binding, so that if a preparation were to react with calf serum, the serum could be replaced with 200 μg of sheep IgG per milliliter[9] or 1 mg of ovalbumin per milliliter. The binding kinetics are defined by analyzing both the time course and effect of antibody concentration in the medium, using cells pretreated with different concentrations of anti-5′-nucleotidase, so that the number of molecules of rabbit antibody on the cell surface is known. Nonspecific binding is determined using cells treated with nonimmune rabbit IgG instead of anti-5′-nucleotidase.

Exchange of 5′-Nucleotidase between the Cell Surface and Cytoplasmic Membranes

In the normal cell, 5′-nucleotidase exchanges continually between the cell surface and the interior, and this exchange is inhibited when pinocytosis is inhibited.[9,10] The exchange can thus provide information on the kinetics of membrane flow from the cell surface to the interior and back. It should be emphasized that there seem to be a variety of intracellular pathways available to membrane that enters the cell from the surface[23-25]; the quantitative importance of each and the nature of the individual membrane proteins involved remains to be clarified. It is thus quite possible that the exchange of 5′-nucleotidase reflects the end result of a very complex pattern of membrane flow.

The exchange is analyzed by two procedures: the internalization of antibody bound initially to the cell surface, and the transfer of enzyme from the interior to the cell surface. Cells are treated with anti-5′-nucleotidase serum so that ~ 85% of the surface activity is inhibited and warmed to room temperature in HEPES–saline containing 10% newborn calf serum as described in the section on antibody binding. One group of cells is analyzed for surface and total 5′-nucleotidase activity and another for surface anti-5′-nucleotidase by the binding of ^{3}H-labeled sheep-anti-rabbit antibody; it is generally advisable to perform the assays in triplicate. Control cells are treated with nonimmune serum instead of anti-5′-nucleotidase serum.

Other groups of cells are returned to culture in complete medium for appropriate times and then analyzed for 5′-nucleotidase activity and surface anti-5′-nucleotidase as at zero time. The time intervals selected depend on the pinocytic rate exhibited by the cells; the exchange of enzyme and

[23] M. G. Farquhar, *in* "Transport of Macromolecules in Cellular Systems" (S. C. Silverstein, ed.), p. 341. Dahlem Konferenzen, Berlin, 1978.

[24] M. C. Willingham and I. Pastan, *Cell* **21,** 66 (1980).

[25] D. A. Wall, G. Wilson, and A. L. Hubbard, *Cell* **21,** 79 (1980).

antibody is essentially complete after 90 min for macrophages,[26] but requires 9 hr for fibroblasts.[9] During the time required for equilibration, the total 5′-nucleotidase activity should remain constant, indicating that no antibody is released from the enzyme during the experiment. It is also necessary to show, in a separate experiment, that [^{3}H]anti-5′-nucleotidase remains associated with the cells; this should be carried out using $F(ab')_2$ fragments of the antibody, since the Fc region can be degraded without releasing the antibody from the enzyme.

Until the intracellular compartments involved in the exchange of 5′-nucleotidase are known, a detailed kinetic analysis of the exchange data is unlikely to reveal meaningful information. The most useful value to be obtained, therefore, is the half-time of the exchange, which may be determined graphically. As has been emphasized earlier, for both macrophages and fibroblasts this value is in close agreement to that for the time required to internalize the entire cell surface, as determined by the more complex and laborious techniques of morphometric analysis.

Acknowledgments

Some of the techniques described here were developed in collaboration with Y.-J. Schneider, P. Baudhuin, and A. Trouet, International Institute of Cellular and Molecular Pathology, Brussels, Belgium. It is a pleasure to recognize their contribution to the work.

[26] W. R. Hitchener and C. C. Widnell, *Eur. J. Cell Biol.* **22,** 206 (1980).

[35] Intracellular Iodination of Lysosome Membrane for Studies of Membrane Composition and Recycling

By WILLIAM A. MULLER, RALPH M. STEINMAN, and ZANVIL A. COHN

Lactoperoxidase (LPO)-catalyzed iodination is a well-established method for labeling membrane proteins.[1-3] We describe here a method whereby LPO selectively iodinates intracellular membranes within living macrophages (Mϕ). The LPO is covalently coupled to carboxylated polystyrene latex spheres (LPO–latex). These particles are readily ingested by Mϕ and rapidly established within the phagolysosomal (PL) compartment. When $^{125}I^-$ and H_2O_2 are added at 4°, the enzyme selectively radiolabels PL

[1] M. Morrison, this series, Vol. 32, p. 103.
[2] A. L. Hubbard and Z. A. Cohn, *J. Cell Biol.* **55,** 390 (1972).
[3] A. L. Hubbard and Z. A. Cohn, *J. Cell Biol.* **64,** 438 (1975).

ISBN 0-12-181998-1

polypeptides. The latter can then be compared to radioiodinated plasma membrane (PM) proteins. In addition the labeled PL membrane proteins rapidly return, or *recycle,* to the cell surface. Therefore, this technique provides a covalent, nonreutilizable label to monitor membrane flow between lysosome and plasma membrane.

Preparation of LPO–Latex

Materials

Microfuge tube, 1.5-ml, plastic (Ulster Scientific, Inc.)
Carboxylate-modified polystyrene latex spheres (CM-latex) $\leq 1\ \mu m$ in diameter (Dow Diagnostics, Inc., Indianapolis, Indiana)
Teflon-coated magnetic stir bar, $\frac{1}{4}$ in.
N-Hydroxysuccinimide (NHS) kept dry (Pierce Chemical Co., Rockford, Illinois)
1-Cyclohexyl-3-(2-morpholinoethyl)carbodiimide metho-*p*-toluenesulfonate (CMC) (Pierce)
Lactoperoxidase (LPO). We routinely used one vial of Calbiochem purified grade, catalog No. 427488 (100 IU per vial) for each batch.
Glycine (crystalline)
Acetate buffer, 0.2 *M*, pH 5.4
Carbonate–bicarbonate buffer, pH 9.6
Phosphate-buffered saline (PBS) (GIBCO, Grand Island, New York)
Glycerol in glass-distilled water, 50% (v/v)
Disposable plastic pipettes are used to handle the latex suspensions, since latex sticks to glass.
Eppendorf microfuge (Brinkmann Instruments, Westbury, New York)
Vortex Genie mixer (Scientific Glass Apparatus)
Branson cell sonifier
Magnetic stirring device

Coupling of Lactoperoxidase (LPO) to Carboxylate-Modified Latex (CM–Latex)

1. Resuspend the CM–latex (10% w/v) by vortexing the stock bottle, and transfer 150 μl to a 1.5-ml microfuge tube.

2. Wash the latex three times in 0.2 *M* acetate buffer, pH 5.4, by sequentially adding 1 ml of buffer, mixing, and centrifuging at 12,000 rpm for 5 min in an Eppendorf microfuge at 4°; carefully decant the supernatant and resuspend the latex pellet in 1 ml of buffer by vortexing or sonication (2–3 sec at setting 2 on a Branson cell Sonifier using the microtip attachment).

3. After the final wash and resuspension, add a clean $\frac{1}{4}$-in. magnetic stirring bar to the microfuge tube.

4. Add 11.5 mg of crystalline *N*-hydroxysuccinimide (NHS) and vortex for several seconds to dissolve (final concentration = 0.1 *M*). Immediately add 42.36 mg of carbodiimide and vortex briefly (final concentration = 0.1 *M*).

5. Stir for 10 hr at room temperature holding the microfuge tube upright in a flat-bottomed glass vial. To insulate the reactants from the heat developed by the magnetic stirrer, place a foam pad under the glass vial.

6. Pellet the activated latex beads in the microfuge as in step 2, rinse the pellet carefully once in the acetate buffer, and resuspend rapidly into 0.5 ml of LPO in carbonate–bicarbonate buffer, pH 9.6. (This solution was prepared by injecting 0.5 ml of buffer into the vial of purified-grade LPO.)

7. Stir magnetically for 30 min at 4°.

8. Add crystalline glycine to a final concentration of 1 *M* to quench ester bonds, and stir for an additional 15–30 min at room temperature.

9. Pellet the latex beads with covalently bound LPO (LPO–latex) and wash four to five times (as in step 2, but using PBS) until two successive supernatants show no LPO activity by the *o*-dianisidine assay (see step 11).

10. Finally, suspend the LPO-latex in 1 ml of 50% glycerol and store at −20°.

11. The enzymic activity of LPO is measured using *o*-dianisidine as described,[4] but at pH 6. To measure the activity of LPO–latex preparations, the stock is first diluted 1:100 to 1:200 in buffer. The LPO–latex settles slowly and does not interfere with the measurements at these dilutions.

12. Relative latex concentrations are determined by light scattering at 500 nm using samples boiled in 2% sodium dodecyl sulfate (SDS). Absorbance is linear with latex concentration from 0.00075 to 0.04% by weight (OD_{500} = 0.03–1.5).

Comments

1. The coupling procedure is an adaptation of the method of Parikh and Cuatrecasas.[5]

2. Aqueous solutions keep the CM–latex insoluble and easy to manipulate by centrifugation. Latex beads should never be dried or frozen, as irreversible clumping will occur.

3. In the first step of the coupling procedure, the carboxyl group of CM–latex is activated with a water-soluble carbodiimide so that it is susceptible to nucleophilic attack by NHS. The latter displaces the carbodii-

[4] R. M. Steinman and Z. A. Cohn, *J. Cell Biol.* **55,** 186 (1972).

[5] P. Cuatrecasas and I. Parikh, *Biochemistry* **11,** 2291 (1972).

mide, forming an ester. The esterified latex is then centrifuged away from the soluble reagents so that LPO is not exposed to the carbodiimide. In the next step, at high pH (9.6), nonprotonated lysine ε-amino groups of the LPO readily attack the ester bond on the latex, displacing NHS and forming a more stable amide linkage (Fig. 1 in Muller *et al.*[9]).

4. This procedure routinely couples 15–20% of the original LPO activity and 15–20% of the original protein to the latex spheres. Thus, the specific activity of the LPO is unchanged. Over the course of 4 years, more than 30 different batches of LPO–latex were made, with enzymic activities 1 ± 0.1 U/ml, by our *o*-dianisidine assay.

5. Enzymic activity is stable for at least 11 months when stored in 50% glycerol at −20°.

6. If LPO is exposed to CM–latex without prior carbodiimide activation, substantial enzymic activity can be bound to the beads. Unlike the covalently coupled LPO, however, enzymic activity continues to elute with successive saline washes. Alternatively, "absorbed" LPO can be rapidly removed by washing the beads with a nonionic detergent, indicating that it is bound by hydrophobic interaction. Detergents do not remove LPO from covalently coupled LPO–latex.

Delivery of LPO–Latex to Lysosomes

Materials

Dulbecco's phosphate-buffered saline with (PBS) and without (PD) calcium and magnesium (GIBCO, Grand Island, New York).

Culture medium 199 (GIBCO) with 10% (v/v) fetal bovine serum (FBS) and 100 U of penicillin G per milliliter

Macrophages (Mϕ). An average of 6×10^6 peritoneal cells can be lavaged from female Swiss mice (25–30 g) with PD; 25–40% adhere to glass or plastic and are >95% Mϕ as described by Cohn and Benson.[6] We culture the cells for 2 days, replacing the medium daily, To achieve nearly confluent monolayers, the following numbers of cells are plated: for routine iodination, 3×10^6 cells in 16-mm-diameter flat-bottom wells (Costar, Data Packaging, Cambridge, Massachusetts); for electron microscopy, 2×10^7 cells in 35-mm plastic dishes (Nunclon Delta, Kamstrup, Roskilde, Denmark); for cell fractionation, 4 to 5×10^7 cells in 60-mm Nunc dishes.

Trypan blue stain: 0.4% (GIBCO) diluted to 0.1% in PBS containing 1% FBS. To test the viability of cells at any step, add this solution to cell monolayers on ice for 3–5 min. Aspirate the stain and wash the

[6] Z. A. Cohn and B. Benson, *J. Exp. Med.* **121,** 153 (1965).

cultures three or four times with cold PBS or until blue color stops eluting from the dish. Examine the cells with an inverted microscope. Dead cells (usually $<1\%$ of the cells) have blue nuclei. If necessary, cultures can be fixed in 1–2% glutaraldehyde for a permanent record.

Method

The following protocol was designed to deliver LPO–latex rapidly and selectively to Mϕ lysosomes.

1. Dilute the LPO–latex stock (see under Preparation of LPO–latex, above) to an $OD_{500} = 0.33$ in PBS. This corresponds to a 1:800 dilution of our stock.

2. Wash the Mϕ monolayers several times with cold PBS and add an appropriate volume of LPO–latex suspension: 5 ml for a 60-mm dish, 1.5 ml for a 35-mm dish, 0.4 ml for a 16-mm Costar well.

3. Since the latex spheres settle very slowly, centrifuge the culture vessels at 1000 *g* for 2 min at 4°. (We have found that placing the culture dishes on Microtiter tray centrifuge carriers works well for this.) At this point latex beads are attached to the surface of all the cells and to the dish. By scanning electron microscopy (EM) the beads are not depressed into the cell.

4. Decant the supernatant and rapidly wash the monolayer twice with warm PBS.

5. Return the cultures to the incubator in PBS or culture medium prewarmed to 37°. A rapid and synchronous wave of phagocytosis ensues, so that by 15–30 min all the latex has been ingested.

6. To be certain that subsequent iodination is entirely intracellular, complete ingestion of all latex particles is crucial. Two methods are employed to eliminate extracellular and dish-bound latex. Usually confluent monolayers are established so that dish-bound latex is effectively cleared by the cells themselves. At lower cell densities, brief trypsinization (200 μg/ml in PBS for 5 min) removes the vast majority of extracellular beads. Trypsin neither inactivates nor releases LPO activity from LPO–latex beads. Similar results are obtained with both protocols, except that there are a few polypeptides on the Mϕ cell surface that are trypsin sensitive.

7. To terminate ingestion, culture dishes are rapidly washed several times with cold (4°) PBS and set on top of an ice-water bath for 10 min.

Comments

1. LPO-latex beads, administered to Mϕ as described, are fully internalized. This has been verified by both scanning and transmission EM.

2. Uptake of LPO–latex is proportional to bead dose over a wide range corresponding to an uptake of 10–100 beads/Mϕ. Higher doses "stuff" the

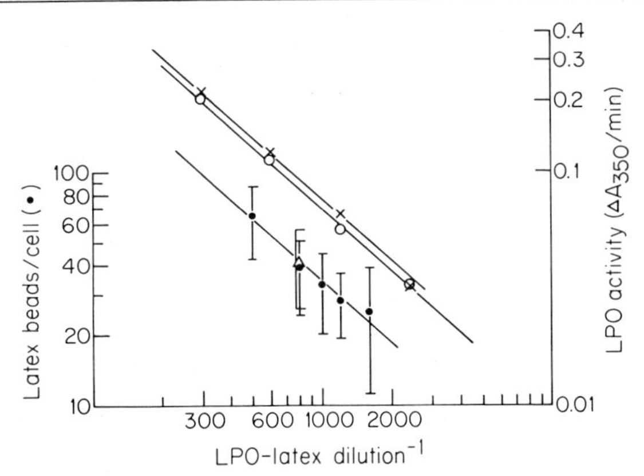

FIG. 1. Ingestion of lactoperoxidase (LPO)–latex as a function of particle dose. Two separate assays are shown here. The LPO enzymic assay was the triiodide assay, performed as described.[7] Direct latex counts (●) were performed on each of 50 cells chosen at random (25 from each of 2 coverslips). Standard deviations are indicated by the vertical bars. The LPO assays were performed on lysates of cell monolayers either immediately after trypsinization to remove extracellular latex (○) or after returning cells to culture for 30 min after trypsinization (×). Triplicate measurements were made at each latex dilution. The standard deviations were all less than 10% of the indicated mean.

cells, making them round and easily dislodged from the culture surface. Uptake can be monitored using LPO enzymic assays[7] or by direct microscopic counts of the refractile latex beads in fixed cells (Fig. 1).

3. Diaminobenzidine (DAB) cytochemistry[8] is used to visualize all cell-associated LPO. Glutaraldehyde fixation should not exceed 10 min because longer fixation preserves an endogenous Mϕ endoplasmic reticulum peroxidase.[9] With fixation for 10 min, the DAB reaction product is restricted to the rim of every latex bead and is not found in any other location (Fig. 2). Latex that has not had LPO coupled to it has no DAB reactivity.

4. The enzymic activity of LPO–latex is not altered by phagocytosis (Fig. 1) and is stable for at least 5 hr as measured by enzymic assay or DAB cytochemistry. The activity falls over the course of 1–2 days. However, DAB reaction product is always bead associated, and exocytosis of ingested beads is not detectable.

5. Under the conditions we employ, all LPO–latex beads are in secondary lysosomes. That is, every bead is surrounded by acid phosphatase;

[7] L. A. Decker (ed.), "Worthington Enzyme Manual," Worthington Biochemical Corporation, Freehold, New Jersey, 1977.

[8] R. C. Graham, Jr. and M. Karnovsky, *J. Histochem. Cytochem.* **14,** 291 (1966).

[9] W. A. Muller, R. M. Steinman, and Z. A. Cohn, *J. Cell Biol.* **86,** 292 (1980).

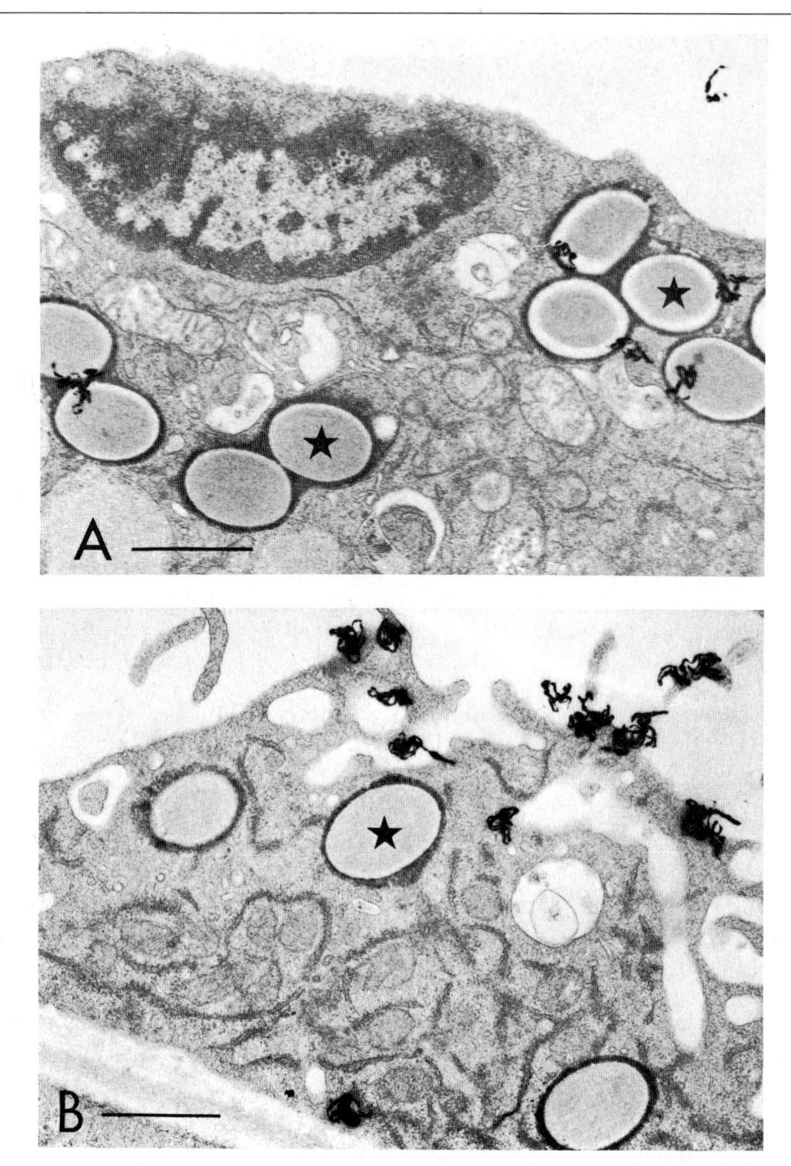

FIG. 2. Recycling of phagolysosome (PL) membrane to the cell surface. Macrophage PL were iodinated using lactoperoxidase (LPO)–latex. After exposure to $^{125}I^-$ for 10 min on ice, cells were either fixed immediately (A) or washed and returned to culture for 5 min before fixation (B). Fixed monolayers were stained with 3,3′-diaminobenzidine to visualize LPO activity. The electron-dense precipitate is restricted to the rims of LPO–latex beads (★). Specimens were then processed for EM autoradiography. At time zero, exposed silver grains were associated with the PL membrane (A). Within 5 min at 37°. exposed grains were seen distributed over the macrophage plasma membrane (B). Bar = 1 μm.

cytochemical reaction product is visualized with cytidine monophosphate as substrate.[9,10]

Intracellular Iodination

Materials

Na^{125}I, carrier free, in solution with 0.1 *N* NaOH. Store at room temperature.

Glucose oxidase (GO) type VI from *Aspergillus niger* (Sigma Chemical Co., St. Louis, Missouri) purchased as a concentrated solution

Ice-water bath: Polystyrene foam boxes work well for this purpose.

Reagents for iodination prepared fresh and kept on ice:

- PBS–glucose: 20 m*M* D-glucose in PBS
- GO dilution tubes: Add PBS–glucose to each of three tubes for subsequent rapid serial dilution of stock GO.
- $^{125}I^-$ tube with a volume of 0.1 *N* HCl equal to the volume of $^{125}I^-$ that will be used, as well as PBS–glucose for dilution
- Stop solution: 0.1 *M* KI, 0.02% NaN_3 in PBS

Cell cultures that have ingested LPO–latex (see section on delivery of LPO–latex, above)

Glass syringe (Hamilton) with needle calibrated from 0 to 50 μl

For recycling experiments, sterile PBS and culture medium that have been warmed to 37°

Procedure

1. After complete ingestion of LPO–latex (see section on delivery of LPO–latex, above), remove the cultures from the incubator, wash six times with PBS, and chill on an ice-water bath for 10 min.

2. Aspirate the PBS and replace with a defined volume of PBS–glucose. The volumes tabulated below work well.

Solution	16-mm wells (ml)	30-mm dish (ml)	60-mm dish (ml)
PBS–glucose	0.350	1.050	3.5
^{125}I mixture	0.050	0.150	0.5
GO	0.100	0.300	1.0
	0.50	1.5	5.0

[10] P. M. Novikoff, A. B. Novikoff, N. Quintana, and J. J. Hauw, *J. Cell Biol.* **50,** 859 (1971).

Iodination should be performed in a well-ventilated, lead-lined fume hood equipped with a vacuum suction flask for aspirating radioactive fluid.

3. Flush the Hamilton syringe once with glass-distilled water, withdraw the proper volume of $Na^{125}I$ from the stock bottle and mix with diluent in the "^{125}I" tube. Note that the concentration of ^{125}I in this tube is 10 times the final concentration

Perform steps 4–6 in rapid succession.

4. Dispense ^{125}I solution into all culture vessels.
5. Rapidly dilute the GO stock solution through three tubes and dispense the proper volume into the culture vessels.
6. Carefully tilt and rotate culture vessels to mix the reagents.
7. Leave on ice for the desired length of time. Iodide incorporation plateaus within 20 min.
8. Aspirate the reagents and wash cells quickly two or three times with stop solution.
9. Wash two or three times with PBS.
10. Test viability by trypan blue dye exclusion as described in section on delivery of LPO–latex, above.

The iodinated cells are now ready for further processing, e.g., lysis for trichloroacetic acid precipitation, homogenization for cell fractionation, fixation for EM autoradiography, or return to culture to monitor recycling. See Muller *et al.*[9] for procedural details. For recycling studies, wash the cells in warm PBS and return them to culture in prewarmed medium.

Comments

1. Iodide is oxidized to volatile iodine at neutral or acid pH and is less stable at low temperatures. For these reasons, store the ^{125}I in base at room temperature, withdraw the stock with a needle through the airtight rubber gasket just prior to use, and work in a ventilated hood until the cells have been rinsed several times. Most of the added $^{125}I^-$ is not incorporated into cells, and >99% of the unreacted label is removed by these washings.

2. Glucose oxidase activity is unstable in aqueous solutions at low concentrations. Therefore, dilute from the stock immediately before addition to the cells.

3. Reserve a separate set of pipetting devices for use in iodinations, since instruments pick up radioactivity with repeated use and cannot be completely decontaminated.

4. Keep the culture vessels in the ice-water bath at all times to inhibit movement of labeled membrane. By EM autoradiography, incorporated ^{125}I recycles from the phagolysosome to the plasma membrane in 5–10 min at

37°; membrane flow proceeds even at reduced temperature at substantial rates. If an iced culture vessel is placed at room temperature, the solutions can warm to 18° within 2 min.

5. To facilitate rapid and multiple washings, PBS and stop solution are dispensed from syringes through large-bore ($\geq$ 18-gauge) needles.

6. The effect of varying the concentration of different reactants has been assessed[9]: (*a*) Iodination increases with increasing numbers of ingested LPO–latex. (*b*) In the absence of exogenous GO, H_2O_2 produced by the Mϕ upon phagocytosis is sufficient to support ^{125}I incorporation into trichloroacetic acid-precipitable counts. Iodination is enhanced three- to sixfold by the addition of exogenous GO. Further increases in exogenous GO above our standard dose of 0.24 mU/ml results in little increase in iodination, and concentrations > 10 mU/ml can be toxic to the Mϕ. (*c*) Iodination varies linearly with the concentration of ^{125}I within the range generally employed (40–400 μCi/ml).

7. After the iodination, most of the cell-associated radiolabel is unreacted $^{125}I^-$. This label elutes slowly from cells. Therefore, all quantitation is performed on samples that have been washed extensively with KI (50–100 m*M* in PBS) before detergent lysis (0.05% Triton X-100 or NP-40) and trichloroacetic acid precipitation. We usually add protease inhibitors to lysates, e.g., phenylmethylsulfonyl fluoride (PMSF) 1 m*M* and aprotinin (0.1–0.2 TIU/ml); however, these are unnecessary if lysis is followed immediately by trichloroacetic acid precipitation or boiling.

8. Radioiodide incorporated by phagocytosed LPO–latex is found exclusively as monoiodotyrosine. There is no detectable lipid labeling. By EM autoradiography, ^{125}I is associated predominantly with PL. For reasons that are unclear, labeling is restricted to the PL membrane and is not detectable in PL contents. We purified labeled PLs on discontinuous sucrose gradients and found that freezing and thawing released >75% of three lysosomal hydrolases but <2% of the ^{125}I. In addition, one-dimensional SDS–polyacrylamide gel electrophoresis indicates that the labeled polypeptides are virtually identical to PM polypeptides iodinated with LPO applied externally at 4° (either soluble LPO or LPO–latex centrifuged onto the Mϕ surface). The labeled proteins are primarily integral membrane proteins, since they cannot be eluted by washing in high salt (6 *M* NaCl or KCl) or alkaline buffer (pH 10.3).

9. Autoradiography of Mϕ labeled with LPO–latex shows that ^{125}I incorporation is heterogeneous. Most beads are labeled, but some (<5%) are very heavily labeled. The two populations can be separated on sucrose gradients, and both have the same spectrum of labeled polypeptides. We do not know the cause for this heterogeneity in labeling.

Application to Membrane Recycling

Labeling of latex PL membrane within intact cells offers several advantages for the study of membrane recycling:

1. The label is covalently incorporated into intrinsic membrane proteins so that the fate of the label directly reflects the fate of that membrane molecule. We have studied the major iodinatable membrane polypeptides,[11] but antibodies can be used to follow specific and minor membrane components.
2. The digestion product of iodinated protein, monoiodotyrosine, cannot be used for the synthesis of new protein.[12] Thus, all label in protein remains on the same protein molecules originally iodinated within the lysosome.
3. LPO remains on the latex beads, and the latter remain within phagolysosomes, marking the site of iodination in EM autoradiograms. PLs are also readily purified by cell fractionation.[9,11]
4. Once the iodination reaction is terminated, any nonincorporated $^{125}I^-$ within the cell diffuses out; there is no residual iodination when cells are returned to culture.
5. Since PL membrane is labeled, one can examine the role of the lysosome in membrane recycling.

Using this method, we have demonstrated that membrane polypeptides return intact from the lysosome to the plasma membrane.[11]

1. Autoradiography of electron microscope specimens shows that radioactivity originally confined to the rims of LPO–latex PLs (Fig. 2A) is randomly distributed over the plasma membrane within minutes of return to culture (Fig. 2B). Recent data demonstrate that the recycling process has reached equilibrium after about 5 min[13] (Fig. 2B).
2. The same spectrum of polypeptides radiolabeled within the PL moves in concert to the PM.
3. These experiments identify the lysosome as an important organelle involved in the recycling pathway.
4. We have demonstrated a reciprocal centripetal flow of membrane from PM to PL via pinocytic vesicles. We believe that the membrane influx by pinocytosis is usually balanced by the recycling efflux, to maintain a lysosome compartment of constant surface area.
5. Iodination of PL membrane in intact cells has also been used to study turnover of lysosome membrane proteins. We found that after an initial

[11] W. A. Muller, R. M. Steinman, and Z. A. Cohn, *J. Cell Biol.* **86,** 304 (1980).
[12] H. J.-P. Ryser, *Biochim. Biophys. Acta* **78,** 759 (1963).
[13] W. A. Muller, R. M. Steinman, and Z. A. Cohn, *J. Cell Biol.* **96,** 29 (1983).

rapid loss of label from PL, the half-life of PL membrane proteins is ~ 30 hr—within the range calculated for PM proteins.

Acknowledgments

This work was supported by U.S. Public Health Service grants to R. M. S. (AI-13013) and Z. A. C. (AI-07012).

[36] Labeling of Plasma Membrane Glycoconjugates by Terminal Glycosylation (Galactosyltransferase and Glycosidase)

by LUTZ THILO

A membrane labeling technique based on the terminal glycosylation of plasma membrane glycoconjugates can be schematically represented as in Fig. 1: ^{3}H- or ^{14}C-labeled galactose is enzymically bound to or released from the plasma membrane. This label provides a biochemical method of measuring internalization and recycling of labeled membrane components,[1-4] viz.:

1. Internalized label is no longer accessible to enzymic release and can therefore be distinguished quantitatively from label remaining on the cell surface.
2. Previously internalized label again becomes accessible to enzymic release when it is recycled back to the cell surface.

Since this labeling technique provides a radioactive marker that is covalently linked to membrane components, it can also be applied for electron microscopic autoradiographic observations of vectorial membrane flow.[2,4,5] In cells of the amoeba *Dictyostelium discoideum* and in cells of a macrophage cell line P388D$_1$ it was found that the label serves as a stable membrane marker, remaining fully membrane bound during at least 40

[1] L. Thilo and G. Vogel, *Proc. Natl. Acad. Sci. U.S.A.* **77,** 1015 (1980).
[2] H. G. Burgert and L. Thilo, *Exp. Cell Res.,* **144,** 127 (1983).
[3] L. Thilo and H. G. Burgert, *Exp. Cell Res.,* in press.
[4] C. de Chastellier, A. Ryter, and L. Thilo, *Eur. J. Cell Biol.,* in press.
[5] H. Schwarz and L. Thilo, *Eur. J. Cell Biol.,* in press.

ISBN 0-12-181998-1

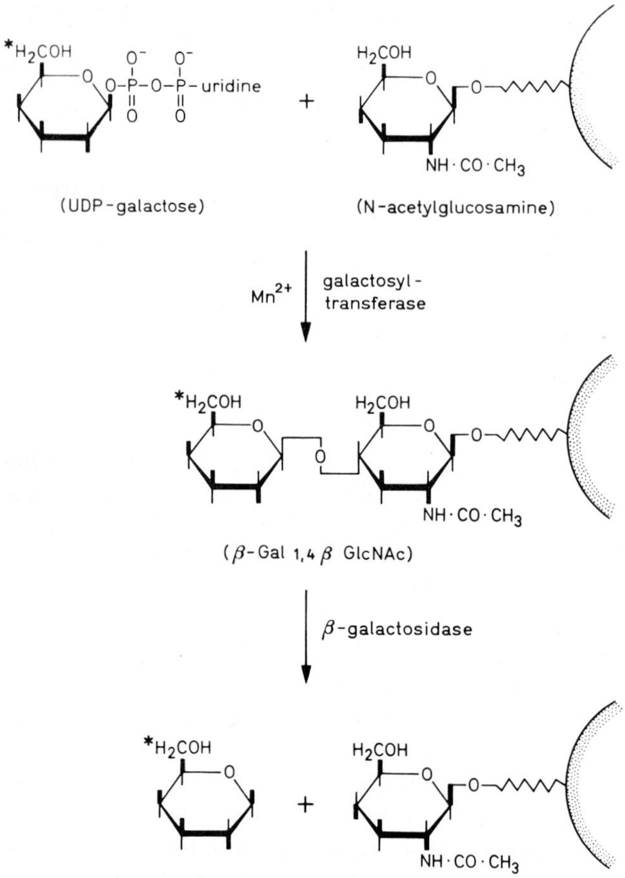

FIG. 1. Schematic representation of reversible labeling technique.

cycles of membrane internalization and recycling.[1,2,5] Previously, this labeling technique was used during structural[6-8] and developmental[9] studies.

Labeling of Plasma Membrane Glycoconjugates

The labeling procedure must be carried out under conditions where membrane internalization is arrested, i.e., <4°. Furthermore, the conditions must be physiological in order not to affect the subsequent behavior of

[6] H. Schenkel-Brunner, *Eur. J. Biochem.* **33,** 30 (1973).

[7] M. Schindler, D. Mirelman, and U. Schwarz, *Eur. J. Biochem.* **71,** 131 (1976).

[8] J. H. Shaper and L. Stryer, *J. Supramol. Struct.* **6,** 291 (1977).

[9] B. Wallenfels, *Proc. Natl. Acad. Sci. U.S.A.* **76,** 3223 (1979).

the cells. On the other hand, a minimum density of label must be attained, depending on the actual experimental requirements. For most biochemical approaches, it will suffice if the labeling density is of the order of 10^4 cpm/10^6 cells. Electron microscopic autoradiographic observations require about 10^5 cpm/10^6 cells in order to yield several grains per cell section per week of exposure time.

The protocol for the labeling procedure described below has been developed for cells of the amoeba *Dictyostelium discoideum*,[1,4,5] but with slight modifications it has equally successfully been applied to cells of a macrophage cell line, $P388D_1$[2,3] and to BHK21 cells.

1. The cell suspension is cooled to $<4°$ in ice to arrest membrane flow and washed free of medium using cold buffer (cf. observation 1 below). This can be done by one centrifugation step (5 min at 500 *g*) or by direct rinsing if cells are attached to a coverslip. The cells are resuspended in buffer at a concentration of about 2.5×10^8 cells/ml.

2. The incubation mixture (10 volumes) consists of cell suspension, 4 volumes (2.5×10^8 cells/ml; see observation 2); UDP[6-^{3}H]Gal, 4 volumes (100 μCi/ml; see observation 3); $MnCl_2$, 1 volume (50 m*M*; see observation 4) Galactosyltransferase, 1 volume (5 units/ml; see observation 5).

The labeling reaction is started by the addition of galactosyltransferase. The final conditions in 10 volumes of incubation mixture are: $<4°$; cell density, 10^8 cells/ml; UDP[6-^{3}H]Gal, 40 μCi/ml; Mn^{2+}, 5 m*M*; and galactosyltransferase, 0.5 unit/ml. During the incubation the reaction mixture should occasionally be agitated to prevent sedimentation of the cells.

3. The reaction is terminated by diluting 10-fold in cold buffer and immediate washing (three times) to remove unbound radioactivity. The radioactivity and cell number are determined and can be expressed as counts per minute of [^{3}H]Gal bound per 10^6 cells.

The kinetics of labeling for a number of different conditions are shown in Fig. 2. The following observations (1-5) were made in the course of developing the present technique.

1. The buffer used in the incubation mixture does not seem to have a significant effect on labeling efficiency. Within the average spread indicated in Fig. 2 (—O—), the same results were obtained when using 20 m*M* HEPES, pH 6; 20 m*M* HEPES, pH 7.4, with or without NaCl at 140 m*M*; 20 m*M* phosphate buffer, pH 6 or pH 7.4. Previously 3-(*N*-morpholino)propanesulfonic acid (MOPS) (125 m*M*, pH 7.4) was used.[9]

2. The labeling density increases with decreasing cell concentrations. Cells can also be labeled when they remain attached to a substratum. BHK21 cells grown on glass coverslips to a density of about 10^6 cells per

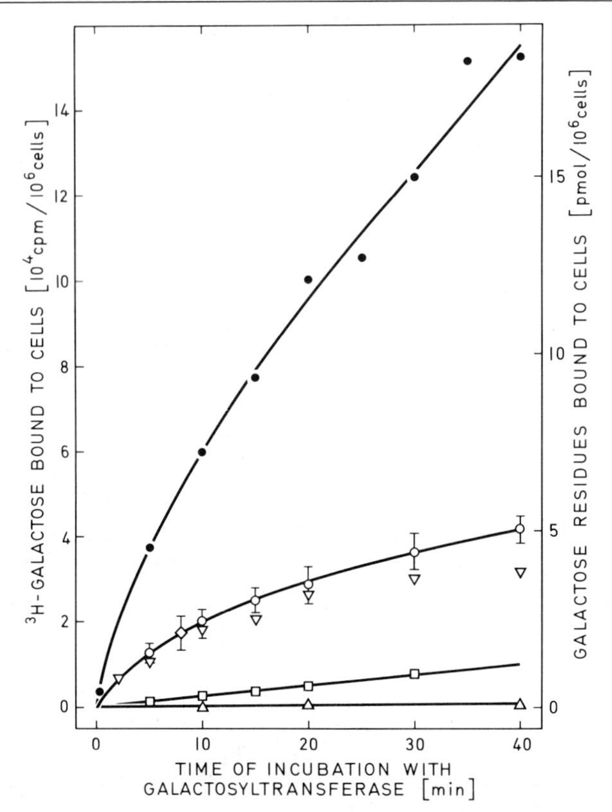

FIG. 2. Kinetics of labeling at <4°. Cells of *Dictyostelium discoideum,* ○: conditions as described in text, bar indicates the average spread between nine independent experiments; ●: tenfold concentration of UDPGal; □: without addition of Mn^{2+}; △: without addition of galactosyltransferase. ▽: $P388D_1$ cells in 10 m*M* HEPES, pH 7.4, 140 m*M* NaCl. ◇: BHK21 cells on coverslips in 10 m*M* HEPES pH 6.8, 100 m*M* NaCl. The ^{3}H counting efficiency is about 0.2.

coverslip, were incubated by submerging each coverslip in 0.1 ml of reaction mixture (Fig. 2, -◇-).

3. UDP[6-^{3}H]Gal (supplied by Amersham as the ammonium salt in 50% aqueous ethanol at about 15 Ci/mmol and 1 mCi/ml) is diluted 10-fold to 100 μCi/ml in buffer prior to use in the reaction mixture. The concentration of UDPGal during the reaction therefore is 2.6 μM, and about 20% of this becomes cell bound after about 40 min (Fig. 2, -○-). For higher densities of label, UDPGal can be used without dilution in buffer. In this case the ethanol must be removed by diluting 2-fold in buffer and by evaporation to the original volume under nitrogen at 4°. A 10-fold increase in the concentration of UDPGal results in a 4-fold increase in labeling density after 40 min (Fig. 2, -●-). Labeling with UDP[U$-^{14}$C]Gal (Amersham, lithium salt, in 2% ethanol, 340 mCi/ mmol, 25 μCi/ml) at a concentration of 1 μCi/ml in

the reaction mixture yields about 3000 dpm of [^{14}C]Gal/10^6 cells after 40 min.

4. The presence of Mn^{2+} is essential.[10] Without the addition of Mn^{2+} the reaction rate is reduced to about 20% of the normal rate (Fig. 2, -□-). No difference in the reaction rate is observed for concentrations of Mn^{2+} between 2 m*M* and 10 m*M*.

5. Galactosyltransferase[11,12] (lactose synthase, EC 2.4.1.22, from bovine milk; Sigma) at a final concentration of 0.5 unit/ml yields the highest reaction rate under the conditions described (Fig. 2, -○-). Higher concentrations up to 2 units/ml do not increase the rate. Below 0.5 unit/ml the reaction rate increases linearly with enzyme concentration.

Release of Label from Plasma Membrane Glycoconjugates

The enzyme β-galactosidase from *Streptococcus pneumoniae* (EC 3.2.1.23)[13-15] can be used to hydrolyze the label off the cell surface.

Membrane flow is arrested by cooling in ice or, permanently, by fixation in 2.5% glutaraldehyde. The cells are then washed free of medium (and of glutaraldehyde) by two washing steps in the 10-fold volume of imidazole buffer (0.1 *M*, pH 6.8). The cells are then resuspended at 2×10^8 cells/ml in this buffer.

The hydrolysis reaction is started by the addition of an equal volume of β-galactosidase solution (1 unit/ml) in imidazole buffer (final concentration 0.5 unit/ml, cell concentration 10^8 cells/ml). The incubation is carried out at $<4°$ (unfixed cells) or at room temperature (fixed cells; for longer incubation times NaN_3 can be added at 5 m*M* to prevent growth of contaminating bacteria, incorporating released galactose).

The hydrolysis reaction is stopped by 10-fold dilution in buffer and immediate separation of cells and supernatant by centrifugation. The fraction of label released is determined by comparing the radioactivity in the supernatant to the total radioactivity in the cell suspension before centrifugation (care must be taken that radioactivity bound to fixed cells is counted with the same counting efficiency as in the supernatant).[16]

[10] J. T. Powell and K. Brew, *J. Biol. Chem.* **251,** 3645 (1976).

[11] J. F. Morrison and K. E. Ebner, *J. Biol. Chem.* **246,** 3977 (1971).

[12] B. S. Khatra, D. G. Herries, and K. Brew, *Eur. J. Biochem.* **44,** 537 (1974).

[13] R. C. Hughes and R. W. Jeanloz, *Biochemistry* **3,** 1535 (1964).

[14] L. R. Glasgow, J. C. Paulson, and R. L. Hill, *J. Biol. Chem.* **252,** 8615 (1977).

[15] β-Galactosidase from *Diplococcus pneumoniae* was kindly supplied by Rudolf Weil (Sandoz Forschungsinstitut, Vienna). One unit is defined as hydrolyzing 1 μmol of *o*-nitrophenyl β-D-galactopyranoside per minute at 20° in 0.1 *M* imidazole buffer, pH 6.8. The enzyme has now become commercially available from Seikagaku Kogyo Co., Ltd., Tokyo.

[16] ^{3}H bound to glutaraldehyde-fixed cells is counted with a lesser counting efficiency. If ^{3}H on free galactose residues is counted with an efficiency of about 20%, then ^{3}H bound to the fixed plasma membrane is counted with about 10%, ^{3}H bound inside vacuoles of fixed cells with

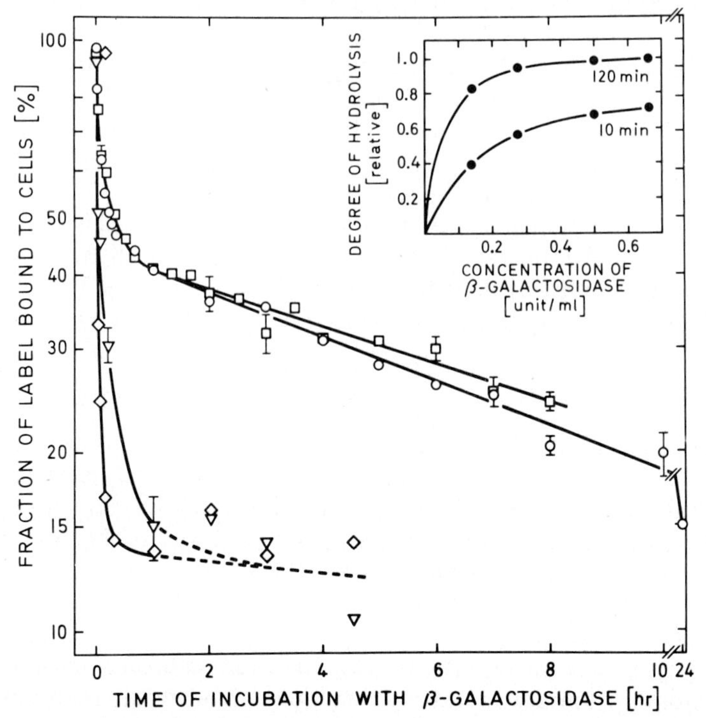

FIG. 3. Kinetics of the release of label from the cell surface by β-galactosidase (0.5 unit/ml).[15] ○: Cells of *Dictyostelium discoideum* fixed in glutaraldehyde, at 20°. □: Unfixed cells of *D. discoideum* at 0°. ▽: Fixed cells of P388D$_1$ at 20°. ◇: Unfixed cells of P388D$_1$ at 0°. Error bars: SEM, $n = 4$. *Inset:* Dependence on β-Gal concentration; 1 △ 62% [^{3}H]Gal released at 20°, fixed cells of *D. discoideum.*

The kinetics of hydrolysis are shown in Fig. 3.[15] There is no significant difference in the reaction rate for fixed and unfixed cells. A much faster release of label is observed for fixed and unfixed macrophages. Depending on the cell type, a fraction of the order of 20% of the label remains cell bound even after long incubation times. The reason for this is not known; it may be due to a higher specificity of the β-galactosidase compared to the galactosyltransferase. As shown in the inset of Fig. 3, increasing the concentration of β-galactosidase results in no significant increase in reaction rate. At lower enzyme concentrations slower reaction rates are observed, especially for the

about 8%, and ^{3}H bound in fixed latex containing phagosomes with > 14% efficiency. This distorting effect can be reduced, but not eliminated, by treating the fixed cells with a solubilizing agent. The difference in relative counting efficiencies can be avoided completely by burning the samples in a sample oxidizer (Packard, Model 306), which has the additional advantage of fully separating ^{3}H and ^{14}C in double-label experiments.[3,4]

initial fast part (<20 min) of the reaction. β-Galactosidase from *Escherichia coli* does not catalyze the hydrolysis reaction.

Acknowledgment

This work was supported by the Deutsche Forschungsgemeinschaft.

[37] *Dictyostelium discoideum* as a Model System to Study Recognition Mechanisms in Phagocytosis

By GÜNTER VOGEL

Phagocytosis is an integrated series of complex events: the first step involves recognition and binding of a particle to the cell surface. Binding seems to initiate the next step, the extension of pseudopodia that embrace the particle. Ultimately the particle is internalized by plasma membrane fusion and pinched off into the cytoplasma.[1]

The cellular slime mold *Dictyostelium discoideum* is particularly well suited as an experimental system to study the mechanism of phagocytosis.[2] In the unicellular phase of the life cycle, it is nutritionally dependent on endocytosis, ingests avidly a variety of substrate particles, and can be grown axenically in large quantities for biochemical investigation. Furthermore, isolation and analysis of mutants with altered phagocytotic properties allow one to dissect the complex process of phagocytosis into individual steps.[3,4]

Strains and Growth Conditions

Strains AX2 (ATCC 24397) and derived mutants are grown in axenic medium[5] by incubation on a rotary shaker (100 rpm; 2.5 cm amplitude). The medium contains, per liter: Oxoid bacteriological peptone, 14.3 g; Oxoid yeast extract, 7.15 g; maltose, 18 g; $Na_2HPO_4 \cdot 2H_2O$, 0.64 g; KH_2PO_4, 0.48 g; and dihydrostreptomycin sulfate, 50 mg. The medium is sterilized by autoclaving at 121° for 20 min. Amoebae are also grown on agar surfaces with *Escherichia coli* or *Klebsiella aerogenes* as the bacterial

[1] S. C. Silverstein, R. M. Steinman, and Z. A. Cohn, *Rev. Biochem.* **46,** 669 (1977).

[2] W. F. Loomis, "*Dictyostelium discoideum:* A Developmental System." Academic Press, New York, 1975.

[3] G. Vogel, L. Thilo, H. Schwarz, and R. Steinhart, *J. Cell Biol.* **86,** 456 (1980).

[4] G. Vogel, *Monogr. Allergy* **17,** 1 (1981).

[5] D. J. Watts and M. J. Ashworth, *Biochem. J.* **119,** 171 (1970).

METHODS IN ENZYMOLOGY, VOL. 98

ISBN 0-12-181998-1

associate, as described in detail elsewhere.[6,7] The cell number is monitored with an electronic particle counter (e.g., Model Z_{BI}, Coulter Electronics). For preservation of strains, clonally derived spores are suspended in axenic medium and stored in liquid nitrogen.

Quantitation of Endocytotic Uptake

Principles. In order to determine initital rates of endocytotic uptake it is essential, first, to separate rapidly and quantitatively cells containing endocytosed material from the bulk of noningested material; and second, to quantitate the small amounts of endocytosed material by sensitive and reliable methods. The first problem is overcome by centrifuging the cell suspension through a column of highly viscous polyethylene glycol (PEG) 6000. Extracellular fluid and small particles with a high surface-to-volume ratio (e.g., bacteria and latex beads with diameters less than 2 μm) remain on top of the column, whereas the large amoebae are found on the bottom. Recovery of cells is about 100%, and amoebae remain fully viable during this procedure. Various types of latex beads are commercially available, and fluorescein-labeled bacteria can easily be prepared. Both can be estimated sensitively by optical methods. For pinocytosis, the most convenient fluid phase markers are fluorescent dextrans of high molecular weight, which are commercially available.[3,8] With appropriate modifications these procedures can be applied to other cell types.[9]

Pinocytosis Assay

Cells are harvested by centrifugation at 200 *g* for 5 min, suspended at a density of 3 to 5 $\times$ 10^6 cells/ml in the chosen medium, and incubated on a rotary shaker (100 rpm) for 15 min to recover. Fluorescein-labeled dextran (e.g., FITC-dextran 60, Pharmacia) is added from a stock solution (20 mg/ml) to a final concentration of 2 mg/ml. To stop pinocytosis, 1-ml samples are taken at various times and diluted into 4 ml of ice-cold 20 m*M* potassium phosphate, pH 6.2. Cells are collected by centrifugation and resuspended in 1 ml of cold phosphate buffer. To remove external FITC-dextran completely, the cell suspension is layered over 10 ml of an aqueous solution of 20% (w/w) PEG 6000 in a centrifuge tube and centrifuged for 10

[6] M. Sussman, *Methods Cell Physiol.* **2,** 397 (1966).
[7] M. J. Krichevsky and L. L. Love, this series, Vol. 39, p. 485.
[8] L. Thilo and G. Vogel, *Proc. Natl. Acad. Sci. U.S.A.* **77,** 1015 (1980).
[9] R. D. Berlin and J. M. Oliver, *J. Cell Biol.* **85,** 660 (1980).

min at 200 g in a swing-out rotor. The supernatant is suctioned off using a Pasteur pipette the tip of which has been bent and to which a small piece of silicon tubing has been attached. This ensures the thorough removal of material sticking to the glass surface.

Pelleted amoebae are washed once by suspension and centrifugation in 3 ml of 50 m*M* Na_2PO_4, pH 9.2, and resuspended in 2 ml of the same buffer. After counting, cells are lysed by addition of Triton X-100 (0.2% final concentration), and the fluorescence intensity of the solution is determined using excitation and emission wavelengths of 470 and 520 nm, respectively. The pinocytosed volume is determined by comparison with a standard curve. Fluorescein fluorescence is very pH sensitive, and therefore it is essential to ensure that the pH is above pH 9, when fluorescence intensity is maximal and constant.

FITC-dextran qualifies as a suitable and convenient fluid-phase marker according to the following criteria. It is commercially available, nontoxic to the cells, and can be sensitively quantified. Uptake is directly proportional to its concentration in the medium over a range 0.5 – 10 mg/ml. This indicates that the dextran is ingested by bulk uptake, since receptor-mediated uptake would exhibit saturation characteristics. Uptake rates are proportional to the cell concentration and proceed linearly with time for about 1 hr. No uptake is observed at 2° or in the presence of metabolic inhibitors or uncouplers of oxidative phosphorylation (e.g., 1 μM carbonyl cyanide-*m*-chlorophenyl hydrazone).

Substrate Particles for Phagocytosis

Fluorescein-labeled bacteria (FITC-bacteria) are prepared by incubating bacteria ($OD_{420} = 20$) in 50 m*M* Na_2HPO_4, pH 9.2, in the presence of 0.1 mg of fluorescein isothiocyanate per milliliter (Isomer 1, Sigma) for 3 hr at 37°. Routinely, a B/r strain of *E. coli* is used but other strains of *E. coli* or *K. aerogenes* work equally well. Bacteria are collected and washed by centrifuging at 5000 *g* for 10 min in 20 m*M* potassium phosphate, pH 6.2, until no fluorescence is detectable in the supernatant. Under these conditions, the dye-to-bacteria ratio has been found to be about the same in different batches of cells.

Polystyrene latex beads, carboxylated latex beads, and amino-containing latex beads are commercially available in different sizes (Polysciences).

Glucosylated latex beads are prepared according to the reaction scheme shown in Fig. 1. 1-Thio-D-glucose is reversibly coupled to amino-containing microspheres (diameter 0.45 μm) by use of the bifunctional reagent *N*-succinimidyl-3-(2-pyridyldithio) propionate (SPDP, Pharmacia). To a suspen-

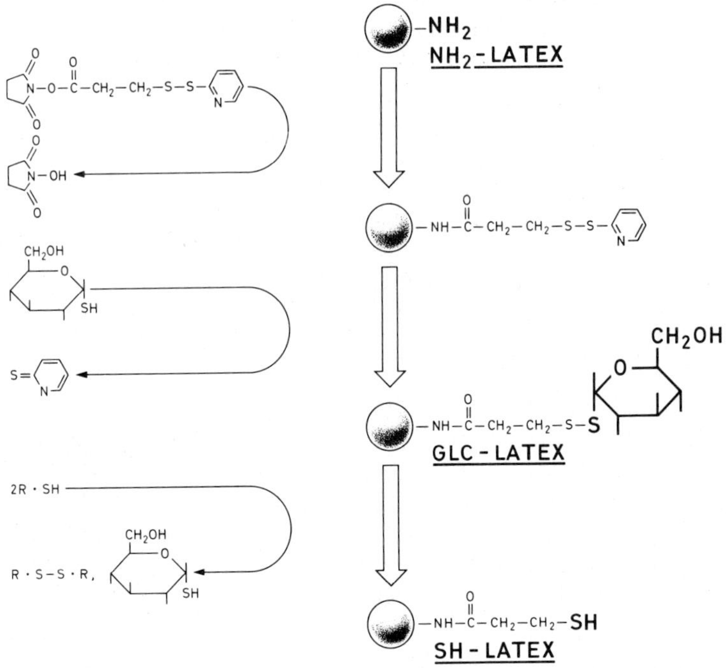

FIG. 1. Schematic presentation of the reaction scheme for reversible glucosylation of amino-containing latex spheres. From Vogel.[4]

sion of latex beads (5% solids) in 100 m*M* potassium phosphate pH 6.5, the same volume of 35 m*M* SPDP dissolved in ethanol is rapidly added, and the mixture is incubated with occasional shaking for 2 hr at room temperature. Excess reagent is removed by centrifuging and washing twice at 1000 *g* for 15 min, and latex beads containing 2-pyridine disulfide groups are suspended to a final concentration of 2.5% solids in the above buffer. 1-Thio-D-glucose (Sigma) is added to a final concentration of 10 m*M*, and the reaction mixture is incubated with gentle shaking at room temperature. The reaction is followed by determining optically the release of pyridine-2-thione ($\epsilon_{343nm} = 8.08 \times 10^3 M^{-1} cm^{-1}$) in aliquots of the solution after centrifuging down the latex beads. When the optical density is maximal and remains constant, latex beads are washed twice by centrifugation and resuspended in 20 m*M* potassium phosphate, pH 6.2. Under these conditions, the degree of substitution is about 1×10^6 glucose molecules on a single sphere. 1-Thio-D-glucose can be subsequently removed by incubation of glucosylated latex beads in the presence of 50 m*M* dithiothreitol. Thereby, thiol-containing latex beads are generated.

Phagocytosis Assays

Phagocytosis in Shaken Suspensions

Amoebae are suspended at a density of 2 to 4×10^6 cells/ml in the chosen medium and incubated on a rotary shaker (routinely at 100 rpm). After 15 min of recovery, substrate particles as required were added at a ratio of particles to amoebae of at least 200 : 1. Phagocytosis is stopped by diluting 1-ml samples at various times into 2 ml of ice-cold potassium phosphate, pH 6.2. To separate amoebae from noningested particles, the cell suspension is centrifuged through a PEG 6000 solution as described for the pinocytosis assay. Noningested particles remain in the top fluid layer, and pelleted amoebae are washed once by centrifuging in 3 ml of 50 m*M* Na_2HPO_4, pH 9.2, and resuspended in the same buffer. After counting, cells are lysed by addition of Triton X-100 (0.2% final concentration).

For FITC-bacteria as substrate particles, the fluorescence intensity is determined as described above. The number of phagocytosed bacteria is determined by comparison with a standard curve. For this, a defined number of bacteria are lysed in a solution of 1% sodium dodecyl sulfate by heating for 2 min at 90° and determining the fluorescence intensity in aliquots of this solution. The additional treatment is necessary, since noningested bacteria, in contrast to ingested ones, are not lysed by 0.2% Triton X-100 and therefore, owing to change in quantum yield, an erroneous relation would otherwise result.

For latex beads as substrate particles, the optical density at 560 nm is measured and the number of ingested particles is determined by comparison with a standard curve.

Phagocytosis on Filters

About 2×10^6 amoebae and 2×10^9 substrate particles are rapidly mixed in 2 ml of the chosen medium. With slight underpressure applied, the suspension is uniformly deposited on filters (0.8 μm pore size, 47 mm in diameter; Millipore, $AABPO_4700$), which then are rested on absorbent support pads presoaked with the same medium. Samples are incubated in petri dishes in a moist atmosphere. After various times, cells are harvested by placing the filter in a centrifuge tube containing 4 ml of ice-cold medium, and cells are resuspended by vigorous shaking. Thereafter, the procedures described above are followed.

Aggregation Assay

Washed cells are suspended in 17 m*M* potassium phosphate, pH 6.2, to a final density of 2 to 3×10^6 cells/ml and incubated on a rotary shaker at 100 rpm. At intervals, 0.1-ml samples are transferred to Accuvette sample beakers by use of a wide-bore plastic pipette tip and aggregation is assessed by determining the decline in particle number with a Coulter counter.

Adhesion Assay

A total of 4×10^6 cells suspended in 4 ml of the medium chosen are placed in tissue culture flasks (plane area, 25 cm^2; volume, 50 ml) and incubated for 40 min without agitation. Subsequently, the flasks are shaken for 3 min on a rotary shaker at 80 rpm, and the percentage of nonadherent cells is estimated by counting cells in the supernatant.

Enrichment of Mutants Defective in Phagocytosis

Principle. Endocytosis is the sole mechanism for nutrient uptake in *Dictyostelium discoideum;* therefore, defective mutant cells are expected to be lethal. Consequently, the selection procedure is devised for isolating temperature-sensitive phagocytosis mutants. Cells that do not phagocytose heavy tungsten beads at 27°, the nonpermissive temperature, are isolated on the basis of their lower density. Since growth on bacteria is dependent on phagocytosis, remaining cells are screened for temperature-sensitive growth on agar plates seeded with bacteria.

Mutagenesis. Amoebae are grown axenically to a density of 2 to 4×10^6 cells/ml, harvested by centrifugation, and washed twice with 20 m*M* potassium phosphate, pH 6.2. Cells are suspended to a concentration of about 3×10^7 cells/ml in the above buffer, and the mutagen *N*-methyl-*N'*-nitro-*N*-nitrosoguanidine, freshly dissolved in dimethyl sulfoxide at 50 mg/ml, is added to give a final concentration of 1 mg/ml. Cells are incubated on a rotary shaker at 120 rpm at 20° for 30 min; subsequently, the reaction is stopped by chilling, and cells are washed twice in phosphate buffer. Viability tests reveal that about 1–5% of cells survive mutagenesis under these conditions. Cells are resuspended in axenic medium at a density of about 1×10^6 cells/ml and immediately subdivided into different batches. Thus mutants derived from different batches are of independent origin. Cells require about 4 days for recovery and are subsequently grown to a density of about 3 to 4×10^6 cells/ml (about five to seven doublings) to allow the expression of mutations.

Selection. Separately grown mutagenized cells (10 ml in a 100-ml Erlenmeyer flask) are shifted from 20° to 27° to a reciprocal waterbath shaker (150 strokes/min; stroke, 2.5 cm) and incubated for 2 hr. Subsequently, 200 mg of tungsten beads (1 μm in diameter, Planseewerke, Austria) are added, and the incubation is continued for another 2 hr. The bulk of noningested tungsten beads is removed by incubating for 5 min without shaking, then decanting the supernatant carefully into centrifuge tubes. The cell suspension is diluted 1 : 3 by addition of axenic medium, and cells containing tungsten beads are mainly precipitated by centrifuging for 2 min at 70 g in a swing-out rotor. The centrifugation procedure is repeated until no cells containing tungsten beads are detectable in the supernatant by microscopic observation. After intermittent growth of the amoebae at 20°, the tungsten treatment is repeated twice. Remaining cells are plated clonally at 20° on nutrient agar plates seeded with bacteria. Clones are transferred with toothpicks in duplicate to agar plates previously spread with bacteria and are examined for temperature-sensitive growth by replica plating at 20° and 27°. About 5 – 10% of cells are usually found to be temperature sensitive for growth. This frequency is about 50 – 100 times higher than that of mutagenized cells before the tungsten selection.

Characterization of Mutants

Most of 100 independently isolated mutants, which were temperature sensitive for growth upon bacteria (which depends upon phagocytosis), were also temperature sensitive for axenic growth, which depends upon pinocytosis. In these mutants, the endocytotic capacity decreases with the decreasing growth rate at the nonpermissive temperature. These strains may be impaired in any essential cellular process either directly or only indirectly participating in endocytosis.

However, three mutant strains, named HV29, HV32, and HV33, carry a mutation that is unequivocally related to the phagocytotic event. These mutants are altered in cell-particle binding in phagocytosis, and, in addition, they are altered in other cohesive properties of the cells. All three have the same phenotype: they grow axenically by pinocytosis like wild-type cells at 20° and 27°, and mutant amoebae pinocytose with at twice the rate of wild-type cells at both temperatures. When incubated in the absence of shear forces on filters, both mutant and wild-type cells phagocytose all the various substrate particles at comparable rates. In contrast to wild-type cells, mutant cells do not phagocytose any of the various substrate particles at any temperature when incubated in axenic medium in agitated suspensions, in which shear forces are generated by shaking (cf. Fig. 3). Detailed analysis of the mutant phenotype reveals the following differences compared to the wild type.

1. Bacteria that contain terminal glucose residues on the surface (e.g., *E. coli* B/r) and glucosylated latex particles are phagocytosed normally (cf. Fig. 2). However, in contrast to wild-type cells, uptake is specifically inhibited by glucose and by molecules containing glucose at the nonreducing terminus.

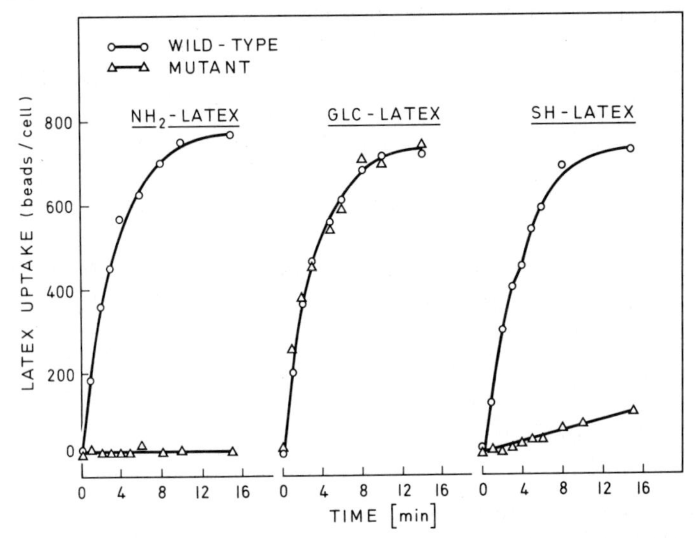

FIG. 2. Comparison of uptake rates of amino-, glucose-, and thiol-containing latex spheres (diameter 0.45 μm) at 20° by wild type (○) and mutant (△) amoebae in shaken cultures in 20 mM phosphate buffer, pH 6.2. From Vogel.[4]

This type of inhibition is competitive with an apparent inhibition constant of about 1 mM.[3,4]

2. Polystyrene latex beads, having very hydrophobic surface properties, are phagocytosed by mutant and wild-type cells at comparable rates. But more hydrophilic particles, such as protein-coated latex beads, amino- or thiol-containing latex beads (Fig. 2), or bacteria without terminal glucose residues,[3] cannot be phagocytosed by mutant cells whereas uptake by wild-type cells remains normal.

3. Mutant strains are also altered in cohesive properties. First, when exponentially growing wild-type cells are starved by incubation in phosphate buffer, they form large, tight aggregates within 15 min. In contrast, mutant amoebae remain as single cells and do not aggregate when incubated under the same conditions (cf. Fig. 4). Second, wild-type cells adhere tightly to the surface of tissue culture plates, and all cells spread in about 20 min. On the other hand, mutant cells adhere only loosely or remain in suspension under the same conditions.

Isolation of Adhesive Revertants

To select for adhesive revertants, mutant cells are suspended in 10 ml of axenic medium at a density of 1 to 2 $\times$ 10^6 cells/ml and placed into a tissue-culture flask (plane area, 75 cm^2; volume, 250 ml). After incubation for 90 min without agitation, the cell suspension is shaken on a reciprocal shaker (140 strokes/min; stroke, 2.5 cm) for 15 min, and cells in the supernatant are sucked off. The bottom of the flask is washed once by flushing with 10 ml of axenic medium; after addition of 10 ml of fresh medium, remaining cells are grown to a density of about 1 to 2×10^6 cells/ml by incubating the tissue culture flasks on a rotary shaker. The selection procedure is repeated about 15–20 times until microscopic observation reveals that most of the cells adhere and spread like wild-type cells. Cells are cloned, and five revertants of each mutant strain are analyzed for phagocytosis and cell aggregation. All the revertants ingest the various substrate particles like wild-type cells when incubated in axenic medium in shaken suspension (cf. Fig. 3). Furthermore, when starved and shaken in phosphate buffer, all the revertants form tight aggregates like those formed by wild-type cells (Fig. 4).

Conclusions

Analysis of the mutant phenotype reveals on the cell surface the presence of two functionally distinct receptors that recognize different surface features on various substrate particles. Wild-type amoebae can bind and internalize a wide variety of substrate particles, either very hydrophobic (e.g., polystyrene latex) or more hydrophilic (e.g., amino- and thiol-containing latex, bacteria) via "nonspecific" receptors. Mere physical forces seem to

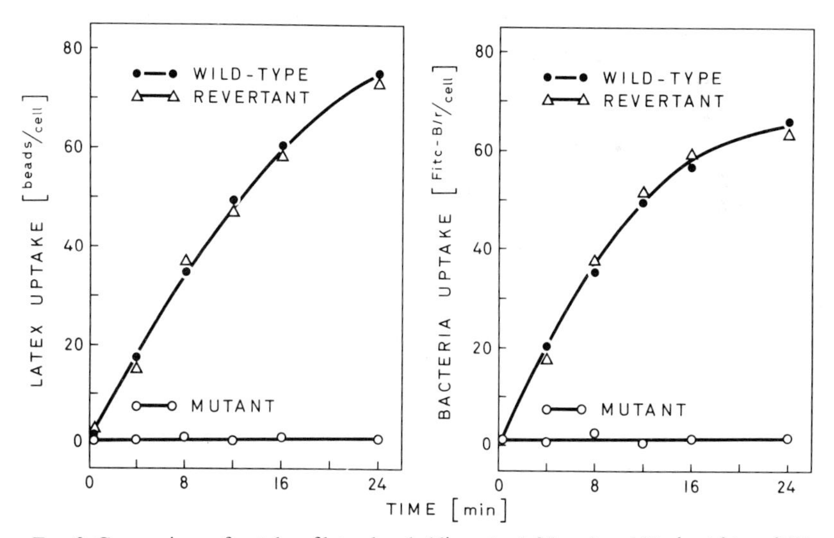

FIG. 3. Comparison of uptake of latex beads (diameter 1.04 μm) and *Escherichia coli* B/r at 20° by wild-type, mutant, and revertant amoebae in shaken cultures in axenic medium.

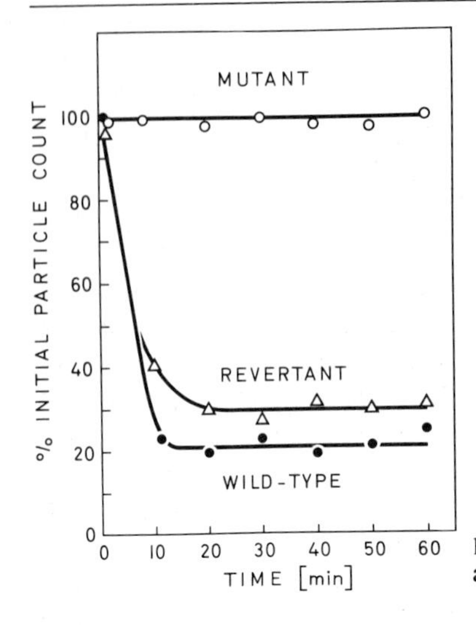

FIG. 4. Comparison of cell aggregation in 20 mM phosphate buffer of wild-type, mutant, and revertant amoebae.

promote adhesion between cells and particles by hydrophobic interaction. In the mutants, nonspecific binding is altered in such a way that only very hydrophobic (e.g., polystyrene beads), but not more hydrophilic (e.g., bacteria), particles can be bound by these receptors. The increased specificity of substrate binding by mutant amoebae enables the identification and characterization of another, specific binding site. Only substrate particles containing terminal glucose residues are bound by this lectin-type receptor, and binding is inhibited specifically and competitively by glucose.

Three independently derived mutants with altered properties in cell–particle binding are additionally altered in other cohesive properties. They do not adhere to plastic surfaces and they do not form tight aggregates when starved in phosphate buffer. Therefore it seems likely that a common structural basis exists for these properties. This is confirmed by isolating revertants that are once more adhesive on plastic surfaces. All the revertants phagocytose like wild-type cells and form aggregates on starvation.

Taken together, the analysis of mutant phenotype makes evident a relationship between the processes of cell–particle recognition in phagocytosis, cell adhesion and spreading on extended surfaces, and cell–cell aggregation.

[38] Recycling of Insulin-Sensitive Glucose Transporter in Rat Adipocytes

By TETSURO KONO

It is well known that insulin lowers the blood sugar concentration by stimulating transport of glucose across the plasma membranes of muscle and fat cells. According to the theory developed in this and other laboratories,[1-5] the glucose transport mechanism in the resting cells that are not stimulated by insulin is mostly localized in an intracellular storage site; however, when cells are exposed to insulin, the glucose transport mechanism is translocated from the storage site to the plasma membrane, and when the hormone is eliminated from the system, the glucose transport activity is retranslocated from the plasma membrane back into the storage site. The translocation, or the recycling, of the glucose transport mechanism can be assessed empirically by the methods described below.[1,2]

Experimental Approaches

In an attempt to study the recycling of the glucose transport mechanism, the first problem was to find a method of assay of glucose transport activity in a cell-free system. This problem has been solved in two laboratories by two different approaches. Thus, Cushman and Wardzala[3] and Karnieli *et al.*[4] estimated the number of the glucose transport carriers by measuring the activity of D-glucose-inhibitable [^{3}H]cytochalasin B binding in the presence of unlabeled cytochalasin E. Cytochalasin B is a potent competitive inhibitor of glucose transport in a number of cell types, and although the binding of cytochalasin B is not specific to the glucose transport carriers, most of the nonspecific binding sites are also bound by cytochalasin E, which does not inhibit glucose transport.[3,4] These investigators found that administration of insulin to fat cells causes an increase in the specific cytochalasin B binding activity in the plasma membrane fraction while decreasing the activity in the low-density microsome fraction. They found also that elimination of insulin from the system initiates the reversal of the above hormone effects. In our

[1] K. Suzuki and T. Kono, *Proc. Natl. Acad. Sci. U.S.A.* **77,** 2542 (1980).

[2] T. Kono, K. Suzuki, L. E. Dansey, F. W. Robinson, and T. L. Blevins, *J. Biol. Chem.* **256,** 6400 (1981).

[3] S. W. Cushman and L. J. Wardzala, *J. Biol. Chem.* **255,** 4758 (1980).

[4] E. Karnieli, M. J. Zarbowski, P. J. Hissin, I. A. Simpson, L. B. Salans, and S. W. Cushman, *J. Biol. Chem.* **256,** 4772 (1981).

[5] L. J. Wardzala and B. Jeanrenaud, *J. Biol. Chem.* **256,** 7090 (1981).

ISBN 0-12-181998-1

laboratory, we solubilized glucose transport activity, reconstituted it into egg lecithin liposomes, and measured the glucose transport activity in the reconstituted system.[1,2] We found that insulin increases the glucose transport activity in the plasma membrane-rich fraction while decreasing the activity in the Golgi-rich fraction, and that elimination of insulin decreases the transport activity in the plasma membrane-rich fraction while increasing the activity in the Golgi-rich fraction. Furthermore, we found that the above insulin effects and their reversal are dependent on metabolic energy and independent of protein synthesis.

In the following section, I describe the experimental procedures that are involved in the reconstitution approach. The advantages of this approach are that the results of the study show the changes in the functional glucose transport activity and that the experiments require a considerably smaller number of animals as compared to the cytochalasin binding approach. On the other hand, the advantage of the latter approach is that the results of the experiment show the actual number of glucose transport carriers.

Treatment of Fat Cells

Principle

The glucose transport activity in isolated rat epididymal fat cells is stimulated by the addition of insulin, and its effect is reversed by elimination of the hormone. In addition, effects of various agents on the development and reversal of the insulin effect are tested. The treated cells are washed with an isotonic sucrose solution and homogenized. Unless otherwise specified, the indicated pH values refer to those determined at room temperature.

Materials

Fat cells: Isolated epididymal fat cells from 1.5 rats (or approximately 1.5 g of adipose tissue) are needed for each test; thus cells from three rats are required to test the basal and plus-insulin activities.

Krebs–Henseleit HEPES buffer (pH 7.4 measured at 37°) containing 2 m*M* glucose and 20 mg/ml fraction V bovine serum albumin. The fraction V albumin we use was purchased from Rheis Chemical Company, a Division of Armour Pharmaceuticals.

Buffer A; 0.25 *M* sucrose containing 1 m*M* EDTA(Na) and 10 m*M* Tris-HCl (pH 7.5). The sucrose we use is the Grade 1 preparation from Sigma.

Pure oxygen

Zinc insulin, 1 μM in 0.154 *M* NaCl containing 3 m*M* HCl. This solution may be stored in a refrigerator for 1 month.

2,4-Dinitrophenol, 50 m*M* Tris salt in 0.154 *M* NaCl
KCN, 100 m*M* in 0.154 *M* NaCl
Puromycin, 10 m*M* in 0.154 *M* NaCl
Cycloheximide, powder
Crude bacterial collagenase. We use type CLS from Worthington; the preparation should contain a minimum amount of trypsin-like protease (see text).

Procedures

Isolated fat cells are prepared by the collagenase method[6] from Sprague–Dawley rats weighing approximately 180–230 g. The cells from 1.5 rats are suspended in 10 ml of Krebs–Henseleit HEPES buffer (pH 7.4 at 37°) containing 2 m*M* D-glucose and 20 mg/ml of fraction V albumin.[7] The cell suspension is first incubated in pure oxygen at 37° for 30 min with gentle shaking (1 stroke/sec). During this preliminary incubation, the basal glucose transport activity declines considerably.[8] The cells are then exposed to 1 n*M* insulin for 10 min in the presence or the absence of various agents. For example, the ATP level in fat cells may be lowered to less than one-tenth of the normal by incubating cells for 5 min with either 1 m*M* 2,4-dinitrophenol (Tris salt) or 1 to 2 m*M* KCN.[9] The protein synthesis is inhibited more than 95% by exposure of cells to either 0.1 m*M* puromycin or 1 m*M* cycloheximide for 90 min.[2] The effect of insulin is reversed by incubation of the insulin-treated cells with 1 mg/ml of crude bacterial collagenase for 45–60 min.[2]

The treated cells are washed twice with 8 ml each of buffer A at approximately 15° by brief centrifugation (e.g., for 30 sec at 700–800 rpm in a bench-top International Clinical Centrifuge set at one-half of the maximum acceleration). The washed cells are suspended in 10 ml of buffer A and homogenized in a Dounce tissue grinder by eight strokes with a type B pestle.[10]

Comments

Contrary to the observation by some investigators, we found that the insulin effect on glucose transport in adipocytes from a 250-g rat is not significantly less than that on the activity in cells obtained from a 150-g rat.

[6] M. Rodbell, *J. Biol. Chem.* **239,** 375 (1964).
[7] T. Kono and F. W. Barham, *J. Biol. Chem.* **246,** 6210 (1971).
[8] F. V. Vega and T. Kono, *Arch. Biochem. Biophys.* **192,** 120 (1979).
[9] T. Kono, F. W. Robinson, J. A. Sarver, F. V. Vega, and R. H. Pointer, *J. Biol. Chem.* **252,** 2226 (1977).
[10] V. Manganiello and M. Vaughan, *J. Biol. Chem.* **248,** 7164 (1973).

Epididymal adipose tissue may be supplemented with perirenal adipose tissue, which is slightly less sensitive to insulin than is epididymal tissue. Crude bacterial collagenase from *Clostridium histolyticum* contains an acidic proteinase[11] that rapidly modifies insulin. Therefore, cells prepared by the collagenase digestion must be washed carefully before the insulin experiment, and the collagenase preparation may be used to eliminate insulin from the incubation mixture.[2] Certain collagenase preparations contain trypsin-like protease, which may either modify the insulin receptor or mimic the action of the hormone depending on the level of contamination.[12,13] We test the quality of a collagenase preparation by measuring the insulin sensitivity of fat cells prepared with the enzyme preparation. Usually fraction V albumin can be used without any purification; however, some preparations contain insulin-like activity that may be eliminated by trypsin treatment.[14] One of the purposes in washing cells with buffer A is to eliminate divalent cations that interfere with the fractionation of the subcellular structures. If the temperature of buffer A is too low, most of the membrane structures would be trapped by partially solidified fat during homogenization and would never be recovered.[15] On the other hand, if the temperature is too high, a part of the insulin effect may be reversed during the washing procedure.

Isolation of the Plasma Membrane-Rich and Golgi-Rich Fractions

Principles

When the fat cell homogenate is centrifuged, subcellular structures are sedimented in the following order: nuclei, mitochondria + lysosomes (major), plasma membrane, endoplasmic reticulum, Golgi apparatus, and lysosomes (minor). The glucose transport activity is associated with the plasma membrane-rich and Golgi-rich fractions, which may be isolated by several different methods. Methods A–C presented below are currently in use in our laboratory.

Materials

Fresh fat cell homogenate
Buffer A: 0.25 *M* sucrose containing 1 m*M* EDTA(Na) and 10 m*M* Tris-HCl (pH 7.5)

[11] T. Kono, *Biochemistry* **7,** 1106 (1968).
[12] T. Kono, *J. Biol. Chem.* **244,** 5777 (1969).
[13] T. Kono and F. W. Barham, *J. Biol. Chem.* **246,** 6204 (1971).
[14] J. E. Jordan and T. Kono, *Anal. Biochem.* **104,** 192 (1980).
[15] T. Kono, F. W. Robinson, and J. A. Sarver, *J. Biol. Chem.* **250,** 7826 (1975).

Sucrose (15%, w/w) containing 1 m*M* EDTA(Na) and 10 m*M* Tris-HCl (pH 7.5). The sucrose concentration is 158.9 g/liter or 0.465 *M*; $d = 1.0592$.

Sucrose (32.5%, w/w) containing 1 m*M* EDTA(Na) and 10 m*M* Tris-HCl (pH 7.5). The sucrose concentration is 370.2 g/liter or 1.082 *M*; $d = 1.1391$.

EDTA(Na), 1 m*M*, in 10 m*M* Tris-HCl (pH 7.5)

Procedures

The methods of fractionation are schematically presented in Fig. 1. In our laboratory, all the centrifugations are carried out in Beckman's J-21 or L8-70 centrifuges set at 2°. The centrifuge time includes the time for acceleration, but not that for deceleration.

Method A. The cell homogenate prepared in buffer A (see Treatment of Fat Cells, above) is immediately centrifuged for 2 min at 7000 rpm in a JA-20 rotor (5900 g_{max}). The infranatant solution (S-1) is withdrawn with a

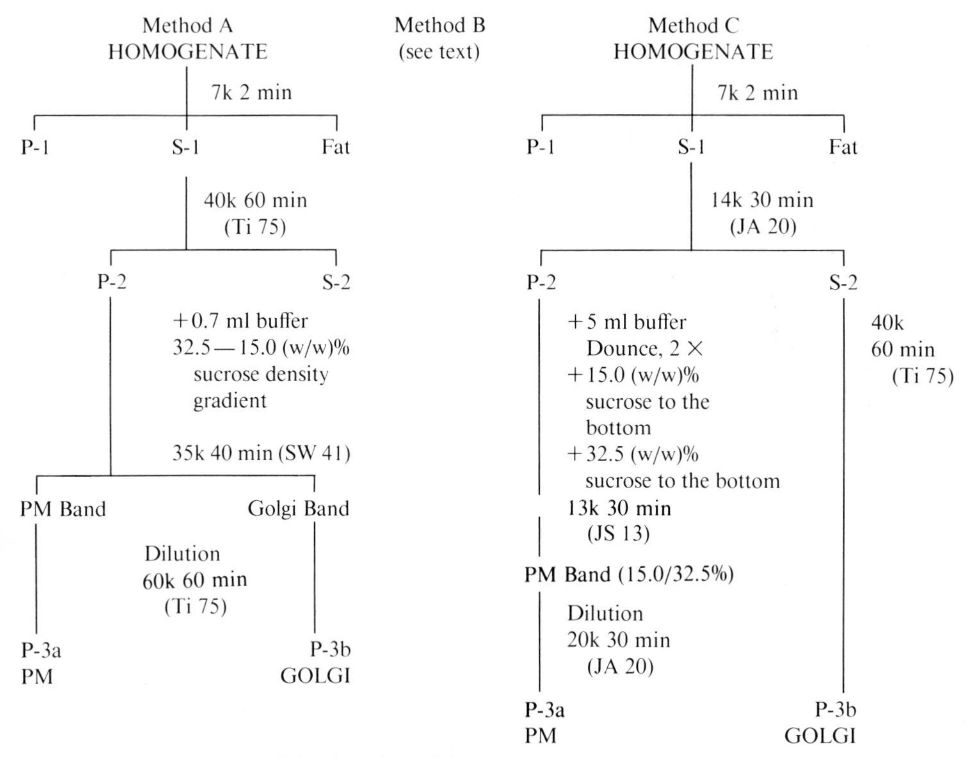

FIG. 1. Methods of fractionation of fat cell homogenate. k = 1000 rpm; PM, plasma membrane rich.

syringe from underneath the partially solidified fat and centrifuged for 60 min at 40,000 rpm in a Ti 75 rotor (150,000 g_{max}, $\omega^2 t = 5.85 \times 10^{10}$). The supernatant solution is discarded, the remaining solution is wiped off, and the pellet (P-2) is suspended in 0.5 ml of ice-cold buffer A using a 500-μl Eppendorf pipette. The suspension is gently placed on top of a sucrose density gradient described below. A fraction of P-2 still remaining in the centrifuge tube is rinsed with 0.2 ml of ice cold buffer A, and the rinse is added to the main fraction on top of the sucrose density gradient.

A linear sucrose density gradient is prepared in a centrifuge tube for an SW 41 rotor (14 $\times$ 89 mm); the limiting sucrose concentrations are 32.5 and 15.0% (w/w). The solution contains 1 m*M* EDTA(Na) and 10 m*M* Tris-HCl (pH 7.5). The centrifugation is carried out for 40 min at 35,000 rpm in an SW 41 rotor (160,000 g_{av}, $\omega^2 t = 2.81 \times 10^{10}$).

After the centrifugation, the sucrose solution, including 0.7 ml of buffer A applied to the top with fraction P-2, is drained from a hole punctured in the bottom of the centrifuge tube and separated into 17 fractions (each 0.75 ml). Fractions 3 through 7 (the plasma membrane-rich fraction) and fractions 13 through 17 (the Golgi-rich fraction) are separately pooled, diluted with 2 ml of 1 m*M* EDTA(Na) in 10 m*M* Tris-HCl (pH 7.5), and centrifuged for 60 min at 60,000 rpm in a Ti 75 rotor (340,000 g_{max}, $\omega^2 t = 1.28 \times 10^{11}$).

Method B. The cell homogenate is centrifuged for 2 min in a centrifuge set at 7000 rpm (5900 g_{max}). Both the pellet (P-1) and the supernatant solution (S-1) are withdrawn from underneath the partially solidified fat and centrifuged for 15 min at 13,000 rpm in a JA-20 rotor (20,200 g_{max}). The pellet (P-2) is saved while the supernatant (S-2) is centrifuged for 30 min at 40,000 rpm in a Ti 75 rotor (150,000 g_{max}). The pellet (P-3, enriched with Golgi vesicles) and that (P-2, enriched with the plasma membrane) saved earlier are separately suspended in 0.7 ml (total) of cold buffer A and subjected to the linear sucrose density gradient centrifugation (35,000 rpm, 40 min) as described in Method A.

Method C. The second centrifugation described in Method B is carried out for 30 min at 14,000 rpm (23,500 g_{max}). The supernatant solution (S-2) is saved for preparation of the Golgi-rich fraction. The pellet (P-2) is suspended in 5 ml of ice-cold buffer A using a Dounce tissue grinder (two strokes with a type B pestle). The suspension is placed in a 15-ml Corex test tube, and 2 ml of 15% (w/w) sucrose solution followed by 3 ml of 32.5% (w/w) sucrose solution both containing 1 m*M* EDTA(Na) and 10 m*M* Tris-HCl (pH 7.5) are introduced to the bottom of the tube using injection syringes equipped with long needles. The test tube is then centrifuged for 30 min at 13,000 rpm in a JS-13 rotor (18,000 g_{av}). After the centrifugation, the plasma membrane-rich band appearing at the interface between 15.0% and 32.5% sucrose solutions is withdrawn along with 2.0 ml of the solution using an injection syringe. The syringe is rinsed with 1 ml of buffer A, and the rinse is mixed with the main fraction. The mixture is diluted with 2 ml of ice-cold 1 m*M* EDTA(Na) in 10 m*M* Tris-HCl (pH 7.5), and the plasma membrane-rich fraction is sedimented by centrifugation for 30 min at 20,000 rpm in a JA-20 rotor (40,000 g_{max}). The Golgi-rich fraction is sedimented by centrifugation of fraction S-2 (see above) for 30 min at 40,000 rpm in a Ti 75 rotor (150,000 g_{max}).

Comments

Methods A and B are modifications of the method described by Kono *et al.*[2] and method C is a modification of the methods described by Kono *et al.*[15a] Method A is useful in preparing reasonably pure subcellular fractions for general characterization of the glucose transport activity. Method B yields the purest subcellular fraction; however, the procedure is tedious as the plasma membrane-rich and Golgi-rich fractions are separately subjected to sucrose density gradient centrifugation, and one extra centrifugation is needed as compared to method A. Method C is considerably simpler than the others and yields membrane preparations almost as pure as those obtained by method A.[15a]

The sucrose density gradient centrifugation involved in methods A and B is not an isopycnic centrifugation, and the results are significantly less satisfactory if the centrifugation is carried our for a longer period of time (e.g., 1 hr) at a higher speed (e.g., at 40,000 rpm). Under the conditions specified above (35,000 rpm for 40 min), both nuclei and mitochondria are pelleted at the bottom of the centrifuge tube, and the plasma membranes form a white band at approximately one-third from the bottom. The bands of the endoplasmic reticulum and Golgi apparatus are invisible.

The plasma membrane-rich fraction is enriched with 5′-nucleotidase and catecholamine-sensitive adenylate cyclase.[12] Contamination of the plasma membrane fraction by mitochondria may be estimated from the content of cytochrome *c* dehydrogenase[12]; and that of lysosomes from the content of *β*-*N*-acetyl-D-glucosaminidase.[16] However, contamination by endoplasmic reticulum is difficult to assess, since both phosphodiesterase and rotenone-insensitive NADH dehydrogenase form a distinct peak in the plasma membrane fraction as well as in the endoplasmic reticulum fraction. The Golgi-rich fraction is enriched with UDPGal:*N*-acetylglucosamine galactosyltransferase.[1,2] However, this enzyme also forms a distinct peak in the plasma membrane-rich fraction.[2]

[15a] T. Kono, M. M. Smith, F. W. Robinson, and T. Watanabe, *Arch. Biochem. Biophys.*, submitted.

[16] A. Horvat, J. Baxandall, and O. Touster, *J. Cell Biol.* **42,** 469 (1969).

Solubilization and Reconstitution of the Glucose Transport Mechanism

Principle

Subcellular membranes are solubilized with detergent, the macromolecular fraction is isolated by gel filtration, and the glucose transporter (along with other lipophilic proteins) is incorporated into egg lecithin liposomes by sonication, freezing, thawing, and a second sonication.[1,2,17]

Materials

The plasma membrane-rich and Golgi-rich fractions of fat cells
Buffer B: 10 m*M* Tris-HCl (pH 7.5)
Sodium cholate 20 m*M* in buffer B
Sephadex G-50 (fine), packed in Econo-Columns from Bio-Rad, 7 × 26-mm (volume = 1 ml), and equilibrated with buffer B at room temperature
Egg lecithin, Sigma type IX-E, P-8640, approximately 60% pure. This waxy preparation is divided into small fractions, placed in glass vials, and stored in nitrogen at −20°.

Procedures

Pellets of the plasma membrane-rich and the Golgi-rich fractions (from 1.5 rats) are each dissolved in 0.22 ml of buffer B. The solubilization is aided by manual mixing for 10 min at 0°, followed by a 5-sec sonication at setting 3 (see below for details of sonication). The solution is then placed in a deep freezer at −70°. Although this freezing is not essential, preparations that are frozen at this step appear to give consistently high transport activity after reconstitution. Also, if the experiment is started in the morning, this is probably the most convenient step to conclude the day's work; the preparations frozen at this step can be stored at −70° at least for 1 week without significantly losing the transport activity.

The frozen preparations are then thawed in water at room temperature. The Golgi-rich fraction should be clear at this stage. The plasma membrane-rich fraction prepared by method A or method B (see the preceding section) also should be clear or almost clear; however, the plasma membrane-rich fraction prepared by method C is considerably turbid. A turbid preparation should be centrifuged for 10 min at 10,000 rpm in a JA-21 rotor (12,000 g_{max}), or it may obstruct gel filtration in the next step.

For the gel filtration, 200 μl of sample in sodium cholate solution is

[17] F. W. Robinson, T. L. Blevins, K. Suzuki, and T. Kono, *Anal. Biochem.* **122,** 10 (1982).

applied to a column of Sephadex G-50 (7 × 26 mm), equilibrated with buffer B at room temperature. The column is washed with buffer B, and the macromolecular fraction is recovered between 0.15 and 0.5 ml of the washing solution. This gel filtration is carried out at room temperature, and the eluates are placed in ice. The protein concentration in the eluate is determined in duplicate (using 25 μl of the solution) by the Bradford method.[18]

For reconstitution of the glucose transport activity, 250 μl of the macromolecular fraction obtained by gel filtration is mixed with 50 μl of 150 mg/ml of egg lecithin in buffer B (see below); this mixing should be done carefully and thoroughly using a Vortex mixer. The mixture is divided into four 75-μl aliquots; and each aliquot is separately sonicated for 5 sec at setting 3, frozen at −70° for 15 min or more, thawed in water at room temperature, and sonicated again for 5 sec at setting 3 (see below for details of sonication). Alternatively, 200 μl of the macromolecular fraction may be mixed with 40 μl of egg lecithin suspension and divided into three 80 μl aliquots.

For preparation of the egg lecithin liposomes used above, a waxy crude egg lecithin preparation (Sigma type IX-E) is scooped with a plastic spatula and placed at the bottom of a polystyrene test tube for sonication (see below). The test tube is weighed before and after the sampling, and a calculated amount of buffer B is added to make the final lecithin concentration 150 mg/ml (we add 0.930 ml buffer to 150 mg lecithin on the assumption that $d = 1.080$). The lecithin is dispersed into the buffer in four steps: (*a*) manual mixing with a plastic spatula; (*b*) sonication for 2 min in nitrogen at setting 4 (the spatula used above should be broken in half, and the lower half covered with lecithin should be sonicated in the test tube); (*c*) vigorous mixing with a Vortex mixer for 30 sec; and (*d*) a second sonication for 3 min at setting 4.[17] The final preparation should be homogeneous and translucent. If necessary, the sonication may be carried out longer (tested up to 9 min). The maximum sample size tested for this dispersal of egg lecithin is 1 ml.[17]

In our laboratory, all the sonication for reconstitution is done in a cup horn from Heat Systems Ultrasonic (catalog No. 431A) filled with water and connected to a Branson sonifier (Model w-185) (Fig. 2). The sample to be sonicated is placed in a polystyrene test tube (Sarstedt, 55-468), 16.8 × 95 mm. The bottom of this test tube is filed smooth, and it is placed 2–3 mm above the sonicator horn using an L-shaped metal strip as a measure (the thickness of the metal strip we use is 2.3 mm, but it can be anywhere between 2 and 3 mm[17]).

The power of the sonic oscillation may be assessed indirectly by sonicat-

[18] M. M. Bradford, *Anal. Biochem.* **72,** 248 (1976).

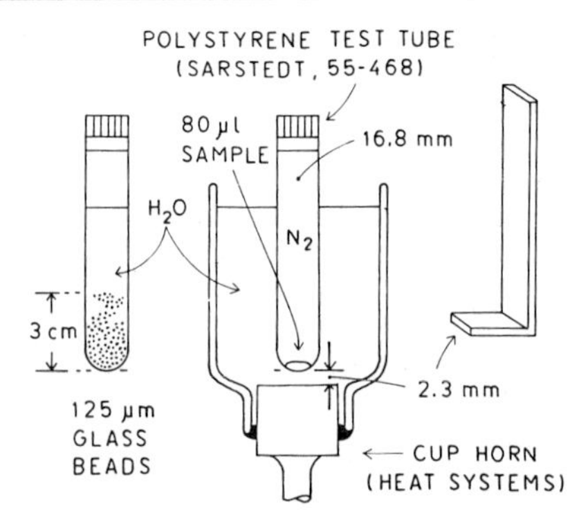

FIG. 2. Sonication apparatus.[17]

ing 0.5 g of glass beads, 125 μm in diameter (from Heat Systems Ultrasonic) in a polystyrene test tube filled with 10 ml of water. At setting 3 of our sonicator, the glass beads are stirred up approximately 3 cm from the bottom.[17] Alternatively, 0.5 g of Spherosil XOA 400 (silica beads, 100–200 μm in diameter, from Supelco) may be sonicated in the polystyrene test tube filled with water; at setting 3, the silica beads form a band approximately between 2 and 3 cm from the bottom.[2]

Comments

For solubilization of the fat cell glucose transporter, the optimum sodium cholate concentration is 15–20 m*M*; this agent is injurious to the transport activity at higher cncentrations.[17] Although the egg lecithin preparation used in this work is not pure, several different batches tested (all, types IX-E from Sigma) gave almost identical results. In contrast, several other phospholipid preparations tested gave poor results.[17] The optimum protein concentration from reconstitution is 200–500 μg/ml, and the optimum sample size is approximately 80 μl. The optimum length of the first sonication is 5–10 sec, and that for the second sonication is 2–8 sec.[17] It is important to use polystyrene test tubes of the indicated size for sonication; otherwise, no reproducible results may be obtained.[17] The sonication may be done in air, although we always sonicate our samples in nitrogen for added safety.[17] While the freezing step is imperative, no special handling (e.g., a rapid freezing in liquid nitrogen) is needed. We simply place the test tube in a deep freezer. The precision of the glucose transport assay in the reconstituted system largely depends on the reproducibility of the sonication steps. For this reason, we routinely divide a mixture of protein and egg lecithin into three or four aliquots prior to the first sonication, and monitor the reproducibility of the reconstitution process.

Glucose Transport Assay in the Reconstituted System

Principle

Liposomes reconstituted with glucose transporter are incubated with a mixture of D-[^{3}H]glucose and L-[^{14}C]glucose. At the end of the incubation, the reaction mixture is diluted with cold $HgCl_2$ solution and immediately filtered with a Millipore filter. Since D-glucose is preferentially transported by the carrier-mediated mechanism, the difference in the amounts of D-[^{3}H]glucose and L-[^{14}C]glucose found on the filter is assumed to represent the carrier-mediated glucose transport activity.[1,2,17]

Materials

Reconstituted liposomes containing glucose transporter
Buffer B; 10 m*M* Tris-HCl, pH 7.5
Labeled glucose. We use D-[6-^{3}H(N)]glucose (NET 100) and L-[1-^{14}C]glucose (NEC 478), both from New England Nuclear.
Buffer C: 2 m*M* $HgCl_2$ in buffer B
Millipore filter, type GSWP 02500 (25-mm disks) or GSWP 304FO (1 ft × 1 ft sheets). The latter should be cut into 1-in. squares.
Scintillation cocktail. We use ACS solution, a xylene base mixture, from Amersham.

Procedures

Preparation of Labeled Glucose Solution. The solution is composed of 5 m*M* D-[^{3}H]glucose (100 μCi/ml) plus 5 m*M* L-[^{14}C]glucose (25 μCi/ml) in buffer B. For preparation of 4 ml of this solution, 400 μCi of D-[^{3}H]glucose and 100 μCi of L-[^{14}C]glucose, both in 90% ethanol, are separately dried in a stream of air at room temperature and redissolved into 2-ml portions of buffer B containing 5 m*M* D-glucose plus 5 m*M* L-glucose. As standards, 5-μl portions of each solution are withdrawn (in triplicate) and diluted with 1-ml portions of buffer B. The rest of the solutions are mixed together, and 5-μl portions of the mixture are also diluted with 1-ml portions of buffer B as standard for the final mixture. It is important to remove ethanol from the glucose preparation, since the glucose uptake by liposomes is inhibited as much as 6–7% by 1% ethanol.[17]

Estimation of the Carrier-Mediated D-Glucose Uptake by Liposomes Containing Glucose Carrier. A 20-μl portion of reconstituted liposomes is placed at the bottom of a small test tube (e.g., 12 × 55-mm polystyrene test tube from Sarstedt, 55-484), stoppered, and placed in ice. The test tube is later transferred to a water bath at 37° approximately 4 min prior to the transport assay. The latter is initiated by mixing liposomes with labeled glucose; in practice, 5 μl of the labeled glucose solution is placed near the bottom of the test tube a few millimeters from the liposome suspension (see Fig. 3). The test tube is then lightly pressed to a Vortex mixer; its vibration will cause the droplet of glucose solution to slide down toward the liposome solution. At the instant when the two solutions are mixed, a stopwatch is started, and the test tube is momentarily pressed harder to the Vortex mixer for better mixing. The test tube is immediately brought back into the 37° water bath, and the incubation is continued for 10–20 sec. At the end of the incubation, the reaction mixture is diluted with 1 ml of ice-cold buffer C and immediately filtered with a piece of Millipore filter, GSWP (0.22 μm in pore size) attached, with its shiny side down, to a filtration apparatus (Millipore, 10-025-00) equipped with a special short funnel (20 mm in height and 16 mm in inner diameter). The test tube is washed once with 1 ml of buffer C, and the funnel and the filter are washed twice with 1-ml portions of buffer C. The filter membrane is then removed from the filtration apparatus, placed in a glass vial for scintillation counting, and dried in a stream of air. The vial is filled with 10 ml of ACS solution and shaken for 1 hr at a rate of 1.5 cycles/sec prior to determination of radioactivity by the liquid scintillation method.

As shown in the table, the final concentration of D-glucose in the incubation mixture (25 μl) is 1 mM (20 μCi/ml or 0.5 μCi in total); that of egg lecithin, 2%; and that of protein, two-thirds of the level determined in the eluate after gel filtration.

A Method for Studying Effects of Potential Inhibitors. Effects of potential transport inhibitors may be tested by a protocol designated as the "alternative method" in the table. In this method, liposomes are reconstituted from 1 volume of protein and one-fifth volume of 200 mg/ml of egg lecithin per milliliter. Then, 15 μl of the reconstituted liposomes is mixed with 5 μl of the inhibitor solution plus 5 μl of the labeled glucose solution. As shown in the table, the final protein concentration is one-half of the level determined after

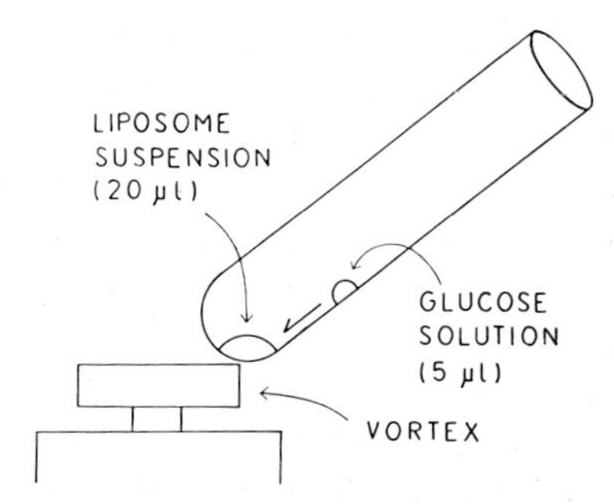

FIG. 3. Mixing of liposomes and labeled glucose.

TWO PROTOCOLS FOR RECONSTITUTION AND TRANSPORT ASSAY

	Standard method	Alternative method[a]
A. Reconstitution		
Protein preparation (a, μg/ml)	250 μl	250 μl
Egg lecithin, 150 mg/ml	50 μl	—
Egg lecithin, 200 mg/ml	—	50 μl
	300 μl	300 μl
Divide into	75 μl × 4	75 μl × 4
B. Transport assay		
Liposomes	20 μl	15 μl
Inhibitors to be tested	—	5 μl
Labeled glucose	5 μl	5μl
	25 μl	25μl
C. Final concentrations		
Protein	$a \times \frac{2}{3}$	$a \times \frac{1}{2}$
Egg lecithin	2%	2%
D-Glucose	1 mM	1 mM

[a] This protocol is designed to test the effects of potential effects of transport inhibitors (see text).

gel filtration. However, the final concentrations of glucose and lecithin are equal to those used in the standard method (see the table).

Determination of Blank Values. During the first 20 sec of incubation, a small amount of D-glucose (15–50 cpm as ^{3}H per total incubation mixture) binds to egg lecithin liposomes.[17] This D-glucose binding can be determined by carrying out the transport assay with liposome preparations free from the transporter protein. The zero-second blank, measured by adding labeled glucose after $HgCl_2$, is usually undetectable.

Calculation of Data. The results of the transport assay are calculated on the assumption that the carrier-mediated D-glucose transport activity is represented by the difference between the amounts of D-[^{3}H]glucose and L-[^{14}C]glucose found on the Millipore filter *minus* the specific binding of D-glucose to liposomes mentioned above. In this calculation, it is most important to know the exact counting efficiencies of ^{3}H and ^{14}C. Therefore, we routinely apply in triplicate 20-μl portions of the standard solution for the labeled glucose mixture (see above) onto a Millipore filter in an empty scintillation counting vial and measure the radioactivity (after being dried in a stream of air) along with the experimental preparations. If the calculated amount of D-glucose and L-glucose in the standard samples are not equal, we make a necessary adjustment (which is usually less than a few percent) on the relative counting efficiences of ^{3}H and ^{14}C.

Comments

We keep reconstituted liposomes in ice until shortly before the transport assay. If the liposomes are kept at 37°, the apparent transport activity increases during 0.5 and 2 hr of incubation and then declines.[15a] We pipette 5 μl of the labeled glucose solution with an Eppendorf pipette; it is suggestd that the precision of the pipetting be ascertained by weighing pipetted water droplets with a chemical balance.

The filtration and washing of liposomes is usually completed in 30–45 sec; however, if the Millipore filter is used upside down (i.e., the shiny side up instead of down), the process will take 5 min or more, and no reproducible results can be obtained.[2,17] The retention of liposomes by GSWP filter (pore size 0.22 μm) is 80–85%, but retention by 0.20-μm Unipore filter is almost zero. These results indicate that the pore size is not the only factor that determines the filtration efficiency. Although we extract radioactivities into ACS solution by mechanical shaking, solubilization of the Millipore membrane may be needed if no efficient shaker is available.

Since L-[^{14}C]glucose costs approximately $2000 per millicurie, attempts were made to use a combination of a less expensive D-[^{14}C]glucose and [^{3}H]sorbitol or [^{3}H]mannitol. However, we found that the latter two compounds are not stable in water. We have not tested a combintion of D-[^{14}C]glucose and L-[^{3}H]glucose, since the latter is commercially available only in its unstable C-1 derivative. The purity and the stability of the labeled compounds are important because less than 1% of the total radioactivity is taken up by liposomes under the experimental conditions described above. The purity of D-[^{3}H]glucose can be ascertained by inhibiting the carrier-mediated transport with 20 *μM* cytochalasin B.[17] The mediated transport activity may also be estimated by incubating reconstituted liposomes with D-[^{3}H]glucose in the presence or absence of 20 *μM* cytochalasin B.[1,2] In this method, the mediated transport activity is calculated on the assumption that it is specifically and completely inhibited by cytochalasin B.[1,2]

Acknowledgments

The methods described here have been developed in our laboratory in collaboration with Dr. Kazuo Suzuki, Miss Lynn E. Dansey, Mr. Robert W. Holloway, Mr. Mark E. Petrick, Mrs. Frances W. Robinson, Miss Teresa L. Blevins, and Miss Melinda M. Smith. Our original work was supported by NIH Grant 5R01 AM 19925 and Juvenile Diabetes Foundation Grant 80-R-340.

Section VI

Transcellular Transport

[39] Preparation of Inside-Out Thyroid Follicles for Studies on Transcellular Transport (Transcytosis)

By VOLKER HERZOG

Thyroglobulin, the macromolecular secretory product of thyroid follicle cells, is released by exocytosis into the follicle lumen; it remains separated from the interstitial space by the tight epithelial monolayer formed by follicular cells. Thyroid hormones appear in the circulation after endocytosis of thyroglobulin and its transfer to lysosomes, where hydrolytic liberation of thyroxine and triiodothyronine from their peptide linkages occurs.[1,2] With the aid of specific and sensitive radioimmunoassays it has been seen that small amounts of intact thyroglobulin are also detectable in the serum of man and several mammalian species.[3,4] It is assumed that this serum thyroglobulin is derived, at least in part, from the thyroglobulin stored in the lumen. However, the mechanism of its transfer through the tight epithelial wall remains obscure.[4]

Studies on the mechanism of transepithelial passage of thyroglobulin can be facilitated by using an *in vitro* system where access to the apical plasma membrane is possible. Such a system is provided by preparations of inside-out follicles from pig thyroid gland.[5] In these follicles, the wall consists of a monolayer of epithelial cells whose polarity is reversed: their apical plasma membrane is directed toward the culture medium and separated from the basolateral cell surface by tight junctions that are impermeable to various tracers.[5] Experiments with inside-out follicles have demonstrated that thyroglobulin can be transferred through the epithelial wall by endocytic vesicles to be released at the basolateral cell surface.[6]

Described here are the conditions that lead to the formation of inside-out follicles and allow one to study the transepithelial vesicular traffic (transcytosis) from the apical to the basolateral cell surfaces.

Preparation of Inside-Out Follicles from Pig Thyroid Gland

Principle. Collagenase digestion of thyroid tissue and application of shearing forces by pipetting result in the dissociation of tissue into intact

[1] S. H. Wollman, *in* "Lysosomes" (J. D. Dingle and H. B. Fell, eds.), Vol. 2, p. 483. North-Holland Publ., Amsterdam, 1969.

[2] L. E. Ericson, *Mol. Cell. Endocrinol.* **22,** 1 (1980).

[3] G. Torrigiani, D. Doniach, and I. M. Roitt, *J. Clin. Endocrinol. Metab.* **29,** 305 (1969).

[4] A. J. van Herle, G. Vassart, and J. E. Dumont, *N. Engl. J. Med.* **301,** 239 (1979).

[5] V. Herzog and F. Miller, *Eur. J. Cell Biol.* **24,** 74 (1981).

[6] V. Herzog, *J. Cell Biol.* **91,** 416a (1981).

ISBN 0-12-181998-1

PREPARATION OF INSIDE-OUT FOLLICLES[a]

Step	Treatment	Composition of medium	Time	Temperature
1	Collection of gland in slaughterhouse	On ice	~20 min	4°
2	Dissection of connective tissue and mincing	Eagle's minimum essential medium (MEM) with Earle's salts	30–40 min	Room
3	Wash several times to remove cell debris	MEM	~10 min	Room
4	Enzymic digestion	MEM containing collagenase	30 min	37°
5	Enzymic digestion and pipetting of tissue fragments	MEM containing collagenase	~90 min	37°
6	Filtration through nylon gauze (200 μm)	MEM containing collagenase	~2 min	Room
7	Centrifugation 30 sec at 70 *g*	MEM containing collagenase	~3 min	Room
8	Resuspension of pellet in MEM and filtration through nylon gauze (150 μm)	MEM	~3 min	Room
9	Centrifugation, 30 sec at 70 *g*	MEM	3 min	Room
10	Wash several times	MEM containing penicillin and streptomycin	10 min	Room
11	Suspension culture	MEM containing fetal calf serum, penicillin, and streptomycin	24 hr	37°

[a] Conditions: Steps 1–9, nonsterile; steps 10 and 11, sterile.

segments of follicles opened during pipetting. Fibroblasts and parafollicular and endothelial cells are separated from follicle segments by low-speed differential centrifugation. Purified follicle segments form closed follicular spheres with a reversed (inside-out) polarity within 24 hr of suspension culture.

Isolation of Follicle Segments. Thyroid glands are removed from pigs (60–80 kg) 10 min after anesthesia by electroshock and killing by bleeding (step 1). Follicles of pig thyroid glands contain large masses of colloid and show the characteristics of a resting gland (Fig. 1). The glands, 5–7 g, are freed of the fibrous capsule and of connective tissue and minced with razor blades into small (1 mm) pieces (step 2). It is important to remove all visible remnants of connective tissue. Thyroid fragments are washed several times (step 3) with Eagle's minimum essential medium (MEM) with Earle's salts until the supernatant medium remains clear. Enzymic digestion is started by incubation of tissue fragments in MEM containing crude collagenase from clostridium histolyticum (Boehringer, Mannheim, Federal Republic of Germany), 150 mU/ml (activity determined with the assay of Wünsch and Heidrich[7]) for 30 min at 37° (step 4). Digestion is performed in a 20-ml Erlenmeyer flask[8] equipped with a pH-electrode connected to a recorder for constant monitoring of the pH in the medium. The medium is equilibrated to pH 7.3–7.4 by gassing with 72% N_2, 20% 0_2, and 8% CO_2. This initial digestion is followed by pipetting of tissue fragments through freshly siliconized pipettes whose tips are of stepwise diminishing diameters,[9] from 1000 to 600 μm (step 5). Pipetting has to be stopped when tissue fragments pass through a pipette with a tip diameter of 600 μm. Further pipetting results in the disintegration of follicle segments into single cells; pipetting for too brief a period reduces the yield. The resulting suspension of follicle segments, nondispersed tissue, fibroblasts, single follicle cells, and cell fragments is filtered (step 6) through 200 μm nylon gauze (Nybolt filters, Schweizerische Seidengaze Fabrik, Zurich, Switzerland). This crude filtrate, containing follicle segments, fibroblasts, single cells, and cell debris, is then centrifuged (step 7) for 30 sec at 70 *g* in polyallomer tubes (Beckman Instruments, Palo Alto, California) in a Labofuge II (Heraeus Christ GmbH, Osterode, Germany). The supernatant, containing fibroblasts, single follicle cells, and cell fragments, is discarded; the loose pellet, made up of follicle segments, is resuspended in MEM, filtered (step 8) through 150-μm nylon gauze, and centrifuged (step 9) for 30 sec at 70 *g*. The pellet is washed several times (step 10) in sterile MEM containing, per milliliter, 100 IU of penicillin and 100 μg of streptomycin. After this washing, colloid is entirely removed from follicle

[7] E. Wünsch and H. Heidrich, *Hoppe Seyler's Z. Physiol. Chem.* **333,** 149 (1963).

[8] All glassware is siliconized by dipping into the undiluted silicone solution (No. 35,130 from Serva Feinbiochemica, Heidelberg, Federal Republic of Germany). The silicone film is cured for 1 hr at 100°.

[9] Pasteur pipettes of defined tip diameter are prepared as follows: The tips of Pasteur pipettes are flamed to soften the tips and to reduce their inner diameters. The inner tip diameter is measured with a lens (magnifying power: 10–15 times) equipped with a graticule scale with 0.1-mm divisions. Pipettes with diameters of 1000, 800, 700, and 600 μm are selected for sequential use during dissociation.

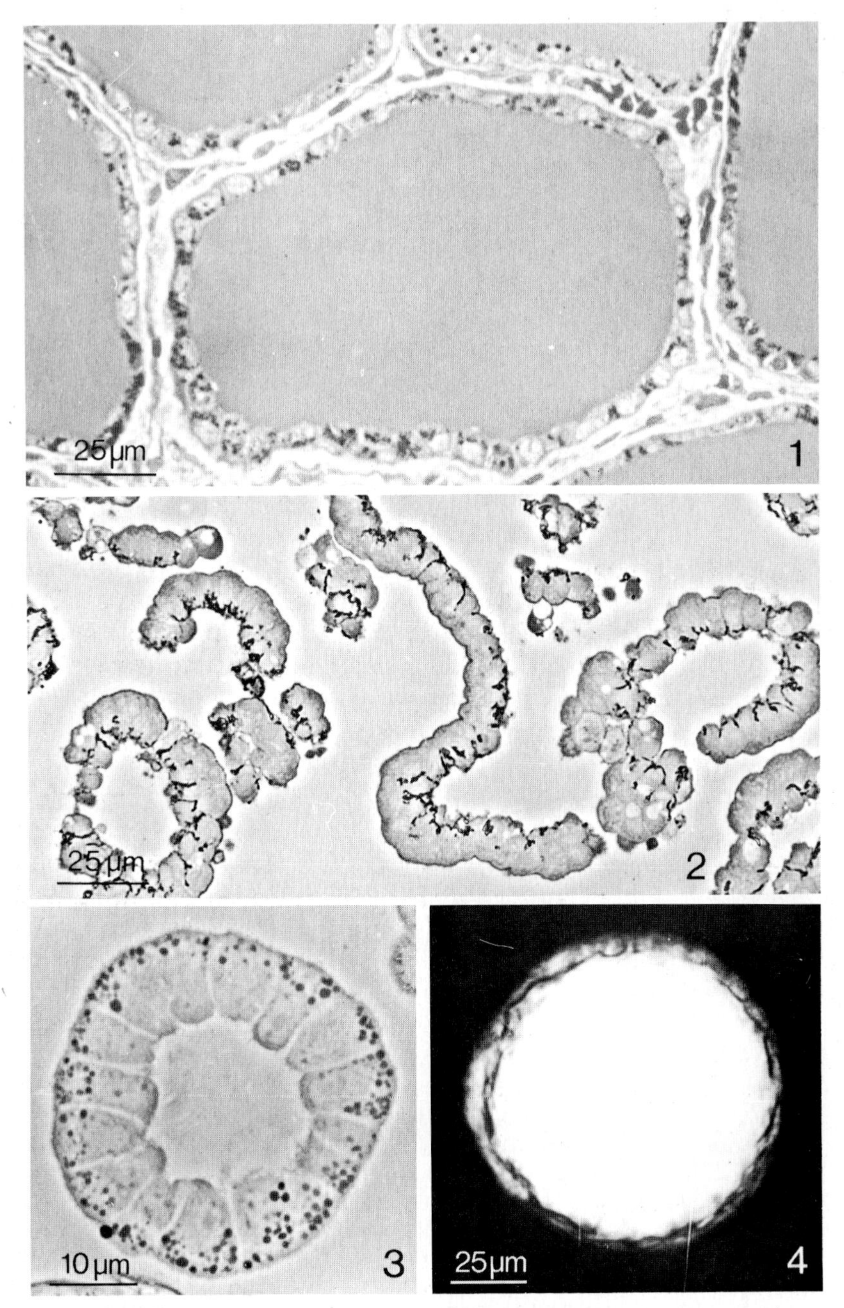

FIGS. 1–4. Light micrographs showing the formation of tight inside-out follicles from pig thyroid gland.

FIG. 1. Follicles *in situ.*

FIG. 2. Isolated follicle segments without colloid.

FIGS. 3 and 4. Reconstituted follicles show a reversed polarity (Fig. 3; see also Fig. 5) and tightness against various tracers such as carbon particles of India ink (Fig. 4). Figure 4 is from Herzog and Miller,[5] reprinted with permission of Wissenschaftliche Verlagsgesellschaft, Stuttgart.

segments (Fig. 2). Usually, the dissociation of 5 g of tissue yields approximately 500 mg of isolated follicle segments, by wet weight.

Formation of Inside-Out Follicles. Freshly isolated follicle segments are resuspended in sterile MEM containing 10% heat-inactivated fetal calf serum, 100 IU of penicillin and 100 μg of streptomycin per milliliter, and transferred to culture dishes (step 11), which are supplied with a hydrophobic membrane (W. C. Heraeus GmbH, Hanau, Germany) to prevent attachment of follicle segments. During suspension culture at 37°, 20% O_2 and 5% CO_2, closed follicular structures are formed. Their closure begins after 4 hr as monitored in a Leitz inverted microscope and is completed within 24 hr of suspension culture (Fig. 3). Resealed follicles are composed of a single layer of epithelial cells and are free of parafollicular cells, fibroblasts, or remnants of capillaries. The epithelial wall of the reconstituted follicles impedes the passage of various tracers,[5] such as ferritin (M_r 500,000, diameter 110 Å; from Miles Laboratories, Inc., Elkhart, Indiana), iron dextran (M_r 200,000, size ~ 210 × 120 Å; from Fisons Ltd., Loughborough, Leicestershire, England), pig thyroglobulin (M_r 670,000; size 300 × 150Å; from Sigma Chemical Co., St. Louis, Missouri) or carbon particles (diameter ~ 250 to 300 nm; India ink No. c11/1413a from Günther Wagner, Hanover, Germany) in a 10% solution as shown in Fig. 4.

Structural and Functional Polarity of Epithelial Cells in Inside-Out Follicles

The polarity of all follicles in suspension is reversed so that the apical plasma membrane is oriented toward the culture medium, and the basal cell surface is directed toward the central cavity (Fig. 5).[10] The apical plasma membrane of each cell bears microvilli and a central cilium and is separated by tight junctions from the smooth lateral plasma membrane. The basement membrane is usually lost during the isolation procedure and is not resynthesized during the period of follicle closure.

The protein to DNA ratio decreases during the isolation procedure from ~ 80 in intact tissue to ~ 13 in follicle segments owing to the loss of colloid.

[10] It is not known why follicle segments always bend in a direction that gives rise to the formation of inside-out follicles. The concentration of negative surface charges on the apical plasma membrane may be of influence. The amount of anionic binding sites is visualized by the preferential binding of the polycationic marker DEAE-dextran on the apical plasma membrane [V. Herzog and F. Miller, *in* "Secretory Mechanisms" (C. R. Hopkins and C. J. Duncan, eds.), p. 101. Cambridge Univ. Press, London and New York, 1979]. The hypothesis that the concentration of anionic charges on the apical cell surface may cause the bending in one direction is supported by the inability of segments to form closed follicles in the presence of cationic molecules.

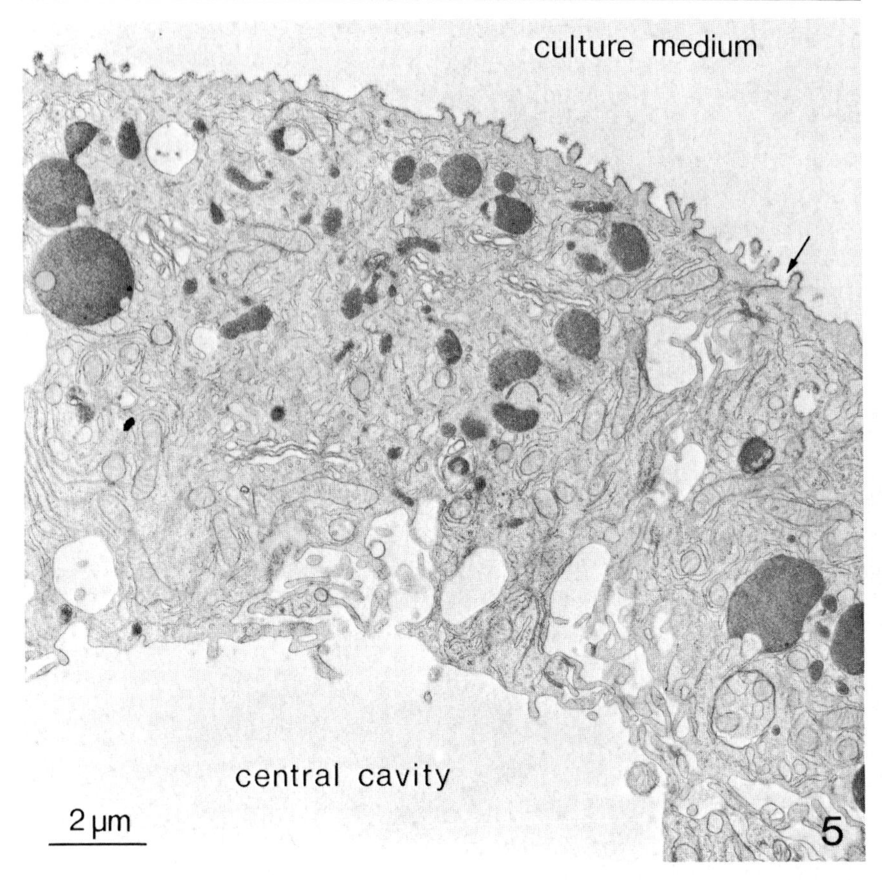

FIG. 5. Inside-out polarity of reconstituted follicles. The apical plasma membrane is directed toward the culture medium and separated from the lateral cell surfaces by tight junctions (arrow). The basement membrane is absent from the basal cell surface in the central cavity.

It remains about the same (~ 14) when inside-out follicles have formed (Fig. 6), indicating that thyroglobulin does not accumulate in the central cavity. Newly synthesized thyroglobulin is released into the culture medium[5] where it becomes detectable after incubation of inside-out follicles with L-[^{3}H]leucine, D-[^{3}H]galactose or ^{125}I (Fig. 7). Since galactose is added to the carbohydrate side chain in the Golgi complex,[11] and iodination of thyroglobulin takes place at the apical cell surface,[12] it is concluded that the vectorial transport of thyroglobulin is maintained in inside-out follicles.

[11] P. Whur, A. Herscovics, and C. P. Leblond, *J. Cell Biol.* **43,** 289 (1969).

[12] R. Ekholm and S. H. Wollman, *Endocrinology* (*Baltimore*) **97,** 1432 (1975).

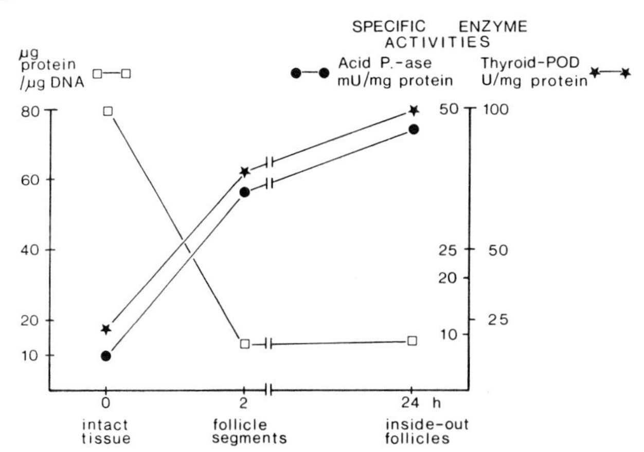

FIG. 6. During dissociation of thyroid tissue, the protein:DNA ratio decreases owing to loss of thyroid colloid. As a consequence, specific activities of thyroid peroxidase (POD; measured with the colorimetric DAB assay of V. Herzog and H. D. Fahimi [*Anal. Biochem.* **55**, 554 (1973)] and of acid phosphatase (P-ase) (measured with the method of M. A. Andersch and A. Szczypinski [*Am. J. Clin. Pathol.* **17**, 571 (1947)] increase. During the formation of inside-out follicles within 24 hr of suspension culture, the protein: DNA ratio remains the same, which indicates that secretion of protein into the newly formed central cavity does not occur.

For several days after follicle closure, inside-out follicles remain single walled and retain their reversed structural and functional orientation. During this period, they are a useful tool for studies on endocytosis from the apical cell surface. When follicles are grown for periods longer than 2 weeks, they have a tendency to lose their characteristic single-walled architecture by the formation of aggregates.

Culture Conditions During the Use of Tracers

Numerous electron-dense tracers interact with proteins in the culture medium. For example, cationized ferritin (p$I \sim 8.5$) forms large aggregates with anionic proteins, such as albumin in the fetal calf serum or thyroglobulin (pI 4.7). Therefore, it is essential to wash follicle preparations extensively in MEM prior to the addition of tracers.

Pathways of Endocytosis from the Apical Cell Surface

The maintenance of the structural and functional polarity and the tightness of the epithelial wall in inside-out thyroid follicles are prerequisites

[13] For terminology of the endosomes, see A. Helenius, M. Marsh, and J. White [*Trends Biochem. Sci.* **5**, 104 (1980)].

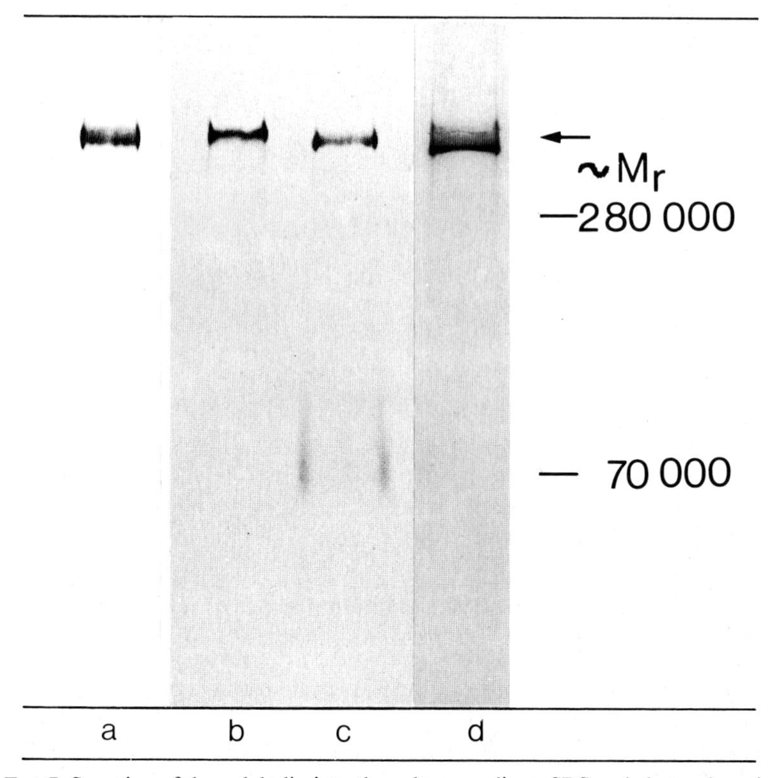

FIG. 7. Secretion of thyroglobulin into the culture medium: SDS–gel electrophoresis in a polyacrylamide gradient, 5 to 15%, and fluorography [W. M. Bonner and R. A. Laskey, *Eur. J. Biochem.* **46,** 83 (1974)] reveal a major band in the L-[^{3}H]leucine- (slot b), the D-[^{3}H]galactose- (slot c), or the ^{125}I- (slot d) labeled secretory products of the same electrophoretic mobility (arrow) as freshly isolated thyroglobulin stained with Coomassie blue (slot a). Single and cross-linked hemocyanin were used as M_r standards.

for studies on endocytosis from the apical plasma membrane. Cationized ferritin (CF), when added to the culture medium at a final concentration of 10–50 μg/ml, adsorbs to anionic sites on the apical plasma membrane is uniform at 4°. At 37°, CF particles collect in coated pits that are located at the bases of microvilli. Coated pits appear to detach from the plasma membrane, to lose their clathrin coat, and to form smooth-surfaced endocytic vesicles (diameter 100 nm).

Route to lysosomes and to the Golgi complex: The endocytic vesicles reach lysosomes within 5 min after having passed through a prelysosomal compartment, the endosome,[13] which is characterized by the absence of acid phosphatase and presumably other lysosomal hydrolases. Cationized ferritin particles are still bound to the inner membrane surface of the endo-

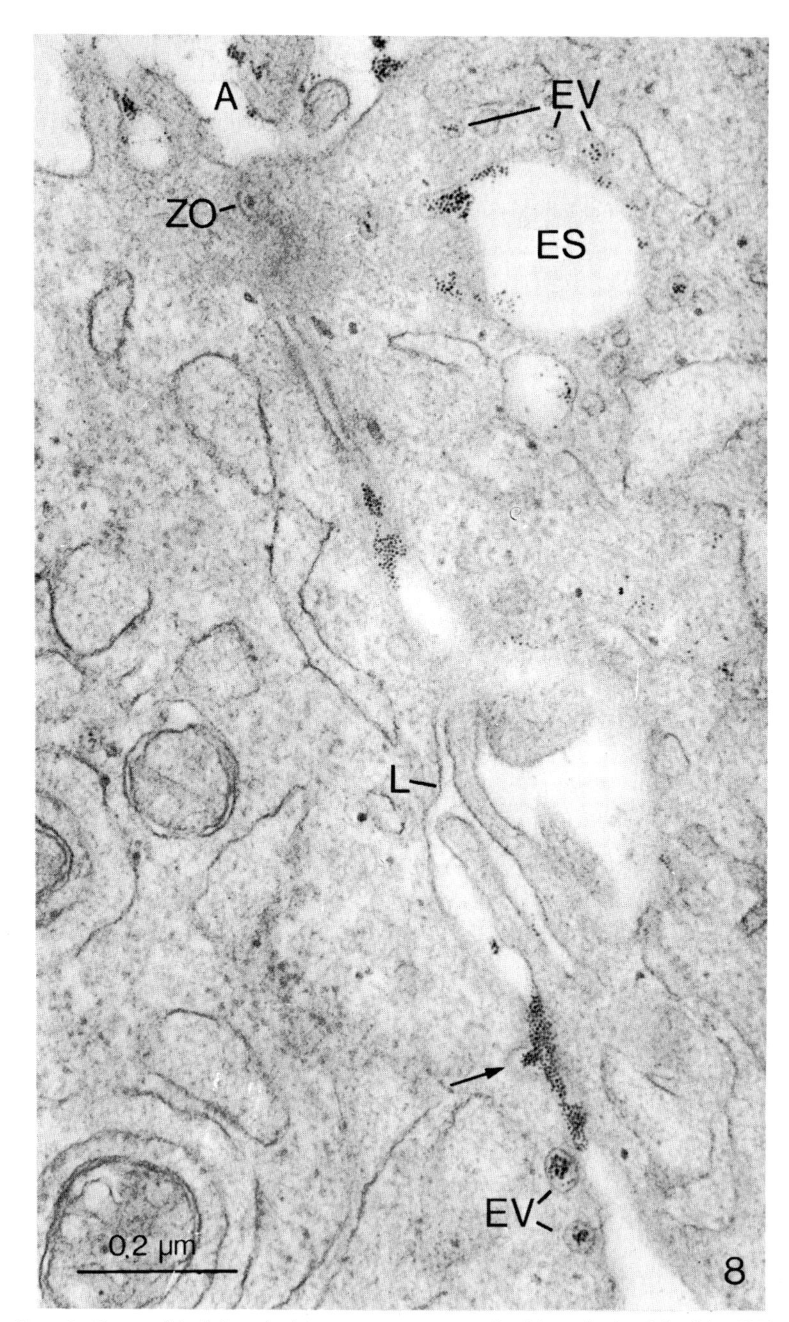

FIG. 8. Transepithelial vesicular transport as traced with cationized ferritin (CF). The CF-labeled endocytic vesicles (EV) deriving from the apical cell surface (A) bypass the tight junctions (ZO) and fuse with the lateral (L, arrow) and basal plasma membranes. After insertion of vesicles, CF particles remain attached to the membrane. Their density on the basolateral cell surface is taken as a measure of transepithelial transport (see Fig. 9). ES, endosome.

somes,[14] but in the secondary lysosomes, CF particles detach from the membrane and collect in the lysosomal matrix.[5] At 30 min, CF particles are also found in the stacked Golgi cisternae. The sequence of the intracellular movement of the surface marker suggests that part of the incoming membrane may detach from the endosomal and lysosomal surfaces to reach the membranes of the Golgi complex for possible reuse during packaging of newly synthesized thyroglobulin.[5,14]

Transcytosis: Endocytic vesicles also fuse with the basolateral plasma membrane, which is reached later than lysosomes, but earlier than the Golgi cisternae: the first CF particles appear at 11 min on the lateral cell surfaces and at 16 min on the basal cell surfaces (Fig. 8). Since CF particles remain attached to the membrane patches inserted into the basolateral cell surface, the transepithelial appearance of CF particles can be quantitated morphometrically by estimation of the particle density (number of CF particles per unit of basolateral membrane length). The results show that transcytosis is increased 6- to 10-fold by stimulation with thyroid-stimulating hormone (TSH) (10 mU/ml), but is arrested at low temperatures (Fig. 9).

Gold particles (diameter 200 Å; prepared after the method of Horisberger and Rosset[15]) when coated with thyroglobulin[16] and added to the medium, are carried by endocytic vesicles mainly to lysosomes. A few endocytic vesicles also release thyroglobulin gold particles into the central cavity of inside-out follicles, indicating a vesicular transepithelial transfer of thyroglobulin.

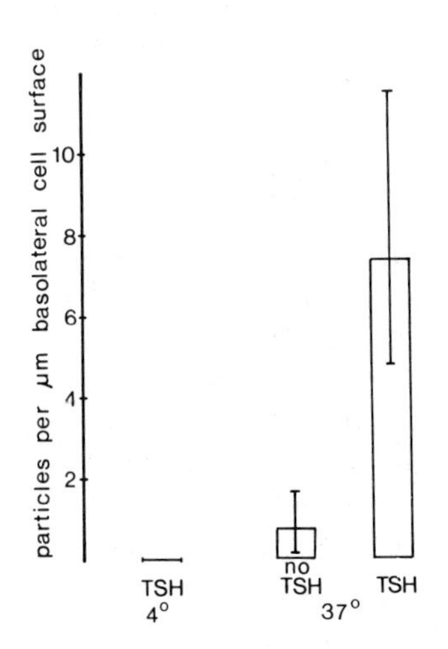

FIG. 9. Morphometric analysis of transcellular transport 60 min after the exposure of inside-out follicles to cationized ferritin. The transcellular vesicular transport is enhanced by stimulation with thyroid-stimulating hormone (TSH) and interrupted at low temperatures. Results are means; bars indicate the range of values from nine cells.

[14] V. Herzog, *Trends Biochem. Sci.* **6,** 319 (1981).

[15] M. Horisberger and J. Rosset, *J. Histochem. Cytochem.* **25,** 295 (1977).

[16] Thyroglobulin from pig thyroid gland is prepared according to the procedure of M. Rolland and S. Lissitzky [*Biochim. Biophys. Acta* **427,** 696 (1976)].

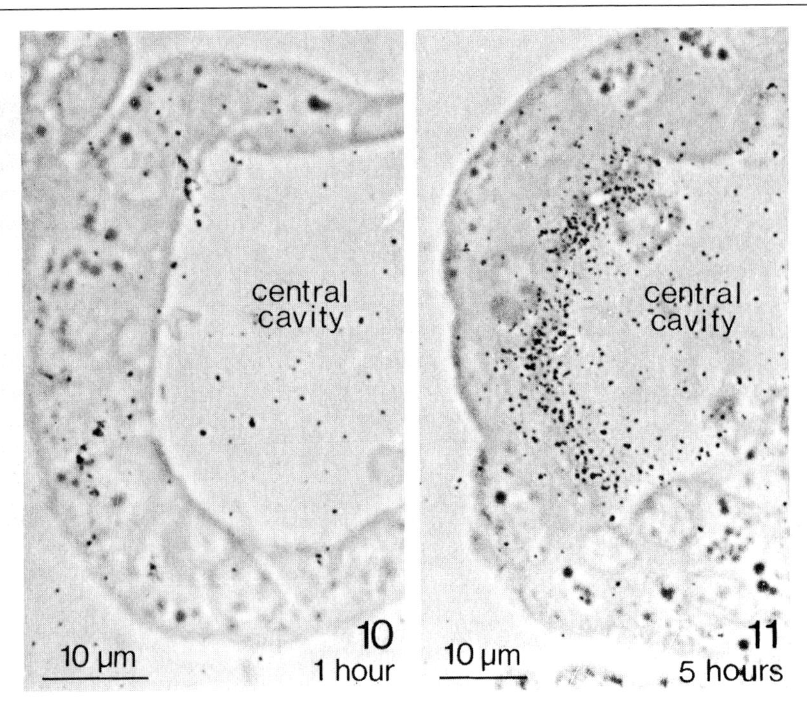

FIGS. 10 and 11. Visualization of transcellular transport of [^{3}H]leucine-labeled thyroglobulin from the culture medium by TSH-stimulated inside-out follicles. Autoradiographic silver grains demonstrate the accumulation of thyroglobulin in the central cavity after 1 hr (Fig. 10) and after 5 hr (Fig. 11). (The extrafollicular radiolabeled thyroglobulin was removed before fixation.)

Evidence for transcytosis of thyroglobulin is also obtained from experiments with biosynthetically radiolabeled thyroglobulin. To prepare radiolabeled thyroglobulin, suspensions of inside-out follicles are incubated in the presence of L-[^{3}H]leucine or ^{125}I for 24 hr at 37°. Radiolabeled thyroglobulin in the culture medium (see Fig. 7) is extensively dialyzed[17] and concentrated. Freshly prepared inside-out thyroid follicles are exposed to [^{3}H]- or [^{125}I]thyroglobulin (1.5 mg/ml) with a specific radioactivity of 200 or 400 μCi/mg, respectively. Samples are washed to remove extrafollicular radiolabeled thyroglobulin, fixed after 1 and 5 hr, and prepared for autoradiographic visualization of internalized thyroglobulin using Ilford L4-emulsion

[17] The culture medium containing [^{3}H]- or [^{125}I]thyroglobulin is extensively dialyzed and concentrated in a microultrafiltration system supplied with a Diaflo PM-10 ultrafiltration membrane to separate radiolabeled thyroglobulin from nonincorporated L-[^{3}H]leucine or ^{125}I. Incubation of freshly prepared follicles with radiolabeled thyroglobulin was performed with an excess of unlabeled L-leucine or iodine.

(Ilford Ltd., Ilford, Essex, England). Part of the internalized radiolabeled thyroglobulin becomes detectable on the transepithelial side, where it accumulates with time in the central cavity of inside-out follicles (Figs. 10 and 11).

Analysis of the transcellularly transported thyroglobulin requires the reopening of inside-out follicles: After 5 hr of accumulation of [^{3}H]- or [^{125}I]thyroglobulin in the central cavity, inside-out follicles are opened again by treatment with 1 m*M* EDTA and gentle pipetting through a Pasteur pipette with a tip diameter of 500 μm. After centrifugation of the reopened follicles in a microfuge (Beckman Instruments, Palo Alto, California), the transcellularly transported, radiolabeled thyroglobulin becomes detectable in the supernatant. It has the same electrophoretic mobility as the thyroglobulin added previously to the culture medium. This indicates that a major part of the thyroglobulin remains intact during its passage through the follicle wall.

Conclusions

Under appropriate culture conditions, follicle segments isolated from pig thyroid gland form tight follicular structures in which the epithelial cells exhibit a stable inside-out polarity. Studies on endocytosis from the apical cell surface of inside-out follicles indicate a TSH-inducible transepithelial traffic of endocytic vesicles. This traffic includes the transfer of thyroglobulin and may explain in part the appearance of thyroglobulin in the circulation, where increased levels are found after stimulation with TSH.

Acknowledgment

I thank S. Fuchs, K. Hrubesch, and U. Reinhardt for technical assistance. This work was supported by Deutsche Forschungsgemeinschaft.

[40] Biosynthesis, Processing, and Function of Secretory Component

By KEITH E. MOSTOV and GÜNTER BLOBEL

Secretory component (SC) was discovered as a glycoprotein associated with polymeric immunoglobulin A and immunoglobulin M (pIg) in a variety of exocrine secretions.[1] The SC is synthesized by glandular epithelial

[1] T. B. Tomasi, E. M. Tan, A. Solomon, and R. A. Prendergast, *J. Exp. Med.* **121,** 101 (1965).

ISBN 0-12-181998-1

cells and has been postulated to be the receptor that mediates the binding and transcellular transport of pIg.[2] By translating mRNA for SC in a cell-free system, we found that SC is initially not made as a secretory protein, but rather as a larger transmembrane precursor.[3] This *in vitro*-synthesized transmembrane precursor can bind IgA. Pulse-chase experiments with cells that synthesize SC have shown that the transmembrane SC precursor is cleaved to a soluble form that is released from cells.[4] Our results, and the cytochemical data of others on the location of SC and IgA in epithelial cells,[2] have led to our proposal of a model of how SC functions in the transcellular transport of pIg[3]: SC is inserted as a transmembrane protein into the rough endoplasmic reticulum membrane. It moves through the Golgi apparatus to the basolateral (or in hepatocytes, sinusoidal) surface of the cell, where a portion of the molecule, the ectoplasmic domain (exposed on the outside of the cell) binds pIg. The transmembrane SC–pIgA complex (or transmembrane SC alone) is endocytosed. The endocytotic vesicle traverses the cell and is exocytosed at the apical (or in hepatocytes, bile canalicular) surface. At some point the transmembrane SC precursor is proteolytically cleaved, releasing mature SC bound to pIg. The fate of the cytoplasmic and membrane spanning portions of the SC precursor is not known. This model explains how SC can be first a membrane receptor and later a soluble protein. The information for guiding SC in its complex pathway through the cell may reside in the cytoplasmic domain of the precursor.

We have studied SC from both rabbits and humans. The rabbit system has the advantage that mRNA for cell-free translation studies may be easily extracted from tissues that make SC, such as liver and lactating mammary gland. However, rabbit SC is a heterogeneous family of at least four polypeptides, each of which is made as a separate primary translation product.[3] This obviously complicates interpretation of data. Also, pulse-chase experiments must utilize primary cultures, since we know of no established rabbit cell line that makes SC. The SC from humans, in contrast, apparently consists of one polypeptide that is made as a single primary translation product.[4] Obtaining mRNA from tissues is problematic, since the tissue must be fresh to prevent degradation of the mRNA. There is, however, a well-differentiated colon adenocarcinoma cell line (HT29) that makes SC[5] and transports IgA from the basolateral to the apical surface of the cells.[6] Pulse-chase experiments can be performed with these cells, and small amounts of mRNA for *in vitro* translation studies can be extracted.

[2] P. Brandtzaeg, *J. Immunol.* **112,** 1553 (1974).
[3] K. E. Mostov, J.-P. Kraehenbuhl, and G. Blobel, *Proc. Natl. Acad. Sci. U.S.A.* **77,** 7257 (1980).
[4] K. E. Mostov and G. Blobel, *J. Biol Chem.* **257,** 11816 (1982).
[5] S. W. Huang, J. Fogh, and R. Hong, *Scand. J. Immunol.* **5,** 263 (1976).
[6] H. Nagura, P. K. Nakane, and W. R. Brown, *J. Immunol.* **123,** 2359 (1979).

Biosynthesis of Rabbit SC

Cell-Free Translation of Rabbit SC

The mRNA is extracted from tissues, such as lactating mammary gland or liver, using the proteinase K–SDS–phenol–chloroform method.[7] If liver is used, contaminating glycogen should be pelleted out after the phenol extractions by centrifuging the solution in a Beckman SW 27 rotor at 25,000 rpm for 45 min. It is generally not necessary to purify poly(A)-containing mRNA by chromatography on oligo(dT)cellulose. It is useful, however, to remove DNA and other contaminants by dissolving the RNA in water at a concentration of 20 A_{260} units/ml, adding LiCl to a final concentration of 2.5 *M*, and letting the RNA precipitate for 12 hr at 4°.

The mRNA is translated cell-free in the wheat germ system.[8] The optimal mRNA concentration must be determined by titration and is usually between 5 and 10 A_{260} units/ml. We have modified the composition of the translation reaction to maximize synthesis of SC. The reaction contains 114 m*M* potassium acetate, 2.7 m*M* magnesium acetate, 0.73 m*M* spermidine–HCl, 40 m*M* HEPES–KOH (pH 7.5), 2.6 m*M* ATP, 0.27 m*M* GTP, 16 m*M* creatine phosphate, 80 μg/ml creatine phosphokinase, 5 m*M* dithiothreitol, 0.75 mCi/ml [^{35}S]cysteine (New England Nuclear, specific activity 700–1000 Ci/mmol), and the 19 amino acids except cysteine at 20 m*M* each. Cysteine is used instead of methionine because SC is rich in cysteine and poor in methionine.[9] Reactions are incubated at 24–26° for 90 min, after which no additional incorporation of [^{35}S]cysteine into protein is observed. Generally, if a 20-μl translation reaction is performed, the immunoprecipitated SC precursor can be visualized after a 2- to 4-day exposure of a fluorographed gel.

To study glycosylation of the SC precursor and its integration into membranes, translations are performed in the presence of dog pancreas rough microsomes as described previously.[10] Preparations of microsomes vary greatly in their glycosylation activity, and the amount of membranes needed must be determined by titration. Generally 1–3 A_{280} units/ml are used. This concentration of microsomes will inhibit the synthesis of SC and other proteins,[10] so it is useful to scale up the translation reaction by a factor of 2–4 to compensate.

To demonstrate that the SC precursor is in fact inserted into the microsomal membrane, the products are treated with trypsin, which digests the cytoplasmic domain of the precursor, exposed on the outside of the micro-

[7] P. M. Lizardi, this series, Vol. 96 [2].

[8] A. H. Erickson and G. Blobel, this series, Vol. 96 [3].

[9] K. Kobayashi, *Immunochemistry* **8,** 785 (1971).

[10] P. Walter and G. Blobel, this series, Vol. 96 [6].

somal vesicles. After translation is complete, the tube is placed on ice and $CaCl_2$ is added from a 100 mM stock to a final concentration of 2 mM. Trypsin (from Worthington) is added to the reaction to a final concentration of 0.3 mg/ml. The trypsin stock is conveniently made up as a 3 mg/ml solution in 10 mM Tris-HCl, pH 7.5, and kept at −80° in single-use aliquots. The trypsinization is performed for 1 hr at 0°. Afterward, Trasylol is added to a final concentration of 1000 units/ml to inhibit the trypsin. Trasylol is purchased as a 10,000 units/ml solution from Mobay Chemical Company, New York. Two controls should be carried out. First, to demonstrate that membrane integration occurs only cotranslationally, the microsomes should be added only after translation is completed, and the sample is then digested with trypsin. Second, to show that the partial protection of the SC fragment from proteolysis is due to its integration into the membrane, translation is first carried out in the presence of microsomes. After translation is completed, the membranes are solubilized by adding Triton X-100 to 1% and then trypsinizing the sample as usual. In both controls, the SC precursor should be completely degraded by the trypsin.

Immunoprecipitation of Cell-Free Synthesized SC

Immunoprecipitation is performed essentially as described by Anderson and Blobel[11] with minor modifications. Sodium dodecyl sulfate (SDS) is added to a final concentration of 2%, and the sample is immediately placed in a boiling water bath for 2 min. After cooling, 4 volumes of Triton dilution buffer [2.5% Triton X-100, 150 mM NaCl, 20 mM triethanolamine–HCl (pH 8.1), 5 mM EDTA, 100 units of Trasylol per milliliter, and 0.02% NaN_3] is added. A pH 8.1 buffer is used because the binding of sheep and goat immunoglobulins to protein A is better at this pH than at pH 7.5. The amount of antiserum needed to immunoprecipitate the SC quantitatively must be determined by titration. The goat anti-rabbit SC antiserum prepared by Kühn and Kraehenbuhl[12] and used in our previous study[3] can be used at a final concentration of 0.05%. Cappel Laboratories (Downington, Pennsylvania) sells both sheep and goat antisera to rabbit secretory IgA. Both of these antisera will immunoprecipitate SC and can be used at a final concentration of 0.2–0.5%. After the antiserum is added, the reactions are incubated at 4° for 8–24 hr. Protein A–Sepharose is then added, and the tubes are incubated for 30–120 min at room temperature with gentle agitation. The beads are washed six times with 1.4- to 1.5-ml portions of 1% Triton X-100, 0.2% SDS, 150 mM NaCl, 20 mM triethanolamine-HCl (pH 8.1), 5 mM EDTA, 0.1% Trasylol, 0.02% NaN_3 and once with the same

[11] D. J. Anderson and G. Blobel, this series, Vol. 96 [8].

[12] L. C. Kühn and J. P. Kraehenbuhl, *J. Biol Chem.* **254,** 11066 (1979).

buffer lacking detergents. Twenty-five microliters of loading buffer (5% SDS, 1 *M* dithiothreitol, 0.2 *M* Tris base, 0.3 *M* sucrose, 5 m*M* EDTA, and 0.002% Bromphenol Blue) is added to the pelleted beads, and the sample is boiled for 2 min. The eluted sample is then subjected to SDS–polyacrylamide gel electrophoresis (PAGE) using the system of Laemmli.[13] Generally, we use gels with a 5 to 10% linear gradient of polyacrylamide. Alkylation of the proteins with iodoacetamide results in poorer separation of the bands of rabbit SC.

Binding of the Transmembrane SC Precursor to IgA

The transmembrane SC precursor synthesized in the cell-free system is capable of binding its ligand, IgA. We have used human IgA because it binds well to rabbit colostral SC[14] and is more readily available than rabbit IgA. Our published studies utilized a human myeloma IgA that was purified by standard techniques[1] and kindly given to us by J. M. McCune of the Rockefeller University. If a myeloma IgA is used, it is important to be sure that it is polymeric, since monomeric IgA does not bind SC. This can be determined by gel electrophoresis under nonreducing conditions.[12] Purified human IgA from pooled serum can also be purchased from Cappel Laboratories. IgA from pooled serum is a heterogeneous mix of molecules and contains an adequate fraction of polymeric species. IgA from secretions should not be used, since it will already contain SC and will not bind additional SC.

The IgA is coupled to CNBr-activated Sepharose (purchased from Pharmacia Fine Chemicals) according to the manufacturer's directions. One milligram of IgA is added to 1 ml of swollen beads and coupled for 2 hr at room temperature. Unreacted amino groups are blocked, and the beads are washed according to the manufacturer's instructions. Generally about 80–90% of the protein is coupled to the beads.

In our preliminary experiments, we were unable to show binding to IgA of the SC precursor made in the cell-free system, apparently because the high concentration of other wheat-germ proteins interfered. Hence, we translated SC in the presence of dog pancreas membranes and reisolated the membranes on a sucrose gradient. The gradient is constructed in a Beckman airfuge tube by first placing 25 μl of 2.0 *M* sucrose in the tube and overlayering with 80 μl of 0.5 *M* sucrose. Both sucrose solutions contain 150 m*M* NaCl and 20 m*M* sodium phosphate, pH 6.5. One hundred microliters of a translation reaction that was performed in the presence of microsomes is

[13] U. K. Laemmli, *Nature* (*London*) **227,** 680 (1970).

[14] D. J. Socken and B. J. Underdown, *Immunochemistry* **15,** 499 (1978).

gently layered on top of the gradient. The tube is centrifuged in the 30°-angle rotor of the Beckman airfuge for 15 min at full speed (160,000 g). The membranes will be visible at the 0.5 M/2.0 M interface and can be collected with a Hamilton syringe in a volume of 10–25 μl. This is diluted to 200 μl with 150 mM NaCl and 20 mM sodium phosphate, pH 6.5. This pH is optimal for binding of SC to IgA. Tririon X-100 is added from a 20% stock to a final concentration of 1%. The detergent solubilizes the microsomal vesicles so that the ectoplasmic domain of the SC precursor, which was in the lumen of the vesicle, becomes accessible to bind IgA. Twenty microliters of IgA–Sepharose is added to the mixture, and the mixture is incubated at 4° for 16 hr with end-over-end mixing. The beads are then washed six times with 1% Triton X-100, 150 mM NaCl, 20 mM sodium phosphate, pH 6.5. Washing of the beads, elution of the bound SC with SDS, and electrophoresis are carried out basically as described above for the protein A–Sepharose beads used in the immunoprecipitation procedure. Unfortunately, there is a good deal of nonspecific absorption of other cell-free translated proteins to the IgA–Sepharose. Two controls can be used to demonstrate that only the binding of the SC precursor is specific. First, an excess of nonradioactive authentic SC (e.g., 10 μg) added to the binding reaction will bind to the IgA–Sepharose and block binding of the radioactive cell-free synthesized SC precursor. Second, if an excess of human IgA (e.g., 50 μg) that is not bound to Sepharose is added, the radioactive SC will mostly bind to this soluble IgA, not to the IgA–Sepharose. In both controls, the binding of nonspecific contaminants to the IgA–Sepharose will be only slightly decreased.

Biosynthesis of Human SC

Cell Culture

The HT29 cell line was derived from a human colon adenocarcinoma by Dr. J. Fogh, Sloan–Kettering Institute for Cancer Research, Rye, New York. We obtained the cell line from him and maintained the cells as described,[5] using McCoys 5a medium, supplemented with 10% fetal bovine serum, 100 units of penicillin, 100 μg of streptomycin, and 2.5 μg of fungizone per milliliter. (All tissue culture reagents were purchased from GIBCO, Grand Island, New York.)

We conducted our initial experiments with the original HT29 cell line. However, the amount of SC precursor that was synthesized by the cells was less than we had hoped, necessitating long (1- to 2-week) exposures of the fluorographs to detect the radioactive SC. We later learned that the HT29 cell line had never been cloned and that not all the cells made SC, as detected by immunofluorescence.[5] It seemed possible that HT29 was a mixed population of cells. We therefore isolated individual clones and screened them for SC synthesis. A single-cell suspension was prepared by trypsinizing a monolayer with 0.25% trypsin (from GIBCO) and 0.4 m*M* EDTA at 37° for 30 min. The cell concentration was determined with a hemacytometer. Cells were diluted to one to two cells per milliliter, and 0.1-ml aliquots were placed in the wells of 96-well Microtiter plates (purchased from Linbro Scientific, McLean, Virginia). Hence, most wells will contain no cells, only about 10–20% of the wells should contain a single cell each, and very few wells will contain two or more cells. This was checked by inspecting the wells with an inverted microscope. The cells were allowed to grow in these wells, and medium was changed every 2–4 weeks. After about 2 months the cells were removed by trypsinization and transferred to 25-cm^2 flasks. After a further 2 months, the cells were removed from the flasks and a portion was allowed to grow in 35-mm dishes. Pulse labeling of the SC produced by the clones was performed as described below. We studied 12 clones by this method. Interestingly, all 12 of them produced at least some SC; however, there was a considerable variation in the amount synthesized. Two clones produced equally large amounts of SC, and one of them, HT29.E10, was selected for further study.

Pulse-Chase Experiments

We generally use 35-mm dishes with confluent monolayers of cells. The medium is changed 12–24 hr before the experiment to ensure that the cells are metabolically active. The medium is removed, and the monolayer is washed twice with phosphate-buffered saline (PBS, from GIBCO). Cells are starved for cysteine for 30 min by adding 0.5 ml of minimal essential medium made up without cysteine, supplemented with 10% dialyzed fetal bovine serum (both from GIBCO). This medium is then replaced with 0.3 ml of the same medium to which [^{35}S]cysteine has been added. Generally we use 50 μCi of label per plate by adding 5 μl of 10 mCi/ml stock to the medium. Cells are generally pulsed for between 5 and 60 min. Short pulses give better temporal resolution in the study of the kinetics of SC processing. Long pulses give proportionately higher levels of incorporation of [^{35}S]cysteine into SC.

If desired, the cells can then be chased with an excess of cold cysteine. The labeling medium is removed, and the cells are washed twice with PBS. McCoys 5a medium (0.8 ml), supplemented with serum and antibiotics, is added. This medium contains a greater than 1000-fold excess of cold cysteine as compared to the concentration of [^{35}S]cysteine used in the labeling, so the chase is effective. We generally chase for 30 min to 72 hr.

Immunoprecipitation of SC Synthesized by Cells

The transmembrane SC precursor is proteolytically cleaved by the cells to yield mature SC. When the cells are lysed to immunoprecipitate the SC precursor, precursor molecules that have not yet been cleaved might be artifactually cleaved by the relevant protease during immunoprecipitation. Additionally, other proteases that normally never see the SC precursor (e.g., lysosomal proteases) may have a chance to attack the molecule. We prevent artifactual proteolysis by lysing the cells in SDS and boiling as quickly as possible, which denatures all proteases. The medium is removed, and 0.5 ml of 0.5% SDS, 100 m*M* NaCl, 50 m*M* triethanolamine–HCl (pH 8.1), 5 m*M* EDTA, and 1% Trasylol are added. The monolayer is scraped off with a rubber policeman, and the lysate is transferred to a 1.5-ml Eppendorf tube. The tube is placed in a boiling water bath for 5 min. After cooling, the clump of DNA in the tube is dispersed by a brief sonication. The lysate is diluted with Triton dilution buffer and immunoprecipitation is performed as described above for cell-free translation products.

The SC secreted into the medium can also be immunoprecipitated. The chase medium from the cells is first centrifuged at 12,000 *g* for 15 min in a Brinkmann microfuge to pellet any dead cells and debris. Then SDS is added to 1%, and the medium is boiled for 5 min. The sample is then diluted with Triton dilution buffer and processed for immunoprecipitation in the same fashion as the cell lysate.

Rabbit anti-human SC antisera are available from Behring Diagnostics, Somerville, New Jersey; Atlantic Antibodies, Scarborough, Maine; Dako, Accurate Chemicals, Hicksville, New York; and Boehringer Mannheim, Indianapolis, Indiana. All of them work fairly well. Generally we use 5–10 μl of antiserum for the lysate or medium of one 35-mm dish. The goat anti-human SC antiserum produced by Miles (Elkhart, Indiana) does not bind well to protein A–Sepharose. We generally add antibody to the reaction and incubate for 4 hr at room temperature. The immune complexes are then absorbed with protein A–Sepharose, washed, and analyzed by SDS–PAGE by the same procedure used for immunoprecipitation from cell-free translation.

Cell-Free Translation of HT29 mRNA

The HT29 cell line, or the cloned line HT29.E10, can also be used as a source of mRNA for cell-free translation studies of SC. Cells are most conveniently grown in 100-mm dishes. We generally harvest confluent monolayers from eight such dishes. As in the pulse-chase experiments, the medium is changed 12–24 hr prior to harvesting. The medium is removed, and cells are scraped off in 2 ml of PBS using a rubber policeman. The cells are transferred to a centrifuge tube and centrifuged at 400 *g* for 2 min in a clinical centrifuge at room temperature. The supernatant is removed, and the pellet is frozen in liquid nitrogen and kept at −80° until use. The mRNA is extracted and translated in the same manner as used for rabbit tissues. Unfortunately, the HT29 cell mRNA is only about one-fourth as active as that from rabbit tissues in terms of the total protein synthesized in the cell-free system. Furthermore, even if the HT29.E10 clone is used, SC is only a very minor component of the total products. Generally, a fluorograph of the SC immunoprecipitated from a 100-μl translation and analyzed by SDS–PAGE requires exposure for 2–3 weeks on Kodak XAR-5 film to yield a reasonably dark band.

[41] Transcellular Transport of Proteins Studied *in Vivo*

By GERHARD ROHR and GEORGE SCHEELE

A new approach to the *in vivo* study of protein transport across epithelial membranes has been introduced.[1,2] This approach follows the distribution of radioactively labeled proteins injected into the blood circulation of unanesthetized rats. Samples of blood, urine, bile, and pancreatic juice are collected over time, and individual radioactively labeled proteins contained in these samples are analyzed after their separation by two-dimensional isoelectric focusing/SDS–gel electrophoresis. We describe here in greater detail the methods used in this analysis and give several examples of data derived from study of the transcellular transport of pancreatic exocrine proteins.

Preparation of Animals

Wistar rats, 250–300 g, are starved overnight prior to ether anesthesia and surgical cannulation.

Cannulation of the Tail Vein.[3] A polyethylene cannula of 1 mm i.d. is heated by brief immersion into water at 70° and reduced to 0.1 mm o.d. by stretching with forceps, being careful to maintain patency of the cannula. An injection needle (0.5 mm i.d.) is filed open as shown in Fig. 1 to provide a channel for introduction of the cannula into the tail vein. After introduction of the cannula, the needle is retracted and the position of the cannula is fixed to the tail with tape. This portal is used for intravenous injection. Patency is

[1] G. Rohr, H. Kern, and G. Scheele, *Nature* (*London*) **292,** 470 (1981).
[2] G. Rohr and G. Scheele, *Gastroenterology* (in press).
[3] G. Pappritz and U. Braüer, *Arzneim-Forsch.* (*Drug Res.*) **27,** 864 (1977).

ISBN 0-12-181998-1

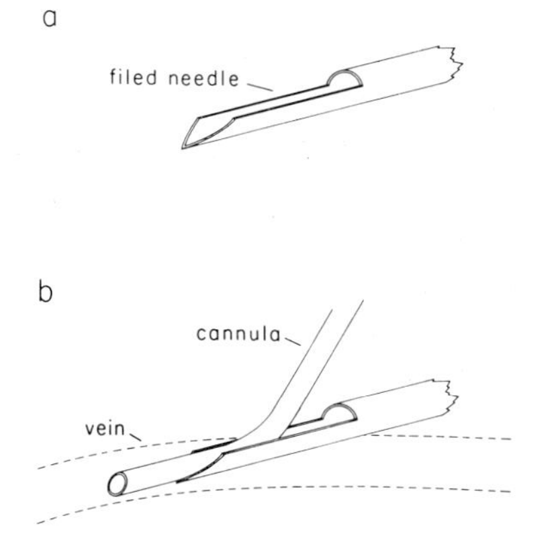

FIG. 1. Insertion of a polyethylene cannula into the tail vein of the rat.

maintained by filling the cannula with heparin at 12 units/ml. An alternative procedure is given by Born and Moller.[4]

Cannulation of the Bile and Pancreatic Ducts. An upper abdominal wall incision is made, and the duodenal C loop is exposed. In the rat, multiple pancreatic ductules drain into the common bile–pancreatic duct as it courses from the liver to the duodenum. The common bile–pancreatic duct is ligated at its point of entry into the duodenum. For the collection of pancreatic juice the common duct is opened immediately proximal to this ligature with a pair of sharp ophthalmic scissors and cannulated with a polyvinyl chloride catheter (i.d. 0.35 mm). The bile duct is then ligated proximal to its entry into the pancreatic tissue. For the collection of bile, a second cannula is inserted proximal to this ligature. Each cannula is secured by ligature, routed subcutaneously with a blunt needle to which the cannula is tied, and exteriorized at the nape of the neck. For experiments in which only the pancreatic juice is collected, the bile duct cannula is introduced, usually near the end of the surgical procedure, into the duodenal lumen near the point of entry of the common duct.

Cannulation of the Inferior Vena Cava or Portal Vein. The distal tip of a syringe needle (0.8 mm o.d.) is cut 1 cm in length, and the blunt end of this tip is inserted 2 mm into a polyvinyl chloride cannula (0.75 mm i.d.). An upper abdominal wall incision is made, the intestinal tract is placed to the side, and the inferior vena cava or portal vein is exposed. The needle tip, grasped by forceps, is inserted 6 mm into the blood vessel. The cannula is fixed to adjacent tissues by ligature, routed subcutaneously, and exteriorized at the nape of the neck. Patency is maintained as described above.

Cannulation of the Ureter. A transverse lower abdominal wall incision is made, and the left and right ureters, in the retroperitoneal space, are exposed

[4] C. T. Born and H. L. Moller, *Lab. Anim. Sci.* **24,** 355 (1974).

by withdrawal of the intestines. The proximal ureter is opened with fine ophthalmic scissors following ligation of the distal ureter. A polyethylene cannula reduced by heat and stretching to 0.4 mm o.d. and 0.2 mm i.d. is inserted into the ureter and secured by ligation. The cannula is routed to the dorsum of the rat, exteriorized, and fixed to the skin by a procedure similar to that described for cannulation of the bile and pancreatic ducts.

Comments. Postoperatively, animals are maintained in restraining cages and allowed to recover from anesthesia and surgery for 24 hr. Restraining cages allow limited excercise and free access to solid food and water. Without intravenous supplementation, rats remain in healthy condition for more than 72 hr after surgery. Throughout our experiments, usually 7 hr, excretory flow rates are generally constant as follows: pancreatic juice, 350–500 μl/hr; bile, 600–800 μl/hr; and urine, 450–900 μl/hr.

Preparation of Radioactive Proteins

Proteins for injection into the blood circulation can be labeled, either exogenously or endogenously, to high specific activities using the following procedures.

Exogenous Radioactive Label. Highest specific radioactivities can be achieved by iodination of proteins with $Na^{125}I$. Iodination of tyrosine residues occurs with either the enzymic (lactoperoxidase) method[5] or the chloramine-T method.[6] Iodination of free amino groups, amino terminal, or ϵ-amino groups of lysine, occurs with the conjugation (Bolton–Hunter) method.[7] After the iodination procedure it is necessary to separate, usually by Sephadex G-25 chromatography, the iodinated protein from the unreacted iodide and to establish that the iodinated protein has retained its biological activity. The enzymic and conjugation methods preserve biological activity to the greatest extent.

Endogenous Radioactive Label. Pancreatic exocrine proteins have been labeled endogenously to levels of highest specific radioactivity by incubation of pancreatic lobules (guinea pig, rat) or tissue slices (dog, human) under physiological conditions in the presence of [^{35}S]methionine or [^{35}S]cysteine.[8] Radioactive proteins, labeled with [^{35}S]methionine and extracted from a crude zymogen granule fraction, contain 1×10^8 cpm/ml and a protein concentration ~ 10 mg/ml.[8] Radioactive proteins are injected into the peripheral blood (tail vein) in volumes $\leqslant$250 μl.

[5] J. I. Thorell and B. G. Johansson, *Biochim. Biophys. Acta* **251**, 363 (1971).

[6] P. J. McConahey and F. Dixon, this series, Vol. 70, p. 210.

[7] A. E. Bolton and W. M. Hunter, *Biochem. J.* **133**, 529 (1973).

[8] G. Scheele, this volume [2].

Analysis of Radioactivity Contained in Body Fluids

Fractionation with Trichloroacetic Acid

Total ^{35}S radioactivity is measured by liquid scintillation spectroscopy after adding 10-μl samples to 5 ml of Liquiscint (National Diagnostics, Somerville, New Jersey). Trichloroacetic acid (TCA)-insoluble radioactivity is measured in 10-μl aliquots applied to Whatman 3MM filter paper disks 25 mm in diameter, which are then immersed in cold 10% TCA, processed by the procedure of Mans and Novelli,[9] washed in 95% ethanol, dried under an infrared lamp, and counted in Liquifluor.

Analysis of TCA-Soluble Radioactivity by Amino Acid Analysis

Sample of serum and bile that contained TCA-soluble ^{35}S are deproteinized with sulfosalicylic acid and neutralized to pH 7 prior to analyses on an amino acid analyzer of the design of Spackman, Moore, and Stein.[10]

Separation of Proteins by Two-Dimensional Isoelectric Focusing Sodium Dodecyl Sulfate (SDS)–Gel Electrophoresis

Proteins contained in serum, bile, and pancreatic juice are separated according to isoelectric point (IEP) and apparent molecular weight (M_r) using the two-dimensional gel procedure developed by Scheele.[11] Proteins are separated in the first dimension by slab gel isoelectric focusing in the presence of 8 *M* urea as described by Bieger and Scheele[12] and in the second dimension by SDS–slab gel electrophoresis in a polyacrylamide gradient (10 to 20%). In addition to urea, the isoelectric focusing gel contains ampholytes in the concentrations pH 3.5–10 (1%), pH 5–8 (0.3%), and pH 4–6 (0.15%) and also the trypsin inhibitors Trasylol (10 KIU/ml; FBA Pharmaceuticals, New York, New York), soybean trypsin inhibitor (2 μg/ml; Worthington Corp., Freehold, New Jersey), and diisopropyl fluorophosphate (0.1 m*M*; Sigma Chemical Co., St. Louis, Missouri). Samples of serum, bile, or pancreatic juice are made 4% in Ampholine, pH 3.5–10, 5 mg/ml in cytochrome *c* (visual marker), and 4–6 *M* in urea to prevent activation of potential proteases. Thirty-microliter samples are applied to the surface of isoelectric focusing gels using 7-mil gel bond (FMC Corp., Rockland, Maine) rectangular frames (10 × 12-mm outside dimensions, 8 × 10-mm inside dimensions) as previously described.[12] Isoelectric focus-

[9] R. J. Mans and G. D. Novelli, *Arch. Biochem. Biophys.* **94,** 48 (1961).

[10] D. H. Spackman, S. Moore, and W. H. Stein, *Anal. Chem.* **30,** 1190 (1958).

[11] G. A. Scheele, *J. Biol. Chem.* **250,** 5375 (1975).

[12] W. Bieger and G. Scheele, *Anal. Biochem.* **109,** 222 (1980).

ing and SDS–gel electrophoresis are carried out as described by Scheele.[11] Proteins are identified with Coomassie Blue R stain.

Analysis of Radioactivity Contained in Two-Dimensional Gel Spots

Fluorography. Radioactive proteins separated in the two-dimensional gel are identified by fluorography after impregnation of the gel with 2,5-diphenyloxazole by the procedure of Bonner and Laskey.[13] Procedural modifications to prevent cracking of gels during vacuum drying at room temperature are given by Scheele.[14]

Radioactivity detected on (preflashed) medical X-ray film have been quantified by two-dimensional spectrophotometric scanning using an Optronics two-dimensional scanner with optical density measurements taken at 100-μm intervals. Data recorded on magnetic tape were analyzed on a Digital Equipment Corporation PDP 11/70 computer using programs that determined the cumulative densities within and fractional densities among the spots.

Liquid Scintillation Spectroscopy. Coomassie Blue-stained spots are excised from the gel and homogenized in 2 ml of distilled water with a motor-driven Teflon pestle. The homogenate is decanted into a scintillation vial followed by two 1-ml washes with distilled water. To a final volume of ~4.5 ml is added 10 ml of scintillation medium 808E from Riechel de Häen. An alternative method is given by Scheele.[11]

Analysis of the Transcellular Transport of Pancreatic Exocrine Proteins

The methods described above have allowed us to study the transport of individual exocrine proteins from the blood circulation across epithelial barriers to separate body fluids.[1,2] Figure 2 shows the kinetics of disappearance of radioactive guinea pig exocrine proteins injected into the blood circulation of the rat and the kinetics of appearance of radioactive proteins in urine, bile, and pancreatic juice. Ninety percent of the radioactive proteins had disappeared from the blood circulation 30 min after injection. Trichloroacetic acid-insoluble radioactivity appeared in urine within 5 min after the radiolabeled proteins were injected into the blood stream, and the kinetics of disappearance of these proteins followed closely those observed in the blood stream. Trichloroacetic acid-insoluble radioactivity appeared in bile just after 5 min, peaked during the 15–30-min period, and declined thereafter to a plateau observed after the 2.5-hr time point. In contrast, pancreatic juice was devoid of radioactivity for the first hour. Trichloroace-

[13] This series, Vol. 96 [15].

[14] G. A. Scheele, this series, Vol. 96 [7].

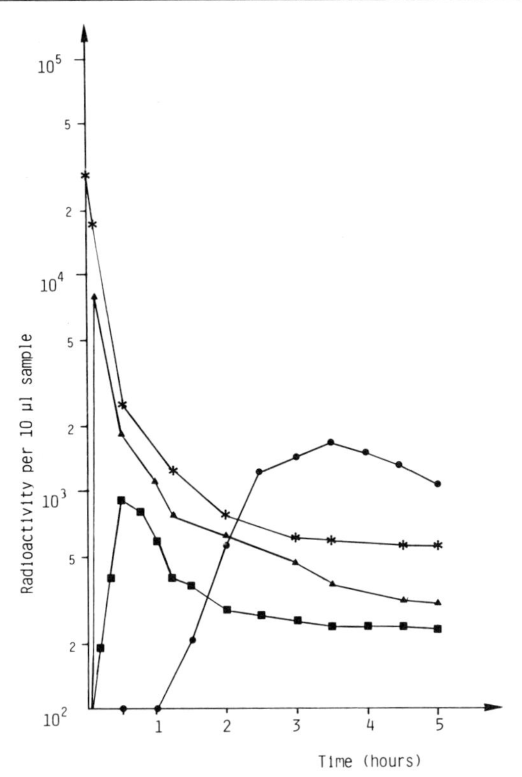

FIG. 2. Kinetics of disappearance of trichloroacetic acid (TCA)-insoluble radioactivity from the blood circulation (*) and kinetics of appearance of TCA-insoluble radioactivity in urine (▲), bile (■), and pancreatic juice (●) following the intravenous injection of radioactive pancreatic proteins. Semilogarithmic plot.

tic acid-insoluble radioactivity first appeared in pancreatic juice in the 60–90-min collection period and thereafter increased, showing a maximal concentration of radioactivity at 3.5–4.0 hr. Radioactive proteins that appeared in urine and bile over the course of 7 hr accounted for 1–2% and 0.3–1.0% of the injected radioactivity, respectively. The majority of pancreatic proteins, approximately 97%, were taken up by a variety of body tissues, particularly kidney, liver, spleen, and lung. Trichloroacetic acid-soluble radioactivity, representing [^{35}S]methionine by amino acid analysis, appeared sequentially in serum, urine, and bile within 5–12 min.

Figures 3–5 show the analysis of individual radioactive proteins contained in these biological fluids. Figure 3 shows the two-dimensional gel pattern of radioactively labeled guinea pig pancreatic proteins injected into the blood circulation of the rat. Figure 4 shows the radioactive proteins that

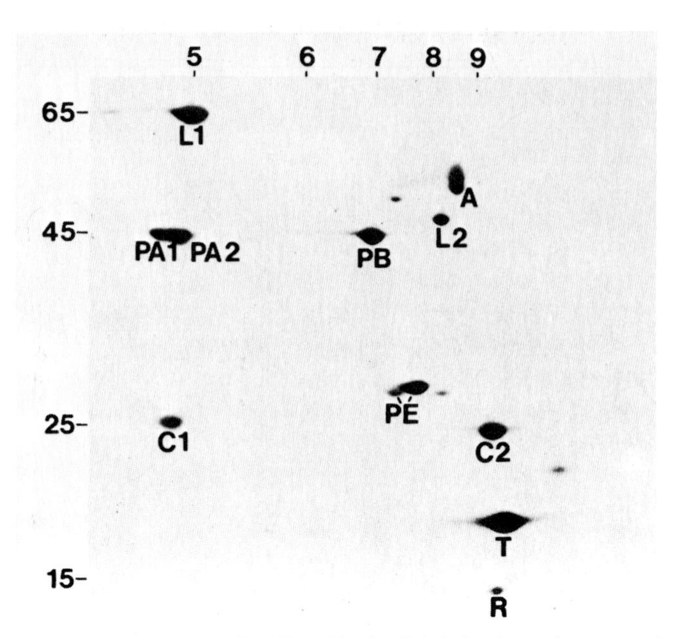

FIG. 3. Fluorographic pattern of [^{35}S]methionine-labeled guinea pig pancreatic proteins separated by two-dimensional isoelectric focusing/SDS–gel electrophoresis. Numbers on the upper abscissa indicate isoelectric points. Numbers on the left ordinate indicate apparent molecular weights $\times 10^{-3}$. Radioactive guinea pig proteins were injected into the blood circulation of the rat. Exocrine pancreatic proteins are labeled according to actual or potential enzyme activity by the following abbreviations: A, amylase; L, lipase; PA, procarboxypeptidase A; PB, procarboxypeptidase B; PE, proelastase; C, chymotrypsinogen; T, trypsinogen; R, ribonuclease. Numbers refer to forms of enzyme or zymogens believed to represent separate gene products.

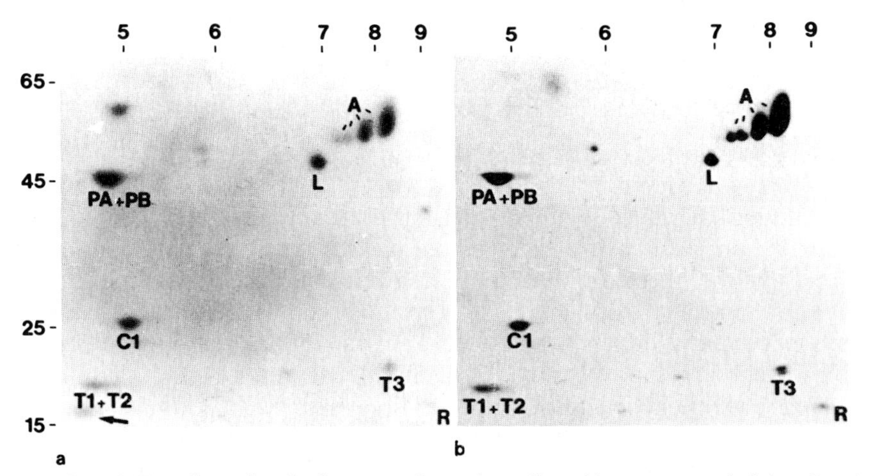

FIG. 4. Two-dimensional gel pattern of proteins collected in rat pancreatic juice after the intravenous injection of radioactively labeled proteins shown in Fig. 3. (a) Coomassie Blue staining pattern. (b) Fluorographic pattern. Numbers on the left ordinate and abscissa and identifications of proteins are given as described in the legend to Fig. 3. Proteins stained with Coomassie Blue represent exocrine pancreatic proteins normally observed in rat pancreatic juice.

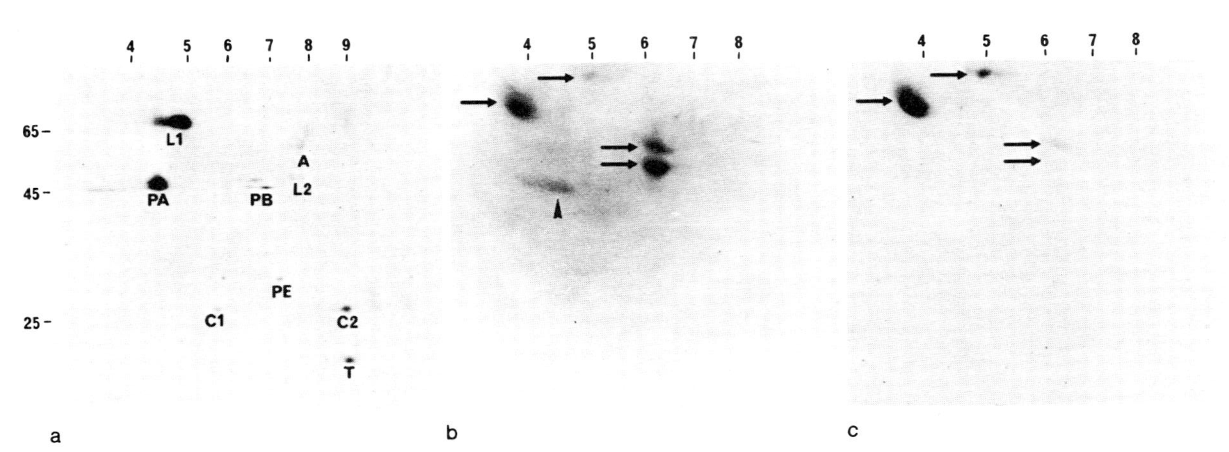

FIG. 5. Two-dimensional gel pattern of proteins collected in rat bile following the intravenous injection of radioactively labeled proteins shown in Fig. 3. (b) Coomassie Blue staining pattern of rat biliary proteins. (a, c) Fluorographic patterns of radioactive proteins that appeared in bile after 0.5 hr (a) and 4.5 hr (c). Numbers on the left ordinate and abscissa and identifications of pancreatic proteins are as described in the legend to Fig. 3. Horizontal arrows indicate bile proteins that are labeled with Coomassie Blue and, during the later collection period, with radioactive methionine. The vertical arrowhead in (b) indicates a bile protein that stains with Coomassie Blue but did not show radioactivity (c) at this level of detection.

appeared in pancreatic juice 3 hr after the injection of radioactive proteins into the blood and compares the pattern of radioactive proteins to the pattern of Coomassie Blue-stained proteins found in the same juice specimen. Although radioactivity was injected in the form of guinea pig pancreatic proteins, radioactivity appeared in pancreatic juice in the form of rat exocrine proteins.

In contrast, radioactive proteins that appeared in bile within the first 30 min represented proteins that were transported intact from blood to bile (Fig. 5). Although radioactive proteins that appeared in bile were composed of the entire group of guinea pig pancreatic proteins, acidic proteins, lipase 1 and procarboxypeptidase A, were considerably enriched over basic proteins (Fig. 5a). These two proteins represented 12% of the radioactivity injected into the bloodstream, but 93% of the radioactive proteins collected in bile. Figure 5c shows the pattern of radioactive proteins that appeared in bile after 4.5 hr and compares their distribution to that of bile proteins stained with Coomassie Blue R (Fig. 5b). At this late time, radioactive proteins that appeared in bile represented bona fide rat biliary proteins. At later experimental time points (>60–90 min) radioactive rat serum proteins appeared in the bloodstream. These studies indicated that pancreatic proteins are removed from the blood circulation by at least three separate pathways: (*a*) uptake and metabolic degradation by a variety of tissues in the body (~97%); (*b*) excretion of intact proteins into the urine (1–2%); and (*c*) transcellular transport of proteins into bile (0.3–1.0%). Transport of exocrine pancreatic proteins from the blood circulation to pancreatic juice could not be demonstrated.

Similar experiments can be carried out *in vivo* to study transcellular transport of radioactively labeled serum proteins, polypeptide hormones, or immunoglobulins. Transcellular transport of these and other proteins can be studied in normal animals or in animals with experimentally induced disease, e.g., hormone-induced pancreatitis or alcohol-induced cirrhosis. Finally, mechanisms for reverse transport may possibly be studied by injection of radioactively labeled proteins, under the proper conditions, into the bile or pancreatic duct and study of the appearance of radioactivity in the blood circulation.

Section VII

Plasma Membrane Differentiation

[42] Polarized Multicellular Structures Designed for the *in Vitro* Study of Thyroid Cell Function and Polarization

By JEAN MAUCHAMP, MARIANNE CHAMBARD, JACQUELINE GABRION, and BERNARD VERRIER

Described here are the experimental systems that have been designed to study *in vitro* thyroid cell function and its regulation with special emphasis on parameters linked to the morphological and functional polarization of the epithelial thyroid cell.

Thyroid Function and Cell Polarization

For an endocrine gland the histological organization of the thyroid gland is rather unusual in that it closely resembles the organization of exocrine glands. The existence of an extracellular store of thyroglobulin, which contains bound tri- and tetraiodothyronine, gives to cell polarization a major role in the sequence of events leading to the synthesis and secretion of thyroid hormones.

Morphology[1-3]

Thyroid epithelial cells are organized into follicles: a monostratified cell layer delineates closed cavities containing thyroglobulin almost exclusively. The apical poles of the cells, which bear numerous microvilli and in many species a cilium, are oriented toward the lumen, whereas the basal surface of the cell layer, lying on a continuous basement membrane, is in contact with the internal milieu, which contains nutrients and stimulators. A tight junction with a variable number of strands forms a continuous seal between adjacent cells and separates the apical and the basolateral domains of the plasma membrane.[4] The height of the cell layer and the size of the colloid depend on the activity of the follicles. When the gland is inactive, the colloid is large and dense and the cells are flattened. In active states the cells are tall and the colloid is small.

The distribution of organelles within the cytoplasm is more or less oriented. The centriole, eventually at the base of the cilium, is close to the

[1] L. W. Tice, *in* "Histology" (L. Weiss and R. O. Greep eds.), p. 1077. McGraw–Hill, 1977.
[2] N. J. Nadler *Handb. Physiol. Sect. 7 Endocrinol.* **3,** 39 (1974).
[3] R. Ekholm, *in* "Endocrinology" (L. J. De Groot, ed.), Vol. 1, p. 305. Grune & Stratton, New York, 1979.
[4] L. W. Tice, S. H. Wollman, and R. C. Carter, *J. Cell Biol.* **66,** 657 (1975).

METHODS IN ENZYMOLOGY, VOL. 98

ISBN 0-12-181998-1

apical pole.[5] The rough endoplasmic reticulum (RER) is usually found throughout the cytoplasm, although in metamorphosing amphibians highly ordered laminar structures are localized at the basal pole.[6] The well-organized Golgi apparatus is mostly supranuclear. Numerous small vesicles are observed in the apical cytoplasm. Some are bristle coated and others, the apical vesicles, appear to discharge their content into the lumen.[7] Actin microfilaments are localized at the apical pole as bundles in the axis of microvilli and as a clear network parallel to the apical surface.[8] Microtubules radiating from the centriole are also more numerous in the apical cytoplasm.

Function

Thyroid hormone synthesis and secretion imply three successive steps.

1. Thyroglobulin synthesized in the RER and iodide concentrated from the circulation are transferred into the follicular cavity. Thyroglobulin travels through the Golgi apparatus and apical vesicles, which discharge the uniodinated protein into the colloid by exocytosis (review in Ekholm[9]). Iodide is actively concentrated at the basal pole and probably enters the colloid by diffusion.[10]
2. A peroxidase localized on the apical plasma membrane iodinates thyroglobulin at the periphery of the lumen.[9] At the same time, thyroid hormone synthesis takes place within the thyroglobulin molecule by intramolecular coupling of iodotyrosyl residues.[11]
3. Thyroglobulin present in the follicular lumen is taken up by the apical pole of the cells by macro- or micropinocytosis.[12] The protein is hydrolyzed, and thyroid hormones liberated from the peptide chain are excreted in the circulation through the basal pole of the cells by an unknown process.

The content of the follicular lumen is a solution of almost pure thyroglobulin (about 100 mg/ml).[13,14] The ionic composition is close to serum except for K^+, which is three to four times more concentrated.[15]

[5] J. D. Zeligs and S. H. Wollman, *J. Ultrastruct Res.* **66,** 97 (1979).
[6] E. Regard and J. Mauchamp, *J. Ultrastruct. Res.* **37,** 664 (1971).
[7] R. Ekholm, G. Engström, L. E. Ericson, and A. Melander, *Endocrinology (Baltimore)* **97,** 337 (1975).
[8] J. Gabrion, F. Travers, Y. Benyamin, P. Sentein, and N. Van Thoai, *Cell Biol. Int. Rep.* **4,** 59 (1980).
[9] R. Ekholm, *Mol. Cell. Endocrinol.* **24,** 141 (1981).
[10] G. Andros and S. H. Wollman, *Am. J. Physiol.* **213,** 198 (1967).
[11] A. Taurog, *Recent Prog. Hormone Res.* **26,** 189 (1970).
[12] L. E. Ericson, *Mol. Cell. Endocrinol.* **22,** 1 (1981).
[13] S. Smeds, *Endocrinology (Baltimore)* **91,** 1288 (1972).
[14] S. Smeds, *Endocrinology (Baltimore)* **91,** 1300 (1972).
[15] J. A. Young, L. J. Hayden, and J. M. Shagrin, *Pfluegers Arch.* **321,** 187 (1970).

Stimulation by Thyrotropin

Thyroid function is stimulated by circulating thyrotropin (TSH) which interacts with a plasma membrane-localized receptor coupled to adenylate cyclase.[16] Although not proved, it is likely that the TSH–receptor–adenylate cyclase complex is localized on the basolateral domain of the plasma membrane. One of the first events following stimulation by TSH occurs at the apical pole of the cells. A rapid discharge of apical vesicles appears to precede the formation of large apical pseudopods that protrude into the lumen.[12] These membrane protrusions engulf a large portion of the colloid content.

The role of cell organization in the normal sequence of events that take place in the gland is obviously important. Thyroglobulin iodination at a normal level and thyroxine synthesis can occur only when thyroglobulin accumulates in the follicular lumen. Therefore, we have not included results obtained with isolated thyroid cells, although such systems can be useful in understanding specific aspects of thyroid cell function.

Membrane Markers of Thyroid Cell Polarity

At present a very small number of molecular markers of the apical and basolateral domains of the plasma membrane of the thyroid cell have been characterized. The thyroperoxidase has been localized on the apical membrane by histochemical methods (see review in Ekholm[9]), but the bulk of the enzyme is present in intracellular organelles (RER, Golgi apparatus, apical vesicles). An aminopeptidase N, similar to the enzyme found on the intestinal brush border, has been observed on the apical thyroid cell surface, using antibodies raised against the intestinal enzyme.[17] The function of this protein in the thyroid is unknown and an intracellular pool was also observed. More recently an Na^+,K^+-ATPase was found exclusively localized on the basolateral portion of the rat thyroid cell plasma membrane using anti-ATPase antibodies.[18]

Consequences of Thyroid Cell Organization

As a result of their integration into follicular structures, thyroid epithelial cells gain the following characteristics: cells are polarized; cells have permanent contact with several neighbors; the apical surface is in contact with the colloid content; the basal surface lies on a basement membrane; the apical

[16] S. Lissitzky, G. Fayet, and B. Verrier, *in* "Methods in Receptor Research" (M. Blecher, ed.), p. 641. Dekker, New York, 1976.

[17] H. Feracci, A. Bernadac, S. Hovsepian, G. Fayet, and S. Maroux, *C. R. Hebd. Seances Acad. Sci. Ser. D* **292,** 271 (1981).

[18] J. Gabrion, unpublished observation.

and basal compartments are separated; and nutrients and stimulators have access to the basal pole.

Experimental procedures have been designed that allow either the isolation and maintenance of isolated follicles or the formation and maintenance of multicellular epithelial structures differing from follicles in various respects: inversion of cell polarity, absence of separation between apical and basolateral compartments, access of nutrients to the apical pole. A comparative study of the expression of thyroid function under these various experimental conditions promotes a better understanding of the consequences of cell organization on the expression of specific differentiated functions and a more complete description of the polarized characters of thyroid cells.

Freshly Isolated Follicles

Complete follicles can be isolated by collagenase treatment of thyroid glands. Various procedures have been described for rat, pig, and human thyroid tissue.[19-21] A mixture of closed and open follicles is usually obtained in various proportions. They can be maintained for short periods of time and display functional characteristics of thyroid cells. The presence of open follicles allows free access to the apical portion of the plasma membrane, but the interpretation of quantitative biochemical studies might be difficult. Such preparations have been used to study apical membrane retrieval,[20,21] localization of the apical iodination site,[9] and production of H_2O_2 at the apical surface of the cells.[22]

Rat thyroid follicles were maintained for about 10 days in suspension culture in agarose-coated dishes.[19] The medium was Coon's modified F_{12} supplemented with 0.5% calf serum.

Formation of Closed Epithelial Structures in Culture (Fig. 1a)

The first type of structure obtained in culture is a closed epithelial structure, in which the cells form continuous polarized cell layers with tight junctions between adjacent cells. A cavity is enclosed by the cells, and only one cell pole, in contact with the culture medium, is accessible. The other cell pole is oriented toward the cavity. Depending on the culture condition, two closed structures that differ in orientation of cell polarity can be obtained: a follicle structure when the apical pole is oriented toward the cavity; a vesicle (or cyst) or monolayer structure when the apical pole is in contact with the culture medium.

[19] L. Nitsch and S. H. Wollman, *Proc. Natl. Acad. Sci. U.S.A.* **77,** 472 (1980).
[20] V. Herzog and F. Miller, *Eur. J. Cell Biol.* **19,** 203 (1979).
[21] J. F. Denef, U. Björkman, and R. Ekholm, *J. Ultrastruct. Res.* **71,** 185 (1980).
[22] U. Björkman, R. Ekholm, and J. F. Denef, *J. Ultrastruct. Res.* **74,** 105 (1981).

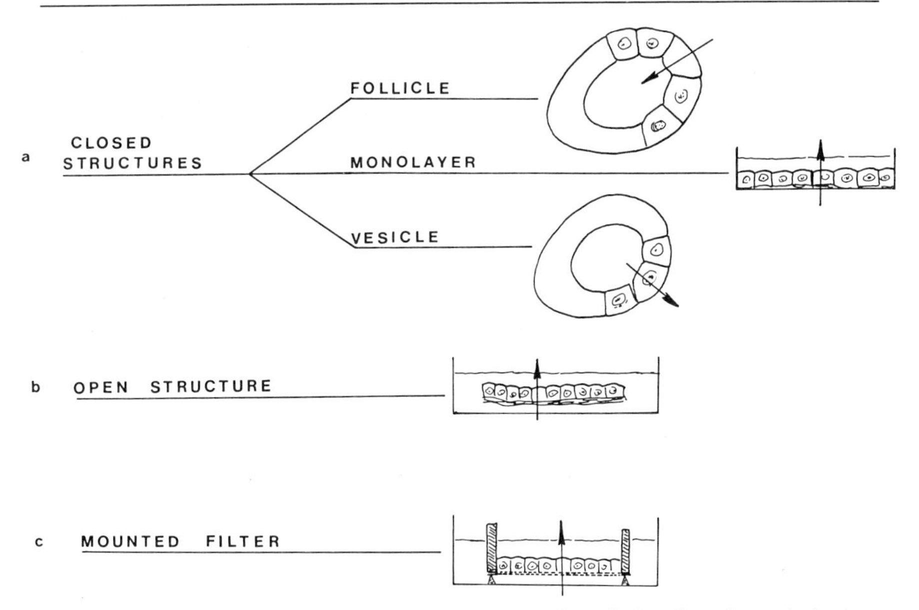

FIG. 1. Various structures formed in culture. Arrows show the basolateral-to-apical orientations.

Reorganization of Isolated Cells into Follicle-Like Structures

Isolated pig, beef, or sheep thyroid cells can be obtained by trypsin or trypsin–EGTA treatment of finely minced thyroid tissue.[23-25] Porcine thyroid cells were used in most cases reported here. When cultured at high cell density, they reorganize into follicle-like structures under a variety of culture conditions. These structures have most of the morphological characteristics of genuine follicles. The apical pole was oriented toward the follicular lumen, and the basal pole was in contact with the culture medium. No basement membrane was observed under these conditions.[24-27] A large proportion of thyroglobulin was found in the culture medium, but not in the follicular lumen.[26] These follicles might have been leaky or an abnormal secretion of thyroglobulin might have occurred at the basal pole of the cells.

[23] W. Tong, P. Kerkof, and I. L. Chaikoff, *Biochim. Biophys. Acta* **60,** 1 (1961).

[24] G. Fayet, H. Pachecot, and T. Tixier, *Bull. Soc. Chim. Biol.* **52,** 299 (1970).

[25] J. Mauchamp, A. Margotat, M. Chambard, B. Charrier, L. Remy, and M. Michel-Bechet, *Cell Tissue Res.* **204,** 417 (1979).

[26] S. Lissitzky, G. Fayet, A. Giraud, B. Verrier, and J. Torresani, *Eur. J. Biochem.* **24,** 88 (1971).

[27] G. Fayet, M. Michel-Bechet, and S. Lissitzky, *Eur. J. Biochem.* **24,** 100 (1971).

Follicles were obtained under four different culture conditions.

1. In NCTC 109 culture medium in the absence of serum, with the culture dish coated with agarose to avoid cell attachment (unpublished observation).
2. In Eagle's minimum essential medium (MEM) or NCTC 109 supplemented with 0.5% to 1% (v/v) calf serum and 0.25% (w/v) gelatin in plain polystyrene culture dishes, which do not support cell attachment.[28]
3. In Eagle's MEM or NCTC 109 supplemented with serum (2–10%) and containing a thyroid stimulator: TSH (50 μU/ml or more) or prostaglandin E_2 (1 μM). The stimulator can be replaced by dibutyryl-cyclic AMP (0.5 or 1 mM). When dishes that do not support cell attachment were used, the follicles remained in suspension. When tissue culture-treated dishes were used, the follicles attached to the plastic substratum.[25,27]
4. Cells were embedded at high density inside a collagen gel reconstituted from acid-soluble rat tail tendon collagen.[29] Culture medium was added after gel polymerization. Follicles were obtained within 3 to 5 days. The basal pole was formed in contact with the surrounding collagen matrix.

When thyroid cells are organized into follicles they concentrate and organify iodide[26,30]; cyclic AMP synthesis responds to TSH stimulation[31] and is inhibited by muscarinic agonists.[32] Thyroglobulin is synthesized at a high rate, and the thyroglobulin–mRNA (TgmRNA) level is elevated.[33,34] A transepithelial potential difference of 30 mV (lumen negative) can be measured.[35]

Formation of Vesicles (*Inside-Out Follicles or Cysts*)[35a]

When isolated porcine thyroid cells[25] or follicle segments[36] were cultured as an unstirred suspension in serum-containing medium (5% or more) in the

[28] J. Mauchamp, M. Chambard, J. Gabrion, and C. Pelassy, *C. R. Seances Soc. Biol. Ses. Fil.* **174,** 241 (1980).

[29] M. Chambard, J. Gabrion, and J. Mauchamp, *J. Cell Biol.* **91,** 157 (1981).

[30] J. Mauchamp, B. Charrier, N. Takasu, A. Margotat, M. Chambard, and D. Dumas, *in* "Hormones and Cell Regulation" (J. Dumont and J. Nunez, eds.), Vol. 3, p. 51. Elsevier/North-Holland, Amsterdam, 1979.

[31] N. Takasu, B. Charrier, J. Mauchamp, and S. Lissitzky *Eur. J. Biochem.* **90,** 139 (1978).

[32] S. Champion and J. Mauchamp, *Mol. Pharmacol.* **21,** 66 (1982).

[33] J. Chebath, O. Chabaud, and J. Mauchamp, *Nucleic Acids Res.* **6,** 3353 (1979).

[34] O. Chabaud, J. Chebath, A. Giraud, and J. Mauchamp, *Biochem. Biophys. Res. Commun.* **93,** 118 (1980).

[35] J. Mauchamp, J. Gabrion, and P. Bernard, *in* "Hormonal Control of Epithelial Transport," *Colloq.-Inst. Natl. Rech. Med.* **85,** p. 43 (1979).

[35a] See Herzog, this volume [39].

[36] V. Herzog and F. Miller, *Eur. J. Cell Biol.* **24,** 74 (1981).

absence of thyroid stimulator, they formed closed epithelial structures that differed from follicles in orientation of cell polarity. The apical pole of the cells was oriented toward the culture medium, and the basal pole faced the cavity encompassed by the cell layer. The cavity contained no electron-dense material. Such structures can also be obtained when rat thyroid follicles maintained in culture with 0.5% serum are further cultured in the presence of 5% serum. An inversion of cell polarity was observed within 3 days.[37] Similarly oriented structures have been obtained with Madin-Darby canine kidney (MDCK) cells in suspension.[38]

When thyroid cells are organized into vesicles they are unable to concentrate and organify iodide,[30] and they display a poor cyclic AMP response to thyrotropin stimulation[30,31]; thyroglobulin, synthesized at a low rate,[34] is excreted into the culture medium.[36] A transepithelial potential difference of about 10–20 mV, cavity positive, can be measured.[35]

Monolayers on an Impermeable Substratum

When thyroid cells were cultured on substrata that allow cell adhesion and migration (glass, tissue culture polystyrene dishes, collagen-coated polystyrene), they formed continuous cobblestone-like monolayers. When cells were seeded at a high cell density (about 3×10^5 cells/cm^2), the monolayer was confluent within 2 days and formed domes. Electron microscopy showed that the cells were polarized with their apical surfaces in contact with the culture medium. The basal pole lay on the culture substratum.[25] Cells organized into monolayers have properties similar to those described for cells cultured as vesicles.

Tranformation of Vesicles into Follicles

Some metabolic differences between cells organized into vesicles and cells forming follicles might be directly related to the inversion of cell polarity. For instance, the inaccessibility of the basal pole of cells forming vesicles or monolayers may indicate that the iodide is not concentrated or that cAMP synthesis is not stimulated by TSH, and the absence of iodide organification may result from the excretion of thyroglobulin into the culture medium. Attempts were made to modify the orientation of cell polarity, thus transforming vesicles into follicles.

Vesicles formed after 4–5 days in culture in suspension were collected by centrifugation, washed once, and embedded inside a collagen gel.[29] After polymerization of collagen, the culture medium was added and incubation continued. Within 2 to 3 days an inversion of cell polarity was observed, and

[37] L. Nitsch and S. H. Wollman, *J. Cell Biol.* **86,** 875 (1980).

[38] J. D. Valentich, R. Tchao, and J. Leighton, *J. Cell. Physiol.* **100,** 291 (1979).

thyroglobulin began to accumulate inside the lumen. During the same period iodide concentration and organification and responsiveness to acute stimulation by TSH were restored.[39] These results show, first, that orientation of cell polarity can be inverted without dissociating the cells, and second, that functions that were not expressed can be recovered upon cell reorientation.

Formation of Open Epithelial Structures (Fig. 1b)

With the previously described structures, follicles, and vesicles, only one cell pole is directly accessible from the culture medium and the content of the cavity cannot be experimentally modified. Simultaneous direct access to both sides of the epithelial structure can be obtained by culturing thyroid cells on a permeable substratum floating in the culture medium. Under such conditions the apical and basal compartments are not separated; the structure is open. We have used floating collagen rafts mainly,[29] as described previously for mammary cells and hepatocytes.[40,41] A layer of collagen gel is polymerized in plastic petri dishes (0.2–0.5-ml/35-mm dishes). Isolated cells are seeded in the collagen-coated dishes at a high density and allowed to form a confluent monolayer. After 3–5 days, the gel supporting the cell layer is released from the dish by rimming and allowed to float in the culture medium. A moderate contraction of the gel is observed, but the cells remain organized in a monolayer with their apical pole facing the culture medium and their basal pole lying on the permeable collagen gel.[29]

Cells cultured in monolayer on attached collagen gel cannot concentrate iodide, and their cAMP content is not modified acutely by TSH. In contrast, immediately after the gel is released from the culture dish, iodide present in the culture medium is concentrated by the cells, with concentration ratios of 7 to 14, and the intracellular cAMP level is increased three to four times by TSH in the absence of a phosphodiesterase inhibitor.[39,42] In addition, large pseudopods similar to those described after *in vivo* stimulation of thyroid glands appear on the apical cell surface. The formation of these membrane protrusions was not observed when cells were grown on plastic or when the collagen gel was attached to the bottom of the culture dishes.[42]

The polarized monolayer formed on collagen gel is able to concentrate iodide or to respond to TSH provided that iodide or TSH can reach the basolateral domain of the plasma membrane.

[39] M. Chambard, J. Gabrion, B. Verrier, and J. Mauchamp, *in* "Membranes in Growth and Development" (G. Giebisch, J. F. Hoffman, and L. Bolis, eds.), p. 403. Liss, New York, 1982.

[40] G. Michalopoulos and H. C. Pitot, *Exp. Cell Res.* **94,** 70 (1975).

[41] J. T. Emerman and D. R. Pitelka, *In Vitro* **13,** 316 (1977).

[42] M. Chambard, B. Verrier, J. Gabrion, and J. Mauchamp, *J. Cell Biol.* **96,** 1172 (1983).

No iodide organification was observed under these conditions, either in the cells or in the culture medium, whereas follicles embedded in collagen synthesized iodinated thyroglobulin.

The impossibility of accumulating thyroglobulin, or of generating a specific ionic environment for the apical pole, might be responsible for the lack of organification. Separation of the apical and basolateral compartments seems required, therefore, for complete expression of the thyroid function.

Closed Structures with Access to Both Sides of the Cell Layer (Fig. 1c)

Access to both sides of the monolayer together with the separation of apical and basolateral compartments can be achieved by culturing the cells on filters glued to a plastic or glass ring. The filter forms the permeable bottom of a cup with the ring as the side. Cells are seeded inside the cup, which is placed in a petri dish containing cell-free culture medium. A similar system has been described for growing bladder or kidney epithelial cell lines.[43] The substratum can be cellulose nitrate (Millipore) or polycarbonate (Nucleopore) coated with collagen. Cells form a monolayer on the filter; the apical pole is oriented toward the medium contained in the ring whereas the basal pole, lying on the substratum, is in contact with the medium present in the dish outside the ring.

Under these conditions iodide added to the medium outside the ring is concentrated by the cells. In contrast, iodide present in the medium inside the ring, in contact with the apical pole, is not transported into the cells. It appears therefore that the iodide concentration mechanism is localized on the basolateral domain of the thyroid cell plasma membrane. A similar conclusion is obtained for the TSH–receptor–adenylate cyclase complex. Indeed increased cAMP levels and pseudopod formation are obtained only when TSH is added to the medium outside the ring, thus reaching the basal surface of the cell layer.[42]

Conclusion

Some of the culture systems that were used to obtain the multicellular structures described here have been used by others to grow various epithelial cells. Generalizing the observations made with porcine thyroid cells to other epithelial cells forming occluding monolayers might, therefore, be possible. This seems to be the case for mammary cells. Conventional culture conditions appear to be inadequate for growing and maintaining in culture this

[43] J. S. Handler, R. E. Steele, M. K. Sahib, J. B. Wade, A. S. Preston, N. L. Lawson, and J. P. Johnson, *Proc. Natl. Acad. Sci. U.S.A.* **76,** 4151 (1979).

type of differentiated cell, since the major characteristic, i.e., cell polarization and asymmetry of the external environment, is strongly disturbed. Our results show that conditions for controlling cell organization experimentally and for finding a correlation between the organization and the expression of a specific function can be found.

[43] Polarized Assembly of Enveloped Viruses from Cultured Epithelial Cells

By ENRIQUE RODRIGUEZ-BOULAN

The well-known structural and functional polarity of secretory and absorptive epithelia is based on the asymmetric distribution of plasmalemmal components between the two opposite cell surfaces, apical and basolateral. The apical or free surface, in contact with the outer medium, may exhibit morphological differentiations like microvilli and cilia; the basolateral surface specializes in adhesion to the basal lamina and to other cells and, hence, displays the different junctional elements; tight junctions (at the border between both surfaces), gap junctions, and desmosomes.[1] Because of their different structure, function, and biochemical composition, the two plasmalemmal regions may be considered as two different organelles. In order to study the basis for this molecular segregation, we are using as a model system polarized epithelial cell lines infected with enveloped RNA viruses, which bud asymmetrically from the cell surface.[1-3] Influenza (a myxovirus), Sendai and simian virus 5 (paramyxoviruses) are assembled from the apical (free) region of the plasmalemma; vesicular stomatitis virus (VSV, a rhabdovirus) buds exclusively from the basolateral surface of Madin-Darby canine kidney (MDCK) cells,[2] a polarized dog kidney epithelial line[1,4,5] (Fig. 1). Other epithelial cell lines are available.[6] Polarized viral budding is preceded, and presumably determined, by the asymmetric insertion of the viral envelope glycoproteins into the surface that the virus utilizes

[1] E. Rodriguez-Boulan, *in* "Modern Cell Biology" (B. Satir, ed). Vol. 1, p. 119. Liss, New York, 1983.

[2] E. Rodriguez-Boulan and D. D. Sabatini, *Proc. Natl. Acad. Sci. U.S.A.* **75,** 5071 (1978).

[3] E. Rodriguez-Boulan and M. Pendergast, *Cell* **20,** 45 (1980).

[4] D. S. Misfeldt, S. T. Hamamoto, and D. R. Pitelka, *Proc. Natl. Acad. Sci. U.S.A.* **73,** 1212 (1976).

[5] M. Cereijido, E. S. Robbins, W. J. Dolan, C. A. Rotunno, and D. D. Sabatini, *J. Cell Biol.* **77,** 853 (1978).

[6] J. S. Handler, F. M. Perkins, and J. P. Johnson, *Am. J. Physiol.* **238,** F_1 (1980).

ISBN 0-12-181998-1

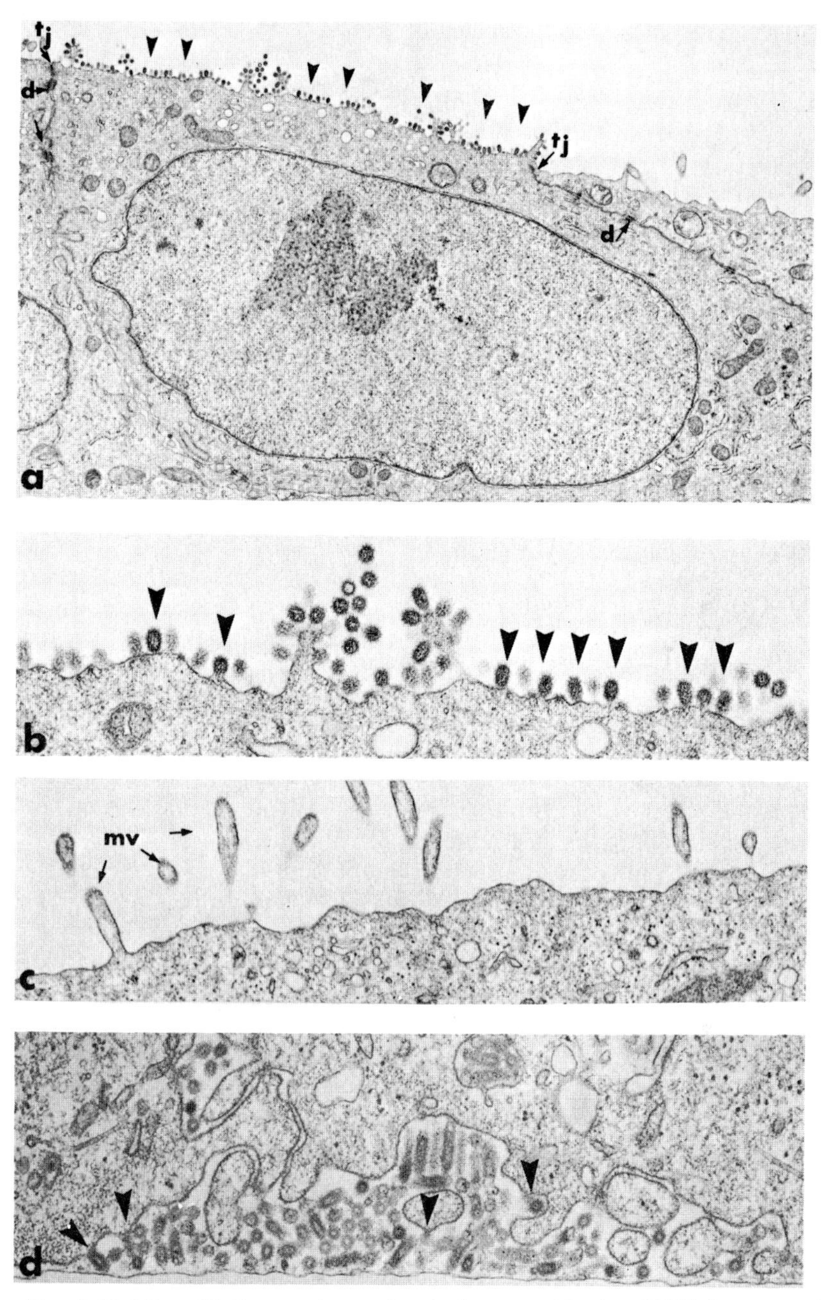

FIG. 1. Budding of influenza virus and vesicular stomatitis virus (VSV) from opposite surfaces in MDCK cells. Confluent monolayers of MDCK cells were infected with influenza virus (a, b) or VSV (c, d), fixed after 8 and 6 hr, respectively, and processed for electron microscopy. (a) Influenza virus buds exclusively from the apical surface (arrowheads). Note the tight junctions (tj) and desmosomes (d) (× 7600). An enlarged portion of the apical surface is represented in (b) (× 34,000). (c) Apical (× 16,000) and (d) basal (× 30,000) surface of MDCK cells infected with VSV. Note microvilli (mv) on the apical surface and virions budding from the basolateral surface (arrowheads).

for budding[3] (Fig. 2). Thus, viral glycoproteins exhibit polarity properties typical of intrinsic epithelial plasma membrane proteins and constitute excellent models to investigate the mechanisms responsible for their segregation.[2,3] The information for the asymmetric surface distribution must reside in special regions of the proteins (sorting or addressing signals).[1,3] The identification of these signals and of the mechanisms involved in their recognition (sorting mechanisms) in epithelial cells should throw light on the nature of the processes controlling the targeting of proteins to specific cellular destinations.

This chapter deals in detail with the methods and procedures utilized in our laboratory to study the origin of epithelial polarity using enveloped RNA viruses. As a corollary, it summarizes the results from our laboratory and other groups that are working with this model system.

Cell Lines

There are several different sublines of MDCK cells being used by various laboratories. We are currently using two of them: one ws obtained from the American Type Culture Collection (MDCK, CCL-34); the other was a gift from Dr. Joseph Leighton (Department of Pathology, Medical College of Pennsylvania, Philadelphia). Both sublines exhibit structural and functional polarity, since they develop hemicysts or domes when cultured on solid substrata, due to unidirectional transport of sodium and water, and a measurable transepithelial electrical resistance, of around 100 Ω cm^2, which correlates with the presence of tight junctions in freeze-fracture replicas.[4,5] The cells are grown at 36.5° in an air – 5% CO_2 atmosphere in plastic tissue culture flasks with 15 ml of Eagle's minimum essential medium (MEM) (Grand Island Biological Co., Grand Island, New York) containing 8% fetal calf serum (FCS), 100 IU of penicillin and 100 μg of streptomycin per milliliter. The cells are harvested weekly by treatment with 0.25% trypsin, 10 m*M* EDTA, replated at a density of 10^4 cells/cm^2, and fed with fresh medium every 3 or 4 days. To prevent changes in the original properties of the cells, new cultures are started every 3 months from a stock frozen in liquid nitrogen. African green monkey kidney cells (Vero, ATCC CCL 81), or baby hamster kidney cells (BHK21, ATCC, CCL 10) utilized for growth and titration of VSV or SV5 are grown in Eagle's MEM containing 4% FCS.

Preparation of Virus Stocks

Work with viruses must be carried out under normal tissue culture sterile conditions. All material contaminated with viruses must be placed in a separate container and autoclaved.

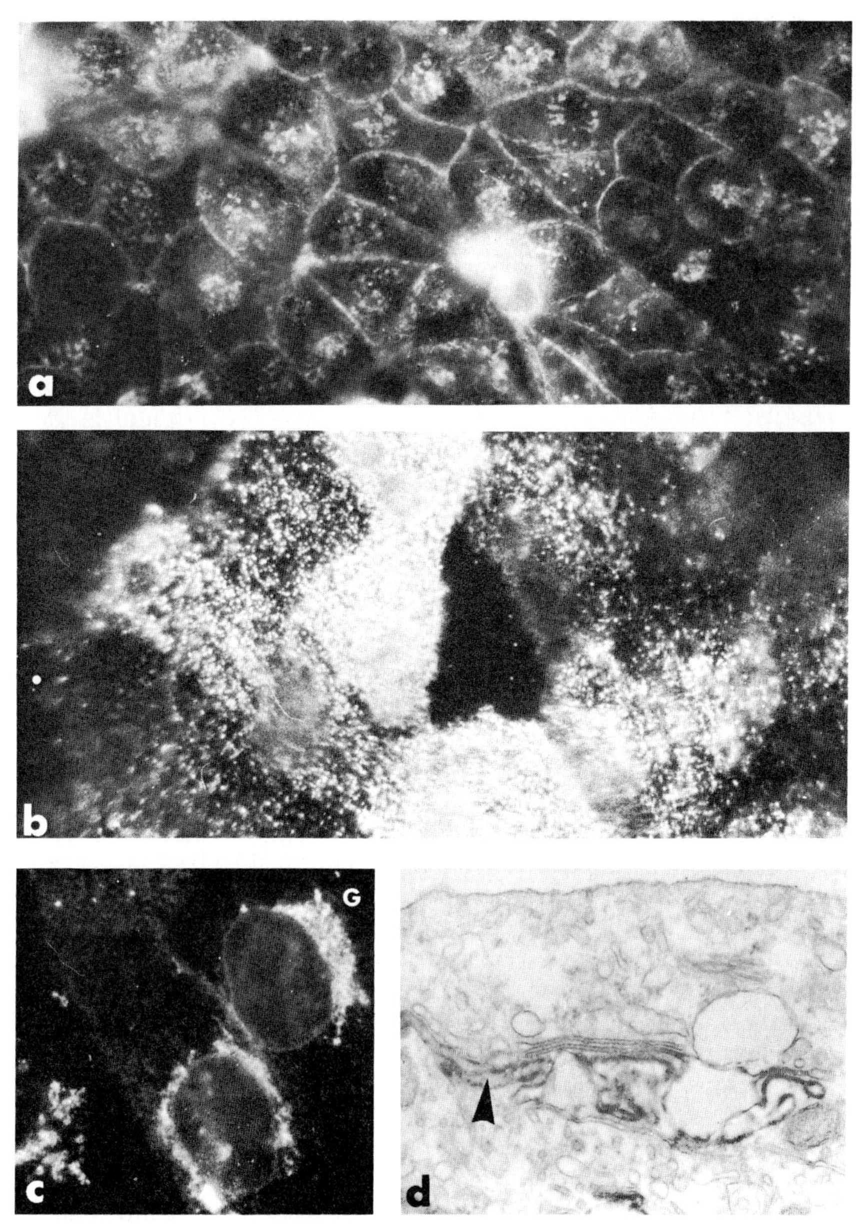

FIG. 2. Migration of influenza virus and VSV envelope proteins to the surface of MDCK cells. MDCK cells were infected with the temperature-sensitive mutants VSV ts 045 (a) or influenza ts 61 (b–d), incubated at the nonpermissive temperature (39.5°) for 4 or 6 hr, respectively, and then transferred to the permissive temperature (32°). The cells were fixed after various times of incubation at this temperature and then processed for immunofluorescence (IF) or immunoperoxidase (IP) electron microscopy. (a) ts 045, 40 min at 32°; IF in cells permeabilized with 0.05% saponin (× 2000). Note staining of the Golgi apparatus and lateral cell surfaces. (b) ts 61, 90 min at 32°; IF without saponin. Note staining of microvilli in the apical surface (× 2700). (c) ts 61, 20 min at 32°; IF with saponin. Note staining of the Golgi apparatus (G) (×2700). (d) ts 61, 20 min at 32°C, IP with saponin. Note staining of some of the Golgi cisternae (arrowhead) (× 18,000).

Vesicular Stomatitis Virus. Confluent monolayers of Vero cells grown in 75 cm^2 (T-75) or 150 cm^2 (T-150) tissue culture flasks are washed once with sterile Moscona's saline, and then inoculated with VSV at a multiplicity of infection (m.o.i.) of 0.01–0.1 plaque-forming units (PFU) per cell), suspended in 2 ml (T-75) or 5 ml (T-150) of MEM containing 2% FCS. Because of these small volumes required to maximize virus adsorption to the monolayers, the flasks must be shaken every 10–15 min during the inoculation, which is carried out at 37° in the incubator for 1 hr. The inoculum medium is replaced by fresh MEM-2% FCS (10 ml for T-75 and 20 ml for T-150 flasks), and the infected cultures are incubated until the virally induced cytopathic effect is complete (usually 24–36 hr). The medium is then collected, the cell debris are removed by centrifugation at 2000 rpm for 15 min in a table-top centrifuge, and the viral suspension is fractionated into 50–100 aliquots and stored at −70°. An aliquot is thawed, and the virus titer is determined by plaque assay or by cytopathic effect 50% (CPE-50%) assay; titers remain constant for at least 6 months.

Influenza. The procedure is the same as that described for VSV, except for the use of MDCK cells, which sustain the growth of influenza much better than do Vero or BHK cells, and for the substitution of Dulbecco's MEM (D-MEM) for MEM and 0.2% bovine serum albumin for the FCS in the inoculum and incubation media (the use of FCS results in lower influenza titers, presumably because of the presence of serum inhibitory factors).

Simian Virus 5 (SV5). This virus has a slower replication cycle than VSV and influenza virus and causes a much smaller cytopathic effect; in fact, cells may remain infected and viable for several days. For virus growth, confluent monolayers of MDCK cells are inoculated at m.o.i. = 1 in D-MEM, as described for VSV, and then incubated at 37° for 2–4 days in 2% *calf* serum D-MEM. It is convenient to check daily the hemagglutinin titers in the medium, usually they stabilize after 3–4 days around values of 320–640 hemagglutination units/ml (HAU/ml).

Temperature-Sensitive Mutants of VSV and Influenza Virus. We are currently using mutants of VSV and influenza virus, ts 045[7] and ts 61,[8] respectively, that exhibit at the nonpermissive temperature (39.5°) defective viral glycoproteins unable to exit from the endoplasmic reticulum (ER). Upon reversal of the temperature to 32°, the proteins start the migration to the cell surface as a synchronous wave; this provides a convenient way to study their intracellular pathway. Since these mutants display a high level of revertancy to the wild-type phenotype, special precautions must be taken to prevent it. They essentially involve growing the virus stocks from cloned virions obtained from single plaques. Plaques and virus stocks are grown at

[7] A. Flamand, *J. Gen Virol.* **8,** 187 (1970).

[8] P. Palese, *Top. Infect. Dis.* **3,** 49 (1978).

the permissive temperature (32°) in the same cell lines as the wild-type viruses. The virus stock must be titered at both permissive and nonpermissive temperature and the ratio between both should exceed 100. Except for the temperature, titrations are carried out as described below for the wild-type viruses.

Titration of Viruses

Hemagglutination

Influenza and Sendai viruses and SV5 can be conveniently titered by hemagglutination assay, which, although not so quantitative as the infectivity assays described below, provides a quick estimate of the amount of virus production.

Procedure

1. Wash chick or human type "O" red blood cells three times in ice-cold PBS.
2. Prepare duplicate, serial 1:2 dilutions of the viral samples in cold PBS in a 96-well Microtiter dish (U-shaped bottom), using 50 μl/well. When many samples are to be titered, a multichannel micropipette (Flow Laboratories, McLean, Virginia) is very convenient.
3. Add 50 μl of a 2% (v/v) suspension of red blood cells in PBS to each well.
4. Leave one well as a control (no virus).
5. Incubate the Microtiter dish at 4° for 90 min.

The hemagglutination titer is the highest dilution of the virus giving positive hemagglutination; positive wells appear homogeneously pink, and negative wells display a distinct pellet of red blood cells. The hemagglutination titer may be expressed as hemagglutination units per milliliter by simply multiplying the inverse of the highest positive dilution by the volume factor 20 (1000 μl/50 μl). Influenza virus-infected MDCK monolayers usually yield titers above 1:256 (or 5120 HAU/ml). Sendai virus and SV5 yield lower titers. A higher precision in the assay can be obtained by preparing 2:3 dilutions instead of 1:2.[9]

Plaque Assay

Materials

MEM or D-MEM, two times concentrated, contaning 4% FCS (VSV), 4% calf serum (SV5), or 0.4% bovine serum albumin (BSA) (influenza)

[9] R. M. Davenport, R. Rott, and W. Schafer, *J. Exp. Med.* **112,** 765 (1960).

Noble agar, 2% (Difco Laboratories, Detroit,Michigan)
Two waterbaths, one at 43° and one at 37°
10 Petri dishes (60 mm) with confluent monolayers of Vero cells (VSV) MDCK cells (influenza virus, VSV) or BHK cells (SV5 or VSV), per sample to be assayed
Sterile disposable 1-ml plastic pipettes and hand pipetter
Sterile disposable plastic tubes (75 × 12 mm)
Vortex mixer

Procedure. Melt the agar in boiling water and then let equilibrate for at least 1 hr at 43°. Meanwhile, incubate the 2× medium (25 ml for every 10 petri dishes) at 37°. Prepare 1 : 10 dilutions of the virus sample up to 10^{-8} in D-MEM (1 ml per dilution). Mix by vortexing, and change the pipette every time. Inoculate duplicate petri dishes with 0.4 ml of each dilution (10^{-4} to 10^{-8} is usually a safe range, but it depends on the amount of virus). Incubate for 1 hr at 37°, shaking the dishes every 15 min. Mix 25 ml of warm 2% agar with 25 ml of warm 2× medium; quickly remove the inocula from the petri dishes and add 5 ml of the 1% agar – 1× medium solution to each petri dish. Let the agar solidify at room temperature; invert the petri dishes on a plastic tray and place them in an incubator at 37°. After 2 days (VSV, influenza virus) or 3–5 days (SV5), plaques have usually developed. Add fixative (3.7% formaldehyde, 0.8% NaCl) to the petri dishes for 10–15 min; carefully remove the agar with a spatula, and stain the monolayers with 1% gentian violet in 20% ethanol.

Plaques are counted at the viral dilution giving 20–100 plaques per petri dish; this number, multipled by 2.5 (volume factor), and by the dilution, gives the viral titer in plaque-forming units per milliliter.

Cytopathic Effect 50% Assay

Like hemagglutination, this is an end-point procedure. Computation of the 50% end point is carried out by the method of Reed and Muench.[10]

Materials

Tissue culture dish, 96-well, with confluent monolyers of Vero (VSV), MDCK (influenza virus), or BHK (SV5) cells
MEM with 2% FCS (VSV), D-MEM with 0.2% BSA (influenza virus) or with 2% calf serum (SV5)
Sterile disposable plastic tubes (75 × 12 mm)
Sterile disposable 1-ml plastic pipettes and a hand pipetter
Vortex mixer

Procedure. Prepare dilutions as described for the plaque assay (1.5 ml per dilution). When all dilutions are ready (usual range 10^{-3} to 10^{-9}), remove by

[10] B. D. Davis, R. Dulbecco, H. Eisen, H. Ginsberg, and W. B. Wood, "Microbiology." Harper & Row, New York, 1973.

suction the medium from one 12-well row and pipette 0.1 ml/well of the highest dilution of the virus. Repeat the same procedure for the second row and the second highest dilution of the virus. Repeat until 7 rows are completed; the last row receives just 0.1 ml of medium per well (uninfected control). Incubate the culture dish at 37° for 2–3 days, and then score the number of wells showing positive cytopathic effect at each dilution of the virus. In order to find the dilution that gives 50% cytopathic effect, use the interpolating procedure of Reed and Muench.[10] An example is given in the table.

In this example, the 50% point falls between 10^{-6} and 10^{-7}. The exact value of the viral titer is calculated with the following formula:

$$\frac{\text{\% at dilution above 50\%} - 50\%}{\text{\% at dilution above 50\%} - \text{\% at dilution below 50\%}}$$

In the example above,

$$\frac{73 - 50}{73 - 19} = -\frac{23}{54} = 0.43 \quad (1)$$

The CPE-50% titer is

$$10^{-6.43} \text{ or } 1/2.7 \times 10^{6} \quad (2)$$

As with hemagglutination, the titer can be expressed as CPE-50% units/ml by taking the inverse of Eq. (2), and correcting for the volume factor (in this case 10, since each well received 0.1 ml). The sample in our example has, then, the following infectivity titer: 2.7×10^7 CPE-50% units/ml.

Infectivity titers obtained with plaque assay are around 70% of the titers obtained with the CPE-50% assay.[10]

Electron Microscopy of Virus-Infected Monolayers

Preparation of MDCK Monolayers

For electron microscopy, monolayers of MDCK cells can be grown on plastic petri dishes, collagen-coated surfaces (glass or plastic), or nitrocellu-

TYPICAL CPE-50% ASSAY

Viral dilution	Positives/ total	Negatives/ total	Sum of positives	Sum of negatives	Positive/ pos. + neg.	%
10^{-5}	12/12	0/12	23	0	23/23	100
10^{-6}	8/12	4/12	11	4	11/15	73
10^{-7}	3/12	9/12	3	13	3/16	19
10^{-8}	0/12	12/12	0	25	0/25	0

lose filters (Millipore). Because the first two methods allow visualization of the monolayer with the inverted microscope, they are the preferred ones in our laboratory. Collagen can be obtained from commercial sources (Vitrogen, available from Flow Laboratories), or may be easily prepared by acetic acid extraction of rat tendons.[11] A convenient method in our hands involves the simultaneous coating of several (6–8) 13-mm round glass coverslips, attached to a glass slide by a water droplet. Pipette 0.2 ml of a 10 mg/ml collagen solution (in water) on one end of the glass slide and smear with another slide. Expose to ammonium hydroxide fumes for 15 min in a plastic box and let dry. Several slides can be processed at the same time and can be stored in 70% ethanol at 4° for at least 2 months. For use, pick individual collagen-coated coverslips with a flame-sterilized fine forceps, place in a 24-well culture dish, and wash several times with sterile phosphte-buffered saline (PBS) to remove the ethanol.

To study the polarity of viral budding, we normally allow monolayers of MDCK cells to be confluent for at least 2 or 3 days; this makes them more resistant to the cell-dissociating action resulting from the cytopathic effect of the viruses (especially VSV). Seeding at densities close to saturation (2×10^5 cells/cm^2) results in a confluent monolayer by the following day.

Viral Infection

Inoculate confluent monolayers of MDCK cells at an m.o.i. of 5–10 PFU/cell for 1 hr at 37° in MEM containing 100 μg of DEAE-dextran (VSV) per milliliter or in D-MEM (influenza virus, SV5). Use the minimum amount needed to cover the monolayer (0.2, 0.4, or 0.8 ml, respectively, for 15 mm-wells or 35- or 60-mm petri dishes). The addition of DEAE–dextran increases the infectivity of VSV in MDCK cells (usually 10–100 times lower than in Vero cells) to levels only somewhat lower (2–5 times) than those obtained in the Vero cells. After inoculation, replace the medium by MEM with 2% FCS (VSV), D-MEM with 0.2% BSA (influenza virus), or D-MEM with 2% calf serum (SV5) and return the samples to the incubator at 37°. The budding of VSV and influenza virus starts after 4–5 hr of infection and continues for several hours; it can be best observed in samples fixed between 6 and 10 hr. The production of SV5, instead, is intense after 24–48 hr.

Processing of Monolayers Grown on Plastic Petri Dishes

To avoid the cell damage caused by scraping, we are currently using amyl acetate to remove the monolayer from the petri dish, according to Revel and Wolken.[12] The following steps are all carried out in the petri dish.

[11] L. C. M. Reid and M. Rojkind, this series, Vol. 58, p. 263.
[12] J.-P. Revel and K. R. Wolken, *Exp. Cell. Res.* **78**, 1 (1973).

Procedure

1. At the desired time of infection, wash the monolayer with cold cacodylate buffer (0.1 *M* sodium cacodylate, pH 7.4) and fix in 2% glutaraldehyde in cacodylate buffer at 4° for 60 min to overnight.
2. Wash with cacodylate buffer, three times for 10 min.
3. Place in 2% OsO_4 in cacodylate buffer for 1 hr at room temperature (RT).
4. Wash with 2× distilled water three times for 5 min each (RT).
5. Keep in 1% uranyl acetate in water overnight at 4°.
6. Treat with 70%, 80%, 90% ethanol, 10 min each at 4°.
7. Dehydrate in 100% ethanol, three times for 15 min each at 4°.
8. Treat with amyl acetate. This step should be carried out in a chemical hood. Remove ethanol and cover monolayer with amyl acetate. Score monolayer with a needle into four regions and let stand for 10–15 min. With a Pasteur pipette, squirt amyl acetate at the corners with a short, quick motion, avoiding the formation of air bubbles. Transfer the freed monolayer pieces to a separate Wheaton jar containing fresh amyl acetate. Incubate in amyl acetate, four times for 5 min, to remove dissolved plastic from the petri dish (test supernatant for the presence of plastic by adding a drop to 100% ethanol; if a precipitate forms, further washing with amyl acetate is needed).
9. Wash in 100% ethanol, three times for 15 min at 4°.
10. Treat with propylene oxide, twice for 10 min at 4°.
11. Add Epon–propylene oxide, 50%:50%, under constant rotation (RT).
12. Warm to room temperature, then embed in fresh Epon mixture.

Processing of Monolayers Grown on Collagen

Remove the medium from the well, wash with cold cacodylate buffer, and add 2% glutaraldehyde in cacodylate buffer. After 5 min, scrape the collagen layer with a razor blade and centrifuge it for 30 sec in a microcentrifuge. Continue the fixation of the pellet for 1 hr to overnight. Process and embed as described above (omitting the amyl acetate step).

Immunofluorescence Localization of Proteins on the Surface of MDCK Cells

Immunofluorescence on MDCK monolayers is a fast and easy method for study of the surface localization of proteins, especially those localized on the apical surface. For proteins localized on the basolateral surface, procedures have to be utilized to make them accessible to the antibodies. These involve the use of calcium chelating agents, such as EGTA, which have been shown to cause a disruption of the tight junctions,[5] or the use of neutral detergents after fixation to permeabilize the cell membrane. With detergent treatment, not only the basolateral antigens are unmasked, but also the intracellular antigens, which may cause some degree of confusion in the interpretation of the fluorescence image (see Fig. 2).

Procedure

1. Seed MDCK cells on autoclave-sterilized 13-mm glass coverslips, placed in a 24-well culture dish, at 10^5 cells/well.
2. After 2–3 days, carry out the viral infection as previously described.
3. Incubate the monolayer in 10 m*M* EGTA for 15–30 min at 37° (when basolateral surfaces are to be exposed).
4. Fix monolayers with ice cold 4% paraformaldehyde in PBS for 15 min (Prepare this solution by adding 2 g of paraformaldehyde to 40 ml of 2× distilled water at 60–70° under constant stirring with a magnetic bar. Add 3–4 drops of 0.1 *N* NaOH, wait until the solution is clear, then add 5× PBS. Add $CaCl_2$ and $MgCl_2$ to 0.1 and 1 m*M*, respectively, and cool on ice.)
5. Wash with PBS–CaMg three times for 5 min each.
6. For intracellular labeling, incubate monolayers in 0.1% Triton X-100 in PBS–CaMg, for 5 min (4°).
7. Incubate with PBS containing 50 m*M* NH_4Cl, to quench free aldehyde groups.
8. Wash with PBS–CaMg containing 0.2% gelatin (PBS-CaMgG) twice for 5 min at 4°.
9. Remove coverslips with fine forceps and place upside down on 50 μl of the first (antiviral envelope) antibody, 50–100 μg/ml in PBS–CaMgG, previously pipetted on a wax plate. Use the same concentration of normal rabbit immunoglobulin G (IgG) for control samples. Incubate for 30–45 min at RT in a humidified environment.
10. Replace coverslips on a 24-well culture dish and wash with PBS–CaMgG, four times for 5 min (RT).
11. Incubate with goat-anti rabbit IgG coupled to rhodamine or fluorescein (Cappel Laboratories, Cochraneville, Pennsylvania) in dilutions of 1:20 to 1:100, as described above. Purification of these antibodies by affinity chromatography on rabbit IgG–Sepharose columns results in much darker backgrounds.
12. Wash with PBS–CaMgG, four times for 5 min.
13. Wash with PBS alone.
14. Pick up with fine forceps, immerse a few times in distilled water, and

mount on 50% glycerol–PBS, or on 20% polyvinyl alcohol in PBS (Gelvatol, Monsanto Laboratories, Indian Orchard, Massachusetts).

Alternatively, a procedure involving fixation of the monolayers with paraformaldehyde and low glutaraldehyde, followed by permeabilization with 0.05% saponin, can be used with excellent results. This method, which gives better preservation of cellular structures, is described in detail below for the immunoperoxidase detection of viral glycoproteins in intracellular structures. Fixation with glutaraldehyde alone, at any concentration, is unsatisfactory, since it yields very high levels of background fluorescence.

Upon observation with an epifluorescence microscope, labeling of apically localized antigens results in a spotty fluorescence pattern, resulting from the staining of the characteristic microvilli at the apical cell surface (Fig. 2b). Basolaterally localized antigens frequently results in a ringlike fluorescence image because of the apparent concentration of antigen at the cell periphery[3] (Fig. 2a).

Fluorescence Experiments with ts Mutants

Inoculation of MDCK cells with the mutants ts 045 (VSV) or ts 61 (influenza virus) is carried out in exactly the same way as for the wild-type viruses except for the temperature (39.5°). After 3–3.5 hr (ts 045) or 5 hr (ts 61) of additional incubation at nonpermissive temperature, no envelope protein can be detected on the surface of the cells or in the Golgi apparatus by immunofluorescence. The cells, however, are synthesizing and accumulating glycoprotein in the endoplasmic reticulum, as shown by cytoplasmic fluorescence and [^{35}S]methionine incorporation followed by sodium dodecyl sulfate–polyacrylamide gel electrophoresis (SDS–PAGE).[13] Upon transfer to 32°, the glycoproteins are detected in the Golgi apparatus after 10–15 min and on the cell surface after 30–40 min, giving the typical images described above (see Fig. 2a,b,c).

Immunoelectron Microscopic Localization of Viral Envelope Glycoproteins

Surface Localization

In order to carry out studies on the localization of viral envelope glycoproteins on the surface of epithelial cells, we use the following procedure (all washes are by centrifugation for 30 sec in a microcentrifuge).

1. Fix infected monolayer, grown on petri dish, with 2% formaldehyde 0.05% glutaraldehyde in PBS for 30 min at 4°.

[13] E. Rodriguez-Boulan, K. T. Paskiet, P. J. I. Salas, and E. Bard, *J. Cell Biol.*, in press (1983).

2. Scrape monolayer with plastic policeman.
3. Incubate for 15 min in PBS containing 50 m*M* NH_4Cl to block free aldehyde groups.
4. Incubate for 45 min at room temperature with PBS containing, per milliliter, 50–100 μg of rabbit IgG fraction containing antibodies to the viral envelope glycoproteins. (We have also been using monoclonal antibodies against the envelope proteins.)
5. Wash with PBS three times for 5 min.
6. Incubate for 45 min at room temperature in PBS containing 500 μg of goat anti-rabbit IgG per milliliter coupled to ferritin (Cappel Laboratories).
7. Wash three times in PBS.
8. Fix with 2% glutaraldehyde in cacodylate buffer, centrifuge for 30 sec in microcentrifuge, while fixing, to form a pellet; continue fixation for 60 min to overnight.
9. Process for electron microscopy as described above.

We have also been successfully utilizing a system in which the first antibody is coupled to biotin and ferritin-avidin is used in the second step.[3] Ferritin-avidin is currently commercially available (Vector Laboratories, Burlingame, California).

Intracellular Localization

Several groups have followed the intracellular pathway of viral glycoproteins from the rough endoplasmic reticulum to the plasma membrane by immunocytochemical procedures.[13–17] We have reported[18] the localization of VSV G protein and influenza HA in the Golgi apparatus of MDCK cells doubly infected with both viruses by ultrathin frozen sections. Another convenient technique, used in our laboratory, that results in a better identification of intracellular structures than frozen sections, involves the use of $F(ab')_2$ antibody fragments coupled to peroxidase on fixed monolayers permeabilized with saponin. The main problem with this method is working out the fixation conditions which allow maximum penetration with minimal disruption of cell structure. The following procedure gives excellent results:

[14] J. E. Bergmann, K. T. Tokuyasu, and S. J. Singer, *Proc. Natl. Acad. Sci. U.S.A.* **78,** 1746 (1981).

[15] J. Green, G. Griffiths, D. Louvard, P. Quinn, and G. Warren, *J. Mol. Biol.* **152,** 663 (1981).

[16] J. Wehland, M. C. Willingham, M. G. Gallo, and I. Pastan, *Cell* **28,** 831 (1982).

[17] J. J. M. Bergeron, G. Kotwal, G. Levine, P. Bilan, R. Rachubinski, M. Hamilton, G. Shore, and H. P. Ghosh, *J. Cell Biol.* **94,** 36 (1982).

[18] M. Rindler, I. Emanuilov Ivanov, E. Rodriguez-Boulan, and D. D. Sabatini, *Membr. Recycling, Ciba Found. Symp.* **92,** 184 (1982).

Procedure

1. Infected monolayers are fixed in 2% formaldehyde, in 0.1 *M* sodium phosphate, pH 7.4, 50 m*M* sucrose (PS) overnight at 4°, and the following morning with 2% formaldehyde, 0.05% glutaraldehyde in PS (for 30 min at 4°).
2. Wash with PS and incubate for 15 min in 50 m*M* $(NH_4)Cl$ in PS.
3. Incubate for 30 min in PS containing 0.2% gelatin and 0.05% saponin (PS–GS). Saponin is kept throughout the rest of the experiment.
4. Incubate for 45 min at room temperature in PS–GS with 50–100 μg of rabbit antibody against the viral glycoprotein.
5. Wash with PS–GS (three times for 10 min).
6. Incubate for 45 min at room temperature with PS–GS containing peroxidase coupled to $F(ab')_2$ fragments of affinity-purified goat anti-rabbit antibody (Cappel Laboratories).
7. Wash with PS containing 0.05% saponin (three times for 10 min each).
8. Postfix with 4% glutaraldehyde in 0.1 *M* cacodylate buffer, pH 7.2 for 30 min at 4°.
9. Wash with 0.1 *M* Tris-HCl, pH 7.6, three times for 5 min each.
10. Incubate in the same buffer containing 0.5% diaminobenzidine and 0.01% H_2O_2 for 30 min. Change to fresh solution at 15 min.
11. Postfix with 2% glutaraldehyde in 0.1 *M* cacodylate, pH 7.4.
12. Flat-embed in the petri dish; observe sections without further staining.

Summary of the Experimental Results Obtained with the Model of Virus-Infected MDCK Cells

Enveloped RNA viruses bud asymmetrically from monolayers of epithelial cells in culture: influenza and Sendai viruses and SV5 from the apical surface; VSV from the basolateral one[1,2] (Fig. 1). Polarized viral budding is preceded by the asymmetric insertion of the viral glycoproteins into the corresponding surface domain.[3] The carbohydrates of viral glycoproteins appear not to be part of the sorting signals that mediate this sorting process, since polarity is preserved (*a*) in lectin-resistant mutant MDCK cells with constitutive changes in glycosylation[19,20]; (*b*) after treatment with tunicamy-

[19] R. F. Green, H. K. Meiss, and E. Rodriguez-Boulan, *J. Cell Biol.* **89**, 230 (1981).
[20] H. K. Meiss, R. F. Green, and E. Rodriguez-Boulan, *Mol. Cell. Biol.* **2**, 1287 (1982).

cin[19,21]; (*c*) in MDCK cells infected with ts mutants of influenza defective in the neuraminidase, which results in the production of abnormal virions containing sialic acid.[19]

Incubation with cytochalasin B or colchicine does not affect the asymmetric viral budding, which suggests that microfilaments and microtubules are not involved in polarity (unpublished results with M. Cereijido). Monensin, a cationic ionophore that blocks protein secretion at the level of the Golgi apparatus in a variety of cells appears selectively to block the production of VSV, but not of influenza virus.[22]

In monolayers doubly infected with SV5 and VSV, or influenza virus and VSV, asymmetric budding is preserved for several hours[18,23]; a significant number of phenotypically mixed virions is detected only late in the infection, when the cytopathic effect of the infection disrupts cell polarity.[23] Simultaneous immunocytochemical localization of G and HA on ultrathin frozen sections of doubly infected cells, using a different size of colloidal gold particle for each glycoprotein, detects both of them in the Golgi apparatus, indicating that the sorting takes place at a Golgi or post-Golgi level.[18] Work in our laboratory has been focused on using mutants of influenza WSN (ts 61) and VSV (ts 045) with ts defects that prevent the exit of the viral glycoproteins from the ER, to synchronize their migration to the cell surface.[13] When MDCK cells are infected at nonpermissive temperature (39.5°) with ts 61 for 6 hr or with ts 045 for 4 hr, no viral glycoprotein is detected in the surface or in the Golgi. After shift to permissive temperature 32°), both proteins can be detected in the Golgi apparatus after 10–15 min and on the cell surface after 30–40 min by immunofluorescence and immunoelectron microscopy (Fig. 2, a–d). Radioimmunoassay (RIA) has confirmed this finding for the HA; we are now attempting to develop a reproducible RIA for the localization of basolateral proteins, a project that presents considerable technical difficulties. Both immunofluorescence and immunoelectron microscopy (using peroxidase) of monolayers infected with ts 61, using a monoclonal antibody against HA, show a considerable number of tiny vesicles becoming labeled 40–50 min after shift to 32°. These vesicles are preferentially located in a supranuclear region, as the Golgi apparatus, are larger than coated vesicles, and do not exhibit a coat or cross react with anticlathrin antibodies; we suggest that they participate in the transport of HA to the apical surface.[13]

Finally, we have shown that polarized viral budding is preserved in isolated MDCK cells attached to a collagen gel and in clusters of cells in

[21] M. G. Roth, J. P. Fitzpatrick, and R. W. Compans, *Proc. Natl. Acad. Sci., U.S.A.* **76,** 6430 (1979).

[22] F. V. Alonso and R. W. Compans, *J. Cell Biol.* **89,** 700 (1981).

[23] M. G. Roth and R. W. Compans. *J. Virol.* **40,** 848 (1981).

suspension, but not in isolated suspended cells or in fibroblasts.[2,24] These results suggest that complete tight junctions are not essential for polarized viral budding; attachment to a substrate may be sufficient signal to trigger the segregation of surface components necessary for the asymmetric assembly of enveloped viruses.

Van Meer and Simons have taken advantage of the fact that the lipid compositions of enveloped viruses reflects that of the original plasma membrane to show, using MDCK cells infected with influenza virus or VSV, that the phospholipids of apical and basolateral surfaces of MDCK cells are considerably different.[25] The basolateral membranes contain significantly higher levels of phosphatidylcholine, phosphatidylinositol, and sphingomyelin, and lower levels of phosphatidylethanolamine, than the apical surface. The same workers have presented evidence that viral glycoproteins inserted in the wrong surface domain can be translocated to the right one[26]; they suggest that this may be the mechanism utilized during the normal biogenesis of influenza virus hemagglutinin.[27] Two separate experimental lines in our laboratory do not support, however, the latter possibility. First, the immunoperoxidase EM experiments with the ts 61 mutant of influenza suggest direct delivery by vesicles that fuse with the apical surface. Second, infection with influenza virus of confluent MDCK monolayers grown on collagen gels that contain antibody against HA results in the production of normal levels of virus, implying that most of the HA is never exposed to the basolateral surface (D. Misek, E. Bard, and E. Rodriguez-Boulan, unpublished results).

Acknowledgments

I thank Maryann Pendergast and Kevin Paskiet for expert technical assistance. This work was supported by grants from the National Science Foundation and the National Institutes of Health and by an Irma T. Hirschl award.

[24] E. Rodriguez-Boulan, K. T. Paskiet, and D. D. Sabatini, *J. Cell Biol.* **96,** 866 (1983).
[25] G. Van Meer and K. Simons, *EMBO* (*Eur. Mol. Biol. Organ.*) *J.* **1,** 847 (1982).
[26] K. S. Matlin, D. F. Bainton, M. Pesonen, D. Louvard, N. Genty, and K. Simons, *J. Cell Biol.,* in press.
[27] K. S. Matlin and K. Simons, *Cell,* in press.

[44] Isolation and Characterization of Liver Gap Junctions

By ELLIOT L. HERTZBERG

The plasma membrane specialization that is responsible for direct communication between contacting cells via the transfer of low-molecular-weight molecules is called the nexus or, more generally, the gap junction.

ISBN 0-12-181998-1

While originally called a tight junction, it was later demonstrated that the tight junction represents an actual point of fusion of the membranes of adjacent cells. In contrast, the membranes from adjacent cells containing a gap junction form a close, regular association in which they are separated by the characteristic 1- to 2-nm "gap."

Most of the studies of gap junctions and gap junctional communication have attempted to examine the properties and significance of this type of intercellular communication. Studies of the regulation of communication, both at the ultrastructural level (presumptive assembly and disassembly of gap junctions) and at the level of regulation of assembled junctions (by intracellular pH[1,2] or intracellular calcium concentration[3], or both[4]) strongly suggest that gap junctional communication, and its regulation, may play a significant role in cell growth control, development, and differentiation. (For entry into the literature, a number of the more recent reviews may be consulted.[5-8])

Because of the regularity of the assembly of the subunits comprising the gap junction, they have proved to be amenable to analysis by a variety of physical and image-processing techniques.[9-12] Although as yet there is not agreement on the details describing the disposition of gap junction polypeptides in the membrane and how this varies as a function of the communication state of the junction, one can anticipate that further studies will permit structure–function correlations to be made at the atomic level of resolution.

Overview on Isolation of Gap Junctions

The basis for the isolation of gap junctions from plasma membranes is their relative resistance to solubilization in detergents. Among those used to solubilize nonjunctional membrane selectively have been Sarkosyl,[13-16]

[1] L. Turin and A. E. Warner, *J. Physiol. (London)* **300,** 489 (1980).
[2] D. C. Spray, A. L. Harris, and M. V. L. Bennett, *Science* **211,** 712 (1981).
[3] B. Rose, I. Simpson, and W. R. Loewenstein, *Nature (London)* **267,** 625 (1977).
[4] B. Rose and R. Rich, *J. Membr. Biol.* **44,** 377 (1978).
[5] M. V. L. Bennett and D. A. Goodenough, *Neurosci. Res. Program Bull.* **16,** 373 (1978).
[6] J. Feldman, N. B. Gilula, and J. D. Pitts, eds. "Intercellular Junctions and Synapses." Chapman & Hall, London, 1978.
[7] W. R. Loewenstein, *Biochem. Biophys. Acta* **560,** 1 (1979).
[8] E. L. Hertzberg, T. S. Lawrence, and N. B. Gilula, *Annu. Rev. Physiol.* **43,** 479 (1981).
[9] D. L. D. Caspar, D. A. Goodenough, L. Makowski, and W. C. Phillips, *J. Cell Biol.* **74,** 605 (1977).
[10] L. Makowski, D. L. D. Caspar, W. C. Phillips, and D. A. Goodenough, *J. Cell Biol.* **74,** 629 (1977).
[11] G. Zampighi and P. N. T. Unwin, *J. Mol. Biol.* **135,** 451 (1979).
[12] P. N. T. Unwin and G. Zampighi, *Nature (London)* **283,** 545 (1980).
[13] E. L. Hertzberg and N. B. Gilula, *J. Biol. Chem.* **254,** 2138 (1979).
[14] E. L. Hertzberg, *In Vitro* **16,** 1057 (1980).
[15] M. Finbow, S. B. Yancey, R. Johnson, and J.-P. Revel, *Proc. Natl. Acad. Sci. U.S.A.* **77,** 970 (1980).
[16] D. A. Goodenough, *J. Cell Biol.* **61,** 557 (1974).

deoxycholate,[17] and Triton X-100.[18] Liver has been the tissue of choice for most of these studies because of the ease of preparation of a plasma membrane fraction. Most early studies utilized proteases, such as trypsin or collagenase, to aid in tissue dissociation, but it is now clear that the gap junctions isolated by such procedures have been partially degraded, with respect to the polypeptide associated with the gap junction fraction, although little or no ultrastructural modification was observed. Procedures developed that omitted exogenous proteolysis as part of the preparative procedure resulted in the identification of a 26,000- or 27,000-dalton polypeptide as the major constituent of the gap junction.[13,18] There now is general agreement that this is the native form of the junction polypeptide.[13-15,17,18] It should be noted that other polypeptides could be associated with the gap junction *in vivo,* which might be selectively removed during gap junction isolation without a loss of ultrastructural definition.

Gap junctions have been isolated from other tissues, notably lens fiber cells (by many laboratories) and mouse heart.[19] It is interesting that, in both cases, apparently nonhomologous polypeptides comprise the gap junction.[19,20] More detailed analysis is necessary, however, before tissue specificity of gap junctional polypeptides can be concluded.

Isolation of Liver Gap Junctions

The isolation procedure for gap junctions can be conveniently considered in two parts—first, the isolation of plasma membranes, and then the isolation of gap junctions from the plasma membrane fraction. Our procedure for the isolation of rat liver plasma membranes is based on upon a modification[21] of that developed by Neville.[22] Unless otherwise noted, all procedures are carried out at 0–4°.

Isolation of Rat Liver Plasma Membranes

Large-Scale Preparation. Fifty rat livers are removed from female, retired breeder Sprague–Dawley rats which are permitted to feed and drink ad libitum. The rats are killed either by decapitation or by placement in a carbon dioxide euthanasia chamber. The livers (approximately 500 g) are cleaned to remove connective tissue and then minced by passage through a Foley food mill. The minced liver is diluted to about 1.5 liters in 1 m*M* $NaHCO_3$ (buffer) and homogenized in 50-ml aliquots using a Dounce

[17] J. C. Erhart, *Cell Biol. Int. Rep.* **5,** 1055 (1981).
[18] D. Henderson, H. Eibl, and K. Weber, *J. Mol. Biol.* **132,** 193 (1979).
[19] R. W. Kensler and D. A. Goodenough, *J. Cell Biol.* **86,** 755 (1980).
[20] E. L. Hertzberg, D. J. Anderson, M. Friedlander, and N. B. Gilula, *J. Cell Biol.* **92,** 53 (1982).
[21] D. A. Goodenough and W. Stoeckenius, *J. Cell Biol.* **54,** 646 (1972).
[22] D. M. Neville, Jr., *Biochem. Biophys. Acta* **154,** 540 (1968).

homogenizer with a loose-fitting pestle. Since great variability has been found among different homogenizers, homogenization must be followed by phase-contrast microscopy. Cell breakage is estimated by the number of nuclei in a field as compared to the number of intact cells, with a target breakage of 95–99%. Excessive homogenization leads to poor recoveries.

The homogenate is then diluted to about 4 liters and filtered first through two layers and then through four layers of cheesecloth. After dilution to 7 liters, initial fractionation is obtained by sedimentation at 6000 rpm (3590 g[23]) through a continuous-flow rotor (Beckman JCF-Z) at a flow rate of 250 ml/min. The pellet is resuspended in buffer using one or two strokes of a loose-fitting Dounce homogenizer, diluted to about 2.5 liters in buffer, and centrifuged for 10 min at 6000 rpm (6370 g) in a JA-10 rotor (Beckman). The supernatant fraction is discarded, and the pellet is resuspended in a minimum volume of buffer to give a volume of 250–300 ml. The volume of resuspended pellet is measured exactly, and two volumes of 67% (w/w) sucrose are added. The crude membrane fraction is now ready for resolution by density gradient fractionation in a zonal rotor.

While rotating at 2000 rpm, the zonal rotor (Ti 15, Beckman) is loaded in order with: 150 ml of 37% (w/w) sucrose, 1 mM $NaHCO_3$, 500 ml 45% (w/w) sucrose, 1 mM $NaHCO_3$, the crude membrane sample, and then 55% (w/w) sucrose to full capacity. The zonal rotor is preloaded with distilled water, and complete filling is indicated by the beginning of sucrose elution from the rotor. The centrifuge chamber is closed, the vacuum system is engaged, and the rotor is accelerated to 29,000 rpm (84,000 g). After centrifugation for 1 hr, the timer is turned off and the rotor is permitted to come to a stop with the brake engaged. During centrifugation, the plasma membranes float up to the 37/45% interface, while the nuclei spin through the gradient and adhere to the outer wall of the rotor. The membrane fraction, as observed when the rotor is opened, appears as tan to reddish sheets floating in the gradient. The sheets are removed by aspiration into a trap bottle, and the material is diluted with an equal volume of buffer. The suspension is centrifuged for 10 min at 10,000 rpm (17,700 g) in a JA-10 rotor (Beckman). The cloudy supernatant fraction is discarded, and the pellet is resuspended in about 150 ml of buffer. The last centrifugation is repeated, then the plasma membrane fraction is resuspended in a small volume of buffer giving a volume of 30–50 ml containing 15–20 mg of protein per milliliter and stored at $-20°$ for later use.

Small-Scale Preparation. The above procedure can be successfully scaled down for no more than 10 rat livers (120–150 g of tissue) with somewhat greater recoveries of plasma membranes. The same general procedure is followed through homogenization and filtration steps and can

[23] The g forces are calculated at r_{max} for the various rotors.

be conveniently carried out without relying on continuous-flow and zonal rotors.

The filtrate is resuspended to about 2.4 liters, then divided into six 500-ml bottles and centrifuged in the JA-10 rotor at 5000 rpm (4420 *g*) for 10 min). The cloudy supernatant fraction is decanted, the pellet is resuspended to 2.4 liters in buffer, and the centrifugation is repeated. Resuspension and centrifugation are repeated until a relatively clear supernatant fraction is obtained (three or four times). More nuclear lysis occurs during these centrifugations than by limited centrifugation in the step before the zonal centrifugation described above, but this has no effect on the quality of the gap junction fraction ultimately obtained.

The pellet is resuspended by gentle homogenization with a Dounce homogenizer in a minimum volume of buffer (60–90 ml). As described above, two volumes of 67% (w/w) sucrose are added and the sample is divided into six aliquots in polycarbonate bottles for the 35 or 45 Ti rotor (Beckman). A solution of 30% (w/w) sucrose is layered on the suspension to the capacity of the tubes. This is conveniently performed using a peristaltic pump. The tubes are placed in the rotor, and centrifugation is carried out at 34,000 rpm (135,000 *g*) for 75 min with the brake engaged at the end of this time. The material at the 30%/sample interface is collected using a syringe fitted with a 6-in., 16-gauge cannula (blunt tip). It is diluted two- to threefold with buffer to give about 250 ml and centrifuged in the JA-10 rotor for 10 min at 10,000 rpm (17,700 *g*). This centrifugation is repeated as a wash, and the final pellet is resuspended in a minimum volume of buffer. The yield is 10–15 ml of suspension with a protein concentration of 15–20 mg/ml.

Notes on Plasma Membrane Isolation. Plasma membrane isolation has been successfully applied to mouse liver as well as liver from male and female rats of varying ages. Retired breeders are preferred because of their relatively low cost.

Homogenization procedures other than that specified above should be approached with caution. Gap junction fractions obtained using a motor-driven Potter–Elvehjem homogenizer have been heavily contaminated with other proteins as detected by SDS–polyacrylamide gel electrophoresis. By electron microscopy this material has an amorphous appearance. We have been able to use a Polytron homogenizer (Brinkmann) at low-power settings for cell disruption. The required power setting varies from instrument to instrument, but, when the guideline for cell breakage described above has been used, this homogenizer has yielded very satisfactory results with much less effort than Dounce homogenization.

The sucrose solutions used for plasma membrane isolation should be filtered through either a 2-μm Millipore filter or a glass-fiber filter of similar cutoff (Whatman GF/B). Reagent grade sucrose need not be used during plasma membrane isolation. Substantial savings in expense can be made by using cane sugar obtained on sale at local groceries, especially when large-scale preparations are routinely carried out.

The plasma membrane fraction obtained by either of the above procedures is heavily contaminated with nucleic acid, which can be readily removed by a salt wash, a step included in the routine procedure for preparation of gap junctions from plasma membranes.

Isolation of Gap Junctions from Rat Liver Plasma Membranes

Preparative Procedure. The isolation procedure described here applies to the material obtained by the "large-scale" procedure described above. It can be modified for the "small-scale" procedure by simply scaling down all steps.

Plasma membranes, either fresh or thawed, are diluted to 80 ml in buffer and centrifuged at 12,000 rpm (17,400 *g*) for 10 min in a JA-20 rotor (Beckman). The pellets are taken up to about 80 ml in 0.1 *M* NaCl in buffer, and the centrifugation is repeated. Salt is washed out by resuspension in 80 ml buffer, and the centrifugation is repeated. The pellet is resuspended to 50 ml in buffer and divided into 10 aliquots. Five milliliters of 0.45% Sarkosyl NL-97 (or *N*-lauryl sarcosine, sodium salt; Sigma Chemical Company) is added to each tube. The contents are sonicated (Branson Model W-185 or Heat Systems Model W-225) for 10 sec at tap setting 5 using the standard tapered microtip, avoiding foaming. After 10 min at room temperature, the sonication is repeated. After an additional 10 min at room temperature, the suspension is centrifuged for 15 min at 20,000 rpm (48,400 *g*) (JA-20 rotor). The supernatant fraction is discarded; the pellets are resuspended by gentle sonication in buffer, pooled into one tube, and washed by dilution to 40 ml in buffer and repeating the centrifugation. The Sarkosyl-resistant pellet is resuspended to 5 ml in 0.1 *M* Na_2CO_3, pH 11, with gentle sonication. (This volume of sodium carbonate is convenient to work with and can be used even with lesser amounts of membrane material.) The 20,000-rpm centrifugation and wash are repeated. The pellet is resuspended by gentle sonication in 1.5 ml of buffer. To this is added an equal volume of 30% (w/w) sucrose, 2 *M* urea, and 1 m*M* $NaHCO_3$. (Fresh urea solutions should be used for this and for the solutions in the sucrose gradients below.)

Sucrose gradients are prepared in cellulose nitrate tubes for the SW40 rotor (Beckman). The gradients contain 60% (w/w) sucrose, 0.5 ml; followed by 4.0 ml each of 45%, 35%, and 30% (w/w) sucrose containing 1 *M* urea and 1 m*M* $NaHCO_3$. (These solutions can be prepared from the 60% sucrose solution by suitable dilution adding 5 *M* urea, 0.1 *M* $NaHCO_3$, and water and should be checked with a refractometer.) One-half milliliter of the sample is placed on top of each gradient, and centrifugation is carried out at 39,000 rpm (271,000 *g*) for 1.5 hr. The white material at the 35%/45% sucrose interface, containing the gap junctions, is removed using a Pasteur pipette[24] and pooled and diluted to 16 ml with buffer. The suspension is divided into two 10-ml polycarbonate Oak Ridge tubes and centrifuged at 20,000 rpm (48,400 *g*) for 20 min in the JA-20 rotor. The pellets are resuspended by gentle sonication in buffer, combined, and centrifuged again in one tube. Finally, they are resuspended in a minimum amount of buffer with gentle sonication. The yield is 50–100 μg of protein. This gap junction fraction can be used immediately or stored at −20° for later use. By electron microscopy and SDS–polyacrylamide gel electrophoresis, no changes are observed upon storage at −20° for at least 1 month. For longer storage, −80° provides better preservation.

Notes on the Isolation of Liver Gap Junctions. The material at the top of the sucrose gradients consists of lipid vesicles and contains no protein that can be detected on SDS–polyacrylamide gels.

Additional material can be salvaged from the sucrose gradient after removal of the gap junctions from the 35%/45% sucrose interface. After the material that fails to enter the gradient is removed, all the sucrose down to the 45%/60% sucrose interface is collected, pooled, diluted with buffer, and centrifuged as described above. Although somewhat more contaminated than the gap junction fraction, this "gradient residue" fraction contains substantial quantities of the gap junction polypeptide (up to 50% as much as is obtained in the gap junction fraction, depending on the efficiency of collection of material at the 35%/45% sucrose interface). Additional junction material sediments to the bottom of the gradient, most likely as a result of trapping by collagen, one of the major contaminants in the sample loaded.

Somewhat better recoveries are obtained by omitting the 35% sucrose and diluting the applied sample proportionately to fill the tubes.[14] In this modification, the pellet obtained after washing out the Na_2CO_3 is resuspended in 13 ml of buffer and is added to 14 ml of 30% (w/w) sucrose, 2 *M* urea, 1 m*M* $NaHCO_3$. This suspension (4.5 ml) is then layered on top of six sucrose gradients containing the 60%, 45%, and 30% sucrose solutions described above. Subsequent to centrifugation, the gap junction fraction is that at the 30%/45% sucrose interface. The increase in yield (500–600 μg of protein) is accompanied by a slight loss in purity but provides a gap junction fraction suitable for many types of experiments.

[24] It is easiest to observe the white, rather diffuse material at the interface by light coming from above and to the rear of the tube, and having a dark background such as by placing a black notebook or piece of construction paper a few inches behind tube.

Characterization of the Gap Junction Fraction

Although the gap junction fraction was originally characterized by morphological criteria, using the electron microscope, correlation of morphological purity with polypeptide profiles on SDS–polyacrylamide gels has been excellent. We, therefore, routinely evaluate the quality of the gap junction preparations by SDS–polyacrylamide gel electrophoresis. The gels are 12.5% gels using the system of Laemmli and Favre,[25] except that solubilization is carried out in 50 m*M* Na_2CO_3, 50 m*M* dithiothreitol, 2% SDS, resulting in more complete solubilization and less aggregation of the 27,000 M_r gap junction polypeptide. A useful diagnostic property of the M_r 27,000 polypeptide is its tendency to aggregate in the solubilization mixture when heated at 100° for 2 min (Fig. 1, lane A).[18] The major aggregate, under these conditions, runs as a broad band with an apparent M_r of 47,000. The unheated sample (Fig. 1, lane B) demonstrates the preponderance of the M_r 27,000 polypeptide.

If care is not taken to keep all solutions cold and to work at 4° through all the steps described above, proteolysis of the M_r 27,000 polypeptide is observed (Fig. 1, lane C). In most cases, this results in the appearance of a number of polypeptide bands with apparent molecular weights of 24,000 to 26,000. Peptide mapping has demonstrated that these bands are derived from the M_r 27,000 polypeptide.[14] Another band, observed in most preparations, has an apparent molecular weight of 22,000. Its relationship to the M_r 27,000 polypeptide is not clear. In cases of extensive endogenous proteolysis, 18,000- and 12,000-M_r bands are observed. If dithiothreitol is omitted from the solubilizing mixture, these latter bands migrate at 24,000–26,000, suggesting that they are held together by intrachain disulfide bonds.

General Comments

The gap junction fraction containing the M_r 27,000 polypeptide has been characterized in a number of laboratories. With the procedure described above, no significant differences were observed in the lipid composition of the junctional membranes and the plasma membrane.[13] Others, using a somewhat different approach to junction isolation, have demonstrated that there is an enrichment for cholesterol in gap junction membranes.[18] No glycosylation of the M_r 27,000 gap junction polypeptide has been detected.[13,18] A partial sequence of this polypeptide has been reported,[26] and the N-terminal amino acid is methionine. The N-terminal analysis has been confirmed by us.

[25] U. K. Laemmli and M. Favre, *J. Mol. Biol.* **80,** 575 (1973).

[26] B. J. Nicholson, M. W. Hunkapillar, L. B. Grim, L. E. Hood, and J.-P. Revel, *Proc. Natl. Acad. Sci. U.S.A.* **78,** 7594 (1981).

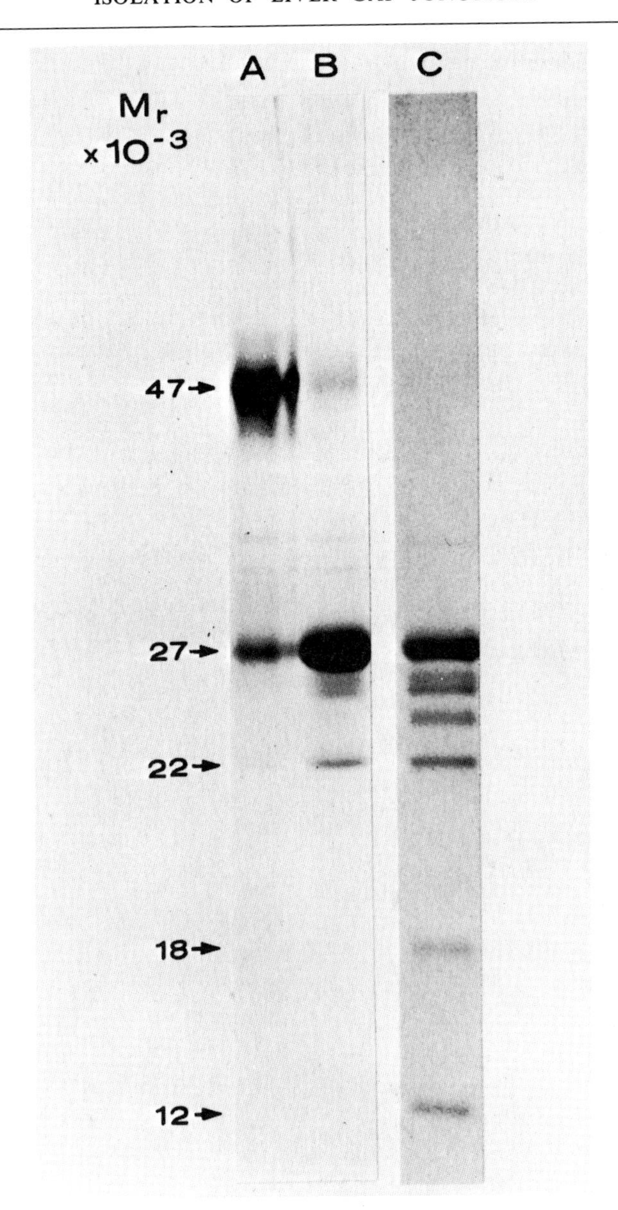

FIG. 1. Sodium dodecyl sulfate (SDS)–polyacrylamide gel of gap junctions. Gap junctions were analyzed on 12.5% gels[25] as described in the text. Lanes A and B are of the same gap junction preparation in which samples were heated (lane A) or unheated (lane B) subsequent to the addition of SDS. Lane C is an unheated sample from a different preparation, in which more proteolysis of the M_r 27,000 polypeptide was observed.

This procedure for the preparation of rat liver gap junctions has worked well when extended to mouse liver. The M_r 27,000 mouse liver gap junction polypeptide is substantially similar to that from rat liver when analyzed by peptide maps.[14,27] It also undergoes the characteristic aggregation when heated in SDS, although the aggregation is not as pronounced as for rat liver. In an effort to further increase the yield of M_r 27,000 polypeptide, we have used this isolation procedure to prepare fractions from bovine and rabbit liver, as well as frozen rabbit, rat, and mouse liver. While an M_r 27,000 polypeptide was found in all cases, that derived from bovine and rabbit liver was not adequately enriched, and recoveries from the frozen rat and mouse liver were too low to encourage their use.

[27] E. L. Hertzberg and N. B. Gilula, *Cold Spring Harbor Symp. Quant. Biol.* **46,** 639 (1982).

[45] Techniques for Studying the Cell-Free Synthesis of the Major Lens Fiber Membrane Polypeptide

By EDWARD H. WILLIAMS, NALIN M. KUMAR, and NORTON B. GILULA

The lens provides a unique system for the study of plasma membrane differentiation. It contains two major cell types—epithelial and fiber cells. The epithelial cells are roughly cuboidal and form a monolayer that covers the anterior surface of the lens. Upon division, they move inward and differentiate into fiber cells. During this process, these cells undergo remarkable elongation, their length sometimes being 3000 times greater than their width. This change in geometry requires an enormous synthesis of plasma membrane from cells that have few remaining cytoplasmic organelles. Concomitant with this increase in plasma membrane is a large increase in a population of junctions that join the adjacent differentiating fiber cells. Morphological studies have revealed that junctions in fiber cells occupy 40–60% of the plasma membrane surface area[1] in contrast to the relative scarcity of these junctions in the outer epithelial layer. Although these junctions appear to be similar to the gap junction that is present in most mammalian tissues, they can be distinguished from the gap junction by biochemical and serological criteria.[2] At present, the possibility exists that two different junction proteins are produced by the lens, i.e., one protein predominantly or solely found in epithelial junctions, and the second one

[1] H. Bloemendal, *Science* **197,** 127 (1977).
[2] E. L. Hertzberg, D. J. Anderson, M. Friedlander, and N. B. Gilula, *J. Cell Biol.* **92,** 53 (1982).

ISBN 0-12-181998-1

expressed during differentiation of fiber cells. Both junctional polypeptides appear to be quite similar in mass; the gap junction polypeptide is about 27,000 daltons,[3,4,4a] and the lens fiber junction polypeptide is about 26,000 daltons.[5,6] In addition, no strong evidence exists that suggests that these polypeptides are glycosylated.

Our purpose is to describe some techniques that can be used for the molecular study of lens plasma membrane differentiation as it applies to the cell junctions. Section I provides details for the fractionation and extraction of lens RNA with the intent of enrichment for mRNA specific for the major fiber junction protein, MP-26. Section II describes the immunoprecipitation procedure that is used to identify MP-26 as a cell-free translation product. Section III describes the procedure for establishing primary cultures of bovine lens epithelial cells.

I. Fractionation and Extraction of Lens RNA

Reagents

TNM buffer: 50 m*M* Tris-HCl (pH 7.5), 50 m*M* NaCl, 5 m*M* $MgCl_2$
SDS, 20% (Pierce)
NaCl, 2 *M*
EDTA, 200 m*M*
Sodium deoxycholate, 15% (Sigma)
Extraction buffer: 1% SDS, 1 m*M* EDTA, 0.1 *M* NaCl, 5 m*M* Tris-HCl (pH 7.5)
Sodium acetate, 4 *M*
Absolute ethanol
Proteinase K (Boehringer Mannheim or E. Merck)
Redistilled phenol
Chloroform
LiCl, 4 *M*
Guanidinium thiocyanate buffer (GT): 4 *M* guanidinium thiocyanate (Fluka, *purum* grade), 0.5% sodium *N*-lauroyl sarcosine (Sigma), 25 m*M* sodium citrate (pH 7.0), 0.1% antifoam (Sigma, type A). This buffer should be filtered and then adjusted to pH 7.0 with 1 *N* NaOH. It can be stored at room temperature for 1 month.
Acetic acid, 1 *M*
Guanidinium hydrochloride, 7.5 *M* (GHCl). This solution should be filtered, neutralized to pH 7 with 1 *N* NaOH, and then buffered with

[3] E. L. Hertzberg and N. B. Gilula, *J. Biol. Chem.* **254,** 2138 (1979).
[4] D. Henderson, H. Eibl, and K. Weber, *J. Mol. Biol.* **132,** 193 (1979).
[4a] B. J. Nicholson and J.-P. Revel, this volume [46].
[5] J. Alcala, N. Leska, and H. Maisel, *Exp. Eye Res.* **21,** 585 (1975).
[6] D. A. Goodenough, *Invest. Ophthalmol. Visual Sci.* **18,** 1104 (1979).

0.025 volume of 1 *M* sodium citrate, pH 7.0. It can be stored at room temperature for 1 month. Just before using, adjust to 5 m*M* dithiothreitol (DTT).

Oligo(dT)cellulose (type 7; P-L Biochemicals, Inc.)

Binding buffer: 0.4 *M* NaCl, 0.01 *M* Tris-HCl (pH 7.5), 1 m*M* EDTA, 0.5% SDS

Elution buffer: 0.01 *M* Tris-HCl (pH 7.5), 1 m*M* EDTA, 0.5% SDS

Ethanol, 70%

Materials

Sorvall Omni-Mixer (Cat. No. 17105)

Eppendorf centrifuge (Cat. No. 5414)

Fractionation and Isolation of "Free" and "Bound" RNA

Fractionation of the lens mRNA into two populations is based on previous observations that the following method can be used as an approach for enriching specific MP-26 mRNA.[7]

A predetermined quantity of frozen calf lenses is pulverized in liquid nitrogen with a ceramic mortar and pestile and then added to 2 volumes (w/v) of TNM buffer. The powdered tissue is allowed to thaw in this buffer. Once thawed, this mixture is homogenized in a Dounce homogenizer with three strokes of a B pestle. The homogenate is then centrifuged at 9000 *g* (r_{max}) for 20 min at 4°. The supernatant containing the so-called "free" polysomes is adjusted to 1% (w/v) SDS, 0.1 *M* NaCl, and 10 m*M* EDTA and stored at 4°, the pellet containing the "bound" polysomes is further fractionated. The pellet is resuspended in 1 volume (relative to original TNM buffer) of TNM buffer and again homogenized in a Dounce homogenizer as described above, then centrifuged at 9000 *g* (r_{max}) for 20 min at 4°. This step is repeated again, and the supernatant is discarded in each case. The final pellet is resuspended and homogenized in a Dounce homogenizer in 1 volume of TNM buffer, adjusted to a concentration of 1.5% (w/v) sodium deoxycholate, then incubated at room temperature with gentle stirring. After 20 min, this fraction is centrifuged at 9000 *g* (r_{max}) for 20 min at 4°. The supernatant, containing the "bound" polysomes released from the pellet, is adjusted to 1% (w/v) SDS and 10 m*M* EDTA.

At this point, a "free" and a "bound" fraction of polysomes exist from which RNA can now be extracted. An equal volume of phenol–chloroform (1 : 1) is added to each supernatant; the mixture is vigorously shaken, then centrifuged at 9000 *g* (r_{max}) at 2° for 15 min. The aqueous phase is collected and stored at 4°. The organic phase is reextracted with 0.3 volume of

[7] F. C. S. Ramaekers, A. M. E. Selten-Versteegen, E. L. Benedetti, I. Dunia, and H. Bloemendal, *Proc. Natl. Acad. Sci. U.S.A.* **77,** 725 (1980).

extraction buffer (volume relative to organic volume), and the two aqueous phases are pooled. The RNA is then precipitated overnight at $-20°$ by adjusting the solution to 0.2 *M* NaOAc and adding 2 volumes of ethanol.

The RNA is spun down and resuspended in an appropriate volume of extraction buffer, proteinase K is added to 5 mg/ml, and the mixture is incubated for 30 min at room temperature. The RNA is extracted with an equal volume of phenol–chloroform (1 : 1) and ethanol precipitated, as described above.

The RNA is then pelleted and resuspended in distilled H_2O, and an equal volume of 4 *M* LiCl is added. This is stored at 4° overnight. The RNA is then pelleted, leaving DNA and tRNA in the supernatant. The RNA pellet is dissolved in an appropriate volume of distilled water if it is to be used for cell-free translation or in oligo(dT) binding buffer (see subsection on affinity chromatography, below) if the RNA is to be further fractionated.

Extraction of Total Lens Fiber RNA

Frozen calf lenses are pulverized in liquid nitrogen and homogenized on ice in 5–6 volumes (w/v) of extraction buffer saturated with phenol, using a Sorvall Omni-Mixer at maximum speed for 1 min at 4°. The procedure for RNA extraction then follows that described above, starting after the addition of phenol–chloroform (1 : 1) to the supernatant.

Extraction of RNA from Cultured Lens Epithelial Cells

The procedure for extraction of RNA follows a previously published guanidinium thiocyanate protocol,[8] which has been modified for application to cell culture systems.

Culture plates are washed three times with ambient temperature phosphate-buffered saline, after which they are carefully drained of excess buffer and 3 ml of 4 *M* guanidinium thiocyanate solution (GT) is added to each plate (6 ml for standard flasks). The GT is evenly distributed over the plate to ensure complete lysis of the cell population. The GT from all plates is pooled and homogenized on ice for 1 min in a Sorvall Omni-Mixer at full power. The homogenate is centrifuged at 10,000 g (r_{max}) for 10 min at 10°. To the supernatant is added 0.025 volume (relative to original GT buffer volume) of 1 *M* acetic acid and then 0.75 volume of absolute ethanol. This mixture is shaken and stored overnight at $-20°$. The RNA is then centrifuged at 6000 g (r_{max}) for 10 min at 10° and resuspended in 0.5 volume (relative to original GT buffer volume) of guanidinium-HCl (GHCl). Then 0.025 volume of 1 *M* acetic acid and 0.5 volume of absolute ethanol are

[8] J. M. Chirgwin, A. E. Przybyla, R. J. MacDonald, and W. J. Rutter, *Biochemistry* **18**, 5294 (1979).

added (both volumes relative to GHCl volume, *not* relative to GT buffer volume), and the mixture is shaken and stored for at least 3 hr at −20°. After centrifugation at 6000 g (r_{max}) for 10 min at 10°, the RNA pellet is resuspended in 0.25 volume of GHCl (relative to GT buffer) and 0.025 volume of 1 M acetic acid and 0.5 volume of absolute ethanol (both relative to GHCl). The pellet is then shaken and stored for at least 3 hr at −20°, at which time centrifugation is repeated at 6000 g (r_{max}) for 10 min at 10°, and the pellet washed in 70% ethanol. The pellet is centrifuged again, the ethanol is removed, and the pellet is dried with a gentle stream of air. Once dried, the RNA can be resuspended in H_2O or oligo(dT) binding buffer (see affinity chromatography, below).

The yields of RNA from the extraction procedures outlined above are given in the table.

Affinity Chromatography for Polyadenylated mRNA

We employ an affinity chromatography system using the oligo(dT)cellulose method modified from a previously reported reference.[9] Because of the relatively small amount of lens mRNA per gram of tissue, we use small quantities of oligo(dT)cellulose to enable us to use small volumes of binding and elution buffers. Rather than building an affinity column, we have chosen to "batch"-purify polyadenylated RNA from lens. The "batch" procedure enables one to elute polyadenylated RNA from the cellulose with small volumes of buffer, thus making it unnecessary to employ a carrier such as tRNA for ethanol precipitation.

An appropriate amount of oligo(dT)cellulose is placed into either a 0.5-ml or 1.5-ml Eppendorf microfuge tube, depending on the quantity of oligo(dT)cellulose required. The oligo(dT)cellulose is then swelled in an excess of binding buffer for 10 min; the tube is then centrifuged, and the buffer is removed and replaced with another volume of binding buffer. This

RNA YIELDS FROM LENS FIBER AND LENS EPITHELIAL CELLS PRIOR TO ISOLATION OF POLYADENYLATED RNA

Source	RNA (μg/g tissue)	RNA (μg/1 × 10^6 cells)
Lens fiber		
Total	50	—
"Bound"	4.0	—
"Free"	34	—
Lens epithelium	—	22

[9] H. Aviv and P. Leder, *Proc. Natl. Acad. Sci. U.S.A.* **69,** 1408 (1972).

change of buffer serves to wash the oligo(dT)cellulose, and the washing is repeated three or four times before RNA is applied. During this washing phase, the RNA, which has been previously suspended in binding buffer, is heated to 65° for 1 min to denature the RNA and then rapidly cooled in ice water. Once the oligo(dT)cellulose is thoroughly washed, the RNA is placed into the tube and the contents are gently agitated for 15 min at room temperature The tube is then centrifuged for 3 min. The supernatant is removed and labeled as the poly(A)$^-$ RNA fraction. The oligo(dT)cellulose is then washed three more times in binding buffer to remove any RNA that is nonspecifically bound. Once the last binding buffer has been removed, an appropriate amount of elution buffer is pipetted into the tube. The oligo(dT)cellulose is resuspended in this buffer and heated at 65° for 2 min prior to centrifugation for 3 min. The supernatant is collected and labeled as the poly(A)$^+$ RNA fraction. Both supernatants, poly(A)$^-$ and poly(A)$^+$, are adjusted to 0.2 *M* NaOAc with a 4 *M* stock solution, and the RNA is ethanol precipitated.

The RNA is pelleted by a 3-min centrifugation in an Eppendorf centrifuge, washed once with 70% ethanol to remove salts, and dried. The RNA is dissolved in an appropriate volume of H_2O for subsequent application of the RNAs. If the RNA is not to be used immediately, it is aliquoted into 0.5-ml Eppendorf tubes and stored at −80°.

II. Immunoprecipitation Procedure

This section describes the detection of the MP-26 polypeptide by immunoprecipitation of cell-free translation products directed by lens RNA.

Reagents

- Dilution buffer: 50 m*M* Tris-HCl (pH 7.4), 190 m*M* NaCl, 6 m*M* EDTA, 1% (v/v) Trasylol (Sigma), 1.25% Triton X-100
- Detergent wash buffer: 50 m*M* Tris-HCl (pH 7.4), 150 m*M* NaCl, 5 m*M* EDTA, 1% (v/v) Trasylol (Sigma), 0.1% Triton X-100 (Polysciences, Inc.), 0.02% SDS (Pierce)
- Nondetergent wash buffer: Same as detergent wash buffer, but without Triton X-100 or SDS
- Sample loading buffer: 200 m*M* Tris-HCl (pH 8.8), 500 m*M* sucrose, 5 m*M* EDTA, 3% SDS, 0.01% (w/v) Bromphenol Blue, 50 m*M* dithiothreitol
- Protein A–Sepharose (Sigma)

Cell-Free Translation

We have employed both the rabbit reticulocyte and wheat germ lysate systems for translating lens RNA. Since our procedures for producing and

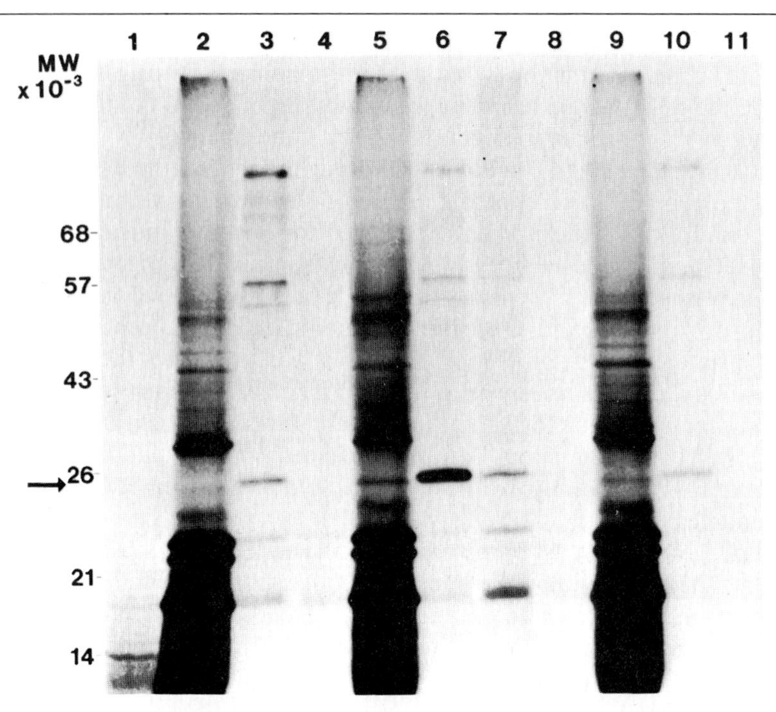

FIG. 1. An autoradiograph of [^{35}S]methionine-labeled polypeptides synthesized in a cell-free wheat germ lysate system directed by lens fiber poly(A)$^+$ RNA and immunoprecipitated products using an anti-MP26 rabbit serum analyzed by 7.5 to 15% gradient SDS–PAGE. Lane 1: H_2O blank; no exogenous RNA added. Lanes 2, 5, and 9: Total products directed by "free," "bound," and total RNA fractions, respectively. Lanes 3, 6, and 10: Polypeptides immunoprecipitated by anti-MP26 rabbit serum from the "free," "bound," and total fractions, respectively. Lanes 4, 8, and 11: Polypeptides precipitated by nonimmune rabbit serum from the "free," "bound," and total fractions, respectively. The arrow in left margin denotes the migration distance of an MP-26 standard electrophoresed on the same gel. Note the relative intensities of the immunoprecipitated products in lanes 3, 6, and 10, which comigrate with the MP-26 standard. the polypeptide products directed by the "bound" RNA (lane 6) are enriched for MP-26 in comparison to the "free" products (lane 3) and total products (lane 10). The relative intensity of the polypeptide in lane 6, which comigrates with the MP-26 standard, can be decreased by the addition of excess MP-26 to the wheat germ system just prior to the addition of antiserum (see lane 7 vs lane 6).

using these lysates follow previously published protocols,[10,11] they are omitted here.

Immunoprecipitation Procedure

The cell-free translation reaction is terminated by adding one-twentieth volume of 20% SDS. The translation reaction tubes are then briefly vortexed

[10] B. E. Roberts and B. M. Paterson, *Proc. Natl. Acad. Sci. U.S.A.* **70,** 2330 (1973).
[11] H. R. B. Pelham and R. L. Jackson, *Eur. J. Biochem.* **67,** 247 (1976).

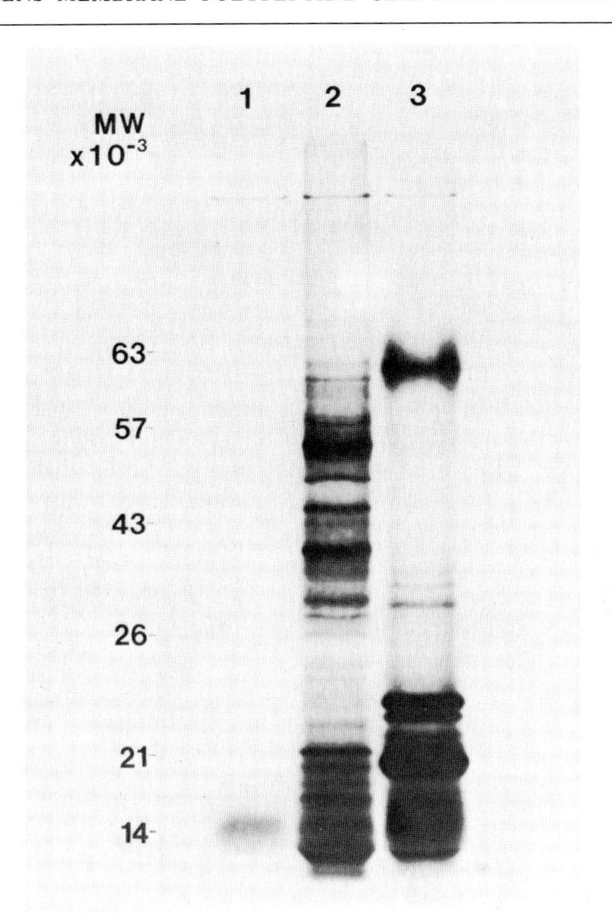

FIG. 2. An autoradiograph of [^{35}S]methionine-labeled polypeptides synthesized in a cell-free rabbit reticulocyte lysate analyzed by 7.5 to 15% gradient SDS–PAGE. Lane 1: H_2O blank; no RNA added. Lane 2: Polypeptides directed by poly(A)$^+$ RNA from lens epithelial cultures. Lane 3: Polypeptides directed by lens fiber total poly(A)$^+$ RNA. The most abundant polypeptides produced by the lens fiber cells, the crystallins, can be seen in lane 3 between 18,000 and 22,500 daltons. These crystallins are significantly reduced or are lacking in the lens epithelial cultures. A comparison of lanes 2 and 3 shows a markedly different synthesis pattern in the lens epithelium and fiber polypeptides. Immunoprecipitation of the lens epithelial polypeptides using anti-MP26 rabbit serum failed to reveal any products that comigrated with an MP-26 standard (data not shown).

and heated at 100° for 4 min. Upon completion of heating, the tubes immediately receive 4 volumes of dilution buffer and are vortexed again. At this time, either antiserum or nonimmune serum is added to the appropriate tubes. The volume of serum used depends upon its titer and relative abundance of antigen in the translation mixture. We routinely use 1–5 μl of crude serum for every 100 μl of cell-free translation volume. The tubes are then placed on a tilting platform at 4° and mixed gently for 12–24 hr. The tubes are then centrifuged, and the supernatants are transferred to new tubes. A small insoluble pellet should be discarded. To the supernatants, 30 μl of a 1 : 1 suspension of protein A–Sepharose in H_2O is added for every 5 μl of serum. The tubes are placed back onto the tilting platform and rotated gently for 1.5–3 hr at room temperature. The tubes are centrifuged, then the supernatant is discarded and the protein A–Sepharose pellets are washed three times with 1 ml of a detergent wash buffer and then once with a nondetergent wash buffer for 5–15 min each at room temperature on the tilting platform. At the end of each wash, the protein A–Sepharose is pelleted by brief centrifugation, and the supernatants are discarded. After the nondetergent wash, sample loading buffer is added to the Sepharose pellets with care taken to suspend the entire pellet in the buffer. The volume of sample loading buffer used will depend on the volume of protein A–Sepharose employed for the immunoprecipitation. Three volumes of sample loading buffer are added for each volume of protein A–Sepharose. The volume of the gel well usually is the limiting factor, since in most cases the entire volume of sample loading buffer is loaded into one gel well in order to provide sufficient amounts of radioactive sample for short autoradiography exposure times. The tubes are centrifuged before loading the appropriate amount of the sample (supernatant) onto an analytical gel.

Total translation products and specific immunoprecipitated polypeptides are shown in Fig. 1. A comparison of the total translation products of lens epithelial and fiber RNA is shown in Fig. 2.

III. Primary Cultures of Bovine Lens Epithelial Cells

Reagents

Culture medium: Dulbecco's modified MEM (GIBCO) with 25 m*M* HEPES and 4500 mg of glucose per liter, supplemented with 10% fetal calf serum (Sterile Systems) plus 1% pen/strep

Trypsin, 0.1% (GIBCO)

Lens Epithelial Culturing Procedure

Freshly excised calf eyes are obtained from a local slaughterhouse, and the lenses are removed by a radial incision along the equator of the eye.

Under sterile conditions, a lens is washed in PBS and the capsule is removed and discarded. With fine forceps, thin strips of the anterior and equatorial portions of the lens are removed and placed in a 100-mm plastic petri dish with 7 ml of culture medium. Each plate will support material from four to six lenses. The plates are incubated at 37°, and 2 ml of additional medium is added to each plate on the second day of culture. On the third day, half of the medium is removed and replaced with fresh medium. For the next 7–10 days, cultures are fed as required. At 13–14 days, the cells are dissociated with 0.1% trypsin in phosphate-buffered saline and removed from the plate. They are then counted and replated into 100-mm dishes at an initial density of 5×10^5 cells/plate. The first-passage plates usually require 4–6 days to reach confluency; thereupon they can undergo a subsequent passage or harvesting for RNA extractions.

[46] Gap Junctions in Liver: Isolation, Morphological Analysis, and Quantitation

By BRUCE J. NICHOLSON and JEAN-PAUL REVEL

Gap junctions are arrays of cell-to-cell channels that permit exchanges of cytoplasmic low-molecular-weight constituents, such as ions and various metabolites (see reviews[1-3]). Gap junctions allow for metabolic cooperation between the cells of a tissue[4] and for electrical coupling of excitable cells,[5] and they are believed to play a major role in the control of growth and differentiation.[6,7] Gap junctions are ubiquitous structures, found in every metazoan phylum and nearly all tissues studied.[8] With few exceptions, all these junctions have a characteristic appearance[9]—a feature of major importance, since it has been the only criterion of purity available for following the isolation of gap junctions.

From a biochemical standpoint, the best-studied tissues are the liver and eye lens, and this chapter focuses attention on the liver gap junctions. Here, the major protein component has been reasonably well characterized,[10-12]

[1] J.-P. Revel, S. B. Yancey, D. J. Meyer, and B. Nicholson, *In Vitro* **16,** 1010 (1980).

[2] M. L. Hooper and J. L. Subak-Sharpe, *Int. Rev. Cytol.* **69,** 45 (1981).

[3] J. Flagg-Newton, I. Simpson, and W. R. Loewenstein, *Science* (*Washington, D.C.*) **205,** 404 (1979).

[4] J. Pitts, *In Vitro* **16,** 1049 (1980).

[5] E. Furshpan and D. Potter, *J. Physiol.* (*London*) **145,** 289 (1959).

[6] W. R. Loewenstein, Y. Kanno, and S. J. Socolar, *Fed. Proc., Fed. Am. Soc. Exp. Biol.* **37,** 89 (1978).

[7] N. B. Gilula, M. L. Epstein, and W. H. Beers, *J. Cell Biol.* **78,** 58 (1978).

[8] C. Peracchia, *Int. Rev. Cytol.* **66,** 81 (1980).

[9] N. B. Gilula, *Int. Cell Biol. Pap. Int. Congr. 1st 1976,* p. 64 (1977).

[10] E. L. Hertzberg and N. B. Gilula, *J. Biol. Chem.* **254,** 2138 (1979).

[11] D. Henderson, M. Eibel, and K. Weber, *J. Mol. Biol.* **132,** 193 (1979).

[12] B. J. Nicholson, M. W. Hunkapiller, I. B. Grim, L. E. Hood, and J.-P. Revel, *Proc. Natl. Acad. Sci. U.S.A.* **78,** 7594 (1981).

ISBN 0-12-181998-1

and a partial amino acid sequence is available.[12] X-Ray diffraction[13] and image reconstruction based on low-dose electron microscopy,[14] in combination with the other techniques, indicate that each gap junctional channel comprises two halves (connexons[15]), one through the membrane of each adjacent cell. Each connexon is composed of six apparently identical polypeptide chains, probably associated with phospholipid. Gap junctions, as usually visualized, consist of large arrays of closely packed connexon pairs, which appear to retain their integrity under a variety of experimental treatments.[16]

A. A Strategy for the Isolation of the Gap Junction Protein

Virtually all the published procedures for the purification of gap junctions involve the isolation of a plasma membrane fraction and its subsequent treatment with a detergent that solubilizes other membrane components but leaves the gap junctions, clearly recognizable by their characteristic lattices, intact. The junctions are then separated on a sucrose gradient from the other detergent-resistant material of the plasma membrane fraction on the basis of density. Additional treatments with reagents such as 8 *M* urea,[17,18] and solutions of high pH (e.g., pH 11)[10] have also been employed in various preparation procedures with little apparent disruption of the gap junction structure. The general protocol outlined above must be specifically modified for each tissue because of differences in membrane or junctional density, the presence of additional components peculiar to that tissue (e.g., urate oxidase in the liver,[10] myosin and actin filaments in cardiac muscle,[19] crystallins of lens[20]), or differences in the sensitivity of gap junctions to detergent treatments.[19] We review here the isolation of gap junctions as applied to a specific tissue, the rat liver, with emphasis on techniques that we have found to provide reproducibly good yields of highly enriched gap junction fractions.

[13] L. Makowski, D. L. D. Caspar, W. C. Phillips, and D. A. Goodenough, *J. Cell Biol.* **74,** 629 (1977).

[14] P. N. T. Unwin and G. Zampighi, *Nature* (*London*) **283,** 545 (1980).

[15] D. A. Goodenough, *Methods Membr. Biol.* **3,** 51 (1975).

[16] D. A. Goodenough and J.-P. Revel, *J. Cell Biol.* **45,** 272 (1970).

[17] J. Alcalá, N. Lieska, and H. Maisel, *Exp. Eye Res.* **21,** 581 (1975).

[18] M. Finbow, S. B. Yancey, R. Johnson, and J.-P. Revel, *Proc. Natl. Acad. Sci. U.S.A.* **77,** 970 (1980).

[19] R. W. Kensler and D. A. Goodenough, *J. Cell Biol.* **86,** 755 (1980).

[20] H. A. Bloemendal, F. Zweers, I. Vermorken, I. Dunia, and E. L. Benede, *Cell Differ.* **1,** 91 (1972).

B. Isolation of Gap Junctions from Rat Liver

1. Isolation of Plasma Membranes

Solutions. The volume is that required for 25 livers, i.e., 200 g of tissue.

Isolation buffer (IB), 20 liters: 2 mM $NaHCO_3$, 0.5 mM $CaCl_2$, pH 7.4 (4°)

Perfusion buffer (PB), 200 ml: IB + 0.9% NaCl (37°)

Two-phase solution: 80 g of dextran (Sigma: $M_{r(av)}$ 500,000), 62 g of polyethylene glycol 6000 (Baker), 11.92 g of $NaH_2PO_4 \cdot H_2O$, 16.22 g of $Na_2HPO_4 \cdot 7H_2O$, 0.3 g of NaN_3, 1486 ml of H_2O (4°). The solution should be made up 1 day before the isolation and allowed to separate in a separatory funnel overnight in a cold room.

Methodology—Available Methods. In the majority of published isolation protocols for gap junctions from rat or mouse liver, the method of Neville,[21] variously modified, has been used to isolate the plasma membrane fraction. The livers are homogenized, and the homogenate is filtered through cheesecloth to remove coarse fibrous material and precipitated nuclear protein. The homogenate is then centrifuged once or twice at relatively low acceleration of gravity (g) to separate the membranes from most of the mitochondria and soluble proteins. The pellets from these spins are then loaded onto a discontinuous sucrose gradient and centrifuged.

This gradient is usually composed of layers of 60, 54, 50, and 43% (w/v) sucrose. The membranes can be collected between the 50% ($d = 1.191$) and 43% ($d = 1.168$) (w/v) sucrose layers (for specific details, see Hertzberg and Gilula[10]), or a crude plasma membrane fraction can be collected instead over a 52% (w/v) sucrose cushion.[22] An alternative route is that developed by Lesko *et al.*,[23] in which the sucrose gradient is supplanted by a two-phase polymer system (Fig. 1). This technique produces fractions similar in purity and yield to those obtained from sucrose gradients.[23] Although the ingredients of the two-phase mixture are somewhat expensive, the method avoids the time-consuming use of sucrose gradients and allows one to handle large quantities of material (~ 120 g wet weight of liver) at one time, even if a zonal rotor is not available. Since the only information available as to the appropriateness of this method in the preparation of plasma membranes for gap junction isolation applies specifically to rat liver plasma membrane,[18] caution should be used in adapting the technique to other tissues and species.

[21] D. Neville, *J. Biophys. Biochem. Cytol.* **8,** 413 (1960).

[22] G. Zampighi and P. N. T. Unwin, *J. Mol. Biol.* **135,** 451 (1979).

[23] L. Lesko, M. Doulin, G. V. Marinetti, and J. D. Hare, *Biochim. Biophys. Acta* **311,** 173 (1973).

Methodology—The Two-Phase Method (Fig. 1). Twenty-five young adult rats (~200 g body weight—if they are much older, contamination with connective tissue becomes a problem) are sacrificed by cervical dislocation. After cutting the vena cava above the liver, each liver (~8 g wet weight) is perfused by injecting 3–5 ml of 37° PB through the spleen. Perfusion is achieved through the portal system. The perfused liver is excised and placed in ice-cold IB within 30 sec of death. All subsequent steps are performed on ice. Each liver is homogenized in 100 ml of IB with a Tissuemizer (Tekmar Ultra-Turrax, SDT-182 EN) at maximum power for 5 sec. This step can also be achieved in a Dounce homogenizer using 25 strokes of a loose-fitting pestle after first dicing the liver with scissors.

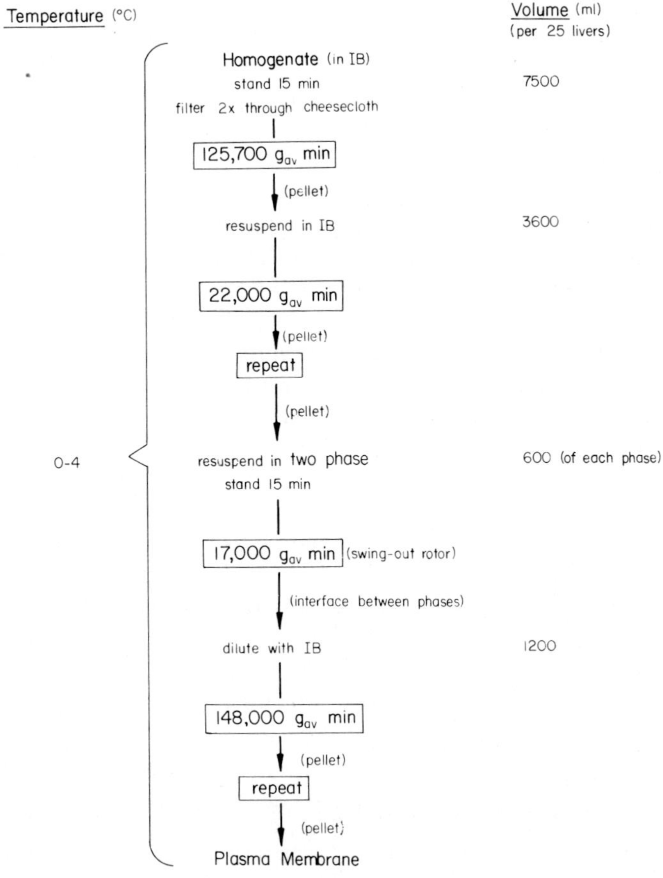

FIG. 1. Protocol for the isolation of plasma membrane by the two-phase method[23] from rat liver. For details, refer to the text.

Homogenates of six livers are pooled, diluted to 1800 ml with IB, and placed on ice for 10–15 min to allow nucleoprotein to precipitate. The homogenate is then filtered twice through four layers of cheesecloth, and the filtrate is centrifuged at 5000 rpm for 30 min (Sorvall HG-4L rotor; RC-3 centrifuge: 125,700 g_{av} min). The pooled pellets are resuspended in 3600 ml of IB with vigorous shaking and centrifuged twice at 3000 rpm for 15 min (HG-4L rotor: 22,000 g_{av} min). The supernatant, containing most of the mitochondria and soluble proteins, is discarded each time. After the second centrifugation at this speed, care must be taken to aspirate *all* of the supernatant, or the subsequent two-phase separation may fail. (If necessary, the pellets obtained after the second spin can be resuspended and compacted at higher g to allow for a more complete aspiration of the supernatant.) These pellets are resuspended by vigorous shaking in 600 ml of each phase of the two-phase system and distributed among four 1-liter bottles (150 ml of each phase per bottle). After shaking, the bottles are allowed to stand for 15 min and then centrifuged for 15 min at 2600 rpm (HG-4L rotor: 17,000 g_{av} min), after which most of the nuclei will be in the pellets. The supernatants containing the separated phases and the plasma membrane-rich interfaces are poured into two 1-liter bottles, shaken vigorously, and recentrifuged as above. The "carpet-like" interface is harvested by aspiration, diluted to 1200 ml with ice-cold IB, and spun down at 10,000 rpm for 15 min (Sorvall GSA rotor; RC-5B centrifuge: 148,000 g_{av} min). This centrifugation is repeated once more out of a volume of 400 ml of IB to yield a plasma membrane fraction. This can be refrigerated overnight for the subsequent isolation of gap junctions.

2. Isolation of "Native" Gap Junctions (Fig. 2)

Solutions. The volume is that required for 25 livers; i.e., 200 g of tissue.

Bicarbonate buffer (BB), 600 ml: 2 mM $NaHCO_3$, pH 7.4 (4°)
NaCl, 0.1 M in BB (4°), 100 ml
Sarkosyl, 1.1% in BB (RT), 50 ml
Na_2CO_3, 0.1 M (4°C), 5 ml
Sucrose, 34% (w/v) in 1 M urea, 2 mM BB, 5 ml
Sucrose, 40% (w/v) in 1 M urea, 2 mM BB, 15 ml
Sucrose, 77% (w/v) in 1 M urea, 2 mM BB, 15 ml
Sucrose, 81% (w/v) in 1.5 M urea ~ 0.1% Sarkosyl, 2 mM BB, 10 ml

Methodology. The first highly purified fractions of gap junctions from liver were obtained by the treatment of plasma membrane fractions with enzymes[24] (e.g., collagenase, hyaluronidase). It soon became apparent that,

[24] D. A. Goodenough, *J. Cell Biol.* **68,** 220 (1976).

although this treatment had little effect on the gap junction's morphological appearance, it did cause partial cleavage of the gap junction protein.[25] While such preparations have their uses (see Section B,3), most current protocols seek to minimize proteolysis in an attempt to isolate the "native" or undegraded protein. In addition to omitting specific proteolysis treatments from the protocol, the inclusion of protease inhibitors, notably 0.5 m*M* phenylmethylsulfonyl fluoride (PMSF), in the original perfusion buffer and all subsequent steps (except the two-phase separations) seems to be useful in preventing cleavage of the junctional protein by endogenous proteases during the isolation procedure.[12]

The procedure we have adopted for the isolation of gap junctions from the plasma membrane fraction (Fig. 2) is very similar to that of Hertzberg and Gilula,[10] but it contains modifications that result in an increased yield without an associated loss in purity.[12] Unless otherwise noted, all solutions are at 4°. The plasma membrane pellets are suspended in 100 ml of 0.1 *M* NaCl in BB with vigorous shaking or use of a Dounce homogenizer. The pellets are allowed to stand on ice for 15 min to release peripheral proteins adhering to the membranes. The salt-washed membranes are pelleted at 12,000 rpm for 15 min (Sorvall SS-34 rotor: RC-5B centrifuge: 168,000 g_{av} min); they are then resuspended in BB, and the centrifugation is repeated. After resuspension in 50 ml of BB at room temperature (RT) by a few strokes of the pestle of a Dounce homogenizer, an equal volume of 1.1% (w/v) Sarkosyl (NL 97) in BB (RT) is added. The mixture is stirred for 40 min, (RT) during which time 15-ml aliquots are removed and sonicated with a microtip for 3–4 sec at setting No. 4 on a Branson sonicator (S-125). The detergent-insoluble material is pelleted at 20,000 rpm for 15 min (SS-34 rotor: 470,000 g_{av} min), washed in ice-cold BB, and recentrifuged. The pellets are next resuspended in 4 ml of 0.1 *M* Na_2CO_3 (pH 11) by brief sonication in order to solubilize the substantial amount of urate oxidase copurifying with the junctions to this point. After exactly 15 min at 4° (longer times cause aggregation of gap junctions with copurifying contaminants), the insoluble material, which includes the gap junctions, is collected by centrifugation at 20,000 rpm for 15 min (SS-34 rotor). This pellet is washed once by resuspension in BB and centrifugation as above. For loading on sucrose gradients, the washed pellets are homogenized by brief sonication into 3 ml of BB to which is added 1 ml of 1.1% Sarkosyl and 8 ml of 81.1% (w/v) sucrose in 1.5 *M* urea and BB. Three sucrose gradients are then poured, each containing 4 ml of 77.2% (w/v) sucrose, 4 ml of sample (final concentrations of 54%, w/v, sucrose, 1 *M* urea, and 0.09% Sarkosyl), 4 ml of 40% (w/v) sucrose, and 1 ml of 34% (w/v) sucrose. All sucrose solutions are made in 1 *M* urea in BB. After centrifugation at 35,000 rpm for at least

[25] J. Duguid and J.-P. Revel, *Cold Spring Harbor Symp. Quant. Biol.* **40,** 45 (1976).

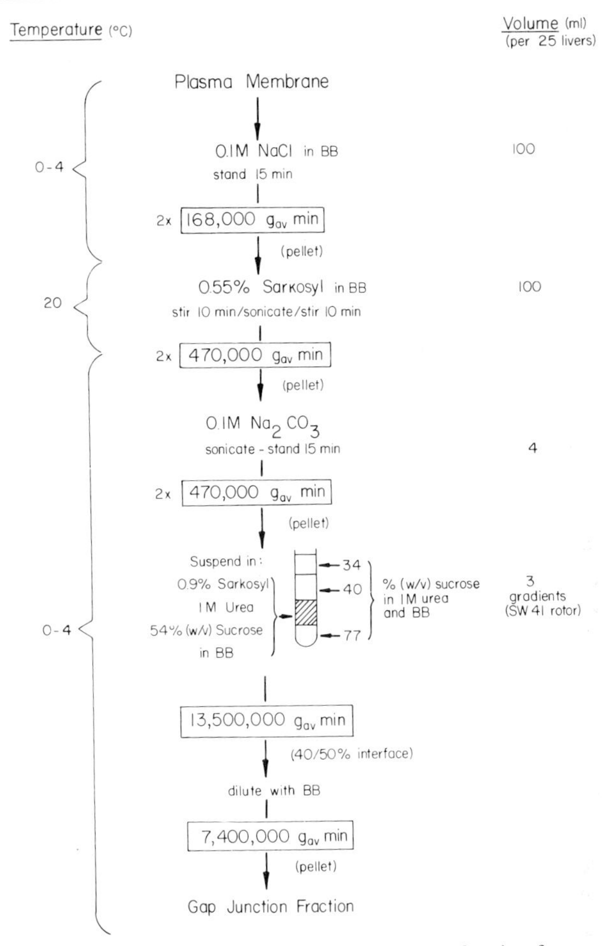

FIG. 2. Protocol for the isolation of a "native" gap junction fraction from rat liver plasma membrane. For details, refer to the text.

1.5 hr at 4° (Beckman SW-41 rotor; L3-50 ultracentrifuge: 13,500,000 g_{av} min), the $\frac{40}{54}$% (w/v) sucrose interfaces are harvested and diluted with BB. The final gap junction fraction is collected by centrifugation at 40,000 rpm for 1 hr (Beckman 42.1 rotor: 7,400,000 g_{av} min). The Sarkosyl and urea present in the sucrose gradient seem to reduce the aggregation of gap junctions with themselves and copurifying contaminants, such as collagen and other fine fibrous material. Loading the sample in the 54% (w/v) sucrose layer also aids in this respect, since the junctions float upward and the denser contaminants sediment to the lowest interface. This avoids the nonspecific trapping of material that occurs when it is spun through the concentrated blanket of material that accumulates at each interface. The interface at which we collect our gap junction fraction (between layers of densities 1.16 and 1.121) is consistent with the reported density of gap junctions on continuous sucrose gradients (i.e., 1.165).[10]

The final junctional fraction can be stored as a pellet at −20° for months without detectable structural or biochemical changes, especially if PMSF is included in the final centrifugation. The fraction shows only small amounts of contaminating amorphous material and collagen (see Section C for a more detailed evaluation) and contains 150–300 μg of gap junction protein (M_r 28,000) which represents a yield of 1–2 μg of junction protein from every gram wet weight of liver. The characteristics of the M_r 28,000 protein and the other components of this fraction are discussed in Section D.

3. Isolation of "Enzyme-Treated" Gap Junctions (Fig. 3)

Solutions. The volume is that required for 25 livers, i.e., 200 g of tissue.

Enzyme buffer (EB), 100 ml: 50 m*M* Tris-HCl, 5 m*M* $CaCl_2$, pH 7.4 (37°)
Bicarbonate buffer (BB), 300 ml: 2 m*M* $NaHCO_3$, pH 7.4 (4°)
Sarkosyl, 1.1% in BB (RT), 75 ml
Sucrose, 32% (w/v) in BB (4°), 24 ml
Sucrose, 54% (w/v) in BB (4°), 24 ml
Triton X-100 wash (for rapid isolation procedure only): 5% Triton X-100, 5 m*M* Tris-acetate, pH 7.4 (RT), ~ 10 ml

Methodology—Standard Isolation Procedure. As mentioned above, gap junctions have previously been isolated from membranes with the aid of proteases, the rationale being that the gap junction ultrastructure appears unscathed even after extensive proteolysis. However, it is now recognized that the intactness of the junctional ultrastructure is not reflected in its major constituent protein (M_r 28,000), which is degraded to two fragments of M_r 10,000[12] (representing 20,000 of the original 28,000 daltons) upon exposure to proteases, specifically trypsin. These polypeptides can be of considerable use in studies on the arrangement of the protein in the lipid bilayer. This method of isolating junctions also has one other major advantage. It produces highly enriched gap-junction fractions from a variety of species whereas the "native" isolation procedure often produces fractions badly contaminated with fibrous material (e.g., collagen) when applied to species other than rat or mouse.

The procedure described here (Fig. 3) is similar to that of Finbow *et al.*[18] The plasma membrane fraction (Section B,1) is homogenized using a Dounce homogenizer with a loose-fitting pestle, into 70 ml of EB, and then 10 ml of a 1.6 mg/ml collagenase (Worthington) solution are added. After gentle agitation at 37° for 25 min, 10 ml of 0.6 mg/ml trypsin (Sigma, type

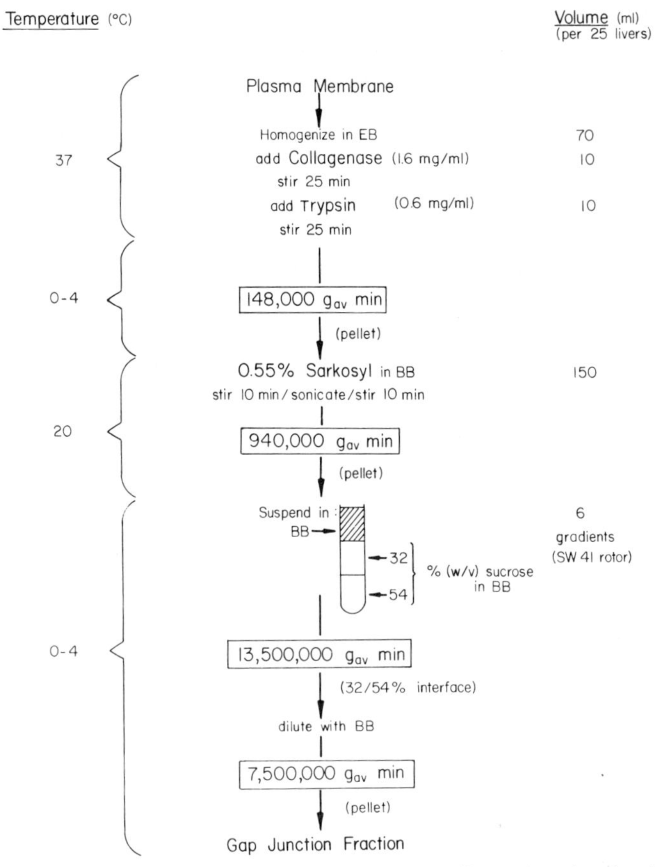

FIG. 3. Protocol for the isolation of an "enzyme-treated" gap junction fraction by the standard protocol from rat liver plasma membrane. For details, refer to the text.

XI) are added, and the solution is shaken gently for an additional 25 min. The enzyme-resistant pellet is collected after centrifugation at 10,000 rpm for 15 min (Sorvall GSA rotor; RC-5B centrifuge: 148,000 g_{av} min) and then treated with Sarkosyl in a total volume of 150 ml as described in Section B,1). A spin of 20,000 rpm for 30 min (Sorvall SS-34 rotor: 940,000 g_{av} min) is used to collect the detergent-insoluble material, which is then resuspended in 30 ml of ice-cold BB and layered on top of six discontinuous gradients containing 4 ml of 32% (w/v) sucrose and 4 ml of 54% (w/v) sucrose in BB. The $\frac{32}{54}$% (w/v) sucrose interface is harvested after centrifugation for at least 1.5 hr at 35,000 rpm (Beckman SW-41 rotor; L3-50 ultracentrifuge: 13,500,000 g_{av} min) and collected after dilution with BB by centrifugation at 40,000 rpm for 1 hr (Beckman, 42.1 rotor: 7,500,000 g_{av} min). This gap junctional fraction, which can also be stored for lengthy periods of time at $-20°$ with no apparent ill effects, looks morphologically very similar to the "native" gap junction preparations, although the yield of junctional protein is usually less (0.5 μg per gram wet weight of liver) and the major (and virtually only) band detectable by Coomassie staining after sodium dodecyl sulfate–polyacrylamide gel electrophoresis (SDS–PAGE) has an M_r of 10,000.

Methodology—Rapid Isolation Procedure (Fig. 4). Since the majority of nonmembranous, but insoluble, components that tend to copurify with the junctional fractions are readily digested by proteases to soluble peptides, a rigorous plasma membrane isolation is not a necessary first step in the isolation of gap junctions when proteases are employed. Consequently, the isolation protocol described in Section B,1 and the first part of this section can be considerably streamlined without great sacrifice in the purity of the final fraction. The advantages of this technique are severalfold: (*a*) the isolation procedure takes only approximately 6 hr (*b*) analyzable quantities (>0.5 μg of protein) of gap junctions can be obtained from a single animal; and (*c*) handling and transfers of the material are minimized. The entire protocol can be performed in only three centrifuge bottles in a total of five centrifugation steps, a great convenience when radioisotopes are employed.[26]

Each rat is sacrificed by cervical dislocation, and the liver is excised and placed in 200 ml of ice-cold IB in a 250-ml centrifuge bottle, where it is homogenized with a Tissuemizer (Tekmar, Ultra-Turrax, SDT-182EN) at maximum power for 6 sec. Homogenates are placed on ice for 15 min before the addition of solid sucrose to yield a final density of 1.16. Centrifugation at 1500 rpm for 5 min (Sorvall HG-4L rotor; RC-3 centrifuge: 2000 g_{av} min) pellets most of the unbroken cells, nuclei, connective tissue, and coagulated nucleoprotein, while leaving the membranes suspended. The supernatant is collected and spun at 10,000 rpm for 20 min (Sorvall SS-34 rotor; RC-5B centrifuge: 260,000 g_{av} min) to yield the crude plasma membrane fraction. This is then treated with 3.2 mg of collagenase (Worthington, type IV) in 40 ml of EB containing 5 mM $MgCl_2$. After stirring for 15 min at 37°, 1 mg of trypsin (Sigma, type XI) is added for an additional 15-min incubation. Upon dilution with 200 ml of ice-cold IB, the protease-treated plasma membrane is centrifuged at 10,000 rpm for 15 min (SS-34 rotor: 190,000 g_{av} min). This pellet is then homogenized and stirred for 15 min in 35 ml of 0.55% Sarkosyl with 10% urea in BB at room temperature to solubilize all the nonjunctional membrane and the small proteolysis products of the previous step, which are released after membrane solubilization. The urea serves to minimize nonspecific aggregation. The remaining

[26] S. B. Yancey, B. J. Nicholson, L. Grim, and J.-P. Revel, *J. Supramol. Struct.* **16,** 221 (1981).

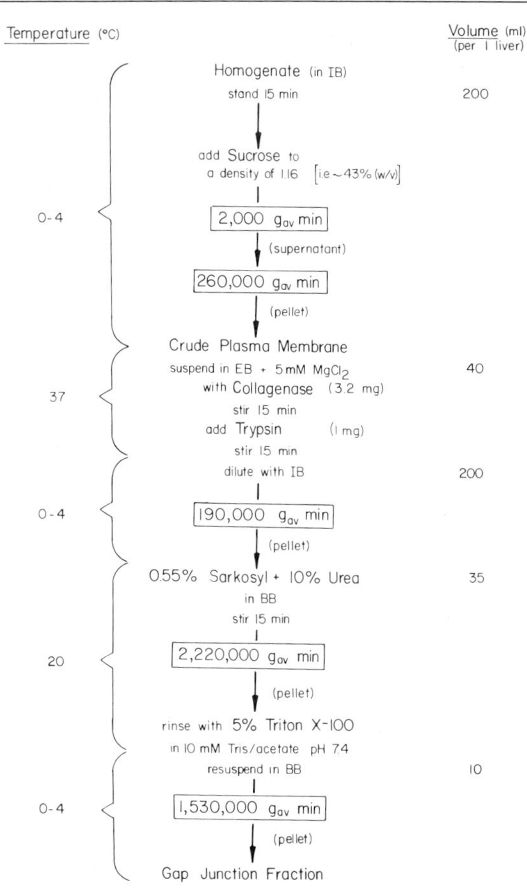

FIG. 4. Protocol for the isolation of an "enzyme-treated" gap junction fraction by a rapid protocol from rat liver. For details, refer to the text.

insoluble structures are collected by centrifugation at 20,000 rpm for 1 hr (Sorvall SS-34 rotor: 2,220,000 g_{av} min). This pellet and the walls of the centrifuge tube are gently rinsed with 5% Triton X-100 in 10 mM Tris–acetate buffer, pH 7.4, to remove excess lipid contaminants. The final gap junction fraction is obtained after the pellet is washed by resuspension in 10 ml of BB and the last centrifugation is repeated, but in a Ty 65 rotor in a Beckman L3-50 centrifuge.

When examined by SDS–PAGE, this final fraction displays a pattern similar to that seen in fractions obtained by the methodology outlined in Section B,1 and the first part of this section (see also Section D below), with a single major band at M_r 10,000. However, additional bands are normally seen near the top of the gel, which indicates an increased level of contaminants (e.g., collagen). The yield is slightly lower than that from the more complete protocols since, with much less starting material, the percentage of nonspecific losses due to absorption onto centrifuge tubes, etc., is much larger. Despite these losses, however, even livers from partially hepatectomized animals (liver wet weight, 2–3 g) yield sufficient protein to be readily analyzed on a microgel SDS–PAGE system (i.e., 0.5–1 μg).[27]

C. Criteria of Purity — Morphological Methods

1. Methodology

Until recently, the only way of estimating the purity of gap junction fractions was to examine them in the electron microscope after negative staining, thin sectioning, or occasionally freeze fracture of the final pellet. For thin sectioning[28] and freeze fracture,[28a] gap junction fractions are first pelleted and then prepared by the same procedures that are used for blocks of tissue. However, by far the most commonly used and simplest technique for examining gap junction fractions is to use negative stain.[29]

After thorough mixing, 1 μl or more of the fraction to be studied is deposited on a dental wax plate and mixed with an equal volume of phosphotungstic acid (PTA) neutralized to pH 7.4. The final concentration of the phosphotungstate is of the order of 0.5%, but should be adjusted upward or downward to give the best results after examination of a sample grid in the electron microscope. A carbon-reinforced Formvar or celloidin-coated grid, held in fine jeweler's tweezers closed by rubber band or paper clip, is touched to the drop. The sample is allowed to stand, drop side up, so that the suspended material can settle on and bind to the surface of the grid. After a few minutes, the excess fluid is removed with a wick (a triangular wedge of Whatman No. 1 filter paper) touched to the edge of the grid. Care must be taken to remove material that might have been aspirated between the tines of the tweezers by capillary action. The grids are allowed to dry thoroughly (5 or 10 min at least) and then examined in an electron microscope operating at 80 kV. The entire grid is surveyed at low magnification, and areas where the junctions are well spread are chosen for further examination and photography. A magnification of 13,000 to 15,000 is usually needed to detect connexons. If the material is very poorly distributed on the surface of the grid, heavily clumped in some areas while others are com-

[27] S. B. Yancey, personal communication, 1980.

[28] A. M. Glauert, "Practical Methods in Electron Microscopy," Vol. 3, Part I. North-Holland Publ., Amsterdam, 1975.

[28a] J. H. Willison and A. J. Rowe, "Practical Methods in Electron Microscopy," Vol. 8. North-Holland Publ., Amsterdam, 1980.

[29] R. J. Haschemeyer and R. J. Myers, *in* "Principles and Techniques of Electron Microscopy" (M. A. Hayat, ed), Vol. 2, p. 99. Van Nostrand–Reinhold, New York, 1970.

pletely bare, fresh grids should be used. Old ones can be made less hydrophobic again by placing them in the refrigerator overnight or by dipping them in 0.1% serum albumin, which is then drained off before addition of the material to be negatively stained.

Superior results in terms of clarity and definition of the images have been published with junctions negatively stained with uranyl salts.[30] The use of uranium salts can be more difficult since they precipitate as hydroxides above pH 6. A possible routine for staining with uranyl acetate or formate would be to deposit a drop of the sample to be examined on the grid and allow it to settle as was described for PTA, but in the absence of any negative stain. The freshly dried sample can then be washed with solutions of appropriate pH (a similar routine is used to get rid of excess sucrose). Washing can consist of dipping the grid in a beaker of appropriate solutions (distilled water, or very dilute buffers, or buffers made from volatile mixtures such as ammonium acetate–acetic acid), draining the excess fluid on the grid and between the tines by touching the edge of the grid to a filter paper, and then applying the appropriate negative stain as a droplet. After a few minutes the excess stain is drained off with a filter paper wick, and the grid is then ready to be examined in the electron microscope.

2. Appearance of Gap Junction Fractions

In the "native" gap junction fractions, the junctions appear as irregularly shaped flat sheets showing the typical double-membrane profile and closely packed, hexagonally arranged particles (connexons) in *en face* views. Occasionally, small gaps in the array of connexons suggest that some portions of the junction may have been lost, perhaps as a result of partial solubilization during the detergent extraction.[31] Of the nonjunctional materials present, the most ubiquitous are clumps of fine fibrous material that appear alone and associated with the surfaces of gap junctions. As yet, the origin of this material has not be identified and it cannot be consistently associated with any specific protein present in the gap junction fraction. Other structures, seen more rarely in the junction fraction, include strands of collagen[12] (these are much more abundant at the $\frac{54}{77}$% (w/v) sucrose interface of the final sucrose gradient) and vesicles of nonjunctional membrane.[18] The "enzyme-treated" gap junctions differ from those in the "native" fraction in that they most frequently appear as closed vesicles or curved sheets.[24] The contaminating material seen in these fractions is similar to that described above with the exception that collagen strands are never seen.

[30] W. J. Carsen, P. M. Heidger, J. C. Herr, and D. A. Goodenough, *J. Cell Biol.* **71,** 333 (1976).
[31] G. Zampighi and J. D. Robertson, *J. Cell Biol.* **56,** 92 (1973).

3. Limitations in Estimating Purity

In general, morphological examination of the purified fractions provides some guidelines as to the nature and degree of contamination in junctional fractions and has played an important part in suggesting improved strategies for the isolation of gap junctions. However, as the purity of junctional fractions has steadily improved, the need for a quantitative estimate of the percentage of the final fraction that is composed of gap junctions has become greater. Electron microscopic examination presents problems in maintaining unbiased sampling both in the selection of a portion of the final pellet for examination and in the selection of the region of an electron microscope grid to be used for analysis. Furthermore, if one attempts to quantitate by measuring the surface area of the grid occupied by junctions, compared to that occupied by other material, the density of protein in each structure is ignored. Particularly troublesome has been the association of an M_r 34,000 protein with gap junctional fractions.[25,32] This is likely to have been due to the presence of small crystals of urate oxidase in the preparation, which are readily overlooked in negatively stained samples. In addition, not all structures visible in the electron microscope can be correlated with the presence of specific proteins recognizable after SDS–PAGE of the fractions (e.g., the fine fibrous material referred to above). This could suggest that the structure contains no protein or that it represents a different form (denatured?) of the gap junctions themselves. The possibility also exists that some of the contaminants in a fraction may not be visualized in preparations examined in the electron microscope.

As a result of these difficulties and the lack of an assay for the gap junction protein other than the presence of morphologically intact gap junctions, it has been difficult to determine reliably the purity of junctional fractions. Although this problem still exists in most instances, characterization of the protein components of liver gap junction fractions, specifically those of rat and mouse,[10–12] has enabled us to assess biochemically the purity of this fraction.

D. Criteria of Purity — A Biochemical Approach

1. Identification and Nature of the Gap Junction Protein

The identity of the protein components of the gap junction has been a major point of contention in the field over the last 10 years. Polypeptides of M_r 10,000–38,000 have been variously proposed as the major protein of gap junctions from rat or mouse liver.[10,11,24,32,33] The problem was the lack of

[32] J. C. Ehrhart and J. Chauveau, *FEBS Lett.* **78,** 295 (1977).

[33] J. G. Culvenor and W. H. Evans, *Biochem. J.* **168,** 475 (1977).

a direct assay for the gap junction protein and only limited success in the production of antibodies. Only in 1982 did preliminary reports of such an antibody appear.[34] The only course was to isolate fractions highly enriched for gap junctions by morphological criteria. The protein components of these fractions were then examined, usually by separation on an SDS-polyacrylamide gel, and the major species were identified as gap junctional proteins. This approach relies heavily on the efficiency of morphological techniques in detecting nonjunctional contaminants, the pitfalls of which have been discussed in Section C,2. Furthermore, the SDS-polyacrylamide gel profiles were complex, not only as a consequence of nonjunctional contaminants, but also because of the susceptibility of the junctional protein to proteolysis and its tendency to aggregate in SDS when heated.[11,12]

A particularly useful tool in identifying junctional protein has been the two-dimensional peptide mapping originally described by Elder *et al.*[35] and modified by Takemoto *et al.*[36] After separation by SDS-PAGE, proteins, labeled either *in vivo* with ^{35}S, or *in vitro* with ^{125}I, are digested exhaustively with a highly specific protease (e.g., trypsin or chymotrypsin). These peptides are then separated in two dimensions on a thin-layer cellulose plate by electrophoresis and chromatography to yield a pattern of labeled peptides (a "fingerprint") that is unique for each protein. By analyzing each of the proteins of the final junctional fraction in this way, relationships between them can be identified. This, in conjunction with other data, has allowed us to conclude that the rat liver gap junction is composed of a single major polypeptide of M_r 28,000 ± 2000.[12]

In "native" gap junction fractions isolated in the presence of protease inhibitors (e.g., 0.5 mM phenylmethyl sulfonyl fluoride) and solubilized in SDS under conditions that minimize aggregation of the junctional protein (i.e., 30-45 min at room temperature in Laemmli solubilization buffer[37]), the M_r 28,000 protein is the only major component detected by SDS-PAGE. The minor components are more evident when samples are overloaded or the fractions are prepared in the absence of specific protease inhibitors. They have been shown[11,12] to result from (*a*) contaminating collagen, which can be identified by peptide mapping (i.e., material remaining at the top of the running gel); (*b*) residual urate oxidase not fully solubilized by the treatment with Na_2CO_3 (i.e., protein of M_r 34,000); (*c*) an unidentified, nonjunctional contaminant that shows a unique fingerprint and is enriched in the denser portions of the final sucrose gradient, where gap

[34] O. Traub and K. Willecke, *Biochem. Biophys. Res. Comm.* **109,** 895 (1982).
[35] J. H. Elder, R. A. Pickett, J. Hampton, and R. A. Lerner, *J. Biol. Chem.* **252,** 6510 (1977).
[36] L. J. Takemoto, J. S. Hansen, and J. Horwitz, *Comp. Biochem. Physiol. B,* **68B** 101 (1981).
[37] U. K. Laemmli, *Nature (London)* **227,** 680 (1970).

junctions are rarely found (i.e., protein of M_r 38,000); (*d*) aggregation of the M_r 28,000 junctional protein in SDS as a result of heating, sitting for too long in SDS before loading on the gel, or too high a concentration of protein in the sample (e.g., highly aggregated material at the top of the running gel and a dimer at M_r 50,000); and (*e*) partial proteolysis of the M_r 28,000 protein (e.g., polypeptides of M_r 26,000 and 24,000). A further minor component of M_r 21,000 (only 4% of the total protein of a rat liver "native" gap junction fraction,[12] but a more major component of mouse gap junction fractions[11]) can also be shown to be related to the M_r 28,000 protein by peptide mapping; yet unlike the other polypeptides, it cannot be explained as a simple serine protease degradation product of the major protein. Whether it is produced by the action of another protease (e.g., the sulfhydryl protease cathepsins of liver) or is a second, but minor, protein component of gap junctions related to the major component remains to be established.

If trypsin is specifically added to a "native" gap junction fraction under conditions in which the overall ultrastructure of the junctional complex remains intact, the final product is two polypeptides, each of M_r 10,000.[12] Peptide mapping has shown that these are the same polypeptides that appear as a single diffuse band of M_r 10,000 on SDS–polyacrylamide gels of "enzyme-treated" gap junction fractions. When chymotrypsin is used instead of trypsin, polypeptides of M_r 14,000 and 10,000 are generated.[38] A consideration of the size of these proteases ($\sim$ 5 nm in diameter[39]) and the dimensions of the gap junction structure suggests that the polypeptides that are protected from further proteolysis represent the portions of the gap junction protein that are embedded in the membrane or located in the 2-nm gap between membranes. Given this premise, it can be inferred from the above observations that the gap junction protein traverses the membrane more than once, as might be expected of a protein forming the walls of a transmembrane channel. N-terminal sequence analysis has further demonstrated that at least the C-terminal 4000 daltons of the protein is exposed at the cytoplasmic surface of the junction.[12]

2. Protein Composition as a Criterion for Purity

Since the identity of the gap junction protein and its behavior under varying conditions has now been defined, it is possible, at least in the case of rat liver, to determine quantitatively both the purity and, consequently, the yield of the various gap junction fractions based solely on a consideration of their protein composition. This can be analyzed by solubilizing a sample of the fraction under consideration in Laemmli solubilization buffer[37] for 30–45 min at room temperature, separating the various protein compo-

[38] Personal observations, 1980.

[39] R. M. Stroud, L. M. Kay, and R. E. Dickerson, *J. Mol. Biol.* **83,** 185 (1974).

nents by SDS–PAGE and staining the gel with Coomassie Blue R-250 (most protocols generally published for staining SDS polyacrylamide gels will suffice). Quantitation of the amount of each polypeptide in the sample can be achieved by scanning the appropriate lane of the gel with a densitometer and integrating the area under each peak. In our laboratory this was achieved by a Joyce Loebl densitometer and a digitizing tablet interfaced to a Tektronix minicomputer (4052), respectively. The total area under all the peaks corresponding to junctional polypeptides [as defined in the preceding section under (*d*) and (*e*)] can then be divided by the total area under the densitometer scan of the gel to obtain an estimate of the purity of the fraction. An estimate of the yield of junctional protein in the fraction requires, in addition, some absolute estimate of protein present. This can be obtained in one of two ways. The protein content of the whole fraction can be determined independently by amino acid analysis or by a Lowry assay,[40] although the latter may be difficult because the junctions are poorly soluble in anything but SDS, which is known to interfere with Lowry assays. This quantity can then be adjusted to include only gap junctional protein by correcting for the purity percentage determined as described above. Alternatively, the total area of the peaks of gap junctional polypeptides determined from the scan of the stained polyacrylamide gel can be standardized against the areas of the peaks in densitometer scans of adjacent lanes on the same gel in which known amounts of various standard proteins have been loaded.

Both the determination of purity and the determination of yield are subject to certain inherent errors, based on limitations of Coomassie staining as a quantitative assay for proteins. In any given gel system it is imperative to determine over what range and under what conditions the Coomassie staining remains linear with respect to protein concentration. However, even within this range, different proteins bind the dye with different efficiencies, so a determination of absolute protein concentration or even relative amounts of two different, unknown proteins can entail substantial errors, especially in the case of glycoproteins. Fortunately, no carbohydrate has been found to be associated with the gap junctional protein.[10,11] An independent check on the values obtained from Coomassie-stained gels can be obtained by radioactively labeling the fraction *in vitro* with ^{125}I, after solubilization in SDS but before separation by PAGE. Autoradiograms of the gel can be scanned in the same way as Coomassie-stained gels. In this instance, differences in the labeling efficiencies of different proteins can be corrected for if their amino acid compositions are known, since iodine is known to label tyrosine residues, and under some conditions histidine, specifically. Figures calculated in our laboratory for the yield of junctional protein in, and purity of, gap junction fractions prepared

[40] O. H. Lowry, N. J. Rosebrough, A. L. Farr, and R. J. Randall, *J. Biol. Chem.* **193,** 265 (1951).

PURITY AND YIELD OF VARIOUS GAP JUNCTION (GJ) FRACTIONS

Isolation procedure	Purity[a] $\left(\frac{\text{GJ protein}}{\text{total protein}} \times 100\right)$	Yield[b] $\left(\frac{\mu\text{g GJ protein}}{\text{g wet weight liver}}\right)$
"Native" gap junction fraction	77 ± 9	0.8–2.0
"Enzyme-treated" gap junction fraction		
Standard protocol	81 ± 6	0.3–0.7
Rapid protocol	33 ± 5[c]	0.17–0.35[d]

[a] The means ± one standard error determined from several (four to nine) different isolations are shown.

[b] Based on an average weight of 8 g for a liver from a 200-g rat.

[c] This estimate does not take into consideration the substantial amount of undissolved material that remains in the tube after treatment with SDS and loading on the gel. The purity of the whole fraction could be lower.

[d] These numbers are based on fractions isolated from 1 liver whereas the two above are based on fractions from 25 livers. Consequently, the lower yield is the result of a higher percentage of nonspecific losses resulting from the adsorption of gap junctions to glassware, etc.

from 25 rat livers by each of the methods described in Section B are presented in the table.

3. Validity of Extrapolating Results to Other Systems

Gap junctions of different origins have many features in common. Only small differences in the size of the connexons or the width of the extracellular gap have been observed,[41] and often they could be ascribed to differences in preparative procedures. There are more differences in terms of packing, but this feature could be under physiological control.[8] Gap junctions can mediate metabolic cooperation or electrotonic coupling between cells in culture.[4] This has proved to be true even in cell combinations derived from physiologically distinct species,[42] which again suggests a kinship extending far and wide. The rather uniform molecular weights ascribed to gap junction proteins (M_r 26,000–34,000[10,11,19,36]) could also imply a high degree of conservation, but it is only recently that this idea has become testable by detailed analysis of the protein(s).

Peptide mapping of the liver gap junction proteins from several mammalian species (rat, mouse, calf, and rabbit) has shown them to be virtually indistinguishable, and preliminary "fingerprints" of the gap junction protein from chicken liver suggest that there is some conservation even between classes.[43] The main intrinsic protein (M_r 26,000) of eye lens fiber cell

[41] L. A. Staehelin, *Int. Rev. Cytol.* **39,** 191 (1974).

[42] M. L. Epstein and N. B. Gilula, *J. Cell Biol.* **75,** 769 (1977).

[43] B. J. Nicholson, L. J. Takemoto, M. W. Hunkapillar, L. E. Hood, and J.-P. Revel, *Cell* **32,** 967 (1983).

junctions, which are similar in appearance and in apparent properties to gap junctions in other tissues, is similarly conserved, with homology demonstrated by both peptide mapping[36] and immunology[44] in species as far apart as man and shark. However, when the gap junction proteins of different tissues are compared, a different story emerges. Major differences in the main intrinsic protein of eye lens and the gap junction protein of liver have now been demonstrated immunologically,[45] by peptide mapping,[43] and in partial sequences of the two proteins purified from rat[43] (the N-terminal 18% of the lens protein and 24% of the liver protein). Indeed, if any homology does exist between the two proteins, it is barely detectable by the methods used to date. We have obtained peptide mapping data that demonstrate that the major protein of gap junction fractions from heart is also very different from the major proteins of liver or lens.[45a] It is, therefore, quite possible that a whole family of tissue-specific junction proteins exist. While there may be homologies between them with respect to both sequence and structure, it is quite clear that results on the properties or biochemistry of gap junctions in one tissue should not, in general, be extrapolated to other systems.

No enzymic activity or even "substrate" specificity has yet been associated with the purified gap junction protein. This is not surprising, since the isolation procedures require treatments with detergents and/or other reagents likely to destroy any activity. The approaches evolved, however, may permit a functional analysis of the gap junction protein. As more becomes known of the structure of the protein, domains that might be involved in functions such as the modulation of junctional permeability, perhaps through specific binding of cations or calmodulin,[8,46,47] should become more clearly defined.

[44] D. Bock, J. Dockstrader, and J. Horwitz, *J. Cell Biol.* **92,** 213 (1982).
[45] E. L. Hertzberg, *In Vitro* **16,** 1057 (1980).
[45a] D. B. Gros, B. J. Nicholson, and J.-P. Revel, **Cell** (submitted).
[46] B. Rose and W. R. Loewenstein, *J. Membr. Biol.* **28,** 87 (1976).
[47] D. C. Spray, A. L. Harris, and M. V. Bennett, *Science* (*Washington, D.C.*) **211,** 712 (1981).

[47] Studies on the Biogenesis of Cell–Cell Channels

By GERHARD DAHL, RUDOLF WERNER, and ROOBIK AZARNIA

The cell–cell channel that connects cells in organized tissues has been the subject of a comprehensive review.[1] It is generally assumed that gap junctions, as observed under the electron microscope, represent cell–cell

[1] W. R. Loewenstein, *Physiol. Rev.* **61,** 829 (1981).

ISBN 0-12-181998-1

membrane channels. However, attempts to isolate the proteins involved in channel formation have given equivocal results so far, although a major component of liver gap junctions appears to exhibit similar properties in several laboratories.[2-4] One difficulty in isolating junctional proteins is that at present no assay other than morphological examination exists for monitoring the purification procedure. Without a functional assay, minor but functionally significant components may be overlooked.

For the study of biogenesis of cell–cell channels and analysis of its protein components, we have developed a complementation test based on the expression of foreign mRNA in channel-deficient cells.[5,6] The mRNA was isolated from uterine smooth muscle cells that were actively involved in new channel formation in response to estrogen treatment.[7] As a vehicle for transporting the mRNA into recipient cells, we used "artificial viruses." These were obtained by encapsulating mRNA in liposomes which, like viruses, interact with cell membranes by fusion (Fig. 1) and probably also to some extent by endocytosis.[8,9] It should be emphasized, however, that at this time we can only assume that the formation of new gap junctions in myometrium is due to the *de novo* synthesis of channel proteins. An alternative hypothesis would be that estrogen induces the synthesis of mRNA coding for a regulatory protein, a possibility that would be no less interesting.

Methodology

Solutions

Phenol: Saturate phenol with water and adjust the pH to 7.0; add 50 mg of 8-hydroxyquinoline per 100 ml and warm to 60° before using.

GT solution: Dissolve 50 g of guanidine thiocyanate (GT) in 80 ml of water; add 2.5 ml of 1 *M* sodium citrate, pH 7.0; adjust to pH 7.0; add 0.5 g of Sarkosyl NL-97 (Geigy); adjust volume to 95 ml; filter. Before use add 5 ml of 2-mercaptoethanol.

Sodium acetate, 1 *M*, pH 5.0

Potassium acetate; 2 *M*, pH 5.0

EDTA, 1 m*M*, pH 7.4

[2] D. Henderson, H. Eibl, and K. Weber, *J. Mol. Biol.* **132,** 193 (1979).

[3] E. L. Hertzberg and N. B. Gilula, *J. Biol. Chem.* **254,** 2138 (1979).

[4] M. Finbow, S. B. Yancey, R. Johnson, and J.-P. Revel, *Proc. Natl. Acad. Sci. U.S.A.* **77,** 970 (1980).

[5] G. Dahl, R. Azarnia, and R. Werner, *Nature* (*London*) **289,** 683 (1981).

[6] G. Dahl, R. Azarnia, and R. Werner, *In Vitro* **16,** 1068 (1980).

[7] G. Dahl, and W. Berger, *Cell Biol. Int. Rep.* **2,** 381 (1978).

[8] S. Dales, *Bacteriol. Rev.* **37,** 103 (1973).

[9] K. S. Matlin, H. Reggio, A. Helenius, and K. Simons, *J. Cell Biol.* **91,** 601 (1981).

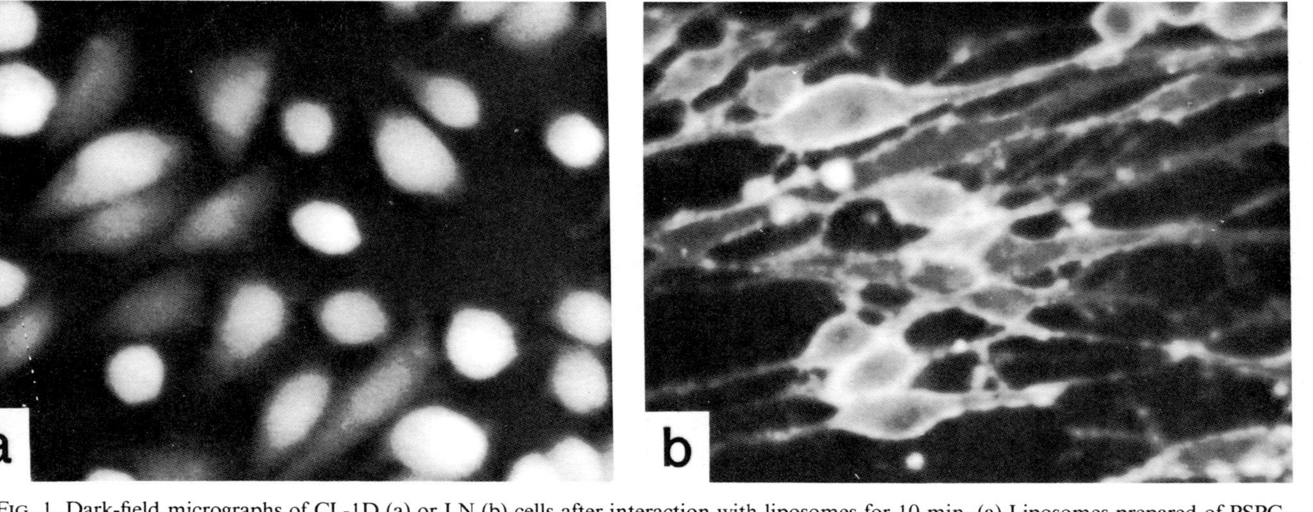

FIG. 1. Dark-field micrographs of CL-1D (a) or LN (b) cells after interaction with liposomes for 10 min. (a) Liposomes prepared of PSPC contained carboxyfluorescein (100 m*M*). The fluorescent tracer is found throughout the cells. The nuclear region is more heavily stained by the fluorescent tracer than the rest of the cytoplasm, as would be observed if the tracer were injected via a microelectrode into the cells. The cell density is lower than in cultures used for mRNA studies in order to show the cell outlines more clearly. (b) Liposomes prepared from phosphatidylserine and fluorescent 1-acyl-2-(*N*-4-nitrobenzo-2-oxa-1,3-diazole)aminocaproyl phosphatidylcholine containing buffer L(nonfluorescent).Fluorescence is restricted to cell boundaries. This indicates that the liposomal content appears intracellularly while the liposomal membranes remain associated with the membrane, presumably by liposome–cell membrane fusion. Some endocytotic uptake of liposomes probably occurs in parallel. Further studies, not shown here, revealed that liposomal contents are free to diffuse within the cytoplasm. [R. Azarnia and G. Dahl, *Fed. Proc., Fed. Am. Soc. Exp. Biol.* **39,** 1787 (1980)].

NaCl–sodium dodecyl sulfate (SDS): 0.5 *M* NaCl, 0.5% SDS, 10 m*M* HEPES, 1 m*M* EDTA, pH 7.4

NaCl solution: 0.5 *M* NaCl, 10 m*M* HEPES, 1 m*M* EDTA, pH 7.4

Buffer A: 25 m*M* sucrose, 1.5 m*M* MgCl, 50 m*M* NaCl, 25 m*M* HEPES, pH 7.2

Buffer L: 150 m*M* KCl, 10 m*M* NaCl, 5 m*M* HEPES, 1 m*M* EGTA, pH 7.4

All solutions, except phenol and GT solution, are treated with 0.2% diethyl pyrocarbonate (DEP) and autoclaved for 90 min to destroy the DEP.

Procedure (Fig. 2)

Step 1. Mature female virgin rats (Sprague–Dawley) are injected intramuscularly on three consecutive days with 5 mg of estrogen (Delestrogen, Squibb) in each hindleg per day. On day 4 the rats are anesthesized and decapitated.

Step 2. For the following procedures all instruments (scissors, forceps, scalpel) are kept in aqueous diethyl pyrocarbonate (0.2%) and rinsed with RNase-free H_2O prior to use. Uteri are excised and washed in ice-cold buffer A. The endometrium is removed by scraping at low temperature. After thorough rinsing, the myometrium is dried by blotting with filter paper and immersed in GT solution.

Step 3. About 15 myometria suspended in 50 ml of GT solution are homogenized (Tekmar, Ultra-Turrax), and the mixture is centrifuged at 5000 *g* for 10 min. After extraction of the pellet with another 20 ml of GT followed by centrifugation, the combined supernatants are adjusted to 80 ml with GT, heated to 60°, and mixed with 80 ml of hot aqueous phenol. After incubation for 10 min at 60°, the mixture is cooled in ice. Then 80 ml of sodium acetate is added, and the mixture is kept at 0° for 10 min, with

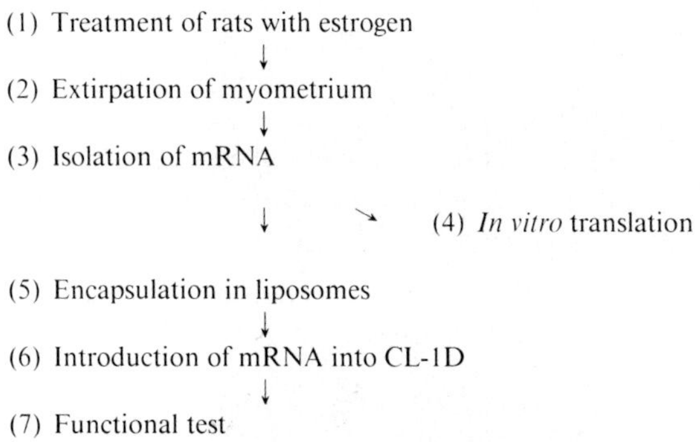

(1) Treatment of rats with estrogen
↓
(2) Extirpation of myometrium
↓
(3) Isolation of mRNA
↓ ↘ (4) *In vitro* translation
(5) Encapsulation in liposomes
↓
(6) Introduction of mRNA into CL-1D
↓
(7) Functional test

FIG. 2. Sequence of preparatory steps.

intermittent shaking. After the addition of 80 ml of chloroform and vigorous shaking, the mixture is centrifuged to separate the phases. The aqueous phase, containing all RNA, is removed and extracted with phenol for a second time. (No further addition of sodium acetate is required.) The final aqueous phase is extracted three times with chloroform and then mixed with 1.5 volumes (about 240 ml) of ethanol. After incubation at −20° overnight, the mixture is centrifuged at 12,000 *g* for 30 min, and the RNA precipitate is washed with ice-cold ethanol and then dried in a lyophilizer. The RNA is dissolved in about 2 ml of 1 m*M* EDTA heated to 70° for 1 min and then chilled in ice. The RNA solution is then adjusted to a concentration of 20 OD_{260} units/ml with NaCl-SDS (about 20 ml), and slowly passed three times through a small column (made from a Pasteur pipette) containing 0.5 g of oligo(dT)cellulose equilibrated with NaCl-SDS. The column is then washed with 10 ml of NaCl-SDS and 10 ml of NaCl, and finally eluted with EDTA. Poly(A)-containing RNA elutes within the first 4-5 ml and can be monitored spectroscopically. After the addition of 0.1 volume of potassium acetate, the mRNA is precipitated with 2 volumes of ethanol. All glassware used throughout this procedure must be heated to 200° to inactivate RNase. Disposable sterile plastic centrifuge tubes are used whenever possible. (Modified from Chirgwin *et al.*[10]; see also other chapters of this volume.)

Step 4. Biological activity of the mRNA is tested by translation in a rabbit reticulocyte lysate system (Bethesda Research Laboratories).[11] Incorporation of [^{35}S]methionine into protein is monitored by scintillation counting and by two-dimensional SDS-polyacrylamide gel electrophoresis (SDS-PAGE).[12]

Step 5. Encapsulation of mRNA into liposomes by reverse-phase evaporation is performed according to a method developed by Szoka and Papahadjopoulos.[13] Specifications: 12 mg of phosphatidylserine and 8 mg of phosphatidycholine (both from Avanti) are taken up in 0.5 ml of chloroform and 1 ml of diethyl ether. After addition of 0.5 ml of aqueous mRNA solution (20-40 μg/ml in buffer L), an emulsion is prepared by repeated pipetting with an Eppendorf pipette (1 ml). (Volumes can be scaled up or down.) The organic phase is removed from the milky white emulsion by rotary evaporation. This procedure results in the formation of large, mostly unilamellar liposomes. The liposome preparation is dialyzed against Ca- and Mg-free Hank's balanced salt solution.

[10] J. M. Chirgwin, A. E. Przybyla, P. J. MacDonald, and W. J. Rutter, *Biochemistry* **18,** 5294 (1979).

[11] H. R. B. Pelham and R. Jackson, *Eur. J. Biochem.* **67,** 247 (1976).

[12] P. H. O'Farrell, *J. Biol. Chem.* **250,** 4007 (1975).

[13] F. Szoka and D. Papahadjopoulos, *Proc. Natl. Acad. Sci. U.S.A.* **15,** 4194 (1978).

Step 6. The communication-defective CL-1D mouse cell lines[14] are grown in Dulbecco's minimum essential medium (DMEM) supplemented with 10% newborn calf serum. Cells are used at confluent state with a cell density of $\geq 2 \times 10^5$ cells/cm^2. Cells are repeatedly washed with serum-free DMEM medium. The liposome preparation is diluted 1:4 (v/v) with serum-free DMEM medium, and 1 ml of this suspension is added to a 9.5-cm^2 dish of CL-1D cells. The incubation period with liposomes is 1 hr at 37°. The liposomes are then removed, and the cells are washed thoroughly and incubated in normal culture medium at 37°.

Step 7. The presence of cell–cell communication is tested by measuring electrical coupling[18] between randomly selected (10 or more) cell pairs. (The very few multinucleated cells (<1%) that can be found after liposome treatment are presumably attributable to cell fusion and are excluded from this scoring procedure.) Two adjacent cells are impaled with a microelectrode each. A bridge circuit allows measurement of membrane potential and simultaneous current injection into either cell (Fig. 3). Current injection into cell 1 displaces its membrane potential. If the cell is connected to its neighbor via low-resistance pathways, part of the current will change the membrane potential of this neighboring cell as well. Usually hyperpolarizing square pulses of electrical current are used, since depolarization can result in artifactual closure of existing cell–cell channels.[19] The ratio of the membrane potentials in neighboring cell (V2) and in the injected cell (V1) is called the coupling ratio; it gives an estimate of the junctional resistance and, assuming uniform cell–cell channels, an estimate of the number of channels. The mRNA-induced electrical coupling observed in CL-1D cells usually exhibits a low coupling ratio (Fig. 3). The coupling incidence, defined as the percentage of cell pairs exhibiting electrical coupling (with any coupling ratio), usually is of the order of 50–70% (on rare occasions 100% coupling incidence can be observed).

[14] The lack of cell–cell coupling and of gap junctions for CL-1D cells has been documented.[15,16] During the course of this work we have learned, however, that there are certain conditions under which this situation is not given. Cell–cell coupling and gap junctions can be expressed in these cells at very low density, and their formation can be induced at any cell density by the addition of exogenous dibutyryl cyclic adenosine monophosphate.[17] The latter requires protein synthesis. There is no positive indication that in the noncoupling state CL-1D cells contain "silent" channels.

[15] R. Azarnia and W. R. Loewenstein, *J. Membr. Biol.* **23,** 1 (1977).

[16] W. J. Larsen, R. Azarnia, and W. R. Loewenstein, *J. Membr. Biol.* **34,** 39 (1977).

[17] R. Azarnia, G. Dahl. and W. R. Loewenstein, *J. Membr. Biol.* **63,** 133 (1981).

[18] For further detail of the technique and interpretation, see S. J. Socolar and W. R. Loewenstein, *in* "Methods in Membrane Biology" (E. D. Korn, ed.), pp. 123–179. Plenum, New York, 1979.

[19] A. L. Obaid and B. Rose, *Biophys. J.* **33,** 106a (1981).

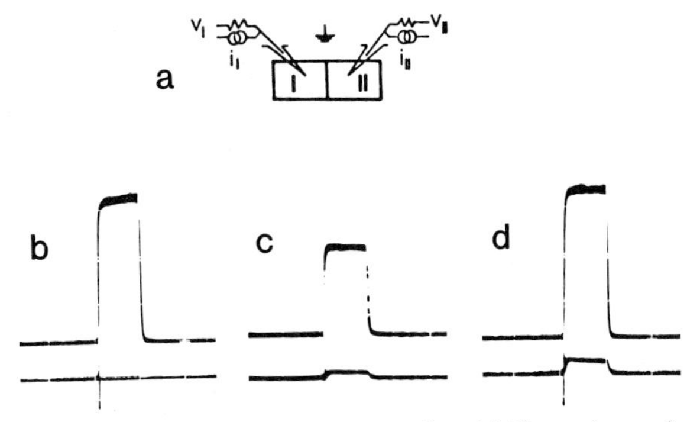

FIG. 3. Measurements of electrical cell–cell coupling. (a) Electrodes are impaled into two adhacent cells. A bridge circuit allows current injection and recording of membrane potential simultaneously. Routinely, the current is sequentially injected into both cell 1 and cell 2 (I, II in figure). For simplicity, the effect of current injection into cell 1 is shown only in the following oscilloscope traces. (b) Membrane potential in control cells 6 hr after treatment wth liposomes containing buffer only. The hyperpolarization is restricted to the membrane potential of the current-injected cell. (c, d) Membrane potentials in cells 6 hr after treatment with liposomes containing the mRNA preparation. Both in the current-injected cell and in the neighboring cell, the membrane potential is displaced by the current pulse. This electrical coupling of the cells usually is of low coupling ratio ($V_2 : V_1$) as seen in (b). Higher coupling ratios, as shown in (c), are rarely observed. Coupling ratios found in normally coupling competent cell types are even higher than those shown in (c). On the average 50–70% of all cell pairs tested show electrical coupling (coupling incidence) if an acceptance criterion of 1-mV membrane potential change in the non-current-injected cell is used.

Comments

The observed induction of cell–cell coupling in CL-1D cells appears to be mediated by mRNA. This conclusion is supported by a series of control experiments including RNase treatment of the mRNA, inhibition by cycloheximide, and study of concentration dependence.[5,6]

The possibility that the observed cell–cell coupling is due to mechanisms other than membrane channels, e.g., pseudocoupling or cell fusion, is remote. Pseudocoupling seems to be ruled out since the anatomical situation providing high external electrical resistance is not given in cultured cells[18] and has not been observed under these conditions. Cell fusion induced by interaction with liposomes has been ruled out by control experiments.

The hypothesis that the mRNA preparation isolated from estrogen-treated myometrium includes sequences that code for channel proteins is compatible with the following observation. The mRNA-induced cell–cell coupling is transient and disappears after about 24 hr.[5,6] An almost identical time course of appearance[20] and disappearance of labeled gap junctional proteins has been observed recently in liver cell membranes.[21]

In order to increase the amount of translation product, one might consider injection of mRNA via microelectrodes. The following considerations make the liposome technique more suitable for our purposes:

1. Liposome-mediated mRNA transfer affects essentially all cells within a culture dish. This facilitates statistical and biochemical analysis.
2. Detection of cell coupling via second- or higher-order neighbors is possible only if more than one cell pair is loaded with mRNA.
3. The functional state of cell–cell channels is controlled by the intracellar Ca^{2+} concentration. Microinjection into such small cells as CL-1D can easily result in a rise of intracellular calcium concentration. This may occur by direct entry of calcium at the site of injury or by a reduction in membrane potential.[19]
4. Cells impaled with large microelectrodes used for the injection of mRNA have a tendency to round up and lose contact with their neighbors.
5. Finally, the liposome-mediated transfer of mRNA is much more convenient than microinjection.

The low coupling ratios observed suggest that cell pairs establish only relatively few functional channels. There are many obvious reasons for this.

1. Under the conditions described above, each cell receives only about 1000 molecules of mRNA.[22]
2. Probably only a small number of these mRNA sequences code for channel components.
3. Even if all channel components are inserted into the membrane in the form of completed hemichannels, the chance for the establishment of functional channels between the membranes of adjacent cells may well be below 100%.

Considering these limitations, it is noteworthy that our assay procedure is extremely sensitive, requiring only a few functional channels per cell pair. More elaborate techniques have been shown to trace even single-channel events.[23] For these reasons it is obvious that less sensitive assays, such as tracer diffusion studies or electron microscopy, are less likely to succeed in detecting mRNA-induced channel formation. These assays probably can be

[20] J.-P. Revel, S. B. Yancey, D. J. Meyer, and B. Nicholson, *In Vitro* **16,** 1010 (1980).
[21] R. F. Fallon and D. A. Goodenough, *J. Cell Biol.* **90,** 521 (1981).
[22] T. Wilson, D. Papahadjopoulos, and R. Taber, *Cell* **17,** 77 (1979).
[23] W. R. Loewenstein, Y. Kanno, and S. J. Socolar, *Nature* (*London*) **274,** 133 (1978).

used more successfully with mRNA preparations that are enriched for channel protein specific sequences.

Perspectives

The functional assay for mRNA described above allows the identification of proteins for which as yet no biochemical assay is available. In order to accomplish this, mRNA sequences responsible for cell–cell channel formation must be isolated in pure form. The most promising method for obtaining single mRNA species involves molecular cloning of cDNA prepared from the mRNA mixture. Identification of clones containing channel-specific sequences may be accomplished by again using the liposome-mediated complementation test in CL-1D cells. DNA isolated from relevant clones is expected to inactivate a mRNA preparation by "hybrid arrest."[24]

This approach is attractive in that, if multiple protein components are required for the formation of a functional channel, these components can be identified individually by testing permutations of mRNA species. Furthermore, the functional contributions of individual components might be assessable by this method.

Application of this technique to the isolation of other types of mRNA of interesting membrane proteins should, in principle, be possible. It should be reemphasized, however, that success will depend on the availability of a comparably sensitive assay.

[24] B. M. Paterson, B. F. Roberts, and E. L. Kuff, *Proc. Natl. Acad. Sci. U.S.A.* **74,** 4370 (1977).

[48] Selective Labeling and Quantitative Analysis of Internalized Plasma Membrane

By IRA MELLMAN and CYNTHIA J. GALLOWAY

Endocytosis generates a rapid and enormous inward flow of plasma membrane (PM)-derived vesicles. This flow is often continuous and can involve amounts of membrane up to twice the cell's surface area every hour. Endocytosis is also balanced by a compensatory centrifugal flow that recycles internalized PM components to the cell surface.[1] This bidirectional traffic is closely regulated and highly specific. Internalized membrane is free to fuse with only a restricted subset of intracellular membranes, i.e., the components of the vacuolar apparatus (PM, endosomes, lysosomes, Golgi).

[1] R. M. Steinman, I. S. Mellman, W. A. Muller, and Z. A. Cohn, *J. Cell Biol.* **96,** 1 (1983).

METHODS IN ENZYMOLOGY, VOL. 98

ISBN 0-12-181998-1

In addition, these organelles are able to maintain their diversity and integrity in spite of extensive communication with the cell surface.

While the general pathways and characteristics of endocytosis and membrane recycling are well known, little information exists regarding the mechanisms that control these processes. Traditionally, the study of endocytosis has concentrated on the contents of endocytic vacuoles as opposed to the composition, interactions, and fate of the vacuole membrane. A more detailed understanding of the PM components that participate in endocytosis, as well as of the composition of membranes throughout the vacuolar system, is likely to be of great importance. Unfortunately, endosomes have yet to be isolated by subcellular fractionation in sufficient yield and purity to permit such analysis. Thus, alternative strategies have been developed.

This chapter summarizes one such approach, which uses iodination catalyzed by lactoperoxidase–glucose oxidase to label endocytic vesicle membrane proteins from within intact cells. This is a versatile technique that has thus far allowed the selective iodination of PM internalized during latex phagocytosis[2] and fluid pinocytosis[3] in mouse macrophages. The latter method is described here. In addition, we illustrate how monoclonal anti-PM antibodies can be employed to obtain a quantitative description of the antigenic composition of labeled (or unlabeled) membrane.

Pinocytosis and Enzymic Determination of Lactoperoxidase

The method we have developed is based on a relatively simple modification of the lactoperoxidase (LPO)–glucose oxidase (GO) technique described by Hubbard and Cohn.[4,5] The principle is to allow cells to internalize the iodination system via pinocytosis at 37°, to remove extracellular LPO–GO by washing at 0°, and then to use the intracellular enzyme to mediate iodination. Most of our work has employed suspension cultures of the mouse macrophage-like cell line J774 (obtained from Dr. Peter Ralph, Sloan–Kettering Institute for Cancer Research, Rye, New York),[6] but the technique should be applicable to many cell types grown either in suspension or in monolayer.

Clearly, the efficacy of this technique depends on the unique localization of LPO to intracellular vacuoles. Thus, before the method can be employed, one should demonstrate that LPO is taken up by the cells of interest without

[2] W. A. Muller, R. M. Steinman, and Z. A. Cohn, *J. Cell Biol.* **86,** 292 (1980).

[3] I. S. Mellman, R. M. Steinman, J. C. Unkeless, and Z. A. Cohn, *J. Cell Biol.* **86,** 712 (1980).

[4] A. L. Hubbard and Z. A. Cohn, *J. Cell Biol.* **55,** 390 (1972).

[5] A. L. Hubbard and Z. A. Cohn, *J. Cell Biol.* **64,** 438 (1975).

[6] P. Ralph, J. Pritchard, and M. Cohn, *J. Immunol.* **114,** 898 (1975).

binding to the PM, i.e., by fluid-phase pinocytosis.[7,8] This is generally determined by both enzymic and cytochemical approaches.

LPO Uptake. The uptake of fluid-phase pinocytic markers, such as sucrose and horseradish peroxidase (HRP),[7,8] proceeds linearly as a function of the concentration of marker in the medium and as a function of time of incubation at 37°. These parameters can easily be determined for LPO by enzymic assay. Confluent monolayers of J774 cells are plated in 35-mm tissue culture dishes (10^6 cells/dish) in Dulbecco's modified Eagle's minimum essential medium (DMEM) containing 5% fetal calf serum (FCS). After three rinses with PBS, the monolayers are refed with medium containing LPO (Calbiochem, purified grade) over a wide concentration range (0.01 – 1.0 mg/ml). After up to 2 hr at 37°, the cells are washed five times on ice over a 15-min period with PBS containing 5 m*M* glucose (PBS-G). Often, cells are then returned to culture at 37° to remove the trace amounts of dish-bound enzyme,[7] but this step may be omitted. Washed cells are lysed in 0.15% (w/v) Triton X-100 and assayed for enzyme activity the same day (see below).

In addition, the amount of surface-adsorbed enzyme can be determined by incubating cells for 2 hr at 0° in 1 mg of LPO per milliliter. Pinocytosis is inhibited by low temperature. For J774 cells, the amount of cell-associated enzyme activity at 0° is <4% that present at 37° and is not usually distinguishable from enzyme-free blanks. Identical results have been obtained using serum-free medium and using cells in suspension washed as described below.

LPO Cytochemistry. The diaminobenzidine (DAB) – H_2O_2 technique of Graham and Karnovsky[9] can be employed for the cytochemical visualization of LPO in the electron microscope using the same procedure that has been employed for detecting HRP activity.[10] The reaction product generated by LPO is significantly weaker than HRP, however. J774 cells are exposed to LPO (2 mg/ml) at 37°, usually for 3 – 5 min, washed extensively at 4° with PBS-G, and fixed for 15 min in 2.5% glutaraldehyde in 0.1 *M* sodium cacodylate, pH 7.4, at room temperature. After four washes with PBS, reaction product is developed using DAB (Sigma; 0.5 mg/ml) and 0.01% H_2O_2 (from 30% solution) in 0.1 *M* Tris-HCl, pH 7.6 (15 min at room temperature). The cells are washed, postfixed in OsO_4 (1% in 0.1 *M* sodium cacodylate, pH 7.4, 60 min, 0°) and stained with 0.5% uranyl acetate (1 hr at room temperature). Cells are then embedded in Epon and sectioned.

Exposure of J774 to LPO for <5 min results in DAB reaction product

[7] R. M. Steinman and Z. A. Cohn, *J. Cell Biol.* **55,** 186 (1972).
[8] R. M. Steinman, J. M. Silver, and Z. A. Cohn, *J. Cell Biol.* **63,** 949 (1974).
[9] R. C. Graham, Jr. and M. J. Karnovsky, *J. Histochem. Cytochem.* **14,** 291 (1966).
[10] R. M. Steinman, S. E. Brodie, and Z. A. Cohn, *J. Cell Biol.* **68,** 665 (1976).

only in electron-lucent vacuoles (usually 0.2–0.5 μm in diameter) in the peripheral cytoplasm. These vacuoles can be termed pinocytic vesicles or endosomes. No reaction product is observed on the cell surface.[3]

LPO Assay. LPO activity can be conveniently determined in solution using the *o*-dianisidine–H_2O_2 method, which has been described for measuring HRP activity.[8] Alternatively, one can avoid the use of the carcinogenic substrate and obtain at least severalfold greater sensitivity using a modification of the triiodide assay developed by Morrison.[11] The reaction mixture is prepared fresh (stable for 2 hr) and consists of 33 mM NaP_i, pH 7.0; 5 mM KI; and 30 μM H_2O_2 (prepared by diluting 90 mM stock, which itself is made by diluting 0.1 ml of 30% H_2O_2 to 10 ml with distilled H_2O). To 1 ml of reaction mixture, $\leq$ 10 μl of cell lysate or 2–10 ng of purified LPO is added, and the increase in absorbance at 350 nm is followed on a recording spectrophotometer at room temperature. The reaction is linear for up to 5 min and can detect as little as 0.5 ng of enzyme.

Using a linear portion of the curve, one can convert activity to units (micromoles of triiodide) using an extinction coefficient of 26. Purified LPO from Calbiochem or the ammonium sulfate salt obtained from Boehringer-Mannheim gives activities of approximately 200 units per milligram of protein. In practice, we have found that the Calbiochem material is very uniform from lot to lot and can usually be assumed to be 1 mg/100 Calbiochem units. The enzyme can be stored in solution (PBS) at $-20°$ indefinitely.

GO Assay. Glucose oxidase from *Aspergillus niger* (type V) is obtained from Sigma. It is stable at 4° for at least 2 years. It is generally unnecessary to assay GO activity. However, a convenient assay (recommended once for every new lot of enzyme) is to use an HRP-linked procedure that employs *o*-dianisidine.[8] The reaction mixture contains 30 ml of 0.1 M NaP_i (pH 6.0), 0.25 ml of *o*-dianisidine (10 mg/ml), 0.02 ml of HRP (20 mg/ml, Sigma); and 2 ml of 1 M glucose. Glucose oxidase is added ($<$ 1 Sigma unit), and the increase in absorbance at 450 nm is followed at room temperature. In practice, GO concentrations are optimized empirically for both surface and intracellular iodination and are usually expressed in terms of the Sigma units ($\sim$ 200 units per milligram of protein).

Iodination Procedure

The general procedure for intracellular iodination of endosomes is given in Table I. For the Metrizamide wash (step 4), 3% Metrizamide (analytical grade; Accurate Scientific, Hicksville, New York) is made up in HEPES-buffered saline (0.12 M NaCl, 6.7 mM KCl, 1.2 mM $CaCl_2$, 10 mM

[11] M. Morrison, this series, Vol. 17A, p. 653.

TABLE I
IODINATION OF ENDOSOME MEMBRANE IN INTACT CELLS[a]

1. J774 macrophages are harvested from suspension culture and washed with PBS–glucose three times.
2. Cells are suspended (~2 × 10^7/ml) in PBS–glucose containing lactoperoxidase (LPO) (0.5–1.5 mg/ml) and glucose oxidase (GO) (50–100 μg/ml).
3. Cell suspension is incubated for 5 min at 37° to permit pinocytosis of LPO and GO.
4. After cooling to 4°, extracellular LPO and GO are washed away by centrifuging cells through a cushion of 3% Metrizamide followed by three sequential washes with PBS–glucose.
5. Cells are suspended (~2 × 10^7/ml) in PBS–glucose on ice, and carrier-free ^{125}I-labeled Na (0.1–1.0 mCi/ml) is added; iodination of endosome membrane is catalyzed by LPO sequestered uniquely within vesicles. Incubation is continued for 15 min at 0°.
6. Iodination is terminated by dilution and repeated washing with serum-free DMEM at 4°.
7. Cell viability is assessed by trypan blue exclusion.

[a] Data are taken from Mellman *et al.*[3]

HEPES, and 1 *M* NaOH to adjust to pH 7.6). Usually, the cell suspension is layered onto a 9-ml cushion and centrifuged at low speed (200 *g*) for 7.5 min at 4°. Cell pellets are resuspended gently using a plastic-tipped automatic pipettor.

For each new cell type, the optimal concentration of GO should be determined. J774 cells are extremely resistant to killing by H_2O_2[12] and therefore can tolerate a relatively large amount of the enzyme. The GO can be reduced or omitted entirely and added during the iodination step (as in cell-surface iodination). However, these alterations may result in decreased iodination. Na^{125}I (carrier-free) is obtained from Amersham/Searle. Iodination of J774 cells results in negligible loss of viability (>95% viable) and negligible incorporation of ^{125}I into cell lipid (<5%, determined by two-dimensional thin-layer chromatography of Folch extracts[2,3]; significant free iodine does partition into the organic phase).

The requirements for incorporation of ^{125}I into trichloroacetic acid (TCA)-precipitable material are shown in Table II. These precipitations are routinely performed by spotting a small aliquot of lysed cells [0.5% Nonidet P-40 (NP-40), centrifuged at 40,000 *g* for 15 min] on Whatman glass-fiber filters (GF/B) and washing the filters on a vacuum manifold with cold 10% TCA containing 0.1 *M* KI (prepared fresh). As indicated, iodination depends on the pinocytic uptake of both LPO and GO. A useful control that estimates the degree to which labeling occurred as a result of surface-adsorbed vs intracellular enzyme is provided by comparing iodination of cells incubated with LPO and GO at 0° vs 37°, respectively (Table II). It is

[12] C. F. Nathan, L. H. Bruckner, S. C. Silverstein, and Z. A. Cohn, *J. Exp. Med.* **149,** 84 (1979).

TABLE II
ENDOSOME IODINATION[a]

Condition	TCA-precipitable ^{125}I [(cpm/10^6 cells) $\times$ 10^5]
LPO + GO[b] 37°, 5 min	4.81
LPO + GO 0°, 5 min	0.13
LPO only 37°, 5 min	0.13
GO only 37°, 5 min	0.05
No addition	0.01
Surface iodination[c]	1–5

[a] Data are taken from Mellman *et al.*[3]

[b] J774 cells were exposed to lactoperoxidase (LPO) (0.75 mg/ml) and glucose oxidase (GO) (0.05 mg/ml) for 5 min at 37°, washed extensively in the cold, and ^{125}I-labeled Na is added (0.4 mCi/ml) at 0°. Iodination was terminated after 12 min as detailed in Table I.

[c] Typical results are given for purposes of comparison.

important to make this comparison several times. If 0° labeling is much more than 10% of 37° labeling, one must suspect that significant adsorption of LPO to the cell surface has occurred, resulting in "background" labeling of the PM in addition to internalized membranes.

PM Iodination. As described by Hubbard and Cohn,[5] LPO-catalyzed cell surface iodination was conducted at 0° (2 to 3 $\times$ 10^7 cells/ml). In terms of the milligram protein estimates given above, the final concentration of LPO is 15 μg/ml ($\sim$ 3 triiodide units), and that of GO is 65 ng/ml (100 Sigma milliunits). Significant loss of viability or labeling of cell lipid is not observed.

Localization of Incorporated ^{125}I

Several methods have been employed to demonstrate that the endosome iodination procedure radiolabels intracellular as opposed to cell surface membrane proteins.[3,13] Among these are quantitative electron microscope autoradiography and quantitative immunoprecipitation. The distribution of incorporated ^{125}I on internally and surface labeled cells is usually compared. For the immunoprecipitation technique, the fraction of total cell ^{125}I-labeled membrane antigen present on the PM was determined. This was accomplished by adding monoclonal anti-macrophage antibodies either to intact cells (which were washed free of unbound antibody before lysis) or

[13] I. Mellman, *Membr. Recycling Ciba Found. Symp.* **92,** 35 (1982).

directly to cell lysates.[3] Accordingly, the amount of labeled antigen detected by immune precipitation represented either surface-associated or total cell (surface plus intracellular) antigen, respectively.

Both of these methods are rather cumbersome and difficult to apply routinely. However, in addition to the 0° incubation control mentioned above, it would be useful to have one additional criterion to determine whether most iodination has occurred intracellularly. One method, which has not been very effective with macrophages, is to treat endosome and surface-labeled cells with protease at 0° and to compare the amounts of ^{125}I released. Trypsin (1 mg/ml), Pronase (20 μg/ml), or proteinase K (0.5 mg/ml) have been used for this purpose (usually 1-hr incubations). On Chinese hamster ovary (CHO) cells. Storrie *et al.*[14] reported that 50% vs <10% of cell-associated ^{125}I was protease releasable following cell surface and intracellular labeling, respectively.

A second approach, which is more quantitative, relies on the use of releasable receptor-bound ligands.[13] We have employed soluble immunoglobulin G (IgG)-containing immune complexes for this purpose. Complexes are formed according to standard techniques[15] by combining dinitrophenylated (DNP) bovine serum albumin (~10 DNP/molecule) with monoclonal mouse anti-DNP or affinity-purified rabbit anti-DNP IgG at a molar ratio of 5:1 (IgG:albumin; 37°, 30 min). The IgG concentration employed is usually 20 μg/ml. These complexes bind specifically to macrophage Fc receptors as indicated by the fact that complex binding is inhibited by simultaneous incubation with the rat anti-mouse Fc receptor monoclonal antibody 2.4G2.[16]

After the binding of immune complexes to J774 cells at 0° (2 hr), the cells are washed and then warmed to 37° for 5 min in the presence or the absence of LPO–GO. The cells are washed in the cold then iodinated intracellularly or on the PM, respectively, as described above. ^{125}I-labeled IgG can be easily detected after both labeling protocols by immune precipitation (see below) with rabbit anti-mouse or goat anti-rabbit IgG (Cappel Laboratories) and protein A–Sepharose (Pharmacia). To determine what fraction of the ^{125}I-labeled IgG was surface bound, cells are incubated at 0° under conditions that remove Fc receptor-bound ligand [50 μg of 2.4G2 IgG per milliliter for 1 hr or 1 mg of subtilisin BPN (Sigma) per milliliter for 2 hr]. The cells are then washed and lysed in 0.5% NP-40. Immunoprecipitation demonstrates that >80% of the cell-associated ^{125}I-labeled IgG can be eluted from surface-labeled cells and <20% can be stripped from internally

[14] B. Storrie, T. D. Dreesen, and K. M. Maurey, *Mol. Cell. Biol.* **1,** 261 (1981).

[15] R. G. Q. Leslie, *Eur. J. Immunol.* **10,** 317 (1980).

[16] I. S. Mellman and J. C. Unkeless, *J. Exp. Med.* **152,** 1048 (1980).

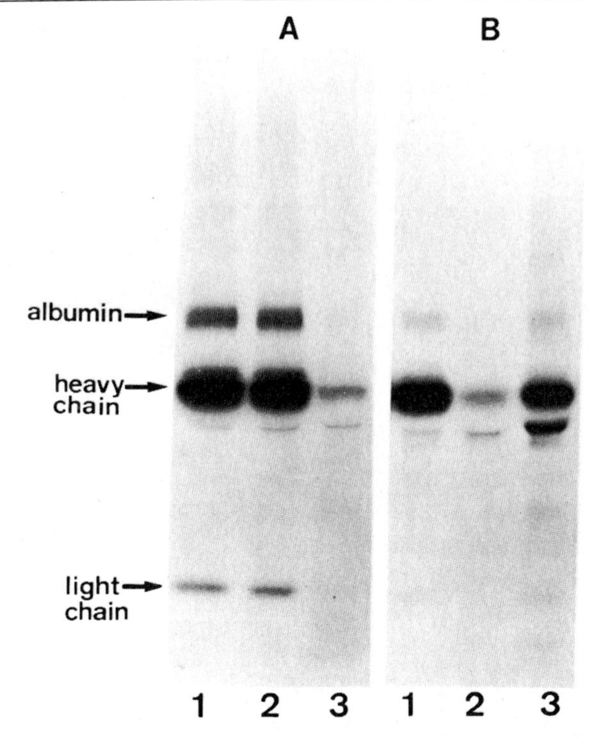

FIG. 1. Iodination and elution of Fc receptor-bound IgG immune complexes from J774 cells. Complexes were formed using a monoclonal mouse anti-DNP IgG and DNP-albumin and detected by immunoprecipitation using a rabbit anti-mouse IgG antiserum (Cappel Laboratories) and protein A–Sepharose (Pharmacia). See the text for details. Panel A: Intracellular iodination. Panel B: Cell surface iodination. Lanes 1: Total cell IgG; Lanes 2: IgG remaining with cells after treatment with 50 μg of 2.4G2 IgG per milliliter for 1 hr at 0°; lanes 3: IgG eluted from cells immunoprecipitated from medium.

labeled cells (Fig. 1). In principle, an analogous approach can be taken using a number of other ligands.

Analysis of Iodinated Polypeptides

The proteins labeled after intracellular and PM iodinations are compared first by SDS–polyacrylamide gel electrophoresis. We usually employ 4–11% acrylamide gradient gels,[17] which are then fixed, dried, and subjected to autoradiography using intensification screens (DuPont Lightning Plus) at −70°. Normally 1 to 4 $\times 10^5$ TCA-precipitable cpm are added per lane. The iodinated bands should represent a restricted subset of total

[17] D. M. Neville, Jr. and H. Glassman, this series, Vol. 32, p. 92.

cellular proteins (i.e., those stained by Coomassie blue or labeled with [^{35}S]methionine). Few if any iodinated species should comigrate with any major cell protein. In J774 cells, the gel profiles from internal and PM labelings are quite similar, except that the presumptive pinocytic vesicle content marker, LPO, is self-iodinated and usually detectable after intracellular labeling.[3]

A more quantitative description of labeled proteins can be obtained by immunoprecipitation. For this purpose, we have used a set of rat anti-mouse macrophge PM monoclonal antibodies, including the anti-Fc receptor antibody 2.4G2.[3,16,18] After cell surface or endosome iodination, cell pellets are lysed in 0.5% NP-40 in PBS containing 1 m*M* phenylmethylsulfonyl fluoride (PMSF) and 0.2 U of aprotinin (Sigma) per milliliter. Lysates are cleared by centrifugation at 45,000 *g* at 4° for 15 min. To 0.2 ml of lysate, 1 μg of purified antibody is added; the preparation is incubated at 0° for 30 min. Antigen–antibody complexes are isolated using F(ab$'$)$_2$ fragments of affinity-purified rabbit anti-rat IgG (Cappel Laboratories) coupled to CNBr-activated Sepharose 4B (2 mg of antibody per milliliter of packed resin).[3] Forty microliters of a 40% (v/v) suspension of the immunoadsorbent is added, incubated for 60 min at 4° with constant agitation (Eppendorf mixer), and collected in a microfuge. The Sepharose pellet is resuspended and washed as follows: 2 × SA buffer (0.3 *M* NaCl, 0.0125 *M* KP_i pH 7.4, 0.02% NaN_3), 3 × 0.1% SDS–0.05% NP-40 (in 10 m*M* Tris–HCl, pH 8.6, 0.3 *M* NaCl, prepared fresh), 2 × SA buffer. Immunoprecipitates are displayed on SDS–polyacrylamide gels and then excised, and ^{125}I is quantitated in a gamma counter. Any immunoprecipitation method that can be shown to remove all detectable ^{125}I-labeled antigen from a lysate should be applicable.

Illustrative data obtained from experiments with J774 cells are summarized in Table III.[3] With one exception (antigen 2.6), each of the eight membrane antigens studied was labeled to the same relative degree on both cell surface and endosome membranes.

Analysis of Unlabeled Membrane

It is also possible to use the anti-PM monoclonal antibodies to measure the amount of antigen present on unlabeled membranes: e.g., detergent lysates of whole cells or subcellular fractions. For this purpose, we are developing (in conjunction with Dr. J. C. Unkeless, Rockefeller University) a sensitive two-site immunoradiometric assay (IRMA) for membrane proteins. This method requires two reagents: a purified, high-affinity monoclonal antibody (labeled with ^{125}I) and a second antibody or crude antiserum

[18] J. C. Unkeless, *J. Exp. Med.* **150,** 580 (1979).

TABLE III
COMPOSITION OF PLASMA MEMBRANE (PM) AND ENDOSOME MEMBRANE AS DETERMINED BY THE RELATIVE LABELING OF EIGHT MEMBRANE ANTIGENS[a,b]

	Relative ^{125}I detected by immune precipitation	
Monoclonal antibody[c]	PM iodination	Endosome iodination
2D2C (90,000)	1.000	1.000
1.21J (94,000, 180,000)	0.552	0.452
2E2A (82,000)	0.150	0.166
F480 (150,000)	0.170	0.092
2F44 (42,000)	0.035	0.024
25-1 (44,000, 12,000)	0.065	0.041
2.4G2 (60,000, 47,000)	0.022	0.026
2.6 (20,000)	0.008	0.017

[a] Data are taken from Mellman *et al.*[3]
[b] ^{125}I was determined by excising and counting radioactive bands identified after SDS–polyacrylamide gel electrophoresis. Antigen 2D2C was always most heavily labeled and was thus assigned a value of 1.0. The labeling of each of the other antigens was then normalized against 2D2C.
[c] Values in parentheses refer to molecular weight.

containing specificities that recognize a determinant(s) on the antigen not seen by the labeled antibody.

To detect macrophage Fc receptors, we have employed ^{125}I-labeled Fab fragments of the anti-receptor monoclonal 2.4G2.[18] The antibody is iodinated using Iodogen (Pierce Chemical Co.).[19] Fifty micrograms of 2.4G2 Fab in 0.1 ml of *M* NaP_i (pH 7.0) and 1.0 mCi of $Na^{125}I$ are transferred to a 13 × 100-mm glass tube coated with 20 μg of Iodogen (dried down from 25 μl of $CHCl_3$). After 5 min at 0°, the reaction mixture is applied to a 0.3-ml column of washed Bio-Rad AG l-X8 resin (Cl^- or OH^- form, 200–400 mesh) and the protein is eluted with 2 × 0.5-ml aliquots of PBS containing 1 mg of bovine serum albumin per milliliter and 0.02% NaN_3. More than 95% of the unincorporated iodine is removed by the column.

Membrane or cell lysates are prepared as above in 0.5% NP-40 containing 1% (v/v) FCS. Then 0.1–0.5 μg of ^{125}I-labeled 2.4G2 Fab is added (~4 × 10^5 cpm, preadsorbed with protein A–Sepharose) to 0.1 ml of lysate in a microfuge tube and incubated at room temperature for 1 hr. Then 10 μl of a 1 : 10 dilution of a rabbit anti-Fc receptor antiserum (that was raised against purified receptor protein[20]) is added, and the sample is incubated for 1 hr at room temperature. To adsorb the antigen–antibody complexes, 50 μl

[19] P. J. Fraker and J. C. Speck, Jr., *Biochem. Biophys. Res. Commun.* **80,** 849 (1978).
[20] I. S. Mellman, H. Plutner, R. M. Steinman, J. C. Unkeless, and Z. A. Cohn *J. Cell Biol.* **96,** 887 (1983).

of a 50% v/v solution of protein A–Sepharose (Pharmacia) in PBS–0.02% NaN_3 is added, and the incubation is continued at room temperature with agitation for 1 hr. After two washes in SA and two washes in 0.1% SDS–0.05% NP-40 (in 10 m*M* Tris-HCl, pH 8.6, 0.3 *M* NaCl), the bottom of the microfuge tube containing the pellet is sliced off and ^{125}I is determined in a gamma counter. Background radioactivity (no lysate added) is usually <150 cpm above machine background.

This method reproducibly allows the detection of Fc receptor present in lysates of $<10^4$ J774 cells. At $\sim 1 \times 10^6$ total Fc receptors per cell, as few as 10^{10} molecules (<20 fmol) are easily detected. The assay exhibits good linearity between 10^9 and 10^{12} molecules (i.e., lysates of 10^3 to 10^6 J774 cells).

Acknowledgments

This work is supported by grants from the National Institutes of Health (GM29075 and RR 05358) and from the Swebilius Cancer Fund. I. M. is a Junior Faculty Research Awardee of the American Cancer Society.

[49] Spontaneous Transfer of Exogenous Epidermal Growth Factor Receptors into Receptor-Negative Mutant Cells[1]

By MANJUSRI DAS, JEFFREY FEINMAN, MARK PITTENGER, HERMAN MICHAEL, and SUBAL BISHAYEE

Epidermal growth factor (EGF), a small single-chain polypeptide of ~ 6000 daltons, is a potent stimulant of DNA replication and cell division in responsive cells.[2] The actions of EGF are mediated through a cell-surface receptor polypeptide of 170,000–180,000 daltons.[3,4] Receptor–EGF interaction produces a variety of rapid events, including activation of a receptor-associated protein kinase (tyrosine specific),[4] and receptor clustering and endocytosis.[3,5,6] One or more of these events leads ultimately to the intracel-

[1] Supported by NIH research grant AM-25819 and Research Career Development Award AM-00693 awarded to Manjusri Das.

[2] G. Carpenter and S. Cohen, *Annu. Rev. Biochem.* **48,** 193–216 (1979).

[3] M. Das and C. F. Fox, *Proc. Natl. Acad. Sci. U.S.A.* **75,** 2644 (1978).

[4] S. Cohen, H. Ushiro, C. Stoscheck, and M. Chinkers, *J. Biol. Chem.* **257,** 1526 (1982).

[5] G. Carpenter and S. Cohen, *J. Cell. Biol.* **71,** 159 (1976).

[6] J. Schlessinger, Y. Schechter, M. C. Willingham, and I. Pasten. *Proc. Natl. Acad. Sci. U.S.A.* **75,** 2659 (1978).

ISBN 0-12-181998-1

lular generation of macromolecular activator(s) of DNA replication.[7,8] Of the few growth factor receptors that have been studied to date, EGF receptor is one of the best characterized, and yet there remains an abyss of ignorance regarding its biochemistry and mechanism of action.

One way to obtain insights into the mechanism of receptor action would be to induce a cell to insert into its plasma membrane a receptor that it normally lacks. Such studies have been performed with other receptor systems using polyethylene glycol-mediated fusion techniques.[9,10] We were interested in performing such studies with a variant cell line NR-6 derived from mouse 3T3, which can neither bind nor biologically respond to EGF. Development of a procedure for receptor transfer could be an important step toward elucidation of the mechanism of receptor action and understanding the mode of integration of receptor proteins in the membrane. During our attempts to develop this procedure, we observed a spontaneous and selective transfer of EGF receptors from donor membranes to recipient cells in a biologically active orientation.[11] The present chapter describes some of the properties of this transfer process.

Methods

EGF receptor is transferred from donor receptor-enriched membranes to recipient receptorless cells in the absence of added fusogenic agents. The cells are assayed for transferred receptor activity (*a*) by ^{125}I-labeled EGF binding, or (*b*) by measuring EGF-induced stimulation of [^{3}H]thymidine incorporation into DNA.

Recipient Cells. Monolayer cultures of EGF receptor-negative mouse NR-6 cells[12,13] are grown and maintained at 37° in a 10% CO_2–90% air atmosphere in Dulbecco's modified Eagle's medium containing 3% fetal calf serum and 7% newborn calf serum. Cells are subcultured by trypsinization. Medium, serum, and other reagents for cell culture are obtained from GIBCO.

Donor Membranes. EGF receptor-enriched plasma membranes are obtained from either mouse (Swiss white) liver or human A-431 carcinoma cells.

[7] M. Das, *Proc. Natl. Acad. Sci. U.S.A.* **77,** 112 (1980).

[8] M. Das, *Proc. Natl. Acad. Sci. U.S.A.* **78,** 5677 (1981).

[9] M. Schramm, *Proc. Natl. Acad. Sci. U.S.A.* **76,** 1174 (1979).

[10] D. Doyle, E. Hou, and R. Warren, *J. Biol. Chem.* **254,** 6853 (1979).

[11] S. Bishayee, J. Feinman, M. Pittenger, H. Michael, and M. Das, *Proc. Natl. Acad. Sci. U.S.A.* **79,** 1893 (1982).

[12] M. Das, T. Miyakawa, C. F. Fox, R. M. Pruss, A. Aharonov, and H. R. Herschman, *Proc. Natl. Acad. Sci. U.S.A.* **74,** 2790 (1977).

[13] R. M. Pruss and H. R. Herschman, *Proc. Natl. Acad. Sci. U.S.A.* **74,** 3918 (1977).

The procedure of Aronson and Touster[14] for isolation of rat liver plasma membranes is used without modification for preparation of liver plasma membranes from Swiss white mice. The pellet (33,000–78,000 *g*) is suspended in 57% (w/w) sucrose–5 m*M* Tris-HCl, pH 8.0, and subjected to sucrose gradient equilibrium centrifugation as described.[14] The fraction P_2 (8.5–34% sucrose) contains about 25% of the input 5′-nucleotidase activity and EGF binding activity and less than 0.1% of the input *N*-acetyl-β-D-glucosaminidase activity present in the original tissue homogenate. The specific EGF binding activity of this fraction varies between 1 and 3 pmol of ^{125}I-labeled EGF bound per milligram of protein under standard incubation conditions (60-min incubation with 10 n*M* ^{125}I-labeled EGF at 20°). This fraction is stored in liquid N_2 until use.

Plasma membranes from A-431 cells are prepared using hypoosmotic borate–EDTA.[15,16] A-431 cells are scraped from culture dishes, and the twice-washed cell pellet is suspended in 2 volumes of harvesting buffer (0.5 *M* boric acid–0.15 *M* NaCl, pH 7.2). The suspension is slowly added to 100 volumes of extraction solution (0.02 *M* boric acid–0.2 m*M* EDTA, pH 10.2) under gentle stirring. During the next 10 min plasma membranes are lysed, and cytoplasmic contents are expelled and gelatinized. Then 8 volumes of borate solution (0.5 *M* boric acid, pH 10.2) are added, and stirring is continued for 5 min. After filtration through glass wool, the suspension is centrifuged at 450 *g* for 10 min at 4°. The supernatant is recentrifuged at 12,000 *g* for 30 min. The pellet is suspended in phosphate-buffered saline (0.137 *M* NaCl–2.7 m*M* KCl–1.47 m*M* KH_2PO_4–0.84 m*M* Na_2HPO_4, pH 7.4) and centrifuged in a SW41 rotor at 15,000 rpm for 1 hr through a 35% (w/w) sucrose solution in phosphate-buffered saline. The membrane fraction accumulating at the interface is collected, washed with phosphate-buffered saline, and stored in liquid N_2 until use.

Before use in experiments with cell monolayers, the membranes are UV sterilized.

Receptor Transfer. Incubation of receptor-negative NR-6 cell monolayers (10^5 cells per 16-mm dish) with donor membranes (~ 140 μg of hepatic membranes; or 15 μg of A-431 membranes) in 0.3 ml of Dulbecco's modified Eagle's medium (containing 2% fetal calf serum plus 100 units of penicillin and 100 μg of streptomycin per milliliter) at 25° for 6 hr under a CO_2 atmosphere results in a spontaneous transfer of ~ 20% of input EGF-receptor activity to NR-6 cells. Under these conditions, ~ 10^5 receptor sites are incorporated per NR-6 cell. Transferred EGF-receptor activity is determined as follows using radioiodinated EGF prepared as described.[12] The

[14] N. N. Aronson and O. Touster, this series, Vol. 31, p. 90.

[15] D. Thom, A. J. Powell, C. W. Lloyd, and D. A. Rees, *Biochem. J.* **168,** 187 (1977).

[16] D. Cassel and L. Glaser, *J. Biol. Chem.* **257,** 9845 (1982).

monolayers are washed free of unbound membranes and then incubated at 20° for 60 min with 10 nM ^{125}I-labeled EGF in Eagle's balanced salt solution containing 10 mM HEPES (pH 7.4) and 0.1% bovine serum albumin. At the end of incubation, unbound radioactivity is removed by rapidly washing the cell monolayers five times with 1 ml of the same salt–albumin solution for each wash. The washed monolayers are solubilized with 0.5 ml of 0.5 M NaOH and assayed in a gamma counter. Specific EGF binding is determined by measuring the difference in cell-bound radioactivity in the presence and in the absence of 2 μg of unlabeled EGF.

Bulk membrane transfer or adhesion is determined using ^{125}I-labeled donor membranes. Membranes (1–3 mg of protein, in 250 μl of 0.1 M sodium phosphate buffer, pH 7.4) are labeled with ^{125}I (0.1–0.5 mCi; Amersham) at 0° for 10 min with Iodogen (50–100 μg; Pierce) as the oxidizing agent.[17] The radioactively labeled membranes (20,000–50,000 cpm per microgram of protein) are pelleted by high-speed centrifugation and washed twice with 0.15 M NaCl–10 mM Tris·HCl, pH 7.4. (Electrophoresis and autoradiography have shown that almost all the Coomassie Blue-stained protein bands are labeled with ^{125}I.) After incubation of NR-6 monolayers with ^{125}I-labeled donor membranes (under conditions identical to those for unlabeled membranes in receptor transfer experiments), the dishes are washed free of unbound radioactivity, and cells containing bound labeled membranes are solubilized with 0.5 ml of 0.5 M NaOH and counted in a gamma counter. Cell-associated membrane radiactivity never exceeded 1.7% of input under all our conditions of receptor transfer.

Selectively of Receptor Transfer. When hepatic membranes (140 μg of protein) are incubated with NR-6 cells (10^5 cells per 16-mm dish) at 26° for 2 hr and then tested for insertion, only about 1–3% of the input EGF receptor and input bulk membrane proteins are transferred from membranes to NR-6 cells. However, when the incubation time is prolonged up to 6 hr there is a transfer of almost 20% of input receptor activity, whereas bulk protein transfer remains near the original 1% level. Receptor insertion after 10 or 12 hr of incubation at 26° was about the same as that observed after 6 hr of incubation, suggesting attainment of equilibrium within 6 hr. The results suggest that during the earlier part of the incubation (2 hr), bulk membrane adhesion occurred, but later a specific mechanism took control, causing a preferential insertion of the EGF receptor over the other hepatic proteins.

A preferential insertion of the EGF receptor is also observed in a temperature-dependence study. Even at 4°, at which there is insertion of only 0.7% of the input bulk membrane proteins, EGF receptor insertion is

[17] P. J. Fraker and J. C. Speck, *Biochem. Biophys. Res. Commun.* **80,** 849 (1978).

up to 7% of the input activity. At 24°, approximately 10% of the input EGF receptors are inserted whereas only about 1% of the bulk membrane proteins are inserted. At 37°, receptor insertion increases, but there is also an increase in bulk membrane incorporation, probably due to an enhanced rate of endocytic uptake. Because the mechanism responsible for preferential receptor insertion operates optimally at 24–26°, with a low degree of nonspecific membrane incorporation, this temperature range is used in all insertion experiments. The average activation energy for receptor transfer in the temperature range of 4–30° is ~9 kcal.

Effects of Various Agents. Preincubation of NR-6 cells with cycloheximide (1 μg/ml) or the presence of cycloheximide (1 μg/ml) during incubation of cells with donor membranes has no effect on the receptor gain by the NR-6 cells, suggesting that endogenous protein synthesis is not involved in this process. Also, the process does not involve glycosylation-induced activation of preexisting aglycoreceptors in NR-6 cells because tunicamycin (1 μg/ml) does not inhibit the receptor gain.

Pretreatment of either donor membranes or recipient cells with low concentrations of phospholipase C (0.5 unit/ml, a concentration that does not affect EGF receptor activity per se) results in a >70% reduction of receptor transfer, which suggests the involvement of phospholipids in insertion. The microfilament disrupting agent, cytochalasin B (~ 10 μ*M*) inhibits >50% receptor transfer, but colchicine, an antimicrotubular agent, has no inhibitory effect even at high concentrations (5–50 μ*M*).

Characteristics of the Inserted Receptor. The inserted receptor can bind ^{125}I-labeled EGF with high specificity and affinity. The apparant dissociation constant of the binding of ^{125}I-labeled EGF with the inserted heptic receptor is about 2.4 n*M*, in agreement with that observed (2.2 n*M*) for binding of ^{125}I-labeled EGF to the donor hepatic receptor in isolated plasma membranes. A large fraction (90%) of the ^{125}I-labeled EGF bound at 23° to NR-6 cells containing inserted receptors could be dissociated from the cells by mild acid treatment (0.2 *M* acetic acid–0.5 *M* NaCl, pH 2.5, 6 min at 4°), which suggests a surface location for the inserted receptors.

The number of EGF receptors that can be inserted into an NR-6 cell during a 6-hr incubation at 26° increased in proportion to the quantity of donor membranes and then gradually plateaued. At maximal insertion, approximately 10^6 receptor sites were incorporated per NR-6 cell. The parent 3T3 cell line (from which the NR-6 line was derived) contains approximately 10^5 EGF-receptor sites per cell.[12]

Pretreatment of the EGF receptors in the donor membranes with a photoreactive derivative of ^{125}I-labeled EGF results in specific radiolabeling of the receptor due to the formation of a M_r 170,000 receptor–^{125}I-labeled EGF covalent complex.[3,12] Transfer of the covalent complex to NR-6 cells also increases in proportion to the quantity of donor membranes. This provides an extra criterion for true transfer of receptor from donor membranes to recipient cells.

Stability of the Inserted Receptor. The inserted receptor is exceptionally stable to dissociation and other losses. When NR-6 containing $\sim 10^5$ receptors/cell are incubated at 37° in Dulbecco's modified Eagle's medium containing 2% fetal calf serum, there is an initial loss of about 40% of the original inserted activity between 10 and 20 hr at 37°, but there is no further loss of remaining receptor up to at least 50 hr of incubation at 37°. A parallel experiment conducted with ^{125}I-labeled membranes showed a rapid loss (80% lost within 5 hr) of bulk inserted proteins.

Transferred Receptor and EGF-Induced DNA Synthesis. In order to determine whether the inserted receptor behaves like a normal functional receptor, it is tested for EGF-induced stimulation of DNA synthesis. Our earlier attempts to test this were foiled by the fact that the membranes without the addition of EGF are active in stimulating [^{3}H]thymidine incorporation into DNA.[18] However, the peak of DNA synthetic activity induced by the membrane-associated mitogen is over at 30 hr after the removal of membranes, and a second peak is not observed at a later point.[18] Because at 30 hr after membrane removal, the NR-6 cells still contain about 60% of the originally inserted EGF receptors, the situation permits a test of EGF receptivity. In these experiments, stationary density-inhibited monolayers of NR-6 cells in 16-mm dishes are incubated at 26° for 6 hr with 160 μg of mouse liver membranes in 0.3 ml of Dulbecco's modified Eagle's medium containing 2% fetal calf serum. At the end of the incubation, the cells are washed free of unbound membranes and incubated at 37° in 0.3 ml of the same medium. After 30 hr of incubation, the medium is removed and the cells are incubated with conditioned Dulbecco's medium containing 2% fetal calf serum and EGF (prepared as described in Savage and Cohen[19]) at different concentrations (0.3 n*M*, 1 n*M*, 3 n*M*). [^{3}H]Thymidine (1 μCi/ml, 0.65 μM; Amersham) is added at 12 hr after EGF addition, and the incubations are continued for an additional 12 hr. At the end of incubation the trichloroacetic acid-insoluble radioactivity is determined as described.[3,11] The EGF was found to stimulate DNA synthesis in these membrane-treated NR-6 cells in a concentration-dependent manner. A three- to fourfold stimulation is observed at optimal concentrations of EGF.[11] In contrast, NR-6 cells not treated with membranes are totally unresponsive to EGF. The dramatic difference in DNA replicative stimulation that is observed with EGF between membrane-treated and untreated cells is not

[18] S. Bishayee and M. Das, *Biochim. Biophys. Acta* **696,** 134 (1982).
[19] C. R. Savage and S. Cohen, *J. Biol. Chem.* **247,** 7609 (1972).

observed with other growth factors such as fibroblast growth factor or insulin.

Concluding Remarks

Spontaneous and selective transfer of integral membrane proteins has been reported to occur between artificial membranes and between artificial and natural membranes.[20,21] Our studies suggest the existence of a natural (affinity-mediated) mechanism for specific EGF receptor transfer between natural membranes. This spontaneous receptor transfer is observed not only with NR-6 cells, but also with other EGF receptor-deficient cells such as lymphocytes, and even with EGF receptor containing cells such as mouse 3T3 cells and human foreskin fibroblasts. Whether similar transfer systems exist for other membrane receptors remains to be tested. Because EGF receptor is an integral membrane protein (detergents are required for its solubilization), it is not easy to visualize a mechanism for its successful insertion in the absence of added fusogens. It is possible that the preferential insertion of EGF receptors over other membrane proteins is due to a specific cell-surface membrane protein that has a high affinity for the receptor. Such a protein with a tendency to associate with the receptor may also be involved in the biological message transmission mechanism. Alternatively, it may have a role in the sorting and integration of membrane proteins after biosynthetic production.

[20] S. L. Cook, R. Bouma, and W. H. Huestis, *Biochemistry* **19,** 4601 (1980).
[21] H. G. Enoch, P. J. Fleming, and P. Strittmatter, *J. Biol. Chem.* **252,** 5656 (1977).

Section VIII

Transfer of Phospholipids between Membranes

[50] Lipid Transfer Proteins: Overview and Applications

By DONALD B. ZILVERSMIT

The observations that phospholipids and cholesterol in plasma lipoproteins are exchangeable and that cholesterol in erythrocytes can exchange as well date back to the early 1950s. These early studies have been reviewed by Bell.[1] It was not until the late 1960s that phospholipid exchange between biological membranes was demonstrated.[2] This exchange of phospholipid appeared to require a cytosolic fraction that, in rat liver, was soon identified as a soluble protein. Since then, several investigators have purified a number of proteins (see Table I) that are capable of accelerating exchange or net transfer of phospholipids among intracellular organelles, plasma membranes, and artificially prepared lipid vesicles.

Nomenclature

Originally, the proteins that accelerated phospholipid exchange between membranes were named "phospholipid exchange proteins." When it became obvious that some of these proteins could also accelerate the net transfer of phospholipids,[3] the name "phospholipid transfer protein" was adopted. When this name also became inadequate, because a few of the isolated proteins were shown to transfer lipids other than phospholipids, the name "lipid transfer protein" came into use.[3,4]

It seems reasonable, therefore, to use the generic name lipid transfer protein and recognize that individual proteins may exhibit such specificities that they might be referred to as phospholipid transfer protein or even phosphatidylcholine transfer protein. Appropriate abbreviations for these proteins are based on naming the known substrate followed by TP, such as PC-TP for the transfer protein that shows a high degree of specificity for phosphatidylcholine[5]; PI-TP or PI/PC-TP for the protein from beef brain[6] or beef heart[7] that shows a preferential transfer of phosphatidylinositol but

[1] F. P. Bell, *Prog. Lipid Res.* **17,** 207 (1978).

[2] K. W. A. Wirtz and D. B. Zilversmit *J. Biol. Chem.* **243,** 3596 (1968).

[3] R. C. Crain and D. B. Zilversmit, *Biochim. Biophys. Acta* **620,** 37 (1980).

[4] K. W. A. Wirtz, *in* "Lipid–Protein Interactions" (P. C. Jost and O. H. Griffith, eds.) p. 151. Wiley (Interscience), New York, 1982.

[5] H. H. Kamp, K. W. A. Wirtz, and L. L. M. Van Deenen, *Biochim. Biophys. Acta* **318,** 313 (1973).

[6] G. M. Helmkamp, Jr., M. S. Harvey, K. W. A. Wirtz, and L. L. M. Van Deenen, *J. Biol. Chem.* **249,** 6382 (1974).

[7] P. E. DiCorleto, J. B. Warach, and D. B. Zilversmit, *J. Biol. Chem.* **254,** 7795 (1979).

ISBN 0-12-181998-1

will also accelerate the transfer of phosphatidylcholine and, to a much lesser extent, sphingomyelin; and ns-TP for the nonspecific lipid transfer protein from liver[3,8] (see Table I).

There is one other distinction worth mentioning. If a relatively nonspecific protein mediates the exchange of lipid between membrane fractions, it may, under certain conditions, accelerate the bidirectional movement of the same lipid and, under different conditions, promote the transfer of one lipid in one direction and that of a different lipid in the opposite direction. We have proposed to use "homo-exchange" for the former and "hetero-exchange" for the latter,[3] leaving the generic name "exchange" for a process that may include both homo- and hetero-exchange.

Properties of Transfer Proteins

Table I shows a listing of various purified lipid transfer proteins, their molecular weights, and their isoelectric points. Both beef and rat liver cytosol contain a transfer protein that is quite specific for the transfer of phosphatidylcholine. Although the molecular weights for the proteins isolated from beef and rat liver are both 28,000, their isoelectric points differ considerably. The primary structure of the phosphatidylcholine-specific transfer protein from beef liver has been determined,[9] and the lipid binding site has been identified,[4]

A phosphatidylinositol–phosphatidylcholine-specific transfer protein has been isolated from beef strain or beef heart cytosol.[6,7] The proteins appear to be quite similar as judged by their molecular weights, isoelectric points, and lipid specificities. It is interesting that two forms of the protein, with slightly different isoelectric points, are present in each tissue.

Nonspecific transfer proteins have been isolated from rat[8] and beef[10] liver as well as from rat hepatomas.[11] The beef liver protein has been most extensively examined for differing lipid substrates and will promote the transfer not only of most phospholipids, but also of a glycosphingolipid,[12] ganglioside,[12] and cholesterol.[10] The protein does not appear to transfer cardiolipin.

Finally, in the $d > 1.21$ density fraction of rabbit and human blood plasma, one or more lipid transfer proteins appear to accelerate the transfer of cholesteryl ester, triglyceride, retinyl ester, and phosphatidylcholine.

[8] B. Bloj and D. B. Zilversmit, *J. Biol. Chem.* **252,** 1613 (1977).

[9] R. Akeroyd, P. Moonen, J. Westerman, W. C. Puyk, and K. W. A. Wirtz, *Eur. J. Biochem.* **114,** 385 (1981).

[10] R. C. Crain and D. B. Zilversmit, *Biochemistry* **19,** 1433 (1980).

[11] E. V. Dyatlovitskaya, N. G. Timofeeva and L. D. Bergelson, *Eur. J. Biochem.* **82,** 463 (1978).

[12] B. Bloj and D. B. Zilversmit, *J. Biol. Chem.* **256,** 5988 (1981).

TABLE I
PURIFIED LIPID TRANSFER PROTEINS

Source	M_r	Isoelectric point	Lipid transferred[a]	References
Brain, beef	I 32,300	5.3	PI,PC	*b,c*
	II 32,800	5.6		
Heart, beef	I 33,500	5.3	PI,PC,SPH	*d*
	II 33,500	5.6		
Hepatoma, rat	11,200	5.2	SPH,PC,PE,PI + PS	*e*
Liver, beef	28,000	5.8	PC	*f,g*
Liver, beef	I 14,500	9.55	PC,PE,SPH,PA,PG, PI,PS,cholesterol, G_{M1},$GbOse_4Cer$	*h,i*
	II 14,500	9.75		
Liver, rat	28,000	8.4	PC	*j–l*
Liver, rat	12,500	8.8	PE,PC,PI,PS,SPH, cholesterol	*m,n*
Plasma, human	66,000	4.8	PL,CE,RE,TG	*o–r*
	58,000			

[a] Abbreviations: CE, cholesteryl ester; $GbOse_4Cer$, globotetraglycosylceramide; G_{M1}, II^3-α-*N*-acetylneuraminosyl-gangliotetraglycosylceramide; PA, phosphatidic acid; PC, phosphatidylcholine; PE, phosphatidylethanolamine; PI, phosphatidylinositol, PG, phosphatidylglycerol; PL, phospholipid; PS, phosphatidylserine; RE, retinyl ester; SPH, sphingomyelin; TG, triglyceride.

[b] G. M. Helmkamp, Jr., M. S. Harvey, K. W. A. Wirtz, and L. L. M. Van Deenen, *J. Biol. Chem.* **249,** 6382 (1974).

[c] R. A. Demel, R. Kalsbeek, K. W. A. Wirtz, and L. L. M. Van Deenen, *Biochim. Biophys. Acta* **446,** 10 (1977).

[d] P. E. DiCorleto, J. B. Warach, and D. B. Zilversmit, *J. Biol. Chem.* **254,** 7795 (1979).

[e] E. V. Dyatlovitskaya, N. G. Timofeeva, and L. D. Bergelson, *Eur. J. Biochem.* **82,** 463 (1978).

[f] H. H. Kamp, K. W. A. Wirtz, and L. L. M. Van Deenen, *Biochim. Biophys. Acta* **318,** 313 (1973).

[g] P. Moonen, R. Akeroyd, J. Westerman, W. C. Puyk, P. Smits, and K. W. A. Wirtz, *Eur. J. Biochem,* **106,** 279 (1980).

[h] R. C. Crain and D. B. Zilversmit, *Biochemistry* **19,** 1433 (1980).

[i] B. Bloj and D. B. Zilversmit, *J. Biol. Chem.* **256,** 5988 (1981).

[j] C. Lutton and D. B. Zilversmit, *Biochim. Biophys. Acta* **441,** 370 1976.

[k] R. H. Lumb, A. D. Kloosterman, K. W. A. Wirtz, and L. L. M. Van Deenen, *Eur. J. Biochem.* **69,** 15 (1976).

[l] B. J. H. M. Poorthuis, T. P. van der Krift, T. Teerlink, R. Akeroyd, K. Y. Hostetler, and K. W. A. Wirtz, *Biochim. Biophys. Acta* **600,** 376 (1980).

[m] B. Bloj and D. B. Zilversmit, *J. Biol. Chem.* **252,** 1613 (1977).

[n] B. J. H. M. Poorthuis, J. F. C. Glatz, R. Akeroyd, and K. W. A. Wirtz, *Biochim. Biophys. Acta* **665,** 256 (1981).

[o] N. M. Pattnaik, A. Montes, L. B. Hughes, and D. B. Zilversmit, *Biochim. Biophys. Acta* **530,** 428 (1978).

[p] R. E. Morton and D. B. Zilversmit, *J. Lipid Res.* **23,** 1058 (1982).

[q] J. Ihm, J. L. Ellsworth, B. Chataing, and J. A. K. Harmony, *J. Biol. Chem.* **257,** 4818 (1982).

[r] A. R. Tall, E. Abreu, and J. Schuman, *J. Biol. Chem.* **258,** 2174 (1983).

Several workers have partially purified these proteins.[13-18] In addition, a protein that inhibits lipid transfer also appears to be present in plasma.[19] Table I shows the lipid transfer properties observed in our own laboratory.

Other Sources of Lipid Transfer Activities

Table II shows additional examples of animal and plant tissues that appear to contain lipid transfer activities. Apparently, lipid transfer activities are widely distributed among cell types. Thus far, there is no indication as to whether a specific type of transfer activity is associated with a particular cell type or with a specialized cell function.

Assay of Transfer Activities

Crude Cytosol Activity

In general, the transfer activity of a purified transfer protein, or of a crude cytosol fraction, is measured by determining the transfer of an isotopically labeled lipid from a donor to an acceptor particle. The donor or acceptor may be an intact cell, an organelle, a plasma membrane, a synthetic vesicle, a lipoprotein, or a monolayer. Since the merits and shortcomings of these assays have been discussed in detail,[4,20] here we shall concern ourselves primarily with general comments.

In using a crude cytosol for the assay, one does not know which transfer protein is being measured, because two or more transfer proteins may be present. In liver, for example. both specific and nonspecific transfer proteins are found. Thus, if one measures the transfer of labeled phosphatidylcholine, one presumably obtains the total of both activities. Some discrimination between transfer activities is still possible, however, by the judicious use of labeled substrate and by choice of certain incubation conditions. If, for example, one uses labeled phosphatidylethanolamine for substrate, only the nonspecific transfer activity is measured. The combined use of labeled phosphatidylcholine and phosphatidylethanolamine could reveal both

[13] D. B. Zilversmit, L. B. Hughes, and J. Balmer, *Biochim. Biophys. Acta* **409**, 393 (1975).

[14] N. M. Pattnaik, A. Montes, L. B. Hughes and D. B. Zilversmit, *Biochim. Biophys. Acta* **530,** 428 (1978).

[15] R. E. Morton and D. B. Zilversmit, *J. Lipid Res.* **23,** 1058 (1982).

[16] J. Ihm, J. L. Ellsworth, B. Chataing, and J. A. K. Harmony, *J. Biol. Chem.* **257,** 4818 (1982).

[17] A. R. Tall, E. Abreu, and J. Schuman, *J. Biol. Chem.* **258,** 2174 (1983).

[18] D. B. Zilversmit, R. E. Morton, L. B. Hughes, and K. H. Thompson, *Biochim. Biophys, Acta,* **712,** 88 (1982).

[19] R. E. Morton and D. B. Zilversmit, *J. Biol. Chem.* **256,** 11992 (1981).

[20] D. B. Zilversmit and M. E. Hughes, *Methods Membr. Biol.* **7,** 211 (1976).

TABLE II
PARTIALLY PURIFIED LIPID TRANSFER PROTEINS

Source	Lipid transferred[a]	References
Animals		
Intestine, rat	PC	*b,c*
Intestine, rat	PI, PC	*c*
Liver, rat	PI, PC	*d*
Lung, rat	PC, PG, PE	*e*
Lung, sheep	PC, PI, PE, PG	*f,g*
Spleen, beef	Glycosylceramide	*h*
Plants		
Endosperm, germinating castor bean	PC	*i,j*
Tuber, potato	PC, PI, PE	*k*
Other		
Rhodopseudomonas sphaeroides	PC, PE, PG	*l*
Yeast	PC	*m*

[a] Abbreviations: PC, phosphatidylcholine; PE, phosphatidylethanolamine, PG, phosphatidylglycerol; PI, phosphatidylinositol.
[b] C. Lutton and D. B. Zilversmit, *Lipids* **11,** 16 (1976).
[c] K. Yamada, T. Sasaki, and T. Sakagami, *J. Biochem.* (*Tokyo*) **84,** 855 (1978)
[d] R. H. Lumb, A. D. Kloosterman, K. W. A. Wirtz, and L. L. M. Van Deenen, *Eur. J. Biochem.* **69,** 15 (1976).
[e] L. M. G. Van Golde, V. Oldenborg, M. Post, and J. J. Batenburg, *J. Biol. Chem.* **255,** 6011 (1980).
[f] M. E. Robinson, L. N. Y. Wu, G. W. Brumley, and R. H. Lumb, *FEBS Lett.* **87,** 41 (1978).
[g] C. D. Whitlow, G. L. Pool, G. W. Brumley, and R. H. Lumb, *FEBS Lett.* **113,** 221 (1980).
[h] R. J. Metz and N. S. Radin, *J. Biol. Chem.* **255,** 4463 (1980).
[i] D. Douady, J. C. Kader, and P. Mazliak, *Plant Sci. Lett.* **17,** 295 (1980).
[j] T. Tanaka and M. Yamada, *Plant Cell Physiol.* **20,** 533 (1979).
[k] J. C. Kader, *Biochim. Biophys, Acta* **380,** 31 (1975).
[l] L. K. Cohen, D. R. Lueking, and S. Kaplan, *J. Biol. Chem.* **254,** 721 (1979).
[m] G. S. Cobon, P. D. Crowfoot, M. Murphy, and A. W. Linnane, *Biochim. Biophys. Acta* **441,** 255 (1976).

activities in a single assay if appropriate standards of the purified lipid transfer proteins are included in the assay. Alternatively, one may increase the ionic strength of the incubation medium. This has a preferential inhibitory effect on the nonspecific transfer activity from beef liver.[10]

An additional problem exists in the use of crude cytosol as the fraction to be assayed. The presence of proteases may lead to rapid deterioration of transfer activity, and the presence of phospholipases may alter the donor and acceptor particles, or solubilize labeled lipids that are used to measure transfer activity.

A similar problem may be encountered with cell organelles that are used as donors or acceptors. Lipase activity of beef heart mitochondria accounts for the transfer of triglyceride fatty acids from sonicated vesicles to mitochondria during prolonged incubations, thus invalidating the use of triglyceride as a nonexchangeable marker (see below). We have avoided this problem by heat treatment of the mitochondria prior to their use as acceptor particles.[20]

Nonexchangeable Markers[20]

The transfer of labeled lipid molecules from one membrane fraction to another is frequently not readily distinguishable from the transfer of this lipid as part of a fusion or binding of the donor lipid particle to the acceptor. In some studies a correction is made for bulk transfer of lipid by subtracting the labeled lipid transferred at 4° from that transferred at the temperature of the experiment. Such a correction may not be valid if this bulk transfer of the labeled lipid species is affected by temperature, as may be the case for fusion. A more satisfactory procedure may be in the incorporation of a nonexchangeable marker in the donor particle. Triglyceride, cholesteryl ester, and cardiolipin have been used for synthetic vesicles,[20] whereas *in vivo* labeled integral membrane proteins may serve as nonexchangeable markers for natural membranes.[21] A third method of measuring bulk transfer of lipid, as part of vesicle sticking or fusion, is by the incorporation of a nonpermeant chemical in the aqueous space of the lipid vesicle.[22] When using this technique, special attention must be given to the development of membrane leaks as time progresses, or as transfer protein is added. There is one more aspect of the corrections for binding or fusion that deserves mention. Even when a nonexchangeable marker is incorporated in the donor particle, it is possible that the correction factor is erroneous if the ratio of nonexchangeable to exchangeable markers in the donor particles is not the same for each particle, i.e., if heterogeneity of labeling is present. Such heterogeneity may occur, for example, if the donor particle population consists of a mixture of unilamellar and multilamellar vesicles. It is sometimes possible to eliminate this heterogeneity by gel filtration of the donor particles before the assay.

[21] D. B. Zilversmit and M. E. Hughes, *Biochim. Biophys, Acta* **469**, 99 (1977).

[22] D. Hoekstra, J. van Renswoude, R. Tomasini, and G. Scherphof, *Membr. Biochem.* **4,** 129 (1981).

Measuring Appearance or Disappearance of Label

In a transfer reaction one can measure either the loss of label from the donor particle or the gain of label in the acceptor. It is good practice, at least when testing new assay conditions, to measure labeled substrate in both donor and acceptor fractions and thus be able to determine the recovery of labeled lipid in the total system. It is not unheard of to lose an appreciable fraction of the donor or acceptor particles by sticking to the surface of the incubation vessel.

If the disappearance of labeled lipid from the donor particle is measured, it is convenient to separate donor and acceptor by centrifugation and choose an acceptor particle that sediments more readily than the donor. If small unilamellar vesicles are used as donors, mitochondria[20] or multilamellar vesicles[23] are suitable acceptors. After centrifugation, one simply determines the ratio of radioactivity due to the exchangeable and nonexchangeable markers in the supernatant fraction. This ratio is directly comparable to a similar ratio in the donor particles at the beginning of incubation.

If one measures the appearance of label in the acceptor particle, it is advantageous to incorporate the nonexchangeable marker in the acceptor particle and arrange the assay so as to leave the acceptor in the supernatant fraction at the end of the incubation and centrifugation. This is accomplished, for example, if rat liver microsomes serve as donors and small unilamellar vesicles as acceptors. These particles are readily separated by lowering the pH followed by centrifugation.[4] In this instance the nonexchangeable marker is used to corrct for the 20–30% of the acceptor particle that sediments during centrifugation.

Procedures other than centrifugation have been used to separate donor and acceptor particles. After incorporation of glycolipid in one of the particles, a separation can be achieved by the addition of a lectin[24] or of antibody.[25] If one of the particles carries a negative charge is may be retained on an anion-exchange column.[26,27]

Certain physical techniques may lend themselves to the measurement of lipid transfer without actually separating the donor and acceptor membranes. An exchangeable lipid species that is either electron spin labeled or fluorescent would exhibit a broadened or quenched spectrum when the labeled lipid is present in high concentrations, such as in the donor particle.

[23] P. E. DiCorleto and D. B. Zilversmit, *Biochemistry* **16,** 2145 (1977).

[24] T. Sasaki and T. Sakagami, *Biochim. Biophys. Acta* **512,** 461 (1978).

[25] C. Ehnholm and D. B. Zilversmit, *Biochim. Biophys. Acta* **274,** 652 (1972).

[26] J. A. Hellings, H. H. Kamp, K. W. A. Wirtz, and L. L. M. Van Deenen, *Eur. J. Biochem.* **47,** 601 (1974).

[27] A. M. H. P. Van den Besselaar, G. M. Helmkamp, Jr., and K. W. A. Wirtz, *Biochemistry* **14,** 1852 (1975).

The transfer to an acceptor membrane would, therefore, be detectable by spectral changes without the necessity of separating the membrane fractions. Studies with electron spin labels have been carried out.[4]

Function of Lipid Transfer Proteins

Little information is available for the role of lipid transfer proteins in the function of the cell or in the metabolism of lipids. Inasmuch as lipids are not water soluble, it would not be unreasonable if the cell possessed proteins that were capable of mediating the transfer of lipids from one site to another. In view of the observation that most lipids are synthesized by the endoplasmic reticulum, and that these lipids ultimately become part of other organelles or of the plasma membrane, a method whereby insoluble lipids are readily transferable seems essential. Although in some instances membrane flow or fusion may account for lipid transfer, the existence of transfer proteins may lend an element of specificity to this transfer and may introduce a step that, in principle at least, can be regulated.

There are several indications that the nonspecific lipid transfer protein of liver may play a role in cholesterol biosynthesis. Nolan *et al.*[28] have purified a protein (SCP_2) that activates one or more steps of cholesterol biosynthesis between lanosterol and cholesterol. They suggest that this protein is identical to the nonspecific lipid transfer protein.

Use of Transfer Proteins as Tools

During the initial stages of the investigation of transfer protein, we observed that phosphatidylcholine of small unilamellar vesicles could be only partially replaced by phospholipids from some other lipid source. The initial rate of phospholipid exchange was, however, directly proportional to the added transfer protein. Moreover, the fraction of exchangeable lipid appeared to be greater for small vesicles and appeared to be identical to the lipid accessible to phospholipase D. It seemed reasonable to conclude, therefore, that the rapidly exchangeable lipid of small unilamellar vesicles was located in the outside monolayer of the membrane.[29]

Measurements of lipid asymmetry by this method confirmed and extended studies by other techniques in showing that individual lipid species may exhibit uneven distributions over the two monolayers. Undoubtedly these asymmetries are determined to some extent by the electric charge and size of polar headgroups and by the chemical nature of the hydrophobic portions of the molecules, as well as by the curvature of the membrane. It is

[28] B. J. Noland, R. E. Arebalo, E. Hansbury, and T. J. Scallen, *J. Biol. Chem.* **255,** 4282 (1980).
[29] L. W. Johnson, M. E. Hughes, and D. B. Zilversmit, *Biochim. Biophys. Acta* **375,** 176 (1975).

of interest that lipid asymmetry is maintained even in membranes in which lipids appear to be transferred relatively rapidly from one side of the membrane to the other.[30] Asymmetry measurements with transfer proteins have been compared to other methods in a review by Etemadi.[31] Van Deenen[32] has reviewed the use of transfer proteins and of phospholipases to asymmetry studies on erythrocytes.

We have also used transfer proteins as modifiers of membrane composition, in order to study the dependence of enzyme activity on membrane lipid composition. Rat liver microsomes were incubated with sonicated lipid vesicles of varying composition, and different transfer proteins were used to accomplish the changes in membrane composition. With the phosphatidylcholine-specific transfer protein, we were able to change the phosphatidylcholine fatty acid composition of the microsomal membranes, whereas with the phosphatidylinositol–phosphatidylcholine-specific and nonspecific lipid transfer proteins we could alter the phospholipid headgroup pattern. Upon measurement of microsomal glucose-6-phosphatase activity, we observed that alterations in phosphatidylethanolamine composition were the most effective in modifying enzyme activity.[33] Similar applications of lipid transfer protein have been proposed by Bergelsen and Barsukov.[34] For example, the incorporation of lysophosphatidylcholine in rat liver microsomes was found to increase activity of glucose 6-phosphatase and decrease cytochrome *P*-450 levels.[35]

The phosphatidylcholine-specific transfer protein has also been used to supply labeled phosphatidylcholine to various membrane fractions to study the conversion of phosphatidyl[^{3}H]choline to sphingomyelin.[36] The authors suggest (personal communication) that the *in vivo* synthesis of sphingomyelin may also be mediated by lipid transfer protein.

Acknowledgment

This work was supported in part by Public Health Service Research Grant HL 10940 from the National Heart, Lung and Blood Institute of the U.S. Public Health Service.

[30] R. C. Crain and D. B. Zilversmit, *Biochemistry* **19,** 1440 (1980).
[31] A-H. Etemadi, *Biochim. Biophys. Acta* **604,** 423 (1980).
[32] L. L. M. Van Deenen, *FEBS Lett.* **123,** 3 (1981).
[33] R. C. Crain and D. B. Zilversmit, *Biochemistry* **20,** 5320 (1981).
[34] L. D. Bergelson and L. I. Barsukov, *Science* **197,** 224 (1977).
[35] E. V. Dyatlovitskaya, A. F. Lemenovskaya, A. T. Valdnietse, E. V. Sinitsyna, and L. D. Bergelson, *Biochemistry* (*Engl. Transl.*) **45,** 1546 (1980).
[36] D. R. Voelker and E. P. Kennedy, *Fed. Proc., Fed. Am. Soc. Exp. Biol.* **40,** 1805 (1981).

[51] Nonspecific Lipid Transfer Protein from Rat and Beef Liver: Purification and Properties

By BERNABÉ BLOJ and DONALD B. ZILVERSMIT

The nonspecific lipid transfer proteins accelerate the intermembrane transfer of phosphatidylcholine, phosphatidylethanolamine, phosphatidylinositol, phosphatidylserine, phosphatidic acid, phosphatidylglycerol, sphingomyelin, cholesterol, neutral glycosphingolipids, and gangliosides.[1-3] They do not appear to accelerate the transfer of diphosphatidylglycerol, cholesteryl ester, or triglycerides. They are also capable of promoting net phospholipid transfer to membranes or modified lipoproteins.[4] The nonspecific lipid transfer proteins differ in various respects from other proteins that promote the transfer of one or more phospholipids between a large variety of membranes, lipoproteins, emulsions, and monolayers. The nonspecific lipid transfer proteins have relatively lower molecular weights, between 13,500 and 14,500, and higher isoelectric points, between 8.6 and 9.75. They are more heat stable and withstand heating at 90° with little loss of activity.[1,2]

The transfer proteins have been useful tools for the study of the distribution and movement of phospholipids in artificial vesicles,[5] microsomes,[6] intact erythrocytes,[7] and Semliki Forest virus[8] and of cholesterol in the brush-border plasma membranes of the rabbit intestine.[9] They have also been employed to study the effect of changes in phospholipid composition on the activity of rat liver microsomal enzymes, e.g., glucose-6-phosphate phosphohydrolase (EC 3.1.3.9),[10] to generate isotopically asymmetric vesicles in studies dealing with vesicle–cell interactions[11] and to introduce spin-labeled phosphatidylcholine into membranes.[12]

The table summarizes the different lipids transferred by the nonspecific

[1] B. Bloj and D. B. Zilversmit, *J. Biol. Chem.* **252,** 1613 (1977).
[2] R. C. Crain and D. B. Zilversmit, *Biochemistry* **19,** 1433 (1980).
[3] B. Bloj and D. B. Zilversmit, *J. Biol. Chem.* **256,** 5988 (1981).
[4] R. C. Crain and D. B. Zilversmit, *Biochim. Biophys. Acta* **620,** 37 (1980).
[5] P. E. DiCorleto and D. B. Zilversmit, *Biochim. Biophys. Acta* **552,** 114 (1979).
[6] D. B. Zilversmit and M. E. Hughes, *Biochim. Biophys. Acta* **469,** 99 (1977).
[7] R. C. Crain and D. B. Zilversmit, *Biochemistry* **19,** 1440 (1980).
[8] G. Van Meer, K. Simons, J. A. F. Op den Kamp, and L. L. M. Van Deenen, *Biochemistry* **20,** 1974 (1981).
[9] B. Bloj and D. B. Zilversmit, *J. Biol. Chem.* **257,** 7608 (1982).
[10] R. C. Crain and D. B. Zilversmit, *Biochemistry* **20,** 5320 (1981).
[11] A. Sandra and R. E. Pagano, *J. Biol. Chem.* **254,** 2244 (1979).
[12] F. M. Megli, C. Landriscina, and E. Quagliarello, *Biochim. Biophys. Acta* **640,** 274 (1981).

METHODS IN ENZYMOLOGY, VOL. 98

ISBN 0-12-181998-1

REFERENCES TO THE LITERATURE ON ASSAY SYSTEMS FOR LIPID TRANSFER BETWEEN DIFFERENT NATURAL AND ARTIFICIAL MEMBRANES IN THE PRESENCE OF NONSPECIFIC LIPID TRANSFER PROTEIN

Lipid	SUV/ MLV[a]	SUV/ mitochondria	SUV/ erythrocyte ghosts	Microsomes/ mitochondria
Phosphatidylcholine	5,2[b]	1	1	6
Phosphatidylethanolamine	2	1	—	6
Phosphatidylinositol	2	1	—	6
Phosphatidylserine	2	—	—	6
Sphingomyelin	2	1	—	6
Phosphatidic acid	2	—	—	—
Phosphatidylglycerol	2	—	—	—
Ceramide tetrahexoside	—	—	3	—
Ganglioside G_{M1}	—	—	3	—
Ceramide	9	9	—	—
Diglyceride	9	9	—	—
Cholesterol	2	1	1	—
Spin-labeled phosphatidylcholine[c]	12	12	—	—

[a] Abbreviations: SUV, small unilamellar vesicles; MLV, multilamellar vesicles.
[b] Numbers in columns 2–5 refer to text footnotes.
[c] 1-Acyl-2,12-(4,4-dimethyloxazolidine-*N*-oxyl)stearoylglycerophosphorylcholine.

lipid transfer protein and some of the transfer systems used for the assay. The nonspecific lipid transfer proteins from rat and beef liver exhibit similar properties.[1,2,13] Although the purification protocol was first developed for the isolation and characterization of the proteins from rat liver,[1] beef liver provides a more convenient starting material when relatively large amounts of protein are required.

Procedures to measure transfer activity and to purify the proteins from rat and beef liver are described in this chapter.

Assay of Transfer Activity

The assay is based on the ability of nonspecific lipid transfer proteins to accelerate the transfer of labeled phosphatidylethanolamine from phosphatidylethanolamine–phosphatidylcholine small unilamellar vesicles (donor particle) to beef heart mitochondria (acceptor particle). A trace of labeled triolein is included in the small unilamellar vesicles during preparation to correct for vesicle recovery. After the incubations, the mitochondria are pelleted and aliquots of the supernatants, containing the vesicles, are counted. A measure of the transfer activity is given by the decrease in the phosphatidylethanolamine/triolein ratio of the supernatants.

[13] B. Bloj, M. E. Hughes, D. B. Wilson, and D. B. Zilversmit, *FEBS Lett.* **96,** 87 (1978).

Lipids

Labeled phospholipids are prepared from rat liver. A rat (250–350 g body weight) is injected intraperitoneally with 10–15 mCi of $^{32}P_i$ 15–18 hr before sacrifice. The liver is removed, minced, and extracted for 3 hr with 20 volumes of chloroform–methanol, 2:1 (v/v). The filtered crude extract is washed by the procedure of Folch *et al.*,[14] and solvent is removed in a rotary evaporator under N_2. The lipid residue is redissolved in 25 ml of chloroform, which contains up to 0.75% of methanol as preservative, and subjected to fractionation by chromatography on alumina. To neutral alumina, activity super I (ICN Pharmaceuticals, Cleveland, Ohio), 10% water is added before the mixture is ground in a mortar. The alumina is then shaken vigorously for 1.5–2 hr to allow for equal distribution of the water. A slurry of 25 g of alumina in chloroform is prepared and packed into a water-jacketed column (1 in. diameter). The lipid extract is applied, and the column is eluted sequentially with 100 ml of chloroform, 100 ml of chloroform–methanol, 2:1 (v/v), 150 ml of chloroform–methanol–water, 25:25:4 (v/v/v), and finally with 325 ml of chloroform–methanol–water, 25:25:7.5 (v/v/v). Fractions of 25 ml are collected. Most of the labeled phosphatidylethanolamine appears in the first 4 or 5 fractions of the last eluent. These fractions are pooled, solvent is evaporated in a rotary evaporator, and the lipids are redissolved in chloroform. Approximately 75 μmol of labeled phosphatidylethanolamine, with specific activity between 2.5 and 3.5×10^6 cpm/μmol, are obtained in one preparation. It is kept in chloroform at $-20°$. The radiopurity of phosphatidylethanolamine is 95–96% as assayed by thin-layer chromatography on silica gel H with chloroform–methanol–acetic acid–water, 25:15:4:2 (v/v/v/v). The partially purified, labeled phosphatidylcholine is eluted in the chloroform–methanol (2:1) fractions. Greater than 98% pure labeled phosphatidylcholine can be obtained from the partially purified phosphatidylcholine by chromatography on a second alumina column containing 1% water. The second column (15 g) is packed in chloroform. The solvent of the partially purified labeled phosphatidylcholine fraction is evaporated in a rotary evaporator, and the lipids are redissolved in 5–10 ml of chloroform–methanol, 19:1 (v/v). The sample is applied to the column and eluted with 100 ml of the same solvent and then with 300 ml of chloroform–methanol, 9:1 (v/v). Fifty-milliliter fractions are collected. Most of the phosphatidylcholine is eluted with the second solvent mixture. Approximately 100 μmol of labeled phosphatidylcholine is obtained in one preparation with a specific activity similar to that of the phosphatidylethanolamine.

[^{3}H]Triolein (Amersham Corp., Arlington Heights, Illinois) is purified

[14] J. Folch, M. Lees, and G. H. Sloane-Stanley, *J. Biol. Chem.* **226**, 497 (1957).

by thin-layer chromatography on silica gel H with hexane–diethyl ether–acetic acid, 60:40:1 (v/v/v). The triolein is eluted with chloroform and stored at −20°. Unlabeled egg phosphatidylcholine (Lipid Products, South Nutfield, England) and butylated hydroxytoluene (Nutritional Biochemical Corporation, Cleveland, Ohio) are used without further purification.

Preparation of Sonicated Unilamellar Vesicles

Appropriate aliquots of the stock solutions, containing 6.45 μmol of [^{32}P]phosphatidylethanolamine, 6.45 μmol of phosphatidylcholine, 0.046 μmol of butylated hydroxytoluene, and a trace amount of [^{3}H]triolein are mixed. The solvent is evaporated under a stream of N_2. The lipids are redissolved in 2–3 ml of diethyl ether, and the solvent is removed again under N_2. After the addition of 2 ml of 250 m*M* sucrose, 1 m*M* Na_2EDTA, 50 m*M* Tris-HCl, 3 m*M* NaN_3, pH 7.4 (SET buffer), the suspension is shaken on a Vortex mixer for 5–10 min and left to swell for 0.5–1 hr at room temperature. The lipid suspension in a sealed test tube is sonicated in a sonicating water bath for 30 min. After sonication, the vesicles are diluted with 8 ml of SET buffer containing 500 mg of bovine serum albumin (fraction V powder, fatty acid poor, Miles Laboratories, Elkhart, Indiana) and stored at 4° under N_2.

Preparation of Mitochondria

Beef heart mitochondria are prepared according to the procedure of Green *et al.*[15] as described in this series.[16] They are stored at −20°. Upon thawing, the mitochondria are heated for 20 min at 80° to inactivate lipases and washed with SET buffer before use.

Assay Procedure

Sonicated vesicles (32 nmol of labeled phosphatidylethanolamine) are incubated with heated mitochondria (1.6 mg of protein, 580 nmol of total phospholipid) and an appropriate aliquot of the transfer protein in a total volume of 0.5 ml of SET buffer for 90 min at 37° with gentle agitation. The transfer is terminated by centrifugation at 8000 *g* for 2 min in an Eppendorf microcentrifuge (Model 3200, Brinkmann Instruments, Inc.). Aliquots (0.3 ml) of the supernatant are counted in 5 ml of ACS scintillation counting solution (Amersham Corp.). The [^{3}H]triolein in the sonicated vesicles serves as a nonexchangeable reference marker[17] so that the decrease in the ^{32}P : ^{3}H ratio in the vesicles after incubation, compared to the ^{32}P : ^{3}H ratio in

[15] D. E. Green, R. L. Lester, and D. M. Ziegler, *Biochim. Biophys. Acta* **23,** 516 (1957).
[16] D. B. Zilversmit and L. W. Johnson, this series, Vol. 35, p. 262.
[17] D. B. Zilversmit and M. E. Hughes, *Methods Membr. Biol.* **7,** 211 (1976).

the vesicles before incubation, measures the transfer of phosphatidylethanolamine. The recovery of sonicated vesicles after the incubations is greater than 95%. Blank incubations, without transfer protein, are performed simultaneously with incubations containing transfer protein. The transfer in these blank incubations is usually less than 3% and is subtracted from the values obtained in the incubations containing transfer protein. The assay is linear up to 15% transfer.

The results can be expressed as percentage of phospholipid transferred per hour, but Zilversmit and Hughes[17] have derived a general expression to calculate the nanomoles of phospholipid transferred per unit of time. For this calculation, in the case of phosphatidylethanolamine–phosphatidylcholine small unilamellar vesicles, one has to take into account that only 53% of the labeled phosphatidylethanolamine in the sonicated vesicles is available for transfer to mitochondria or other acceptor particles.

Comments

Although mitochondria are routinely used in the standard assay, other acceptor particles, such as erythrocyte ghosts[1] or multilamellar vesicles,[2] have proved to be appropriate for some purposes. In the case of multilamellar vesicles, only about 7% of the phospholipids are available for exchange, but the amount of phospholipid transferred in the absence of transfer protein is considerably lower than with mitochondria as acceptor particles. Acceptor particles with purified phospholipid must be used when studies are carried out in which the same phospholipid composition is desired in the donor and acceptor particles.

The transfer activity of a protein preparation will vary according to the method of assay, so that the nanomoles of phospholipid transferred per unit of time and per unit weight of protein are not easily comparable between laboratories. In the case of the standard sonicated vesicle–mitochondria system, a 10-fold increase in the concentration of acceptor and donor particles yields a linear increase in the nanomoles of phospholipid transferred per milligram of protein. At the highest concentrations used, transfers of about 40 μmol hr^{-1} mg $protein^{-1}$ have been obtained with no indications of having reached saturation.

Purification of Nonspecific Lipid Transfer Proteins from Rat Liver

pH 5.1 Supernatant. Livers (200–220 g) from rats fasted overnight are minced, and the pieces are rinsed with cold SET buffer. Unless noted otherwise, all procedures are performed at 4°. A 20% (w/v) homogenate is prepared in a Waring blender in the same buffer and is centrifuged subsequently for 10 min at 15,000 *g*. The supernatant is adjusted to pH 5.1 with

1 *N* HCl, allowed to stand for 15 min, then centrifuged for 15 min at 15,000 *g*. The clear supernatant is adjusted to pH 7.2–7.4 and processed immediately.

Ammonium Sulfate Step. Solid ammonium sulfate is gradually added to 50% saturation. After 15 hr the precipitate is removed by centrifugation (30,000 *g* for 15 min) and discarded. The supernatant is adjusted to 90% saturation with ammonium sulfate, and after 2–4 hr the precipitate is collected by centrifugation. The precipitate is redissolved in 80–100 ml of 42 m*M* Tris–acetate buffer, 5 m*M* 2-mercaptoethanol, 3 m*M* NaN_3, pH 7.4. A small insoluble residue is removed by centrifugation.

Sephadex G-75 Step. The protein solution (80–100 ml, 50 mg of protein per milliliter) is loaded onto a Sephadex G-75 column (5 × 80 cm, Pharmacia Fine Chemicals, Piscataway, New Jersey), which is connected in series with a second Sephadex G-75 column (5 × 45 cm). The columns, which have been equilibrated with Tris–acetate buffer, are eluted at 3.5 ml cm^{-2} hr^{-1} and 15-ml fractions are collected. Fractions containing transfer activity are pooled and equilibrated overnight by dialysis against 1 m*M* 2-mercaptoethanol in distilled water to a buffer concentration of 10 m*M* Tris–acetate, 1 m*M* 2-mercaptoethanol and adjusted to pH 7.9.

(Carboxymethyl)cellulose Step. The pooled fractions of the Sephadex column are applied to a 2.8 × 30 cm column of (carboxymethyl)cellulose (CM-52, Whatman, Ltd., Springfield Mill, England) that has been equilibrated with 10 m*M* Tris–acetate, 1 m*M* 2-mercaptoethanol, pH 7.9. The sample is applied to the column, and the column is washed with 120 ml of buffer at 29 ml cm^{-2} hr^{-1}. The active fractions are eluted with a linear gradient formed by 600 ml of 10 m*M* and 600 ml of 50 m*M* Tris–acetate buffer, 1 m*M* 2-mercaptoethanol, pH 7.9. Flow rates are 9 ml cm^{-2} hr^{-1}, and 15-ml fractions are collected. Two activity peaks are obtained, CM_1 and CM_2, at Tris concentrations of 20 and 45 m*M*, respectively. They are concentrated by ultrafiltration through a Diaflo UM-2 membrane (Amicon Corp, Lexington, Massachusetts) in an Amicon ultrafiltration cell and dialyzed overnight against 100 volumes of 42 m*M* Tris–acetate buffer, 5 m*M* 2-mercaptoethanol, 3 m*M* NaN_3, pH 7.4.

Heat-Treatment Step. The concentrated fractions CM_1 and CM_2 are heated in a water bath at 90° for 5–10 min. Denatured proteins are discarded after centrifugation (10 min, 10,000 *g*), and the active supernatants are stored at 4°. At this stage, the CM_2 fraction is purified about 800-fold compared to the pH 5.1 supernatant. An alternative method for preparing CM_1 and CM_2 has been published.[18] The authors reported both high yields and purity.

[18] B. J. H. M. Poorthuis, J. F. C. Glatz, R. Akeroyd, and K. W. A. Wirtz, *Biochim. Biophys. Acta* **665,** 256 (1981).

Purification of Nonspecific Lipid Transfer Proteins from Beef Liver

pH 5.1 Supernatant. One liver (4–5.5 kg) is trimmed of fat, cut into 1-in. cubes, and rinsed with cold 0.25 *M* sucrose. The following steps are carried out at 4°. A 35% homogenate is prepared in a Waring blender and then centrifuged for 30 min at 13,000 *g*. The supernatant is adjusted to pH 5.1 with 3 *N* HCl; after 30 min it is centrifuged for 20 min at 13,000 *g*. The clear supernatant is adjusted to pH 7.4 with 3 *N* NaOH and stored in 900-ml fractions at −20°. The transfer activity is stable for several months.

Ammonium Sulfate Step. Two 900-ml fractions are thawed, and solid ammonium sulfate is added to 40% saturation. After 1 hr the precipitate is removed by centrifugation (30 min at 13,000 *g*). The supernatant is adjusted to 90% saturation with solid ammonium sulfate, and after 1 hr the precipitate is collected by centrifugation. The pellet is suspended in 100 ml of 25 m*M* sodium phosphate, 10 m*M* 2-mercaptoethanol, 3 m*M* NaN_3, pH 7.4. The suspension is dialyzed extensively against 5 m*M* sodium phosphate, 5 m*M* 2-mercaptoethanol, 3 m*M* NaN_3, pH 8.0.

(Carboxymethyl)cellulose Step. The dialyzed solution is applied to a 5 × 10-cm column of (carboxymethyl)cellulose, equilibrated with 5 m*M* sodium phosphate, 5 m*M* 2-mercaptoethanol, and 3 m*M* NaN_3, pH 8.0. The column is loaded at 30 ml cm^{-2} hr^{-1} and washed with 500 ml of the same buffer. The transfer activity is eluted with 500 ml of 25 m*M* sodium phosphate, 45 m*M* NaCl, 5 m*M* 2-mercaptoethanol, 3 m*M* NaN_3, pH 8.0. In the original procedure,[2] this step was carried out at pH 7.4.

Heat Treatment Step. The active fraction eluted from the ion-exchange column is heated at 90° for 5 min. The precipitated proteins are discarded after centrifugation, and the supernatant is concentrated to 50 ml by ultrafiltration through a Diaflo UM-10 membrane.

Octylagarose Column Chromatography Step. This step is performed at room temperature. The protein solution is applied to an octylagarose column (1.0 × 30 cm) that had been equilibrated with 25 m*M* sodium phosphate, 45 m*M* NaCl, 5 m*M* 2-mercaptoethanol, 3 m*M* NaN_3, pH 7.4. The flow rate is 150 ml cm^{-2} hr^{-1}. The column is washed with 50 ml of this buffer and 50 ml of 5 m*M* sodium phosphate, 5 m*M* 2-mercaptoethanol, 3 m*M* NaN_3, pH 7.4. The transfer activity is eluted with 50 ml of this solution adjusted to pH 3 and collected in 7-ml fractions. The pH is increased immediately by collecting the fractions in an equal volume of 25 m*M* sodium phosphate, 45 m*M* NaCl, 5 m*M* 2-mercaptoethanol, 3 m*M* NaN_3, pH 7.4. The protein solution is stored at 4° with little loss of activity during 2 months. If desired, CM_1 and CM_2 can be separated by (carboxymethyl)cellulose ion-exchange chromatogoraphy.[2] From 10 preparations with four different livers, we obtained 5.7 ± 0.6 (average ± SEM) mg of protein purified 1704 ± 265-fold compared to the pH 5.1 supernatant.

Inhibitors of the Nonspecific Lipid Transfer Protein from Beef Liver. The transfer activity of the beef liver protein decreases linearly with the ionic strength of the medium. With NaCl at an ionic strength of 0.20, one-fourth of the transfer activity was present compared to the activity at an ionic strength of 0.05.[2] This contrasts with the characteristics of the phosphatidylcholine-specific transfer protein from beef liver[19] and the transfer protein from beef heart.[20] *N*-Ethylmaleimide (1 m*M*) inhibits the transfer activity of the nonspecific transfer protein 48%.[21]

Acknowledgment

This work was supported in part by Public Health Service Research Grant HL 10940 from the National Heart, Lung and Blood Institute of the U.S. Public Health Service.

[19] K. W. A. Wirtz, W. S. M. Geurts van Kessel, H. H. Kamp, and R. A. Demel, *Eur. J. Biochem.* **61,** 515 (1976).

[20] L. W. Johnson and D. B. Zilversmit, *Biochim. Biophys. Acta* **375,** 165 (1975).

[21] R. C. Crain and D. B. Zilversmit, unpublished results.

[52] Phosphatidylcholine Transfer Protein from Bovine Liver

By JAN WESTERMAN, H. H. KAMP, and K. W. A. WIRTZ

Phosphatidylcholine transfer protein (PC-TP) is one of a series of soluble proteins that facilitate the intracellular transfer of lipids between membranes.[1] The PC-TP from bovine liver functions as a specific carrier of phosphatidylcholine by forming a 1 : 1 molecular complex.[2,3] The protein catalyzes a strict exchange reaction between membranes that contain phosphatidylcholine.[4] Under conditions where one of the membranes lacks phosphatidylcholine, PC-TP has been found to catalyze a net transfer reaction.[5,6] Inherent to its mode of action PC-TP only transfers phosphati-

[1] K. W. A. Wirtz, *in* "Lipid–Protein Interactions" (P. C. Jost and O. H. Griffith, eds.), Vol. 1, p. 151. Wiley (Interscience), New York, 1982.

[2] H. H. Kamp, K. W. A. Wirtz, and L. L. M. van Deenen, *Biochim. Biophys. Acta* **318,** 313 (1973).

[3] R. A. Demel, K. W. A. Wirtz, H. H. Kamp, W. S. M. Geurts van Kessel, and L. L. M. van Deenen, *Nature (London) New Biol.* **246,** 102 (1973).

[4] G. M. Helmkamp, *Biochem. Biophys. Res. Commun.* **97,** 1091 (1980).

[5] D. B. Wilson, J. L. Ellsworth, and R. L. Jackson, *Biochim. Biophys. Acta* **620,** 560 (1980).

[6] K. W. A. Wirtz, P. F. Devaux, and A. Bienvenue, *Biochemistry* **19,** 3395 (1980).

ISBN 0-12-181998-1

dylcholine that is present in the outer monolayer of membranes.[7,8] This property has made PC-TP a very useful tool in studies on the distribution and transbilayer movement of phosphatidylcholine in membranes.[9,10] This chapter describes the large-scale purification of PC-TP from bovine liver.

Assay Method

Microsome–Vesicle Assay. The purification of PC-TP is followed by measuring the transfer of [^{14}C]phosphatidylcholine from rat liver microsomes to phospholipid vesicles according to a slight modification of a method previously described.[11] The microsomes are specifically labeled with [^{14}C]phosphatidylcholine, whereas the vesicles consist of phosphatidylcholine (98 mol %), phosphatidic acid (2 mol %), and a trace of [^{3}H]cholesteryl oleate as a nontransferable internal standard. In brief, ^{14}C-labeled microsomes (1.25 mg of protein) are incubated with ^{3}H-labeled vesicles (1 μmol of lipid phosphorus) and PC-TP activity for 20 min at 25° in a total volume of 2.5 ml of 0.25 *M* sucrose–1 m*M* EDTA–10 m*M* Tris-HCl, pH 7.4. At the end of incubation the microsomes are quantitatively sedimented (10 min at 10,000 *g*) by adjusting the pH of the medium to 5.1 by addition of 0.5 ml of 0.2 *M* sodium acetate–acetic acid (pH 5.0). Controls without PC-TP activity are carried through the entire procedure. Lipids are extracted from the supernatant fractions, and the ^{14}C and ^{3}H radioactivity is determined. The ^{14}C : ^{3}H ratio corrected for the ratio of the control incubation is a measure for transfer activity. The percentage of microsomal [^{14}C]phosphatidylcholine transferred to the vesicles equals $A \times a/b \times 100\%$ in which A is the corrected ^{14}C : ^{3}H ratio, a (cpm) the ^{3}H label in the vesicles before incubation, and b (cpm) the ^{14}C label in the microsomes before incubation. From the percentage figure and the pool of microsomal PC involved in the transfer reaction (i.e., 0.65 μmol), units of transfer activity are calculated. One unit of activity is defined as the amount of protein required to transfer 1×10^{-3} μmol of PC from microsomes to vesicles per minute at 25°.

Other Assays. PC-TP has also been purified by the use of the vesicle–mitochondrion assay.[12] Typically, phospholipid vesicles containing [^{32}P]phosphatidylcholine and a trace of [^{14}C]triolein as nonexchangeable

[7] J. E. Rothman and E. A. Dawidowicz, *Biochemistry* **14,** 2809 (1975).
[8] B. de Kruijff and K. W. A. Wirtz, *Biochim. Biophys. Acta* **468** 318 (1977).
[9] J. M. Shaw, N. F. Moore, E. J. Patzer, M. C. Correa-Freire, R. R. Wagner, and T. E. Thompson, *Biochemistry* **18,** 538 (1979).
[10] G. van Meer, B. J. H. M. Poorthuis, K. W. A. Wirtz, J. A. F. Op den Kamp, and L. L. M. van Deenen, *Eur. J. Biochem.* **103,** 283 (1980).
[11] H. H. Kamp and K. W. A. Wirtz, this series, Vol. 32, p. 140.
[12] P. E. DiCorleto and D. B. Zilversmit, *Biochemistry* **16,** 2145 (1977).

marker are incubated with beef heart mitochondria as acceptor membrane. After sedimentation of the mitochondria the $^{32}P:^{14}C$ ratio of the supernatant is determined. The decrease of this ratio relative to its value before incubation is a measure of PC-TP transfer activity. A continuous spectroscopic assay has been developed that makes use of fluorescent 2-parinaroylphosphatidylcholine.[13] Donor vesicles of this phospholipid show a low level of fluorescence due to self-quenching. When PC-TP and acceptor vesicles of egg phosphatidylcholine are added, transfer of fluorescent phosphatidylcholine to the acceptor vesicles is accompanied by a marked increase of fluorescence. The initial rate of fluorescence enhancement is a measure for transfer activity. This assay has been found useful to monitor transfer activity in column eluents.

Purification Procedure

The table summarizes the various steps of the purification procedure based on a modification of a previously described method.[2,11] The data presented derive from the processing of 40,000 g of bovine liver. In practice, two times 20,000 g of liver are processed on different days for steps 1–4 of the procedure. The active fractions from the DEAE-cellulose column (step 4) are pooled and used for steps 5 and 6. All manipulations are performed at 4°.

Step 1. First pH Adjustment. Fresh bovine liver (20,000 g) is cut in small pieces that are washed with 20 liters of 0.25 *M* sucrose. A 30% homogenate in 0.25 *M* sucrose is prepared with an Ystral homogenizer (type 40/34,

PURIFICATION OF PHOSPHATIDYLCHOLINE TRANSFER PROTEIN FROM BOVINE LIVER

Step	Volume (ml)	Protein[a] (mg)	Specific activity[b]	Recovery (%)	Purification factor
1. First pH adjustment	80,200	1,636,000	0.5	100	—
2. Second pH adjustment	80,200	533,000	2	130	4
3. $(NH_4)_2SO_4$ precipitation	8,850	197,000	4	91	8
4. DEAE-cellulose	6,050	13,400	40	66	80
5. CM-cellulose	620	900	254	28	508
6. Sephadex G-50	950	97	2108	25	4216

[a] The protein content is determined as before[11] except for step 6, which is calculated from $\epsilon_{280}{}^{1\%} = 24.4$.

[b] Specific activity is expressed as units of phosphatidylcholine transfer protein activity per milligram of protein.

[13] P. Somerharju, H. Brockerhoff, and K. W. A. Wirtz, *Biochim. Biophys. Acta* **649**, 521 (1981).

Dottingen, FRG), for 1 min at high speed. The pH of the homogenate is adjusted to 5.1 with 4 *M* HCl under continuous stirring. A massive precipitate is formed that is removed by centrifugation in a WKF Model G 50 K centrifuge (15 min at 14,000 *g*) within 1 hr after pH adjustment. Transfer activity is present in the supernatant.

Step 2. Second pH Adjustment. The supernatant is adjusted to pH 3 with 4 *M* HCl under continuous stirring. After approximately 15 min the pH of the solution is gradually readjusted to pH 6.2 by addition of 1 *M* Tris-HCl (pH 8.2). A heavy precipitate forms that is removed by centrifugation in the WKF centrifuge (10 min at 14,000 *g*). Activity is recovered in the supernatant.

Step 3. Ammonium Sulfate Precipitation. The active protein is precipitated with ammonium sulfate (45 g/100 ml), left overnight, and collected by centrifugation in the WKF centrifuge (15 min at 13,000 *g*). The precipitate is dissolved in 5 m*M* sodium phosphate, 10 m*M* 2-mercaptoethanol (pH 7.2) and dialyzed for 2 days against three times 15 liters of this buffer until the pH and ionic strength are correct. The dialyzate is centrifuged in the WKF centrifuge (15 hr at 14,000 *g*) to sediment denatured protein and the bulk of the glycogen that is still present. Removal of glycogen is very important to assure a uniform binding of protein to DEAE-cellulose.

Step 4. Chromatography on DEAE-Cellulose. The total dialyzate is applied to a column (84 × 6.5 cm) of DEAE-cellulose (Whatman DE-52 microgranular) in 5 m*M* sodium phosphate, 10 m*M* 2-mercaptoethanol (pH 7.2) at a flow rate of 150 ml/hr. Unbound protein is eluted by rinsing the column with 8 liters of buffer. The transfer activity is eluted by increasing the ionic strength of the buffer with 15 m*M* sodium chloride. Aliquots of 0.1 ml are assayed and compared with the transfer activity of 0.1 ml of pH 5.1 supernatant. Fractions with an activity exceeding that of the pH 5.1 supernatant are pooled.

Step 5. Chromatography on CM-Cellulose. Protein is precipitated from the pooled fractions with ammonium sulfate (65 g/100 ml), left overnight, and collected by centrifugation in the WKF centrifuge (15 min at 13,000 *g*). The pellets are dissolved in 10 m*M* citric acid–20 m*M* Na_2HPO_4–10 m*M* 2-mercaptoethanol (pH 5.0) and dialyzed against two times 15 liters of this buffer. The dialyzate is applied to a column (53 × 4 cm) of CM-cellulose (Whatman, CM-52 microgranular), which is rinsed with buffer until the unbound protein has eluted. The elution is continued with a linear gradient of 3500 ml of eluting buffer and 3500 ml of 40 m*M* citric acid–80 m*M* Na_2HPO_4–10 m*M* 2-mercaptoethanol (pH 5.0), at a flow rate of 100 ml/hr. Transfer activity appears in the effluent at about 25 m*M* citric acid–50 m*M* phosphate. Fractions (0.025-ml aliquot) with an activity exceeding that of the pH 5.1 supernatant (0.1 ml aliquot) are pooled.

Step 6. Fractionation on Sephadex G-50. Aliquots (100 ml) of the pooled fractions are applied directly to a column (77 × 7 cm) of Sephadex G-50 (Pharmacia). Protein is eluted with 20 m*M* citric acid–40 m*M* Na_2HPO_4–10% (v/v) glycerol (pH 5.0) at a flow rate of 100 ml/hr. The elution pattern shows two peaks; the second peak contains PC-TP (purity > 90%). The PC-TP in 10% glycerol is concentrated by dialysis against an appropriate volume of 60% (v/v) glycerol in water, so that the final glycerol concentration is 50%; the PC-TP concentration varies from 100 to 250 μg/ml.

Properties

Storage and Stability. PC-TP in 50% glycerol is stored at −20°, a condition under which it retains its full activity for at least a year. The protein loses its activity upon freezing and lyophilization. At concentrations above approximately 250 μg/ml, PC-TP tends to aggregate.

Physical Properties. PC-TP has an isoelectric point of 5.8, a molecular weight of 24,681 (calculated from the primary structure), and a molar absorbance at 280 nm of 60,365 M^{-1} cm^{-1}.[14] It has a fluorescence emission maximum at 327 nm[15] and an α-helix content of 29% (based on circular dichroism).[16] The complex formed between the protein and vesicles of egg yolk phosphatidylcholine has an apparent dissociation constant of 13 m*M*.[17] This constant decreases to 1.4 m*M* and 0.014 m*M* with negatively charged vesicles consisting of phosphatidylcholine and 5 and 20 mol % phosphatidic acid, respectively.[18]

Chemical Properties. The protein contains one molecule of noncovalently bound phosphatidylcholine.[3] It follows from the primary structure that PC-TP consists of a single polypeptide chain of 213 amino acid residues including two disulfide bonds.[14] *N*-Acetylmethionine forms the blocked NH_2 terminus; and threonine, the COOH terminus. The protein has a hydrophobicity index of 1356 calories per residue, which reflects the high content of aromatic (15 Tyr, 9 Phe, 5 Trp) and apolar amino acid residues (19 Val, 17 Leu, 15 Ala, 6 Ile). Activity is inhibited by the disulfide bond modifying reagent dithiothreitol (10 m*M*) and the arginine modifying reagents butanedione (50 m*M*) and phenylglyoxal (10 m*M*).

Substrate Specificity. The specificity of PC-TP for phosphatidylcholine

[14] R. Akeroyd, P. Moonen, J. Westerman, W. C. Puijk, and K. W. A. Wirtz, *Eur. J. Biochem.* **114,** 385 (1981).

[15] K. W. A. Wirtz and P. Moonen, *Eur. J. Biochem.* **77,** 437 (1977).

[16] R. Akeroyd, J. A. Lenstra, J. Westerman, G. Vriend, K. W. A. Wirtz, and L. L. M. van Deenen, *Eur. J. Biochem.* **121,** 391 (1982).

[17] K. Machida and S. I. Ohnishi, *Biochim. Biophys. Acta* **507,** 156 (1978).

[18] K. W. A. Wirtz, G. Vriend, and J. Westerman, *Eur. J. Biochem.* **94,** 215 (1979).

has been demonstrated for both natural[2,19] and artificial membranes.[3,20] The protein is sensitive to variations in the glycerol-backbone region and shows a preference in the order: diester- > diether- ≫ dialkylphosphatidylcholine.[13] It transfers both unsaturated and disaturated species of phosphatidylcholine.[21]

[19] J. E. Rothman, D. K. Tsai, E. A. Dawidowicz, and J. Lenard, *Biochemistry* **15,** 2631 (1976).
[20] H. H. Kamp, K. W. A. Wirtz, P. R. Baer, A. J. Slotboom, A. F. Rosenthal, F. Paltauf, and L. L. M. van Deenen, *Biochemistry* **16,** 1310 (1977).
[21] L. G. Lange, G. van Meer, J. A. F. Op den Kamp, and L. L. M. van Deenen, *Eur. J. Biochem.* **110,** 115 (1980).

[53] Phosphatidylcholine Transfer Protein from Rat Liver: Purification and Radioimmunoassay

By T. TEERLINK, B. J. H. M. POORTHUIS, and K. W. A. WIRTZ

Three different phospholipid transferring proteins have been identified in the membrane-free cytosol of rat tissues, i.e., the phosphatidylcholine transfer protein (PC-TP), the phosphatidylinositol transfer protein (PI-TP), and the nonspecific lipid transfer protein (nsL-TP). PC-TP specifically accelerates the transfer of phosphatidylcholine between membranes,[1-3], PI-TP has a preference for phosphatidylinositol but also transfers phosphatidylcholine[2,3] and nsL-TP accelerates the transfer of all common phospholipids (except for cardiolipin) and cholesterol.[4] A transfer protein for phosphatidylglycerol has been detected in rat lung.[5]

Although we presume that these proteins play an important role in the intracellular transfer of lipids, direct evidence for their physiological significance remains to be established. In order to advance our understanding, sensitive and specific radioimmunoassays need be developed to determine the intracellular levels of these proteins under various physiological conditions. This chapter describes the radioimmunoassay for PC-TP from rat

[1] C. Lutton and D. B. Zilversmit, *Biochim. Biophys. Acta* **441,** 370 (1976).
[2] R. H. Lumb, A. D. Kloosterman, K. W. A. Wirtz, and L. L. M. van Deenen, *Eur. J. Biochem.* **69,** 15 (1976).
[3] K. Yamada, T. Sasaki, and T. Sakagami, *J. Biochem.* (*Tokyo*) **84,** 855 (1978).
[4] B. Bloj and D. B. Zilversmit, *J. Biol. Chem.* **252,** 1613 (1977).
[5] L. M. G. van Golde, V. Oldenborg, M. Post, J. J. Batenburg, B. J. H. M. Poorthuis, and K. W. A. Wirtz, *J. Biol. Chem.* **255,** 6011 (1980).

ISBN 0-12-181998-1

PURIFICATION OF PHOSPHATIDYLCHOLINE TRANSFER PROTEIN FROM RAT LIVER

Step	Volume (ml)	Protein (mg)[a]	Specific activity[b]	Recovery (%)	Purification factor
1. pH 5.1 supernatant	2870	48,800	1.1	100	—
2. $(NH_4)_2SO_4$ precipitation	460	21,480	1.7	69	1.5
3. DEAE-cellulose	205	4,920	5.4	54	5
4. CM-cellulose	400	1,240	16	38	15
5. Sephadex G-50	180	16.2	559	17	508
6. Hydroxyapatite	52	1.5	5633	17	5121

[a] Protein is measured by the method of O. H. Lowry, N. J. Rosebrough, A. L. Farr, and R. J. Randall, *J. Biol. Chem.* **193,** 265 (1951).

[b] Specific activity is expressed as units of phosphatidylcholine transfer protein activity per milligram of protein.

liver[6] including the purification of this protein.[7] Antisera against PC-TP from bovine liver were available[8] but were of no use because of lack of cross-reactivity with rat PC-TP.

Purification Procedure

The table summarizes the various steps of the purification procedure based on a modification of a method described by Lumb *et al.*[2] The assay for PC-TP transfer activity was exactly as described.[9] All manipulations are performed at 4°.

Step 1. Preparation of pH 5.1 Supernatant. Rat livers (1500 g) stored at −20° are thawed and rinsed with 0.25 *M* sucrose–1 m*M* EDTA–10 m*M* Tris-HCl, pH 7.4 (SET) to remove blood. A 30% homogenate in SET is prepared with an Ystral homogenizer (type 40/34, Dottingen, FRG) for 1 min at high speed. The homogenate is centrifuged in a WKF model G 50 K centrifuge (20 min at 14,000 *g*) to sediment cell debris, nuclei, mitochondria, and lysosomes. The supernatant is adjusted to pH 5.1 with 3 *N* HCl and allowed to stand for 30 min with stirring, after which the aggregated proteins and microsomes are removed by centrifugation as above. Transfer activity is present in the supernatant.

Step 2. Ammonium Sulfate Precipitation. The supernatant is adjusted to pH 7.2 with 1 *M* Tris and taken to 60% saturation with solid ammonium

[6] T. Teerlink, B. J. H. M. Poorthuis, T. P. van der Krift, and K. W. A. Wirtz, *Biochim. Biophys. Acta* **665,** 74 (1981).

[7] B. J. H. M. Poorthuis, T. P. van der Krift, T. Teerlink, R. Akeroyd, K. Y. Hostetler, and K. W. A. Wirtz, *Biochim. Biophys. Acta* **600,** 376 (1980).

[8] G. M. Helmkamp, S. A. Nelemans, and K. W. A. Wirtz, *Biochim. Biophys. Acta* **424,** 168 (1976).

[9] J. Westerman, H. H. Kamp, and K. W. A. Wirtz, this volume [52].

sulfate. After standing overnight with stirring, the precipitated proteins are sedimented by centrifugation of the solution in the WKF centrifuge (20 min at 14,000 *g*). The pellet containing PC-TP is dissolved in a small volume of 10 m*M* Tris-HCl, pH 7.2, and extensively dialyzed against this buffer. Denaturated proteins are removed from the dialyzate by centrifugation.

Step 3. Chromatography on DEAE-Cellulose. The dialyzate is applied to a column (34 × 4 cm) of DEAE-cellulose (Whatman DE-52, microgranular) in 10 m*M* Tris-HCl, pH 7.2. The column is rinsed with this buffer; the active protein is recovered in the unbound protein fraction. This fraction is adjusted to 90% saturation with solid ammonium sulfate and allowed to stand overnight, and the precipitated protein is collected by centrifugation as in step 2. The pellet is dissolved in a small volume of 20 m*M* citrate–40 m*M* phosphate, pH 5.0, and dialyzed extensively against this buffer.

Step 4. Chromatography on CM-Cellulose. After removal of precipitated protein the dialyzate is applied to a column (3.4 × 36 cm) of CM-cellulose (Whatman CM-52, microgranular) equilibrated with the citrate–phosphate buffer. The protein is eluted with a linear gradient of 20 m*M* citrate–40 m*M* phosphate, pH 5.0, to 80 m*M* citrate–160 m*M* phosphate, pH 5.0 (2 × 5 column volumes). The active protein elutes at a buffer strength of about 35 m*M* citrate–70 m*M* phosphate and is pooled.

Step 5. Fractionation on Sephadex G-50. The pooled fraction (volume of 400 ml) is adjusted to pH 6.8 with 3 *N* NaOH, put in a dialysis bag, and concentrated to one-fourth of the volume by equilibration against Aquacide (Calbiochem). The protein is applied to a column (80 × 6 cm) of Sephadex G-50 and eluted with 10 m*M* potassium phosphate, pH 6.8; active protein elutes at $V_e/V_o = 1.8$.

Step 6. Chromatography on Hydroxyapatite. The pooled active fraction is applied directly to a hydroxyapatite column (13 × 1 cm) equilibrated with the potassium phosphate buffer. The protein is eluted with a linear gradient of 10 m*M* to 300 m*M* potassium phosphate, pH 6.8 (2 × 5 column volumes). Pure PC-TP elutes at about 180 m*M* potassium phosphate. After addition of glycerol to a concentration of 10% (v/v) the PC-TP solution is threefold concentrated by dialysis against three volumes of 65% (v/v) glycerol–100 m*M* potassium phosphate, pH 6.8. The PC-TP in a final concentration of 50% (v/v) glycerol is stored at −20°.

Properties

In other laboratories PC-TP has been partially purified from rat liver[1] and rat small intestinal mucosa.[3] The protein has a molecular weight of 28,000 estimated from sodium dodecyl sulfate–gel electrophoresis[7] and an isoelectric point of 8.4.[1,2] It accelerates the transfer of phosphatidylcholine between membranes, but not the transfer of phosphatidylinositol and phosphatidylethanolamine.[1-3] This suggests a specificity similar to that of bovine PC-TP.[10] Antisera raised against rat PC-TP in rabbits are not cross-reactive with bovine PC-TP.[7]

Radioimmunoassay Procedure

Immunization Procedure. Antisera are raised in New Zealand white rabbits (weight 2500 g). Prior to immunization PC-TP in 50% (v/v) glycerol is dialyzed against isotonic saline and subsequently concentrated against Aquacide to a concentration of approximately 0.6 mg/ml. An emulsion of 1 ml of PC-TP solution and 1 ml of Freund's complete adjuvant (Miles Laboratories, UK) is used for each immunization and injected in portions of 0.2 ml intradermally in the back and the shoulders. A booster injection of 0.3 mg of PC-TP in isotonic saline is given subcutaneously after 7 weeks and, if necessary, repeated after an additional 2 weeks. The rabbits are exsanguinated 2–2½ months after the primary immunization. Preimmune and immune sera are obtained by allowing the blood to clot for 1 hr at 37° and the clot to retract at 4°, followed by centrifugation to remove the residual red blood cells.

Preparation of Immunoadsorbents. Immunoadsorbents used for affinity chromatography are prepared by covalent coupling of PC-TP and total IgG fraction to Sepharose 4B.[11] Sepharose 4B is activated with cyanogen bromide according to an established procedure.[12] The protein solution is dialyzed overnight against 0.1 *M* $NaHCO_3$ and mixed with activated Sepharose 4B (1–2 g wet weight per milligram of protein). The suspension is gently stirred overnight at 4° and washed once with 0.1 *M* $NaHCO_3$. Remaining active groups are blocked by suspending the immunoadsorbent in 1 *M* ethanolamine, pH 8.0, for 2 hr followed by extensive washing with 0.15 *M* sodium chloride–0.01 *M* sodium phosphate, pH 7.1 (phosphate-buffered saline).

Isolation of Specific Immunoglobulin G (IgG). The antisera are dialyzed against 0.015 *M* potassium phosphate, pH 8.0, and applied to a DEAE-cellulose column (1 ml of serum/5 ml of DEAE-cellulose) equilibrated with the same buffer. The crude IgG fraction elutes in the void volume and is stored at −20° in small aliquots. Specific IgG is isolated from this crude fraction by affinity chromatography. Rat liver PC-TP (1 mg) coupled to

[10] H. H. Kamp, K. W. A. Wirtz, P. R. Baer, A. J. Slotboom, A. F. Rosenthal, F. Paltauf, and L. L. M. van Deenen, *Biochemistry* **16,** 1310 (1977).

[11] S. Fuchs and M. Sela, *in* "Handbook of Experimental Immunology" (D. M. Weir, ed.), Vol. 1, Chapter 10. Blackwell, Oxford, 1978.

[12] S. C. March, I. Parikh, and P. Cuatrecasas, *Anal. Biochem.* **60,** 149 (1974).

Sepharose 4B (2 g) is packed in a small column and washed with 5 column volumes each of phosphate-buffered saline, 0.53 *M* formic acid (pH 2.05), and phosphate-buffered saline. Then, 30 mg of the crude IgG fraction in 0.015 *M* potassium phosphate, pH 8.0, is applied to the column. The column is washed with phosphate-buffered saline until no more protein is detected in the eluate. Bound specific IgG is eluted with 0.53 *M* formic acid (pH 2.05) as described before.[13] The IgG-containing fractions are immediately neutralized with 1 *M* Tris-HCl, pH 8.0, dialyzed overnight against phosphate-buffered saline and stored at $-20°$ in small aliquots. The column is regenerated by washing with phosphate-buffered saline. All operations are performed at 4°.

Preparation of ^{125}*I-Labeled PC-TP.* Transfer protein is labeled with ^{125}I using a solid-phase lactoperoxidase–glucose oxidase enzyme system (Enzymobeads, Bio-Rad Laboratories, Richmond, California). To a 1-ml Pyrex tube are added, in this order, 50 μl of the Enzymobead suspension, 50 μl of 0.2 *M* sodium phosphate (pH 7.4), 40 μl (12 μg) of PC-TP, 10 μl (1 mCi) of carrier-free $Na^{125}I$ solution, and finally 50 μl of 2% D-glucose solution. The reaction is allowed to proceed for 30 min at ambient temperature and is then terminated by sedimentation of the Enzymobeads at 2000 *g* for 5 min. The supernatant is immediately transferred to a 1-ml affinity column containing 1 mg of crude IgG coupled to Sepharose 4B. The column is washed with 0.05 *M* sodium phosphate (pH 7.4), containing 0.5% bovine serum albumin, until the eluate is free of radioactivity. The bound ^{125}I-labeled PC-TP is eluted with 0.1 *N* glycine–HCl (pH 2.8) containing 0.5% bovine serum albumin. The eluate is immediately neutralized with 1 *M* Tris-HCl (pH 8.0) and dialyzed overnight against 0.05 *M* sodium phosphate (pH 7.4). After dialysis, 87% glycerol is added to a final concentration of 55% (v/v), and the labeled protein is stored at $-20°$. Under these conditions, the protein retains its immunoreactivity for several months.

Radioimmunoassay. The assay buffer consists of 0.5 *M* sodium phosphate, pH 7.4–0.5% (w/v) bovine serum albumin–0.02% (w/v) sodium azide–1% (v/v)Triton X-100. All components used in the assay are diluted to the desired concentration with this buffer. The assay is carried out in 3-ml disposable polystyrene tubes. A sample in 0.2 ml of buffer is mixed with 0.1 ml of specific IgG and 0.1 ml of ^{125}I-labeled PC-TP (4000–12,000 cpm). A standard curve is prepared by using pure PC-TP. The amount of IgG chosen (60–100 ng) is sufficient to bind approximately 35% of the labeled PC-TP in the absence of unlabeled protein. Incubation is carried out overnight at 4°. Separation of bound and free tracer is accomplished by the use of protein A-bearing *Staphylococcus aureus* cells (Calbiochem, San

[13] C. C. Young, M. F. Lin, and C. C. Chang, *Toxicon* **15,** 51 (1977).

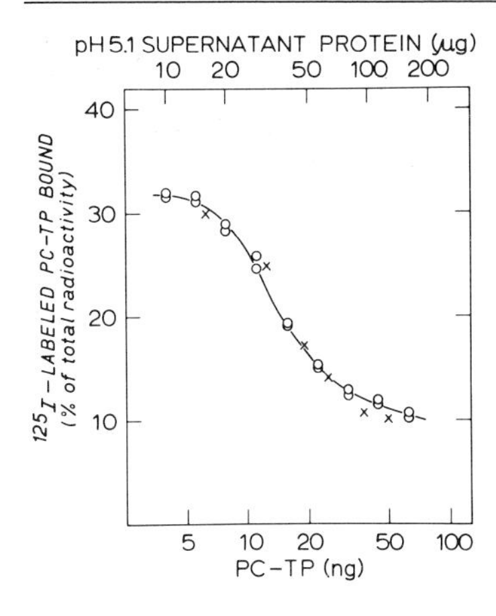

FIG. 1. Radioimmunoassay standard curve for rat liver PC-TP (O), comparing the results obtained with serial dilutions of the pH 5.1 supernatant from rat liver (×).

Diego, California), which provide a satisfactory immunoadsorbent.[14,15] Shortly before use the cells are washed by the following procedure. To 1 ml of a 10% (w/v) suspension of cells, 9 ml of assay buffer is added and the suspension is centrifuged for 10 min at 12,000 *g*. The bacterial pellet is resuspended in 10 ml of assay buffer by careful agitation on a Vortex mixer. This procedure is repeated twice and then the bacteria are diluted with assay buffer to a 1% (w/v) suspension. This suspension (0.05 ml) is added to each assay tube and incubated for 15–30 min at room temperature. Then 2 ml of the assay buffer is added and the tubes are centrifuged for 10 min at 3000 *g*. The supernatants are carefully decanted and the tubes are counted directly in a Packard autogammacounter. Data are expressed as a percentage of the total radioactivity added. In each assay a standard curve is prepared in duplicate (Fig. 1). For rat tissues it is generally observed that serial dilutions of 105,000 *g* supernatant fractions produce displacement curves parallel to the standard curve (see Fig. 1). For calculation of the PC-TP content, only values in the steep part of the curve are considered. The operating range of the assay is from 5 to 50 ng of PC-TP.

Applications. The assay has been used to determine the content of PC-TP in the 105,000 *g* supernatant of rat liver and Morris hepatomas.[6] In this study, it was not necessary to add inhibitors of protease activity. This was generally found to hold for other tissues (i.e., lung, kidney, heart, brain, spleen, and adrenals) with the exception of intestinal mucosa, in which protease activity interfered with the assay (T. Teerlink, unpublished obser-

[14] S. W. Kessler, *J. Immunol.* **115,** 1617 (1975).
[15] E. O'Keefe and V. Bennett, *J. Biol. Chem.* **255,** 561 (1980).

vations). The presence of Triton X-100 in the assay buffer is advantageous because it makes possible measurement of the PC-TP associated with the membrane fractions.

[54] Nonspecific Lipid Transfer Protein from Rat Liver

By B. J. H. M. POORTHUIS and K. W. A. WIRTZ

The membrane-free cytosol of rat liver contains a nonspecific lipid transfer protein (nsL-TP) which *in vitro* stimulates the transfer of all common diacylphospholipids as well as cholesterol between membranes.[1] This protein has been purified to homogeneity from this source.[1-4] nsL-TP appears identical to sterol carrier protein 2, which modulates cholesterol biosynthesis.[5] Both nsL-TP and sterol carrier protein 2 have been shown to stimulate microsomal cholesterol ester formation.[4,6,7] Morris hepatomas have strongly reduced levels of nsL-TP.[4,8] This deficiency may be related to the lack of regulation of cholesterol biosynthesis in these malignant tissues.[4,8,9] This chapter describes a high-yield purification of nsL-TP from rat liver.[10]

Assay Method

Principle. Since nsL-TP is the only transfer protein known to transfer phosphatidylethanolamine, an assay based on this property is used in the purification. The assay is a modification of a method previously described.[1] It involves measurement of transfer of [^{14}C]phosphatidylethanolamine from sonicated unilamellar vesicles consisting of phosphatidylethanolamine and phosphatidylcholine to beef heart mitochondria. At the end of

[1] B. Bloj and D. B. Zilversmit, *J. Biol. Chem.* **252,** 1613 (1977).
[2] B. Bloj, M. E. Hughes, D. B. Wilson, and D. B. Zilversmit, *FEBS Lett.* **96,** 87 (1978).
[3] B. J. H. M. Poorthuis, J. F. C. Glatz, R. Akeroyd, and K. W. A. Wirtz, *Biochim. Biophys. Acta* **665,** 256 (1981).
[4] J. M. Trzaskos and J. L. Gaylor, *Biochim. Biophys. Acta* **751,** 51 (1983).
[5] B. J. Noland, R. E. Arebalo, E. Hansbury, and T. J. Scallen, *J. Biol. Chem.* **255,** 4182 (1980).
[6] K. L. Gavey, B. J. Noland, and T. J. Scallen, *J. Biol. Chem.* **256,** 2993 (1981).
[7] B. J. H. M. Poorthuis and K. W. A. Wirtz, *Biochim. Biophys. Acta* **710,** 99 (1982).
[8] B. J. H. M. Poorthuis, T. P. van der Krift, T. Teerlink, R. Akeroyd, K. Y. Hostetler, and K. W. A. Wirtz, *Biochim. Biophys. Acta* **600,** 376 (1980).
[9] L. A. Bricker, H. P. Morris, and M. D. Siperstein, *J. Clin. Invest.* **51,** 206 (1972).
[10] Formerly this protein was called "nonspecific phospholipid transfer protein."

ISBN 0-12-181998-1

incubation the mitochondria are quantitatively separated from the vesicles by centrifugation, and the amount of radioactivity in the mitochondrial pellet is determined.

Preparation of Vesicles. Single bilayer vesicles are prepared of egg yolk phosphatidylcholine (60 mol%) and rat liver [^{14}C]phosphatidylethanolamine (40 mol%). A stock solution of this lipid mixture is stored at $-10°$ in chloroform. An aliquot of 10 μmol of phospholipid is pipetted into an extraction tube, the chloroform is removed *in vacuo* in a rotary evaporator, and 5 ml of 0.25 *M* sucrose–1 m*M* EDTA–10 m*M* Tris-HCl, pH 7.4 (SET) buffer is added. After gentle agitation in the presence of two glass beads, the milky suspension is irradiated ultrasonically with a Branson sonifier for 10 min under nitrogen at 0°. The clear vesicle suspension is used the same day without further treatment. [^{14}C]Phosphatidylethanolamine is prepared from rat livers as described.[8]

Preparation of Mitochondria. Beef heart mitochondria are prepared according to an established procedure.[11] The mitochondria are suspended in SET (50 mg of protein, i.e., 25 μmol of phospholipid phosphorus per milliliter) and stored at $-20°$. Before use the mitochondria are washed once by diluting 1 ml of suspension with 30 ml of SET. After sedimentation (10 min at 8600 rpm) the mitochondrial pellet is resuspended in 1.8 ml of SET containing bovine serum albumin (31 mg/ml) and used for the assay.

Assay Procedure. nsL-TP activity is determined by incubating aliquots of the appropriate protein fractions with ^{14}C-labeled vesicles (100 nmol of phospholipid) and mitochondria (500 nmol of phospholipid) for 45 min at 37° in a total volume of 0.5 ml of SET buffer, which also contains 0.5% fatty acid-poor bovine serum albumin (Sigma, St. Louis, Missouri).

The tubes are swirled every 15 min to prevent settling of the mitochondria. The reaction is terminated by sedimentation (10 min at 11,000 *g*) of the mitochondria through a cushion of 0.5 ml of 14% (w/w) sucrose. The supernatant is carefully removed by aspiration; the mitochondrial pellet is dissolved in 0.1 ml of 10% (w/v) sodium dodecyl sulfate by heating at 60°, and the radioactivity is determined by liquid scintillation counting. Percentage of transfer is calculated from the radioactivity in the mitochondrial pellet and the vesicles before incubation. Blank incubations are carried through the entire procedure to correct for noncatalyzed transfers (4–6% of the total ^{14}C radioactivity). Specific transfer activity is expressed as nanomoles of phosphatidylethanolamine transferred per hour per milligram of protein.

[11] P. V. Blair, this series, Vol. 22, p. 565.

Purification Procedure

The table summarizes the various steps in the purification procedure developed in our laboratory.[3] In this method nsL-TP is purified to homogeneity with a yield of 50%. All manipulations are performed at 4°.

Step 1. Preparation of pH 5.1 Supernatant. Fresh rat livers (500 g) are rinsed with SET to remove blood, and a 30% homogenate is prepared. The homogenate is centrifuged in a WKF Model G 50K (12,000 *g* for 20 min). The pH of the supernatant is adjusted to 5.1 with 3 *M* HCl and allowed to stand for 30 min with stirring, after which the precipitated protein is removed by centrifugation as above.

Step 2. Batch Treatment with DEAE-Cellulose. Prior to treatment the pH 5.1 supernatant is adjusted to pH 7.2 with 1 *M* Tris, and solid ammonium sulfate is added to 90% saturation. After stirring for 3–4 hr the precipitated proteins are sedimented in a Sorvall GSA rotor (16,000 *g* for 20 min). The pellet is dissolved in a small volume of 10 m*M* Tris-HCl, pH 7.6, and extensively dialyzed against the same buffer. Packed wet DEAE-cellulose, previously equilibrated with the buffer, is added to the dialyzate at a ratio of 1 ml of DEAE-cellulose to 2 ml of dialyzate. After gentle stirring overnight the DEAE-cellulose is removed by centrifugation in the Sorvall GSA rotor; activity is recovered in the supernatant. In order to stabilize the activity 2-mercaptoethanol and glycerol are added to the supernatant to a final concentration of 5 m*M* and 10% (v/v), respectively.

Step 3. Fractionation on Sephadex G-50. After a six-fold concentration to approximately 70 ml with Ficoll (Pharmacia) or Aquacide (Calbiochem), the supernatant is applied to a Sephadex G-50 column (6 × 80 cm) equilibrated with 10 m*M* potassium phosphate–5 m*M* 2-mercaptoethanol–10% glycerol. The column is eluted with the same buffer at a flow rate of 60 ml/hr. Fractions active in phosphatidylethanolamine transfer elute at $V_e/V_o = 1.7$.

Step 4. Chromatography on Hydroxyapatite. The active fractions from

PURIFICATION OF THE NONSPECIFIC LIPID TRANSFER PROTEIN FROM RAT LIVER

Step	Volume (ml)	Protein[a] (mg)	Specific activity[b]	Recovery (%)	Purification factor
1. pH 5.1 supernatant	780	21,000	1.4	100	1
2. DEAE-cellulose	425	2,760	10	100	7
3. Sephadex G-50	203	340	32	38	23
4. Hydroxyapatite	36	6.8	2080	50	1486

[a] Protein is measured by the method of O. H. Lowry, N. J. Rosebrough, A. L. Farr, and R. J. Randall, *J. Biol. Chem.* **193,** 265 (1951).

[b] Specific activity is given in nanomoles of phosphatidylethanolamine transferred per milligram of protein per hour.

the Sephadex G-50 are pooled and applied directly to a hydroxyapatite column (2 × 10 cm) equilibrated with the same buffer. The protein is eluted at a flow rate of 15 ml/hr with a linear gradient (2 × 150 ml) of 10–300 m*M* potassium phosphate–5 m*M* 2-mercaptoethanol–10% glycerol (v/v). Homogeneous nsL-TP elutes at about 150 m*M* potassium phosphate.

Properties

Stability. Both glycerol and 2-mercaptoethanol are required to stabilize the protein during purification. Omission of 2-mercaptoethanol in step 4 lowers the yield to 10% and the purification factor to 910. Omission of glycerol leads to a very rapid loss of activity. In the absence of both compounds pure nsL-TP has a high tendency for aggregation.[2] Storage of nsL-TP in 100 m*M* potassium phosphate, pH 6.8–5 m*M* 2-mercaptoethanol–50% glycerol (v/v/v) at −20° preserves the activity for months.

Physical Properties. The protein has an isoelectric point between pH 8.6 and 9.0[1,4] in accordance with its failure to bind to DEAE-cellulose at pH 7.2 (step 2). A molecular weight of 14,800 has been estimated from its behavior on sodium dodecyl sulfate–polyacrylamide gel electrophoresis. Protein dimers may be detected under conditions where the concentratin of 2-mercaptoethanol in the sample buffer is not carefully controlled (B. J. H. M. Poorthuis, unpublished observation). The molar absorbance at 280 nm calculated from the amino acid composition is 12,900 M^{-1} cm^{-1}. It has a fluorescence emission maximum of 335 nm. The protein has a substantial heat stability (90° for 5 min).[1,5]

Chemical Properties. The protein has a low content of aromatic amino acid residues (1 Trp, 1 Tyr, 5 Phe), which accounts for the low absorbance at 280 nm. Its activity is inhibited by 10 m*M* *N*-ethylmaleimide and 1 m*M* *p*-hydroxymercuribenzoate. This strongly suggests an essential role for the one free sulfhydryl group present.

Specificity. In addition to phosphatidylethanolamine, nsL-TP accelerates the transfer of phosphatidylcholine, phosphatidylinositol, sphingomyelin, and cholesterol between membranes.[1] The protein stimulates microsomal cholesterol ester formation[4,6,7] which is probably related to transfer of cholesterol to the site of esterification. Moreover, it activates the microsomal conversion of both 4,4-dimethyl-Δ^8-cholestenol to C_{27} sterol and 7-dehydrocholesterol to cholesterol.[5]

Reaction Mechanism. The mode of action of nsL-TP remains to be established. A protein probably identical to nsL-TP has been found to bind phosphatidylserine.[12] The formation of a lipid–protein complex supports the notion of a carrier mechanism.

[12] J. Baranska and Z. Grabarek, *FEBS Lett.* **104,** 253 (1979).

Similar Proteins

A nonspecific transfer protein has been isolated from a fast-growing rat hepatoma that could not be detected in rat liver.[13] This protein has an estimated molecular weight of 11,200 and an isoelectric point of 5.2. Two species of nsL-TP have been purified from bovine liver.[14] Both proteins have a molecular weight of 14,500 but differ in isoelectric point (pH 9.55 and 9.75).

[13] E. V. Dyatlovitskaya, N. G. Timofeeva, and L. D. Bergelson, *Eur. J. Biochem.* **82,** 463 (1978).

[14] R. C. Crain and D. B. Zilversmit, *Biochemistry* **19,** 1433 (1980).

[55] Phospholipid Exchange Protein-Dependent Synthesis of Sphingomyelin

By DENNIS R. VOELKER and EUGENE P. KENNEDY

$$\text{Phosphatidylcholine} + \text{ceramide} \rightleftharpoons \text{sphingomyelin} + \text{diacylglycerol}$$

Cellular,[1] enzymic,[1-3] and genetic[4] experiments have now shown convincingly that the proximal donor of the phosphocholine moiety of sphingomyelin is phosphatidylcholine. The plasma membrane appears to be the principal subcellular location of this enzyme.[1,3] As with other membrane-bound enzymes of phospholipid metabolism, the presentation of water-insoluble substrates to this enzyme is a difficult problem. The problem is accentuated by the fact that the phosphocholine transferase is inactivated by low levels of detergents. Phospholipid exchange proteins[5,6] provide a mechanism for introducing the phosphatidylcholine substrate into the membrane domain of the enzyme without disrupting native membrane structure.

Assay Method

Principle. The transfer of phospho[^{3}H]choline from phosphatidyl[^{3}H]choline to endogenous ceramide in plasma membrane vesicles is measured.

[1] D. R. Voelker and E. P. Kennedy, *Biochemistry* **21,** 2753 (1982).

[2] M. D. Ullman and N. S. Radin, *J. Biol. Chem.* **249,** 1506 (1974).

[3] W. D. Marggraf, F. A. Anderer, and J. N. Kanfer, *Biochim. Biophys. Acta* **664,** 61 (1981).

[4] J. D. Esko and C. R. H. Raetz, *Proc. Natl. Acad. Sci. U.S.A.* **77,** 5192 (1980).

[5] K. W. A. Wirtz and D. B. Zilversmit, *Biochim. Biophys. Acta* **193,** 105 (1969).

[6] D. B. Zilversmit and M. E. Hughes, *Methods Membr. Biol.* **7,** 211 (1976).

METHODS IN ENZYMOLOGY, VOL. 98

ISBN 0-12-181998-1

Phosphatidyl[^{3}H]choline is incorporated into plasma membranes by the action of phosphatidylcholine exchange protein. The reaction is terminated by extraction of the lipids. The amount of radiolabeled sphingomyelin in the extract is determined by acid degradation and extraction of sphingosylphosphocholine or by alkaline methanolysis and thin-layer chromatography.

Reagents

Plasma membrane fraction from rat liver[7]
Phosphatidylcholine exchange protein
Unilamellar phosphatidyl[^{3}H]choline liposomes
Chloroform
Methanol
6 *N* HCl–butanol (1 : 1)
HCl, 0.1 *N*, saturated with butanol
NH_4OH
0.3 *N* NaOH–methanol
Silica gel H plates

Partially purified as well as purified phosphatidylcholine exchange proteins will support the reaction equally well.[8] The crudest source of exchange protein used has been a pH 5.1 supernatant fraction derived from rat liver cytosol.[6] Radiolabeled phosphatidylcholine can be prepared from BHK21 cells as described elsewhere,[1] although phosphatidyl[^{3}H]choline of high specific activity has become available from New England Nuclear. Small unilamellar liposomes are prepared by sonicating the labeled phosphatidylcholine for 15 min in 1-min bursts followed by cooling. During sonication the liposome-containing vessel is maintained in an ice-water bath and under a nitrogen atmosphere. Multilamellar liposomes are removed by centrifugation at 100,000 *g* for 1 hr.

Procedure. Reactions are conducted in a final volume of 250 μl and contain 10 m*M* HEPES (pH 7.2), 50 m*M* NaCl, 0.5 μCi (1 nmol) of phosphatidylcholine vesicles, 50–100 μg of plasma membrane–protein, and saturating amounts (usually greater than 3 units[9]) of purified or partially purified phosphatidylcholine exchange protein. After incubation at 37° for 60–90 min, the reactions are terminated by the addition of 1.5 ml of methanol–chloroform (2 : 1, v/v). Lipid extraction is completed by the addition of 0.5 ml of chloroform and 0.6 ml of water, followed by centrifugation and removal of the chloroform phase.[10] The lipids are dried under nitrogen, resuspended in 1 ml of 6 *N* HCl–butanol (1 : 1, v/v) and heated at

[7] S. Fleischer and M. Kervina, this series, Vol. 31, p. 6.

[8] The purified phosphatidylcholine exchange protein used in the development of this method was the generous gift of Donald B. Zilversmit , Cornell University, Ithaca, New York.

[9] P. E. DiCorleto, J. B. Warrach, and D. B. Zilversmit, *J. Biol. Chem.* **245,** 7795 (1979).

[10] E. G. Bligh and W. J. Dyer, *Can. J. Biochem. Physiol.* **37,** 911 (1959).

100° for 1 hr.[11] The solution is cooled and 0.5 ml of 3.5 N NH_4OH is added, followed by 0.5 ml of butanol. The resultant lower phase is removed and discarded, and the upper phase, is washed twice with 1 ml of 0.1 N HCl saturated with butanol. A 0.5-ml portion of the upper phase is counted in a scintillation counter. Alternatively, the lipid extract can be hydrolyzed with 0.3 N NaOH–methanol at 37° for 1 hr. After acidification and extraction, the alkali-stable lipids (sphingomyelin and ether-linked lipids) are chromatographed on silica gel H plates developed in chloroform–methanol–acetic acid–water (50 : 25 : 8 : 2.5, by volume). In this solvent system sphingomyelin has an R_f of 0.2.

Comments

Under the conditions described, the synthesis of sphingomyelin is completely dependent upon the addition of phosphatidylcholine exchange protein. The maximum rate of sphingomyelin synthesis observed has been 3.5 nmol hr^{-1} mg $protein.^{-1}$ As shown by Rothman and Dawidowicz,[12] only about 65% of the phosphatidyl[^{3}H]choline in the liposomes is available for exchange with the plasma membranes. During the initial stages of the incubation, the specific activity of plasma membrane phosphatidylcholine increases rapidly. It is practical to assume that all of the exchangeable phospholipid present in the liposomes is transferred to the plasma membrane instantaneously. If both saturating amounts of exchange protein and a 15- to 20-fold excess of plasma membrane phosphatidylcholine are used this assumption is reasonable. It will, however, always give a conservative underestimate of the subsequent catalytic event.

Future Applications

The introduction of phospholipid substrates into membranes via phospholipid exchange proteins should be useful in studying additional aspects of phospholipid metabolism. The principal advantage of the technique is that it does not measurably alter native membrane structure. Some areas for which the technique is applicable are regulation of membrane-bound phospholipases, substrate preference in arachidonic acid release from phospholipids, the biosynthesis of platelet-activating factor, and the biosynthesis of pulmonary surfactant phospholipids. In addition, the different specificities of the phospholipid exchange proteins can be used to modify subcellular membranes compositionally to determine the effect such perturbations have upon the activities of specific membrane-bound enzymes.

[11] H. Kaller, *Biochem. Z.* **334,** 451 (1961).

[12] J. E. Rothman and E. A. Dawidowicz, *Biochemistry* **14,** 2809 (1975).

[56] Use of a Nonspecific Lipid Transfer Protein to Modify the Cholesterol Content of Synaptic Membranes[1]

By PAULA NORTH and SIDNEY FLEISCHER

The use of a nonspecific lipid transfer protein to vary the cholesterol-to-phospholipid (Ch/PL)[2] ratio of membranes is described. The method is illustrated with synaptosomes and synaptic plasma membranes, where we correlate loss of sodium-dependent γ-aminobutyric acid uptake with decreased Ch/PL ratio in the membrane.[3] The method consists of incubating the membranes with the transfer protein in the presence of lipid vesicles containing either cholesterol and phosphatidylcholine (PC) or PC alone, resulting in net transfer of cholesterol to the membranes or net removal of cholesterol from the membranes, respectively.

Previous approaches toward study of the role of cholesterol in membrane function have employed manipulation of cholesterol content by dietary means[4] or by sterol partitioning between liposomes and isolated membranes *in vitro*.[5-8] These methods have severe limitations. Dietary manipulation is not effective for all cell types, and it also leads to changes in the fatty acid moiety of phospholipids.[9,10] The liposome sterol partitioning method[11] has been effective for cells of the circulatory system, but changes in other types of membranes are difficult to accomplish by this method. In our experience, incubation for several hours with liposomes in the absence of transfer protein achieves only small changes (10–20%) in the Ch/PL ratio of synaptic plasma membranes, when correction is made for sticking of liposomes to the membranes. The Ch/PL ratio of membranes can be varied over

[1] Supported in part by National Institutes of Health Grant AM 14632.

[2] Abbreviations used: Ch/PL, cholesterol/phospholipid (molar ratio); GABA, γ-aminobutyric acid; HEPES, *N*-2-hydroxyethylpiperazine-*N*′-2-ethanesulfonic acid; KH medium, Krebs–HEPES, pH 7.4 (130 m*M* NaCl, 5 m*M* KCl, 1.3 m*M* $MgCl_2$, 1.2 m*M* sodium phosphate, 0.5 m*M* EGTA, 10 m*M* glucose, 10 m*M* HEPES); KHS medium, KH medium + 0.2 *M* sucrose; PC, phosphatidylcholine; SH medium, 0.32 *M* sucrose, 2.5 m*M* HEPES, pH 7.4; Tris, tris(hydroxymethyl)aminomethane.

[3] P. North and S. Fleischer, *J. Biol. Chem.* **258,** 1242 (1983).

[4] J. Kroes and R. Ostwald, *Biochim. Biophys. Acta* **249,** 647 (1971).

[5] S. J. Masiak and P. G. LeFevre, *Arch. Biochem. Biophys.* **162,** 442 (1974).

[6] S. Shattil and R. A. Cooper, *Biochemistry* **15,** 4832 (1976).

[7] H. Borochov and M. Shinitzky, *Proc. Natl. Acad. Sci. U.S.A.* **73,** 4526 (1976).

[8] M. Shinitzky and B. Rivnay, *Biochemistry* **16,** 982 (1977).

[9] R. Ostwald, W. Yamanaka, and M. Light, *Proc. Soc. Exp. Biol. Med.* **134,** 814 (1970).

[10] J. J. Baldassare, Y. Saito, and D. F. Silbert, *J. Biol. Chem.* **254,** 1108 (1979).

[11] K. R. Bruckdorfer, P. A. Edwards, and C. Green, *Eur. J. Biochem.* **4,** 506 (1968).

ISBN 0-12-181998-1

a wide range, rapidly and reversibly, with the use of the nonspecific lipid transfer protein.

Procedures

A. Reagents

The sucrose used in all solutions is of density gradient grade, obtained from EM Laboratories (Elmsford, New York). HEPES is obtained from Calbiochem-Behring (La Jolla, California), and Tris (reagent grade) is obtained from Sigma Chemical Company (St. Louis, Missouri). EGTA and 2-mercaptoethanol are also from Sigma. All other reagents are from Fisher Scientific Company (Pittsburgh, Pennsylvania), certified ACS grade. Sucrose concentrations were adjusted using a refractometer (Bausch & Lomb, Abbé 3L, or equilvalent) at room temperature, and, unless otherwise indicated, pH values were adjusted at room temperature.

B. Isolation of the Transfer Protein

The nonspecific lipid transfer protein is purified from beef liver as described by Crain and Zilversmit.[12] In our studies, the transfer protein is prepared without the heat treatment and octylagarose column chromatography steps. This shortened procedure is adequate for our experiments and increases the yield of transfer activity. Normally, 5-10 kg of liver obtained fresh from a local slaughterhouse is used in the 4-day purification procedure. (The steps are as outlined by Crain and Zilversmit.[12]) The pH 5.1 supernatant (~ 1.7 liters per kilogram of liver) is prepared on the first day and divided into 900-ml portions. Two to six of these portions are processed at a time for further purification. The remainder can be stored frozen without loss of activity at $-20°$. The pH 5.1 supernatant (fresh or stored frozen) is processed through ammonium sulfate precipitation and dialysis and then applied to a CM-cellulose column (5 × 30-cm bed). The cholesterol transfer activity is eluted with 25 m*M* sodium phosphate, 45 m*M* NaCl, 5 m*M* 2-mercaptoethanol, 0.02% sodium azide, pH 7.4.[13] The peak of exchange activity immediately follows a band of dark red material eluted by the buffer. The pooled active fractions (excluding the red material) can be stored at 0-4° in the elution buffer with little loss of activity for 1-2 months. Each 900-ml portion of the pH 5.1 supernatant yields after CM-cellulose chromatography about 1800 units (see below for definition) of cholesterol transfer

[12] R. C. Crain and D. B. Zilversmit, *Biochemistry* **19**, 1433 (1980).

[13] The transfer protein was eluted from the CM-cellulose column at pH 7.4 as described in the original purification procedure of Crain and Zilversmit.[12] Bloj and Zilversmit have reported[14] that purification of the protein is improved by performing this step at pH 8.0.

activity, which is sufficient for extensive cholesterol loading or depletion of about 100 mg (protein) of synaptic plasma membranes under the conditions described below. Prior to use, an aliquot (usually 20–50 ml) of the transfer protein preparation is concentrated 5- to 10-fold (to 75–250 units/ml) without significant loss of activity in an Amicon ultrafiltration cell with a PM-10 membrane under N_2 pressure. For application to synaptosomes where function is measured, the transfer protein should be dialyzed to remove the azide (against 25 m*M* sodium phosphate, 45 m*M* NaCl, 5 m*M* 2-mercaptoethanol, pH 7.4) directly before or after concentration. Whether dialyzed or not, the concentrate is centrifuged to sediment aggregated material.

The transfer activity is stable to freezing at −20° or in liquid N_2 in small (2 ml) aliquots without added cryogenic agents. There is variable loss of activity in freezing larger volumes under these conditions.

C. Preparation of Liposomes

Source of Lipids

Cholesterol (+99%; Eastman Organic Chemicals, Rochester, New York)

Egg yolk PC (type V-E; Sigma Chemical Company, St. Louis, Missouri)

Dioleoyl PC (approximately 98%; Sigma)

[4-^{14}C]Cholesterol (57 mCi/mmol; Amersham, Arlington, Illinois)

[2-^{3}H]Glycerol trioleate (triolein) (2 Ci/mmol; ICN Pharmaceuticals, Irvine, California)

[*Carboxyl*-^{14}C]Triolein (95 mCi/mmol; New England Nuclear, Boston, Massachusetts)

Solutions

Sucrose, 0.25 *M*; HEPES, 50 m*M*; EDTA, 1 m*M*; pH 7.4

Sucrose, 0.32 *M*; HEPES, 10 m*M*; pH 7.4

Procedure. For assay of cholesterol transfer activity, equimolar cholesterol/egg PC liposomes containing trace amounts of [4-^{14}C]cholesterol and [2-^{3}H]triolein are prepared. The lipids (1–2 mg of lipid) are mixed in chloroform in acid-cleaned glass-stoppered test tubes (~ 10 ml) and evaporated under a stream of N_2. The lipid films are dispersed in 0.25 *M* sucrose, 50 m*M* HEPES, 1 m*M* EDTA, pH 7.4 at 1 mg of PC/ml, and then sonicated to clarity (10–20 min) in a bath sonicator (Laboratory Supplies Company, Inc., Hicksville, New York) at 25–35° under N_2 atmosphere.

For use in cholesterol loading or depletion experiments, egg or dioleoyl PC (30–100 mg) with or without 1.7–1.9 mol of cholesterol per mole of PC is mixed with a trace (2000–10,000 dpm/μg P) of [^{3}H]triolein or [^{14}C]triolein as described above, in 25 to 100-ml round-bottomed flasks. After evaporation of the solvent under a stream of N_2, the lipids are further dried under vacuum in a desiccator preflushed with N_2. The lipid film is dispersed in 0.32 *M* sucrose, 10 m*M* HEPES, pH 7.4, by swirling the buffer across the film's surface and briefly submerging the bottom of the flask in the bath sonicator. The concentrations of PC in the dispersions should be 5–10 mg/ml for the PC liposomes and 3–5 mg/ml for the cholesterol/PC liposomes. After hydration for about 1 hr, the PC liposomes are sonicated in glass tubes, under N_2, until translucent (about 20 min) in the bath sonicator at 25–35°. Mixtures of PC and cholesterol are sonicated at 90 W using a Branson sonicator (Heat Systems Ultrasonics, Model W-350 sonifier) with at $\frac{1}{2}$-inch tipped horn, 80% duty cycle, under a gentle stream of N_2. Three 4-min periods of sonication, alternated with cooling periods of equal duration, are sufficient for production of translucent dispersions. Samples with volumes of 3–5 ml are conveniently and effectively sonicated in 20-ml glass scintillation vials. Larger samples (6–20 ml) can be sonicated in 25- or 50-ml glass beakers. After probe sonication, liposomes are centrifuged at 20,000 g_{av} for about 60 min at 25° to remove titanium fragments and undispersed lipid. The aqueous microdispersed lipids are unilamellar vesicles (liposomes). In general, the liposomes have been used within a few hours after preparation.

D. Assay of Nonspecific Lipid Transfer Activity

Assay of nonspecific lipid transfer activity is based on measurement of transfer of radiolabeled cholesterol to heat-treated beef heart mitochondria. A trace of radiolabeled triolein is included in the liposomes as a nontransferable marker. The assay is an adaption of that described by Bloj and Zilversmit[14] in which [^{14}C]cholesterol transfer is studied instead of radiolabeled phosphatidylethanolamine transfer. [^{3}H]Triolein is used as the nontransferable marker. Liposomes (see Section C, above) containing egg PC (3.8 μg of phosphorus), [^{14}C]cholesterol (47 μg of cholesterol), and a trace amount of [^{3}H]triolein are added to a mixture containing heat-treated mitochondria (69 μg of phosphorus, prepared as described elsewhere[14]), and 1.0–2.0 units (defined below) of transfer protein in 0.25 *M* sucrose, 50 m*M* HEPES, pH 7.4 (preincubated at 32° for 10 min) in a final volume of 1.0 ml, in a polycarbonate centrifuge tube (Beckman, 16.1 × 76.2 mm). A linear rate of [^{14}C]cholesterol transfer can be observed for about 10 min. An incubation is also carried out in the absence of transfer protein so that cholesterol transfer not catalyzed by the protein can be subtracted. The reaction is terminated by fivefold dilution with ice-cold buffer. Two milliliters of cold 0.65 *M* sucrose is then pipetted to the bottom of the tube to form a lower layer, and the sample is promptly centrifuged at 35,000 rpm for

[14] B. Bloj and D. B. Zilversmit, this volume [51].

20 min at 2° in a Beckman 50 Ti rotor. The pellet is surface rinsed with water, solubilized in 1 ml of Protosol (New England Nuclear), then assayed for [^{14}C]cholesterol and [^{3}H]triolein by double-isotope liquid scintillation counting. The counts of [^{14}C]cholesterol transferred to the pellet are corrected for sticking of liposomes by subtracting the counts of [^{3}H]triolein in the pellet multiplied by the $^{14}C/^{3}H$ ratio of the initial liposomes. The specific activity of the transfer protein prepared as described in Section B above, is about 30 units per milligram of protein at 32°, where 1 unit is the number of nanomoles of cholesterol transferred to the mitochondria per minute under these conditions.

E. Preparation of Synaptosomes and Synaptic Plasma Membranes

Reagents

SH medium: 0.32 *M* sucrose, 2.5 m*M* HEPES, pH 7.4
Sucrose, 0.8 *M*; HEPES, 2.5 m*M*; pH 7.4
KH medium (Krebs–HEPES) pH 7.4: 130 m*M* NaCl, 5 m*M* KCl, 1.3 m*M* $MgCl_2$, 1.2 m*M* sodium phosphate, 0.5 m*M* EGTA, 10 m*M* glucose, 10 m*M* HEPES
KHS medium: KH medium + 0.2 *M* sucrose
Tris-HCl, 5 m*M*, pH 8.1 (pH adjusted at 0–4°)
Sucrose, 48% (w/w)
Sucrose, 28.5% (w/w), pH 8.1

Preparation of Synaptosomes. Rat forebrain synaptosomes are prepared from a crude mitochondrial pellet essentially by the method of Gray and Whittaker[15] as modified by Hajos.[16] All steps subsequent to decapitation of the rats and removal of the forebrains are performed at 0–4°. The forebrains are homogenized in a Potter–Elvehjem-type glass–Teflon homogenizer (1.000-in. inner diameter) by three strokes with a loose-fitting pestle[17] (0.974-in. inner diameter) followed by six strokes with a tight-fitting pestle [0.991-in. diameter, available from Arthur Thomas Company (Philadelphia, Pennsylvania)] at 800 rpm. SH medium is used for the homogenization, at a final volume of 5 ml per gram of forebrain. The homogenate is diluted with an equal volume of SH medium and centrifuged at 1000 g_{av} (3500 rpm) for 10 min in a Beckman JA-20 rotor (Beckman, Palo Alto, California). The supernatant (S_1) is saved while the pellet (P_1) is resuspended in SH medium to the original volume and centrifuged again at 1000 g_{av} for 10 min. The supernatant is combined with S_1 and centrifuged at 11,000 g_{av} (12,000 rpm) for 15 min at speed in the JA-20 rotor. The resulting crude "mitochondrial" pellet is suspended in 20 ml of SH medium per gram of

[15] E. G. Gray and V. P. Whittaker, *J. Anat.* (*London*) **96,** 79 (1962).
[16] F. Hajos, *Brain Res.* **93,** 485 (1975).
[17] S. Fleischer and M. Kervina, this series, Vol. 31, p. 6.

forebrain and recentrifuged at 11,000 g_{av} for 15 min. This washed pellet is suspended in 4–7 ml of SH medium per gram of forebrain and layered, in approximately 20-ml aliquots, over 35 ml of 0.8 *M* sucrose, 2.5 m*M* HEPES, pH 7.4. The gradients are centrifuged at 10,000 g_{av} (9000 rpm) for 25 min at speed in a Beckman SW 25.2 rotor. The 0.8 *M* layer containing synaptosomes is collected and diluted over a 1-hr period with 3 volumes of KH medium with gentle stirring. The diluted synaptosomes are then centrifuged at 9900 g_{av} (10,000 rpm) for 20 min in a Beckman JA-14 rotor and resuspended in KHS medium to 10–20 mg of protein per milliliter. The synaptosomes are best used soon after preparation for lipid exchange.

Preparation of Synaptic Plasma Membranes. Synaptic plasma membranes from rat forebrain are prepared from the washed mitochondrial pellet by a procedure similar to that of Jones and Matus.[18] The washed mitochondrial pellet (prepared as described above for synaptosomes) is suspended in 5 m*M* Tris-HCl, pH 8.1, 2 ml per gram of forebrain, using six hand-driven strokes of a tight-fitting Teflon pestle.[17] The suspension is maintained in an ice-water bath for 60 min, then rehomogenized with six strokes of the pestle, diluted with 2 volumes of cold 48% (w/w) sucrose (per volume of suspension) to approximately 34% (w/w) sucrose, and transferred in 30-ml aliquots to the bottom of SW 25.2 centrifuge tubes (Beckman). These are overlaid with 20 ml of 28.5% (w/w) sucrose, pH 8.1, followed by 5 ml of 10% sucrose. The gradients are centrifuged at 49,000 g_{av} (20,000 rpm) for 4 hr in the SW 25.2 rotor. The band at the 28.5–34% sucrose interface is collected and diluted twofold with cold water. The synaptic plasma membranes are then recovered by centrifugation at 87,000 g_{av} (35,000 rpm) for 2 hr in a Beckman 60 Ti rotor, forming a pellet with a white outer rim and a slightly darker center underneath. The outer rim, which is most enriched in synaptic plasma membranes, is mechanically teased away from the center region with a spatula and suspended in 0.32 *M* sucrose, 2.5 m*M* HEPES, pH 7.4, at 10–30 mg of protein per milliliter. This fraction has a Ch/PL molar ratio of 0.52 ± 0.01 and a phosphorus-to-protein ratio of 33–37 μg of P per milligram of protein. The membranes can be quick-frozen in liquid nitrogen and stored at $-85°$.

F. Procedures for Lipid Transfer

Materials

Synaptic plasma membranes in 0.32 *M* sucrose, 2.5 m*M* HEPES, pH 7.4 (10–30 mg of protein per milliliter), or synaptosomes in KHS medium (10–20 mg of protein per milliliter) (cf. Section E, above)

Egg PC or dioleoyl PC liposomes in 0.32 *M* sucrose, 10 m*M* HEPES,

[18] D. H. Jones and A. I. Matus, *Biochim. Biophys. Acta* **356,** 276 (1974).

pH 7.4 (5–10 mg of PC per milliliter, containing a trace of [^{14}C]triolein, 2000–10,000 dpm per microgram of P) (cf. Section C, above)
Cholesterol/PC liposomes in 0.32 *M* sucrose, 10 m*M* HEPES, pH 7.4 (3–5 mg of PC per milliliter; 1.7–1.9 mol of cholesterol per mole of PC, containing a trace of [^{14}C]triolein) (e.g., Section C)
Elution buffer: 25 m*M* sodium phosphate, 45 m*M* NaCl, 5 m*M* 2-mercaptoethanol, 0.02% sodium azide, pH 7.4
Transfer protein in elution buffer (75–250 units/ml), concentrated and precentrifuged (cf. Section B, above)
Sucrose, 0.32 *M*, 2.5 m*M* HEPES; pH 7.4
Sucrose, 0.5 *M*
Sucrose 0.5 *M*, containing 40% of KH medium
Sucrose, 1.8 *M*
KH medium
KHS medium
Glucose, 1 *M*

Synaptic Plasma Membranes. The volume of transfer protein that can be added to the transfer mixture is limited because the buffer in which it is stored is relatively high in ionic strength, and high ionic strength is inhibitory to transfer activity.[12] In a typical reaction mixture with a final volume of 1.0 ml, up to 0.2 ml of concentrated transfer protein (in elution buffer) is mixed with 0.08 ml of 1.8 *M* sucrose and 0.17 ml of deionized H_2O to give a sucrose concentration of 0.32 *M*. A 0.05-ml aliquot of synaptic plasma membranes, in 0.32 *M* sucrose at 20 mg of protein per milliliter, is then added to give a final membrane protein concentration of 1 mg/ml. The amount of transfer activity present is varied between 0 and 45 units per milligram of membrane protein as a means of controlling the extent of lipid transfer (cf. Table I), with the volume of elution buffer held constant within a given set of samples by appropriate addition of the buffer without transfer protein (0–0.2 ml). After preincubation of the above mixture at 32° for 10 min, a 0.5-ml volume of PC or cholesterol/PC liposomes (for cholesterol depletion or loading, respectively) is added to start lipid transfer. Controls without liposomes are prepared by addition of 0.5 ml of 0.32 *M* sucrose, 10 m*M* HEPES, pH 7.4, instead of the liposomes. For cholesterol depletion, the molar ratio of liposomal phospholipid to membrane phospholipid is generally 4–6 (on the basis of phosphorus). For cholesterol loading, the liposomal cholesterol to membrane cholesterol molar ratio is usually about 10. The transfer is terminated, usually after 1–2 hr, by centrifuging the membranes through a layer of 0.5 *M* sucrose at 2° for 40 min at 100,000 g_{av} (in a Beckman 50 Ti rotor at 40,000 rpm). The pellets are surface rinsed with, and then suspended in, cold 0.32 *M* sucrose (pH 7.4) at 2–4 mg of protein per milliliter.

Synaptosomes. Lipid transfer with synaptosomes is accomplished as described above except that (*a*) the azide is removed from the transfer protein preparation by dialysis prior to use; (*b*) 10 m*M* glucose (an aliquot from the 1 *M* stock) and 15% of KH medium concentration (0.15 ml of KH medium per milliliter final volume) are added to the transfer mixture to preserve synaptosomal viability; (*c*) the synaptosomes are recovered after lipid transfer by centrifugation at 60,000 g_{av} (30,000 rpm in a Beckman 50 Ti rotor) for 30 min at 2° through a layer of 0.5 *M* sucrose containing 40% of KH medium; and (*d*) the synaptosomes are resuspended in KHS medium at 5–10 mg of protein per milliliter.

The use of 15% of the concentration of physiological saline (KH medium) during the lipid transfer incubation is a compromise between inhibition of the transfer protein by salt[12] and preservation of synaptosomal viability. After lipid transfer, the concentration with respect to KH medium is adjusted to 100% in two steps (*c* and *d* above) in order to make exposure of the synaptosomes to saline more gradual. Assays for function are performed without delay.

Secondary Lipid Transfer to Restore Membrane Composition. A second lipid transfer can be carried out to see whether reversal of the modification in membrane Ch/PL ratio can restore original membrane function. The second transfer is performed directly after the first and differs only in the type of liposome present (PC or cholesterol/PC) and its nontransferable marker ([^{3}H]triolein rather than [^{14}C]triolein). The primary cholesterol-modified membranes and their controls are divided into two portions; one portion (a control for sample handling) is incubated with transfer protein only and therefore without further change in membrane composition, whereas the sample for reversal is treated with the transfer protein and the type of liposomes appropriate for restoration of the original membrane Ch/PL ratio. The incubations are terminated as before by centrifugation through 0.5 *M* sucrose, and the membrane pellets are resuspended as described previously.

G. *Membrane Analysis*

The resuspended membranes are assayed for phosphorus, cholesterol, and protein as described below. The [^{14}C]triolein content is used to correct the phosphorus and cholesterol content of the samples for the presence of cosedimented liposomes (see Section H).

Cholesterol. Cholesterol is determined by a cholesterol oxidase assay in which cholesterol is oxidized to cholest-4-en-3-one with release of H_2O_2. The H_2O_2 that is formed reacts with 4-aminoantipyrine and phenol in the presence of horseradish peroxidase to form a product absorbing at 500 nm. The reaction is carried out as a single-step addition to an assay reagent mixture (Sigma kit 350). Membranes (0.05–0.3 mg of protein) or cholesterol/PC liposomes in a volume of 0.1 ml or less are added directly to 1.0 ml of the enzyme–reagent mix containing cholate, without prior lipid extraction, and the absorbance at 500 nm is allowed to develop fully (15 min). Turbidity is normally responsible for 4–10% of the total absorbance for the membrane samples and is corrected for by remeasuring the absorbance after decolorizing with one drop of 0.5 *M* ascorbic acid (provided in a dropping bottle with the kit).

Phosphorus. Phosphorus is measured by the method of Chen *et al.*[19] as described in Rouser and Fleischer.[20]

Protein. Protein is determined according to Lowry[21] or, when 2-mercaptoethanol is present, as described by Ross and Schatz.[22] Contribution of cosedimented transfer protein to the protein content of the membrane pellets can be a problem after incubations with high ratios of transfer protein to membrane protein. Such errors in the membrane protein content can be estimated by comparison of the lipid-to-protein ratios of membranes incubated with or without transfer protein, in the absence of liposomes. In our experience, contribution of transfer protein (precentrifuged before use) to final membrane protein is not significant at transfer protein-to-membrane protein ratios of 1 or below, which give near-maximal transfer rates.

Measurement of Radioactivity. Aliquots of the membrane pellets are solubilized in 1 ml of Protosol, then partially neutralized, to avoid chemiluminescence, with several drops of glacial acetic acid prior to addition of 10 ml of ACS scintillation fluid (Amersham, Arlington, Illinois). Counting is performed in a liquid scintillation counter programmed for determination of ^{3}H and ^{14}C counting efficiency, based on quenched standards, and for double-isotope analysis.

H. Calculation of Lipid Transfer

During the lipid transfer process, some "sticking" of liposomes to the membranes occurs that should be subtracted in order to determine the cholesterol and phospholipid content referable to the membrane. This is achieved by including a trace amount of ^{14}C-labeled (or ^{3}H-labeled) triolein, considered to be a nontransferable marker,[12] in the liposomes to be incubated with the membranes.[23] The presence of [^{14}C]triolein in the recovered membrane pellet, in conjunction with the known ratios of [^{14}C]triolein to cholesterol and lipid phosphorus in the initial liposomes, allows estimation of the amount of each of these liposomal components present in the pellet

[19] P. S. Chen, T. Toribara, and H. Warner, *Anal. Chem.* **28,** 1756 (1956).
[20] G. Rouser and S. Fleischer, this series, Vol. 10, p. 385.
[21] O. H. Lowry, N. J. Rosebrough, A. L. Farr, and R. J. Randall, *J. Biol. Chem.* **193,** 265 (1951).
[22] E. Ross and G. Schatz, *Anal. Biochem.* **54,** 304 (1973).

due to liposome sticking rather than molecular transfer into the membrane bilayer. Calculations involved in the correction for liposome sticking are illustrated below for sample No. 3 in Table I, which contains the primary data.

Calculations. The [^{14}C]triolein content in the sample (No. 3, Table I) is used to estimate the amount of phospholipid phosphorus and cholesterol in the pellets referable to sticking, using the measured radioactivity of [^{14}C]triolein in the initial liposomes, expressed per microgram of phosphorus (4311 dpm [^{14}C]triolein/μg P) and per microgram of cholesterol (185 dpm [^{14}C]triolein/μg cholesterol). Then, for sample No. 3, the amount of phosphorus referable to sticking is

$$(695 \text{ dpm } ^{14}\text{C})/(4311 \text{ dpm } ^{14}\text{C}/\mu\text{g P}) = 0.16\ \mu\text{g P}$$

The amount of cholesterol referable to sticking is

$$(695 \text{ dpm } ^{14}\text{C})/(185 \text{ dpm } ^{14}\text{C}/\mu\text{g cholesterol}) = 3.76\ \mu\text{g cholesterol}$$

The phospholipid P and cholesterol referable to sticking are then subtracted from the total P and cholesterol (see Table I) to give the actual membrane P and cholesterol:

$$\text{Membrane phospholipid P} = 1.18 - 0.16 = 1.02\ \mu\text{g P}$$

$$\text{Membrane cholesterol} = 17.83 - 3.76 = 14.07\ \mu\text{g cholesterol}$$

These values, normalized per milligram of protein (divided by 0.04 mg of protein per aliquot), are 25.5 μg of P per milligram of protein, and 352 μg of cholesterol per milligram of protein. The molar ratio of membrane choles-

[23] "Sticking" is defined operationally using radiolabeled triolein as a tracer in the phospholipid vesicles, based on the indication that the triolein is not transferred to the membranes by the transfer protein. The triolein is used to correct for "sticking" of the comparable amount of phospholipid and cholesterol. Liposome sticking was confirmed by freeze–fracture analysis of a synaptic plasma membrane preparation that had undergone lipid transfer. Small lipid vesicles were observed associated with and distinct from the membrane. Such associated lipid is subtracted so that the composition of the synaptic plasma membrane can be correlated with its function. However, some degree of fusion cannot be precluded. If fusion of lipid vesicles with the membrane had occurred in part, the lipid composition of the membrane would have been modified, and the correction for "sticking" would be inappropriate. Hence, it is instructive to estimate an upper limit of error in our analysis that would occur in the event of fusion. We observed the largest amounts of "sticking" in cholesterol-depletion experiments, where it represented as much as 25% of the sample phosphorus. For such cholesterol depletion, if the "sticking" were due entirely to fusion, the Ch/PL ratio of the membrane would be lower by 25% than that estimated by the sticking correction. It should be noted that for cholesterol loading the error due to fusion would be in the opposite direction.

TABLE I
TYPICAL PRIMARY LIPID TRANSFER DATA OBTAINED USING THE NONSPECIFIC TRANSFER PROTEIN[a]

	Transfer conditions		Analysis				
Sample	Type of liposome[b]	Transfer protein (units)[c]	Membrane protein (mg)	[^{14}C]Triolein (dpm)	Cholesterol (μg)	Phospholipid (μg P)	Ch/PL (mole/mole)
1. Control	None	0	0.045	—	9.6	1.5	0.51
2. Cholesterol-depleted	Egg PC	26.1	0.063	5762	9.01	3.92	0.18
3. Cholesterol-loaded	Cholesterol/egg PC	26.1	0.040	695	17.83	1.18	1.21

[a] Synaptic plasma membranes were incubated (1) in the absence of both transfer protein and liposomes; (2) with egg PC liposomes and transfer protein; or (3) with cholesterol/egg PC liposomes and transfer protein. The three examples are taken from Table II, where the reaction conditions are described. The amounts of protein, lipid, and radioactivity given above were measured in 0.025-ml aliquots of the resuspended membrane pellets recovered by centrifugation after the transfer incubation (see text).

[b] The egg PC liposomes contained 7025 dpm of [^{14}C]triolein per microgram of P. The cholesterol/egg PC liposomes (molar ratio = 1.9) contained 4311 dpm [^{14}C]triolein per microgram of P (185 dpm [^{14}C]triolein per microgram of cholesterol).

[c] The lipid transfer was catalyzed by addition of transfer protein, expressed here in units added to a final volume of 1.75 ml (cf. Table II).

terol ($M_r = 387$) to phospholipid [based on phosphorus (atomic weight = 31)] can then be calculated:

$$\text{Ch/PL} = \frac{352\ \mu\text{g cholesterol/mg protein}}{25.5\ \mu\text{g P/mg protein}} \times \frac{31\ \mu\text{g P}/\mu\text{mol P}}{387\ \mu\text{g cholesterol}/\mu\text{mole cholesterol}}$$

$$\text{Ch/PL} = 1.11$$

(molar ratio)

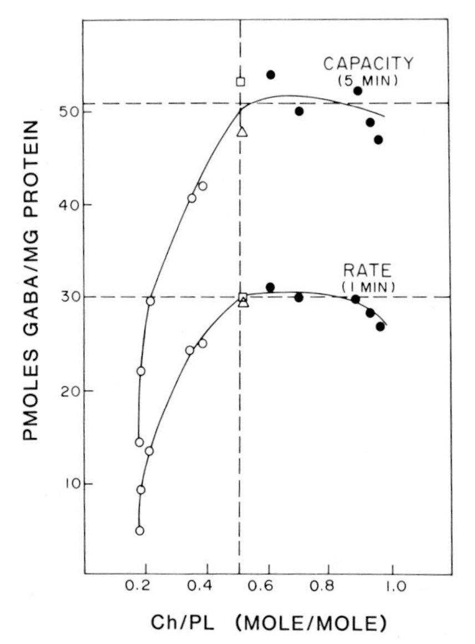

FIG. 1. Sodium gradient-dependent γ-aminobutyric acid (GABA) uptake by synaptic plasma membrane vesicles as a function of Ch/PL ratio. Synaptic plasma membrane vesicles (233 μg of cholesterol and 35 μg of P per milliliter) were incubated in the presence of varying amounts of transfer protein (2–45 units/ml) with egg PC liposomes (150 μg of P per milliliter, ○) or cholesterol/egg PC liposomes (2105 μg of cholesterol and 91 μg of P per milliliter, ●) for 90 min at 32° in order to modify the Ch/PL ratio. Control membranes were incubated with transfer protein only (□) or with neither transfer protein nor liposomes (△). Uptake of GABA was measured at 25° in 0.15 μM [2,3-^{3}H]GABA, 0.16 mg of membrane protein per milliliter, in the presence of a sodium gradient (0.1 M K phosphate, 1 mM $MgCl_2$ inside and 0.1 M NaCl, 1 mM $MgCl_2$ outside) immediately following reisolation of the membranes after lipid transfer.[3] The uptake rate was approximately linear at 1 min, and the uptake was maximal at 5 min (capacity). The phosphorus/protein ratios (μg/mg) for the 12 samples shown, reading from the lowest to the highest Ch/PL ratio, were 68.6, 67.2, 60.1, 43.4, 40.3, 36.7 (□), 36.0 (△), 35.6, 36.4, 32.8, 34.4, and 32.2. The dashed horizontal and vertical lines denote the Ch/PL ratio and GABA uptake rate and capacity in the control synaptic plasma membranes.

TABLE II
CHOLESTEROL DEPLETION AND LOADING OF SYNAPTIC PLASMA MEMBRANES[a]

		Analysis of lipid-modified synaptic plasma membranes			Net transfer	
Type of liposome used[b]	Transfer units	Cholesterol/protein (μg/mg)	Phospholipid/protein (μg P/mg)	Ch/PL (mole/mole)	ΔCholesterol (nmoles/mg protein)[c]	ΔPL (nmoles/mg protein)[c]
None[d]	0	213	33.3	0.51	0	0
None	26.1	214	33.0	0.52	+3	−10
Egg PC	0	200	31.1	0.52	−34	−71
Egg PC	1.9	175	35.7	0.39	−98	+77
Egg PC	3.7	166	37.1	0.36	−121	+123
Egg PC	11.2	146	37.4	0.31	−173	+132
Egg PC[d]	26.1	143	49.2	0.23	−181	+513
Egg PC	80.7	149	58.0	0.21	−165	+797
Chol/Egg PC	0	217	33.6	0.52	+10	+10
Chol/Egg PC	0.8	244	31.3	0.62	+80	−65
Chol/Egg PC	1.9	259	29.3	0.71	+119	−129
Chol/Egg PC	3.7	250	27.0	0.74	+96	−203
Chol/Egg PC	11.2	318	25.7	0.99	+271	−245
Chol/Egg PC[d]	26.1	352	25.5	1.11	+357	−255
Chol/Egg PC	80.7	371	24.9	1.19	+408	−271

[a] Synaptic plasma membranes (63 μg of P, 0.4 mg of cholesterol, 1.9 mg of protein) were incubated with or without egg PC liposomes (267 μg of P) or cholesterol/egg PC liposomes (3.41 mg cholesterol, 146 μg of P) in the presence of varying amounts of transfer protein for 60 min at 32° in a final volume of 1.75 ml. The membranes then were separated from the liposomes and transfer protein by centrifugation (see Section F). Correction was made for liposome-sticking using a nontransferable marker, [^{14}C]triolein (see Section H). The total sample phosphorus referable to sticking of liposomes, which has already been subtracted, varied from 18 to 25% in the case of cholesterol depletion and from 10 to 15% in the case of cholesterol loading.

[b] The egg PC liposomes contained 7025 dpm of [^{14}C]triolein per microgram of P. The cholesterol/egg PC liposomes (molar ratio = 1.9) contained 4311 dpm of [^{14}C]triolein per microgram of P.

[c] Nanomoles of cholesterol or phospholipid gained or lost per milligram of membrane protein during lipid transfer compared to the membranes incubated with neither the transfer protein nor liposomes.

[d] Samples for which primary data is given in Table I.

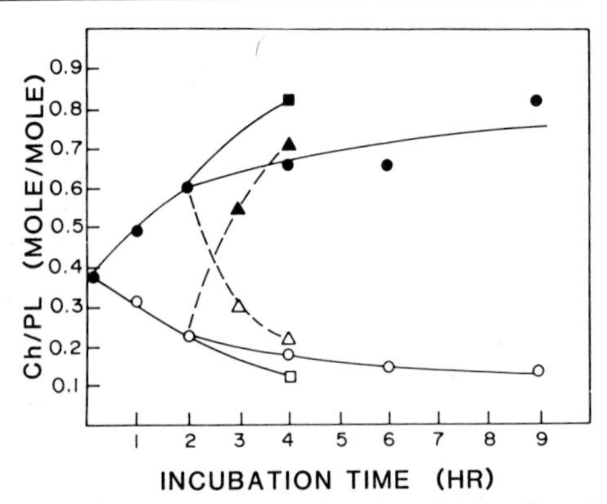

FIG. 2. Cholesterol depletion or loading as a function of time of incubation and the use of a secondary lipid transfer. Synaptosomes (0.75 mg of protein per milliliter) were incubated with dioleoyl PC liposomes (○), or cholesterol-dioleoyl PC liposomes (●), and 25 units of transfer protein per milliliter at 32° for 9 hr as described in Section F. During the incubation, aliquots were taken for analysis. After 2 hr of loading or depletion, aliquots of the membranes were recovered by centrifugation and incubated a second time with transfer protein and either the same type of liposomes as during the first incubation (□ and ■) or with the other type of liposomes (△ and ▲) (see Section F). Liposome "sticking," which increased the phosphorus content of the samples by 12–20%, was corrected for using [^{3}H]triolein (see Section H).

Application: Correlation of Lipid Composition with Function

The lipid transfer protein is an effective and gentle means of varying the Ch/PL ratio in synaptic membranes. With this methodology, the Ch/PL ratio of synaptic plasma membranes has been varied in a 1-hr incubation to as low as 0.21 or as high as 1.19 from a normal value of 0.52 (Table II). By a similar treatment, the Ch/PL ratio of synaptosomes has been varied from 0.16 to 0.81 compared with a normal value of 0.38. The effect of the altered Ch/PL ratio on sodium gradient-dependent γ-aminobutyric acid (GABA) uptake by synaptic plasma membrane vesicles is shown in Fig. 1. Lowering of the Ch/PL ratio caused a marked reduction in GABA uptake, whereas increased membrane cholesterol and incubation with liposomes or transfer protein alone were without effect. A second transfer incubation can be used to increase or reverse induced alterations in membrane Ch/PL ratio (Fig. 2). The loss of GABA uptake observed in synaptic plasma membranes upon lowering of Ch/PL ratio is reversible, strengthening the correlation between reduced membrane cholesterol content and loss of transport activity. The decreased uptake was correlated with a lowering of the number of available GABA binding sites, not to a change in the sodium permeability of the vesicles.[3]

Added Comments

Treatment with transfer protein to alter membrane cholesterol content also results in changes in the phospholipid/protein and total lipid/protein ratios of the membrane as well as the Ch/PL ratio. Cholesterol loading or depletion of synaptosomes and synaptic plasma membranes is accompanied by a decrease or increase, respectively, in the phospholipid-to-protein ratio (Table II). The total lipid content of the membranes increases with large amounts of cholesterol loading or depletion, although the absolute change in the lipid content of the membranes varies somewhat with the experiment. During cholesterol depletion, transfer of radiolabeled liposomal PC to the membrane is largely accounted for by the increase in the measured phospholipid/protein ratio of the membrane.[3] Thus, exchange of membrane and liposomal phospholipid appears to be minimal, since such exchange would transfer labeled PC to the membrane without affecting its phospholipid/protein ratio.

[57] Cerebroside Transfer Protein

By NORMAN S. RADIN and RAYMOND J. METZ

Glucosylceramide, galactosylceramide, and lactosylceramide are transferred by "cerebroside uptake protein" (CUP) from membrane to membrane.[1,2] The transfer protein is actually a pair of similar proteins, differing slightly in isoelectric point. The procedure described below yields only one of the two from bovine spleen.

Materials

DEAE-Sepharose CL-6B
Sephadex G-75, dry bead diameter 10–40 μm
Sephadex G-75, dry bead diameter 40–120 μm
CM-Sepharose CL-6B
Affi-Gel 501 (Bio-Rad)
Ampholines, pH 7–9 and 8–9.5 (LKB)
Glucosylceramide (palmitoyl glucosyl sphingosine), about 250,000 cpm/nmol
Lecithin, from eggs
Cholesterol
Triton X-100, peroxide-free,[3] M_r asumed to be 636

[1] R. J. Metz and N. S. Radin, *J. Biol. Chem.* **255,** 4463 (1980).
[2] R. J. Metz and N. S. Radin, *J. Biol. Chem.* **257,** 12901 (1982).
[3] Y. Ashani and G. N. Catravas, *Anal. Biochem.* **109,** 55 (1980).

ISBN 0-12-181998-1

Dithioerythritol (DTE)
Phosphate-buffered saline (PBS): 20 m*M* sodium phosphate, pH 7.4, 127 m*M* NaCl
Buffer A: 5 m*M* sodium phosphate, 1 m*M* $MgSO_4$, pH 8
Buffer B: PBS containing 1 m*M* DTE, 3.1 m*M* NaN_3 (0.2 mg/ml)
Buffer C: 0.25 *M* sucrose, 10 m*M* Tris- Cl, 1m*M* DTE, pH 7.4 at 4°
Buffer D: 50 m*M* Tris- Cl, pH 7.8 at 4°
Buffer E: 1 m*M* DTE in buffer D
Buffer F: 20 m*M* sodium phosphate, pH 6.2, 3.1 m*M* azide, 1 m*M* DTE

Assay Method

Glucosylceramide is transferred from cholesterol–lecithin liposomes containing labeled cerebroside, to acceptor membranes prepared from human red cells. The liposomes are made by strong sonication of a mixture of the lipids, and the membranes are made by hemolyzing the cells under conditions that yield resealed ghosts. Incubation of transfer protein with the two accelerates the uptake of cerebroside by the ghosts, which are isolated subsequent to the incubation, washed, and analyzed for cerebroside uptake by liquid scintillation counting of the membrane lipids. A single time point is adequate for monitoring purification procedures, but the initial incubation period (about 20 min) produces a temporarily rapid uptake, so more accurate rate measurements are made by carrying out two measurements, at 20 and at 80 min. This system yields excellent separation of donor vesicles from acceptor membranes, thus a nonexchangeable labeled marker is unnecessary. The method is probably useful for assaying other lipid transfer processes.

Membrane Preparation

Outdated human blood from a blood bank is centrifuged at low speed and about 35 ml of cells are washed several times with PBS, until the supernatant liquid is colorless. The cells are resuspended in an equal volume of PBS, and a portion is centrifuged to determine the hematocrit. Seventy milliliters of the suspension are stirred for 15 min with 1400 ml of cold buffer A to lyse the cells. The suspension is centrifuged in six 250-ml bottles for 30 min at 25,000 g_{max}, then the combined pellets are washed twice (only 15 min now) with 200 ml of PBS containing 1 m*M* $MgSO_4$ and 3.1 m*M* NaN_3. The final pellet, which retains a good deal of red color, is suspended in the washing buffer to a final volume equal to twice that of the packed, washed cells and stored in the cold.

The ghosts are useful for at least 1 month, after which they yield smaller and smaller pellets. Before each assay, the ghosts are resuspended by gentle end-over-end rotation of the storage flask. Aliquots of 0.2 ml (equivalent to 0.1 ml of packed red cells) are pipetted into 1.5-ml Eppendorf tubes (conical polypropylene) held in ice.

Liposome Preparation

Radioactive glucocerebroside can be prepared in several ways. It can be made from labeled glucose by coupling to 3-acyl ceramide,[4] by catalytic reduction with tritium,[5,6] by acylation of glucosyl sphingosine with labeled acyl chloride,[7] or by a series of steps starting with unlabeled glucosylceramide and ending with tritium in the 6-position of the glucose.[8] The reductive method is not recommended, since the saturated compound may behave atypically. For the same reason, it seems unwise to start with completely synthetic glucosyl sphingosine or cerebroside (from Miles Laboratories), since these are the saturated DL compounds.

Phosphatidylcholine (8.8 mg, about 11.3 μmol), cholesterol (3.2 mg, 8.2 μmol), and labeled glucosylceramide (14 μg, 20 nmol, about 5,000,000 cpm) are evaporated to dryness from their mixed stock solutions in chloroform–methanol as a thin film inside a 15-ml Corex test tube. Several such tubes are prepared at the same time and stored at room temperature in an evacuated desiccator. On the day of the assay, 10 ml of PBS are added and the tube is covered with Parafilm and vortexed for 15 min (done conveniently with the Big Vortexer, Kraft Industries). Nitrogen is carefully bubbled through the coarse suspension for 5 min to protect the lecithin, and the material is sonicated with a tapered microtip probe at maximum power for 45 min (Branson Sonifier, Model S-75). The tube is carefully centered around the tip and supported in an ice bath. To keep out oxygen somewhat, the annular space between the probe and test tube is covered with Teflon tape.

The tube contents are now transferred to a polycarbonate ultracentrifuge tube and centrifuged at 200,000 *g* for 1 hr to remove metal particles and large liposomes. About 20% of the lipid is lost in the pellet (but is recoverable for further use). The slightly opalescent suspension of liposomes can be stored at 4° under nitrogen up to a week if recentrifuged before each use.

Assay Procedure

The Eppendorf tubes containing 0.2 ml of ghost suspension are held in ice, and 0.2 ml of PBS (for the blanks) or of putative CUP is added to each.

[4] R. O. Brady, J. N. Kanfer, and D. Shapiro, *J. Biol. Chem.* **240,** 39 (1965).
[5] N. W. Barton and A. Rosenberg, *J. Biol. Chem.* **250,** 3966 (1975).
[6] G. Schwartzmann, *Biochim. Biophys. Acta* **529,** 106 (1978).
[7] N. S. Radin, this series, Vol. 28, p. 300.
[8] M. C. McMaster, Jr. and N. S. Radin, *J. Labelled Comp. Radiopharm.* **13,** 353 (1977).

When all the tubes are loaded, but not mixed, 0.2 ml of liposome suspension is added. The tubes are then capped, vortexed briefly, and incubated at 37° in a shaker set for 120 oscillations per minute. After 80 min, the tubes are chilled in ice briefly and centrifuged in the Eppendorf centrifuge for 2.5 min at 15,600 *g*. (Since our centrifuge holds only 12 tubes, we stagger the incubations to allow for the handling time needed for each group of 12.)

The supernatant fluid, containing nontransferred lipid, is removed with a tube attached to a suction flask, and each pellet is washed with 3 × 1 ml of cold PBS. The radioactive cerebroside associated with the membranes is then determined by transferring the pellet to a 20-ml counting vial, washing the incubation tube with 2 × 0.3 ml of water, and adding 10 ml of water-miscible scintillation fluid. Care is needed in squeezing the rubber bulb of the Pasteur pipette to avoid getting part of the pellet too high in the glass tube.

Labeled cerebroside can be recovered from the supernatants by extraction with hexane–isopropanol and silica gel chromatography in the usual way.

The amount of labeled cerebroside taken up by the cell membranes is calculated, after subtracting the activity found in the blanks, from knowledge of the specific activity of the labeled lipid. We generally normalize the uptake values to obtain picomoles per hour per milliliter of membranes.

The rate of uptake is rapid at first, then constant between 20 and 80 min. This is true for the blank incubations too, so the increase between 20 and 80 min must be measured for both the blank and samples. Separate incubation tubes are used for each time point. However, the 20-min value can be overlooked when one wants to know only which column fractions to pool.

A problem with this method is that there is some day-to-day variation in uptake rates, making comparisons between samples imprecise unless they are saved for assay on the same day. The samples obtained from the various purification steps are dialyzed against buffer B and stored at −20°, except for the fractions obtained after the third ammonium sulfate precipitation, which are unstable in the frozen state and must be kept at 4°.

Preparation of Spleen Cytosol

Bovine spleens from a local abattoir are collected in ice and used promptly. These spleens contain tough connective tissue and outer membrane, so we slit them longitudinally and scrape out the pulp with a mastic trowel (a notched steel plate with handle, sold for spreading floor tile cement). The cold pulp (440 g) is homogenized for 2 min with 1027 ml of buffer C using a Polytron at maximum speed.

The homogenate is centrifuged in six 250-ml bottles at 18,000*g* for 30 min in the cold, yielding about 960 ml of cloudy supernatant fluid. To this is added $CaCl_2$, 888 mg/liter, to flocculate the microsomal membranes. The mixture is stirred for 15 min, then centrifuged for 15 min at 25,000 *g*. The supernatant is frozen quickly with Dry Ice–acetone and stored at $-20°$. Because of the large volumes involved, several runs are made and pooled.

At this point, the cytosolic extract contains 38 g of protein and 3.3 million CUP units (picomoles of glucosylceramide transferred during the 1-hr constant-uptake period).

The Coomassie blue dye-binding method for assaying protein was used.[9]

Purification Scheme

Concentration with Ammonium Sulfate. About 2560 ml of cytosol, thawed rapidly by shaking in a 37° bath, are mixed with 420 g of ammonium sulfate, added over a period of 30 min. The mixture is stirred 30 min more and centrifuged at 25,000 *g* for 15 min. The pooled supernatant liquids are now brought to 60% saturation by adding 480 g of ammonium sulfate (181 mg/ml) as above. The pellet now obtained by centrifugation as above is taken up in 700 ml of buffer E and dialyzed against 6×6000 ml of the same buffer. We use three dialysis bags (4 cm flat diameter), each suspended in a 2000-ml cylinder containing a stirring bar, in the cold room for 3 days. However, it seems likely that the time could be shortened somewhat. The retentate is centrifuged for 30 min at 25,000 *g* to clarify it.

At this stage, the protein content is about 12 g and the total activity is 2.1 million units.

DEAE Chromatography. A column of diethylaminoethyl-Sepharose, 4.7×43 cm (750 ml bed volume), is washed with 6000 ml of 1 *M* NaCl and 9000 ml of buffer D, then with 1000 ml of buffer E. The above dialyzed sample (980 ml) is added to the column at a flow rate of 477 ml/hr, followed by 1500 ml of buffer E. Fractions of 16 ml each are collected, and 0.2-ml portions are assayed for CUP activity by the standard method, but with the inclusion of 50 μl of 557 m*M* NaCl, 57.7 m*M* Tris base to adjust the mixture to pH 7.4 and 310 mOsM.

The ion exchanger is washed with 1500 ml of buffer E containing 1 *M* NaCl to remove bound proteins. The effluent from this wash is dialyzed against buffer B before assaying for transfer activity, but little activity should be seen.

This step acts primarily to retard a large proportion of the proteins and the CUP elutes quickly. A curiously shaped elution curve for transfer activity is obtained, apparently because of the presence of interfering material that comes off the column early.

Size Exclusion Chromatography with Coarse Gel. The active fractions

[9] M. M. Bradford, Anal. Biochem. **72,** 248 (1976).

from the DEAE column (30 through 130; 3.8 g of protein, 2 million units) are pooled and treated with 632 g of ammonium sulfate (70% saturation, 436 mg/ml) as before to concentrate the CUP as a pellet. This is taken up in about 200 ml of buffer B, dialyzed against the same buffer, and centrifuged for 15 min at 25,000 *g*; the soluble portion is brought to a volume of 370 ml. The protein content is now 3.6 g, but only 860,000 CUP units are present. Four chromatographic runs are made, each with 90 ml, through a column of Sephadex G-75 (40–120 μm beads). The packing is 5 × 90 cm, or 1800 ml, and elution is carried out with buffer B at 95 ml/hr, with collection of 8-min fractions. Portions (0.1 ml) are assayed for CUP, which appears in fractions 76 through 108. Each run takes 1 day.

At this stage the crude CUP contains 780 mg of protein and 900,000 CUP units. The gel permeation step is thus an effective one for purification.

Ion Exchange with Carboxymethyl Resin. The active fractions from the Sephadex column are pooled (1640 ml) and the CUP is concentrated as above, by adding 1136 g of ammonium sulfate, centrifuging, and dialyzing, this time against buffer F. This preparation contains 690 mg of protein and 735,000 CUP units. The supernatant volume is adjusted to 155 ml and applied to a column of CM-Sepharose (1.3 × 35 cm, 46 ml bed volume), previously washed with 2000 ml of 1 *M* NaCl and 1000 ml of buffer F. The sample and a 46-ml wash are applied at 49 ml/hr, then a linear gradient solution is pumped through at 25 ml/hr. The gradient is formed from 460 ml of buffer F and 460 ml of a similar buffer containing 60 m*M* phosphate. Fractions are collected at 30-min intervals. At the end of the run, the column is washed with similar buffer, but more concentrated (310 m*M*).

Transfer activity is determined in 0.1 ml of each fraction in the usual way, but with addition of 0.1 ml of concentrated sodium phosphate sufficient to adjust the osmolality to 310 mOsM. This is done in an approximate way by adding (to fractions 1–10) a mixture of 26 m*M* monobasic phosphate, 175 m*M* dibasic phosphate. For fractions 11–40, 19 m*M* monobasic, 173 dibasic phosphate; for fractions 41–86, 2 m*M* monobasic, 169 dibasic sodium phosphate is added.

This column yields two relatively small, but distinct, peaks of transfer activity near the beginning of the separation, then a major peak in fractions 44–60. The ion-exchange step yields only 36 mg of protein, with 360,000 activity units in the major peak. It is evident that the two earlier peaks contain an appreciable fraction of the transfer activity and that the degree of removal of extraneous protein is quite good.

Ampholyte Displacement Chromatography. A column of carboxymethyl-Sepharose (1 × 6.4 cm, 5 ml) is equilibrated with buffer F at 1 ml/min. One-half of the CUP sample (100 ml) is mixed with an equal volume of 1 m*M* DTE to lower the ionic strength and is applied to the column at the same flow rate. The packing is now rinsed with 5 ml of 1 m*M* DTE and 25 ml of 0.5% Ampholines, pH 7–9 (0.31 ml of the stock 40% solution diluted to 25 ml with 1 m*M* DTE). Fractions of 1 ml are collected. Elution is now continued with 5 ml of water, then the more alkaline ampholyte solution (0.31 ml of Ampholines, pH 8–9.5, diluted to 25 ml with *water*). The protective mercaptan is omitted at this step in order to allow binding to the next, mercurial column. Aliquots of 15 μl are assayed for CUP activity in 0.2 ml of buffer B in the usual way. The active fractions are pooled, adjusted to pH 7.4 with 0.1 *M* HOAc, and used for the next step. A 0.2-ml portion is saved for more precise assay by adding 20 μl of 0.33 m*M* Triton X-100, 11 m*M* DTE, 34 m*M* azide.

The column is washed out with 310 m*M* sodium phosphate, pH 7.4, containing azide and DTE.

Monitoring of the column effluent at 280 nm shows a major early peak, which largely precedes two active peaks similar in size. The first activity peak reaches its maximum during the elution with water, at effluent pH 8.8, and the second peak reaches its maximum at effluent pH 9.1. Previous workers, with some other lipid transfer proteins, have observed the existence of "twins" differing slightly in apparent p*I*. For lack of time, and because the second peak seemed to contain less protein, we studied only this one. At this stage the protein content of peak 2 is 8 mg, and its total activity is 83,000 units.

Chromatography with a Mercurial Column. A column of Affi-Gel (0.5 × 20 cm, 4 ml) is prepared by washing with 200 ml of PBS and 200 ml of 4 m*M* $HgCl_2$ in PBS at 30 ml/hr. Just before use it is washed with 200 ml of 1 m*M* Na_2EDTA in PBS to remove mercury. The CUP sample is now added to the column at 5.3 ml/hr, followed by 100 ml of PBS–EDTA to remove the ampholytes and some protein. Now the CUP is eluted with 10 ml of the same buffer containing 5 m*M* DTE. The thiol groups of the DTE displace the protein from the organomercurial cation group. At this point the fractions (1.1 ml at 15-min intervals) are collected in tubes preloaded with 0.11 ml of the above-mentioned Triton–azide–DTE solution, which stabilizes the very dilute solution. The protein content of the active material is 3 mg, with good recovery of activity (70,000 units).

A nice feature of the mercurial column is that it removes ampholytes from the CUP effluent. Attempts at removing these buffers by standard procedures had not been effective.

Samples of 5 μl are removed for CUP assay in the usual way.

Chromatography with Small Bead Size Exclusion Gel. At this stage of purification, at least five bands can be seen after silver staining of a polyacrylamide gel electrophoresis slab. Electrophoresis was run with 0.75 mm-thick gels containing 15% acrylamide and 0.136% methylenebisacrylamide at room temperature, using samples mixed with sodium dodecyl sulfate–Tris–glycerol–mercaptoethanol.[10,11] Staining was done with Coomassie blue, using larger samples, or with silver, using 0.2–2 μg of protein.[12]

Further purification can be obtained with a column of Sephadex G-75 (10–40 μm beads, 2.5 $\times$ 41 cm, 200 ml) using 30 μM Triton in buffer B. The active fraction peaks at around 122 ml, and electrophoresis now reveals only three bands with silver. Judging by band intensities in the different fractions, the active one is the slowest. Standardization of this column with thyroglobulin, ovalbumin, chymotrypsinogen A, and ribonuclease A, run in the same buffer, indicates an M_r for CUP of about 20,3000. However, the M_r indicated for the slowest band in dodecyl sulfate electrophoresis gels is about 24,400. This type of discrepancy has been noted for some other lipid transfer proteins.

Properties

Stability. Crude CUP is rather stable if stored in 1 mM DTE; it is inactivated by incubation with *N*-ethylmaleimide. The presence of liposomes does not protect against this inactivation. The material obtained from the CM-Sepharose salt gradient is stable for several months, but loses about half its activity when heated for 5 min at 60°. The material obtained after ampholyte displacement chromatography is quite unstable but can be stabilized by the addition of 50 μg of bovine serum albumin per milliliter or 30 μgM Triton X-100. Triton should probably be added to the assay system, as it enhanced cerebroside transfer due to CUP, not in the blanks. This effect is probably due to increased membrane fluidity produced by the detergent. Stabilization of CUP was produced also by 10% l-propanol, cerebroside-containing liposomes, and various hydrophobic protein.

Kinetics. A few studies showed that the rate of cerebroside transfer is a function of both cerebroside and red cell concentrations. The rate did not reach a maximal value under the conditions used. The suspension of membranes would become rather viscous at higher concentrations.

Lipid Transfer Specificity. When liposomes were prepared with labeled galactosyl- or lactosylceramide in the lecithin–cholesterol mixture (the glycolipids being maintained at 6 μM in the incubation tube), the galactolipid was found to transfer 60% faster than glucocerebroside and the lactolipid, 39% more slowly. Labeled globotriaosylceramide transferred slowly (14% as fast as glucocerebroside), and ceramide and globotetraosylceramide were transferred at negligible rates. Transfer of lecithin or cholesterol was

[10] K. Weber and M. Osborn, *J. Biol. Chem.* **244,** 4406 (1969).

[11] U. K. Laemmli and M. Favre, *J. Mol. Biol.* **80,** 575 (1973).

[12] C. R. Merril, D. Goldman, S. A. Sedman, and M. H. Ebert, *Science* **211,** 1437 (1981).

not enhanced by CUP. Judging from these results, the assay procedure would be improved by the use of galactocerebroside instead of the standard lipid. Galactosylceramide is readily labeled with the galactose oxidase–borotritide method.[13]

Tests with crude splenic cytosol, after dialysis against buffer B, showed that material was present with the ability to enhance ceramide transfer.[2] A preparation of "nonspecific transfer protein" from bovine liver exhibited enhancement of transfer of glycosphingolipids, especially of the globotetraosyl compound.[14] A crude extract of brain cytosol enhanced transfer of galactocerebroside, sulfatide, and ganglioside G_{MI}.[15] It remains to be seen how many sphingolipid transfer proteins exist.

Nature of the Acceptor Membrane. Intact red cells were found to accept cerebroside from liposomes, and chemical or enzymic modifications of the cell surface did not make much difference in their effectiveness. Proteases, periodate, glutaraldehyde, detergents (at low concentrations), and thiols had little effect. Even ghosts treated with glutaraldehyde and heated at 90° for 5 min were useful. (It is possible that such membranes would be more stable than the membranes we used, allowing longer storage and less day-to-day variation.) Partial removal of cell lipids with isopropyl ether–butanol was not harmful. Unsealed red cell ghosts were the best acceptors, but we used resealed ghosts because they centrifuged more rapidly and were superior to intact cells.

It was shown that red cells, previously labeled with glucosylceramide with the aid of CUP, could serve as a donor to unlabeled liposomes containing or lacking cerebroside. In addition, CUP facilitated transfer of cerebroside between liposomes and multilamellar vesicles.

It would appear that CUP is a true "transfer" protein, rather than an "exchange" protein, and that the only requirement by way of acceptor or donor membranes is that they constitute a distinct nonaqueous phase. Presumably *some* lipid must be present in the membranes.

Mode of CUP Action. Unlike lecithin transfer protein, CUP could not be shown to form a firm complex with glucocerebroside. Some binding was demonstrated by use of a cerebroside-containing column and by dialysis between two cell compartments. Because of the relatively weak binding between the lipid and protein, we suggest that CUP acts by helping the glycolipid leave its donor membrane and enter the aqueous phase in true solution (dispersed as single molecules or small aggregates). The dissolved lipid molecules then diffuse to the acceptor membrane, where the CUP assists them in penetrating the outer regions of the membrane.

[13] N. S. Radin and G. P. Evangelatos, *J. Lipid Res.* **22,** 536 (1981).

[14] B. Bloj and D. B. Zilversmit, *J. Biol. Chem.* **256,** 5988 (1981).

[15] S. Sanyal and F. B. Jungalwala, *Trans. Am. Soc. Neurochem.* **13,** 129 (1982).

This hypothesis explains the initially rapid uptake of cerebroside by the membranes. The liposomal preparation presumably exists in equilibrium between liposomes and molecularly dispersed cerebroside, so that the concentration of cerebroside in the aqueous phase is maximal. After these molecules are taken up, the rate of dissolving becomes limiting. When CUP is present, the rate of dissolving is accelerated.

Comments

The method of preparation described here is in need of improvement to reduce losses, apparently the result of inactivation during several steps. The most serious losses occurred during the ammonium sulfate concentration steps. These were not apparent when small batches of extract were used, so it may be that the longer times involved in the large-scale operation are responsible. A more rapid system of dialysis might help, as might the inclusion of Triton, bovine serum albumin, or *n*-propanol. Possibly ammonium sulfate should be replaced by batch binding to the Affi-Gel mercurial column, from which CUP is readily eluted. In addition, the ampholyte displacement step might yield better recovery of activity if serum albumin or propanol were to be included. The more recently introduced ampholyte methods might improve the step.

The physiological roles for CUP are presently unknown. Since the lipids it can transfer occur in various membranes, it may be involved in the formation of membranes (by bringing newly synthesized molecules to growing membrane fragments), in the modification of membranes to suit changing physiological states, or in the destruction of membrane regions (by bringing the lipids to the lysosomes). Judging by recent reports for other lipid transfer proteins, CUP might be involved in bringing cerebrosides to the active regions of enzymes that utilize them as substrates. Potential applications are for the transferases that make sialyl lactocerebroside (ganglioside G_7), trihexosylceramide, lactocerebroside, galactosylceramide, and galactocerebroside sulfate.

Our chromatograhic steps showed that several forms of CUP exist. They could be called "isopheric" forms, or "isopheres," after the Greek word *pherein* ("to carry"). The proteins may be derived from different cell types within the spleen or may have differing, but overlapping, lipid specificities.

A protein very similar to CUP has been isolated from pig brain and rat liver by a procedure not too different from ours.[16,17]

[16] A. Abe, K. Yamada, and T. Sasaki, *Biochem. Biophys. Res. Commun.* **104,** 1386 (1982).
[17] K. Yamada and T. Sasaki, *J. Biochem.* (*Tokyo*) **92,** 457 (1982).

[58] Synthesis of Labeled Phospholipids in High Yield[1]

By HANSJORG EIBL, J. OLIVER MCINTYRE, EDUARD A. M. FLEER, and SIDNEY FLEISCHER

There is a general need for labeled phospholipids in almost all fields of membrane biochemistry and membrane biophysics. The method described here is based on the observation that bromoethyl esters of phosphatidic acids are readily converted to the respective lyso compounds by phospholipase A_2.[2] The resulting lysophospholipids are then reesterified with the labeled free fatty acid in the presence of dicyclohexylcarbodiimide and 4-dimethylaminopyridine. The polar moiety is then converted to any of several phospholipid classes.

The methodology is divided into three sections. Section A describes the synthesis of 1,2-dipalmitoyl-3-benzyl-*sn*-glycerol (~0.09 mol) from 1,2-isopropylidene-*sn*-glycerol. An intermediate (3-benzyl-*sn*-glycerol) can routinely be prepared in larger quantity (~1 mol) for later use, but it is convenient to acylate only a portion (0.1 mol) of this material. The product of this series of reactions, 1,2-dipalmitoyl-3-benzyl-*sn*-glycerol, is stable and can be further subdivided into 20 mmol aliquots to prepare—by sequential hydrogenolysis, phosphorylation, and alkylation—several different 1-palmitoyl-*sn*-glycero-3-phosphoric acid alkyl esters (Section B, see Fig. 2). Section C (see Fig. 3) describes the methodology for incorporation of the labeled fatty acid and the subsequent reactions to obtain the desired polar moiety. These later reactions, described in Section C, use 1 mmol of the alkyl esters so that 10 or more different labeled phospholipids can be prepared from each 1-palmitoyl-*sn*-glycero-3-phosphoric acid alkyl ester intermediate.

Materials

1,2-Isopropylidene-*sn*-glycerol was prepared as described earlier.[3] The optical rotation is $[\alpha]_D^{20} = 15.20$. Phospholipase A_2 from porcine prancreas was a product of Boehringer (Mannheim, FRG).

[1] These studies were supported in part by the Deutsche Forschungsgemeinschaft through SFB_{33}; by the Stiftung Volkswagenwerk, Hanover; and by National Institutes of Health Grant AM-21987.

[2] H. Eibl, *Chem. Phys. Lipids* **26,** 239 (1980).

[3] H. Eibl, *Chem. Phys. Lipids* **28,** 1 (1981).

ISBN 0-12-181998-1

Palmitic acid was obtained from Nu Chek Prep, Inc. (P.O. Box 172, Elysian, Minnesota). Labeled fatty acids, incorporated into the *sn*-2 position of the phospholipids were either radiolabeled or spin-labeled. The radiolabeled fatty acids ([9,10-^{3}H]oleic acid, NET-289; [1-^{14}C]oleic acid, NEC-317) were obtained from New England Nuclear (Boston, Massachusetts). Spin-labeled fatty acids can be obtained from a variety of sources as follows: (*a*) Syva Research Chemicals (3181 Porter Drive, Palo Alto, California): 2-(3-carboxypropyl)-2-tridecyl-4,4-dimethyl-3-oxazolidinyloxyl; 2-(10-carboxydecyl)-2-hexyl-4,4-dimethyl-3-oxazolidinyloxyl; 2-(14-carboxytetradecyl)-2-ethyl-4,4-dimethyl-3-oxazolidinyloxyl (abbreviated as 5-, 12-, and 16-doxylstearic acids, respectively). (*b*) Molecular Probes, Inc. (24750 Lawrence Rd., Junction City, Oregon): 5-, 7-, 10-, 12-, or 16-doxylstearic acids; and (*c*) J. Stefan Institut (61001, Ljubljana, Jamova 39, P.O.B 199-4 Yugoslavia):a variety of doxyl-labeled palmitic acids and other labeled fatty acids. Effective acylation (Section C,1) was obtained with spin-labeled fatty acids from either of these three sources. The purity of the fatty acids was >99%.

Other reagents required were the best grade available (reagent grade or higher purity). Benzyl chloride and hydrogen gas (cylinder) were from Fisher Scientific (Pittsburg, Pennsylvania). Paladium on carbon (10% Pd/C) catalyst, potassium *tert*-butoxide and phosphorus oxychloride were from Aldrich Chemical Co. (Milwaukee, Wisconsin). 2-Bromoethanol and 4-dimethylaminopyridine were from Sigma. Dicyclohexylcarbodiimide (X7298), trimethylamine, dimethylamine, and methylamine were from Eastman Kodak Co. (Rochester, New York). Thin-layer plates (Redi-Coat-2D) were from Supelco (Bellefonte, Pennsylvania). Chromatographic separations were carried out using columns of either silica gel 60 (35–70 mesh SSTM, E. Merck, Darmstadt, Germany) or silicic acid (Bio-Sil A, 100–200 mesh, Bio-Rad Laboratories, Richmond, California).

Solvents. All solvents were reagent grade (Fisher Scientific, Pittsburgh, Pennsylvania) with the exception of tetrahydrofuran, which was UV grade, distilled in glass (Burdick and Jackson Laboratories, Inc., Muskegan Michigan). After use, many of the solvents can be recovered by redistillation. The following solvents are required: acetone, ammonia (28% in water), chloroform, dichloromethane, diisopropyl ether, distilled water, ether, hexanes (boiling point 65.6–69.0°), methanol, 2-propanol, *p*-xylene, tetrachloromethane, tetrahydrofuran, and toluene.

Equipment. All reactions should be carried out in glass vessels. Only standard laboratory glassware is required. A three- or four-fraction "udder" collector (adapter and distribution receiver, e.g., K535750 or K544000 from Kontes, Vineland, New Jersey) is useful for the vacuum distillation (Section A,1).

Silica Gel or Silicic Acid Column Chromatography

When indicated, reaction products are purified by chromatography on silica gel or silicic acid. Of the two materials, silica gel 60 is by far the less expensive and is recommended for the purification of 1,2-dipalmitoyl-*sn*-glycerol (Section A,2). The silicic acid gives better resolution and is recommended for purification of final products (Section C,2). The elution solvents selected are based on the relative mobility of the desired product and the by-products in these solvents, as tested by thin-layer chromatography (TLC). The column is first eluted with a solvent of lower polarity in which the desired product has an R_f value of zero. Elution continues until no further material elutes from the column, as tested by TLC. The next solvent is applied only after all by-products, which run faster on TLC than the desired product, have been eluted. Then the polarity of the solvent is increased sufficiently to elute the product. The solvent systems given (Sections B,1 C,1 and C,2) result in quantitative elution of pure product in the last solvent system applied to the column. More polar contaminants, if present, remain on the column.

A. Synthesis of 1,2-Dipalmitoyl-3-benzyl-*sn*-glycerol (Fig. 1)

1. Preparation of 3-Benzyl-sn-glycerol from 1,2-Isopropylidene-sn-glycerol

1,2-Isopropylidene-*sn*-glycerol (Compound I, Fig. 1) (312 g, 1 mol) is dissolved in 1 liter of *p*-xylene. After the addition of potassium-*tert*-butoxide (157 g, 1.4 mol), the mixture is stirred with a mechanical stirrer and heated under reflux. Benzylchloride (152 g, 1.2 mol) in 500 ml of *p*-xylene is added dropwise within 1 hr, and stirring is continued until completion of the reaction, as indicated by TLC from the disappearance of 1,2-isopropylidene-*sn*-glycerol (R_f values: 0.3 for 1,2-isopropylidene-*sn*-glycerol and 0.6

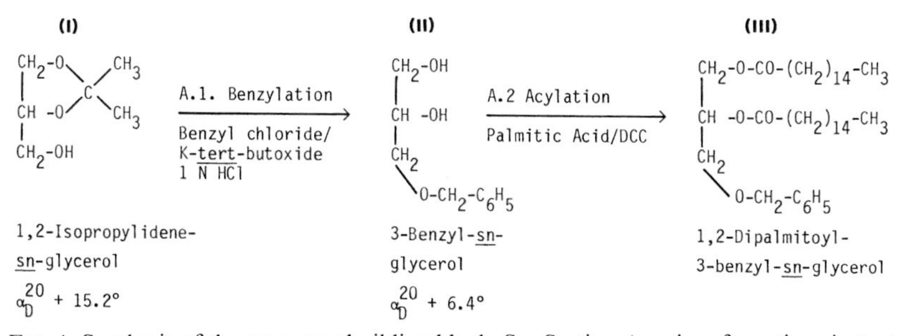

FIG. 1. Synthesis of the precursor building block. See Section A series of reactions in text.

for 1,2-isopropylidene-3-benzyl-*sn*-glycerol in diisopropyl ether–hexane, 1 : 1 by volume).

The reaction mixture is cooled to 20°; 1 liter of diisopropyl ether and 1 liter of water are added. After the separation of two phases in a separating funnel, the upper phase is collected and the lower phase is reextracted with additional diisopropyl ether (500 ml). The combined upper phases are evaporated. The oily residue is dissolved in 1 liter of methanol and heated under reflux in the presence of 80 ml of 1 *N* HCl. The reaction is followed by TLC. After complete conversion of 1,2-isopropylidene-3-benzyl-*sn*-glycerol to 3-benzyl-*sn*-glycerol (R_f values: 0.6 for 1,2-isopropylidene-3-benzyl-*sn*-glycerol and 0.0 for 3-benzyl-*sn*-glycerol in diisopropyl ether–hexane, 1 : 1 by volume), the reaction mixture is cooled to 20° and extracted with 1 liter of hexane to remove, in the upper hexane phase, traces of the starting product and by-products formed during the reaction, such as dibenzyl ether.

The lower methanol–water phase is diluted with a 10% solution of NaCl in water (1 liter). Extraction with 800 ml of chloroform is repeated twice, and the three lower phases are combined and dried over 100 g of Na_2SO_4 with stirring for 6 hr. Filtration and evaporation of chloroform from the filtrate results in an oil that is purified by distillation (bp_{1mm} 145–147°; $[\alpha]_D^{20}$ + 6.4°). The yield of pure product, 3-benzyl-*sn*-glycerol (Compound II, Fig. 1), is 155 g (85% based on 1,2-isopropylidene-*sn*-glycerol).[4]

2. 1,2-Dipalmitoyl-3-benzyl-sn-glycerol[8]

3-Benzyl-*sn*-glycerol (18 g, 0.1 mol; $[\alpha]_D^{20}$ + 6.4) and palmitic acid (56 g, 0.22 mol) are dissolved in 300 ml of CCl_4. The solution is cooled and kept in a water bath at 20°. 4-Diethylaminopyridine (2.4 g, 0.02 mol) is added while stirring with a magnetic stirrer. Dicyclohexylcarbodiimide (50 g, 0.25 mol) is dissolved in 100 ml of CCl_4 and added dropwise with continuous stirring to

[4] The optical rotation of 3-benzyl-*sn*-glycerol, $[\alpha]_D^{20}$ + 6.4° is higher than that reported in the literature [values of $[\alpha]_D^{20}$ + 5.5° (Kates *et al.*[5]) and $[\alpha]_D^{20}$ + 5.8° (Berchthold[6])]. The lower values indicate the presence of a considerable amount of the isomer, 1-benzyl-*sn*-glycerol, which usually results from the use of 1,2-ispropylidene-*sn*-glycerol of low optical rotation as starting material. Consequently, the phospholipids formed from starting materials that contain 3-benzyl-*sn*-glycerol and 1-benzyl-*sn*-glycerol will result in mixtures of natural and unnatural configuration. This aspect is treated in more detail elsewhere.[7]

[5] M. Kates, T. H. Chan, and M. Z. Stanacev, *Biochemistry* **2,** 394 (1963).

[6] R. Berchthold, *Chem. Phys. Lipids* **30,** 389 (1982).

[7] H. Eibl, *Chem. Phys. Lipids* **26,** 405 (1980).

[8] Other saturated fatty acids, e.g., myristic or stearic acids, can be substituted for the palmitic acid to prepare other 1,2-diacyl-3-benzyl-*sn*-glycerols. The time for hydrolysis with phospholipase A_2 (cf. Section B,3) will vary somewhat with different acyl chains. Phospholipids with the spin-labeled fatty acid incorporated into either *sn*-1 or *sn*-2 of the glycerol moiety, with either saturated or unsaturated fatty acids in the other position, can be prepared by synthetic routes that use selective protection of the hydroxyl groups of the glycerol.[2,7]

the solution of the reactants, taking care that the temperature of the reaction mixture does not exceed 25°. After completion of the reaction, as indicated by the disappearance of 3-benzyl-*sn*-glycerol (R_f values; 0.1 for 3-benzyl-*sn*-glycerol and 1.0 for 1,2-dipalmitoyl-3-benzyl-*sn*-glycerol in diisopropyl ether–ether, 1 : 1 by volume), the precipitated dicyclohexylurea is removed by filtration, and the filtrate is freed from solvent by evaporation. The residue, about 70 g, is dissolved in 400 ml of hexane and transferred to a column (7 × 50 cm) packed with 500 g of silica gel in hexane. The product is eluted with solvents of increasing polarity—diisopropyl ether–hexane 1 : 40; 1 : 20, and 1 : 10 by volume. The pure fractions (R_f value of 0.2 in diisopropyl ether–hexane, 1 : 9) are combined and evaporated to dryness. The residue is dissolved in hexane and cooled in an ice bath.

The mixture is filtered by suction, and a white precipitate, 1,2-dipalmitoyl-3-benzyl-*sn*-glycerol[9] (Compound III, Fig. 1), is obtained. The yield is 60 g, which corresponds to 91% based on 3-benzyl-*sn*-glycerol (calculated for $C_{42}H_{74}O_5$, M_r 659.05: C, 76.54%; H, 11.32%; found: C, 76.45%; H, 11.29%).

B. Synthesis of 1-Palmitoyl-*sn*-glycero-3-phosphoric Acid Alkyl Esters

1. 1,2-Dipalmitoyl-sn-glycerol

1,2-Dipalmitoyl-3-benzyl-*sn*-glycerol (13.2 g, 0.02 mol) is dissolved in 40 ml of tetrahydrofuran. Catalytic hydrogenolysis of the benzyl group is achieved in the presence of 2 g of catalyst (10% Pd/C) within 2 hr.[10] Completion of the reaction is indicated by the termination of the hydrogen uptake and confirmed by TLC (R_f values: 0.6 for 1,2-dipalmitoyl-*sn*-glycerol, 0.75 for the 1.3-isomer, and 1.0 for 1.2-dipalmitoyl-3-benzyl-*sn*-glycerol in diisoproyl ether–ether, 1 : 1 by volume). The starting material is

[9] 1,2-Dipalmitoyl-3-benzyl-*sn*-glycerol is a stable and general intermediate for the synthesis of labeled phospholipids. Silica gel chromatography, in our experience, is necessary to achieve a rapid removal of the benzyl group by catalytic hydrogenolysis in the next step of the procedure.

[10] Catalytic hydrogenolysis is carried out in a sealed system in the absence of oxygen to minimize the danger of explosion of the hydrogen gas. The system should be set up in a fume hood. A reservoir is first filled with nitrogen gas and the system is flushed out with nitrogen to displace any oxygen. The reaction mixture is stirred, and the reservoir is then filled with hydrogen (at least 0.5 liter). The reaction mixture is stirred vigorously, and the volume of hydrogen taken up is monitored. For 0.02 mol of 1,2-dipalmitoyl-3-benzyl-*sn*-glycerol, 0.02 mol of hydrogen gas (~ 0.4 liter) should be taken up when the reaction is complete and hydrogen uptake ceases. The system is then flushed out with nitrogen gas before opening the reaction chamber.

(III)

CH_2-O-CO-$(CH_2)_{14}$-CH_3
|
CH -O-CO-$(CH_2)_{14}$-CH_3
|
CH_2
 \ O-CH_2-C_6H_5

1,2-Dipalmitoyl-3-benzyl-*sn*-glycerol

B.1. Hydrogenolysis → H_2 ; Pd/C

(IV)

CH_2-O-CO-$(CH_2)_{14}$-CH_3
|
CH -O-CO-$(CH_2)_{14}$-CH_3
|
CH_2-OH

1,2-Dipalmitoyl-*sn*-glycerol

B.2 Phosphorylation → $POCl_3/N(C_2H_5)_3$

B.2 Alkylation (a,b or c) → B.2 Hydrolysis (H_2O/Na_2CO_3)

(V a–c)

CH_2-O-CO-$(CH_2)_{14}$-CH_3
|
CH -O-CO-$(CH_2)_{14}$-CH_3
| O
| ‖
CH_2-O-P -O-R
      |
      ONa

1,2-Dipalmitoyl-*sn*-glycero-3-phosphoric acid alkyl ester

B.3. Hydrolysis → Phospholipase A_2

(VI a–c)

CH_2-O-CO-$(CH_2)_{14}$-CH_3
|
CH -OH
| O
| ‖
CH_2-O-P -O-R
      |
      ONa

1-Palmitoyl-*sn*-glycero-3-phosphoric acid alkyl ester

FIG. 2. Phosphorylation, alkylation, and hydrolysis of the building block to obtain the precursor for incorporation of labeled fatty acid. See Section B series of reactions in text. The alkylation (Step B,2) is carried out with either: (*a*) bromoethanol to obtain the bromoethyl ester precursor for phosphatidylcholine (Section B,2,a); or (*b*) (*N*-butoxycarbonyl)ethanolamine to obtain the alkyl ester precursor for phosphatidylethanolamines (Section B,2,b); or (*c*) (*N*-butoxycarbonyl)-*tert*-butylserine to obtain the alkyl ester precursor for phosphatidylserine (Section B,2,c). The alkyl ester is abbreviated as R.

completely converted to 1,2-dipalmitoyl-*sn*-glycerol (Compound IV, Fig. 2); the 1.3-isomer is not formed. The catalyst is removed by filtration, and the filtrate directly used for phosphorylation.

2. *Preparation of Dipalmitoyl Esters*

a. 1,2-Dipalmitoyl-sn-glycero-3-phosphoric Acid Bromoethyl Ester, Sodium Salt. Phosphorus oxychloride (3.3 g, 0.022 mol) is stirred with a magnetic stirrer and cooled in an ice bath. After the addition of triethylamine (2.5 g, 0.025 mol) in 10 ml of tetrahydrofuran, stirring is continued and a solution of 1,2-dipalmitoyl-*sn*-glycerol (0.02 mol) in tetrahydrofuran (40 ml) is added dropwise over 60 min. Thin-layer chromatography shows complete conversion of the diacylglycerol to 1,2-dipalmitoyl-*sn*-glycero-3-phosphoric acid dichloride (R_f values: 0.6 for 1,2-dipalmitoyl-*sn*-glycerol and 0.0 for the glycero-3-phosphoric acid dichloride in diisopropyl ether–ether, 1 : 1 by volume).

Triethylamine (5 g, 0.05 mol) in 20 ml of tetrahydrofuran is added with continuous stirring followed by the dropwise addition of bromoethanol (3.2 g, 0.025 mol) in 30 ml of tetrahydrofuran. The temperature is raised to 25°. After 2 hr the main product is 1,2-dipalmitoyl-*sn*-glycero-3-phosphoric acid bromoethyl ester monochloride (R_f value: 0.6 in diisopropyl ether–ether, 1 : 1 by volume) with only little formation of a by-product, the respective glycero-3-phosphoric acid bisbromoethyl ester.

The reaction mixture is filtered by suction to remove the precipitated triethylamine hydrochloride. Hydrolysis of the phosphorus monochloride is achieved as follows: The mixture is cooled with an ice bath, and 50 ml of distilled water is added. After 2 hr, 50 ml of 1 *M* sodium carbonate and 100 ml of hexane are added with shaking. After phase separation, the upper phase is collected and the lower phase is reextracted with 50 ml of hexane. The combined hexane phases contains the product, 1,2-dipalmitoyl-*sn*-glycero-3-phosphoric acid bromoethyl ester, sodium salt (Compound Va, Fig. 2).

The solvent is removed by evaporation, and the residue, about 15 g, is dissolved in 50 ml of chloroform–methanol (10 : 1, v/v) and transferred to a column (4 × 60 cm) packed with 300 g of silica gel in chloroform. Mixtures of chloroform, methanol, and ammonia (10% solution in water) are used for elution, starting with 200 : 20 : 1 (v/v/v) followed by 65 : 15 : 1 (v/v/v) for the elution of the product. The yield of 1,2-dipalmitoyl-*sn*-glycero-3-phosphoric acid bromoethyl ester, sodium salt (Compound Va, Fig. 2), is 11.5 to 12.5 g; this corresponds to 72–84% based on 1,2-dipalmitoyl-3-benzyl-*sn*-glycerol (calculated for $C_{37}H_{71}BrNaO_8P \cdot 1H_2O$, M_r 795.87: C, 55.84%; H, 9.25%; Br, 10.04%; P, 3.89%; found: 55.98%; H, 9.11%; Br, 9.54%; P, 3.87%).

b. 1,2-Dipalmitoyl-sn-glycero-3-phosphoric Acid (N-Butoxycarbonyl)-ethanolamine Ester, Sodium Salt.[11] The procedure was followed as described for the preparation of the 1,2-dipalmitoyl-*sn*-glycero-3-phosphoric acid bromoethyl ester. However, (*N*-butoxy carbonyl)ethanolamine is used for the formation of the second phosphate ester bond instead of bromoethanol. The yield of 1,2-dipalmitoyl-*sn*-glycero-3-phosphoric acid-(*N*-butoxycarbonyl)ethanolamine ester sodium salt (Compound Vb. Fig. 2) is 11–12 g; this corresponds to 68–74% based on 1,2-dipalmitoyl-3-benzyl-*sn*-glycerol (calculated for $C_{42}H_{81}NNaO_{10}P \cdot 1H_2O$, M_r 832.10: C, 60.63%; H, 10.06%; N, 1.68%; Na, 2.76%; O, 21.15%; P, 3.72%; found: C, 60.70%; H, 9.98%; N, 1.65%; P, 3.76%)

c. 1,2-Dipalmitoyl-sn-glycero-3-phosphoric Acid (N-Butoxycarbonyl)-tert-butylserine Ester[11]. The procedure followed is as described for the preparation of the 1,2-dipalmitoyl-*sn*-glycero-3-phosphoric acid bromoethyl ester. However, (*N*-butoxycarbonyl)-*tert*-butylserine is used for the formation of the second phosphate ester bond instead of bromoethanol. The

[11] Care has to be taken to avoid acidic conditions in any subsequent step of the synthesis. Otherwise the yield of the desired product is low owing to loss of material by removal of the protecting groups.

yield of 1,2-dipalmitoyl-*sn*-glycero-3-phosphoric acid-(*N*-butoxycarbonyl)-*tert*-butylserine ester, sodium salt (Compound Vc, Fig. 2), was 10.5–11.5 g; this corresponds to 56–62% based on 1,2-dipalmitoyl-3-benzyl-*sn*-glycerol (calculated for $C_{47}H_{89}NNaO_{12}P \cdot 1H_2O$, M_r 932.22: C, 660.56%; H, 9.84%; N, 1.50%; Na, 2.47%; O, 22.31%; P, 3.32%; found: C, 60.47%; H, 9.73%; N, 1.40%; P, 3.29%).

3. Hydrolysis with Phospholipase A_2

Solutions[12]

Palitzsch buffer A: 50 m*M* borax

Palitzsch buffer B: 200 m*M* boric acid

Procedure. The different 1,2-dipalmitoyl-*sn*-glycero-3-phosphoric acid esters of bromoethanol, of (*N*-butyloxycarbonyl)ethanolamine, or of (*N*-butyloxycarbonyl)-*tert*-butylserine, 0.01 mol each, are dissolved in a mixture of 300 ml of diethyl ether and 300 ml of distilled water, which contains $CaCl_2 \cdot 2H_2O$ (2.3 g; 0.016 mol). Palitzsch buffer A[12] (285 ml) is added followed by the addition of 15 ml of Palitzsch buffer B,[12] giving a pH of the resulting emulsion of 7.5. Phospholipase A_2 (10 mg) is added, and the mixture is stirred with a magnetic stirrer for ~60 min at 35°, after which time the starting material is completely transformed to the respective 1-palmitoyl-*sn*-glycero-3-phosphoric acid esters and fatty acid. Stirring is stopped, and the reaction mixture separated in two phases. The upper phase containing the product, as shown by TLC, is mixed with 100 ml of toluene and evaporated to dryness. The residue forms white crystals with acetone (100 ml). The mixture is filtered by suction; the filtrate contains the palmitic acid. The precipitate (1-palmitoyl-*sn*-glycero-3-phosphoric acid alkyl ester; Compounds VIa–c, Fig. 2) is pure, as shown by TLC, and is obtained in yields of more than 95% based on the respective dipalmitoylglycerophosphoric acid esters.

C. Synthesis of 1-Palmitoyl-2-acyl*-*sn*-phospholipids

1. Reacylation of the 1-Palmitoyl-sn-glycero-3-phosphoric Acid Esters with Labeled Fatty Acids[13] *and Dicyclohexycarbodiimide*

The respective (1-palmitoyl-*sn*-glycero-3-phosphoric acid esters of bromoethanol, of (*N*-butoxycarbonyl)ethanolamine, or of (*N*-butoxycarbonyl)-*tert*-butylserine, 1 mmol each, are dissolved in 20 ml of tetrachloromethane. After the addition of 1.2 mmol of labeled fatty acid,[13] the reaction

[12] S. Palitzsch, *Biochem. Z.* **70,** 333 (1915).

[13] The labeled fatty acids that have been used successfully in these syntheses include spin-labeled and radiolabeled fatty acids. Specific sources are given under Materials. Henceforth the incorporated labeled fatty acyl chain will be designated acyl* or R*.

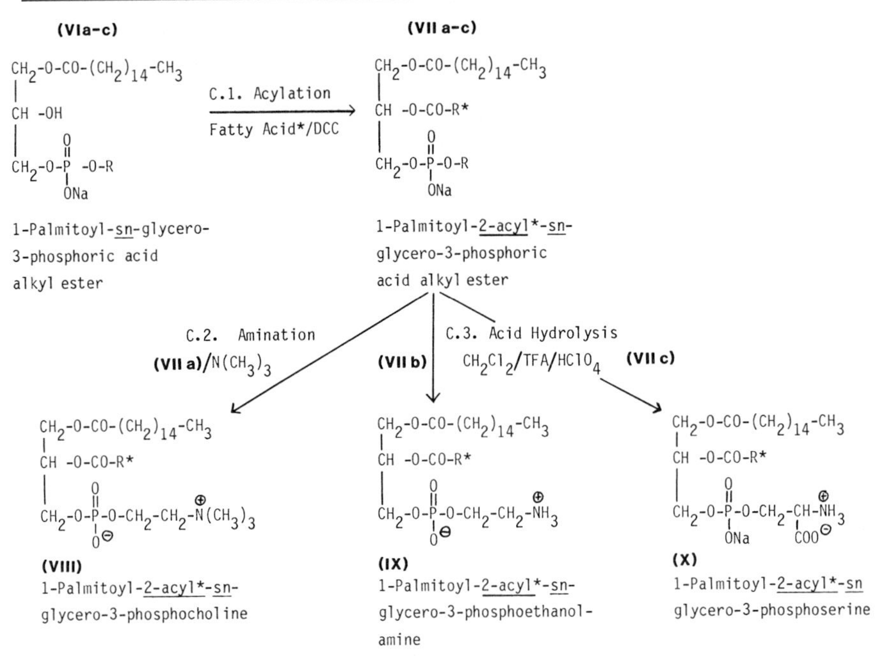

FIG. 3. Incorporation of labeled fatty acid into phosphorylated and alkylated precursors followed by modification of the polar moieties. See Section C series of reactions in text. The labeled fatty acid is indicated by R* or acyl*. The three different 1-palmitoyl-2-acyl*-*sn*-glycero-3-phosphoric acid alkyl esters (Compounds VIIa–c) (cf. Fig. 2) are converted to the three phospholipid classes, phosphatidylcholine, ethanolamine, or serine, respectively. In addition, the bromoethyl ester (VIIIa) can be converted to dimethyl- or monomethylethanolamine by using either dimethylamine or methylamine (cf. Section C,2).

mixture is stirred at 20° with a magnetic stirrer. 4-Dimethylaminopyridine (20 mg) is added, followed by the dropwise addition of 300 mg of dicyclohexylcarbodiimide in 5 ml of tetrachloromethane. The reaction is completed within 2 hr. The respective 1-palmitoyl-2-acyl*-*sn*-glycero-3-phosphoric acid esters of bromoethanol, of (*N*-butoxycarbonyl)ethanolamine, and of (*N*-butoxycarbonyl)-*tert*-butylserine (Compounds VIIa–c, respectively, in Fig. 3) are purified by chromatography (cf. Section B,1) and obtained in yields of 95%.

2. Phosphatidylcholines, (N,N-Dimethyl)phosphatidylethanolamines and (N-Methyl)phosphatidylethanolamines by Amination of 1-Palmitoyl-2-acyl-sn-glycero-3-phosphobromoethyl Ester, Sodium Salt*

The respective *sn*-glycero-3-phosphobromoethyl ester (~ 1 mmol) is dissolved in 5 ml of chloroform;1 g of 2-propanol–trimethylamine(1:1 by weight), 2-propanol–dimethylamine (1 : 1 by weight) or 2-propanol–methylamine (1 : 1 by weight) is added. The test tube is heated to 40°, closed carefully, and kept at 50° in a water bath for 12 hr. The solvents are evaporated by flushing with nitrogen. The residue is dissolved in 3 ml of chloroform and extracted with 3.3 ml of methanol water. After phase separation, the lower chloroform layer contains the product. The solvent is evaporated with nitrogen, and the residue purified by chromatography on 10 g of silicic acid starting with chloroform–methanol–water (200 : 15 : 1, v/v/v) and increasing the polarity stepwise, e.g., 100 : 15 : 1, 65 : 15 : 1, and 65 : 30 : 1. The yield of product (Compound VIII, Fig, 3) is more than 70% based on the starting *sn*-glycero-3-phosphoric acid bromoethyl ester.

Phosphatidylethanolamines and Phosphatidylserines

The respective *sn*-glycero-3-phosphoric acid (*N*-butoxycarbonyl)ethanolamine ester or (*N*-butoxycarbonyl)-*tert*-butylserine ester (1 mmol) is dissolved in 20 ml of CH_2Cl_2, 10 ml of trifluoroacetic acid, and 5 ml of 70% $HClO_4$. The mixture is kept cold in an ice bath and stirred for 30 min with a magnetic stirrer. The solution is then diluted with 40 ml of H_2O, 20 ml of chloroform, and 40 ml of methanol. After mixing and phase separation, the lower chloroform phase is extracted with 20 ml of 0.5 *M* Na_2CO_3. The combined upper phases are reextracted with 40 ml of chloroform. The lower phases are combined, filtered, and evaporated to dryness. Addition of 50 ml of methanol gives crystals that are collected. The respective phosphatidylethanolamines and phosphatidylserines (Compounds IX and X, respectively, Fig. 3) are obtained in quantitative yields.[14] No decomposition is observed during the removal of the protecting groups.

[14] The polar moieties can also be interconverted to prepare phosphatidylcholines, phosphatidylethanolamines, and phosphatidic acids via transphosphatidylation or hydrolysis with phospholipase D [H. Eibl and S. Kovatchev, this series, Vol. 72, p. 632 (1981)]. However, the yields of the reactions are generally lower (40–60%) than the chemical methods described here.

Addendum

Addendum to Article [21]

By MARK C. WILLINGHAM and IRA H. PASTAN

1. An improved image intensification video camera manufactured by Venus Scientific, Inc. (model TV3M) is now available from Carl Zeiss, Inc., expressly designed for use with light microscopy. In our experience, this camera (3-stage intensifier) provides resolution and gain comparable to, and in some cases better than, the EMI direct acceleration device described in this article. Because of its simpler installation, ease of operation, and fail-safe features, we would recommend this new camera as the most practical commercially available system for use with fluorescence microscopy presently available. It is possible that future developments with microchannel plate and other devices may make these systems obsolete in a few years. One such device (Intensicon 8 Camera) is offered by Lenzar Optics Corporation, Graflex, Inc. (Riviera Beach, Florida), but its efficacy with microscopy systems has not yet been reported.

2. Recently, reagents that reduce the photobleaching of fluorochromes have been reported including *n*-propyl gallate [H. Giloh and M. Sedat, *Science* **217**, 1252 (1982)] and *p*-phenylenediamine [J. L. Platt and A. F. Michael, *J. Histochem. Cytochem.* **31**, 840 (1983)]. In our hands, these reagents have proved to be exceptionally useful in experiments on fixed cells, making double-label experiments with fluorescein much easier to photograph.

3. Another recent development is the availability from Molecular Probes, Inc., of purified phycobiliproteins, a family of fluorescent proteins from algae that have advantages over rhodamine and fluorescein in their hydrophilic nature, high quantum efficiency, and large fluorescence emission per molecule [V. T. Oi, A. N. Glazer, and L. Stryer, *J. Cell Biol.* **93**, 981 (1982)]. One of these, β-phycoerythrin, has spectral characteristics similar to tetramethylrhodamine. An evaluation of the usefulness of these reagents is in progress.

ISBN 0-12-181998-1

Author Index

Numbers in parentheses are footnote reference numbers and indicate that an author's work is referred to although his name is not cited in the text.

C

D

E

G

H

I

J

L

M

N

O

P

Q

R

S

T

U

V

W

Y

Z

Subject Index

A

B

C

D

G

H

I

L

M

R

S

T

U

V

X

Y

Z